ELSEVIER SCIENTIFIC PUBLISHING COMPANY
335 JAN VAN GALENSTRAAT
P.O. BOX 211, AMSTERDAM, THE NETHERLANDS

DISTRIBUTORS FOR THE U.S.A. AND CANADA

ELSEVIER NORTH-HOLLAND INC.
52 VANDERBILT AVENUE, NEW YORK, N.Y. 10017

WITH 47 TABLES

ISBN: 0-444-40664-6 (SERIES)
ISBN: 0-444-41575-0 (VOL. IVH)

LIBRARY OF CONGRESS CARD CATALOG NUMBER 64-4605

PRINTED IN THE NETHERLANDS

RODD'S CHEMISTRY OF CARBON COMPOUNDS

ADVISORS

The late Professor Sir ROBERT ROBINSON, O.M., M.A. (Oxon.), D.SC. (Manc.),
HON.D.SC. (Lond., Liv., Wales, Dunelm, Sheff., Belfast, Bris., Oxon., Nott.,
Strath., Delhi, Sydney, Zagreb), HON. SC.D. (Cantab.), HON. LL.D. (Manc.,
Edin., Birm., St. Andrews, Glas., Liv.). HON. D. PHARM. (Madrid and Paris),
HON. F.R.S.E., F.R.S., *London*

Professor A. R. BATTERSBY, M.SC. (Manc.), PH.D. (St. Andrews), D.SC.
(Bris.), M.A., SC.D. (Cantab.), F.R.S., *Cambridge*

Professor R. N. HASZELDINE, M.A., PH.D., SC.D. (Cantab.), PH.D., D.SC.
(Birm.), C.CHEM., F.R.I.C., F.R.S., *Manchester*

Professor R. D. HAWORTH, D.SC., PH.D. (Manc.), B.SC. (Oxon.), C.CHEM.,
F.R.I.C., F.R.S., *Sheffield*

The late Professor Sir EDMUND HIRST, M.A., PH.D. (St. Andrews), D.SC.
(Birm.), HON.LL.D. (St. Andrews, Aberdeen, Birm., Strath.), HON. D.SC.
(Dublin), F.R.I.C., F.R.S., *Edinburgh*

Professor Lord TODD, O.M., M.A. (Cantab.), D.SC. (Glas.), D.PHIL. (Oxon.),
DR.PHIL. NAT. (Frankfurt), HON.LL.D. (Glas., Edin., Melb., Calif.), HON.
DR. RER. NAT. (Kiel), HON. D. MET. (Sheff.), HON. D.SC. (Oxon., Dunelm,
Lond., Exe., Leic., Liv., Adel., Alig., Madrid, Stras., Wales, Strath.),
C.CHEM., F.R.I.C., F.R.S., *Cambridge*

RODD'S CHEMISTRY
OF CARBON COMPOUNDS

VOLUME I

GENERAL INTRODUCTION

ALIPHATIC COMPOUNDS

*

VOLUME II

ALICYCLIC COMPOUNDS

*

VOLUME III

AROMATIC COMPOUNDS

*

VOLUME IV

HETEROCYCLIC COMPOUNDS

*

VOLUME V

MISCELLANEOUS

GENERAL INDEX

*

RODD'S CHEMISTRY
OF CARBON COMPOUNDS

A modern comprehensive treatise

SECOND EDITION

Edited by
S. COFFEY

M.Sc. (London), D.Sc. (Leyden), C.Chem., F.R.I.C.
formerly of
I.C.I. Dyestuffs Division, Blackley, Manchester

VOLUME IV PART H

HETEROCYCLIC COMPOUNDS

Six-membered heterocyclic compounds with (*a*) a nitrogen atom common
to two or more fused rings; (*b*) one hetero-atom in each of two fused rings.
Six-membered ring compounds with two hetero-atoms from Groups VI B,
or V B and VI B of the Periodic Table, respectively. Isoquinoline, lupinane
and quinolizidine alkaloids

ELSEVIER SCIENTIFIC PUBLISHING COMPANY

AMSTERDAM   OXFORD   NEW YORK

1978

CONTRIBUTORS TO THIS VOLUME

N. Campbell, o.b.e., ph.d., d.sc., f.r.s.e.
Department of Chemistry, The University, Edinburgh EH9 3JJ

S. F. Dyke, d.sc., ph.d., b.sc., f.r.i.c.
School of Chemistry, University of Bath, Bath BA2 7AY

M. Sainsbury, a.c.t.(brist.), ph.d., f.r.i.c.
School of Chemistry, University of Bath, Bath BA2 7AY

H. C. S. Wood, b.sc., ph.d., f.r.i.c., f.r.s.e.
Department of Pure and Applied Chemistry, University of Strathclyde, Glasgow G1 1XL

R. Wrigglesworth, b.sc., d.phil.
Chemical Research Laboratory, Wellcome Research Laboratories, Langley Court, Beckenham, Kent BR3 3BS

R. E. Fairbairn, b.sc., ph.d., c.chem., f.r.i.c.
formerly of Research Department, Dyestuffs Division, I.C.I. Ltd., Manchester 9 *(Index)*

PREFACE TO VOLUME IV H

The first four sub-volumes (Parts) of Volume IV of this revised edition of *Rodd's Chemistry of Carbon Compounds* are concerned with the chemistry of organic compounds possessing in their structure an isolated three-, four- or five-membered ring in which a single hetero-atom is present. Parts E, F and G deal with the chemistry of the wider and more varied types of compounds based on six-membered monoheterocyclic systems.

The first four chapters of the present sub-volume, IV H, are a continuation of this theme, the last two, however, are the opening chapters of the series dealing with the next major group of compounds, namely, those based on a six-membered heterocycle with two hetero-atoms in the ring.

Chapter 36, contributed by Dr. S. F. DYKE, is — although its length may appear to belie the statement — a concise account of the isoquinoline alkaloids comprising a very wide range of compounds of great interest from structural organic chemical and biochemical points of view. Chapter 37, by Professor NEIL CAMPBELL, is concerned with a further basic type of heterocyclic system to those already considered, in which a nitrogen atom is situated at the junction of a system of two or more fused-ring structures having the hetero-atom in common, or in which the hetero-atom forms the bridge in a bicyclic structure. The following chapter by Professor H. C. S. WOOD and Dr. R. WRIGGLESWORTH, describes the group of related natural products, the quinolizidine alkaloids. The systematic treatment of six-membered monoheterocyclic compounds containing a hetero-atom from Groups V or VI B, is concluded with Chapter 39 by Professor CAMPBELL in which he discusses fused-ring compounds containing two heterocyclic rings each possessing a single hetero-atom.

The last two chapters in this sub-volume, by Dr. MALCOLM SAINSBURY, are devoted, respectively, to the chemistry of six-membered heterocyclic compounds containing in the ring two hetero-atoms from Group VI B of the Periodic Table, *i.e.* oxygen, sulphur, selenium and tellurium, or one element from this Group along with nitrogen in Group V.

Our knowledge of the chemistry of each of the classes of compounds referred to above has greatly increased since the original edition of this book was published; the volume of literature the contributors have needed to take into consideration in preparing their scripts has been truly formidable. The editor takes this opportunity to congratulate them, on the completion of their tasks, on the way they have presented the results of their labours and to thank them for their help and cooperation in preparing the scripts for publication.

March 1978   S. COFFEY

CONTENTS

VOLUME IV H

Heterocyclic Compounds: Six-membered heterocyclic compounds with (*a*) a nitrogen atom common to two or more fused rings; (*b*) one hetero-atom in each of two fused rings. Six-membered ring compounds with two hetero-atoms from Groups VI B, or V B and VI B of the Periodic Table, respectively. Isoquinoline, lupinane and quinolizidine alkaloids.

Chapter 36. The Isoquinoline Alkaloids
by S. F. DYKE

Chapter 37. *Fused Heterocyclic Systems having a Nitrogen Atom in Common to Two or More Rings*
by NEIL CAMPBELL

Chapter 38. *Lupinane and Quinolizidine Alkaloids*
by H. C. S. WOOD AND R. WRIGGLESWORTH

Chapter 39. *Compounds Containing Two Fused Five- and Six-Membered Heterocyclic Rings each with One Hetero Atom*
by NEIL CAMPBELL

Chapter 40. *Compounds Containing a Six-Membered Ring having two Hetero-atoms from
Group VI B of the Periodic Table: Dioxanes, Oxathianes and Dithianes*
by MALCOLM SAINSBURY

Chapter 41. Compounds Containing a Six-Membered Ring with Two Hetero Atoms from Groups V and VI respectively, of the Periodic Table. Oxazines, Thiazines and their Analogues by MALCOLM SAINSBURY

CONTENTS

OFFICIAL PUBLICATIONS

B.P.	British (United Kingdom) Patent
F.P.	French Patent
G.P.	German Patent
Ger. Offen.	German Patent Application, open for inspection
Sw. P.	Swiss Patent
U.S.P.	United States Patent
U.S.S.R.P.	Russian Patent
B.I.O.S.	British Intelligence Objectives Sub-Committee Reports, H.M. Stationery Office, London.
C.I.O.S.	Combined Intelligence Objectives Sub-Committee Reports
F.I.A.T.	Field Information Agency, Technical Reports of U.S. Group Control Council for Germany
B.S.	British Standards Specification
A.S.T.M.	American Society for Testing and Materials
A.P.I.	American Petroleum Institute Projects
C.I.	Colour Index Number of Dyestuffs and Pigments

SCIENTIFIC JOURNALS AND PERIODICALS

With few obvious and self-explanatory modifications the abbreviations used in references to journal and periodicals comprising the extensive literature on organic chemistry, are those used in the World List of Scientific Periodicals.

LIST OF ABBREVIATED NAMES OF CHEMICAL FIRMS
MENTIONED IN PATENT REFERENCES

A.G.F.A., Agfa A.G.	Aktiengesellschaft für Anilinfabrikation (Berlin)
B.A.S.F.	Badische Anilin- und Soda-Fabrik (Ludwigshafen)
Bayer	Farbenfabriken vorm. Friedrich Bayer und Co. (Leverkusen)
Cassella	Leopold Cassella und Co. (Frankfurt am Main)
C.F.M.	Compagnie française des Matières Colorantes (Paris)
CIBA	Gesellschaft für chemische Industrie (Basel)
Du Pont	E.I. Du Pont de Nemours and Co. (U.S.A.)
G.A.F.	General Anilin and Film Corporation (U.S.A.)
Geigy A.G.	J. R. Geigy S.A. (Basel)
Hoechst	Hoechst A.G. (see M.L.B.)
I.C.I.	Imperial Chemical Industries, Ltd. (London)
I.G.	(= Interessen Gemeinschaft Farbenindustrie) of the principal dyestuffs manufacturers in Germany
Kalle	Kalle und Co., A.G. (Biebrich am Rhein)
M.L.B.	Farbwerke vormals Meister, Lucius und Brüning (Hoechst)
Sandoz	Sandoz A.G. Chemische Fabrik (Basel)

LIST OF COMMON ABBREVIATIONS AND SYMBOLS USED

A	acid
Å	Ångström units
Ac	acetyl
a	axial
as, asymm.	asymmetrical
at.	atmosphere
B	base
Bu	butyl
b.p.	boiling point
C, mC and μC	curie, millicurie and microcurie
c, C	concentration
c.d.	circular dichroism
conc.	concentrated
crit.	critical
D	Debye unit, 1×10^{-18} e.s.u.
D	dissociation energy
D	dextro-rotatory; dextro configuration
DL	optically inactive (externally compensated)
d	density
dec. or decomp.	with decomposition
deriv.	derivative
E	energy; extinction; electromeric effect
E1,E2	uni- and bi-molecular elimination mechanisms
E1cB	unimolecular elimination in conjugate base
e.s.r.	electron spin resonance
Et	ethyl
e	nuclear charge; equatorial
f	oscillator strength
f.p.	freezing point
G	free energy
g.l.c.	gas liquid chromatography
g	spectroscopic splitting factor, 2.0023
H	applied magnetic field; heat content
h	Planck's constant
Hz	hertz
I	spin quantum number; intensity; inductive effect
i.r.	infrared
J	coupling constant in n.m.r. spectra
K	dissociation constant
k	Boltzmann constant; velocity constant
kcal.	kilocalories
L	laevorotatory; laevo configuration
M	molecular weight; molar; mesomeric effect
Me	methyl

m	mass; mole; molecule; *meta-*
ml	millilitre
m.p.	melting point
Ms	mesyl (methanesulphonyl)
[M]	molecular rotation
N	Avogadro number; normal
n.m.r.	nuclear magnetic resonance
N.O.E.	Nuclear Overhauser Effect
n	normal; refractive index; principal quantum number
o	*ortho-*
o.r.d.	optical rotatory dispersion
P	polarisation; probability; orbital state
Pr	propyl
Ph	phenyl
p	*para-;* orbital
p.m.r.	proton magnetic resonance
R	clockwise configuration
S	counterclockwise config.; entropy; net spin of incompleted electronic shells; orbital state
S_N1, S_N2	uni- and bi-molecular nucleophilic substitution mechanisms
S_Ni	internal nucleophilic substitution mechanisms
s	symmetrical; orbital
sec	secondary
soln.	solution
symm.	symmetrical
T	absolute temperature
Tosyl	*p*-toluenesulphonyl
Trityl	triphenylmethyl
t	time
temp.	temperature (in degrees centigrade)
tert	tertiary
U	potential energy
u.v.	ultraviolet
v	velocity
α	optical rotation (in water unless otherwise stated)
$[\alpha]$	specific optical rotation
α_A	atomic susceptibility
α_E	electronic susceptibility
ε	dielectric constant; extinction coefficient
μ	microns (10^{-4} cm); dipole moment; magnetic moment
μ_B	Bohr magneton
μg	microgram (10^{-6} g)
λ	wavelength
v	frequency; wave number
χ, χ_d, χ_μ	magnetic, diamagnetic and paramagnetic susceptibilities

$\sim$	about
$(+)$	dextrorotatory
$(-)$	laevorotatory
$\ominus$	negative charge
$\oplus$	positive charge

Chapter 36

The Isoquinoline Alkaloids

S. F. DYKE

Alkaloids containing an isoquinoline or reduced isoquinoline ring-system are very numerous; they are usually classified into the groups shown in Scheme I, which also emphasises the biogenetic interrelationships. All of these groups will be described in this chapter, *except* the erythrina bases, alkaloids of the lycorine (*cf.* C.C.C., Vol. IV B, pp. 184 *et seq.*) and morphine (*cf.* C.C.C. Vol. IV G, pp. 267 *et seq.*) groups and the protostephanine (*cf. ibid.*, p. 297) types. All dimeric alkaloids, apart from the bisbenzylisoquinolines, are also excluded. Minor groups such as the isoquinolones, phenanthrenes and azafluoranthenes are not covered.

SCHEME 1

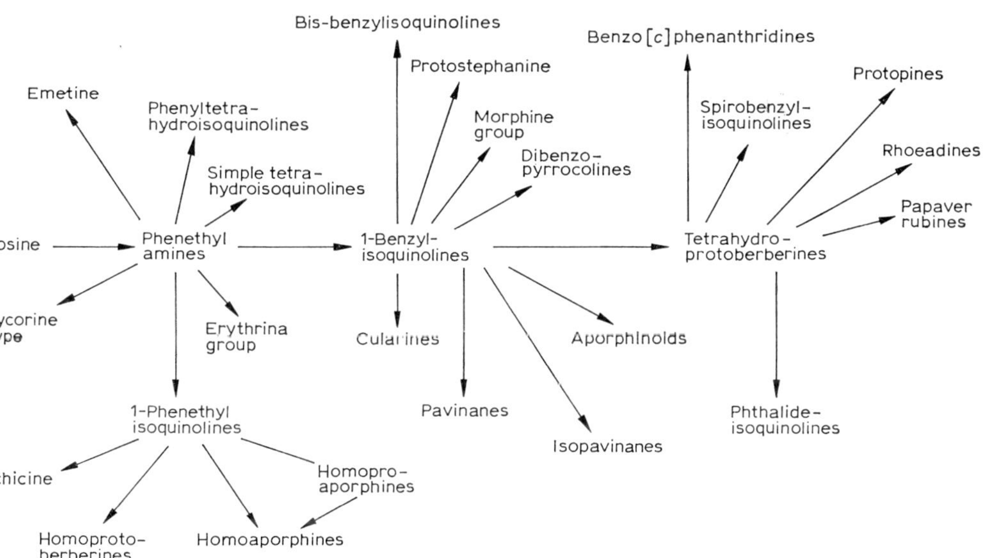

Numerous reviews of the isoquinoline alkaloids have appeared, the more important being the following:

(*1*) "The alkaloids", Academic Press, New York, Volumes 1–15, 1950 to 1975. Volumes 1–4 edited by *R. H. F. Manske* and *H. L. Holmes*; Volumes 5–15 edited by *Manske*. Vol. 4 is devoted to the isoquinoline alkaloids and various chapters, scattered throughout the remainder of the series, up-dates this material. Referred to in the sequel as "The Alkaloids".

(*2*) *T. Kametani*, "The Chemistry of the Isoquinoline Alkaloids", Vol. I, 1969; Vol. II, 1974, Elsevier, Amsterdam. Despite the title, these volumes contain an excellent tabulated summary of the isoquinoline alkaloids, giving the structures and physical data, together with references to isolation, structure determination, synthesis and biosynthesis.

(*3*) *M. Shamma*, "The Isoquinoline Alkaloids: Chemistry and Pharmacology", Academic Press, New York, 1972. The literature up to 1971 is assessed in this excellent standard text on the subject.

(*4*) Specialist Periodical Reports of the Chemical Society: "The Alkaloids". The literature coverage of each volume is indicated: Vol. 1 (Jan. 1969–June 1970); Vol. 2 (July 1970–June 1971); Vol. 3 (July 1971–June 1972); Vol. 4 (July 1972–June 1973); Vol. 5 (July 1973–June 1974); Vol. 6 (July 1974–June 1975). Senior Reporter, Vols. 1–5, *J. E. Saxton*; Vol. 6, *M. F. Grundon*. These works are referred to in the sequel as SPR: The Alkaloids.

(*5*) *S. McLean* and *J. Whelan*, "Alkaloids" ed. *K. Wiesner* (MTP International Review of Science) Organic Chemistry Series One, Vol. 9, Butterworths, London, 1973, p. 161. A very brief survey of benzyl- and homobenzyl- isoquinolines and spirobenzyl-isoquinolines is included.

(*6*) *T. R. Govindachari* and *N. Viswanathan*, in *S. Rangaswami* and *N. V. Subba Rao* (eds.), "Some Recent Developments in the Chemistry of Natural Products", Prentice-Hall of India, New Delhi, 1972.

(*7*) *Govindachari* and *Viswanathan*, "Recent Developments in the Chemistry of Isoquinoline Alkaloids", J. Sci. Ind. Res., New Delhi, 1972, **31**, 244.

(*8*) The Synthesis of Isoquinoline Alkaloids: (*a*) by Phenolic Cyclisation, *Kametani* and *K. Fukumoto*, Accounts Chemical Research, 1972, **5**, 212; Heterocycles, 1975, **3**, 311; (*b*) by Biogenetic-type Reactions, *Kametani, Fukumoto* and *F. Satoh*, Bio-organic Chemistry, 1974, **3**, 430; (*c*) by Thermolysis of Benzcyclobutanes, *Kametani* and *Fukumoto*, Heterocycles, 1975, **3**, 29; (*d*) by Coupling of Phenols, *S. Tobinaga*, Bioorganic Chemistry, 1975, **4**, 110; (*e*) by Systematic Design, *Kametani* and *Fukumoto*, Heterocycles, 1973, **1**, 129; (*f*) by use of Reissert Compounds, *F. D. Popp*, Heterocycles, 1973, **1**, 165; (*g*) from 1,2-Dihydroisoquinolines, *S. F. Dyke*, in *A. R. Katritsky* and *A. J. Boulton* (eds.), "Advances in Heterocyclic Chemistry", Vol. 14, Academic Press, New York, 1972, p. 279.

1. Derivatives of isoquinoline and tetrahydroisoquinoline

(*a*) *The simple tetrahydroisoquinolines*

Many of the alkaloids of this group have been isolated from Cactaceae, and some have also been found in members of the Papaveraceae and Fumariaceae (*S. Agurell*, Lloydia, 1969, **32**, 206; *G. J. Kapadia* and *M. B. E. Fayez*, J. pharm. Sci., 1970, **59**, 1699; *K. M. K. Hornemann, J. M. Neal* and *J. L. McLaughlin*, *ibid.*, 1972, **61**, 41). They all possess at least two oxygen functions (hydroxyl, methoxyl or methylenedioxy) at $C_{(6)} + C_{(7)}$ or at

$C_{(7)} + C_{(8)}$; some members of the group possess three such functions [at $C_{(5)} + C_{(6)} + C_{(7)}$ or at $C_{(6)} + C_{(7)} + C_{(8)}$]. Most of the alkaloids are tertiary bases with an *N*-methyl substituent, although **peyophorine,** unusually, has an *N*-ethyl group (*Kapadia* and *H. M. Fales, ibid.,* 1968, **57,** 2017). Some *N,N*-dimethyl quaternary salts are also known. In many cases there is a substituent at $C_{(1)}$ (Me, CH_2OH or Bu^i). The 1-phenyl- and 1-benzyl-isoquinolines will be considered separately. The 4-phenyl-1,2,3,4-tetrahydroisoquinoline, **cherylline,** will also be included. The lactams **peyoglutam** and **mescalotam** are among some recently isolated amides, and the acids peyoxylic and peyoruvic are of significance in the biosynthesis of the alkaloids of this group.

Peyophorine

Cherylline

Peyoglutam   (R = H)
Mescalotam   (R = Me)

Peyoxylic acid   (R = H)
Peyoruvic acid   (R = Me)

The chemistry of the peyote alkaloids has been reviewed (*Kapadia* and *Fayez*, Lloydia, 1973, **36,** 9).

(*i*) *Structure determination and synthesis*

In early work the usual procedure was to *O*-ethylate any phenolic hydroxyl groups, then subject the compound to oxidative degradation. Inter-relationships and synthesis are by the usual methods involving the Bischler–Napieralski, Pictet–Spengler or Pomeranz–Fritsch reactions (see *S. F. Dyke*, C.C.C. 2nd edn., Vol. IV F, pp. 359 *et seq.*). Thus, **anhalamine** (I), after ethylation and oxidation with potassium permanganate yields 3-ethoxy-4,5-dimethoxyphthalic acid (II):

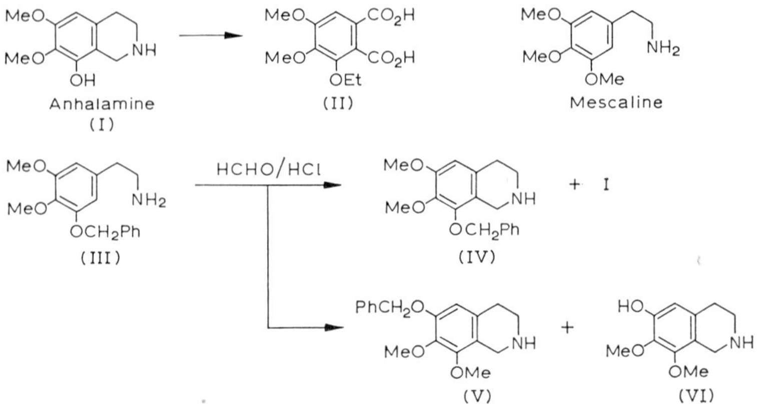

MeO / MeO / OH — Anhalamine (I)

MeO / MeO / CO$_2$H / CO$_2$H / OEt (II)

MeO / MeO / OMe — Mescaline

MeO / MeO / OCH$_2$Ph (III) → HCHO/HCl → MeO / MeO / OCH$_2$Ph (IV) + I

PhCH$_2$O / MeO / OMe (V) + HO / MeO / OMe (VI)

Condensation of mescaline with formaldehyde, followed by quaternisation gave a methiodide identical with O,N,-dimethylanhalaminium iodide, hence solving the structure (*E. Spath*, Monatsh., 1921, **42**, 97; 1922, **43**, 93; Ber., 1934, **67**, 2100). Cyclisation of 2-(3-benzyloxy-4,5-dimethoxyphenyl)ethylamine (III) in a Pictet–Spengler reaction gave the required anhalamine (I) together with the corresponding isomeric isoquinoline IV and, by cyclisation *para* to the benzyloxy function, a mixture of V and VI (*A. Brossi et al.*, Helv., 1966, **49**, 403; for a brief review of synthetic methods see *M. Shamma*, "The Isoquinoline Alkaloids", Academic Press, New York, 1972, p. 7).

More recently spectroscopic methods have been used in structure determination. In the mass spectrometer the loss of a $C_{(1)}$-substituent occurs so readily *e.g.* VII → VIII that the molecular ion is seldom observed:

MeO / MeO / MeO / Me / NMe (VII) → MeO / MeO / OMe / $\overset{\oplus}{N}$Me (VIII)

(IX) → (X) + CH$_2$ $\overset{\oplus}{N}$Me

However, a chemical ionisation mass spectral measurement overcomes this problem. Very frequently a strong peak is observed corresponding to the retro-Diels–Alder reaction *e.g.* IX → X.

The u.v. spectra of simple tetrahydroisoquinolines is essentially benzenoid, and the n.m.r. of such bases has been discussed (*idem, ibid.*, p. 38). Nevertheless errors have been made; the most notable one probably was the assignment of structure XI to gigantine (*J. E. Hodgkins, S. D. Brown* and *J. L. Massingill*, Tetrahedron Letters, 1967, 1321) on spectral evidence:

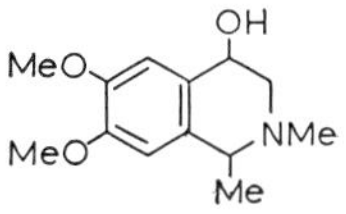

This was subsequently corrected (*Kapadia et al.*, Chem. Comm., 1970, 856) to XII, which was confirmed by synthesis involving the Bischler–Napieralski cyclisation of the amide XIII.

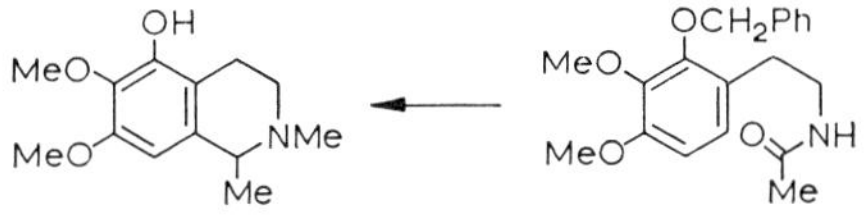

The absolute configuration of (−)**salsolidine** was established (*A. R. Battersby* and *T. P. Edwards*, J. chem. Soc., 1960, 1214) by oxidising the *N*-formyl derivative (R = CHO) first with ozone, then with peracetic acid to the amino acid XIV:

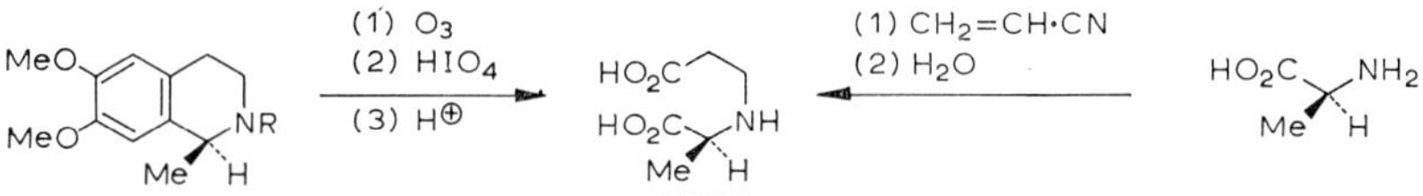

This was correlated with L-alanine by a synthesis involving Michael addition to acrylonitrile, followed by hydrolysis. It has been established that, for the simple monomeric tetrahydroisoquinoline alkaloids, the molecular rotations will show a characteristic shift with change of polarity of the solvent, so that absolute configurations can be assigned (*Battersby* and *Edwards, loc. cit.*). An X-ray crystallographic study of the hydrobromides of (−)**anhalonine** and (+)*O*-**methylanhalonidine** has confirmed the *S*-configuration for these bases and also shown that the $C_{(1)}$-methyl group is pseudo-equatorial in the former but pseudoaxial in the latter (*Brossi et al.*, J. Amer. chem. Soc., 1971, **93**, 6248):

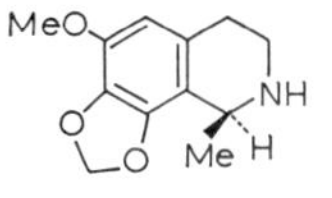

(−)Anhalonine 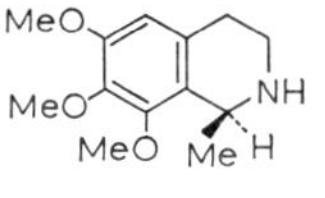   (+)*O*-Methylanhalonidine

(ii) Biosynthesis

The biosynthesis of alkaloids of the simple tetrahydroisoquinoline type has attracted considerable attention, and on the basis of the feeding of labelled precursors, its sequence is now quite well understood (see *S. A. Broun* in SPR Biosynthesis, Vol. I, 1972, p. 1; *A. G. Paul*, Lloydia, 1973, **36**, 36; *R. B. Herbert*, SPR Alkaloids, Vol. I, 1971, p. 16; *J. Staunton, ibid.*, Vol. II, 1972, p. 10). The routes from tyrosine are summarised in Scheme 2.

SCHEME 2

THE BIOSYNTHESIS OF SIMPLE TETRAHYDROISOQUINOLINES

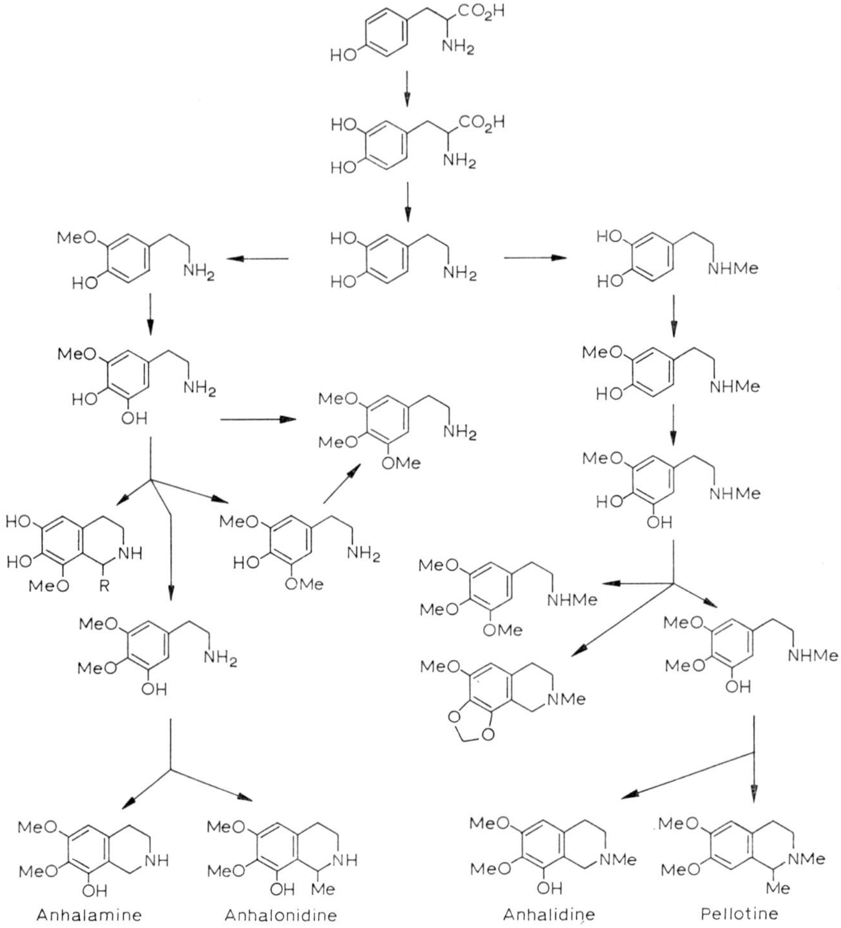

Anhalamine

Anhalonidine

Anhalidine

Pellotine

SCHEME 3

THE ORIGIN OF THE $C_{(1)}$-CARBON ATOM

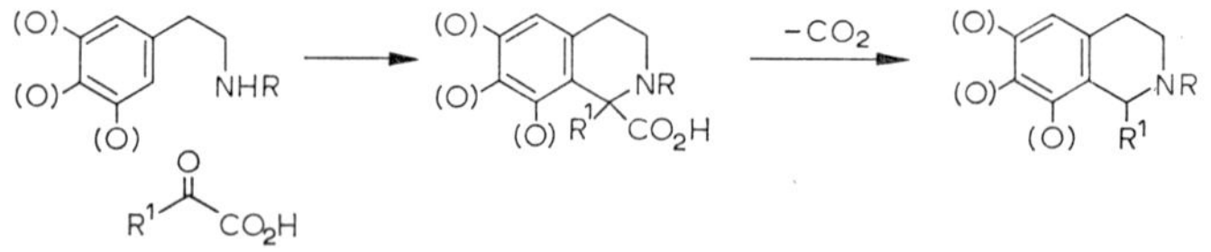

R = H or Me
R^1 = H or Me

Until recently some uncertainty existed concerning the origin of the $C_{(1)}$-side-chain carbon atom. The suggestion that the isoquinoline ring is formed between a β-arylethylamine and an α-oxo acid (Scheme 3) was supported by the observation (*E. Leete et al.*, J. Amer. chem. Soc., 1970, **92**, 6943) that peyoxylic and peyoruvic acids are effectively converted into anhalamine and anhalonidine, respectively:

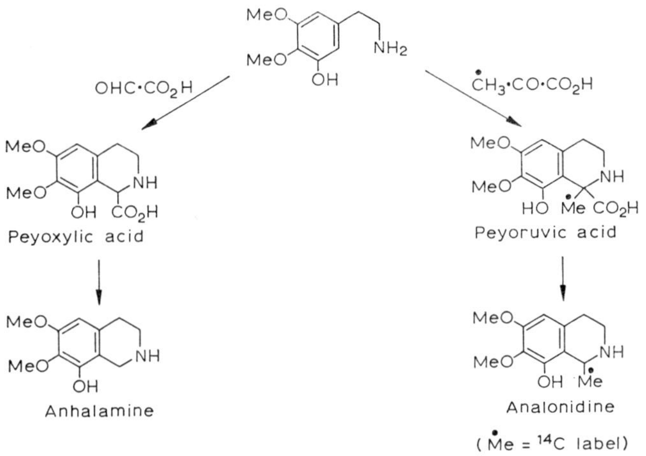

OHC·CO₂H   CH₃·CO·CO₂H

Peyoxylic acid

Peyoruvic acid

Anhalamine

Analonidine

(Me = ¹⁴C label)

(b) 1-Phenyltetrahydroisoquinolines

(i) The cryptostylines

Three closely related alkaloids have been isolated from *Cryptostylis pulva* Schltr. and named (+)**cryptostyline I, II** and **III** (*K. Leander* and *B. Luning*, Tetrahedron Letters, 1968, 1393; *Leander, Luning* and *E. Runsa*, Acta Chem. Scand., 1969, **23**, 244). The first one, $C_{19}H_{21}NO_4$, exhibits a benzenoid-type u.v. spectrum typical of a tetrahydroisoquinoline. In the n.m.r. spectrum the $C_{(1)}$-H absorption appears as a singlet at 4.2 δ, which is shifted to 5.9 δ in the methiodide. Additionally the spectrum exhibits signals due to two methoxyl groups, one methylenedioxy and an *N*-methyl group. In the mass spectrum the base peak, at *m/e* 206, arises from the molecular ion by loss of a methylenedioxyphenyl residue. These facts are accommodated in structure XV for cryptostyline-I, which has been confirmed by a synthesis of the racemate XVI → XVII → XV (*Shamma*, "The Isoquinoline Alkaloids", Academic Press, New York, 1972, p. 490):

Similar work established the structures for cryptostylines-II and -III.

Resolution of the synthetic bases yielded the natural (+)-alkaloids and the unnatural (−)-isomers. A single crystal X-ray analysis of the unnatural (−)cryptostyline-II hydrobromide showed that the molecule has the R-configuration and that the aromatic ring is oriented as shown in XVIII (*Brossi* and *S. Teitel*, Helv., 1971, **54**, 1564). It follows that the natural (+)-alkaloids possess the S-configuration.

(ii) Cherylline

The phenolic alkaloid **cherylline**, $C_{17}H_{19}NO_3$, was isolated from various species in the Amaryllidaceae and its structure was elucidated by spectroscopic methods, coupled with a synthesis (*Brossi et al.*, J. org. Chem., 1970, **35**, 1100). The n.m.r. spectrum is summarised in the structural proposal XIX. The aromatic one-proton singlet at 6.23 δ was assigned to $C_{(5)}$ since

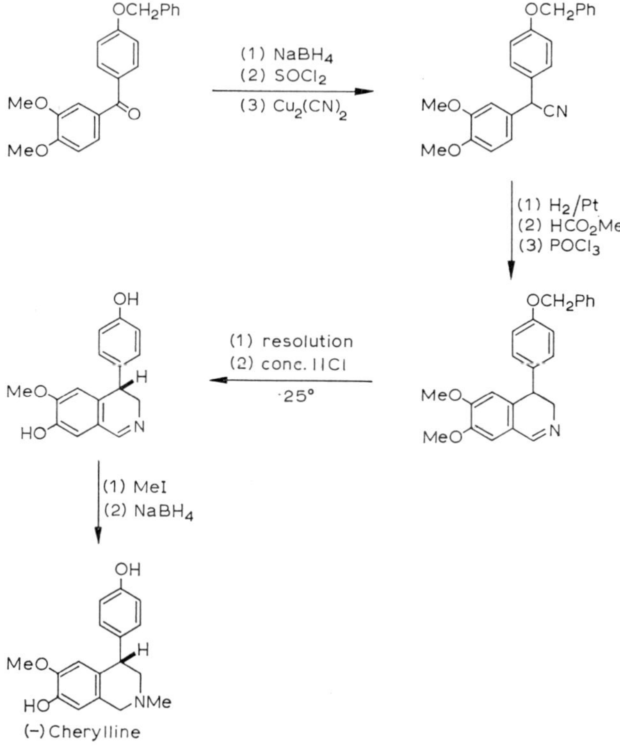

it must be shielded by the aromatic ring at $C_{(4)}$. Hence the absorption at 6.49 δ is due to the $C_{(8)}$-H. When the spectrum was re-determined after the addition of NaOD, the $C_{(8)}$-H underwent a larger upfield shift than the absorption due to $C_{(5)}$-H, and hence a phenolic hydroxyl group must be located *ortho* to it (at $C_{(7)}$). This evidence alone is unsufficient to deduce

SCHEME 4

A SYNTHESIS OF CHERYLLINE

a unique structure, but a synthesis of the alkaloid confirmed XIX (*Brossi* and *Teitel*, Tetrahedron Letters, 1970, 417) (Scheme 4).

The absolute configuration was assigned by a comparison of the o.r.d. curve of cherylline with the curves of 4-aryltetralins, and it was confirmed by an X-ray analysis of the *O,O*-dimethyl-4-bromobenzamide derivative (XX).

It has been suggested (*M. A. Schwartz* and *S. W. Scott*, J. org. Chem., 1971, **36**, 1827) that the alkaloid is biosynthesised as shown in XXI → XXII → XIX. A laboratory synthesis based upon this route has also been reported by these investigators:

(XXI) (XXII) (XIX)

(c) The 1-benzylisoquinolines

The 1-benzylisoquinoline alkaloids fall naturally into the fully aromatic compounds such as **papaverine** (XXIII), and the 1,2,3,4-tetrahydro derivatives, *e.g.* **laudanosine**. In each subdivision quaternary salts, as well as bases, have been characterised, *e.g.* **takatonine** and **magnocurarine.** In the tetrahydro series additional variation is possible by virtue of the asymmetric centre at $C_{(1)}$ and because the bases may be secondary or tertiary (in which case the nitrogen carries a methyl group). Ring A in both the fully aromatic and the 1,2,3,4-tetrahydro states of oxidation may have two or three oxygen functions, whereas ring C possesses one or two such substituents [at $C_{(4)}{}'$

Papaverine
(XXIII)

Laudanosine

Takatonine

Magnocurarine

or $C_{(3)}' + C_{(4)}'$]. No alkaloids of this group have yet been isolated that have an oxygen function at $C_{(4)}$, although it is highly probable that such structures exist in nature. In those examples that are disubstituted in ring A, all but petaline, have these oxygen functions at $C_{(6)} + C_{(7)}$ (*L. Reti*, in "The Alkaloids", Vol. 4, 1954, pp. 7 and 23; *V. Deulofeu, J. Comin* and *M. J. Vernengo, ibid.*, Vol. 10), 1968, p. 401):

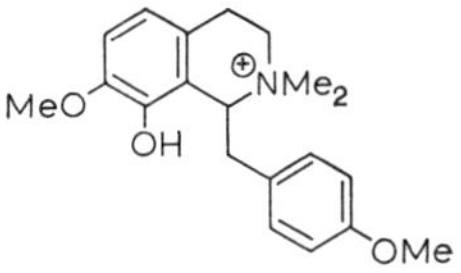

Petaline

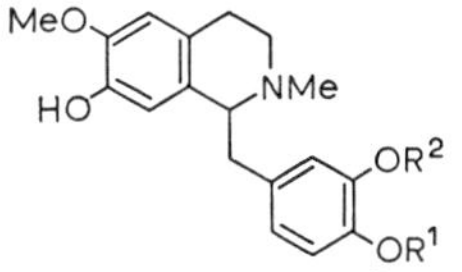

Reticuline ; R^1 = Me , R^2 = H
Orientaline ; R^1 = H , R^2 = Me

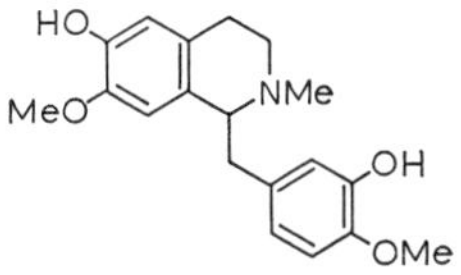

Protosinomenine

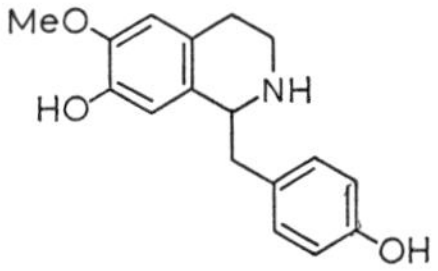

Coclaurine

The 1-benzylisoquinoline alkaloids occupy a position of crucial importance in the biosynthesis of alkaloids of the morphine, protoberberine, protopine, benzo[*c*]phenanthridine, bisbenzylisoquinoline, cularine, erythrina, rhoediine, phthalide and spirobenzylisoquinoline groups. In this context **reticuline, orientaline, protosinomenine** and **coclaurine** are particularly significant.

(*i*) *Structure determination and synthesis*

Papaverine (XXIII) is one of the principal alkaloids of opium, and is probably the first alkaloid to have its structure determined (*G. Goldschmiedt*, Monatsh., 1885, **6**, 954; 1886, **7**, 485). A fascinating account of this early work has been presented by *D. Ginsberg*. "The Opium Alkaloids", Interscience, New York, 1962). A brief summary of the evidence appears in Scheme 5. This structure for papaverine has been confirmed by numerous syntheses (*A. V. Lukyanov, V. S. Onoprienko* and *V. A. Zasosov*, Khim. farm. Zhur., 1972, **6**, 14; C.A., 1972, **77**, 101940b). The approach utilising the Bischler–Napieralski reaction is summarised in Scheme 6, a method still in use, *e.g.* for takatonine (*S. Kubota et al.*, J. org. Chem., 1966, **31**, 516) (Scheme 7).

A very useful modification involves the conversion of the nitrostyrene derivative into the β-aryl-β-methoxynitroethane derivative, followed by the reduction, amide formation and Bischler–Napieralski ring-closure. Under the conditions of the cyclisation, thermal loss of methanol occurs to yield

SCHEME 5

THE STRUCTURE OF PAPAVERINE

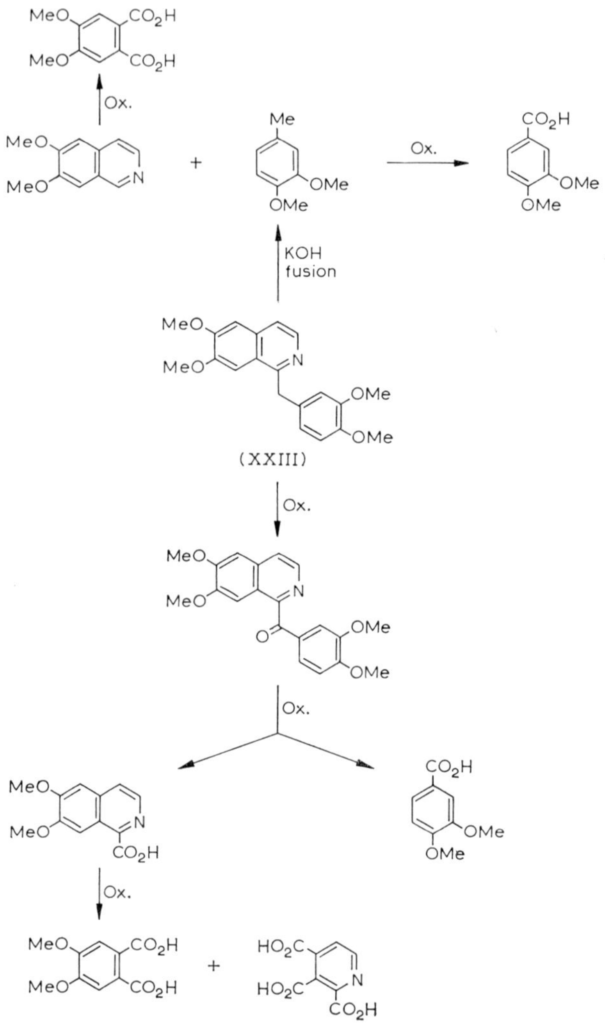

SCHEME 6

A SYNTHESIS OF PAPAVERINE

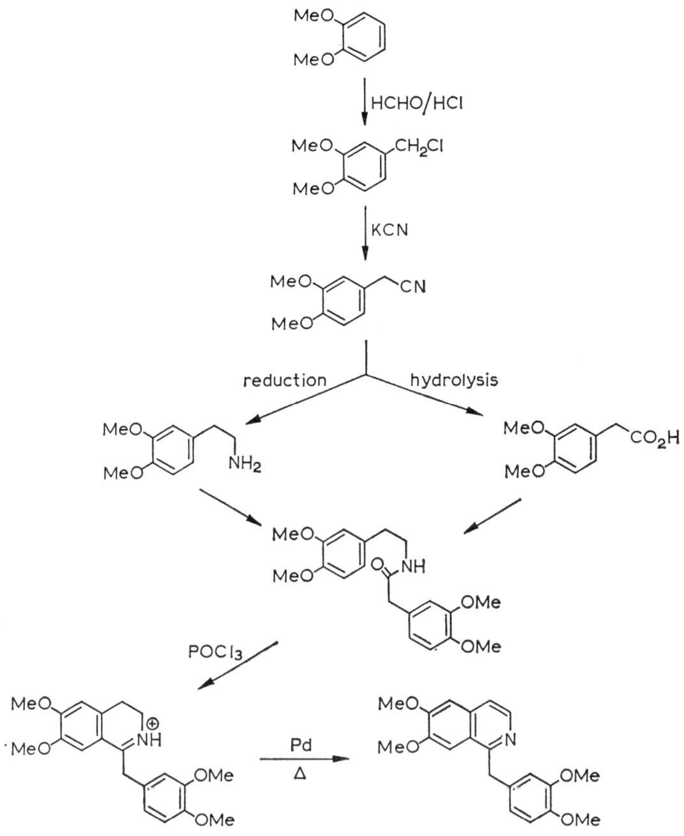

Papaverine
(XXIII)

SCHEME 7
THE SYNTHESIS OF TAKATONINE

Takatonine

SCHEME 8
A MODIFIED SYNTHESIS OF PAPAVERINE

Papaverine
(XXIII)

the fully aromatic isoquinoline directly (Scheme 8) (*Brossi* and *Teitel*, Helv., 1966, **49**, 1757). A recent synthesis of papaverine involves the condensation between 1-chloro-6,7-dimethoxyisoquinoline and the Wittig reagent XXIV (*E. C. Taylor* and *S. F. Martin*, J. Amer. chem. Soc., 1972, **94**, 2874):

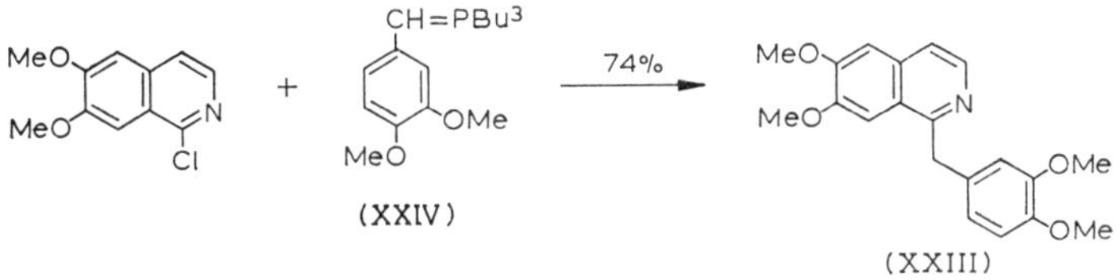

Reissert compounds have provided starting materials for an alternative synthesis of 1-benzylisoquinolines (Scheme 9) (*F. D. Popp*, Heterocycles, 1973, **1**, 165).

The Pomeranz–Fritsch reaction has not proved to be very satisfactory for the synthesis of 1-benzylisoquinoline derivatives. In an attempted synthesis of papaverine by this method, cyclisation of the Schiff's base XXV with sulphuric acid gave very little papaverine (*P. Fritsch*, Ann., 1903, **329**, 37); the major product proved to be XXVI (*S. F. Dyke et al.*, Tetrahedron, 1969, **25**, 1881). A by-product of a different type XXIX was produced, together with the required isoquinoline XXVIII when the acetal XXVII was treated with polyphosphoric acid (*T. Kametani et al.*, J. chem. Soc., C, 1966, 715).

In the Bobbitt modification of the Pomeranz–Fritsch synthesis (*J. M. Bobbitt*, Advances in heterocycl. Chem., 1973, **15**, 99; *Dyke*, C.C.C. 2nd

SCHEME 9

USE OF THE REISSERT REACTION

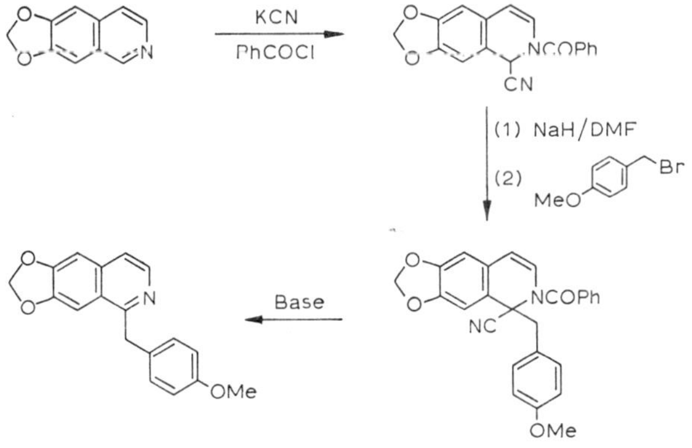

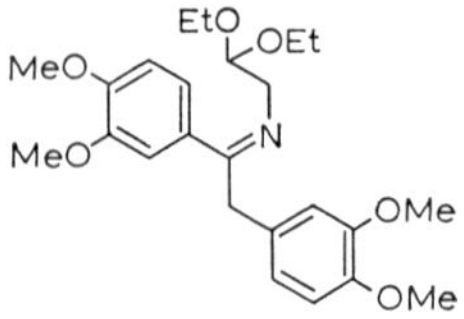

(XXV)

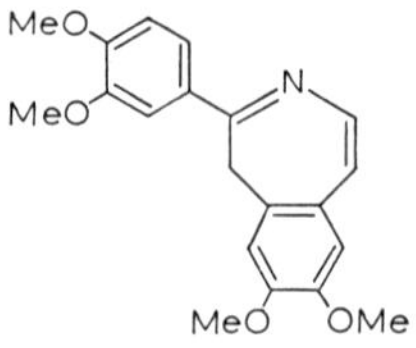

(XXVI)

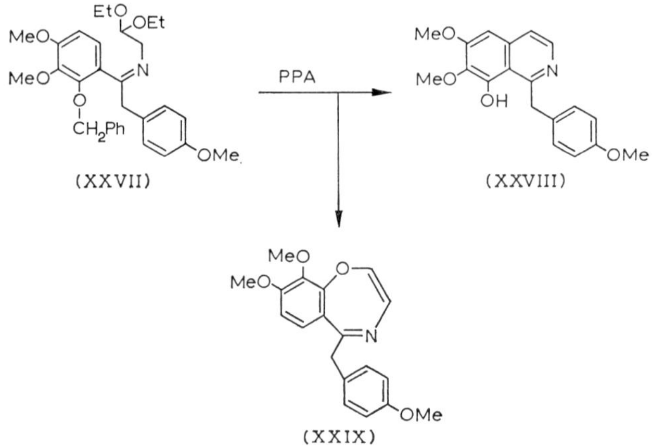

(XXVII)

PPA

(XXVIII)

(XXIX)

SCHEME 10

THE MODIFIED POMERANZ–FRITSCH SYNTHESIS

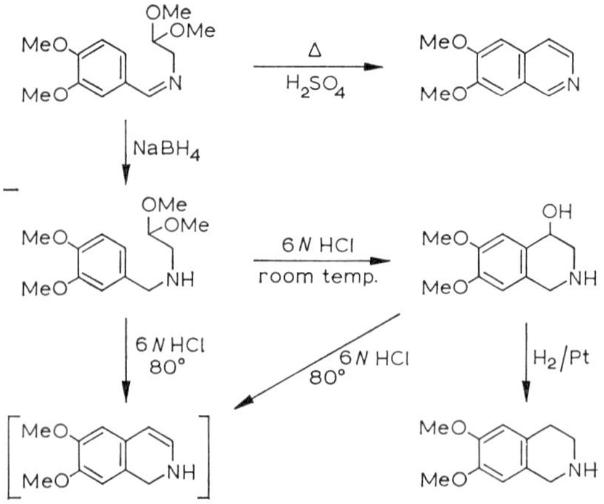

edn., Vol. IV F, p. 380) the Schiff's base is reduced to the benzylaminoacetal which is then cyclised with $6N$ hydrochloric acid. At room temperature the 4-hydroxy-1,2,3,4-tetrahydroisoquinoline is probably formed which can undergo hydrogenolysis to remove the benzylic hydroxyl group. However, if the cyclisation is carried out at 70–80°, or if the 4-hydroxytetrahydro-isoquinoline is heated, a 1,2-dihydroisoquinoline is formed which may then undergo a variety of reactions (*Dyke*, Advances in heterocycl. Chem., 1972, **14**, 279) (Scheme 10). However, this sequence is not very successful for the preparation of 1-benzylisoquinoline derivatives because when benzylamino-acetals such as XXX are treated with mineral acid, the initially formed 1-benzyl-4-hydroxytetrahydroisoquinoline XXXI rapidly cyclises to yield the isopavinane XXXII (*Dyke* and *A. C. Ellis*, Tetrahedron, 1971, **27**, 3803):

(XXX)

6 N HCl

(XXXI)

H⁺

(XXXII)

The fully aromatic 1-benzylisoquinoline skeleton is quite rare in nature. Structural elucidation is achieved nowadays largely by spectroscopic methods, coupled with comparisons with known compounds. In some cases (*e.g.* takatonine, *Kubota et al., loc. cit.*) reduction to the 1,2,3,4-tetrahydro derivative, followed by Hofmann degradation, may be required.

A new alkaloid was isolated from opium, and named **palaudine**, $C_{19}H_{19}NO_4$ (*E. Brockmann-Hanssen* and *K. Hirai*, J. pharm. Sci., 1968, **57**, 940). The parent ion at m/e 325 in the mass spectrum is weaker than the M-1 peak, and the M-15 peak is also strong, whereas the M-31 peak is weak. The u.v. spectrum, with λ_{max}^{MeOH} at 239, 280, 314, 326 nm is very similar to that of papaverine. The i.r. spectrum, with absorption at 3430 cm^{-1} was interpreted in terms of an –OH group present in the molecule. The n.m.r. spectrum (CDCl$_3$ solution) revealed the presence of three methoxyl groups (at 3.99, 3.88 and 3.81 δ), which correspond very closely to the resonances of three of the four methoxyl groups of papaverine. A two proton singlet at 4.49 δ

also corresponded to that due to the methylene group at $C_{(1)}$ in papaverine. Hence, palaudine is probably papaverine with one methoxyl group replaced by hydroxyl. A positive Gibbs test suggested that the position *para* to the phenolic hydroxyl is unsubstituted. The structure deduced from this evidence was confirmed by synthesis involving the Bischler–Napieralski cyclisation of the amide XXXIII:

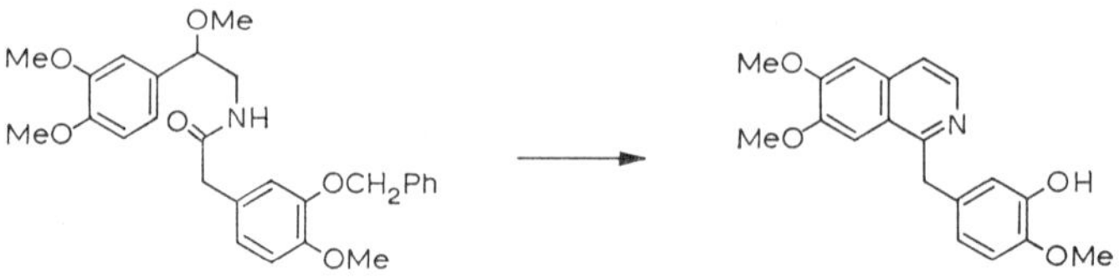

(XXXIII) Palaudine

In early work the structures of the 1-benzyl-1,2,3,4-tetrahydroisoquinoline alkaloids were elucidated by the application of the Hofmann degradation, followed by oxidation to a mixture of carboxylic acids. Laudanosine is an optically active base, $C_{21}H_{27}NO_4$, which also was first isolated from opium (*O. Hesse*, Ber., 1871, **4,** 693; Ann., 1875, **176,** 189). The structure was deduced by the application of the Hofmann exhaustive methylation procedure. These reactions have been re-investigated (*A. R. Battersby* and *B. J. T. Harper*, J. chem. Soc., 1962, 3526) in connection with biosynthetic studies (Scheme 11). It is unusual for a mixture of the *cis-* and *trans-*stilbenes to be reported. In principle an alternative methine XXXIV, can also be produced in the first stage of the Hofmann degradation but only in rare cases has it been reported.

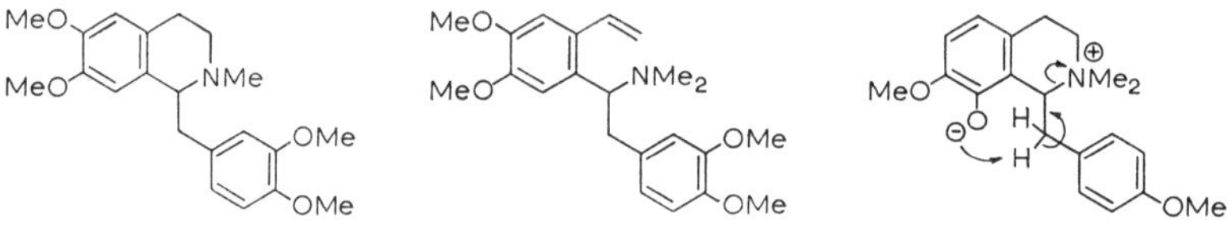

Laudanosine (XXXIV) Petaline (XXXV)

Petaline (XXXV) undergoes Hofmann degradation very easily (passage of the quaternary salt through a hydroxide ion-exchange column) and this has been explained (*N. J. McCorkindale et al.*, Tetrahedron Letters, 1964, 3841) in terms of the phenolate anion at $C_{(8)}$ being close enough to the benzylic methylene group protons to initiate elimination as shown in XXXV.

Most of the 1-benzyltetrahydroisoquinoline alkaloids have been synthesised by use of the Bischler–Napieralski reaction. The 3,4-dihydroisoquinoline derivative produced can either be reduced to the secondary base,

SCHEME 11

DEGRADATION OF LAUDANOSINE

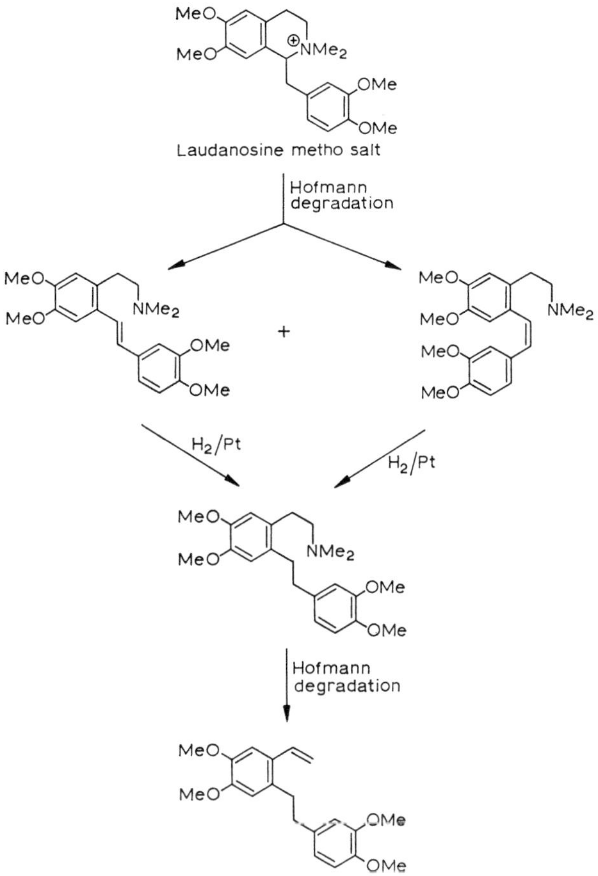

or *N*-methylated to the quaternary salt, then reduced to the *N*-methyltetra-hydroisoquinoline (Scheme 12). The phenolic hydroxyl group may be protected as the benzyl ether, so that the final stage in the synthesis, after reduction, is hydrolysis (*M. Tomita* and *H. Yamaguchi*, Pharm. Bull., Tokyo, 1953, **1**, 10). If a benzoyl or *p*-toluenesulphonyl group is used in the synthesis of armepavine, then reduction of the 3,4-dihydroisoquinolinium salt with sodium tetrahydridoborate causes hydrolysis of the phenolic ester as well to give (±)-**armepavine** (*D. B. Gurfinkel*, Ph.D. Thesis, University of Buenos Aires, 1965, quoted by *Deulofeu, Comin* and *Vernengo, loc. cit.*).

SCHEME 12

THE SYNTHESIS OF 1-BENZYLTETRAHYDROISOQUINOLINES

An improved synthesis of polyhydroxybenzylisoquinolines, utilising the Bischler–Napieralski reaction, but without protection of phenolic hydroxyl groups has been developed, involving cyclisation with phosphoryl chloride in acetonitrile solution (*N. Whittaker*, J. chem. Soc., C, 1969, 85, 94). A biogenetic-type of synthesis of $S(+)$**laudanosine,** involving asymmetric induction, is summarised in Scheme 13 (*S. Yamada, M. Konda* and *T. Shioiri*, Tetrahedron Letters, 1972, 2215).

In more recent structural studies, use has been made of mass and n.m.r. spectral data. The main fragmentation of the molecular ion is at the $C_{(1)}$–CH$_2$–bond, *e.g.* XXXVI → XXXVII. N.m.r. spectral studies are useful for the detection of *N*-Me, OMe and CH$_2$O$_2$ groups. In some cases it is also possible to deduce the site of substitution by an examination of the aromatic proton absorption region, or by noting the chemical shift of the $C_{(8)}$–H and the $C_{(7)}$–OMe group. If the nitrogen carries a methyl group, then the preferred conformation of the molecule XXXVIII is such that these two absorptions are shifted up-field by the shielding effect of the aromatic ring of the benzyl group (*Tomita et al.*, Chem. and pharm. Bull., Japan, 1965, **13**, 921; J. pharm. Soc., Japan, 1966, **86**, 373; *D. R. Dalton, M. P. Cava* and *K. T. Buck*, Tetrahedron Letters, 1965, 2687). The effect is not observed in secondary amines of the 1-benzyltetrahydroisoquinoline group, and the $C_{(8)}$–H is not displaced upfield if the $C_{(7)}$–OMe group is replaced by OH (*V. St. Georgiev* and *N. M. Mollov*, Compt. rend. Acad. bulg. Sci., 1971, **24**, 1329).

An alternative method of synthesis of 1-benzyltetrahydroisoquinolines is illustrated by the first synthesis of petaline (*G. Grethe, M. Uskokovic*

SCHEME 13

A BIOGENETIC-TYPE SYNTHESIS OF (*S*)(+)LAUDANOSINE

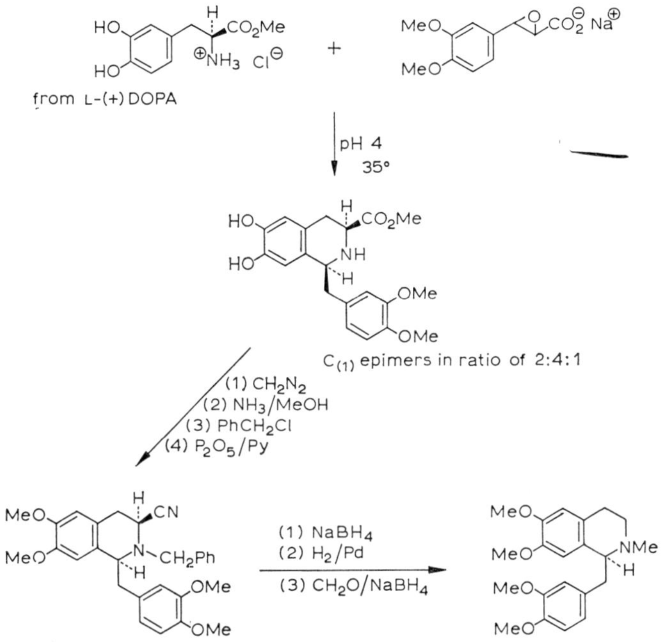

and *A. Brossi*, J. org. Chem., 1968, **33**, 2500). Here a 3,4-dihydroisoquinolinium salt was subjected to reaction with *p*-methoxybenzylmagnesium chloride (XXXIX → XL). A more interesting synthesis of the same alkaloid involves a Stevens rearrangement (XLI → XLII → petaline) (*Brossi et al.*, Helv., 1970, **53**, 874).

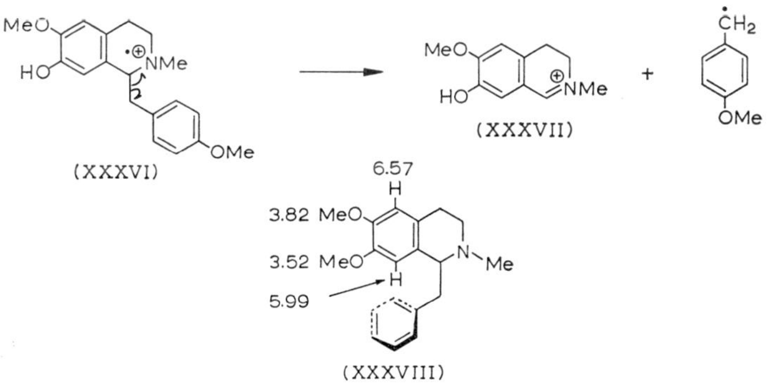

[4]MeOC6H4·CH2MgCl
MeO
PhCH2O
NMe
MeO
(XL)
MeO
CH2Ph
(XXXIX)
PhLi
MeO
OMe
NMe
OMe
(XLII)
MeO
OMe
NMe
OMe
(XLI)
(1) HBr
(2) MeI
MeO
OH
NMe
OMe
Petaline

2 x
MeO
MeO
CO2H
PPA
MeO
MeO
CO2H
O
OMe
OMe
NH4OAc
MeO
MeO
O
N
OMe
OMe
Decalin
190°
MeO
MeO
O
O
OMe
OMe
(1) MeNH2
(2) H2/Pt
(3) B2H6
(1) H2/Pt
(2) B2H6/THF
MeO
MeO
NMe
OMe
OMe
Laudanosine
MeO
MeO
NH
OMe
OMe
Tetrahydropapaverine

SCHEME 14

A simple synthesis of ($\pm$)laudanosine is summarised in Scheme 14 (*G. N. Dorofeenko* and *V. G. Korobkova*, Zhur. obshcheĭ Khim., 1970, **40**, 249; *3. W. Elliot*, J. heterocycl. Chem., 1970, **7**, 1229; 1972, **9**, 853).

Yet another method utilised the Reissert reaction (*B. C. Uff, J. R. Kershaw* and *S. R. Chhabra*, J. chem. Soc., Perkin I, 1972, 479).

An *N*-benzyl-1,2,3,4-tetrahydroisoquinoline alkaloid, **sendeverine**, is known (*T. Kametani* and *K. Ohkubo*, Tetrahedron Letters, 1965, 4317); its structure was elucidated by spectral, oxidative and synthetic evidence:

Sendeverine

(*ii*) *Absolute configuration*

The absolute configuration of ($-$)**norlaudanosine** has been established by ozonolysis to the known amino acid XLIII (*H. Corrodi* and *E. Hardegger*, Helv., 1956, **39**, 889). The absolute configuration of other members of the

Norlaudanosine   (XLIII)

group depends upon comparison or interrelationship with laudanosine. It has been established that o.r.d. curves may be used to determine absolute configuration; the shape of the curve at 200–225 nm and, more usefully, the sign of the Cotton effect at 270–290 and 240–255 nm, is studied. Tetrahydrobenzylisoquinolines of the D-series (1*R*) show negative Cotton effects at these wavelengths, while members of the L-series (1*S*) exhibit positive Cotton effects. Quaternisation of the nitrogen atom has little effect on the o.r.d. curve (*G. Grethe et al., ibid.*, 1970, **53**, 874; *M. Shamma et al.*, Tetra-

($\pm$)Coclaurine

hedron, 1966, **22**, 1467; *Battersby et al.*, J. chem. Soc., 1965, 2239; *J. C. Craig et al.*, Tetrahedron, 1966, **22**, 1335). It must be emphasised that the specific rotation measured at the sodium D line is not a reliable criterion for the allocation of absolute configuration. Deductions made from rotation measurements have been confirmed for (+)**coclaurine** by a crystal structure analysis (δ. *Fridrichsons* and *A. McL. Mathieson, ibid.*, 1968, **24**, 5785).

(iii) Biosynthesis

As long ago as 1910, *E. Winterstein* and *G. Trier* suggested that the 1-benzylisoquinoline alkaloids are formed in nature as indicated in Scheme 15 ("Die Alkaloide", Bornträger, Berlin, 1910, p. 307).

SCHEME 15

WINTERSTEIN AND TRIER'S SCHEME FOR BIOSYNTHESIS

Tyrosine

DOPA

$-CO_2$

$-NH_2$
$-CO_2$

Dopamine

(XLIV)

Papaverine

[44]

[48]

Norlaudanosoline

[Figures are % relative specific activities; intact alkaloid = 100%]

The essential correctness of this scheme was verified by feeding [2-^{14}C]-DL-tyrosine to opium poppies (*Battersby* and *Harper, loc. cit.*; *Battersby, R. Binks* and *Harper*, J. chem. Soc., 1962, 3534; *Battersby et al., ibid.*, 1964, 3600). If the scheme is correct, the papaverine isolated from the opium should be labelled only at $C_{(1)}$ and $C_{(3)}$. By reduction and Hofmann degradation, this was shown to be so, and that the level of activity is about equal. *N*-Norlaudanosoline has been shown to be an effective precursor of papaverine in *Papaver somniferum* (*Battersby et al.*, Tetrahedron Letters, 1965, 1275).

It has been established by careful degradation on ^{14}C-labelled alkaloids of the 1-benzylisoquinoline group in general, that the two aromatic units that originate from tyrosine, are not identical (*H. Rapoport, N. Levy* and *F. R. Stermitz*, J. Amer. chem. Soc., 1961, **83**, 4298; *J. R. Gear* and *I. D. Spenser*, Nature, 1961, **191**, 1393). It has been suggested that DOPA is converted into dopamine and into 3,4-dihydroxyphenylpyruvic acid (XLV) instead of the phenylacetaldehyde XLIV (Scheme 15) (*Battersby*, Proc. chem. Soc., 1963, 189). The product of the Pictet–Spengler type of condensation is then XLVI which undergoes decarboxylation to norlaudanosoline. This proposal has the great merit that both dopamine and XLV are the normal products of amino-acid metabolism (of DOPA). Some recent results using *P. orientale* (*M. L. Wilson* and *C. J. Coscia*, J. Amer. chem. Soc., 1975, **97**, 431) and with *P. somniferum* (*Battersby, R. C. F. Jones* and *R. Kazlauskas*, Tetrahedron Letters, 1975, 1873) have shown conclusively that XLVI is formed from DOPA and that XLVII is probably an intermediate between XLVI and norlaudanosoline.

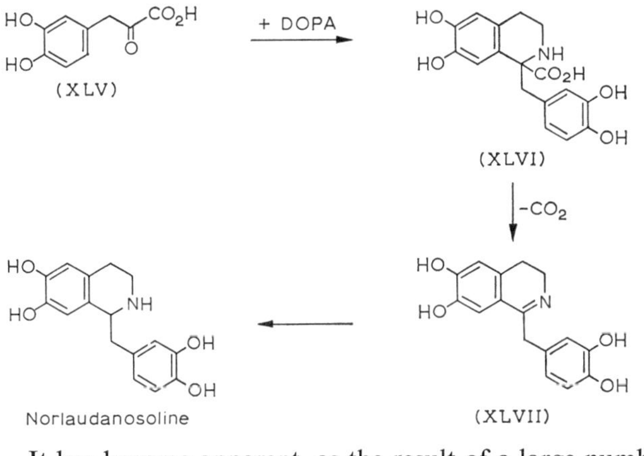

It has become apparent, as the result of a large number of studies on the biosynthesis of the more highly condensed alkaloids of isoquinoline (proto-berberins, protopines, aporphines, etc.) that norlaudanosoline may be regarded as the common precursor of all of them. The actual pathway taken will depend upon the actual enzyme system present in various plant species, and also upon the extent and position of *O*- and *N*-methylation.

As far as the simple 1-benzylisoquinolines are concerned, most work has involved the alkaloids of *P. somniferum*, and a much clearer picture is emerging of the sequence of events (*E. Brockmann-Hanssen et al.*, J. pharm.

SCHEME 16

(-)Norreticuline

(-)Nororientaline

(+)Orientaline

(+)Reticuline

(+)Laudanidine

(-) Reticuline

(-) Laudanidine

Sci., 1971, **60**, 1673; J. chem. Soc., Perkin I, 1975, 1531). The initially formed (−)norlaudanosoline undergoes *O*-methylation by catechol *O*-methyltransferase to (−)**norreticuline** and (−)**nororientaline,** each of which may be *N*-methylated to (+)reticuline and (+)orientaline, respectively. The former can undergo isomerisation, *via* 1,2-dehydroreticuline, to (+) reticuline—the precursor of morphine (Scheme 16). It seems that *O*-methylation of the $C_{(7)}$–OH group in reticuline can occur in either stereo-isomer, giving rise to the laudanidines. However, only the $C_{(3)}'$–OH group of (−)norreticuline can be *O*-methylated (to give (−)norcodamine) and not in the enantiomorph. (−)**Norcodamine** can be *N*-methylated to (+) codamine, which may also arise from (+)orientaline or (+)reticuline (Scheme 17).

SCHEME 17

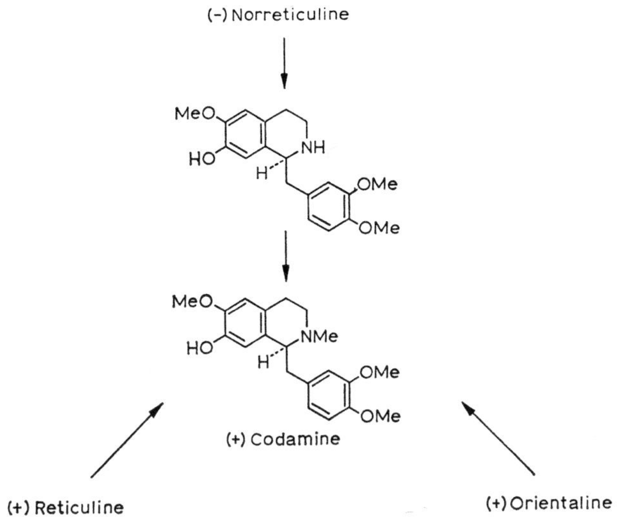
(-) Norreticuline
MeO
HO
NH
H
OMe
OMe
MeO
HO
NMe
H
OMe
OMe
(+) Codamine
(+) Reticuline
(+) Orientaline

SCHEME 18

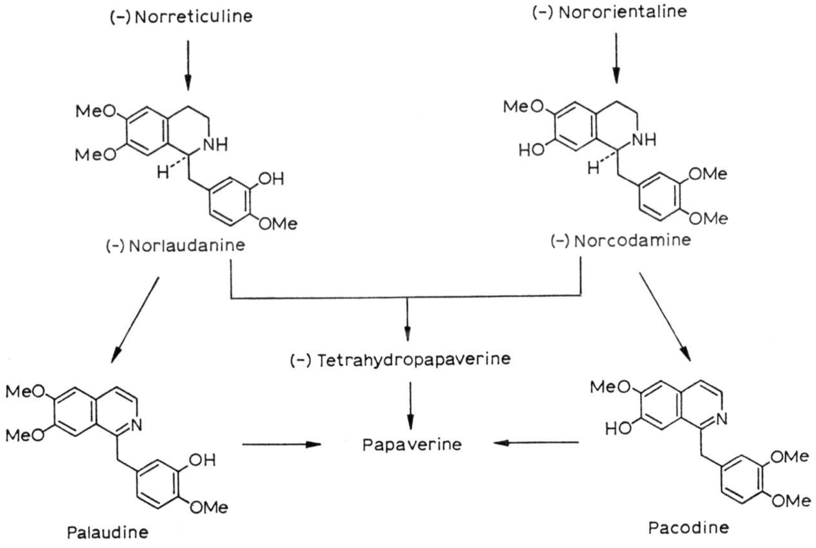
(-) Norreticuline
(-) Nororientaline
MeO
MeO
NH
H
OH
OMe
(-) Norlaudanine
MeO
HO
NH
H
OMe
OMe
(-) Norcodamine
(-) Tetrahydropapaverine
MeO
MeO
N
OH
OMe
Palaudine
Papaverine
MeO
HO
N
OMe
OMe
Pacodine

As far as papaverine biosynthesis is concerned, it has been established (*Brockmann-Hanssen et al., loc. cit.*) that dehydrogenation of the nitrogen-containing ring does not occur at the norreticuline stage. In view of its excellent incorporation (17.6%) into papaverine, (−)tetrahydropapaverine must be the principal immediate precursor. The major pathways to papaverine are then from (−)norreticuline to (−)norlaudanine to (−)tetrahydropapaverine or alternatively (−)nororientaline to (−)norcodamine to (−)tetrahydropapaverine. A minor pathway involves **palaudine** and **pacodine** (Scheme 18). The di-*O*-methyltetrahydroisoquinolines (−)**norprotosinomenine** and XLVIII are not incorporated into tetrahydropapaverine, so it seems that the opium poppy is not capable of methylating at $C_{(6)}$–OMe group in the presence of a $C_{(7)}$–OMe group. So the $C_{(6)}$–OH of norlaudanosoline must be methylated *before* the $C_{(7)}$–OH group:

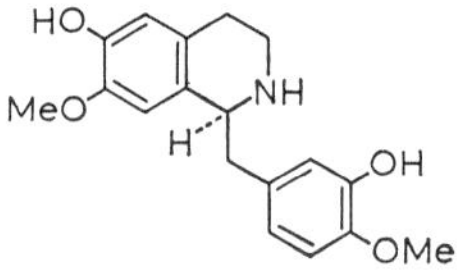

(−)Norprotosinomenine

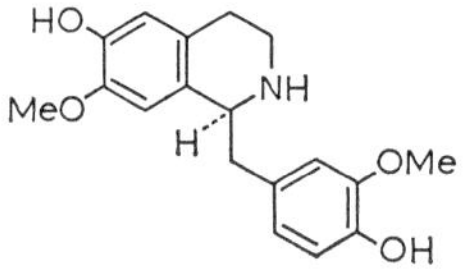

(XLVIII)

(d) The cularines

All four known cularine-type alkaloids have been isolated from the Dicentra and Corydalis genera.

(i) Structure determination

Although **cularine**, $C_{20}H_{23}NO_4$, the first member of the group, was first isolated in 1938 (*R. H. F. Manske*, Canad. J. Res. Sec. B, 1938, **16**, 81) it was 1950 before its structure was established (*idem.* J. Amer. chem. Soc., 1950, **72**, 55). Classical analytical and degradative methods revealed one *N*-methyl and three *O*-methyl groups. The fourth, inert oxygen atom was assigned to a diphenyl ether linkage. Hofmann degradation and reductive cleavage with sodium in liquid ammonia followed by degradation (Scheme 19) enabled a structural proposal XLIX to be made.

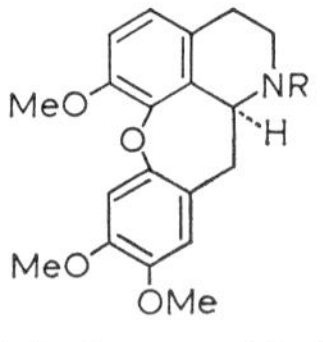

(+)Cularine   (R = Me)
(+)Cularimine   (R = H)

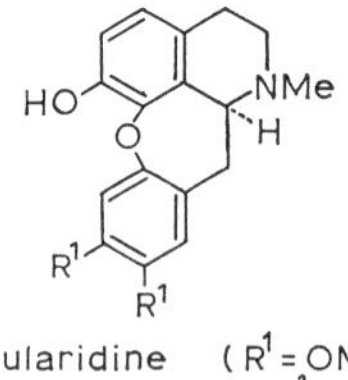

(+)Cularidine   (R^1 = OMe)
(+)Cularicine   (2 R^1 = OCH$_2$O−)

SCHEME 19

STRUCTURE DETERMINATION OF CULARINE

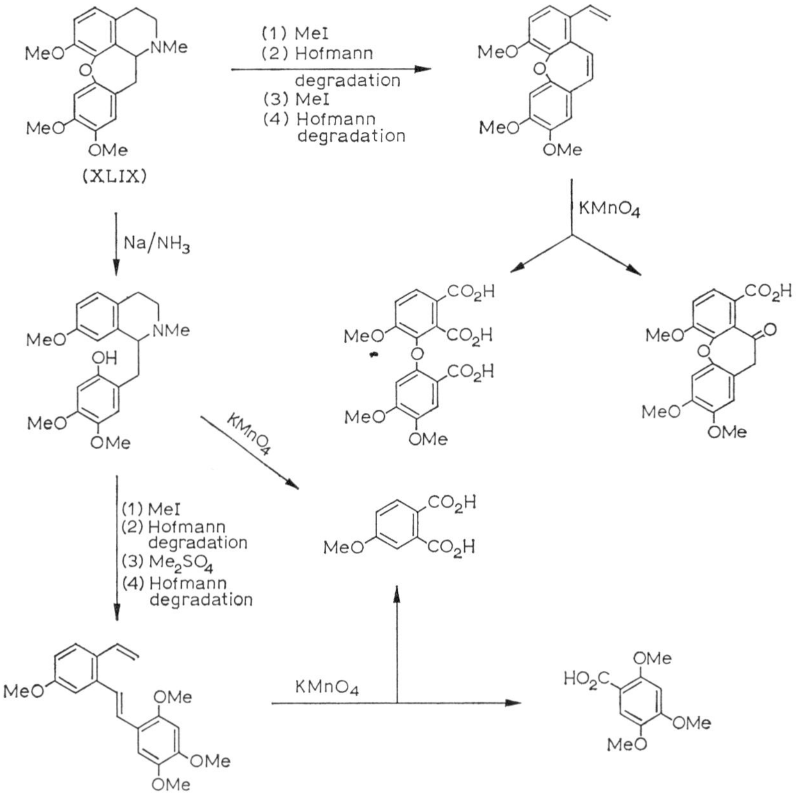

From a detailed study of the proton-n.m.r. spectrum, the geometry shown in L was deduced, especially from the values of J_{AB} (16 Hz), J_{AX} (4 Hz) and J_{BX} (12 Hz) (*N. S. Bhacca et al.*, Tetrahedron, 1966, **22,** 1467). The mass spectrum of cularine shows a strong $M^{\oplus}$ peak, but the base peak at M-15 has been attributed to LI (*M. Ohashi et al.*, J. Amer. chem. Soc., 1963, **85,** 2807).

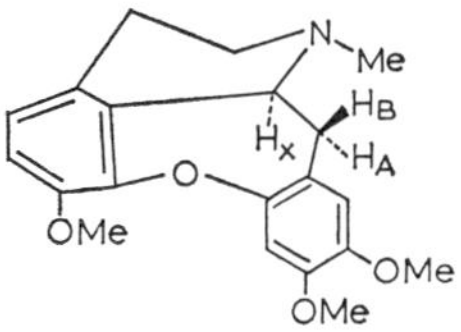

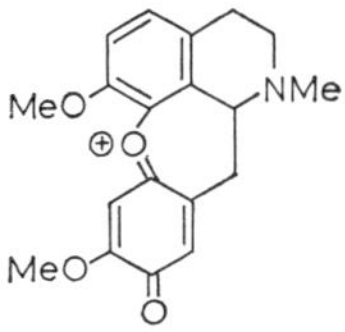

(ii) *Absolute configuration*

The absolute configuration of (+)cularine was deduced by interrelating the alkaloid with L(*S*)-**romneine** of known absolute configuration (*J. Kunitomo et al.*, Tetrahedron, 1971, **19**, 2197). Cleavage of cularine with sodium in liquid ammonia gave a phenolic 1-(hydroxydimethoxy-1-benzyl)-tetrahydroisoquinoline which was shown to be identical with the product

SCHEME 20

THE ABSOLUTE CONFIGURATION OF CULARINE

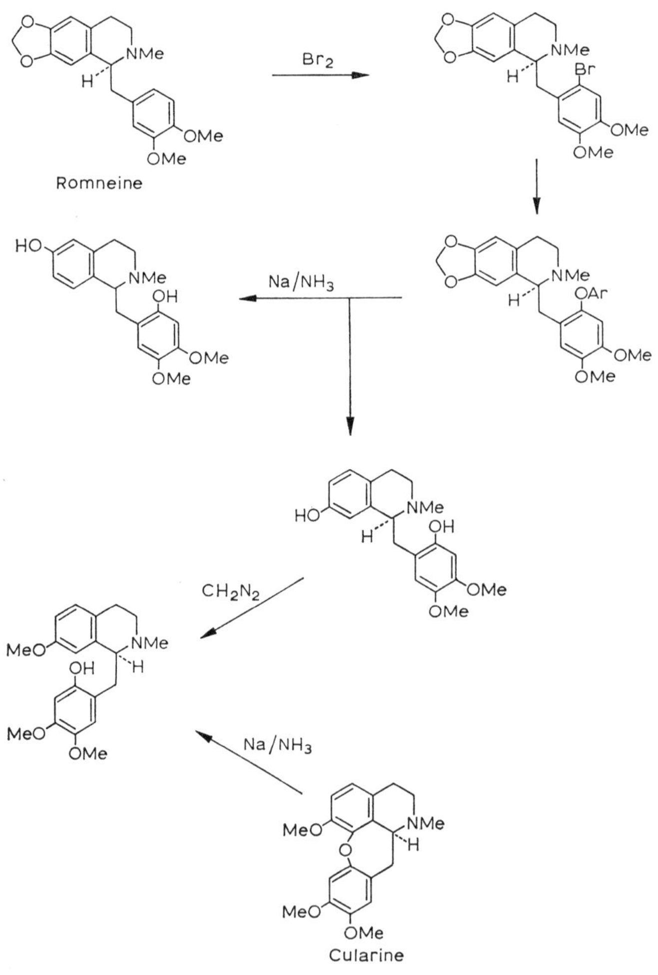

formed from romneine, as shown in Scheme 20. This assignment was confirmed by an X-ray analysis of the methiodide (*Kametani et al.*, Chem. Comm., 1972, 1072).

(*iii*) *Biosynthesis*

The biosynthesis of cularine has not yet been elucidated, but several proposals have been considered, based upon phenolic coupling involving the formation of carbon–oxygen bond present in the diaryl ether linkage.

(*iv*) *Synthesis*

Several syntheses of cularine, or close analogues, have been reported, some inspired by biosynthetic considerations. The first synthesis of cularine, however, involved the preparation of the ketone LII *via* an Ullman reaction, followed by a Pomeranz–Fritsch synthesis of the isoquinoline ring (*Kametani and K. Fukumoto*, J. chem. Soc., 1963, 4289) (Scheme 21).

SCHEME 21

THE SYNTHESIS OF CULARINE

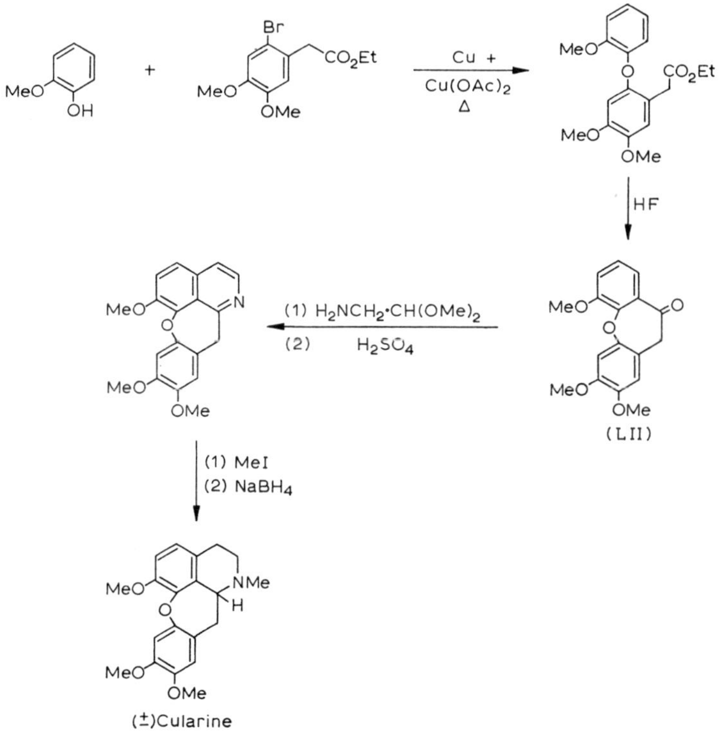

SCHEME 22

IIDA'S SYNTHESIS OF A CULARINE ANALOGUE

(LIII)

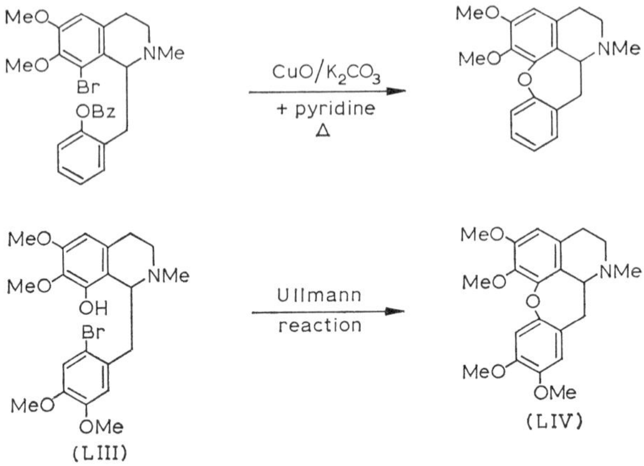

In an alternative approach the Ullmann reaction was deferred until a late stage in the synthesis (Scheme 22) (*H. Iida et al.*, J. pharm. Soc., Japan, 1969, **89**, 645). A slight modification of this method involved the Ullmann reaction with LIII to yield the cularine analogue LIV (*idem, ibid.*, 1969, **89**, 1169). This strategy has been applied to the synthesis of (±)cularine, (±)**cularimine** and to (±)**cularicine** (*idem*, Yakugaku Zasshi, 1972, **92**, 1030, 1242; C.A., 1972, **77**, 140365n; 1973, **78**, 16343; *S. Ishiwata et al.*, Chem. pharm. Bull., Japan, 1970, **18**, 1850).

SCHEME 23

PHENOLIC COUPLING IN THE SYNTHESIS OF A CULARINE ANALOGUE

(LV)   (LVI)

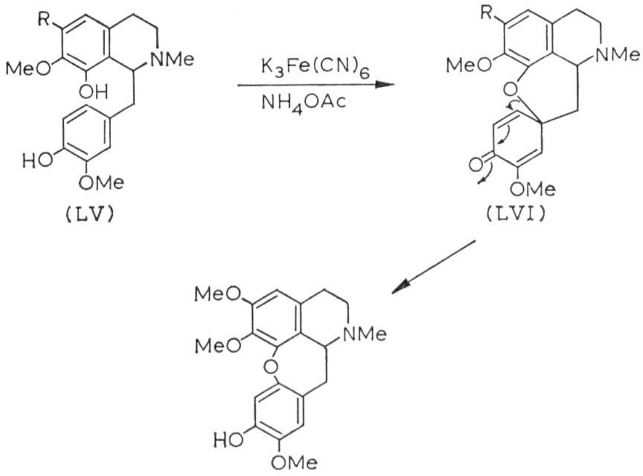

Inspired by biogenetic considerations *Kametani, T. Kikuchi* and *Fukumoto* (*ibid.*, 1968, **16,** 1003; see Scheme 23) reported the synthesis of a cularine analogue by phenolic coupling. The dienone LVI (R = OMe), rearranged by the required phenolate migration, rather than by a benzyl migration.

When the sequence was repeated with LV (R = H), the dienone LVI (R = H) failed to rearrange to the cularine skeleton (*Kametani, Fukumoto* and *M. Fujihara*, Chem. Comm., 1971, 352). However, cularine has been synthesised by phenolic coupling from the alternative diphenol LVII (*A. H. Jackson* and *G. W. Stewart, ibid.*, 1971, 149; *A. J. Birch, Jackson* and *P. V. R. Shannon*, J. chem. Soc., Perkin I, 1974, 2185; *Kametani, Fukumoto* and *Fujihara*, Bio-org. Chem., 1971, **1,** 40). The latter group reported both the required coupled product LVIII and the isomeric LIX.

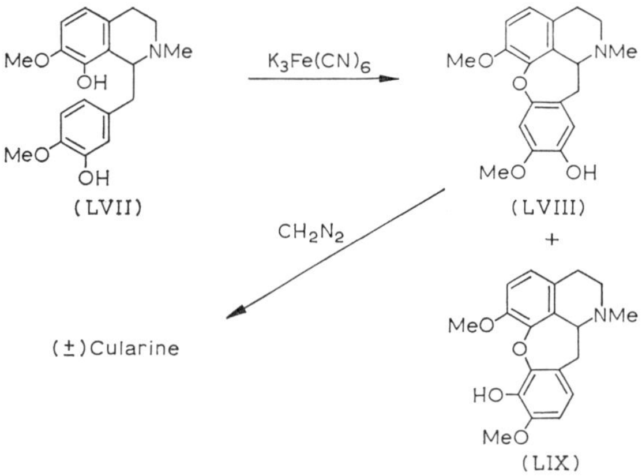

A synthesis of cularicine has been achieved from the ketone LX (*I. Noguchi* and *D. B. MacLean*, Canad. J. Chem., 1975, **53,** 125) and **cularidine** has also been synthesised conventionally (*Iida, H.-C. Hsu* and *Kikuchi*, Chem. and pharm. Bull., Japan, 1973, **21,** 1001).

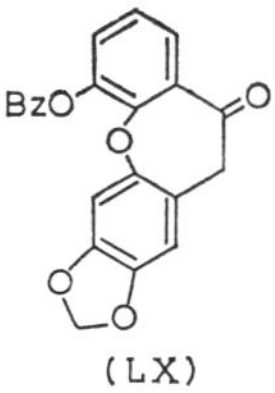

SCHEME 24

DEGRADATION OF CRYPTAUSTOLINE

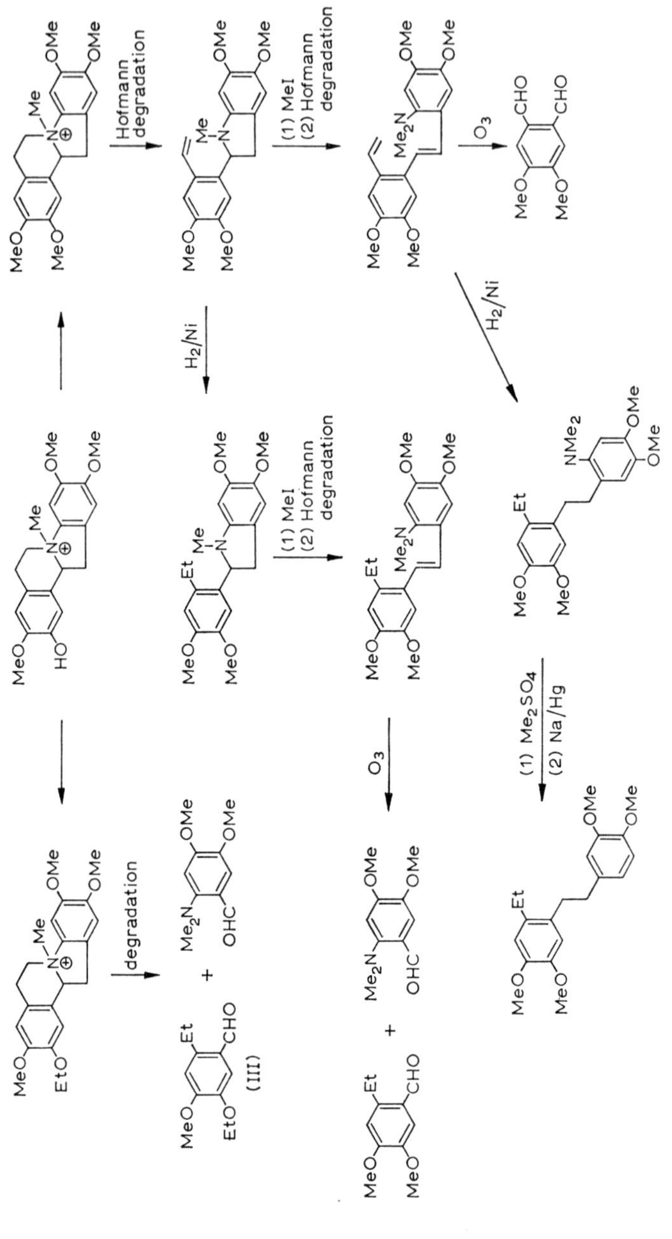

2. The dibenzopyrrocoline alkaloids

Only two alkaloids are known based upon the dibenzopyrrocoline ring system, (−)**cryptaustoline** (I) and (−)**cryptowoline** (II). Nevertheless they have some historical interest in considerations of the biosynthesis of alkaloids of the aporphine and morphine groups. Both cryptaustoline and cryptowoline were isolated from the Queensland shrub *Cryptocaria bowiei* (Hook) Druce (*E. Ritchie et al.*, Austral. J. Chem., 1953, **6**, 78).

(−)Cryptaustoline
(I)

(−)Cryptowoline
(II)

(*i*) *Structure determination*

By the usual methods the molecules of cryptaustoline were found to possess one phenolic hydroxyl, one *N*-methyl group and three *O*-methyl groups. Hofmann degradation of the *O*-methyl ether gave firstly a methine, then a bismethine that still contained basic nitrogen, thus indicating that the alkaloid contains a bridgehead nitrogen atom. The further degradations summarised in Scheme 24 were adequate to establish the structure of the *O*-methyl ether. The position of the phenolic hydroxyl group of cryptaustoline was established when the *O*-ethyl ether was degraded to the known aldehyde III.

(III) + (IV)

Degradation of cryptowoline by the methods indicated in Scheme 24 gave 5-ethoxy-2-ethyl-4-methoxybenzaldehyde (III) and 6-dimethylamino-piperonal (IV), thus establishing the structure.

The n.m.r. spectrum of de-*N*-methylcryptaustoline *O*-methyl ether has been reported. In the mass spectrum of cryptaustoline itself, loss of the *N*-methyl group occurs to give the ion V at *m/e* 327, which then undergoes further fragmentation *via* VI to the base peak VII at *m/e* 325 (*Kametani* and *K. Ogasawara*, Chem. and pharm. Bull., Japan, 1968, **16**, 1498).

(*ii*) *Synthesis*

The ring-system of cryptaustoline and cryptowoline was originally obtained some 30 years before the characterisation of the alkaloids in the course of some studies inspired by biogenetic considerations in the morphine and aporphine alkaloids (*R. Robinson* and *S. Sugasawa*, J. chem. Soc., 1932, 789; *C. Schöpf* and *K. Thierfelder*, Ann., 1932, **497**, 22). Oxidation of laudanosoline with chloranil, ferric chloride or potassium ferricyanide gave the dibenzopyrrocoline VIII instead of the expected aporphine. This quaternary salt was characterised as the tetra-*O*-methyl ether (*O*-methyl-cryptaustoline) (IX).

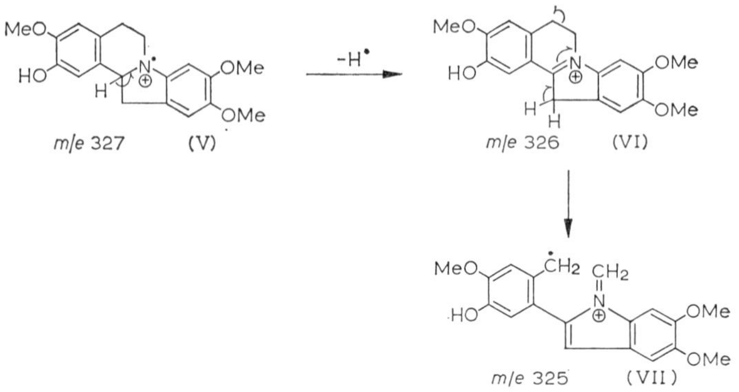

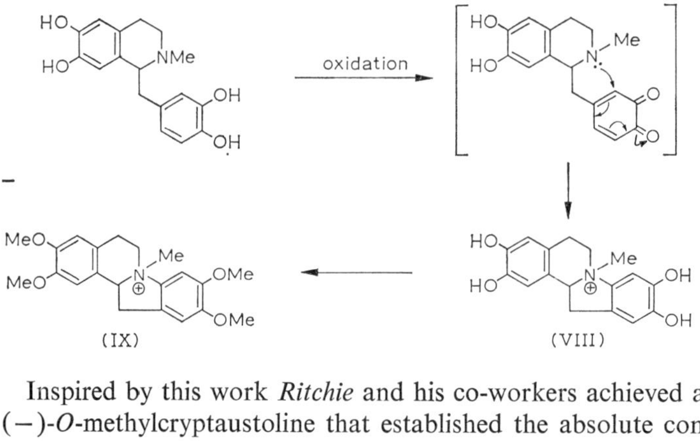

Inspired by this work *Ritchie* and his co-workers achieved a synthesis of (−)-*O*-methylcryptaustoline that established the absolute configuration at $C_{(13)}$ (Scheme 25) (*G. K. Hughes, Ritchie* and *Taylor*, Austral. J. Chem., 1953, **6**, 315). The stereochemistry at the nitrogen remained unknown.

(±)Cryptaustoline and (±)cryptowoline have been synthesised by treating the halogenotetrahydroisoquinolines X and XI, respectively, with strong base. An aryne intermediate was implicated in each case (*Kametani* and *Ogasawara*, J. chem. Soc., C, 1967, 2208) (Scheme 26).

SCHEME 25

THE SYNTHESIS OF (–)CRYPTAUSTOLINE *O*-METHYL ETHER

(–) Laudanosine

(1) Chloranil
(2) *O*-Methylation

(–) Cryptaustoline *O*-methyl ether
(IX)

SCHEME 26

A SYNTHESIS OF CRYPTAUSTOLINE AND CRYPTOWOLINE *VIA* ARYNE INTERMEDIATES

(X)   R = OMe; X = Br
(XI)   2 R = –OCH$_2$O–; X = Cl

(1) MeI
(2) H$^\oplus$
(3) KI

It has been reported that when the 1-benzylisoquinoline derivative XII was reacted with dimsyl sodium the product was XIII, resulting from a Stevens rearrangement (*S. Kano et al.,* Chem. and Pharm. Bull., Japan, 1974, **22**, 1607):

(XII)

(XIII)

(iii) *Biosynthesis*

Nothing is known about the biosynthesis of these two alkaloids but there is little doubt that they arise by oxidative cyclisation of reticuline.

3. Pavinanes and isopavinanes

When papaverine (I) is reduced with tin and hydrochloric acid, the major product is the expected 1,2,3,4-tetrahydropapaverine (II), but a second base, later called **pavine** (*G. Goldschmiedt*, Monatsh., 1886, **7**, 485; *F. L. Pyman* and *W. C. Reynolds*, J. chem. Soc., 1910, **97**, 1320; *Pyman, ibid.*, 1909, **95**, 1610) is also produced in about 5% yield. This new secondary base, $C_{20}H_{23}NO_4$, is resolvable (*W. J. Pope* and *S. J. Peachey, ibid.*, 1898, **73**, 893). During his investigation of the structure *P. Schöpf* (Experientia, 1949,

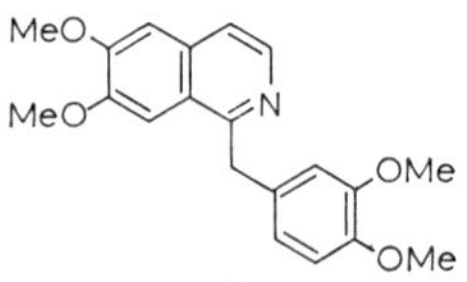

(I)

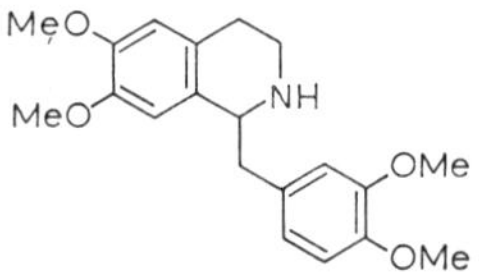

(II)

5, 201) proposed the structures III (R = H) and IV (R = H), but was unable to distinguish between them. He reported that *N*-methylpavine could be formed either by *N*-methylation of pavine or by treating 2-methyl-

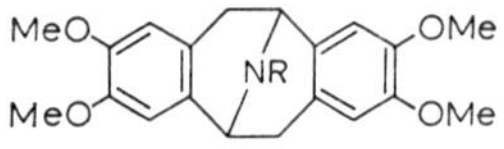

Pavine   (R = H)

(III)

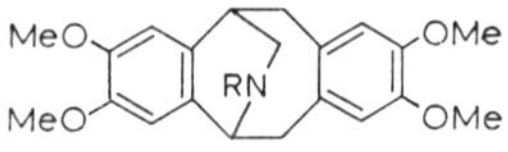

Isopavine   (R = H)

(IV)

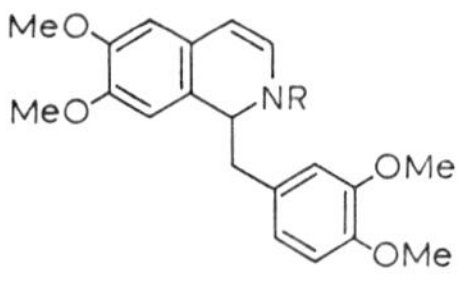

(V)

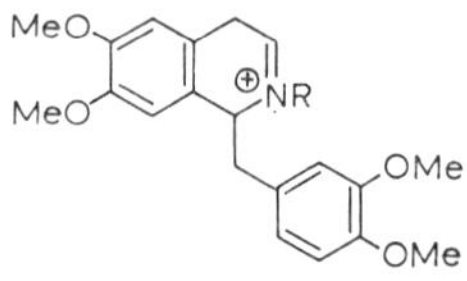

(VI)

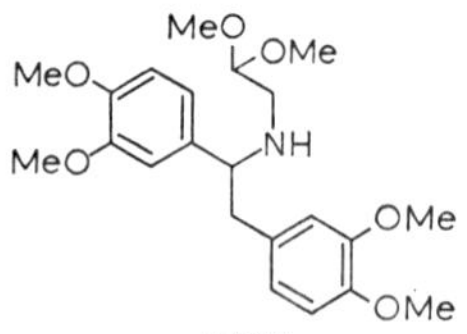

(VII)

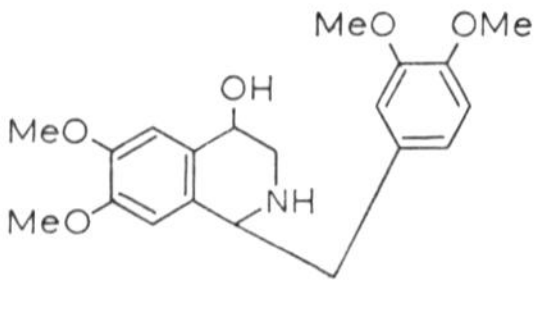

(VIII)

SCHEME 27

DEGRADATIONS OF PAVINE

1,2-dihydropapaverine (V, R $=$ Me) with acids. *A. R. Battersby* and *R. Binks* (J. chem. Soc., 1955, 2888) proved that III (R $=$ H) is the structure of pavine by degradation and synthesis (Scheme 27). They also provided a rationale for the formation of the base, suggesting that papaverine is initially reduced by the tin and hydrochloric acid to 1,2-dihydropapaverine (V, R $=$ H) which, as an enamine, is protonated at $C_{(4)}$ to form the iminium ion VI (R $=$ H). This may then be reduced further to II or undergo an intramolecular nucleophilic attack by the 3,4-dimethoxybenzyl group at $C_{(3)}$, thus leading to pavine (III, R $=$ H). A similar mechanism was postulated for the formation of III (R $=$ Me) from V (R $=$ Me). The isomeric base IV (R $=$ H) was given the trivial name **isopavine** (*Battersby* and *D. A. Yeowell, ibid.*, 1958, 1988) and it was shown that, when the aminoacetal derivative VII is treated with mineral acid, the product is isopavine (IV, R = H), thus correcting some earlier work (*D. A. Guthrie, A. W. Frank* and *C. B. Purves*, Canad. J. Chem., 1955, **33**, 729). It is very likely that VII is first cyclised to the 4-hydroxy-1,2,3,4-tetrahydroisoquinoline (VIII), which then undergoes an intramolecular nucleophilic displacement of the $C_{(4)}$-hydroxyl group by the 3,4-dimethoxybenzyl group at $C_{(1)}$, thus leading to IV (R = H) (*S. F. Dyke et al.*, Tetrahedron Letters, 1969, 1515). The degradations of isopavine are summarised in Scheme 28.

SCHEME 28

DEGRADATION OF ISOPAVINE

Isopavine (IV)   (1) MeI / (2) Hofmann degradation   →   (XIII)

+

(XIV)

H_2/Pt

When the methine base IX from pavine is subjected to Hofmann degradation, the product is X, which, with hydrochloric acid undergoes a Wagner–Meerwein rearrangement to give XI (*M. M. Abdel-Monen* and *T. O. Soine*, J. pharm. Sci., 1967, **56**, 976; *J. Slavik, L. Slavikova* and *K. Haisova*, Coll. Czech. chem. Comm., 1967, **32**, 4420). The same bis-methine XII can be obtained from XI or from isopavine methines XIII and XIV (Scheme 28). Herein lies some of the earlier confusion relating to the structure of pavine. Of the two possible methines XIII and XIV from isopavine, only XIII was isolated and characterised.

(a) *Pavinane alkaloids*

The alkaloid **argemonine**, which was isolated from various species of Argemone plants, was initially thought to be an aporphine type of alkaloid (*L. B. Kier* and *T. O. Soine*, J. Amer. pharm. Assoc. Sci. Ed., 1960, **49**, 187; *J. W. Schermerhorn* and *Soine*, *ibid.*, 1951, **40**, 19; *M. Shamma*, Experientia, 1962, **18**, 34; *Soine* and *Kier*, J. Amer. pharm. Assoc. Sci. Ed., 1962, **51**, 1196), but it was subsequently realised (*F. R. Stermitz, S.-Y. Lwo* and *G. Kallos*, J. Amer. chem. Soc., 1963, **85**, 1553; *M. J. Martell, Soine*

TABLE 1

THE PAVINANE ALKALOIDS

Name	Substituents								m.p. (°C)	$[\alpha]_D$	Configuration	References
	1	2	3	4	7	8	9	10				
Argemonine	H	OMe	OMe	H	H	OMe	OMe	H	155	−188(C)	5S,11S	1–6
Norargemonine	H	OMe	OMe	H	H	OMe	OH	H	238	−154(C)	5S,11S	1,3,7
Isonorargemonine	H	OH	OMe	H	H	OMe	OMe	H	177		5S,11S	7,8
Bisnorargemonine	H	HO	OMe	H	H	OMe	OH	H	254	−266(M)	5S,11S	1,3,9,10
Eschscholtzine	H	O–CH₂–O		H	H	O–CH₂–O		H	128	−202(M)	5S,11S	1,3,11
Eschscholtzidine	H	OMe	OMe	H	H	O–CH₂–O		H	Oil	−194(M)	5S,11S	1,3,12
Caryachine	H	O–CH₂–O		H	H	OH	OMe	H	175	−270	5S,11S	1,13,14
Munitagine	H	H	OMe	OH	H	OMe	OH	H	169	−239(C)	5S,11S	1,15,16
Platycerine	H	H	OMe	OH	H	OMe	OMe	H	132	−267(C)	—	17,18
Unnamed	H	OH	OMe	H	H	OMe	OH	H	197	−254(M)	—	19
Californidine (methiodide of eschscholtzine)	H	O–CH₂–O		H	H	O–CH₂–O		H	286	−212	—	20

References to Table 1

1 *A. C. Barker* and *A. R. Battersby*, J. chem. Soc., C, 1967, 1317; Tetrahedron Letters, 1967, 135.
2 *Battersby* and *R. Binks*, J. chem. Soc., 1955, 2888.
3 *R. H. F. Manske et al.*, Tetrahedron, 1967, **23**, 4209.
4 *M. J. Martell, T. O. Soine* and *L. B. Kier*, J. Amer. chem. Soc., 1963, **85**, 1022.
5 *S. F. Mason et al.*, Tetrahedron Letters, 1967, 137.
6 *Mason, G. W. Vane* and *J. S. Whitehurst*, Tetrahedron, 1967, **23**, 4087.
7 *F. R. Stermitz* and *J. N. Seiber*, Tetrahedron Letters, 1966, 1177.
8 *Stermitz* and *K. D. McMurtrey*, J. org. Chem., 1969, **34**, 555.
9 *Soine* and *Kier*, J. pharm. Sci., 1963, **52**, 1013.
10 *Soine, C. N. Chen* and *K. H. Lee, ibid.*, 1970, **59**, 1529.
11 *Manske et al.*, Canad. J. Chem., 1965, **43**, 2183.
12 *Manske* and *K. H. Shin, ibid.*, 1966, **44**, 1259.
13 *S. T. Li* and *P. K. Lan*, J. pharm. Soc. Japan, 1966, **83**, 177.
14 *S. Natarajan* and *B. R. Pai*, Indian J. Chem., 1972, **10**, 451.
15 *F. Giral* and *A. Sotelo*, Ciencia (Mex.), 1959, **19**, 67.
16 *Stermitz* and *Seiber*, J. org. Chem., 1966, **31**, 2925.
17 *J. Slavik* and *L. Slavikova*, Coll. Czech. chem. Comm., 1963, **28**, 2530.
18 *Stermitz* and *D. K. Williams*, J. org. Chem., 1973, **38**, 1761.
19 *Stermitz et al., ibid.*, 1973, **38**, 3701.
20 *Slavik, Slavikova* and *K. Haisova*, Coll. Czech. chem. Comm., 1967, **32**, 4420.

and *Kier, ibid.*, 1963, **85**, 1022) to be $(-)$-*N*-methylpavine. Since that time a number of pavinane alkaloids have been isolated and identified (Table 1). Recently it was suggested by *C.-H. Chen* and *Soine* (J. pharm. Sci., 1972, **61**, 55, quoting a private communication from *Stermitz*) that the basic ring skeleton should be given the trivial name **pavinane** and numbered as shown in Table 1.

The structures of pavinane alkaloids have been elucidated by Hofmann degradation, interrelations, n.m.r. and mass spectral studies and by synthesis. A set of empirical rules has been proposed (*Chen* and *Soine, ibid.*, 1972, **61**, 55) to predict the chemical shift positions of the aromatic protons (Table 2). Data for **eschscholtzine** and **eschscholtzidine** have been added to show that the rules also apply to those pavinanes containing a methylene-dioxy group (*Dyke* and *A. W. C. White*, unpublished). In argemonine the aromatic region exhibits two singlets at 6.54 and 6.78 δ. It is argued that $H_{(4)}$ and $H_{(10)}$ appear together as the lower field singlet due to deshielding by the inductive effect of the C–N bridge. This argument is not consistent with the observed spectrum of argemonine methohydroxide (*M. P. Cava* and *M. Srinivasan*, J. org. Chem., 1972, **37**, 330) which would be expected to exert a larger inductive effect, but actually exhibits aromatic absorptions very similar to argemonine itself. *Chen* and *Soine* also argue that the methoxyl groups at $C_{(2)}$ and $C_{(8)}$ appear 0.1 ppm upfield from the absorption due to the $C_{(3)}$ and $C_{(9)}$ methoxyl groups. It seems more likely (*White*, Ph.D. Thesis, University of Bath, 1975) that anisotropic shielding, rather than deshielding is predominant and a study of molecular models shows

TABLE 2

N.M.R. OF AROMATIC REGION OF PAVINANES

Name	δ-values				
	$H_{(1)}$	$H_{(2)}$	$H_{(4)}$	$H_{(7)}$	$H_{(10)}$
Argemonine	6.70	—	6.48	6.70	6.48
Norargemonine	6.56	—	6.45	6.73	6.50
Isonorargemonine	6.70	—	6.36	6.71	6.51
Bisnorargemonine	6.53	—	6.45	6.68	6.33
Eschscholtzine	6.58	—	6.40	6.58	6.40
	or	—	or	or	or
	6.40	—	6.58	6.40	6.58
Munitagine	6.48	6.62	—	6.53	6.46
O,O-Dimethylmunitagine	6.72	6.72	—	6.65	6.51
Caryachine	6.51	—	6.75	6.39	6.72

that shielding by the aromatic rings is such that $H_{(1)}$ and $H_{(7)}$ are more shielded than $H_{(4)}$ and $H_{(10)}$. This explains not only the nonequivalence of the methoxyl groups but also that of the hydrogen atoms of the methylene-dioxy functions which appear typically as two doublets in the 5.9 δ region.

The mass spectrum of argemonine, which is summarised in Scheme 29, exhibits an intense M-1 peak, but the base peak of the spectrum at m/e 204 is attributed to the ion XV. The mass spectrum of an unsymmetrically substituted pavinane is typified by that of **eschscholtzidine** (Scheme 30). As in the case of argemonine, the $M_{\oplus}$ and $(M - 1)_{\oplus}$ peaks are strong, and again there is a peak at m/e 204 attributable to the same ion. However, the base peak, at m/e 188 is due to ion XVI.

SCHEME 29

MASS-SPECTRAL FRAGMENTATIONS OF ARGEMONINE

SCHEME 30

MASS-SPECTRAL FRAGMENTATIONS OF ESCHOLTZIDINE

U.v. spectroscopy is of little value for structural elucidation (λ_{max}^{EtOH} broad 287–295 nm) but the spectra measured in cyclohexane show considerable fine structure (*Stermitz* and *J. N. Seiber*, J. org. Chem., 1966, **31**, 2925) thus enabling a distinction to be made between the isomeric pavinane and isopavinane structures.

(*i*) *Configuration*

The absolute configuration of argemonine has been established (*A. C. Baker* and *A. R. Battersby*, J. chem. Soc., C, 1967, 1317) by degradation and correlation with aspartic acid (Scheme 31). Most of the other pavinane alkaloids have been correlated with (−)argemonine by partial synthesis, or by comparison of c.d. and/or o.r.d. spectra, and have been found to have the same absolute configuration as (−)argemonine.

SCHEME 31

THE DEGRADATION OF ARGEMONINE

Hofmann degradation

H₂/Pd/C

HCONH₂ Δ

(1) O₃
(2) HCO₃H

(Known configuration)

(ii) Synthesis

The synthesis of alkaloids of the pavinane group has, almost without exception, involved the preparation and acid-catalysed cyclisation of the appropriate 1-benzyl-1,2-dihydroisoquinoline derivative. The conditions of the acid treatment (*e.g.* polyphosphoric acid and phosphoryl chloride at 150°) are such that yields are often very low, due probably to the cleavage of ether functions (especially of methylenedioxy groups). The synthesis of **platycerine** is illustrative of the method, where the main problem is the preparation of the required 1-benzylisoquinoline derivative (*Stermitz* and *D. K. Williams*, J. org. Chem., 1973, **38**, 1761) (Scheme 32). Recently it was found that when substituted acetals such as XVII are treated with mineral acid, a mixture of the *N*-methylpavinane and the *N*-methyliso-

SCHEME 32

THE SYNTHESIS OF (±)PLATYCERINE

Platycerine

pavinane are produced. The ratio of these products depends largely upon the conditions of the reaction (*Dyke et al.*, Tetrahedron, 1974, **30**, 1193):

(XVII)

SCHEME 33

SYNTHESIS OF *O*,*O*-DIMETHYLMUNITAGINE

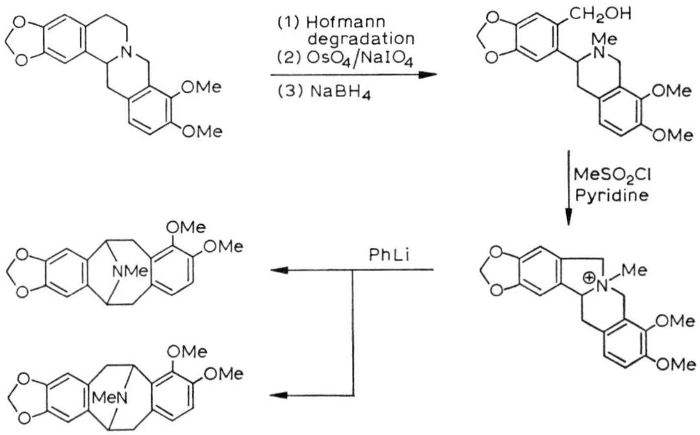

A novel synthesis of *O*,*O*-**dimethylmunitagine** starts from tetrahydroberberine and involves a Stevens rearrangement (*K. Ito et al.*, Chem. Comm., 1974, 1037) (Scheme 33).

(*b*) *Isopavinane alkaloids*

The demonstration that the isopavine ring system XVIII (10,11-dihydro-10,5-iminomethane-5*H*-dibenzo[*a,d*]cycloheptene) occurs in nature, came with the identification of **amurensine** and **amurensinine** (*F. Santavy, L. Hruban* and *M. Maturova*, Coll. Czech. chem. Comm., 1966, **31**, 4286). Since that time several other isopavinane alkaloids have been isolated, characterised and synthesised, and they are summarised in Table 3. Amurensine was isolated in 1962 by *Santavy et al.* (Planta Med., 1962, **10**, 345) from *Papaver anomalum* and *P. nudicaule*, var. *xanthopetalum* and was names xanthopetaline. However, it was subsequently shown to be identical with amurensine isolated previously (*H. G. Boit* and *H. Flentje*, Naturwiss., 1959, **46**, 514) from *P. nudicante* var. *amurense*. The location of the hydroxyl group in amurensine, **reframoline** and **thalisopavine** were established by synthesis (*Dyke et al.*, Tetrahedron, 1972, **28**, 3999; 1974, **30**, 1193; *S. M.*

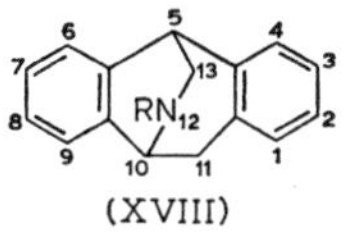

(XVIII)

TABLE 3

THE ISOPAVINANE ALKALOIDS

Name	R^4	R^3	R^2	R^1	m.p.(°C)	[α]	Ref.
Amurensinine	O–CH₂–O		OMe	OMe	164	−162(C)	1,2,3
Amurensine	O–CH₂–O		OMe	OH	215	−178(M)	1,2,4,12
Reframidine	O–CH₂–O		O–CH₂–O		amorphous	−123(M)	5,6,7
Reframine	OMe	OMe	O–CH₂–O		amorphous	−146(M)	5,6,7
Remrefine	method salt of reframine				242ᵃ	−147(W)	5,6,7, 8,9,10
Thalisopavine	OMe	OMe	OMe	OH			11
Reframoline	OH	OMe	O–CH₂–O		amorphous	−140(M)	5,13

ᵃ chloride.

References
1 *F. Santavy, M. Maturova* and *L. Hruban*, Chem. Comm., 1966, 36.
2 *Santavy, Hruban* and *Maturova*, Coll. Czech. chem. Comm., 1966, **31**, 4286.
3 *S. F. Dyke et al.*, Tetrahedron, 1969, **25**, 1881.
4 *H. G. Boit* and *H. Flentje*, Naturwiss., 1959, **46**, 514; 1960, **47**, 180.
5 *J. Slavik, L. Slavikova* and *L. Dolejs*, Coll. Czech. chem. Comm., 1966, **33**, 4066.
6 *Dolejs* and *Slavik, ibid.*, 1968, **33**, 3917.
7 *Dyke* and *A. C. Ellis*, Tetrahedron, 1971, **27**, 3803.
8 *M. S. Yunusov, S. T. Akramov* and *S. J. Yunusov*, Doklady Akad. Nauk. Uz. S.S.S.R., 1966, **23**, 38.
9 *Idem*, Khim. prir. Soedin, 1967, **3**, 68.
10 *Idem, ibid.*, 1968, **4**, 225.
11 *S. M. Kupchan* and *A. Yoshitake*, J. org. Chem., 1969, **34**, 1062.
12 *Dyke* and *Ellis*, Tetrahedron, 1972, **28**, 3999,
13 *Dyke et al., ibid.*, 1974, **30**, 1193.

Kupchan and *A. Yoshitake*, J. org. Chem., 1969, **34**, 1062). The original claims (*M. S. Yunusov, S. T. Akramov* and *S. J. Yunusov*, Doklady Akad. Nauk, Uz. U.S.S.R., 1966, **23**, 38; Khim. prir. Soedin, 1967, **3**, 68; 1968, **4**, 225) concerning the structure of the quaternary alkaloid called **roemrefine** are erroneous. The compound is identical with **remrefine** (*J. Slavik, L. Slavikova* and *L. Dolejs*, Coll. Czech. chem. Comm., 1968, **33**, 4066) which itself has been shown (*Dolejs* and *Slavik, ibid.*, 1968, **33**, 3917; *Dyke* and *A. C. Ellis*, Tetrahedron, 1971, **27**, 3803) to be identical with the metho salt of reframine.

The structures of the isopavinane alkaloids have been deduced by a combination of oxidative degradation, Hofmann degradation, n.m.r. and mass spectral studies. Some interconversions between the alkaloids have also been reported.

Amurensine is converted by diazomethane into amurensinine, which with potassium permanganate, is degraded to 3,4-methylenedioxyphthalic acid. The Hofmann degradations are summarised in Scheme 34. Although both

SCHEME 34

THE STRUCTURAL ELUCIDATION OF AMURENSINE

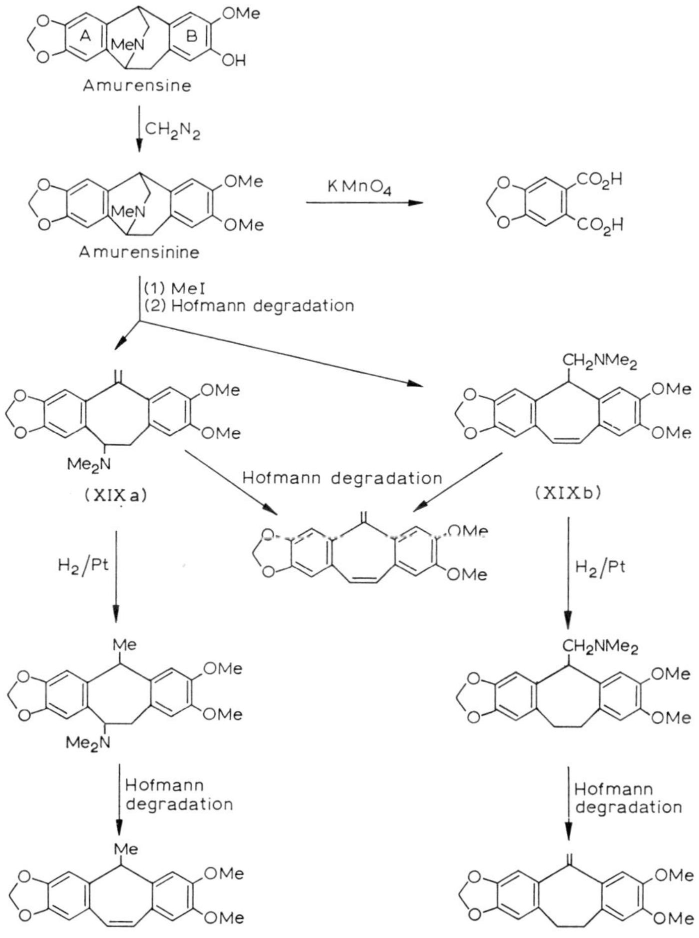

of the possible methines XIXa and XIXb are obtained in this case, only the methine of type XIXb has been reported for other members of the group.

Because 3,4-methylenedioxyphthalic acid is the product of oxidation, the methylenedioxy group must be located on ring A rather than ring B. The location of the two oxygen functions *ortho* to each other on each ring, as shown, follows from the n.m.r. spectra of the two alkaloids which exhibit four one-hydrogen singlets in the aromatic region. The results of Hofmann degradation can be ambiguous *e.g.* amurensinine and reframine give the same methine structure XX (although of opposite configurations, consistent with the same absolute configurations for the alkaloids).

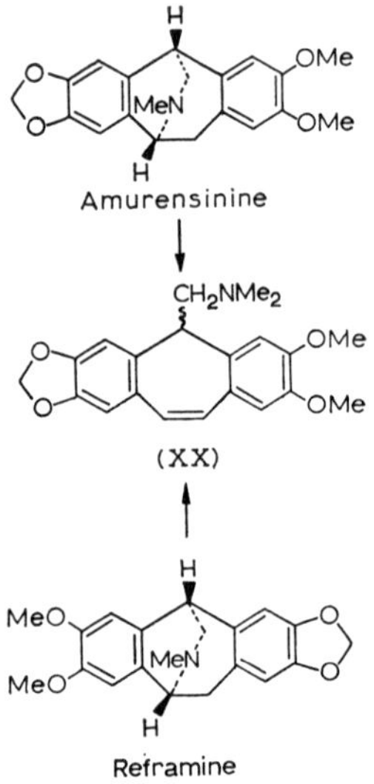

The mass-spectral fragmentations postulated for these alkaloids (*Dolejs* and *V. Hanus*, Coll. Czech. chem. Comm., 1968, **33**, 600; *Dolejs* and *Slavik*, *ibid.*, 1968, **33**, 3917) is summarised in Scheme 35. The base peak in the spectrum of amurensinine is at *m*/*e* 188 and is attributed to the ion XXII which is formed directly from the parent ion, probably *via* XXI (metastable at *m*/*e* 104.3). The accompanying nitrogen-free fragment appears as the tropylium ion at *m*/*e* 151. The $M^{\oplus}$ and $(M-1)^{\oplus}$ ions are strong; the latter can be represented as XXIII or XXIV. A prominent ion at $(M-43)^{\oplus}$ can be attributed to XXV and formed from $M^{\oplus}$ by a retro-Diels–Alder process. The formation of the ion XXII is important since it enables a distinction to be made between the isomeric structures of amurensinine and reframine. The appearance of a $(M+14)^{\oplus}$ peak in the mass spectrum of reframidine is explained by partial loss of a methyl radical from the molecular ion and

SCHEME 35

THE MASS-SPECTRAL FRAGMENTATIONS OF AMURENSININE

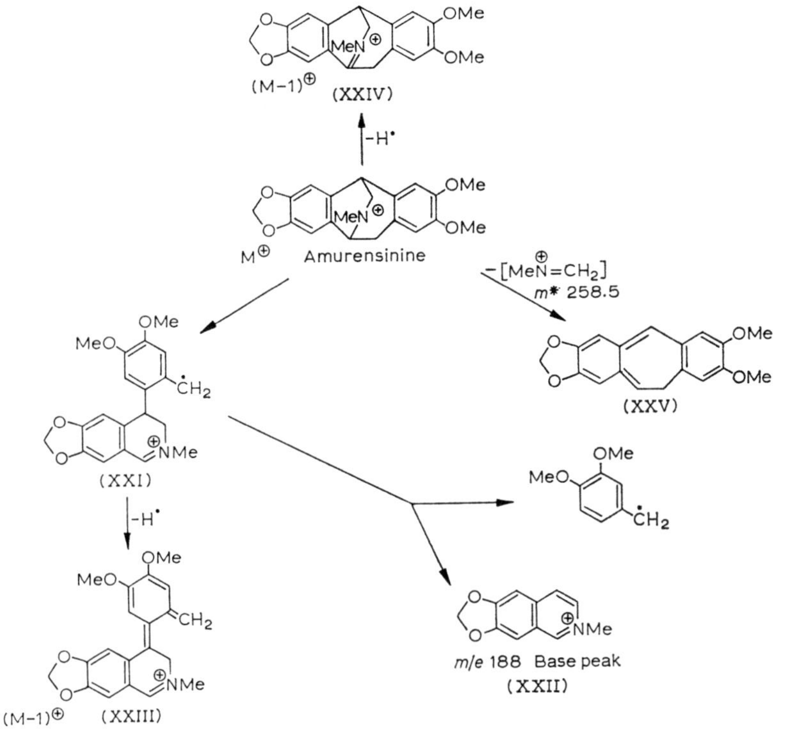

TABLE 4

N.M.R. CHEMICAL SHIFTS FOR THE ISOPAVINANE ALKALOIDS

Name	δ for the aromatic protons				Ref.
Amurensine	6.53	6.58	6.70	6.70	1,2
	6.40[a]	6.72[a]	6.80[a]	6.80[a]	
Amurensinine	6.53	6.63	6.63	6.73	2
Reframine	6.55	6.70	6.83	6.90	3
Reframidine	6.55	6.65	6.75	6.75	3
Thalisopavine	6.54	6.61	6.75	6.75	4

[a] in DMSO.

References

1 *S. F. Dyke* and *A. C. Ellis*, unpublished.
2 *F. Santavy*, in The Alkaloids, Vol. 12, 1970, p. 333.
3 *S. F. Dyke* and *A. C. Ellis*, Tetrahedron, 1971, **27**, 3803.
4 *S. M. Kupchan* and *A. Yoshitake*, J. org. Chem., 1969, **34**, 1062.

recapture of it by another molecular ion, followed by loss of a hydrogen radical (*Dyke* and *Ellis*, Tetrahedron, 1971, **27**, 3803). This explanation was confirmed by a study of appropriately deuterated compounds (*idem*, unpublished).

The n.m.r. spectra of the isopavinanes have been recorded (Table 4). In most cases the aromatic region comprise three or four singlets but it has not yet proved possible to make firm assignments. The methylenedioxy groups appear as two doublets because of the asymmetry of the ring system.

(*i*) *Configuration*

The absolute configuration of (−)amurensine was deduced (*Shamma et al.*, Tetrahedron Letters, 1971, 3415) by c.d. studies, and is shown as XXVI. This has now been confirmed by an unambiguous synthesis of reframoline (*Dyke et al.*, Tetrahedron, in press). Both the pavinane and isopavinanes have the nitrogen bridge below the V plane of the molecules, a point of significance in biosynthetic discussions (see later).

(XXVI)

(*ii*) *Synthesis*

The principal method of synthesis of the isopavinane alkaloids has been based upon the original preparation of isopavine itself. A typical example, reframine (*Dyke* and *Ellis*, Tetrahedron, 1971, **27**, 3803) is summarised in Scheme 36.

SCHEME 36

THE SYNTHESIS OF (±)REFRAMINE

SCHEME 37

THE SYNTHESIS OF (±)*O*-METHYLTHALISOPAVINE

O-Methylthalisopavine

In an alternative approach *(idem, loc. cit.)*, 2-methyl-1,2-dihydropapaverine was converted (Scheme 37) into (±)*O*-methylthalisopavine (*N*-methylisopavine) *via* the 4-hydroxy-2-methyl-1,2,3,4-tetrahydropapaverine.

In a further development of the oxidation of phenolic isoquinoline derivatives with lead tetraacetate (*O. Hoshino, M. Tata* and *B. Umezawa,* Heterocycles, 1973, **1,** 223), syntheses of (±)*O*-methylthalisopavine and (±)reframine have been reported in which the substrates were XXVII and XXVIII, respectively.

A novel synthesis of the isopavinane ring system is shown in Scheme 38 (*T. Kametani* and *K. Ogasawara,* Chem. pharm. Bull., Japan, 1973, **21,** 893).

(XXVII)   R = OMe
(XXVIII)   2R = –OCH$_2$O –

(iii) Biosynthesis of pavinanes and isopavinanes

Only one study has been made on the biosynthesis of pavinanes (*D. H. R. Barton, R. H. Hesse* and *G. W. Kirby,* J. chem. Soc., 1965, 6379) in which labelled (+)reticuline was fed to the mixed plants of *Argemone mexicana* L. and *A. hispida* Gray. An insignificant uptake of the label into (−)argemonine was found. This may be due to the very small amounts of argemonine present in these plants (*Stermitz* and *Seiber,* J. org. Chem., 1966, **31,** 2925).

SCHEME 38

A NOVEL SYNTHESIS OF ISOPAVINANES

It has been suggested by *Stermitz* and *Seiber* and by *A. C. Barker* and *Battersby* (J. chem. Soc., C, 1967, 1317) that (+)reticuline is oxidised to the 1,4-dihydroisoquinolinium ion XXIX, which can then be cyclised to the pavinane ring-system XXX. The phenolic hydroxyl group on the benzyl portion of XXIX provides the activation for the cyclisation under physiological conditions. Subsequent alteration of the oxygen functions then provides most of the observed alkaloids. Cyclisation of XXIX can also occur *ortho* to the benzylic hydroxyl group to yield munitagine:

No work has yet been reported on the biosynthesis of the isopavinane alkaloids, but it seems reasonable to postulate (*Dyke et al.*, Tetrahedron Letters, 1969, 1515; *Dyke* and *Ellis*, Tetrahedron, 1971, **27**, 3803) that a 1-benzyl-4-hydroxy-1,2,3,4-tetrahydroisoquinoline is the immediate precursor—probably 4-hydroxyreticuline (XXXII) which itself is derivable from noradrenaline (XXXI):

(XXXI)

(XXXII)

It is also possible that the pavinanes are derived from XXXII by simple dehydration to 1,2-dihydroreticuline, followed by $C_{(4)}$-protonation and cyclisation. The fact that the absolute stereochemistry of the pavinanes and isopavinanes is the same, lends support to this view (*Shamma et al.*, Tetrahedron Letters, 1971, 3425). Although 1-benzyl-4-hydroxyisoquinoline alkaloids have not yet been isolated from plant sources, certain other alkaloids, *e.g.* **imenine** and **erythristemine** are known which may well arise from such intermediates. It has already been established that **berberastine** is biosynthesised from noradrenaline (XXXI):

Imenine

Erythristemine

Berberastine

4. The bisbenzylisoquinolines

The bisbenzylisoquinoline alkaloids, which are widely distributed in plants of the Menispermaceae family growing in South America and the Far East, and in Daphnandra, Berbis, Magnolia and Nectandra families, is one of the largest sub-groups of the isoquinoline alkaloids. The subject has been reviewed several times (*M. Kulka*, "The Alkaloids", 1954, **4**, 199; 1960, **7**, 439; *M. Curcumelli-Rodostamo* and *Kulka, ibid.*, 1967, **9**, 133; *Curcumelli-Rodostamo, ibid.*, 1971, **13**, 303; *M. F. Grundon*, Progr. in org. Chem., 1964, **6**, 38; *M. Shamma*, "The Isoquinoline Alkaloids", Academic Press, New York, 1972, p. 115). They are all dimers with, in the simplest cases, two 1-benzyltetrahydroisoquinoline residues linked together by one, two or three diphenyl ether linkages. Some examples are known that contain a carbon–carbon linkage between the residues as well as diphenyl ether linkages. The two 1-benzyltetrahydroisoquinoline components may be secondary or tertiary amines with the same or opposite configurations, and carrying the usual oxygen functions.

A classification of bisbenzylisoquinolines based upon biogenetic proposals was made by *F. Faltis et al.* (Ber., 1941, **74B**, 79); with modifications this gained wide acceptance. The basic concept is that the bisbenzylisoquinolines are derived from two molecules of coclaurine (see p. 11) (or norclaurine) by carbon–oxygen–carbon phenolic oxidative coupling, preceded or succeeded by *O*- and/or *N*-methylation. A few alkaloids, *e.g.* **magnolamine** and **tenuipine,** that contain one more oxygen function per 1-benzylisoquinoline

SCHEME 39

BISBENZYLISOQUINOLINES WITH ONE DIPHENYL ETHER LINKAGE

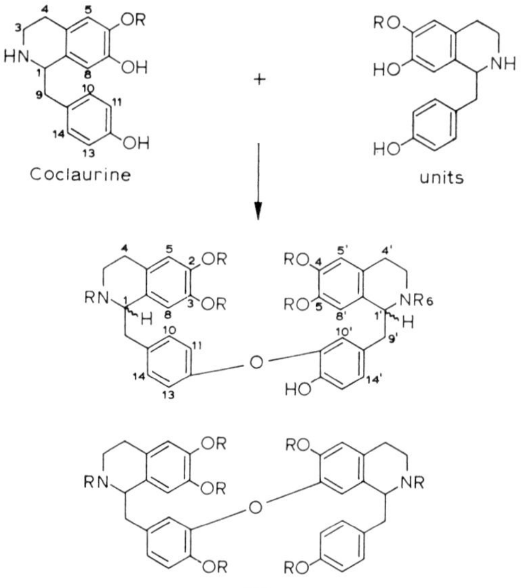

unit than coclaurine are considered to arise by hydroxylation, possibly after the phenolic coupling has occurred.

Thus, alkaloids with one diphenyl ether linkage are considered to be formed by the linkage of the $C_{(12)}$–OH group of one coclaurine molecule with the $C_{(11)}$-position of a second unit (Scheme 39). *O*-Methylation etc., then provides the distinction between individual members. Linkage between coclaurine is not confined to $C_{(12)}$–O–$C_{(11')}$; examples of the $C_{(7)}$–O–$C_{(11')}$ type (I) are known.

SCHEME 40

SOME POSSIBILITIES CONTAINING
TWO DIPHENYL ETHER LINKAGES

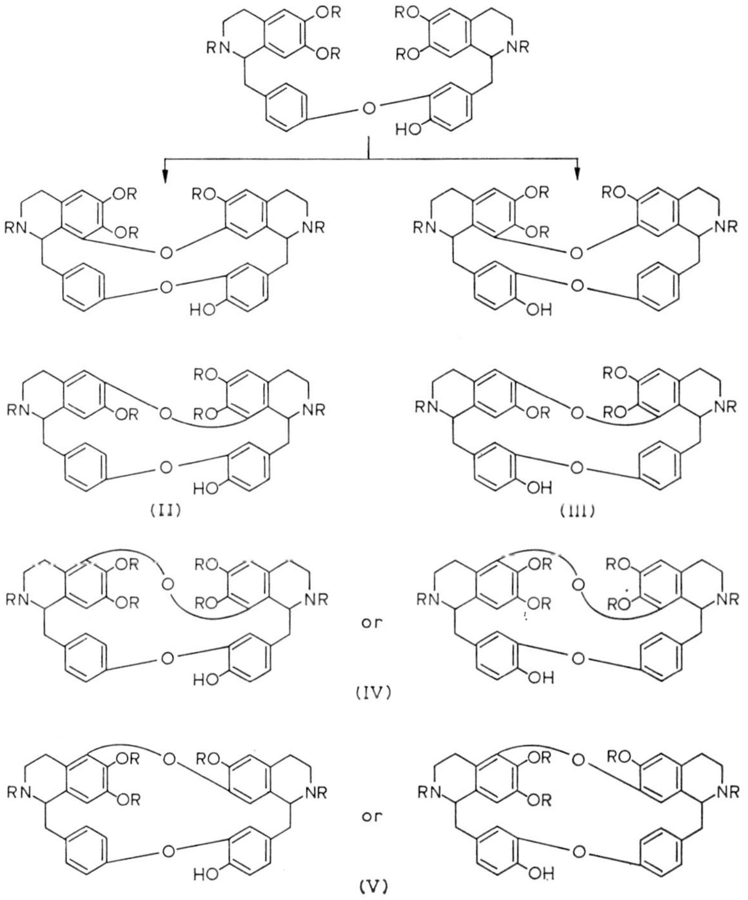

In those compounds that contain two diphenyl ether linkages, several variations are possible. If it is assumed (for the purposes of classification) that the first linkage, $C_{(12)}$–O–$C_{(11')}$, is formed as above, there are two ways in which a $C_{(7)}$–O–$C_{(8')}$ linkage can become established (Scheme 40). Some alkaloids are known in which the second diphenyl ether bond is formed between $C_{(6')}$ and $C_{(8')}$; there are two ways of doing this (II) and (III). Again on the basis of a $C_{(12)}$–O–$C_{(11')}$ linkage, the second one can be $C_{(5)}$–O–$C_{(8')}$ (IV) or $C_{(7)}$–O–$C_{(5')}$ (V).

Alternatively, two diphenyl ether linkages can be formed between $C_{(12)}$ and $C_{(8)}$ to give the symmetrical structure VI, whilst yet another naturally occurring variation VII involves a $C_{(12)}$–O–$C_{(8')}$ together with a $C_{(7)}$–O–$C_{(11')}$ link (Scheme 41).

SCHEME 41

ALTERNATIVE DIPHENYL ETHER LINKAGES

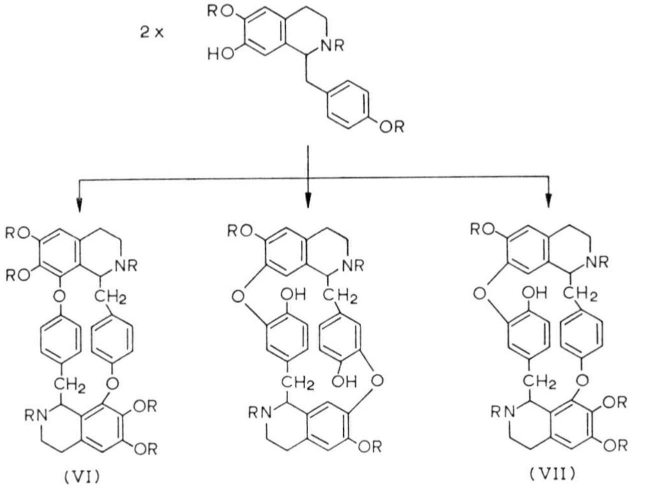

When three diphenyl ether linkages are present there is a large number of possibilities; the main naturally occurring ones are derivatives of VIII.

In addition to these categories some alkaloids are known that contain a carbon–carbon as well as at least one carbon–oxygen–carbon linkage, *e.g.* IX and X. Some bisbenzylisoquinolines are now known that contain a benzyl–O–aryl bond, *e.g.* XI.

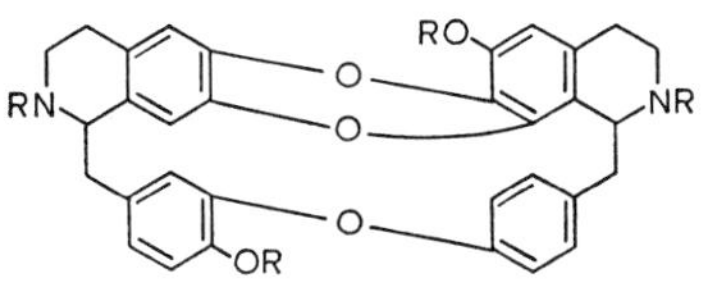

(VIII) 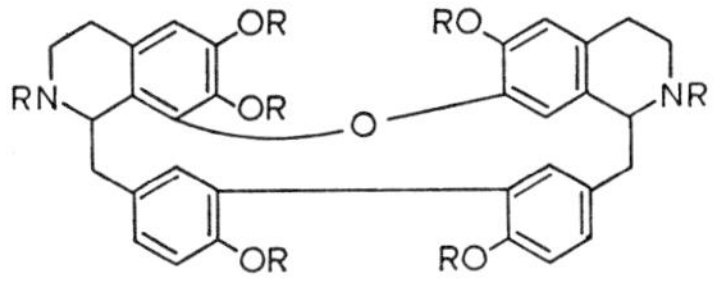　(IX)

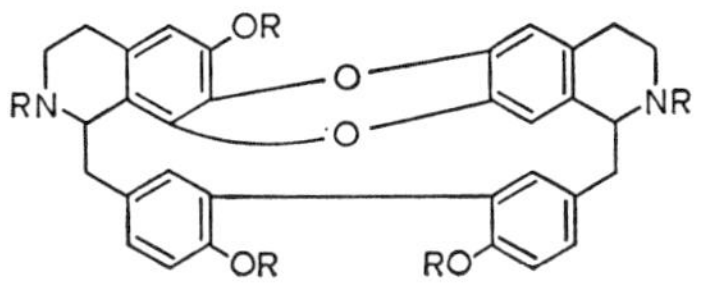

(X) 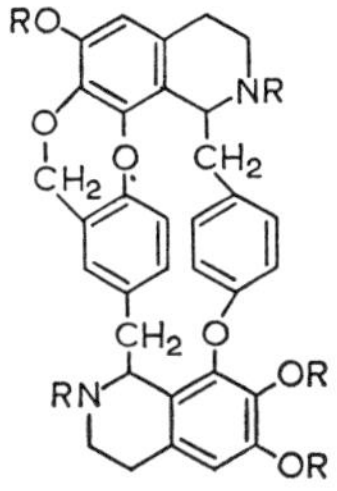　(XI)

(*a*) *Structure determination*

In early investigations, the main techniques depended upon oxidation, Hofmann degradation and oxidative cleavage of the methines so formed. An important development occurred when diphenyl ether cleavage with sodium in liquid ammonia was introduced. More recently mass and proton magnetic resonance spectroscopy have been utilised with some success.

Magnoline is the simplest member of the bisbenzylisoquinoline series, containing only one diphenyl ether linkage. Its structure was elucidated in 1940 (*N. F. Proskurnina* and *A. P. Orekhov*, J. gen. Chem., U.S.S.R., 1940, **10,** 707; C.A., 1941, **35,** 2520). The usual tests established a molecular formula of $C_{36}H_{40}N_2O_6$, containing two *N*-methyl, two *O*-methyl and three phenolic hydroxyl groups. Methylation (CH_2N_2) gave a tri-*O*-methyl ether which was oxidised with potassium permanganate. The important fragments XII and XIII were recognised by comparison with authentic samples. Repetition of the degradation with the tri-*O*-ethyl ether gave XIV and XV. Hence, magnoline must be represented by XVI.

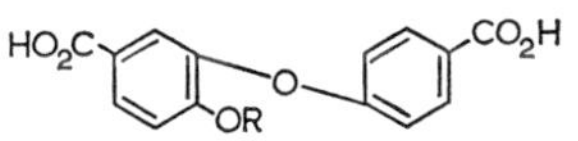

(XII):　R = Me
(XIV):　R = Et

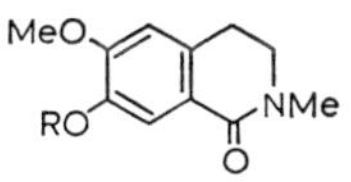

(XIII):　R = Me
(XV):　R = Et

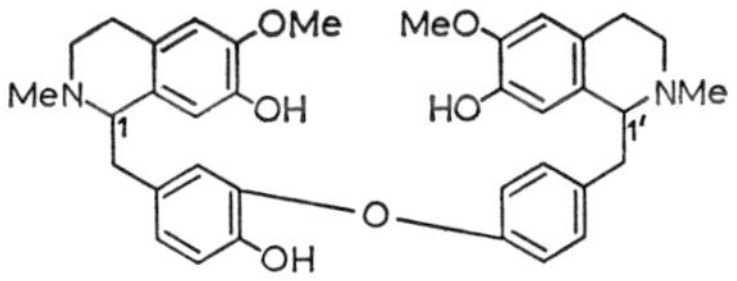

Magnoline
(XVI)

The remaining problem is the configurations at $C_{(1)}$ and $C_{(1')}$, which was solved by application of the sodium/liquid ammonia cleavage.

It was reported in 1937 that diphenyl ethers can be cleaved by this reagent (*P. A. Sartoretto* and *F. J. Sowa*, J. Amer. chem. Soc., 1937, **59**, 603; *A. L. Kranzfelder, J. J. Verbanc* and *Sowa, ibid.*, 1937, **59**, 1488; *F. C. Weber* and *Sowa, ibid.*, 1938, **60**, 94). In those cases where an *ortho* methoxy group is present, fission occurs preferentially *ortho* to this group (*e.g.* XVII and XVIII):

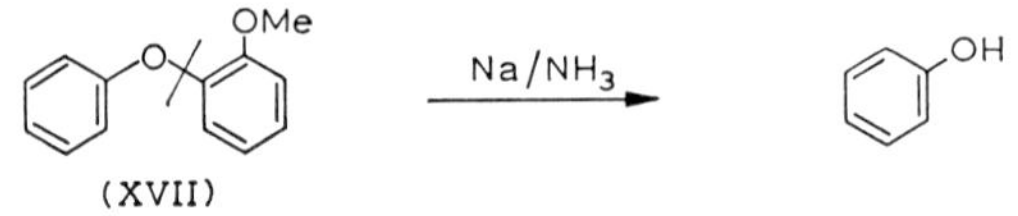

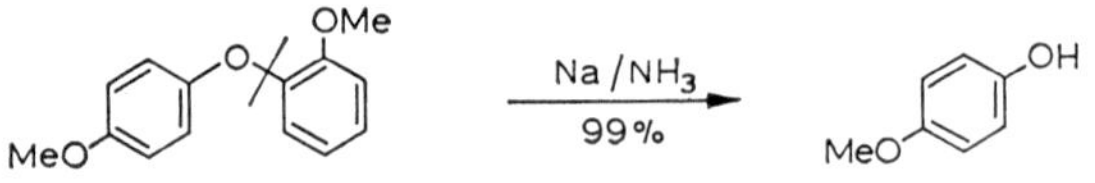

The reaction was first applied to bisbenzylisoquinolines in 1951 (*M. Tomita, E. Fujita* and *F. Murai*, J. pharm. Soc., Japan, 1951, **71**, 226) and has been used widely ever since. Usually a complex mixture of 1-benzyltetrahydroisoquinolines is produced, which can be separated by chromatography. The structures and configurations of the components can be established by standard procedures and the position(s) of ether link(s) in the original alkaloid deduced. Ambiguity can arise, and this is illustrated by the case of **thalisopine**. Cleavage of the one diphenyl ether linkage present gave armepavine (XIX) and the 1-benzyltetrahydroisoquinoline (XX) (*Z. F. Ismailov, A. U. Rakhmakariev* and *S. Y. Yunusov*, Uzb. Khim. Zh.,

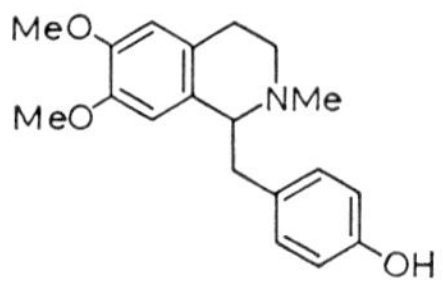

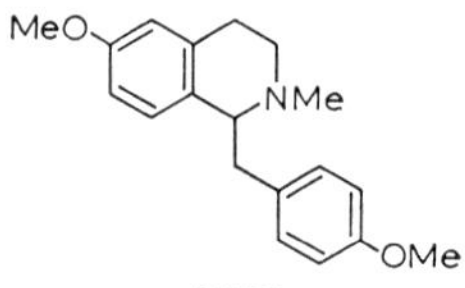

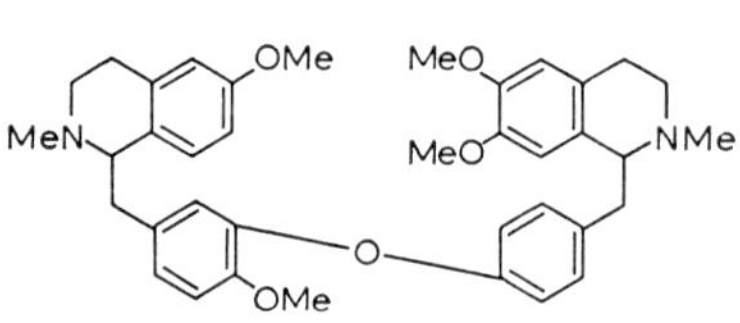

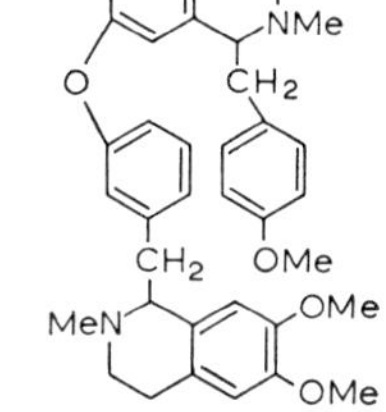

1961, 56; C.A., 1963, **58**, 3469). Structure XXI was then proposed for the alkaloid, but the alternative structure XXII is probably correct, thus retaining the biogenetically more acceptable 6,7-di-oxygenation pattern of each unit (*Curcumelli-Rodostamo* and *Kulka*, "The Alkaloids", 1967, **9**, 133). From the results of sodium/ammonia cleavage of a wide variety of diaryl ethers, it seems that in the bisbenzylisoquinoline series, fission occurs at specific carbon–oxygen bonds to give derivatives of 6,7-dihydroxy-1-(4-hydroxybenzyl)-1,2,3,4-tetrahydroisoquinolines as the products. A useful method for labelling the site of a diphenyl ether linkage has been developed that involves cleavage with Na/ND_3 (*I. R. C. Bick et al.*, Chem. Comm., 1971, 1056). Alternatively the *O*-methylated dimer is treated with 3% DCl in D_2O at 120° in a sealed tube. Only protons *ortho* to methoxyl groups are exchanged under these conditions. Subsequent cleavage with sodium/ammonia, isolation and characterisation of the tetrahydroisoquinolines is achieved in the usual way and an aromatic *hydrogen ortho* to methoxyl indicates the site of the original diphenyl ether linkage (*Y. Inubushi et al.*, Tetrahedron Letters, 1972, 423).

A new degradative procedure with non-phenolic bisbenzylisoquinoline alkaloids involves ultraviolet irradiation in the presence of oxygen when oxidative cleavage occurs at the benzylic centres (XXIII → XXIV + XXV) (*Bick, J. B. Bremner* and *J. Wiriyachitra, ibid.*, 1971, 4795).

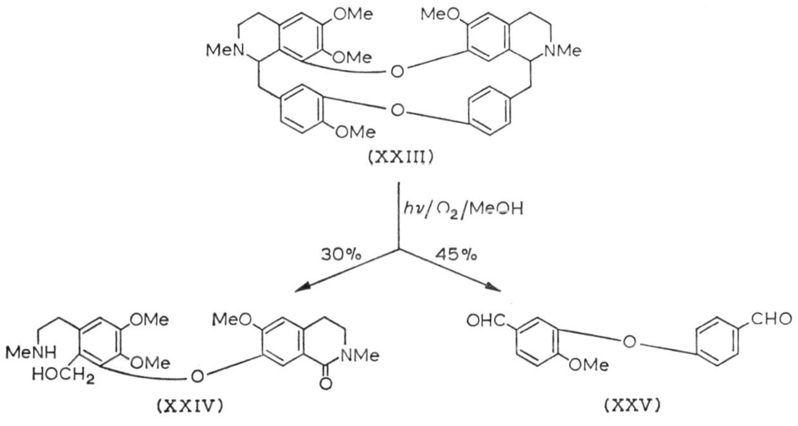

The correlation of (+)laudanosine (XXVI) (of known absolute configuration—see p. 23) with the trimethoxy-1-benzylisoquinoline derivative,

XXVII, has been achieved in at least two ways. In one, XXVI was de-*O*-methylated to **laudanine** which, after conversion to the *O*-phenyl ether, was reduced to XXVII with sodium/ammonia (*M. Tomita* and *J. Kunimoto*, J. pharm. Soc. Japan, 1962, **82,** 734). Hence, the absolute configurations of a large number of bisbenzylisoquinolines were deduced (*Tomita* and *Kunitoma, ibid.*, 1962, **82,** 741). By this means, the absolute configuration of magnoline was found to be that shown.

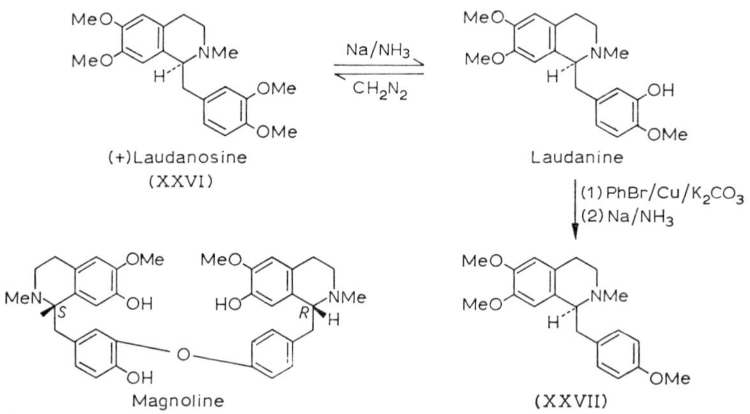

Oxyacanthine (XXVIII). The correct molecular formula, $C_{37}H_{40}N_2O_6$, was established by *E. Spath* and *A. Kolbe* (Ber., 1925, **58,** 2280), who also found that the alkaloid contains three methoxyl and two *N*-methyl groups, and one phenolic hydroxyl function. *O*-Methylation, followed by Hofmann degradation gave the bis-methine XXIX, which was oxidised with potassium permanganate to the dicarboxylic acid XXX (R = Me); this was synthesised by use of the Ullmann reaction. The degradative sequence was repeated using *O*-ethyloxyacanthine, when XXX (R = Et) was produced (*Spath* and *J. Pike, ibid.*, 1929, **62,** 2251). Ozonolysis of the bis-methine XXIX gave the nitrogen-containing dialdehyde XXXI which, when subjected to further Hofmann degradation, followed by catalytic hydrogenation, gave XXXII (*F. von Bruckhausen* and *P. H. Gericke*, Arch. Pharm., 1931, **269,** 115; *von Bruckhausen, H. Oberembt* and *A. Feldhaus*, Ann., 1933, **507,** 144). Clemmenson reduction of XXXII yielded XXXIII, which was identified

by Ulmann reaction as indicated in Scheme 42. These data are consistent
with XXVIII or XXXIV for the structure of oxyacanthene:

SCHEME 42

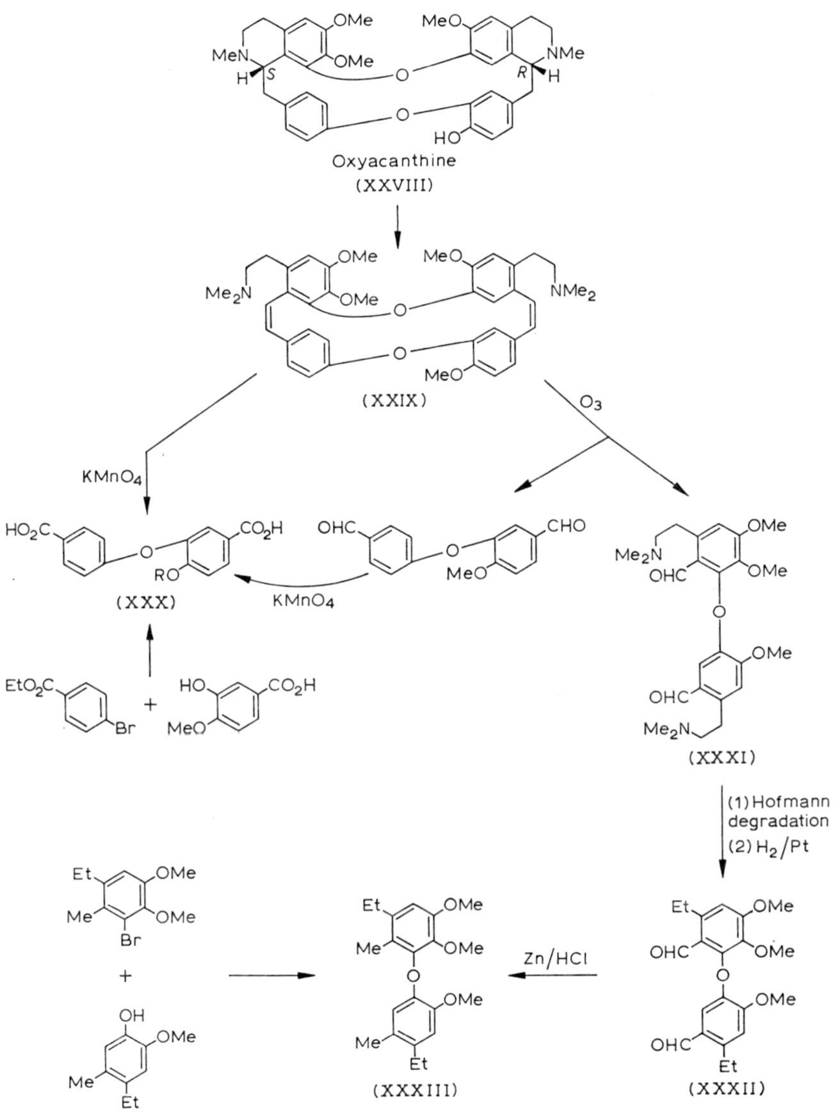

The alkaloid **berbamine** was known to be a structural isomer of oxyacanthine. When berbamine-*O*-methyl ether was cleaved with sodium/ammonia, the two 1-benzylisoquinolines XXXV and XXXVI were produced (*Tomita et al.*, J. pharm. Soc. Japan, 1951, **71**, 226, 301; C.A., 1952, **46**, 4554, 4555). Hence, berbamine is represented by structure XXXIV and oxyacanthine must be XXVIII. The configurations at $C_{(1)}$ and $C_{(1')}$ were found to be (*S*), (*R*) in oxyacanthine and (*R*), (*S*) in berbamine (*Tomita* and *J. Kunitomo*, Yakagaku Zasshi, 1962, **82**, 741; C.A., 1963, **58**, 4613).

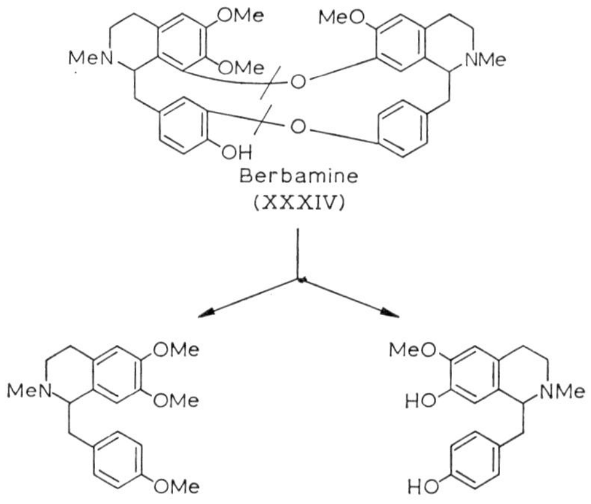

Berbamine
(XXXIV)

(XXXV)

(XXXVI)

Some useful generalisations have been made concerning proton n.m.r. chemical shifts of OMe and NMe groups in the oxyacanthine type of structure (*I. R. C. Bick et al.*, J. chem. Soc., 1961, 1896). From the data summarised in XXXVII and XXXVIII it would seem that the value of the methoxyl resonance gives information concerning the position of methoxyl groups and the stereochemistry at $C_{(1)}$ and $C_{(1')}$. In the oxyacanthine series the two *N*-methyl groups resonate at 2.25 δ, whereas in the berbamine types two separate signals are observed.

The application of mass spectrometry is becoming increasingly important in structural elucidation in bisbenzylisoquinolines (*Shamma et al.*, Chem. Comm., 1966, 7; *J. Baldis et al.*, *ibid.*, 1971, 132; Tetrahedron Letters, 1966, 2059; Austral. J. Chem., 1968, **21**, 2305; J. chem. Soc., Perkin I, 1972, 592, 597, 599).

Tetrandrine (XXXIX), like most bisbenzylisoquinolines, undergoes fission readily at the $C_{(1)}$ and $C_{(1')}$ positions to give (in this case) the ion *m/e* 198

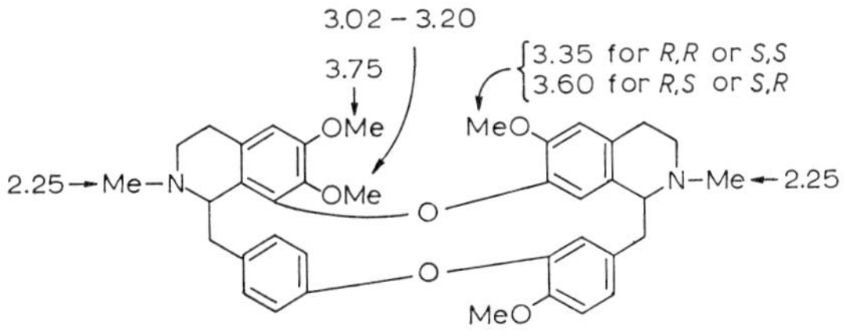

3.02 – 3.20

3.75

{3.35 for *R,R* or *S,S*
3.60 for *R,S* or *S,R*}

2.25 → Me–N   OMe   MeO   N–Me ← 2.25

OMe   O   O   MeO

(XXXVII)

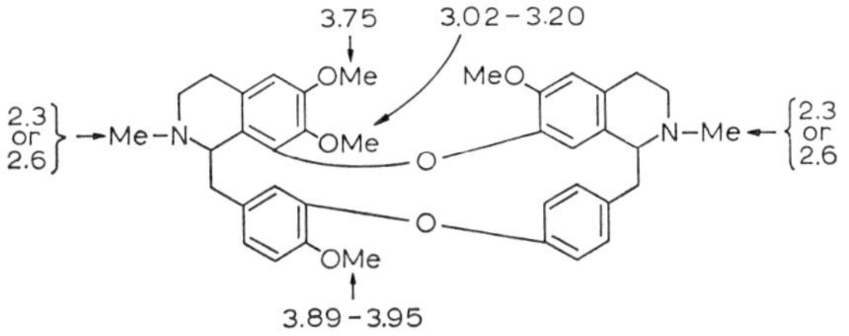

3.75   3.02 – 3.20

2.3 or 2.6 → Me–N   OMe   MeO   N–Me ← {2.3 or 2.6}

OMe   O   O

OMe

3.89 – 3.95

(XXXVIII)

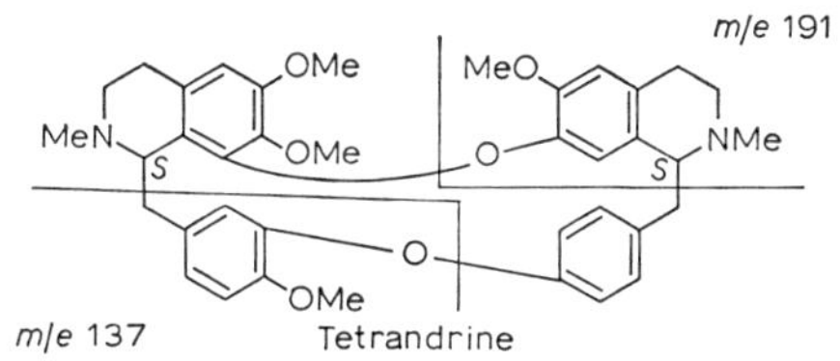

m/e 191

MeN   OMe   MeO   NMe

S   OMe   O   *S*

O

OMe

m/e 137   Tetrandrine

(XXXIX)

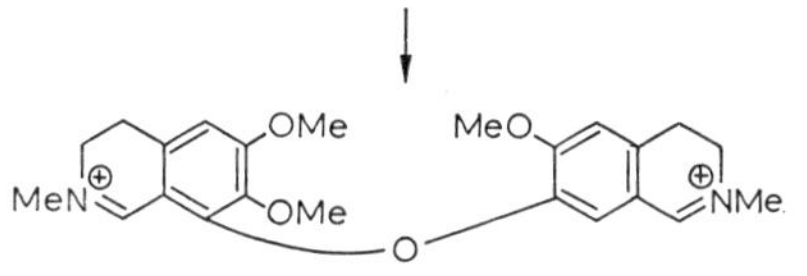

MeN⊕   OMe   MeO   ⊕NMe

OMe   O

(XL)

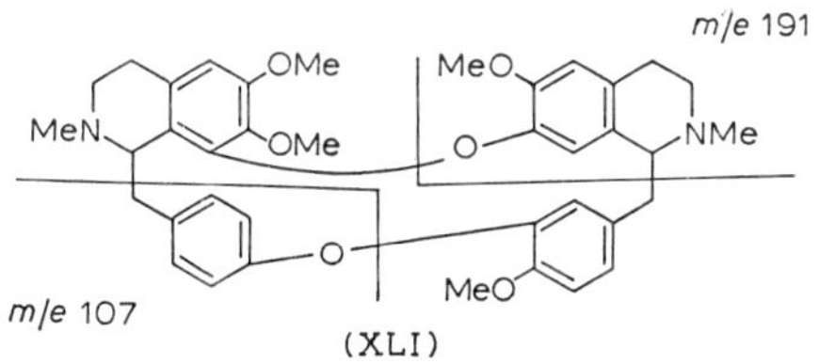

m/e 191

MeN   OMe   MeO   NMe

O   O

O   MeO

m/e 107   (XLI)

(XL) as the base peak. A number of other singly and doubly charged ions in the spectrum have been rationalised, but two ions of importance in structure determination are at *m/e* M-137 and *m/e* M-191, and these are shown by the dissection lines XXXIX. In the oxyacanthine type of structure a similar cleavage, as indicated in XLI, gives ions at *m/e* M-191 and *m/e* M-107.

Trilobine is an example of an alkaloid that contains three diphenyl ether links. The molecular formula, $C_{36}H_{36}N_2O_5$, was established by *H. Kondo* and *T. Nakazata* (J. pharm. Soc. Japan, 1924, **511**, 691; C.A., 1925, **19**, 1708), who showed that there are two methoxyls and two *N*-methyl groups present.

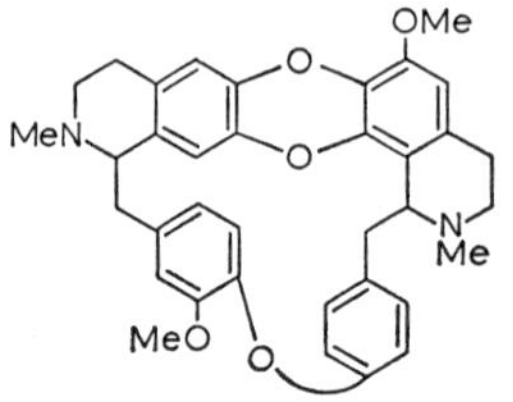

Trilobine

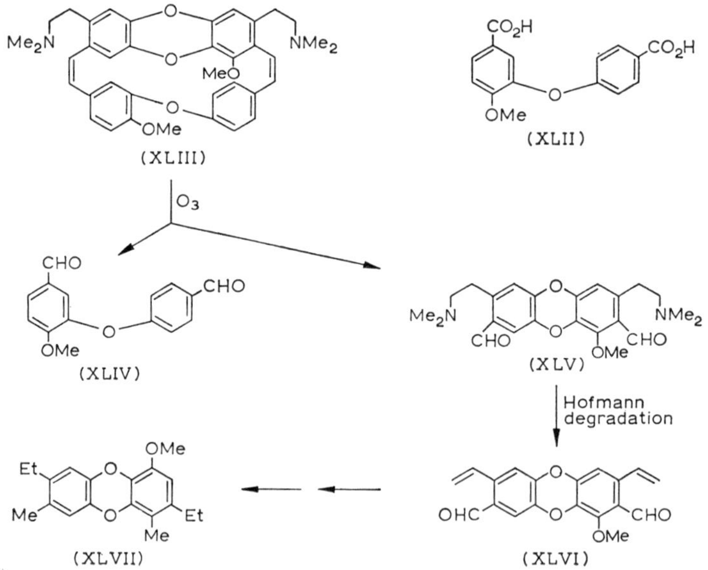

Oxidation of the alkaloid with potassium permanganate gave the known compound XLII. Hofmann degradation of trilobine yielded a bismethine formulated as XLIII, which, upon ozonolysis, gave XLIV and a nitrogen-containing dialdehyde, to which structure XLV was assigned (*Kondo* and *Tomita*, Ann., 1932, **497**, 104). A second Hofmann degradation on the latter provided a divinyl dialdehyde, allotted structure XLVI. However, *F. Faltis et al.* pointed out that on the basis of this and further degradative evidence the structures XLIII–XLVI could be re-drawn with a change in the position of the methoxyl group (Ber., 1941, **74B,** 79; *Faltis,* Ann., 1932, **499,** 301). The divinyl dialdehyde was reduced catalytically, then again by Clemmenson method, to a compound XLVII, which was claimed to be identical with a synthetic specimen (*Tomita* and *C. Tani,* J. pharm. Soc. Japan, 1942, **62,** 468, 481; C.A., 1951, **45,** 4728, 4729). Since the reductive cleavage of the dibenzo-1,4-dioxin type of structure is complex, it is not a very useful method in structural elucidation of alkaloids of the trilobine type.

A correlation between the trilobine and oxyacanthine types of alkaloids has been found and the method is summarised in Scheme 43 (*Y. Inubushi* and *M. Kozuba,* Pharm. Bull., Japan, 1954, **2,** 215; C.A., 1956, **50,** 1052).

SCHEME 43

THE CORRELATION OF OXYACANTHINE WITH TRILOBINE

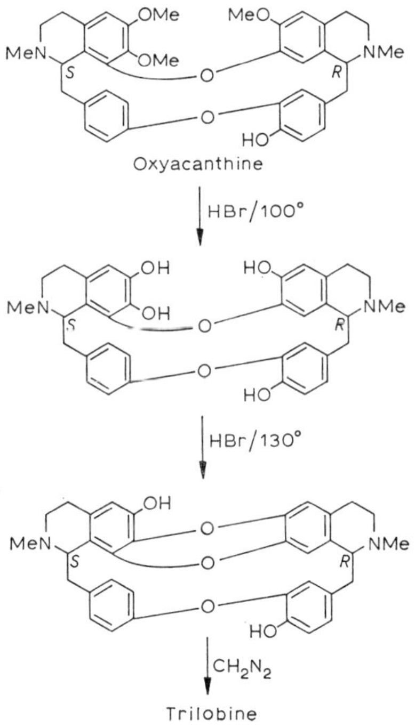

(b) Biosynthesis

The bisbenzylisoquinoline alkaloids are formed in nature by phenolic oxidative coupling between 1-benzyltetrahydroisoquinoline units. Coclaurine and N-methylcoclaurine are the key building blocks, but the sequence of formation of the second and third diphenyl ether linkage is unknown.

Little tracer work has been reported. Coclaurine with a ^{14}C label in the N-methyl group (XLVIII), is incorporated into ($\pm$)**epistephanine** (L), and ($-$)N-methylcoclaurine, labelled with ^{14}C as indicated in XLIX, is incorporated into only one half of L (*D. H. R. Barton, G. W. Kirby* and *A. Wilchers*, Chem. Comm., 1966, 266).

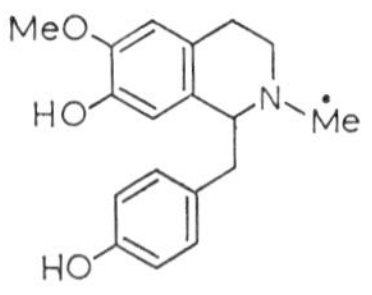

(XLVIII)

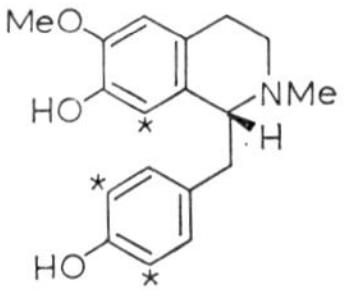

(XLIX)

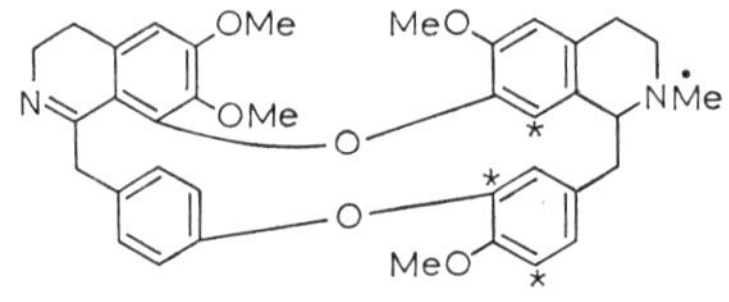

Epistephanine
(L)

The biosynthesis of those alkaloids that possess three diphenyl ether linkages has been discussed (*A. R. Battersby*, in "Oxidative Coupling of Phenols", ed. *W. I. Taylor* and *Battersby*, Dekker, New York, 1967, p. 142; *Barton* and *T. Cohen*, Festsch. A. Stoll, 1957, p. 117).

(c) Chemical synthesis

Although several attempts have been made to stimulate the supposed biosynthetic pathway *in vitro*, little success has been achieved in the synthesis of the actual alkaloids; however, some analogues have been described. The

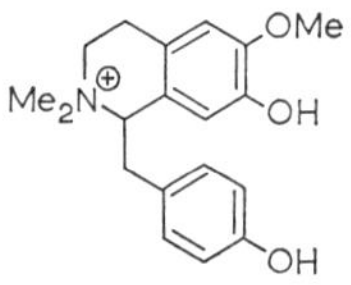

(LI)

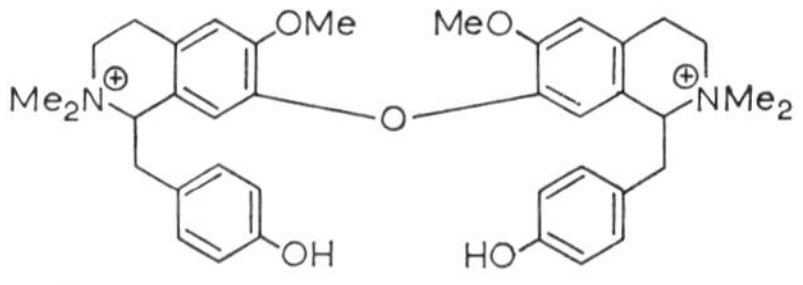

(LII)

SCHEME 44

THE SYNTHESIS OF ISOTETRANDRINE

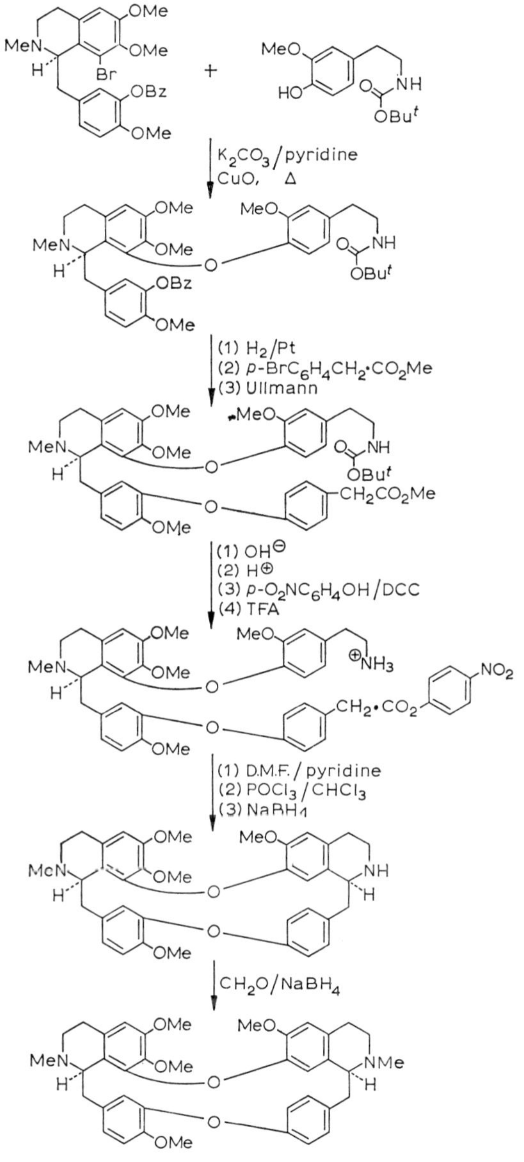

(+) Isotetrandrine

first success of this type involved the formation of LII by the oxidation of LI with potassium ferricyanide in basic solution (*B. Franck, G. Blaschke* and *G. Schlingloff*, Angew. Chem., intern. Edn., 1964, **3**, 192).

Most of the successful syntheses of the alkaloids of the bisbenzylisoquinoline group have depended upon the Ullmann reaction to form the diphenyl ether link or links and the Bischler–Napieralski reaction to form the isoquinoline rings. The proper use of protecting groups for phenolic hydroxyl groups has been a notable feature of this approach, which is illustrated in Scheme 44 for (+)**isotetrandrine** (*Y. Inubushi et al.*, J. chem. Soc., C, 1969, 1547) and in Scheme 45 for **magnoline** (*T. Kametani et al.*, J. heterocycl. Chem., 1966, **3**, 239; Chem. and pharm. Bull., Japan, 1967, **15**, 56).

SCHEME 45

SYNTHESIS OF MAGNOLINE

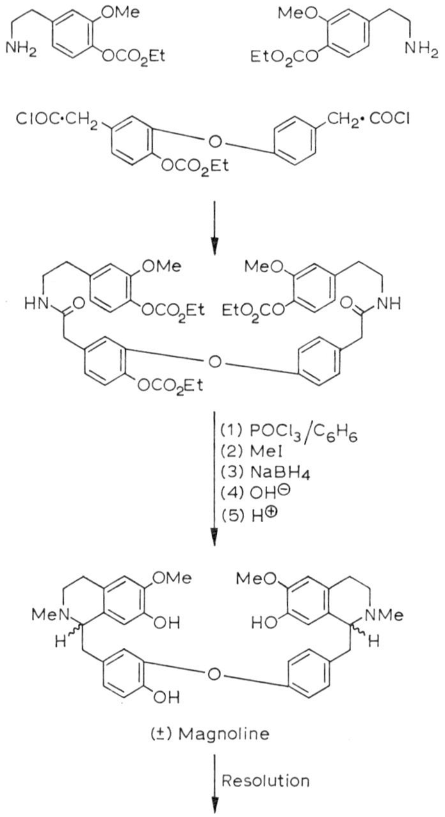

SCHEME 46

THE SYNTHESIS OF *O*-METHYLTILIACORINE

Other alkaloids that have been synthesised include **liensinine** (*idem, ibid,* 1966, **14,** 67; J. chem. Soc., C, 1969, 298), **cepharanthine** (*Tomita, K. Fujitani and Y. Aoyagi,* Tetrahedron Letters, 1967, 1201), and *N*-methyldihydro-menisarine (*S. Ueda,* J. pharm. Soc. Japan, 1962, **82,** 714; *Tomita, Ueda and A. Teraoka,* Tetrahedron Letters, 1962, 635).

The synthesis of (±)*O*-**methyltiliacorine,** which contains a carbon–carbon linkage, is summarised in Scheme 46 (*B. Anjaneyulu, T. R. Govindachari and N. Viswanathan,* Tetrahedron, 1971, **27,** 439).

5. Aporphinoids

Aporphinoid is the term used by *M. Shamma* (in SPR "The Alkaloids", Vol. 6, p. 170), to cover alkaloids of the aporphine, proaporphine, neoproaporphine and oxoaporphine groups. Together they constitute the largest group of isoquinoline alkaloids (over 100 are known). Aporphines are most abundant in the Papaveraceae, but are also found in Araceae, Lauraceae and Monimiaceae, among other families. All are derivatives of the 4*H*-dibenzo[*d,e–g*]quinoline (I) system, with R = H or Me. The currently accepted numbering system is shown, although some confusion may arise when earlier literature is consulted since different numbering systems have been in use from time to time.

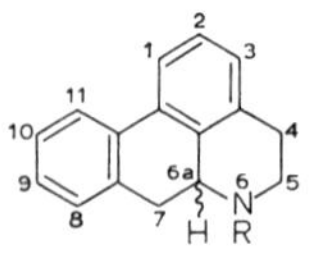

(I) 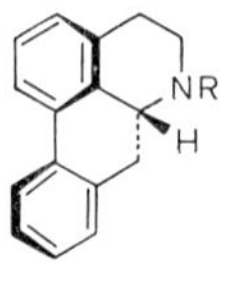   (II)
D-*(R)*-configuration   (III)
L-*(S)*-configuration

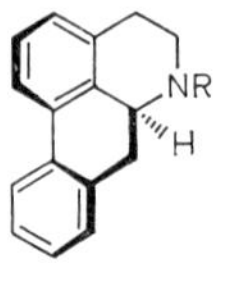

The aporphines can be divided into three major groups, depending upon the extent of methylation of the nitrogen atom. The noraporphines are secondary bases with R = H in I, whereas the aporphines proper (I, R = Me) are tertiary bases. Finally, some quaternary salts have been isolated from plant material in which the nitrogen carries two methyl groups. All naturally occurring aporphines possess oxygen functions at $C_{(1)}$ and $C_{(2)}$ (*K. W. Bentley* and *S. F. Dyke*, J. org. Chem., 1957, **22**, 429). Trioxyaporphines have the additional oxygen function usually at $C_{(9)}$, $C_{(10)}$ or $C_{(11)}$, although a few $C_{(4)}$-, $C_{(7)}$- or $C_{(8)}$-substituted compounds are also known. The majority of the alkaloids of this group possess four oxygen functions, and these are usually to be found at $C_{(1)}$, $C_{(2)}$, $C_{(9)}$, $C_{(10)}$ or at $C_{(1)}$, $C_{(2)}$, $C_{(10)}$, $C_{(11)}$. Finally, some pentaoxyaporphines are known in which the additional oxygen is placed at $C_{(3)}$ or $C_{(4)}$. The oxoaporphines constitute that subgroup of the aporphines with an oxygen function at $C_{(7)}$. These oxygen functions may be OH, OMe, OCH_2O or ketonic oxygen. Aporphines with the (*R*)-configuration II* and with the *S*-configuration III are known. A further chiral centre is created if the molecule possesses a hydroxyl group at $C_{(4)}$ or $C_{(7)}$.

---

* To emphasise the relationship of these alkaloids to isoquinoline, the relevant formulae are usually drawn as represented in II and III.

(a) Aporphines

The chemistry of aporphines has been reviewed on several occasions (*Shamma* and *W. A. Slusarchyk*, Chem. Reviews, 1964, **64**, 59; *R. H. F. Manske*, "The Alkaloids", 1954, **4**, 119; *Shamma*, "The Alkaloids", 1967, **9**, 1; "The Isoquinoline Alkaloids", Academic Press, New York, 1972, p. 194; *M. P. Cava* and *A. Venkateswarlu*, Ann. Rpts. of med. Chem., 1968, ed. *C. K. Cain*, Academic Press, New York, 1969, p. 331; *Shamma* and *S. S. Salgar*, in SPR "The Alkaloids", **4**, 1974, 197; *K. L. Stuart* and *Cava*, Chem. Reviews, 1968, **68**, 321; *Shamma*, SPR "The Alkaloids", **6**, 170).

It is convenient to discuss structural elucidation by chemical and by physical methods separately, but in order to compare the efficacy of each, the same example is discussed under both headings.

(i) Structure determination

(1) By chemical methods. Two principal methods were developed during the classical era—oxidation and Hofmann degradation—although it was realised that reaction of aporphines with acetic anhydride or with benzoyl chloride could also be informative. The von Braun degradation was also used occasionally.

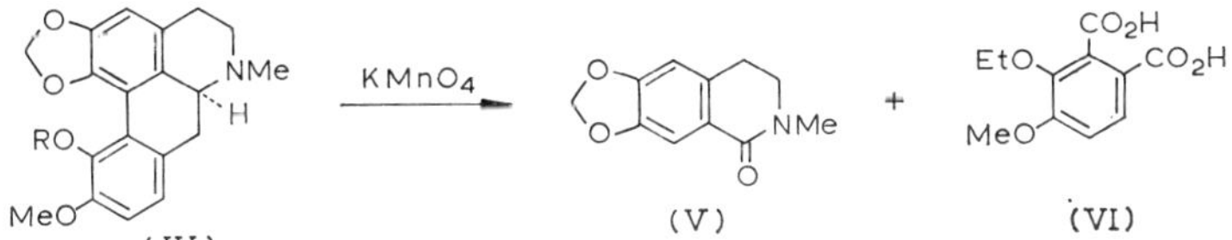

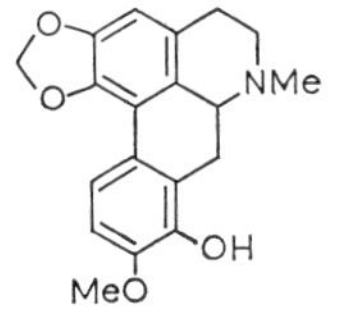

Direct oxidation of aporphines with concentrated nitric acid under forcing conditions yields mettophanic acid (benzene-1,2,3,4-tetracarboxylic acid) and this is now accepted as sufficient evidence that the compound under investigation is an aporphine. Under milder conditions and using potassium permanganate, a 1-oxo-1,2,3,4-tetrahydroisoquinoline is formed, together with a substituted phthalic acid. Phenolic aporphines are usually converted into their *O*-ethyl ethers before oxidation, both to protect the phenolic ring from degradation, and to act as a marker for the position of the hydroxyl group. Thus, $(+)$**bulbocapnine,** $C_{19}H_{19}NO_4$, (IV, R $=$ H) was oxidised, after O-ethylation, to IV (R $=$ Et), to a mixture of V and VI. The structure of bulbocapnine therefore can be either IV (R $=$ H) or VII.

methine

isomethine

(1) MeI
(2) Ag_2O
(3) heat

(1) $KMnO_4$
(2) $-CO_2$

(VIII)

Hofmann degradation of the derived methohydroxide of an aporphine may proceed in two ways to yield an optically inactive methine, and an optically active isomethine. Further Hofmann degradation of either of these bases yields the same vinylphenanthrene (Scheme 47). Since in earlier studies of aporphines it was not possible to recognise the structure of the alkoxy-substituted 1-vinylphenanthrene, it was degraded further by oxidation to a

SCHEME 47

HOFMANN DEGRATION OF APORPHINES

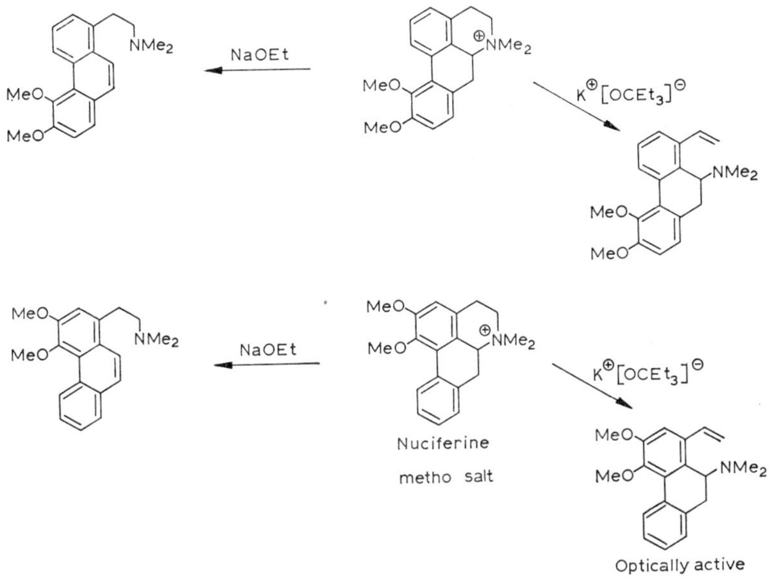

phenanthrene-1-carboxylic acid, which was decarboxylated. The task of identification of the resultant polyalkoxyphenanthrene (VIII) was easier since independent synthesis by the Pschorr method could be employed. Small changes in reaction conditions can affect the course of the Hofmann degradation. Some control is possible by using a bulky base (*J. G. Cannon et al., J. med. Chcm.*, 1973, **16**, 219; *L. D. Yakhontova, O. N. Tolkachev* and *D. A. Pakaln*, Khim. prir. Soedin., 1973, 684) (Scheme 47).

On the basis of the isolation of V in the oxidative degradation and of VIII in the exhaustive methylation procedure, bulbocapnine must be IV (R = H) but with stereochemistry as yet undefined (*J. Gadamer* and *F. Kuntzo*, Arch. Pharm., 1911, **249**, 598; *E. Spath, H. Holter* and *R. Posega*, Ber., 1928, **61**, 322).

The Hofmann degradation to alkoxyphenanthrenes can be ambiguous (*Bentley* and *Dyke, loc. cit.*). Thus, isothebaine methyl ether (IX, R = Me), upon degradation by the method used with bulbocapnine, gave 3,4,5-trimethoxyphenanthrene (X). This could arise from IX (R = Me) or from XI:

(IX)

III

(XI)

(X)

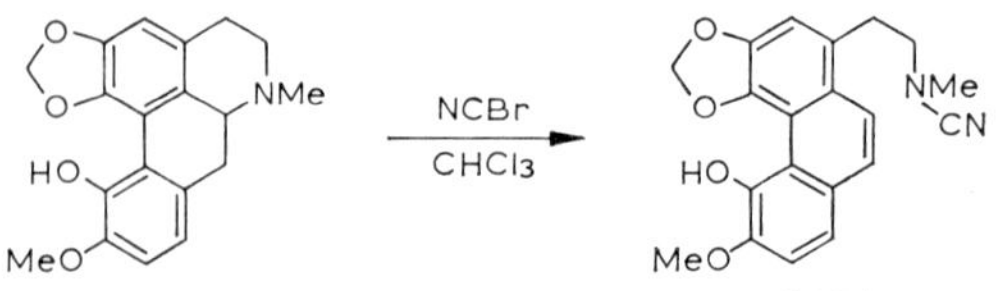

Bulbocapnine

NCBr / CHCl3

(XII)

SCHEME 48

THE ABSOLUTE CONFIGURATION OF GLAUCINE

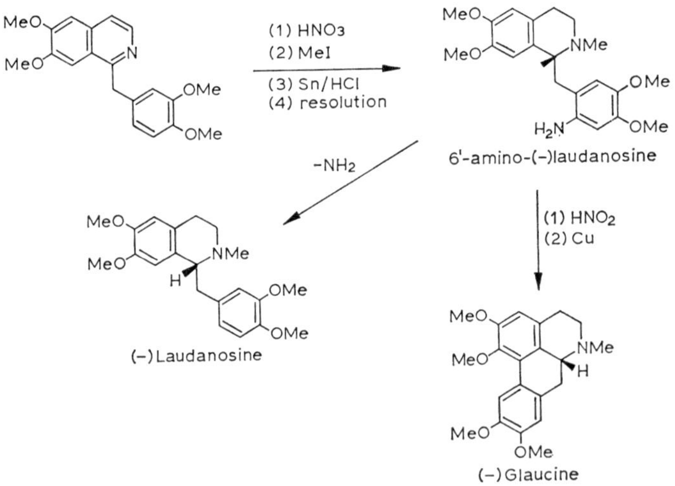

(1) HNO3
(2) MeI
(3) Sn/HCl
(4) resolution

6'-amino-(−)laudanosine

−NH2

(−)Laudanosine

(1) HNO2
(2) Cu

(−)Glaucine

Further information is required in a situation like this—usually an un-ambiguous synthesis.

Von Braun degradation (treatment of a tertiary base with cyanogen bromide) causes cleavage of the $C_{(6a)}$–$C_{(7)}$ bond with the formation of the aromatic methine—for example bulbocapnine (IV, R = H) is converted into XII (*E. E. Smissman et al.*, J. med. Chem., 1970, **13**, 640).

The absolute configuration of (−)laudanosine is known, and since it has been converted into (−)**glaucine** (*F. Faltis* and *E. Adler*, Arch. Pharm., 1951, **284**, 281) (Scheme 48) by a series of reactions which does not affect the asymmetric centre, (−)glaucine must be represented as the (*R*)-configuration shown.

Several aporphines have been inter-related by *O*-methylation, *e.g.* (+)**iso-boldine** and (+)**glaucine** or by cleavage of methylenedioxy groups, followed, by methylation, *e.g.* (+)**dicentrine** and (+)glaucine.

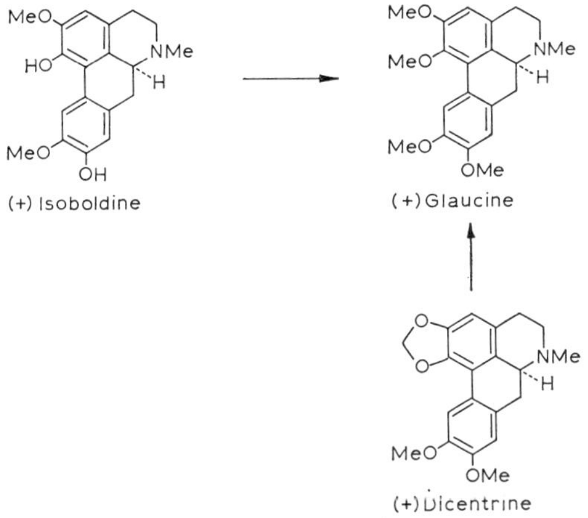

(+) Isoboldine

(+) Glaucine

(+) Dicentrine

It was suggested some time ago that all aporphines that are appreciably dextrorotatory at the sodium D line belong to the L-series [(*S*)-configuration] and that laevorotatory aporphines should be assigned to the D-(*R*)-con-figurational series (*Bentley* and *H. M. E. Cardwell*, J. chem. Soc., 1955, 3252). This rule has stood the test of time (*Shamma* and *M. J. Hillman*, Experientia, 1969, **25**, 544). Further, it has been found that with tetra-oxyaporphines members of the 1,2,9,10-series have $[\alpha]_D$ + 119° or less, whereas those in the isomeric 1,2,10,11-series have $[\alpha]_D$ of +139° or more for the (*S*)-configuration. This useful correlation seems to be independent of the nature of the substituents.

(+)Bulbocapnine fits into this rotation rule, but the absolute configura-

SCHEME 49

THE ABSOLUTE CONFIGURATION OF (+)BULBOCAPNINE

Thebaine
(known configuration)

$H^{\oplus}$

(−)Morphothebaine

(+)Bulbocapnine

Na/NH$_3$

(+)Morphothebaine

tion has been independently deduced by a correlation with **morphothebaine** of established absolute configuration (*W. A. Ayer* and *W. I. Taylor*, J. chem. Soc., 1956, 472) (Scheme 49). An X-ray analysis of bulbocapnine methiodide has also been carried out (*T. Ashida, R. Pepinsky* and *Y. Okaya*, unpublished, quoted by *M. Shamma* and *Slusarchyk, loc. cit.*) confirming the assignment. The analysis also showed that the biphenyl chromophore of the alkaloid is not planar, but has an angle of twist of almost 30°. It was *Shamma* (Experientia, 1960, **16**, 484) who pointed out earlier that aporphines are not planar, and that the conformations are as shown in Scheme 48.

As a result of the studies on o.r.d. curves of a number of aporphines (*C. Djerassi, K. Mislow* and *Shamma, ibid.*, 1962, **18**, 53; *M. J. Vernengo, ibid.*, 1961, **17**, 420; *J. C. Craig* and *S. K. Roy*, Tetrahedron, 1965, **21**, 395) it has been found that aporphines exhibited a Cotton effect of high amplitude between 235 and 245 nm, which seems to be independent of substituents at 1-, 2-, 3-, 9-, 10- and 11-positions. It is of diagnostic value; if the Cotton effect is positive, the alkaloid belongs to the L-(*S*) series, whereas if the Cotton effect is negative, the D-(*R*) series is indicated. (+)*Bulbocapnine*, $[\alpha]_D$ +237° (CHCl$_3$), has maxima specific rotations of $[\alpha]_{242.5}$ 44,900 and $[\alpha]_{227}$ −37,000 and hence belongs to the L-(*S*) series.

(XIII): (R = H)
(XIV): (R = Et)

Corydine
= Glaucentrine

The structures of some aporphines have been assigned on the basis of inter-relationships with other members of supposedly known structure. **Glaucentrine** was originally isolated over 40 years ago (*R. H. F. Manske*, in "The Alkaloids", 1954, **4**, 119) and structure XIII allocated to it on the basis that the melting point of its *O*-ethyl ether tartrate and methiodide were undepressed by the corresponding salts of the synthetic compound XIV. No degradations of **glaucentrine** were examined, and the u.v. spectrum was not recorded. The synthesis of 1-hydroxy-2,9,10-trimethoxyaporphine (XIII; R = H) was described in 1965 (*Shamma* and *Slusarchyk*, Tetrahedron Letters, 1965, 1509) and it was found *not* to be identical with glaucentrine. Further investigation then showed that glaucentrine is identical with natural **corydine.**

(2) *By spectroscopic methods. Ultraviolet spectroscopy* is a very useful method for structural elucidations because the spectra are derived from the biphenyl chromophore which is influenced by the number and position of oxygen functions and not by their nature (*Shamma*, Experientia, 1960, **16**, 484; *Shamma* and *Slusarchyk, loc. cit.*; *A. W. Sangster* and *Stuart, ibid.*, 1965, **65**, 69; *Shamma* and *Salgar*, SPR "The Alkaloids", **4**, 197). Thus, for tetra-oxyaporphines it is a simple matter to distinguish between the 1,2,9,10- and 1,2,10,11-tetraoxygenation patterns. Similarly, the spectra of 1,2,9- and 1,2,10-trioxyaporphines are distinguishable, and those aporphines with only 1,2-dioxygenation also provide characteristic spectra. When the u.v. spectra of monophenolic aporphines in basic solution are examined (*Shamma et al.*, J. org. Chem., 1971, **36**, 3253) it is possible to deduce the presence of a $C_{(9)}$–OH group.

Proton magnetic spectra of aporphines have been closely studied and found to provide a powerful tool for the elucidation of structure (*I. R. C. Bick et al.*, J. chem. Soc., 1961, 1896; *W. H. Baarschers et al., ibid.*, 1964, 4778; *T. H. Yang*, J. pharm. Soc. Japan, 1962, **82**, 804; *A. H. Jackson* and *J. A. Martin*, J. chem. Soc., C, 1966, 2061, 2222; *D. S. Bhakuni, S. Tewari* and *M. M. Dhar*, Phytochemistry, 1972, **11**, 1819). The δ values discussed below refer to chloroform as solvent and TMS as internal standard.

3.42 -3.63 { CH₃O 2 ... CH₃O 1 } 3 4 6a NR 7

3.65-3.72 → CH₃O 11

3.72-3.89 → CH₃O 10 8 9

(a) ortho-*Methoxyl groups.* Thus a $C_{(1)}$-OMe is the most shielded and appears upfield from methoxyl groups at $C_{(2)}$, $C_{(9)}$ or $C_{(10)}$, whereas a $C_{(11)}$-methoxyl resonance has an intermediate chemical shift.

(*b*) *Methylenedioxy groups*. Because the biphenyl system is twisted, the two hydrogen atoms of a methylenedioxy group spanning $C_{(1)}$, $C_{(2)}$ will be in different magnetic environments and will appear as a double doublet. The same, presumably applies to a $C_{(10)}$, $C_{(11)}$-methylenedioxy group. Such a grouping at $C_{(9)}$–$C_{(10)}$, however, absorbs as a two-hydrogen singlet.

(*c*) *Aromatic protons*. The $C_{(3)}$-H is almost always at highest field and a $C_{(11)}$-H test is the most deshielded. Other aromatic hydrogen resonances are intermediate in value. However, for a $C_{(9)}$, $C_{(10)}$-disubstituted ring D, the hydrogens at $C_{(8)}$ and $C_{(11)}$ will appear as singlets, whereas $C_{(10)}$, $C_{(11)}$-disubstitution should allow an AB quartet for $C_{(8)}$ and $C_{(9)}$ absorptions, and this is usually observed if the $C_{(10)}$ and $C_{(11)}$ functions are methoxyl groups. However, if a phenolic hydroxyl is attached to $C_{(11)}$, the $C_{(8)}$ + $C_{(9)}$ protons appear as a singlet absorption. Recently shift-reagent studies have shown (*R. V. Smith* and *A. W. Stocklinski*, Tetrahedron Letters, 1973, 1819) that $Eu(fod)_3$ enables all aromatic protons to be resolved.

The n.m.r. spectra of phenolic aporphines may be simplified by deuterium exchange of *ortho* and *para* protons (*D. H. R. Barton et al.*, Proc. chem. Soc., 1964, 261). The influence of alkali on aromatic proton resonances of phenolic aporphines has also been studied (*K. G. R. Pachler, R. R. Arndt* and *W. H. Baarschers*, Tetrahedron, 1965, **21**, 2159). In the NaOH–DMSO solvent system, characteristic upfield shifts of aromatic protons occur. Usually the shifts observed are *ortho:* 0.47–0.54 ppm., *meta:* 0.25–0.42 ppm and *para:* 0.86 ppm. Aromatic hydrogens attached to the non-phenolic ring undergo an unfield shift of 0.11–0.26 ppm. However, when a phenolic hydroxyl group is at $C_{(1)}$ or at $C_{(11)}$, the proton at $C_{(11)}$ or at $C_{(1)}$, respectively, undergoes a *downfield* shift of almost 1.0 ppm.

The *mass spectra* of a number of aporphines have been recorded and analysed (*C. Djerassi et al.*, J. Amer. chem. Soc., 1963, **85**, 2807; *Jackson* and

SCHEME 50
MASS SPECTRAL FRAGMENTATIONS OF APORPHINES

$$(M-1)^{\oplus}$$

$-OMe \rightarrow (M-31)^{\oplus}$

$-OH \rightarrow (M-17)^{\oplus}$

$-Me \rightarrow (M-15)^{\oplus}$

retro-Diels-Alder

$+\ CH_2=\overset{\oplus}{N}Me$

$(M-43)^{\oplus}$

$-OMe \rightarrow (M-74)^{\oplus}$

$-Me$

$(M-58)^{\oplus}$

Martin, J. chem. Soc., C, 1966, 2181) and the main fragmentations are summarised in Scheme 50.

In 1,2,9,10-tetraoxyaporphines, the $(M-1)^{\oplus}$ peak is the base peak, whereas in the isomeric 1,2,10,11-series this ion has only 50% or less of intensity of the $M^{\oplus}$ ion, which is often the base peak. Also the intensities

SCHEME 51

$(M-31)^{\oplus}$ $(M-15)^{\oplus}$

of the $(M - 15)^{\oplus}$, $(M - 17)^{\oplus}$ and $(M - 31)^{\oplus}$ ions are usually greater in the 1,2,10,11-series than in the 1,2,9,10-tetraoxygenated aporphines. In the secondary amines (NH group present) the $(M - 43)^{\oplus}$ ion is replaced by one at $(M - 29)^{\oplus}$.

If a $C_{(1)}$-OMe group is absent the $(M - 31)^{\oplus}$ ion is weak, but is otherwise strong in intensity. This has been interpreted by the fragmentations shown in Scheme 51.

SCHEME 52

SYNTHESIS OF APORPHINES

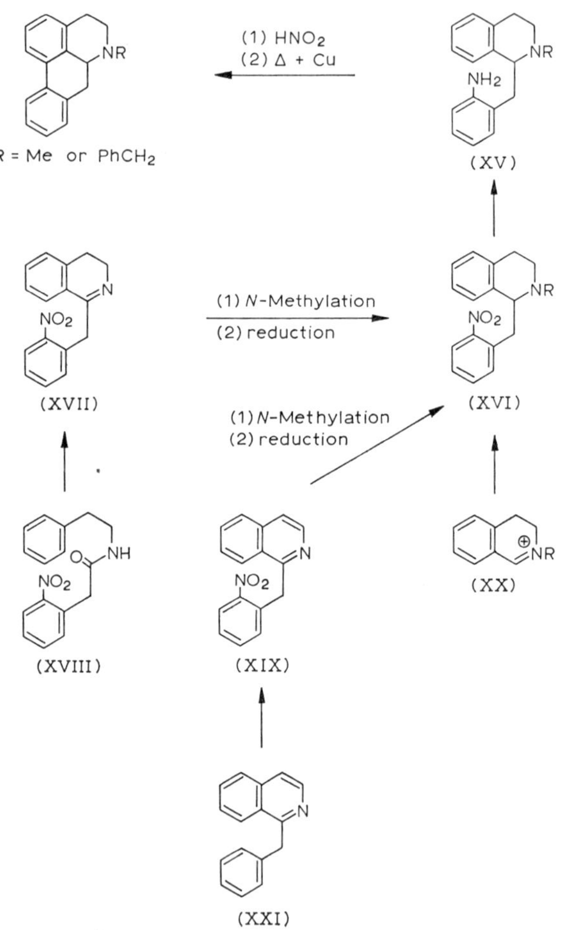

SCHEME 53

A SYNTHESIS OF (±)BULBOCAPNINE

(XXII)

(1) MeI
(2) NaBH$_4$

(1) HNO$_2$
(2) Δ + Cu

H$_2$/Pt

(ii) Chemical synthesis

For many years the only practicable syntheses of aporphines depended upon the Pschorr ring closure of an appropriately substituted 6′-amino-1,2,3,4-tetrahydroisoquinoline, XV (Scheme 52). This precursor is usually derived from the corresponding nitro derivative XVI, which in turn is synthesised by the Bischler–Napieralski ring closure of the required amide, XVIII, to the 3,4-dihydroisoquinoline, XVII, followed by reduction. A synthesis of (±)bulbocapnine by this general method is summarised in Scheme 53; the phenolic hydroxyl group was protected by a benzyl group which was removed in the final step.

Often the major problem is preparing the starting materials required for the Bischler–Napieralski reaction *e.g.* XXII, where, in any case, yields are often low, due in part to the presence of the nitro group. The poorest step in the entire sequence is the Pschorr cyclisation (0–15% are common-place). The yield decreases as the number of oxygen substituents in the 1-benzyl-tetrahydroisoquinoline increases (*D. F. De Tar*, Org. Reactions, 1957, **9**, 490; *J. A. Weisbach et al.*, J. med. Chem., 1963, **6**, 91). However, a C$_{(7)}$-

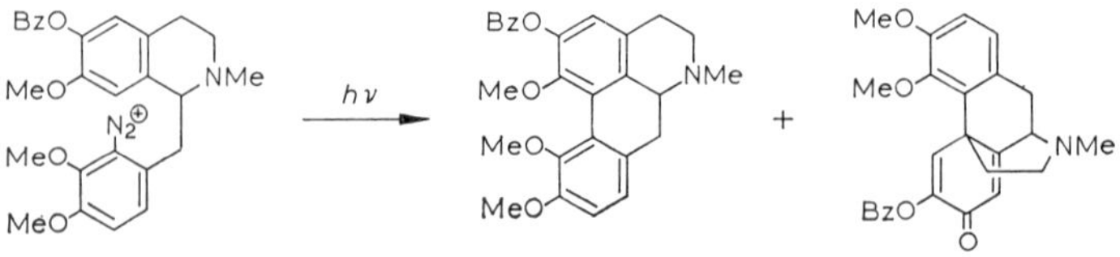

(XXIII)   (XXIV)   (XXV)

hydroxyl group facilitates the Pschorr ring closure (*S. M. Kupchan, V. Kameswaran* and *J. W. A. Findlay*, J. org. Chem., 1973, **38,** 406). The yield of aporphine is also increased by increasing the steric size of the group attached to nitrogen. Some improvements have also been made by photolysing the intermediate diazonium salt, *e.g.* XXIII → XXIV, but the dienone such as XXV is also obtained (*Kametani* and *K. Fukumoto*, Accounts Chem. Res., 1972, **5,** 212).

Alternative methods have been developed to obtain the intermediate of type XIX (Scheme 52). In one method the fully aromatic 1-benzylisoquinoline (XXI) can be nitrated at the 6′-position provided that an alkoxy group is present at $C_{(3')}$ (*para* to the position of substitution). Glaucine was the first aporphine to be synthesised, and papaverine was the starting material (Scheme 54) (*J. Gadamer*, Arch. Pharm., 1911, **249,** 680). 1-Benzyl-

SCHEME 54

THE FIRST SYNTHESIS OF GLAUCINE

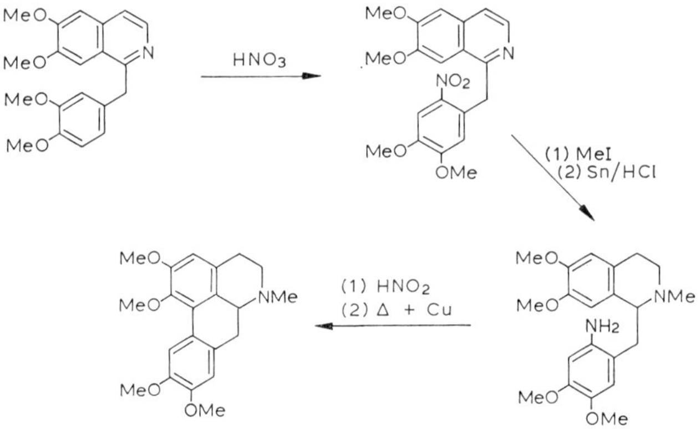

tetrahydroisoquinolines can also be nitrated at $C_{(6')}$ and this provides a useful route to 1,2,9,10-tetraoxyaporphines. Thus **predicentrine, thalicsimidine** and **thaliporphine,** among others, have been synthesised in this way (*Kametani et al.*, J. chem. Soc., C, 1971, 1032; *Shamma* and *Slusarchyk*, Tetrahedron, 1967, **23,** 2503):

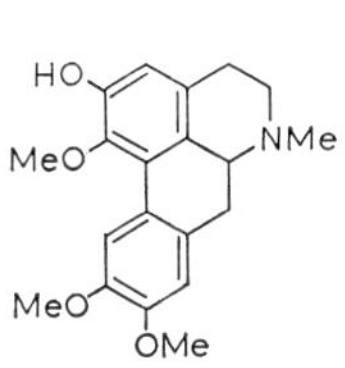

Predicentrine

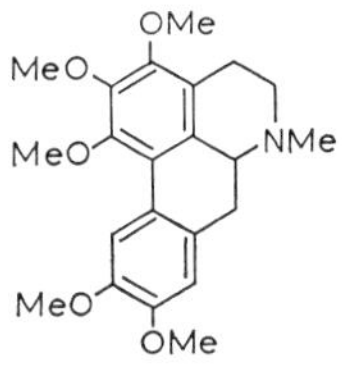

Thalicsimidine

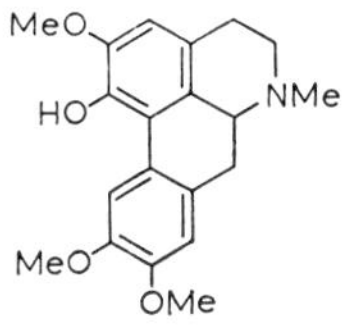

Thaliporphine

Pschorr ring closure is usually effected on the 1,2,3,4-tetrahydroisoquinoline
XV (Scheme 52) but studies have also been made with the fully aromatic
6′-amino derivatives. In the case of 6′-aminopapaverine, a recent reinvestiga-
tion of the Pschorr reaction has revealed that in addition to a low yield of
the oxidised aporphine XXVI, the major product is XXVII. Under different
conditions a 30% yield of XXVIII was obtained, which rearranged in acid
solution to XXIX (*M. P. Cava, I. Noguchi* and *K. T. Buck*, J. org. Chem.,
1973, **38**, 2394). Pschorr conditions, when applied to XXX gave the oxoaporphine XXXI (*I. Ribas, J. Saa* and *L. Castedo*, Tetrahedron Letters, 1973,
3617).

The intermediate XVI (Scheme 52) can also be prepared by the addition
of an *o*-nitrotoluene derivative to a 3,4-dihydroisoquinolinium salt in the

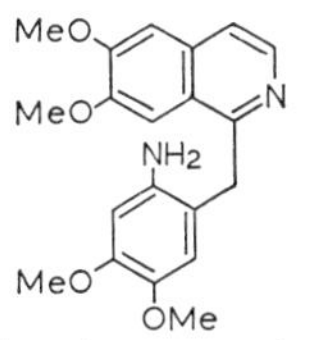

6′-Aminopapaverine

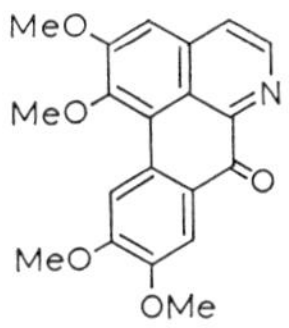

(XXVI)

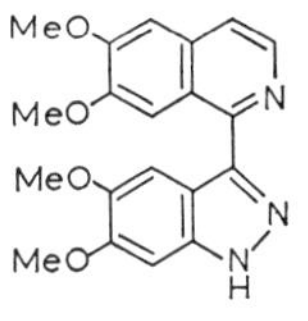

(XXVII)

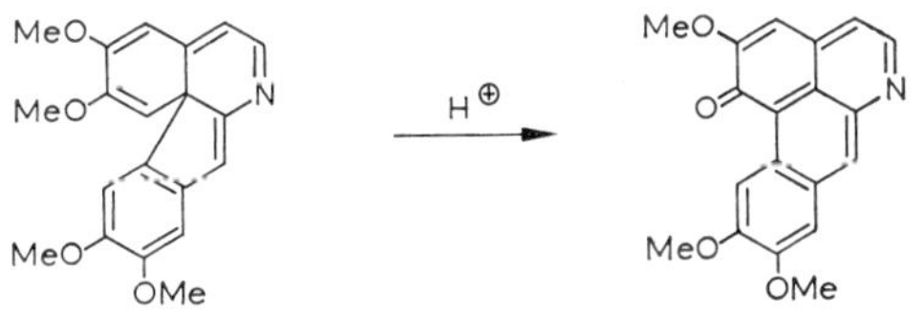

(XXVIII)

H⊕ →

(XXIX)

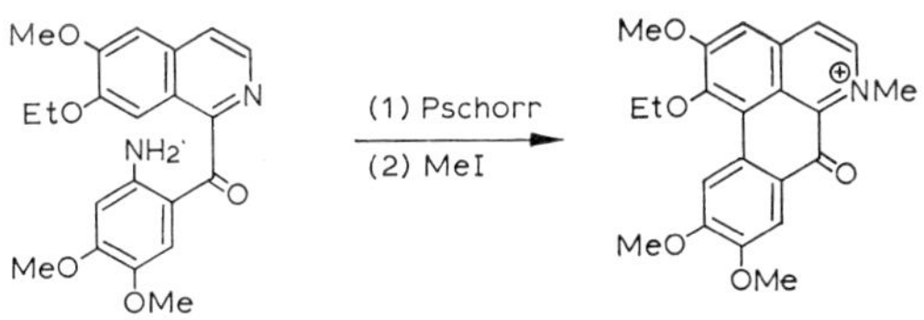

(XXX)

(1) Pschorr
(2) MeI →

(XXXI)

SCHEME 55

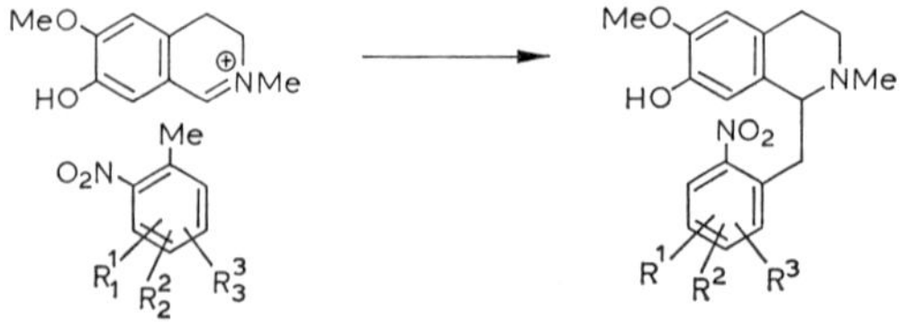

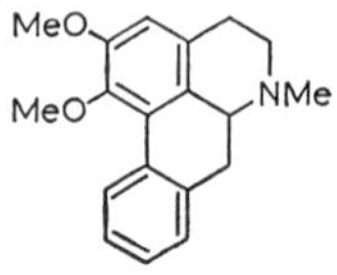

Nuciferine

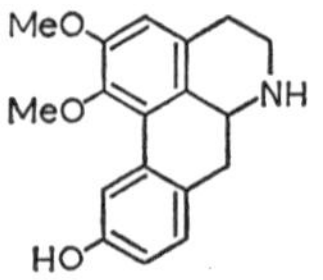

Tuduramine

presence of base (Scheme 55). By this method the alkaloids **thaliporphine, nuciferine, glaucine** and **tuduramine,** among others, have been synthesised (*J. A. Weisbach et al.*, J. med. Chem., 1963, **6,** 91; *Kupchan, Kameswaran* and *Findlay*, J. org. Chem., 1972, **37,** 405; 1973, **38,** 405; *S. Narayanaswami, S. Prabhakar* and *B. R. Pai*, Indian J. Chem., 1969, **7,** 945).

An alternative sequence involves the use of the Reissert reaction (*J. L. Neumeyer et al.*, J. org. Chem., 1969, **34,** 3786; J. pharm. Sci., 1970, **59,** 1850; *Cava* and *M. V. Lakshmikantham*, J. org. Chem., 1970, **35,** 1867):

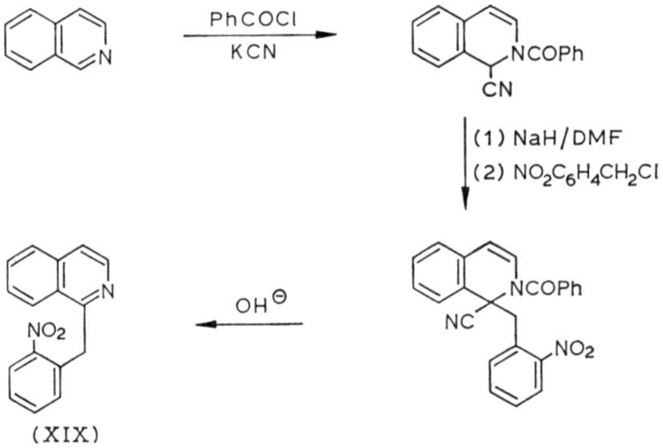

(XIX)

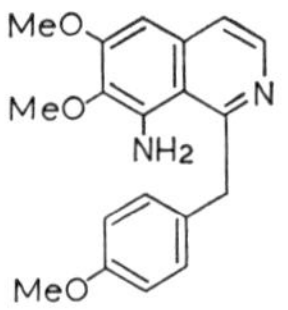

(XXXII)

Another variation on Scheme 52 involves the Pschorr reaction on an 8-amino-1-benzyltetrahydroisoquinoline derivative; the compound XXXII has been successfully ring closed, but 8-aminoisoquinolines are not readily accessible (*S. Ishiwata* and *K. Itakura*, Chem. and Pharm. Bull., Japan, 1969, **17**, 1298; *A. R. Battersby et al.*, Chem. Comm., 1968, 1214).

Yet another method for the formation of the 6′-aminotetrahydroisoquinoline of type XV (Scheme 52) involves the intermediate generation of an aryne (*M. S. Gibson, G. W. Prenton* and *J. W. Walthew*, J. chem. Soc., 1970, 2234). The intermediate aryne itself will cyclise to an aporphine (*S. V. Kessar, S. Batra* and *S. S. Gandhi*, Indian J. Chem., 1970, **8**, 468; *T. Kametani et al.*, J. chem. Soc., C, 1971, 2712). A typical side-reaction, however, is formation of the quaternary salts XXXIII. Yields are often very poor due to a variety of side-reactions (*R. J. Spangler, D. C. Boop* and *J. H. Kim*, J. org. Chem., 1974, **39**, 1368), however, better yields are obtained if the 3,4-dihydroisoquinolinium salt is used, for example XXXIV → XXXV (*Kametani et al.*, J. chem. Soc., Perkin I, 1973, 1212; Chem. and Pharm. Bull., Japan, 1973, **21**, 768).

R = OMe or 2 R = -OCH₂O -

A modification of the Pschorr reaction has led to a synthesis of 7-hydroxyaporphines (Scheme 56) (*F. E. Granchetti* and *Neumeyer*, Tetrahedron,

SCHEME 56

THE SYNTHESIS OF 7-HYDROXYAPORPHINES

1974, **30,** 3701). An extremely useful stereospecific reduction of XXXVI has enabled a synthesis of ($\pm$)norushinsunine to be achieved (*J. Kunitomo et al.,* Chem. and Pharm. Bull., Japan, 1971, **19,** 1502.

Since 1966 photochemical routes to aporphines have become increasingly important (*Kametani* and *Fukumoto,* Accounts Chem. Res., 1972, **5,** 212). Nuciferine was one of the first alkaloids to be synthesised by this means (*Cava et al.,* J. org. Chem., 1970, **35,** 175). The 3,4-dihydroisoquinoline

(XXXVI)

($\pm$) Norushinsunine

SCHEME 57

THE PHOTOCHEMICAL SYNTHESIS OF NUCIFERINE

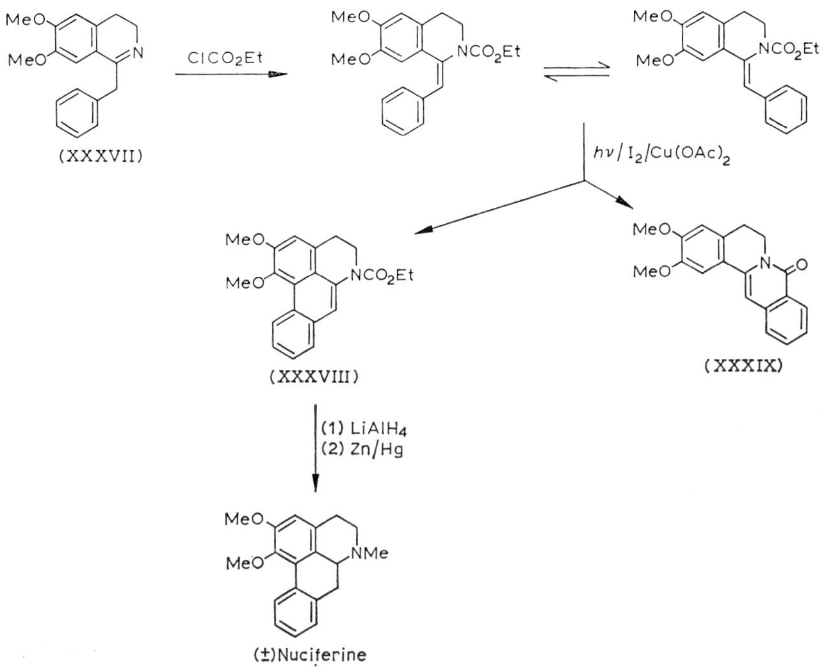

XXXVII (Scheme 57) was reacted with ethyl choloroformate, and the resulting carbamate ester was irradiated in the presence of iodine and cupric acetate. The dihydroaporphine XXXVIII was then reduced, first with lithium tetrahydridoaluminate, then with amalgamated zinc in aqueous alcoholic hydrochloric acid to give ($\pm$)nuciferine. A by-product in the photolysis is the protoberberine derivative XXXIX. Improved yields of aporphines are realised if the 1-benzyl-6′-halogenoisoquinoline derivative is photolysed (*Kupchan* and *Konoji*, Tetrahedron Letters, 1966, 5353). In a minor modification of this method, the halogen is placed at $C_{(8)}$ (*Kametani et al.*, J. chem. Soc., Perkin I, 1972, 1435). A further variation involves photolysing the halogeno compound in the presence of base, *e.g.* XL → XLI (*Cava, P. Stern* and *K. Wakisako*, Tetrahedron, 1973, **29**, 2245).

It is not necessary to protect the nitrogen in these photocyclisations, *e.g.* as in XLII → XLIII (*R. J. Spangler* and *D. C. Boop*, Tetrahedron Letters, 1971, 4851).

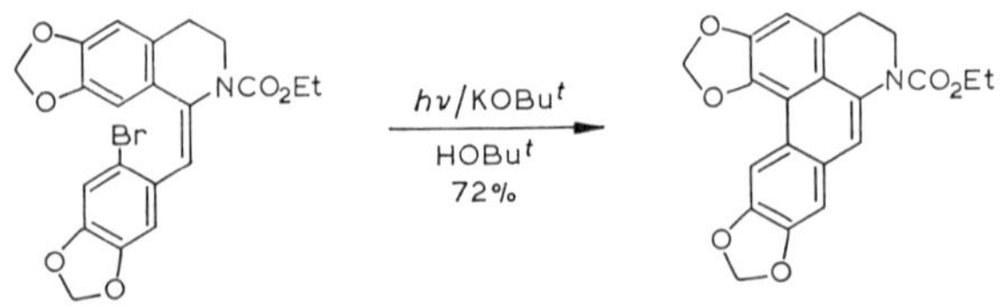

(XL) → (XLI)

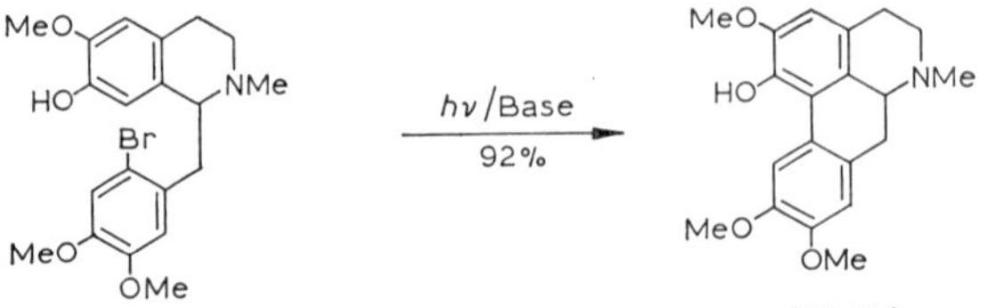

$h\nu/\mathrm{KOBu}^t$

HOBu^t

72%

(XLII) → (XLIII)

$h\nu/$Base

92%

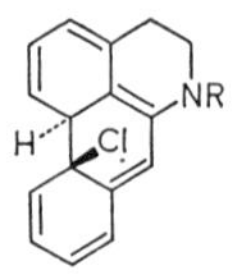

(XLIV)

In these base-catalysed photocyclisations, it is supposed that the first step is reversible formation of an intermediate of the type XLIV, which then loses hydrogen chloride in an irreversible process.

A further method of synthesis of aporphines — phenolic coupling — will be considered in the section on biosynthesis (p. 92).

Steporphine and **cataline** are examples of the rare subgroup of aporphines that are hydroxylated at $C_{(4)}$ (*J. Kunitomo et al.*, Tetrahedron Letters, 1969, 3287; *Ribas, J. Sueiras* and *Castedo, ibid.*, 1972, 2033); **imenine** is also a 7-oxo-aporphine (*Cava et al.*, Chem. Comm., 1969, 1217). The latter has been synthesised by the method shown in Scheme 58 (*Cava* and *I. Noguchi*, J. org. Chem., 1973, **38**, 60). The key reaction, catalytic dehydrogenation of XLV to XLVI is the second example of a reaction first reported in 1966 (*H. Nakao, Y. Yura* and *M. Ito*, Ann. Rep. Sankyo Res. Lab., 1966, **18**, 48).

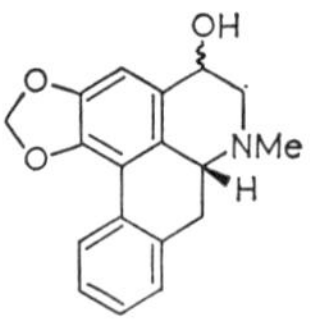

Steporphine

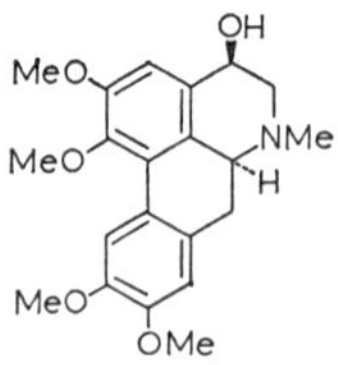

Cataline

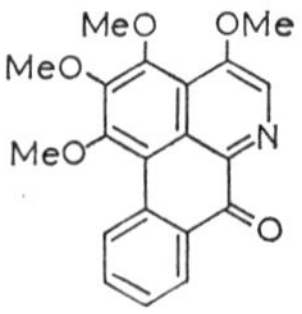

Imenine

SCHEME 58

THE SYNTHESIS OF IMENINE

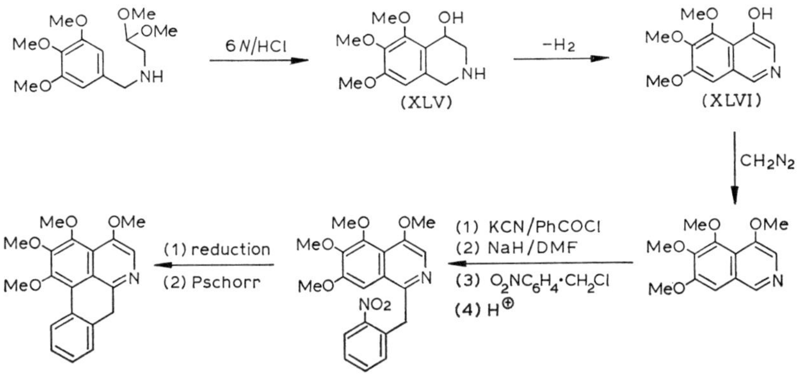

When the 1-benzyltetrahydroisoquinoline XLVII is treated with lead tetra-acetate in acetic acid at room temperature, the 4-acetoxyaporphine XLVIII is among the products (*O. Hoshino, T. Toshioka* and *B. Umezawa, J. chem. Soc., D*, 1971, 1533). The yield (6%) probably compares favourably with that to be expected from the multi-step route involving the Pschorr ring closure. In a further development, the phenolic aporphine XLIX has been converted stereospecifically into L (*Umezawa et al.*, Chem. Comm., 1975, 306).

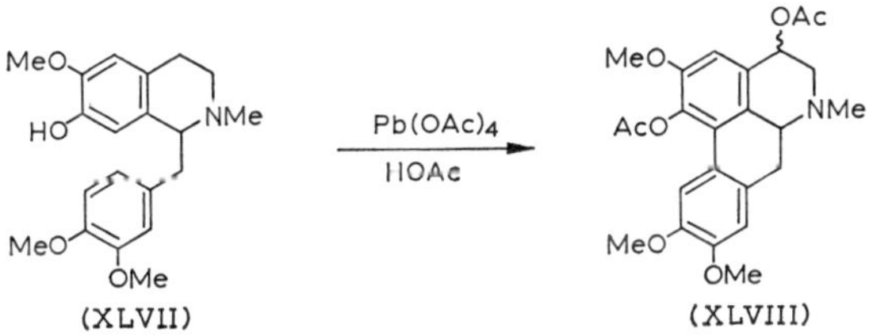

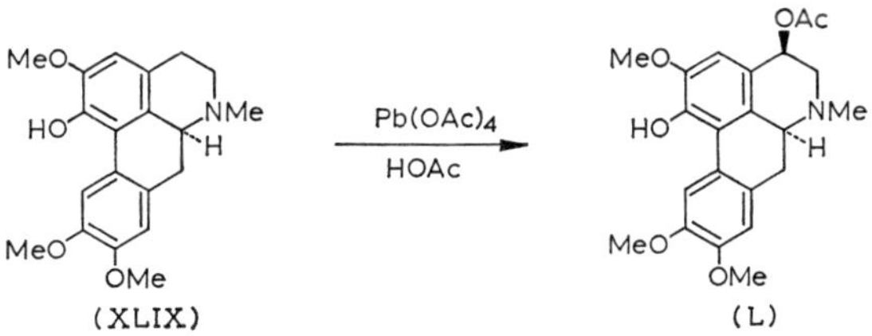

SCHEME 59

WINTERSTEIN AND TRIER'S BIOSYNTHETIC POSTULATE

Norlaudanosoline

≡

Glaucine-type

Corytuberine-type

(iii) *Biosynthesis of aporphines*

As long ago as 1910, *Winterstein* and *Trier* postulated that the aporphine alkaloids could arise in nature from nor-laudanosoline by aryl–aryl bond formation (Scheme 59).

(LI)

(LIII)

(LII)

(LIV)

Cryptowaline

The processes of cyclisation and *O*- and *N*-methylation, and the order of these events were not detailed. The major attraction of the scheme was its ready explanation of the then known oxygenation patterns of aporphines. There are, however, some exceptional substitution patterns, in addition to which, the tri- and di-oxyaporphines are not easily accommodated in this scheme. Nevertheless attempts were made to duplicate these ideas *in vitro*. (For a review see *T. Kametani* and *K. Fukumoto*, Synthesis, 1972, 657). Oxidation of laudanosoline (LI) might yield the aporphine LII, but the substance actually formed was the quaternary salt LIV (*R. Robinson et al.*, J. chem. Soc., 1932, 789; *C. Schöpff et al.*, Ann., 1939, **537**, 143), probably formed *via* the ortho-quinone LIII. Some 25 years later, alkaloids possessing the ring skeleton of LIV were isolated from a North Queensland plant, *e.g.* (−)**Cryptowoline** (see p. 34). It was not until the 1960's that the seemingly obvious modification to LII was made (*B. Frank* and *G. Blaschke, ibid.*, 1962, **659**, 123; 1963, **668**, 145; 1966, **695**, 144; Angew. Chem., intern. Edn., 1964, **3**, 192; *Frank, Blaschke* and *G. Schlingloff*, Tetrahedron Letters, 1962, 439). Oxidation of LV with ferric chloride gave the aporphine LVII in 60% yield. A radical-pairing mechanism is shown in LV → LVIII → LIX → LVII, but the quinone LVI may be the intermediate.

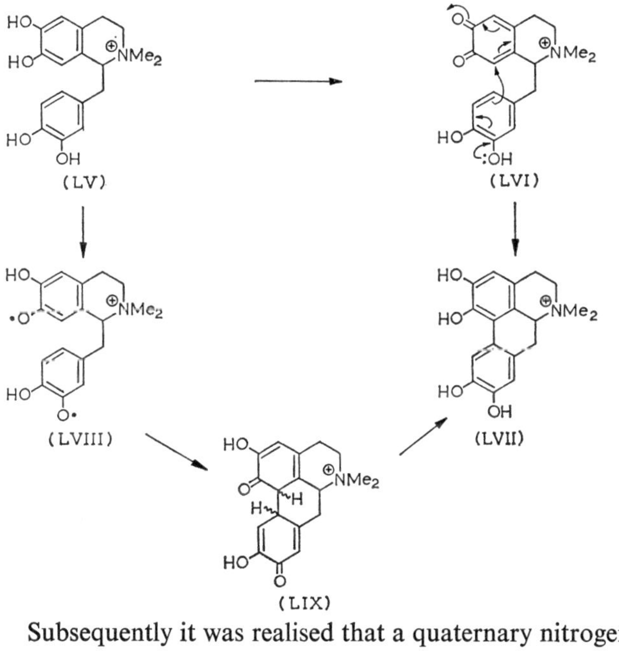

(LV)   (LVI)   (LVIII)   (LIX)   (LVII)

Subsequently it was realised that a quaternary nitrogen is not necessary if *o*-quinone formation can be suppressed. Ferric chloride will form a complex,

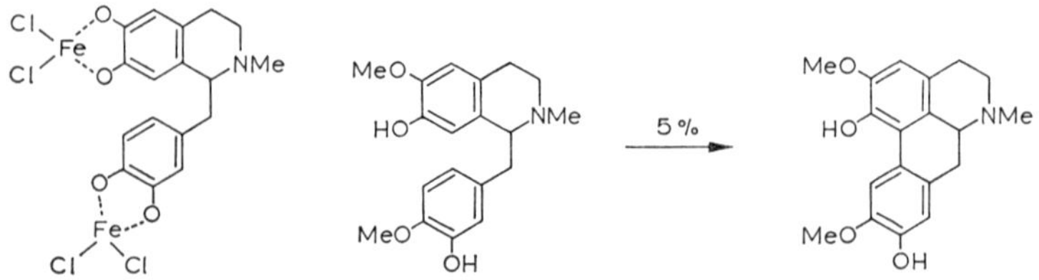

(LX) (LXI) Isoboldine

possibly as LX, if a concentrated solution is used (*Frank* and *L. F. Tietze*, Angew. Chem., intern. Edn., 1967, **6**, 799; *Frank*, *G. Dunkelmann* and *H. J. Lubs*, ibid., p. 1075). A closer analogy to a possible *in vivo* reaction was realised when it was found that oxidation of LXI with potassium ferricyanide lead to **isoboldine** (*A. H. Jackson* and *J. A. Martin*, J. chem. Soc., C, 1966, 2061, 2222). A number of phenolic oxidations have now been reported, and a number of new oxidising agents have been described in syntheses of aporphines. These include vanadyl chloride ($VOCl_3$) (*M. A. Schwartz*, *R. A. Holton* and *S. W. Scott*, J. Amer. chem. Soc., 1969, **91**, 2800; *Schwartz* and *Holton*, ibid., 1970, **92**, 1090; *Schwartz*, Synthetic Comm., 1973, **3**, 33), vanadyl fluoride (*S. M. Kupchan* and *A. J. Liepa*, J. Amer. chem. Soc., 1973, **95**, 4062; *Schwartz*, *B. F. Rose* and *B. Vishnuvajjala*, ibid., 1973, **95**, 612), vanadium tetrachloride (*W. L. Carrick*, *G. L. Karapinka* and *G. T. Kwiatkowski*, J. org. Chem., 1969, **34**, 2388), ferric chloride in dimethylformamide (*S. Tobinga* and *E. Kotani*, J. Amer. chem. Soc., 1972, **94**, 390) and lead tetra-acetate (*O. Hoshino*, *T. Toshioka* and *Umezawa*, Chem. Comm., 1971, 1533; Chem. and Pharm. Bull., Japan, 1974, **22**, 1302).

In 1957 *Barton* and *Cohen* discussed the mechanism of phenolic oxidation (*D. H. R. Barton* and *T. Cohen*, Festschr. A. Stoll Siebzigsten Geburtstag 1957, 117) and applied these ideas to a variety of types of natural products. In the case of aporphine biosynthesis they postulated that the *in vivo* oxidation of a suitably constituted diphenol would involve the initial formation of a cyclohexadienone which could then, by a dienone–phenol rearrangement, be transformed into the observed aporphine bases. These ideas are illustrated in Scheme 60, starting with orientaline. Since the dienone LXII is unsymmetrically substituted, there are FOUR theoretically possible ways for the dienone–phenol rearrangement to occur, thereby giving rise to the observed 1,2,10,11-tetraoxygenation (LXIII) and the isomeric 1,2,9,10-substitution pattern LXIV or LXV. The scheme has the great merit that it also easily explains the unusual oxygenation pattern found in **crebinine.**

Barton and *Cohen* suggested further that the trioxyaporphines could be derived similarly from trioxy-1-benzylisoquinoline derivatives. These ideas are illustrated in Scheme 61, using *N*-methylcoclaurine as precursor, when

SCHEME 60

THE BIOSYNTHESIS OF TETRAOXYAPORPHINES

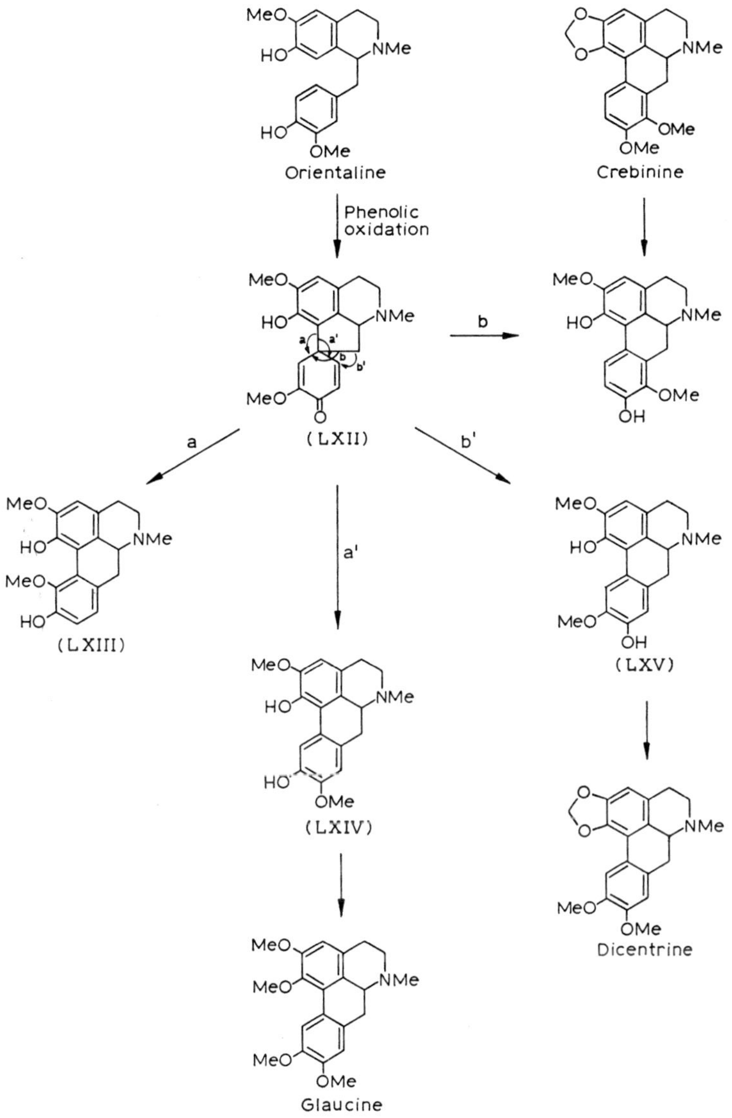

SCHEME 61

A POSSIBLE BIOSYNTHETIC PATHWAY TO TRIOXYAPORPHINES

N-Methylcoclaurine

(LXVI)

a

b

Laureline

N-Methylanaboline

the formation of **laureline** and *N*-**methylanalobine** can arise from the two possible rearrangement pathways of the symmetrical dienone LXVI.

However, to explain the formation of trioxyaporphines like isothebaine, the precursor would need to be LXVII which is very unlikely, and an orthodienone LXVIII would have to be formed (Scheme 62).

The biosynthesis of dioxyaporphines cannot be accommodated at all on this scheme.

SCHEME 62

A POSSIBLE BIOSYNTHETIC ROUTE TO ISOTHEBAINE

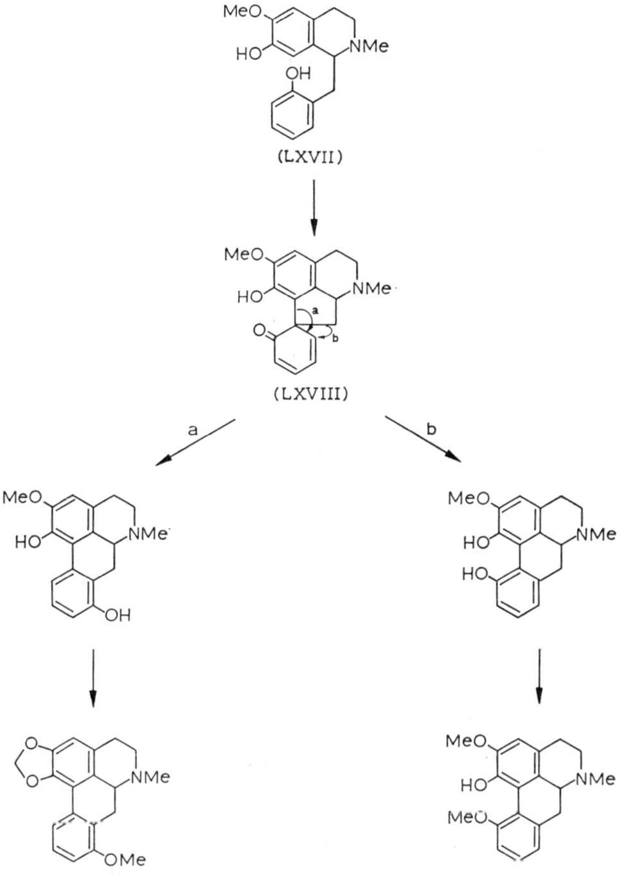

Stephanine Isothebaine

A more satisfactory proposal (*Battersby*, Proc. chem. Soc., 1963, 199) involves genesis of the trioxyaporphines from *tetra*oxy-1-benzylisoquinolines as outlined in Scheme 63. Thus, the dienone LXII formed from orientaline undergoes a reduction to LXIX, which may then undergo a dienol–benzene type of rearrangement. FOUR products are possible for LXIX, but only two are shown, leading to isothebaine and the stephanine-type oxygenation pattern LXX. This modified scheme has the great merit that the biosynthesis of the dioxyaporphines are easily explained (Scheme 64) from a trioxybenzylisoquinoline precursor.

SCHEME 63

THE BIOSYNTHESIS OF TRIOXYAPORPHINES

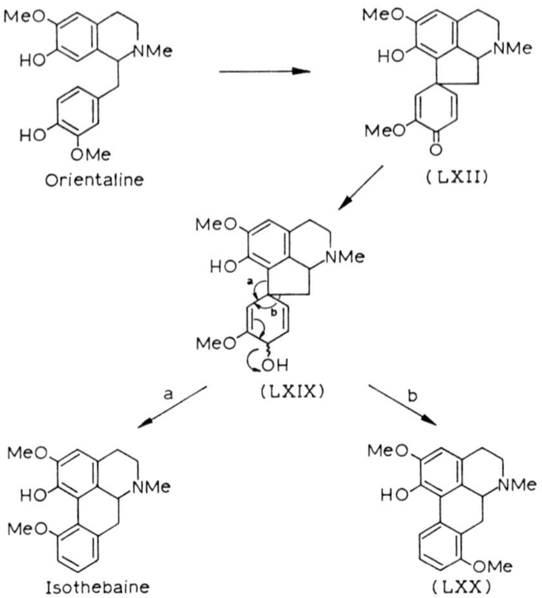

SCHEME 64

THE BIOSYNTHESIS OF DIOXYAPORPHINES

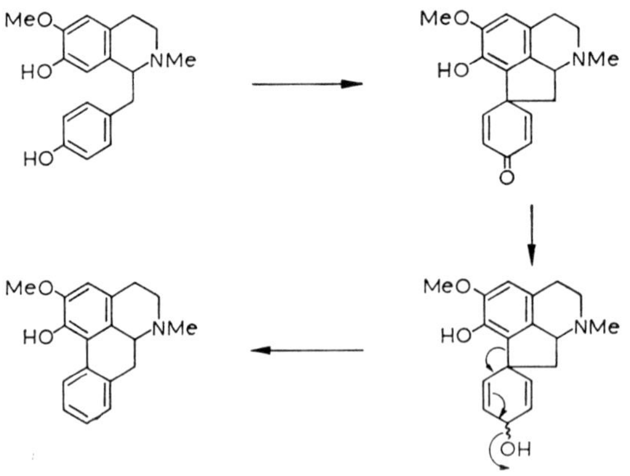

The ideas concerning the biosynthesis of aporphines have been convincingly substantiated in two main ways. Firstly, some very elegant labelling experiments have been carried out and secondly, and most excitingly perhaps, compounds possessing the cyclohexadienone ring system have been isolated from plant materials — the group of alkaloids now known as proaporphines. They are often accompanied by aporphines, into which they are very easily converted upon treatment with dilute acids under mild conditions. In addition several *in vitro* syntheses of proaporphines and aporphines have been achieved. Each of these will be discussed below. There have been also some surprises in the results obtained for *in vivo* studies, and these too will be mentioned.

[^{14}C]Orientaline, labelled as shown in LXXI, is incorporated into isothebaine in *P. orientale* (*Battersby et al.*, Chem. Comm., 1965, 230; 1969, 464). The ratio of the ^{14}C-labelled atoms remains constant, thus showing that the labelled OMe group of orientaline remains intact. (+)Orientaline, ^{14}C-labelled at $C_{(3)}$ was incorporated twentyeight times more efficiently into (+)isothebaine than the (−)isomer. The synthesis of (±)isothebaine *in vitro* along the route summarised in Scheme 63 has been achieved (*Battersby*, *T. H. Brown* and *J. H. Clements*, J. chem. Soc., 1965, 4550). If orientalinone is treated with acid, instead of being reduced with sodium tetrahydridoborate, the product is the tetraoxyaporphine, **isocorytuberine** (*Jackson* and *Martin*, ibid., 1966, 2222).

It has been demonstrated that when tritium-labelled (+)*N*-methylcoclaurine is fed to *P. dubium*, it is incorporated stereospecifically into (+)**roemerine** (*Barton et al.*, ibid., 1967, 2134). However, when *N*-methylcoclaurine ^{14}C-labelled in both methyl groups was administered, the derived roemerine showed a significant loss of the activity associated with the [^{14}C-OMe] group, indicating that the methylenedioxy group formation is preceded by significant *O*-demethylation, followed by re-methylation. It was found also that (±)coclaurine is the precursor of **anonaine** and of **mecambrine**; the latter is transformed *in vivo* into **mecambroline**.

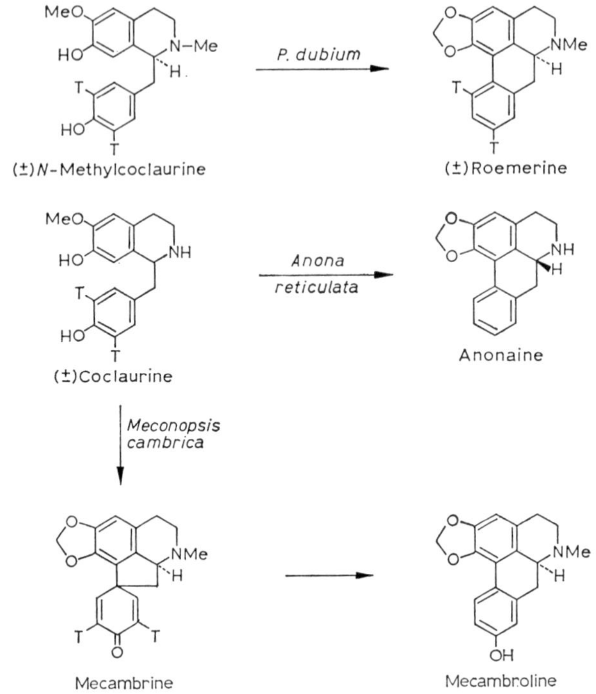

(±)N-Methylcoclaurine → *P. dubium* → (±)Roemerine

(±)Coclaurine → *Anona reticulata* → Anonaine

(±)Coclaurine → *Meconopsis cambrica* → Mecambrine → Mecambroline

(b) The proaporphines

The first two members of this rapidly growing group of alkaloids to be characterised were **pronuciferine** (LXXIIa) (*K. Bernauer*, Helv., 1963, **46**, 1783; 1964, **47**, 2122) and **crotonosine** (LXXIIb) (*Barton et al.*, Proc. chem. Soc., 1963, 280).

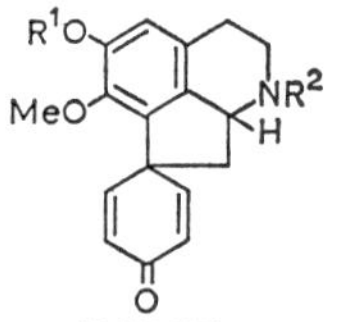

(LXXII)

(a): $R^1 = R^2 = Me$, (+)Pronuciferine
(b): $R^1 = R^2 = H$, (+)Crotonosine
(c): $R^1 = H$; $R^2 = Me$, (+)Homolinearisine

Several are now known, together with partially reduced derivatives based upon the following structures:

(For reviews see *K. L. Stuart* and *M. P. Cava*, Chem. Reviews, 1968, **68,** 321; *Bernauer* and *W. Hofheinz*, Fortschr. Chem. org. Naturstoffe, 1968, **26**, 245; *F. Santavy*, in "The Alkaloids", 1970, **12, 349**.)

A characteristic reaction of the proaporphines, which has aided structural studies, is of course the acid-catalysed rearrangement to aporphines. Reductive cleavage with sodium in liquid ammonia on the other hand, yields a 1-benzylisoquinoline alkaloid, *e.g.* (+)pronuciferine (LXXIIa) gave (−)armepavine, a reaction of considerable importance in assigning absolute configuration at $C_{(6a)}$ of the proaporphines. The spectral properties have been summarised by *Shamma* in "The Alkaloids", Academic Press, New York, 1972, p. 191.

Armepavine

The structures of the reduced proaporphines are based upon interrelationships with the proaporphines proper, *e.g.* (+)linearisine can be hydrogenated to a dihydro compound, LXXIII, which is the enantiomer of the reduction product of (+)crotonosine (LXXIIb) (*Barton et al.*, J. chem. Soc., C, 1966, 1676). The configuration at $C_{(7a)}$ was deduced from c.d. measurements (*G. Snatzke* and *G. Wollenberg*, *ibid.*, 1966, 1681). Several syntheses of proaporphines have now been recorded (*Shamma*, *ibid.*, p. 186). The first synthesis of a proaporphine by phenolic coupling involved the formation

Linearisine   (LXXIII)   Crotonosine
                                            (LXXII b)

(LXXIV)

(±)Glaziovine

of LXXIV, together with orientalinone, (LXII), p. 97, when orientaline was oxidised with potassium ferricyanide (*Battersby* and *T. H. Brown,* Chem. Comm., 1966, 1970). In a second example, similar oxidation of *N*-methylcoclaurine gave (±)**glaziovine,** which, with diazomethane was converted into (±)pronuciferine (LXXIIa) (*Kametani* and *H. Yagi,* J. chem. Soc., C, 1967, 2182).

An alternative method, first used in a synthesis of (+)pronuciferine is summarised in Scheme 65 (*Bernauer,* Helv., 1968, **51,** 1119).

SCHEME 65

THE SYNTHESIS OF (±)PRONUCIFERINE

Pronuciferine

When the Pschorr reaction was applied to LXXV, the products were the aporphine (LXXVI) and (±)**homolinearisine** (LXXVII) (*S. Ishiwata* and *K. Ilakura,* Chem. and Pharm. Bull., Japan, 1970, **18,** 416, 1841). Proa-

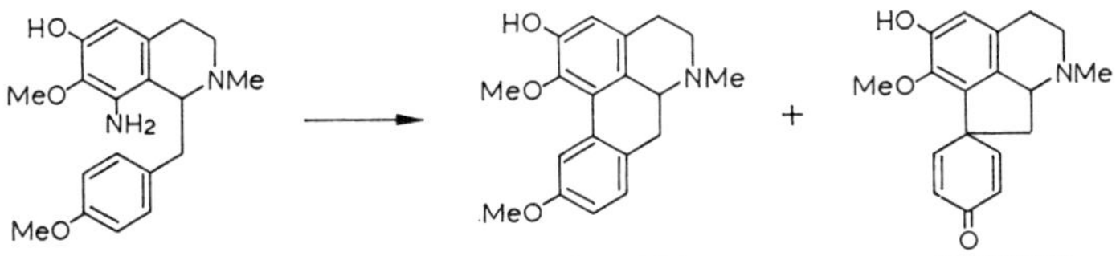

(LXXV) (LXXVI) (LXXVII)

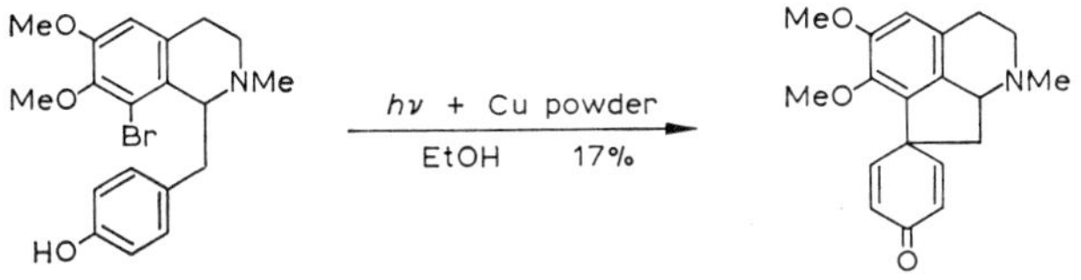

(LXXVIII) (LXXIIa)

porphines may also be obtained in the course of photochemical reactions designed to prepare aporphines, *e.g.* LXXVIII → LXXIIa (*Z. Horii*, *Y. Nakashita* and *C. Iwata*, Tetrahedron Letters, 1971, 1167; *Kametani et al.*, Chem. Comm., 1971, 724).

The biosynthesis of the proaporphine alkaloids has also been studied by the use of tracer-labelled precursors. Thus, it was found that (+)nor-orientaline, labelled as shown in LXXX was incorporated into (+)crotonosine,

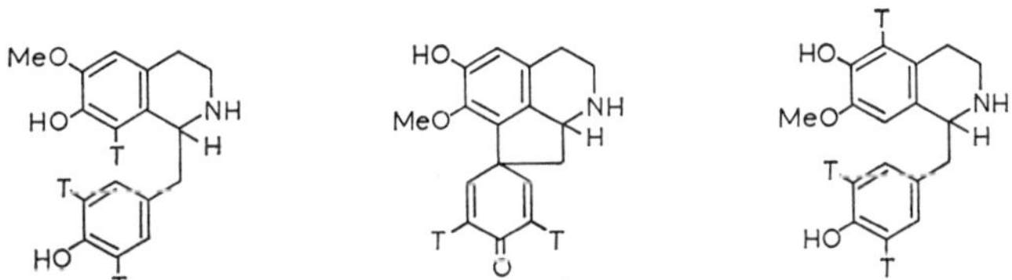

(LXXX) (+)Crotonosine (LXXIX) (LXXXI)

as LXXIX in *Croton linearis* (*Barton et al.*, Chem. Comm., 1965, 141). Because the isomeric precursor LXXXI was not incorporated, it means that a de-*O*-methylation-remethylation sequence is involved.

In principle some aporphine alkaloids, *e.g.* bulbocapnine, may be biosynthesised by direct phenolic coupling of reticuline or from orientaline *via* the dienone (Scheme 66).

SCHEME 66

POSSIBLE BIOSYNTHETIC PATHWAYS TO BULBOCAPNINE

Reticuline

Bulbocapnine

Labelling experiments involving the feeding of (±)reticuline to *Corydalis cava* suggests the direct pathway (*Blaschke*, Arch. Pharm., 1968, **301,** 432; 1970, **303,** 358). This type of pathway is supported by the observation that (±)[$^{14}C_{(3)}$]reticuline is incorporated into isoboldine in *P. somniferum* (*E. Brochmann-Hanssen, C.-C. Fu* and *L. Y. Musconi*, J. pharm. Sci., 1971, **60,** 1880). The same precursor, labelled at the *N*-methyl group is incorporated very efficiently into **magnoflorine** (*Brochmann-Hanssen et al.*, Chem. Comm., 1972, 1269). (±)Orientaline was not incorporated into isoboldine; neither was (±)**norprotosinomenine** (*Idem*, J. pharm. Sci., 1973, **62,** 1291).

Unexpected results were obtained during a study of the biosynthesis of (+)corydine, (+)glaucine and (+)dicentrine in *Dicentra eximia* (*Battersby et al.*, Chem. Comm., 1971, 985). It was eventually found that the key

Isobaldine

Magnoflorine

Norprotosinomenine

SCHEME 67

THE INVOLVEMENT OF NEOPROAPORPHINES IN BIOSYNTHESIS

Norprotosinomenine   (LXXXII)

Corydine

Boldine

2 R = CH$_2$: Dicentrine
R = OMe: Glaucine

intermediate to all three alkaloids is norprotosinomenine. A new type of dienone was postulated to account for the results (Scheme 67). ($\pm$)Laudanosoline, ($\pm$)reticuline and ($\pm$)orientaline were not incorporated.

A compound, LXXXIII, structurally related to neoproaporphines such as LXXXII has been obtained by diazotising 6'-aminopapaverine (*Cava, I. Noguchi* and *K. T. Buck*, J. org. Chem., 1973, **38**, 2394; *Kametani et al.*, Chem. and Pharm. Bull., Japan, 1972, **20**, 1973). In acid solution, this dienone rearranges to the aporphine LXXXIV. The same substance has been prepared by oxidising LXXXV with ceric sulphate in acid solution (*Kupchan* and *Liepa*, J. Amer. chem. Soc., 1973, **95**, 4062).

$H^{\oplus}$

ceric
sulphate

(LXXXIII)   (LXXXIV)   (LXXXV)

(c) *The oxoaporphines*

The isolation of the so-called oxoaporphines from plants in the Anonaceae, Araceae, Hernandiaceae, Lauraceae, Magnoliaceae, Menispermaceae, Papaveraceae and Ranunculaceae families is a fairly recent development in the chemistry of aporphines (*Shamma* and *R. L. Castenson*, "The Alkaloids", 1973, **14**, 225). Although only about 20 are known at present, this number will undoubtedly increase sharply. They are probably derived *in vivo* by oxidation of the corresponding aporphine bases. All of the oxoaporphines characterised so far have a fully aromatic nitrogen-containing ring together with a ketone carbonyl group at $C_{(7)}$. Other oxygen functions are found in those positions expected for aporphines themselves. In one case, **imenine,** an additional oxygen function is present at $C_{(4)}$, whereas **pontevedrine** contains an amide-carbonyl group at $C_{(5)}$. All are characterised by deep colours.

(*i*) *Structure determination*

The first member of the group to be examined was **lidiodenine**. It analyses for $C_{17}H_{19}NO_3$, is optically inactive, contains one methylenedioxy group, no phenolic hydroxyl, methoxyl or *N*-methyl gioups (*W. I. Taylor*, Tetrahedron, 1961, **14**, 42). Although lidiodenine forms an oxime, the carbonyl band is at 1639 cm^{-1} in the infrared, suggesting a highly conjugated carbonyl group. Oxidation of the alkaloid with chromic acid leads to 1-aza-anthraquinone-4-carboxylic acid, which is decarboxylated to the known 1-aza-anthraquinone. On the basis of this evidence, together with the knowledge that aporphine alkaloids occur in members of the Magnoliaceae family from whence **lidiodenine** came, the following structure was proposed for the alkaloid:

Lidiodenine　$\xrightarrow{\text{CrO}_3}$

This was confirmed by a synthesis (Scheme 68) utilising a Pschorr phenanthrene ring closure. The alkaloid was obtained subsequently by oxidising ushinsunine or roemerine with chromium trioxide (*T.-H. Yang*, J. pharm. Soc. Japan, 1962, **82**, 794; *M. Tomita et al.*, *ibid.*, 1962, **82**, 1574). The n.m.r. spectrum of lidiodenine has been described (*Kupchan, M. I. Suffness* and *E. M. Gordon*, J. org. Chem., 1970, **35**, 1682; *D. Warthen, E. L. Gooden* and *M. Jacobson*, J. pharm. Sci., 1969, **58**, 637; *I. R. C. Bick* and *G. K.*

SCHEME 68

A SYNTHESIS OF LIDIODENINE

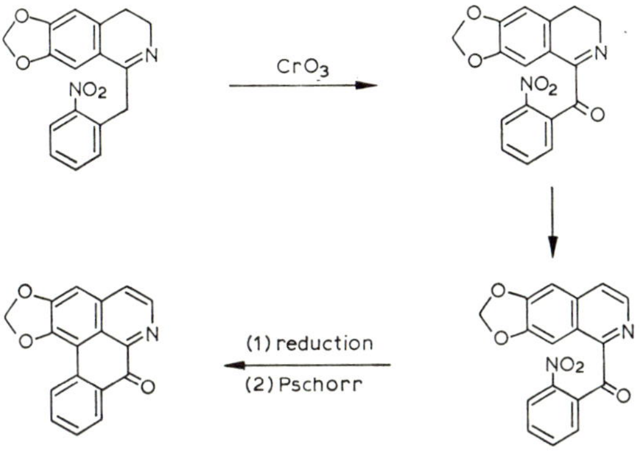

Douglas, Tetrahedron Letters, 1964, 1629). Lidiodenine is of considerable current interest because of its significant cytotoxic inhibitory activity against cells derived from human carcinoma of the nasopharynx (*Warthen, Gooden and Jacobson, loc. cit.*).

Alkaloids of the oxoaporphine group isolated after lidiodenine have been allocated structures principally as a result of u.v. and n.m.r. evidence, together with a correlation with the corresponding aporphine base — usually the latter are oxidised to the oxoaporphine with chromium trioxide in pyridine. In some cases the structures have been supported by a study of mass spectral fragmentations and confirmed by total synthesis.

In the u.v. spectrum a band at about 280 nm is held to be characteristic of a 1,2-methylenedioxy-3-methoxy-, or of a 1,2,3-trimethoxy-oxoaporphine that is unsubstituted at $C_{(4)}$. A band at 222–226 nm is associated with a 1,2,10,11-tetraoxygenated oxoaporphine. The n.m.r. spectral data on oxoaporphines have been summarised by *Shamma and Castenson (loc. cit.)*. In general the $C_{(3)}$-H atom resonates at high field and the $C_{(5)}$-H and $C_{(11)}$-H atoms appear furthest downfield. A $C_{(1)}$-OMe group is usually at higher field (4.0–4.2δ) than other methoxyl groups. A methylenedioxy group at $(C_{(2)} + C_{(3)})$ resonates at a lower field (6.6–6.8δ) than a similar group at $(C_{(9)} + C_{(10)})$.

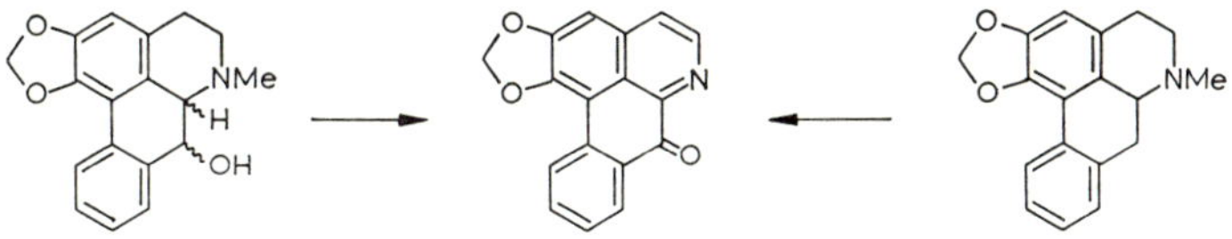

Ushinsunine   Lidiodenine   Roemerine

SCHEME 69

FRAGMENTATIONS OF OXOAPORPHINES

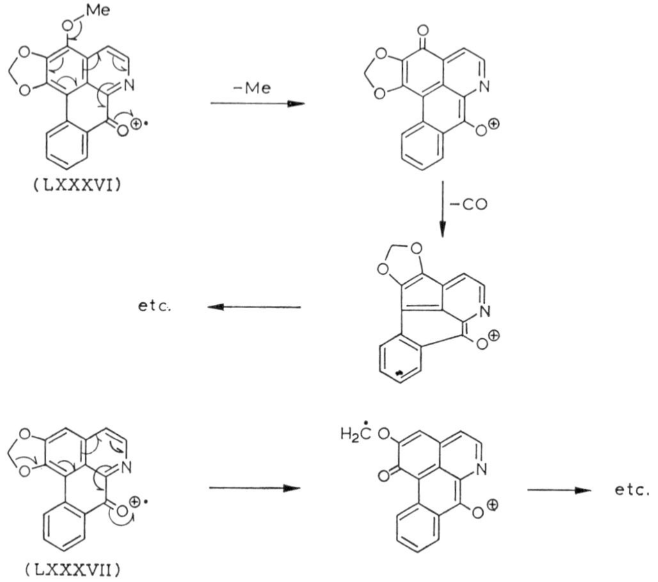

(LXXXVI)

$-Me$

$-CO$

etc.

(LXXXVII)

etc.

A study has been made of the mass spectral fragmentations of representative members of the oxoaporphine alkaloids (*Bick, J. H. Bowie* and *Douglas*, Austral. J. Chem., 1967, **20**, 1403). When a $C_{(3)}$-oxygen function is present (*e.g.* in **atherospermidine,** LXXXVI), the fragmentations shown in Scheme 69 occur, whereas lidiodenine (LXXXVII) is thought to undergo the initial cleavage shown.

The position of the OH group in **moschatoline** was deduced by an examination of the fragmentation pattern of the derived *O*-acetate (*Bick et al., loc. cit.*).

Alkaloid PO-3, a green crystalline salt, is an example of the so far rare variation in which there is a quaternary nitrogen and a phenolic hydroxyl group. This compound was isolated from *P. orientale*, the major alkaloid

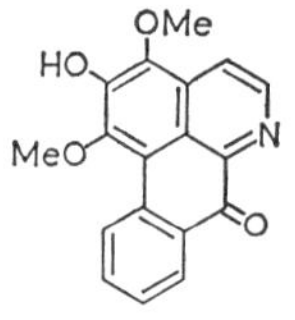

Moschatoline

of which is isothebaine (*V. Preininger* and *F. Santavy*, Acta Uni. Palacki. Olomuc. Fac. Med., 1966, **43**, 5; *Preininger et al.*, Arch. Pharm., 1969, **302**, 808). Alkaloid PO-3 analysed for $C_{19}H_{16}NO_4ClO_4$ and exhibited an i.r. band at 1650 cm^{-1}, indicating a highly conjugated carbonyl group. Since reduction of PO-3 with zinc in acetic acid gave *rac.*-isothebaine, alkaloid PO-3 was allotted the structure shown:

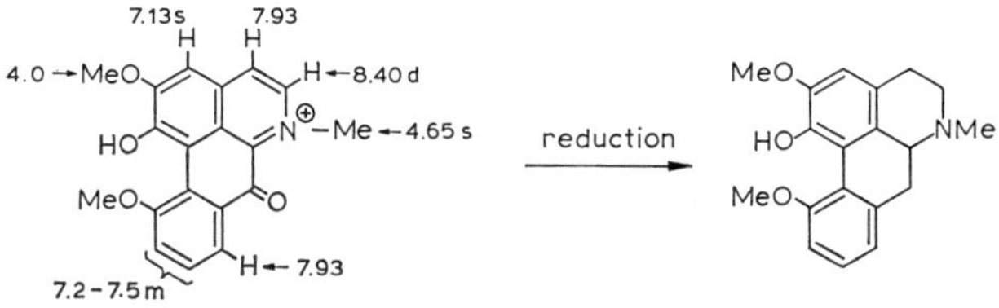

Alkaloid PO-3

Pontevedrine, $C_{21}H_{19}NO_6$, is an *N*-methylamide. From the i.r. spectrum (band at 1660 cm^{-1}) it was concluded that the alkaloid has a highly conjugated ketonic chromophore, together with an amide-carbonyl function. The u.v. spectrum was unchanged on addition of acid or base. The n.m.r. spectrum contained signals for four methoxyl groups and one *N*-methyl group and four aromatic one-proton singlets. The structure was confirmed when pontevedrine was formed in the oxidation of glaucine with a large excess of chromium trioxide in pyridine (*I. Ribas, J. Sueiras* and *L. Castedo*, Tetrahedron Letters, 1971, 3093; *H. Furukawa et al.*, J. pharm. Soc., Japan, 1972, **92**, 150).

An unusual base, **glauvine,** has been allocated the structure shown (*L. D. Yakhontova, V. I. Sheichenko* and *O. N. Tolkachev*, Khim. prir. Soedin, 1972, 214; C.A., 1972, **77**, 48675).

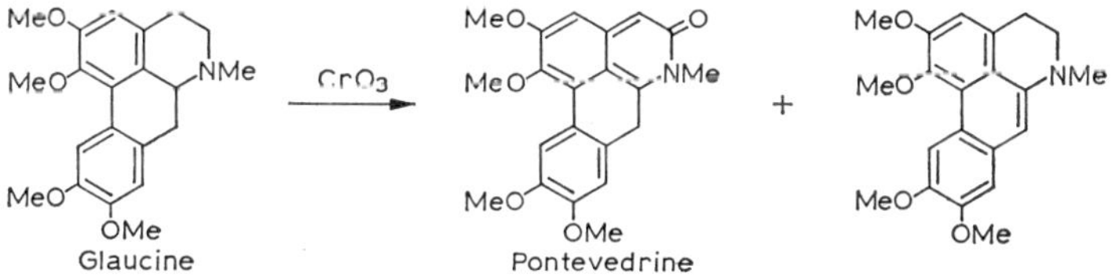

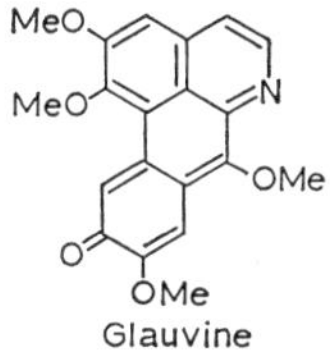

Glaucine

Pontevedrine

Glauvine

6. The protoberberines

The system of nomenclature of this group has not been standardised and can often be confusing, partly because there are in fact two series of alkaloids, one based upon the quaternary ring system I and the other upon the tertiary base II. The numbering system most commonly used now is shown in II and is quite different from earlier systems. Position $C_{(14)}$ is sometimes numbered as $C_{(13a)}$.

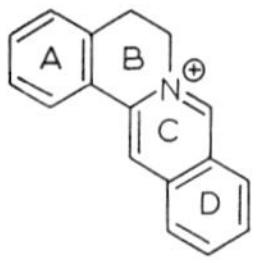

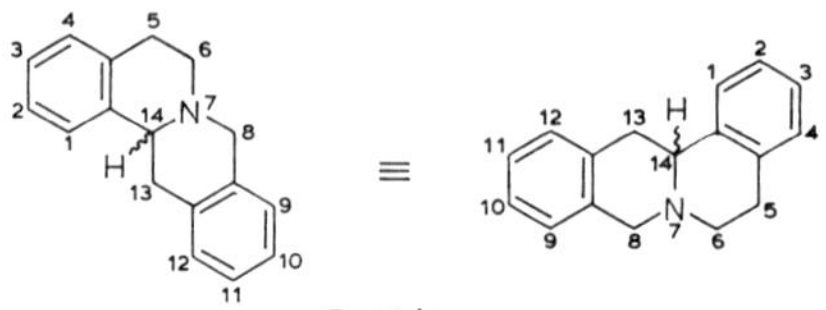

Berbine
(I) (II)

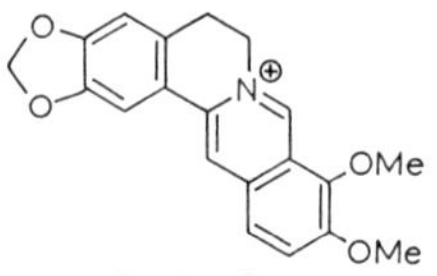

Berberine
(III) (IV)

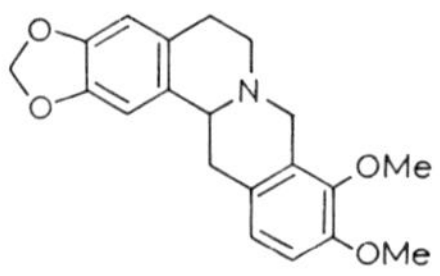

The first member of the group to be isolated and characterised was **berberine** (III) and this led to the entire group being called the berberine alkaloids. However, it was *W. H. Perkin* who suggested the name "protoberberine" for derivatives of the quaternary salt I, and this has been adopted. Most of the naturally occurring derivatives of I or II have been given trivial names upon isolation, and these are still used. When a derivative of I is known, current practice is to name the related racemic form of II as the tetrahydro derivative of the protoberberine, but if the tetrahydro form is optically active, or has no related quaternary form, a trivial name is used. Thus, if berberine (III) is reduced, the product is ($\pm$)tetrahydroberberine (IV). However, the ($-$)form occurs naturally as ($-$)**canadine**. The reduced ring system II has also been referred to as berbine. This system of nomenclature is often used for non-naturally occurring compounds.

Reviews of the protoberberine alkaloids are to be found in *M. Shamma*, "The Alkaloids", Chemistry and Pharmacology, Academic Press, New York, 1972, p. 269 and also in *P. W. Jeffs*, "The Alkaloids", 1967, **9**, 41.

(*a*) *Occurrence and structural variations*

The protoberberines as a group are widely distributed among the Papaveraceae (about 60% of the known alkaloids), Berberidaceae (10%), Ranunculaceae (10%), Menispermaceae (10%) and Rutaceae (10%). Since the first isolation of berberine from *Berberis vulgaris* in 1837, over one hundred members of this family have been characterised.

All known protoberberines (and tetrahydroprotoberberines) contain at least four oxygen functions (OH, OMe, OCH_2O) and most have these groups attached at $C_{(2)}$, $C_{(3)}$, $C_{(9)}$ and $C_{(10)}$. A small number have the $C_{(2)}$, $C_{(3)}$, $C_{(10)}$, $C_{(11)}$-oxygenation pattern. The $C_{(1)}$, $C_{(2)}$, $C_{(9)}$, $C_{(10)}$-pattern is very unusual but is represented in (−)**caseanadine.** Some alkaloids are known with one or more additional oxygen functions attached to the aromatic rings A and/or D; (−)**capaurine** and **stepharotine** are illustrative (*M. Tomita, M. Kozuka* and *S. Uyeo*, J. pharm. Soc., Japan, 1966, **86,** 460). An alcoholic OH group may be present at $C_{(13)}$ (*e.g.* (−)**ophiocarpine,** or at

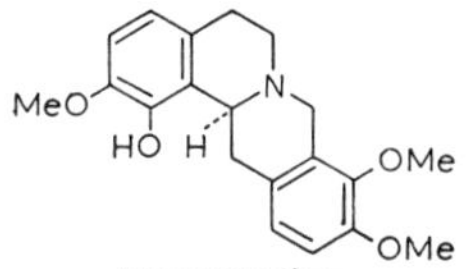

Caseanadine

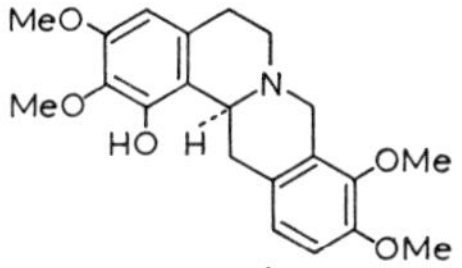

Capaurine

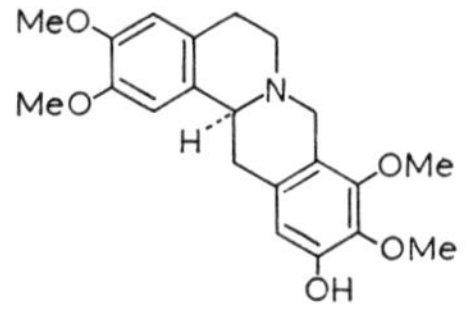

Stepharotine

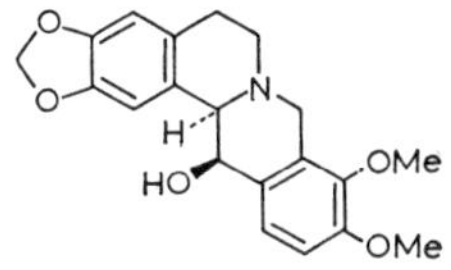

(−)Ophiocarpine

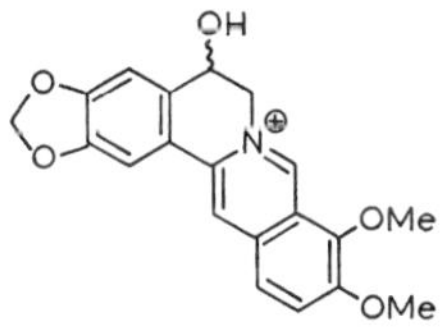

Berberastine

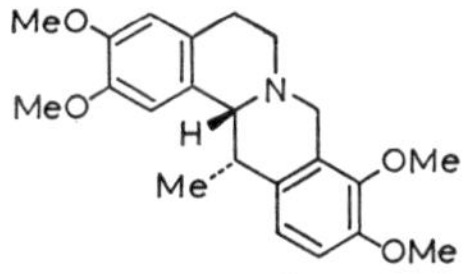

Corydaline

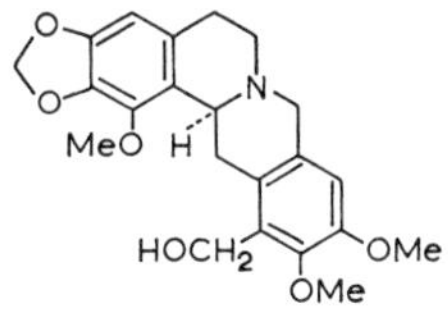

(−)Mecambridine

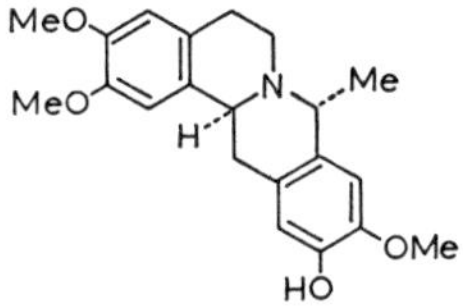

Corytenchirine

$C_{(5)}$ as in **berberastine.** A few tetrahydroprotoberberines are known with a methyl group at $C_{(13)}$ [(−)**corydaline**] or a hydroxymethyl group [(−)**mecambridine**]. The first example of an 8-methyltetrahydroprotoberberine was described recently (*T. Kametani et al.*, J. chem. Soc., Perkin I, 1976, 63). The structure of **corytenchirine** was elucidated by spectral methods and two syntheses have confirmed it (*idem, ibid.*, 1976, 1218).

SCHEME 70

OXIDATIVE DEGRADATION OF BERBERINE

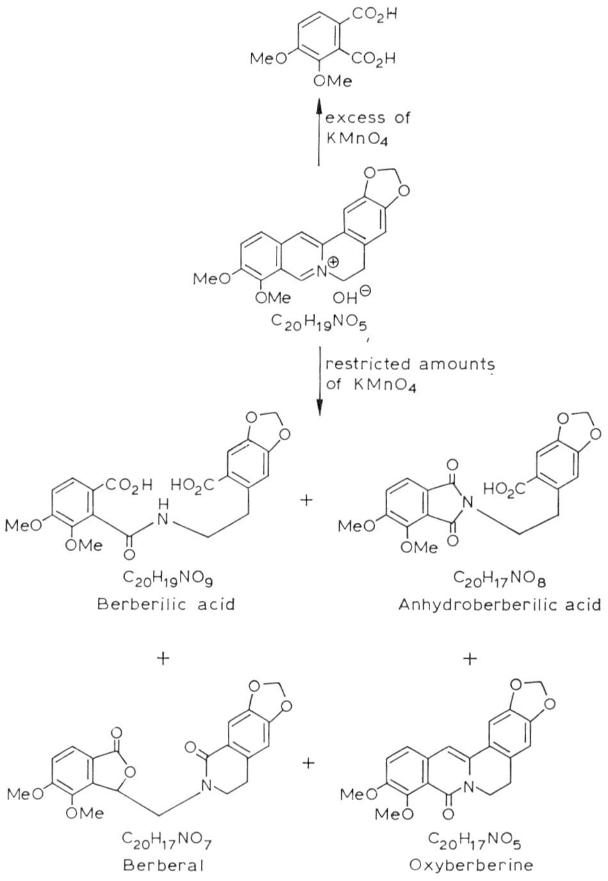

(*b*) *Structure determination*

The structure of berberine itself was elucidated by oxidative degradation in a classical research by *W. H. Perkin* (J. chem. Soc., 1889, **55,** 63; 1890, **57,** 992; *Perkin* and *R. Robinson, ibid.*, 1910, **97,** 305). This is summarised in Scheme 70. Oxidation of berberine with an excess of potassium permanganate yields hemipinic acid, but by using restricted amounts of oxidant, Perkin isolated several compounds, each of which still contained all twenty carbon atoms of the original alkaloid. The structures of these were elucidated in a series of investigations culminating in the structure of berberine itself.

It was *J. Gadamar* who first recognised that berberine is a quaternary hydroxide (pK_a 11.8) and who pointed out that three structures V–VII can be written for it (Arch. Pharm., 1905, **243,** 12). He proposed that in alkaline solution an equilibrium exists between the quaternary hydroxide form V, the pseudobase VI and the aldehyde form VII. An enormous amount of work has been reported over the years on this problem, which is summarised by *D. Beke*, who also sets out the modern view in Advances in heterocycl. Chem., ed. by *A. R. Katritzky*, Academic Press, Vol. I, 1963, p. 167 *et seq.* Thus, although the forms V and VI are undoubtedly present in equilibrium, evidence for the aldehyde form VII is lacking. Reactions such as disproportionation to dihydroberberine (VIII) and oxoberberine (IX) need not proceed through the aldehyde form, but can involve hydride ion transfer between the quaternary and pseudobase structures. An extensive study, using i.r., u.v., n.m.r. and polarographic methods has shown that the position of the equilibrium is strongly influenced by the position of oxygen functions in ring D (*F. Santavy et al.*, Coll. Czech. chem. Comm., 1972, **37,** 2746).

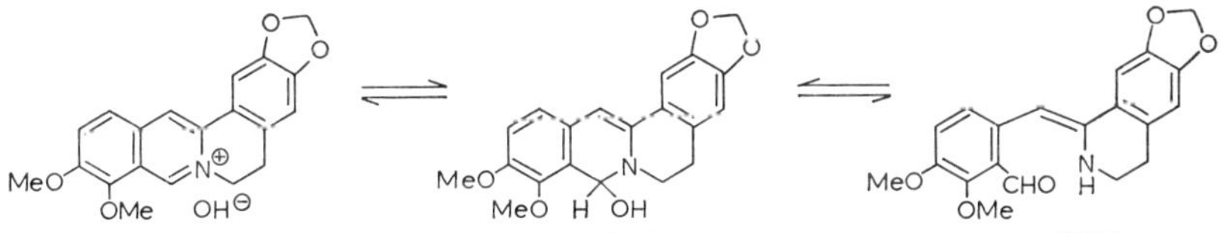

(V) (VI) (VII)

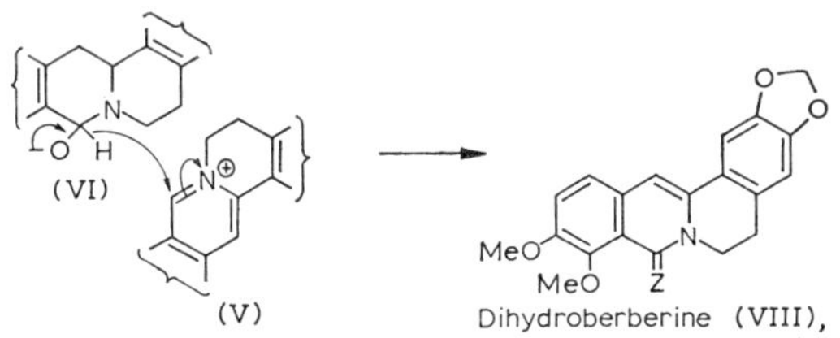

Dihydroberberine (VIII), Z = 2H
Oxoberberine (IX), Z = O

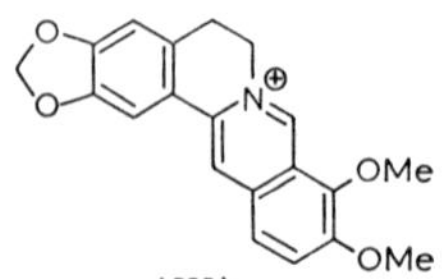

(III)

λ_{max}^{EtOH} 263 345 423

log ε_{max} 4.4 4.4 3.7

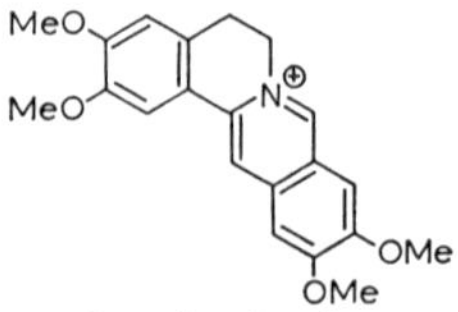

Pseudopalmatine

265 287 310 345 375

4.3 4.7 4.5 4.3 4.0

The u.v. spectra of protoberberinium salts are characteristic of the ring system, with significant differences apparent when a 2,3,9,10- and a 2,3,10,11-tetraoxy substitution pattern are compared, *e.g.* berberine (III) and **pseudopalmatine** iodide. Thus, the former type exhibit a minimum of absorption at 300–310 nm, whereas the latter show strong absorption over this region. Since tetrahydroprotoberberines are easily dehydrogenated to protoberberines (*e.g.* with iodine in ethanol containing sodium acetate), u.v. spectroscopy can be valuable for locating substituents in ring D of the bases.

Nowadays chemical degradations in the tetrahydro series are preferred to *Perkin's* original oxidative degradation methods, so any new protoberberines are reduced (*e.g.* with H_2/Pt or $NaBH_4$/EtOH/H_2O) to the racemic tetrahydroprotoberberines. The structure (and stereochemistry where appropriate) of the reduced compounds is then elucidated by a combination of chemical degradation, inter-relationships with known structures, and physical methods, especially i.r., n.m.r., mass spectral and o.r.d./c.d. measurements.

SCHEME 71

OXIDATIVE DEGRADATION OF TETRAHYDROPROTOBERBERINES

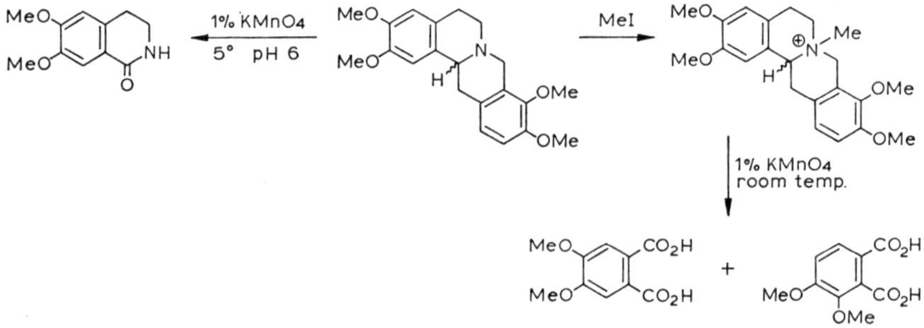

Oxidation of tetrahydroprotoberberines with potassium permanganate occurs easily to yield mixtures of dihydroisocarbostyrils and phthalic acids, either as bases or as metho salts (Scheme 71). However, it seems that if the alkaloid possesses a methoxyl group at $C_{(11)}$, the dihydroisocarbostyril derivative is not obtained (*Santavy*, in "The Alkaloids", 1970, **12**, 383).

Hofmann degradation usually leads to a mixture of the two possible methines, *e.g.* X and XI from (−)canadine methiodide with XI usually predominating:

The isomer X may be partially or fully racemised because XI may undergo a transannular reaction to regenerate (racemic) **canadine,** which can then in turn be degraded once more to X + XI. Some interesting results have been obtained from a study of the Hofmann degradation of phenolic tetrahydro-protoberberines (*Kametani et al.*, Heterocycles, 1974, **2**, 433, 653; J. chem. Soc., Perkin I, 1974, 2678; 1975, 1012). Degradation of **discritine** metho salt gave the expected methine XII, but similar treatment of **schefferine** (XIII), **nandinine** (XIV) and the metho salt XV, yielded 1-benzylisoquinoline derivatives of the type XVI in addition to the methines of type XII. Hofmann degradation of **coreximine metho salt** gave the XVI-type of structure, and the unusual product XVII:

Hofmann degradation

Discritine
metho salt

(XII)

(XIII): $R^1 = R^2 = Me$; $R^3 = H$
(XIV): $R^1 + R^2 = CH_2O_2$; $R^3 = H$
(XV): $R^1 + R^2 = CH_2O_2$; $R^3 = OH$

(XVI)

Coreximine
metho salt

(XVII)

A tetrahydroprotoberberine that possesses a $C_{(13)}$-methyl group is best examined by a combination of Hofmann and Emde degradations. This procedure is illustrated with (+)**thalictricavine** (Scheme 72) (*C. Tani et al.*, J. pharm. Soc. Japan, 1962, **82**, 751), where the $C_{(13)}$ and $C_{(14)}$ hydrogen atoms are *cis* to each other. In the alternative *trans* geometry XVIII, Hofmann degradation proceeds in the alternative way to yield the methine XIX (*H. W. Bersch*, Arch. Pharm., 1958, **291**, 595; *Jeffs*, Experientia, 1965, **21**, 690).

Tetrahydroprotoberberines are easily degraded by cyanogen bromide (Van Braun reaction) to yield two products — for example, canadine gives a mixture of XX and XXI (*I. Salley* and *R. H. Ayers*, Tetrahedron, 1963, **19**, 1397). The formation of XXI in preference to XXII, the expected product in this type of reaction, was attributed to steric hindrance by the $C_{(9)}$-methoxyl group in the transition state leading to XXII. In another examination of this reaction, the product isolated under somewhat different

SCHEME 72

DEGRADATION OF $C_{(13)}$-METHYLTETRAHYDROPROTOBERBERINES

O₃

(1) MeI
(2) Emde degradation

(+) Thalictricavine

Hofmann degradation

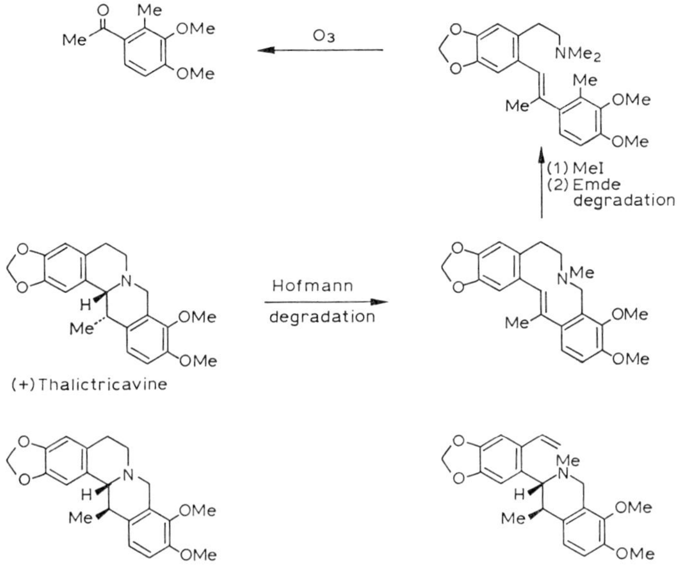

(XVIII)

(XIX)

conditions proved to be **XXIII** (*J. D. Albright* and *L. Goldman*, J. Amer. chem. Soc., 1969, **91,** 4317):

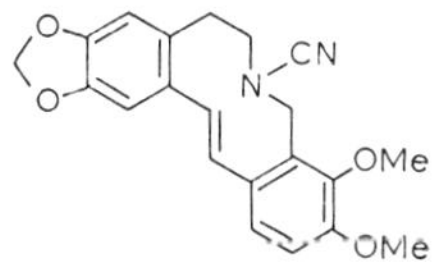

(XX)

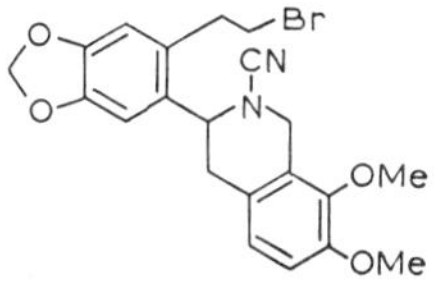

(XXI)

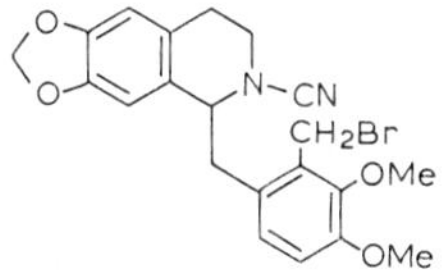

(XXII)

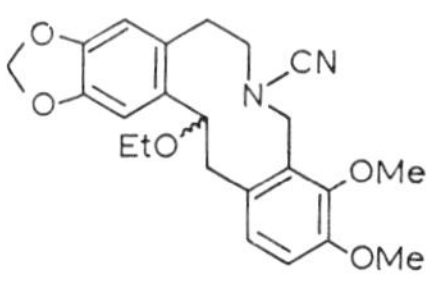

(XXIII)

The u.v. spectra of tetrahydroprotoberberines are sufficiently characteristic (*A. W. Sangster* and *K. L. Stuart*, Chem. Reviews, 1965, **65,** 69) for the ring system to be differentiated from many others; it is not possible, however, to decide between a 2,3,9,10- and a 2,3,10,11-tetraoxygenation pattern. The

m/e 339
(XXIV)

(XXV)

m/e 164
(XXVI)

−H•

m/e 174
(XXVII)

principal use of i.r. spectroscopy of this group is to determine whether the rings B/C are *trans*-fused (more usual) or *cis*-fused. The former geometry gives rise to characteristic absorption in the region of 2800 cm^{-1} (the Bohlmann bands). Mass spectrometry is now a powerful tool in structural elucidation since fragmentation patterns are well understood. The key fragmentation is a retro-Diels–Alder reaction of the molecular ion, *e.g.* XXIV, to provide the ions XXV, XXVI and XXVII, so that it is possible (*C. Djerassi et al.*, J. Amer. chem. Soc., 1963, **85**, 2807) to differentiate between substituents on rings A and D. Further, it is possible to distinguish between a 9-hydroxy-10-methoxytetrahydroprotoberberine and its 9-methoxy-10-hydroxy isomer. This is because in the former case a methyl

m/e 150
(XXVIII)

− Me•

m/e 135
(XXIX)

(XXX)

−H•

m/e 149
(XXXI)

(XXXII)

m/e 178
(XXXIII)

group is preferentially expelled from the ion XXVIII corresponding to the ring-D fragment to give an ion at m/e 135, formulated as XXIX. The ion XXX, derived from the 9-methoxy-10-hydroxy derivative, fragments further by loss of hydrogen to give XXXI at m/e 149. The mass spectral fragmentation of $C_{(13)}$-methyltetrahydroprotoberberines have also been examined and the expected pattern is observed *e.g.* XXXII → XXXIII.

There is no doubt that n.m.r. spectroscopy provides the most informative physical method of structure determination (*C.-Y. Chen* and *D. B. Maclean*, Canad. J. Chem., 1968, **46,** 2501), especially when it is combined with selective deuteration and Nuclear Overhauser Effect studies. The oxygenation pattern in ring D may be deduced by examining the signals due to the methylene group at $C_{(8)}$ (*R. H. F. Manske et al., ibid.*, 1970, **48,** 3673; *T. R. Govindachari et al.*, Indian J. Chem., 1970, **8,** 769; *Chen* and *Maclean, loc. cit.*). Thus, with $C_{(9)}$, $C_{(10)}$-disubstitution an AB quartet ($J_{AB} \simeq 16$ Hz) at δ 3.65 and δ 4.35 occurs, whereas only a broadened singlet at δ 4.05 appears with the alternative $C_{(10)}$, $C_{(11)}$-disubstitution pattern. It may be possible to deduce the position of a phenolic hydroxyl group by exchanging aromatic hydrogen atoms *ortho* or *para* to it with deuterium (warming the phenol briefly with 12 N deuterium chloride effects this reaction). Thus, with $(-)$**caseadine** it was established by mass spectral studies that a part-structure XXXIV can be written (*Chen, Maclean* and *Manske*, Tetrahedron Letters, 1968, 349) for the alkaloid. A $C_{(2)} + C_{(3)}$ oxygenation of ring A is ruled out since caseadine methyl ether is not identical with norcoralydine, and a $C_{(3)} + C_{(4)}$-disubstitution is unlikely on biogenetic grounds. The unusual structure XXXV was deduced by a closer study of the aromatic region of the n.m.r. spectrum. This consisted of an AB quartet ($J = 8.4$ Hz) and two singlets. One half of the quartet and both singlets were broadened by benzylic coupling. After treatment with 12 N deuterium chloride solution, the signal due to H_B in XXXV had disappeared and the H_A absorption had collapsed to a singlet:

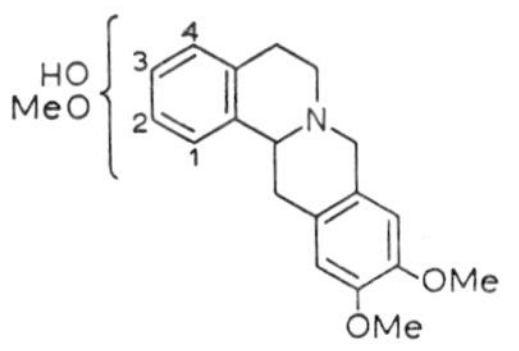

(XXXIV) 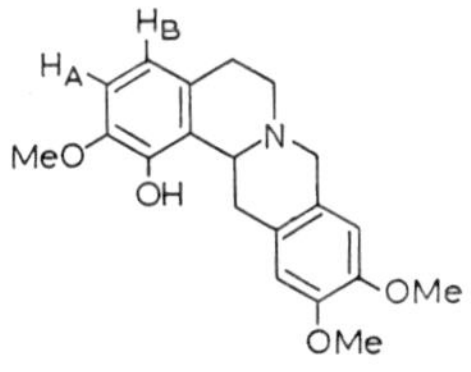 (XXXV)

The elucidation of the structure of $(\pm)$**apocavidine** illustrates the use of the NOE (*Manske et al., loc. cit.*). Since $J_{13,\,14} = 3$ Hz, these protons are *cis* orientated ($J_{trans} \simeq 8$ Hz). Irradiation of $C_{(14)}$-H at δ 3.71 caused a sharpening of the aromatic hydrogen singlet at δ 6.78, which is thus at $C_{(1)}$. Irradiation of the $C_{(5)}$ methylene group at δ 2.68 sharpened the signal at δ 6.58 which is thus due to $C_{(4)}$-H. Irradiation of the OCH_3 resonance at δ 3.87 caused a 23% enhancement of the signal at δ 6.58, so that the OMe group must be at $C_{(3)}$.

The stereochemistry and configuration of members of the tetrahydro-protoberberine group have been studied by both chemical and physical methods.

Apocavidine

$H.$ *Corrodi* and $E.$ *Hardegger* (Helv., 1956, **39**, 889), who determined the absolute configuration of $(-)$tetrahydropapaverine by degradation to N-β-carboxyethyl-L-aspartic acid of known configuration (see p. 23), then condensed the isoquinoline with formaldehyde to give $(-)$**norcoralydine** (Scheme 73). This configurational correlation is valid since it has been shown (*Kametani* and $M.$ *Ihara*, J. chem. Soc., 1968, 1305) by the use of deuterium labelling that configuration is retained in this Mannich reaction. With the configuration of $(-)$norcoralydine established as $14(R)$, it was assumed that all tetrahydroprotoberberines which are laevorotatory have the $14(R)$ configuration, since it was not possible to achieve a direct correlation, using this method, with a 2,3,9,10-tetraoxy substituted base. However, it is known that changes in the nature of substituents have only a minor effect on the specific rotation. The configuration is, however, now firmly established ($A. R.$ *Battersby et al.*, *ibid.*, C, 1966, 1052) since $(-)$norreticuline of known absolute configuration was converted into a separable mixture of $(-)$**scoulerine** and $(-)$**coreximine** by a Mannich reaction (Scheme 74).

SCHEME 73

THE CONFIGURATION OF $(-)$NORCORALYDINE

SCHEME 74

(−)Norreticuline

Mannich reaction

(−)Scoulerine

(−)Coreximine

Optical rotatory dispersion and circular dichroism spectra provide the most useful methods for establishing configuration. By comparing the negative Cotton effect curve of (−)tetrahydropalmatine in the 300-nm region with the plain positive curve of (−)1,2-bisphenylethylamine (XXXVI) it was concluded that the alkaloid has the 14(R) configuration (*G. Lyle*, J. org. Chem., 1960, **25**, 1779). However, this is not a good model because of the greater degree of flexibility in XXXVI. Laevorotatory tetrahydro-protoberberines which exhibit a second negative Cotton effect in the region 240–280 nm have been examined (*Kametani* and *Ihara, loc. cit.*), and it was found that 1-benzyltetrahydroisoquinolines of the L-series show positive absorptions in this region. So, for the same absolute configuration, the two series have curves of opposite sign. It was also found that the amplitude of

(XXXVI)

this second Cotton effect is always greater in the spectra of 2,3,10,11-tetra-oxytetrahydroprotoberberines than in the 2,3,9,10-tetraoxy analogues. Other studies have shown (*M. Legrand* and *R. Viennet*, Bull. Soc. chim., 1966, 2798) that a direct correlation of configuration is possible from the Cotton effect in 220–200 nm region.

For a tetrahydroprotoberberine without a $C_{(1)}$ or a $C_{(13)}$-substituent the *trans*-quinolizidine conformation, XXXVII with rings B and C as half-

(XXXVII)

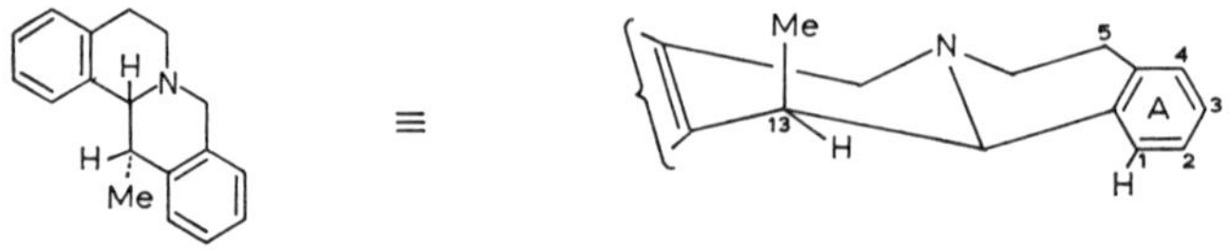

(XXXVIII)

chairs, is the most stable. Such compounds will exhibit Bohlman bands in the i.r. region at about 2800 cm^{-1}. When a methyl group is placed at $C_{(13)}$ such that the hydrogen atoms at $C_{(13)}$ and $C_{(14)}$ are *cis*, the most stable conformation remains the same as that above with the $C_{(13)}$-methyl group axial XXXVIII. However, when the $C_{(13)}$, $C_{(14)}$ hydrogen atoms are *trans* to each other, so that the methyl group at $C_{(13)}$ is equatorial, there is the possibility of steric interaction between it and the $C_{(1)}$-position (XXXIX) and one of the two alternative *cis*-quinolizidine conformations may become more stable; this would certainly be true if a $C_{(1)}$-oxygen function were present as shown in XL:

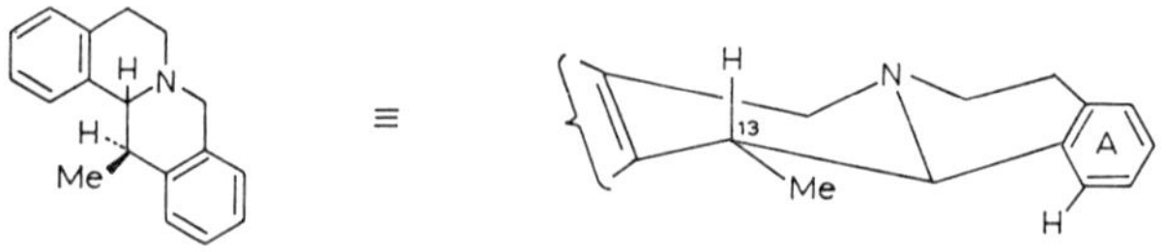

(XXXIX)

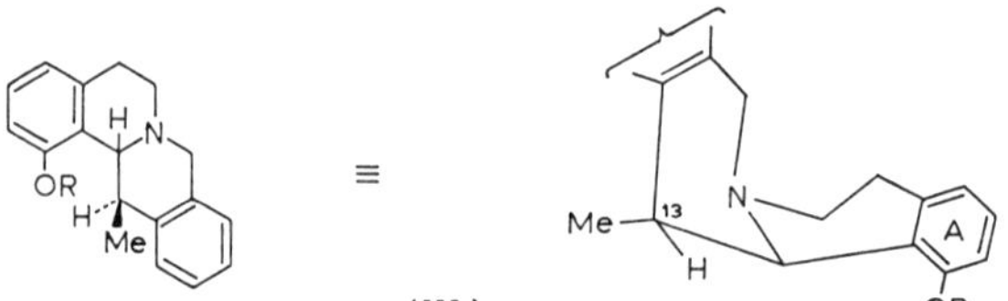

(XL)

By a study of the rates of methiodide formation (*M. Shamma, C. D. Jones* and *J. A. Weiss*, Tetrahedron, 1969, **25**, 4347) it was concluded that XLI has the *trans*-quinolizidine conformation, but that XLII has the *cis*-fused B/C function. The alkaloids **capaurine** (XLIII, R = Me) and **capaurimine**

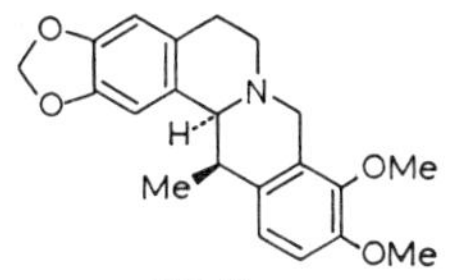

(XLI)

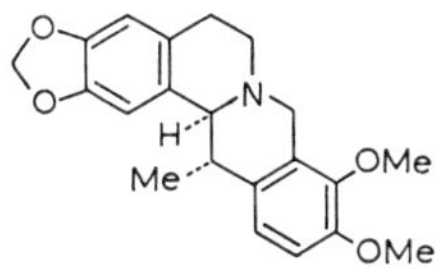

(XLII)

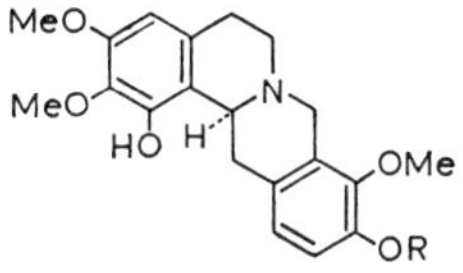

Capaurine: (R = Me)
Capaurimine: (R = H)

(XLIII)

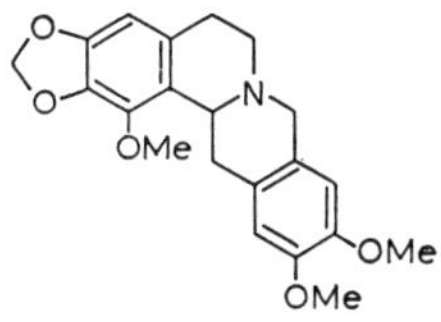

(XLIV)

(XLIII, R = H) have been shown by X-ray analysis (*Kametani et al.*, Tetrahedron Letters, 1968, 4251; Chem. Comm., 1970, 1241; J. chem. Soc., C, 1971, 2541) to have the rings B/C *cis*-fused.

The preferred conformation can also be assigned to tetrahydroprotoberberines from the ^{13}C-chemical shift values for $C_{(6)}$ (*Kametani*, Heterocycles, 1975, **3**, 371). If the δ-value is upfield of 48.3 ppm, the preferred conformation is *cis*, *e.g.* in XLIV. In compounds without a $C_{(1)}$-substituent the *trans* B/C ring function is indicated if δ for C_6 is at about 51.3 ppm.

The introduction of an additional centre of assymmetry at $C_{(13)}$ has no effect on the sign of rotation and makes little contribution to the molecular rotation of tetrahydroprotoberberines.

Hydrogenolysis of (—)**ophiocarpine** acetate (XLV, R = OAc) gave (−)tetrahydroberberine thus establishing the configuration at $C_{(14)}$ as (*R*) (*M. Ohta, Tani* and *S. Morozumi*, Chem. and Pharm. Bull., Japan, 1964, **12**, 1072). This is supported by the fact that (−)ophiocarpine (XLV,

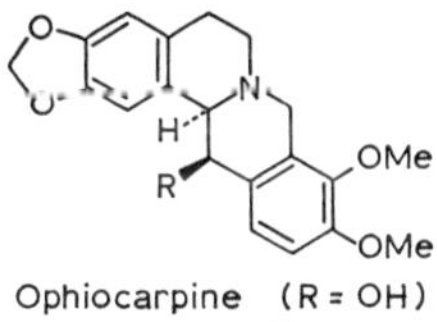

Ophiocarpine (R = OH)

(XLV)

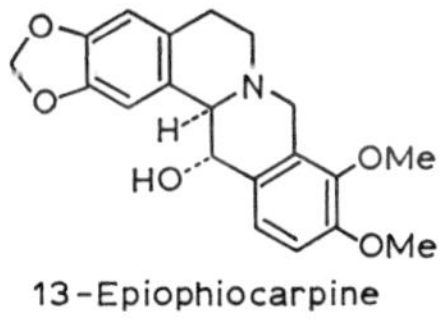

13-Epiophiocarpine

(XLVI)

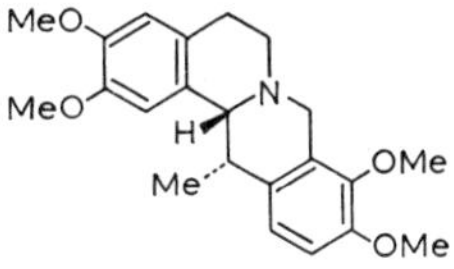

Corydaline

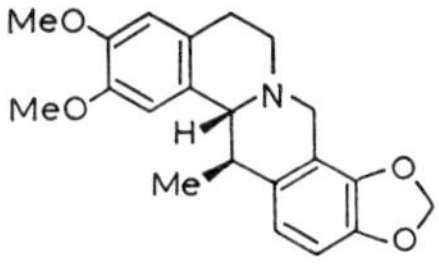

Thalictrifoline

R = OH) exhibits a negative o.r.d. curve. The i.r. spectra of both ophio-carpine (XLV, R = OH) and 13-epiophiocarpine (XLVI) contain Bohlmann bands indicating that both compounds have the *trans*-quinolizidine con-formation. However, ophiocarpine has OH absorption characteristic of strong intramolecular hydrogen bonding, whereas the epimer XLVI exhibits a sharp OH band. It is concluded that the hydroxyl group of ophiocarpine must be axial, thus leading to structure XLV (R = OH), whereas in the synthetic epiophiocarpine (XLVI), the OH group is equatorial. These assignments are confirmed by the n.m.r. spectra of the compounds.

The assignment of relative configuration of 13-methyltetrahydroproto-berberine has been clarified by *P. W. Jeffs* (Experientia, 1965, **21**, 690). It was established that **corydaline** has the stereochemistry shown and so too have all the alkaloids except **thalictrifoline** which does not exhibit Bohlmann bands in the i.r. spectrum, and consequently has the *cis*-quinolizidine conformation. In general the $C_{(13)}$-methyltetrahydroprotoberberines have positive Cotton effect curves so that the configuration at $C_{(14)}$ is (*S*).

Examination of the n.m.r. spectra of a number of 13-methyltetrahydro-berberines of known structure and stereochemistry has established (*Manske et al.*, Canad. J. Chem., 1970, **48**, 3673) it to be a useful technique for deducing relative stereochemistry. Thus, if the $C_{(13)}$-H is *cis* to the $C_{(14)}$-H, the absorption due to the $C_{(8)}$-methylene group is an AB quartet with $J \simeq 16$ Hz; a large (0.6–0.7 ppm) chemical-shift difference is also apparent. If, however, $C_{(13)}$-H and $C_{(14)}$-H are *trans* to each other, the $C_{(8)}$-methylene protons absorb with only a small (0.1–0.2 ppm) chemical-shift difference.

Berberastine (XLVII, R = Me) and **thalidastine** (XLVII, R = H), are two 5-hydroxyberberines, but the configuration at $C_{(5)}$ is not known:

Berberastine (R = Me)
Thalidastine   (R = H)
(XLVII)

(c) Some reactions of protoberberine derivatives

If a quaternary berberinium salt, for example XLVIII (Scheme 75), is reduced with lithium tetrahydridoaluminate to the dihydro derivative XLIX (R = H), or is reacted with acetone to yield XLIX (R = CH$_2$·COMe), the resultant enamine can be alkylated directly at $C_{(13)}$. In the resultant iminium ion L,

SCHEME 75

ENAMINE REACTIONS IN DIHYDROBERBERINES

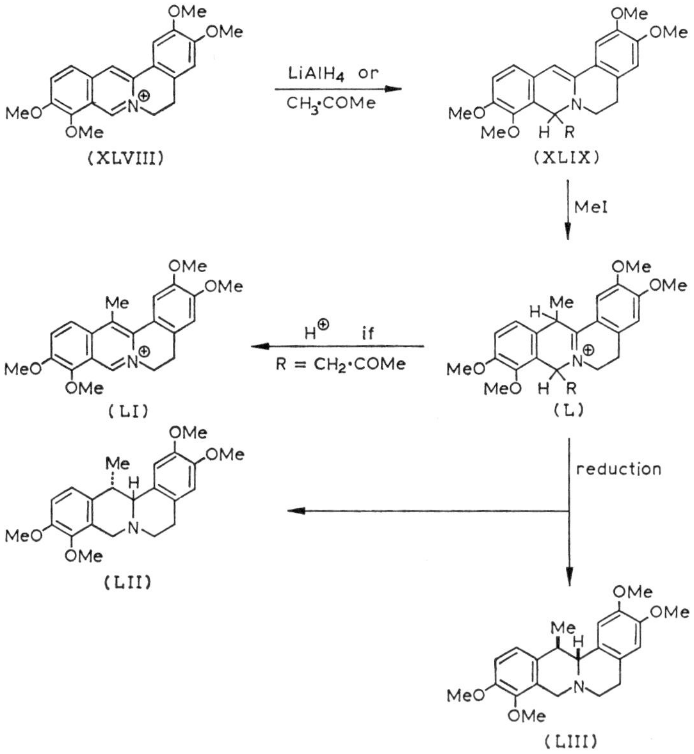

if R = $CH_2 \cdot COMe$, mild acid treatment generates the quaternary salt LI. However, if R = H in L, reduction gives a mixture of LII and LIII (*M. Freund* and *K. Fleischer*, Ann., 1915, **409**, 188; *J. Gadamer*, Arch. Pharm., 1910, **248**, 681; *Gadamer* and *W. Klee*, ibid., 1916, **254**, 295; *Tani, N. Takao* and *S. Takao*, J. pharm. Soc., Japan, 1962, **82**, 748; *T. Takemoto* and *Y. Kondo, ibid.*, p. 1408). The relative proportions of LII and LIII depend upon the reducing agent used.

Dihydroberberine has been converted into ophiocarpine (XLV) and 13-epiophiocarpine by hydroboration of the enamine system, followed by treatment with hydrogen peroxide in sodium hydroxide solution (*I. W. Elliot*, J. heterocycl. Chem., 1967, **4**, 639):

Dihydroberberine

Ophiocarpine
(XLV)

Berberine acetone (LIV) yields a series of products when oxidised (*J. Iwasa* and *S. Naruto*, J. pharm. Soc., Japan, 1966, **86,** 534). With osmium tetroxide, berberine phenol betaine (LV) is produced (Scheme 76), whereas with potassium permanganate (LV), neoxyberberine acetone (LVI) and LVII are formed. Neoxyberberine acetone yields berberine phenol, the protonated form of LV when treated with acid.

SCHEME 76

OXIDATION OF BERBERINE ACETONE

MeO
MeO CH₂·COMe
Berberineacetone
(LIV)

OsO₄

KMnO₄

MeO
OMe
(LV)

OH
MeO
OMe
(LVI)

+

HO₂C
OMe
OMe
(LVII)

HO⊖ H⊕

H⊕

OH
MeO
OMe

Oxidation of 13-methylberberine acetone (LVIII) with potassium permanganate gives LIX as the major product, together with small amounts of

LX and LXI (*Naruto, H. Nishimura* and *H. Kaneko*, Tetrahedron Letters, 1972, 2127; Chem. and Pharm. Bull., Japan, 1975, **23**, 1276):

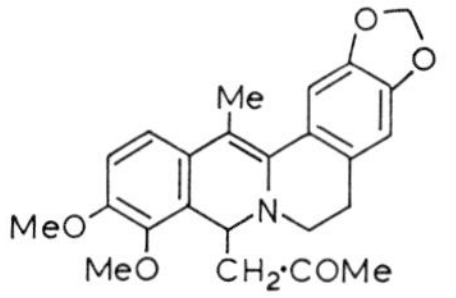

(LVIII)

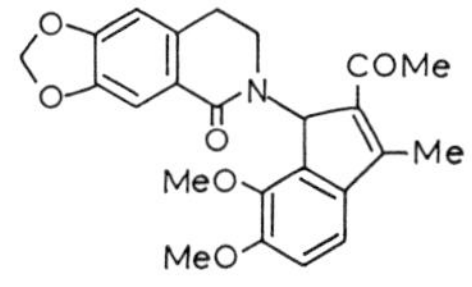

(LIX)

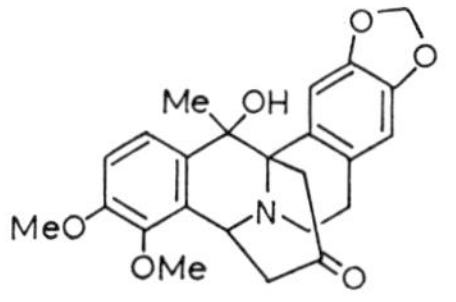

(LX)

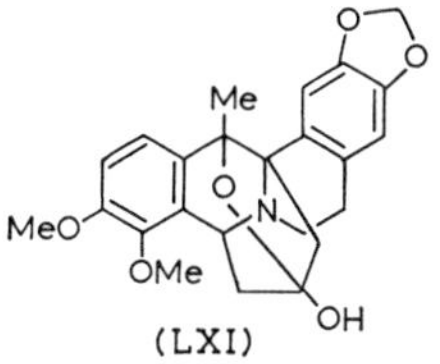

(LXI)

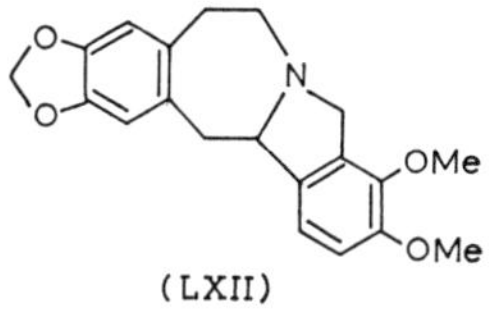

(LXII)

Two products are formed when berberine phenol (LV) is reduced with zinc and acetic acid (*Takemoto, Y. Kondo* and *K. Kondo*, J. pharm. Sci., Japan, 1963, **83**, 162). The expected tetrahydroberberine is accompanied by a substance later proved to be LXII (*C. Schopf* and *M. Schweickert*, Chem. Ber., 1965, **93**, 2566).

(d) Synthesis

The more important methods for the synthesis of the protoberberine or tetrahydroprotoberberine, ring systems can be classified into five types (Scheme 77). All require a pre-formed isoquinoline nucleus. Route 5, from the protopine type of structure, is of academic interest only since in practice the protopine alkaloids are usually synthesised from the corresponding protoberberines. Syntheses of Type 1 have been the most popular and, with modifications, are still used widely. Practical syntheses based upon Type 3

SCHEME 77

THE MAIN SYNTHETIC ROUTES TO PROTOBERBERINES

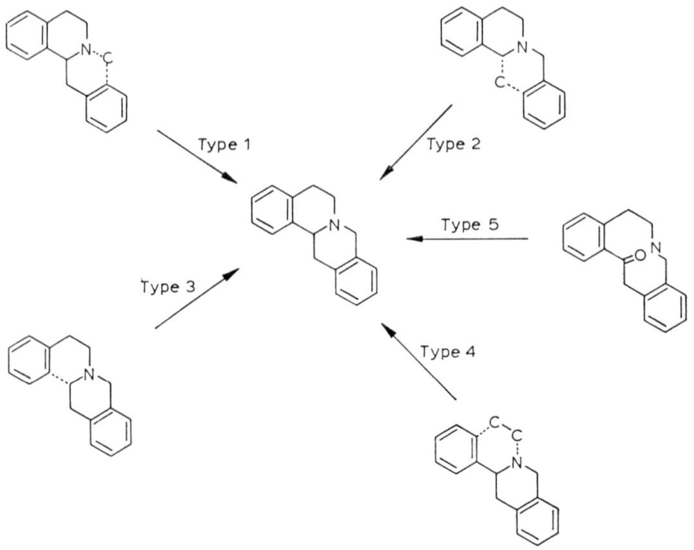

Type 1

Type 2

Type 5

Type 3

Type 4

and Type 4 approaches have been realised only recently. In addition to these routes, a number of interconversions have been reported which involve manipulation of the oxygen functions, together with the preparation of 13-methyl- and 13-hydroxy-tetrahydroprotoberberines described above. The early literature has been reviewed (*K. Pelz*, Chem. Listy, 1963, **57**, 1107).

(i) Type 1 syntheses

The first successful synthesis of a berbine was achieved by *A. Pictet* and *A. Gams* (Ber., 1911, **44**, 2480). The 1-benzyltetrahydroisoquinoline (LXIII) was treated with methylal and concentrated hydrochloric acid to form the berbine derivative, **pseudocanadine** (LXV), presumably by way if LXIV. The reaction may be regarded as a Pictet–Spengler reaction — an example of the Mannich reaction. Formaldehyde in place of methylal is now preferred for the introduction of the "berberine bridge" carbon atom. By using acetal in place of methylal, 8-methyltetrahydroprotoberberines, *e.g.* LXVI, can be obtained (*Pictet* and *S. Malinouski, ibid.*, 1913, **46**, 2688):

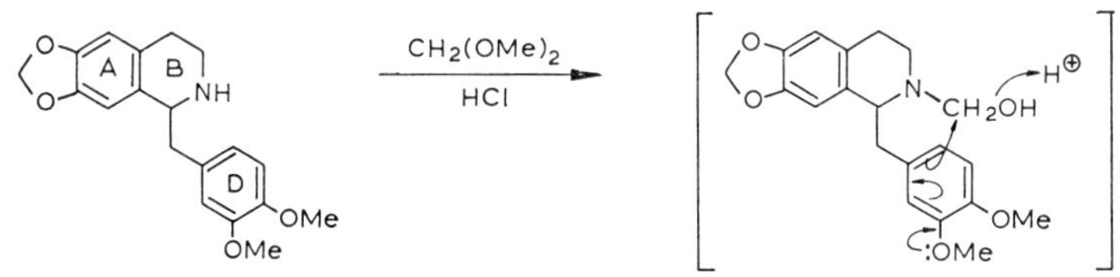

(LXIII).

(LXIV)

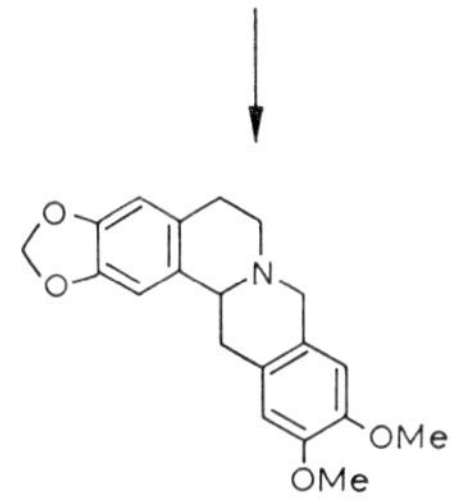

(LXVI)

Pseudocanadine
(LXV)

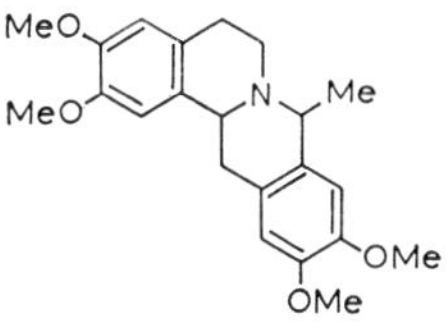

In principle the cyclisation of a 1-benzyltetrahydroisoquinoline such as
LXIII, with formaldehyde/hydrochloric acid, can occur either *ortho* or *para*
to an oxygen function in the potential ring D. In practice, cyclisation *para*
to an ether oxygen invariably occurs leading to the 2,3,10,11-tetraoxyberbine
derivatives. Since the majority of the alkaloids exhibit the 2,3,9,10-tetraoxy-
genation pattern, various attempts have been made to modify this cyclisation
reaction. Early attempts to block the 6′-position with bromine failed; when
LXVII and LXVIII were treated with formaldehyde/hydrochloric acid, or
when the derived *N*-formyl compounds were heated with phosphoryl
chloride, cyclisation occurred with elimination of the bromine to give the
2,3,10,11-tetraoxyberbines once more (*E. Spath* and *N. Lang, ibid.*, 1921,
54, 3064; *R. D. Haworth* and *W. H. Perkin Jr.*, J. chem. Soc., 1925, **127**,
1448). However, it has been found more recently that provided one or more
phenolic groups are present, cyclisation will occur with retention of the
bromine atom (*Kametani* and *M. Ihara, ibid.*, C, 1967, 530; 1968, 1179;
C. Tani et al., Yakugaku Zasshi, 1973, **93**, 268; C.A., 1973, **79**, 5478c)
(Scheme 78).

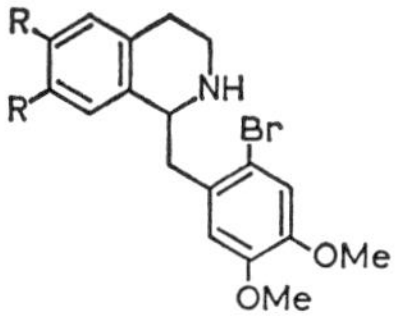

(LXVII): R = OMe
(LXVIII): 2R = CH$_2$O$_2$

SCHEME 78

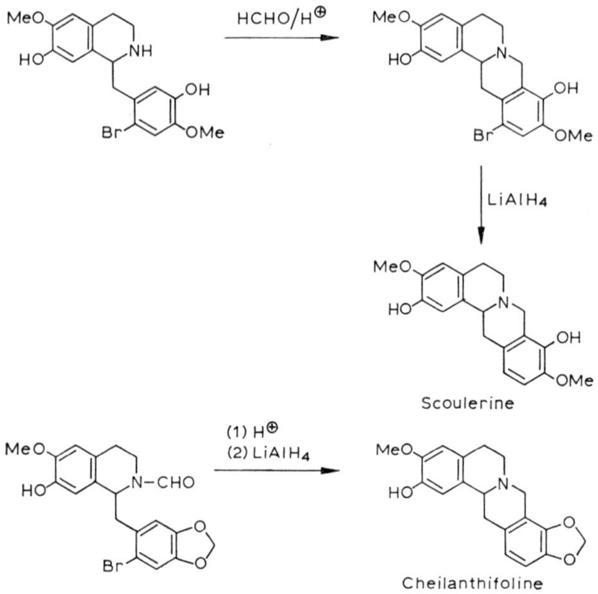

MeO
HO
NH
OH
Br
OMe
HCHO/H⊕
MeO
HO
N
OH
Br
OMe
LiAlH4
MeO
HO
N
OH
OMe
Scoulerine
MeO
HO
N-CHO
O
O
Br
(1) H⊕
(2) LiAlH4
MeO
HO
N
O
O
Cheilanthifoline

Some time ago it was found that if LXIX is reacted with formaldehyde in hydrochloric acid, then with diazomethane, a separable mixture of **norcoralydine** and **tetrahydropalmatine** (LXX) was obtained (*Spath* and *E. Kruta*, Monatsh., 1928, **50**, 341). By carrying out the reaction under "physiological conditions" (aqueous solution, pH 7, room temperature, atmospheric pressure), *C. Schopf* (Angew. Chem., 1937, **50**, 797) was able to isolate

SCHEME 79

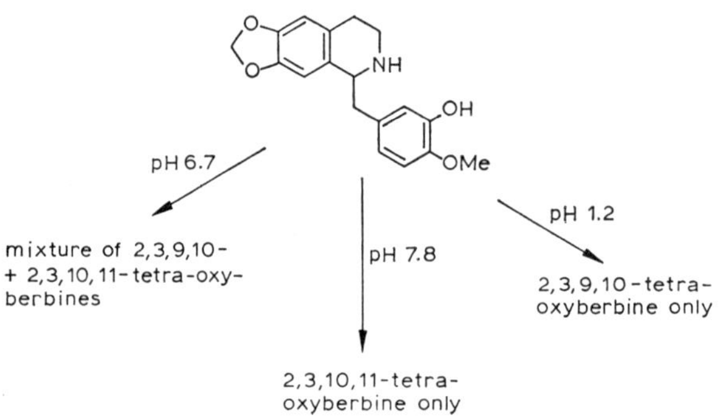

O
O
NH
OH
OMe
pH 6.7
mixture of 2,3,9,10-
+ 2,3,10,11-tetra-oxy-
berbines
pH 7.8
2,3,10,11-tetra-
oxyberbine only
pH 1.2
2,3,9,10-tetra-
oxyberbine only

LXXI in 80% yield. This type of approach to naturally occurring berbines has proved to be popular, especially when it was realised that pH may be of critical importance (Scheme 79) (*Kametani et al.*, J. chem. Soc., C, 1971, 2709). However, steric factors are considered to be responsible for the observation that LXXII is cyclised to LXXIII in a sequence where pH control is not important (*Kametani, T. Honda* and *Ihara, ibid.*, 1971, 2396). Some examples of successful cyclisations illustrate the method (Scheme 80) (*Battersby et al.*, *ibid.*, 1966, 1052; *Kametani et al.*, *ibid.*, 1971, 2709; *Kametani, Honda* and *Ihara, ibid.*, 1971, 3318; *Kametani, Ida* and *T. Kikuchi*, Yakagaku Zasshi, 1968, **88**, 1185).

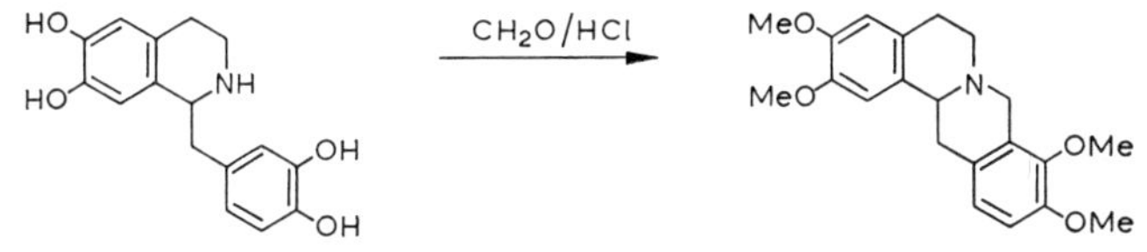

CH₂O/HCl

(LXIX)   (LXX)

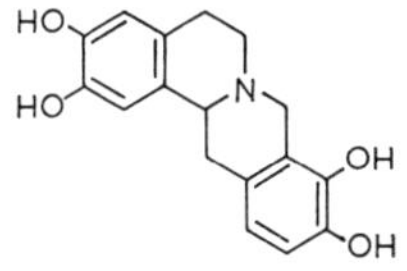

(LXXI)

HCHO/HCl

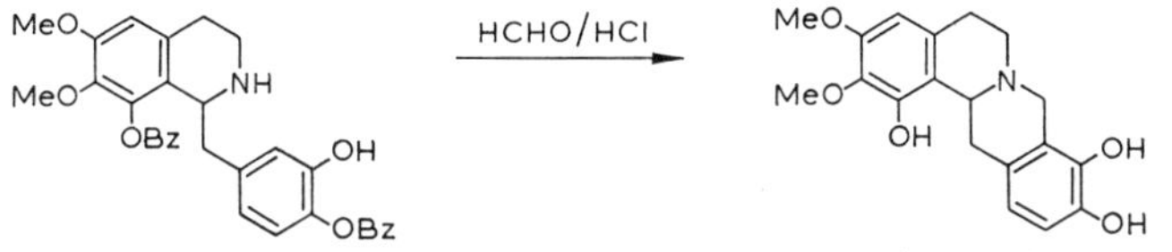

(LXXII)   (LXXIII)

A modification that leads directly to the protoberberine salt involves the *N*-acylation of a 1-benzyl-3,4-dihydroisoquinoline derivative, followed by acid-catalysed cyclisation — for example, LXXIV → LXXVI. If the amides, of type LXXV are photolysed, high yields of protoberberines can be obtained (*G. R. Lenz* and *N. C. Yang*, Chem. Comm., 1967, 1136; *Yang, A. Shani* and *Lenz*, J. Amer. chem. Soc., 1966, **88**, 5369; *M. P. Cava* and *S. C. Havlicek*, Tetrahedron Letters, 1967, 2625). If the carbamate ester is used, for example LXXV (R = OEt), 8-oxoberbines are produced, for example LXXVII. (For reviews of these photocyclisations, see *Kametani* and *K. Fukumoto*, Accounts Chemical Research, 1972, **5**, 212; *Y. Kondo*, Heterocycles, 1976, **4**, 197).

SCHEME 80

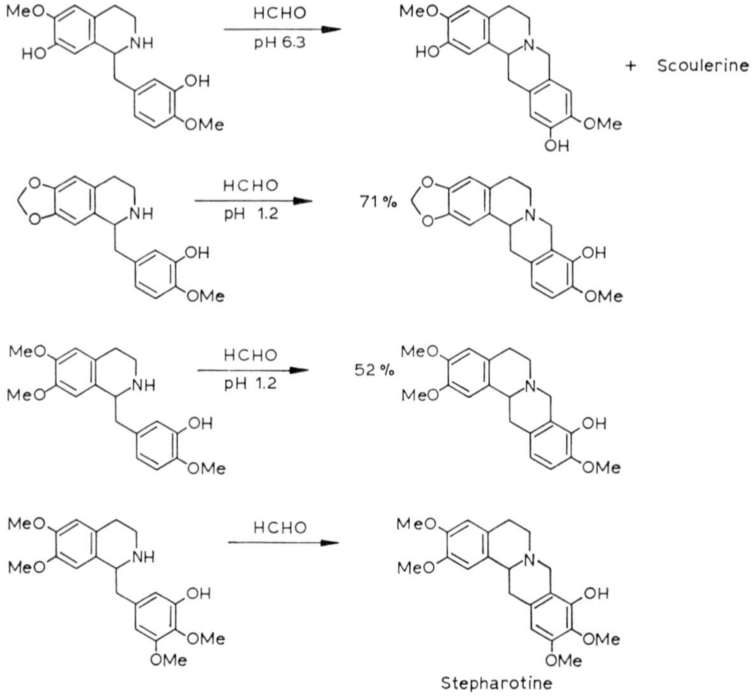
MeO
HO
NH
OH
OMe
HCHO
pH 6.3
MeO
HO
N
OMe
OH
+ Scoulerine
NH
OH
OMe
HCHO
pH 1.2
71 %
N
OH
OMe
MeO
MeO
NH
OH
OMe
HCHO
pH 1.2
52 %
MeO
MeO
N
OH
OMe
MeO
MeO
NH
OH
OMe
OMe
HCHO
MeO
MeO
N
OH
OMe
OMe
Stepharotine

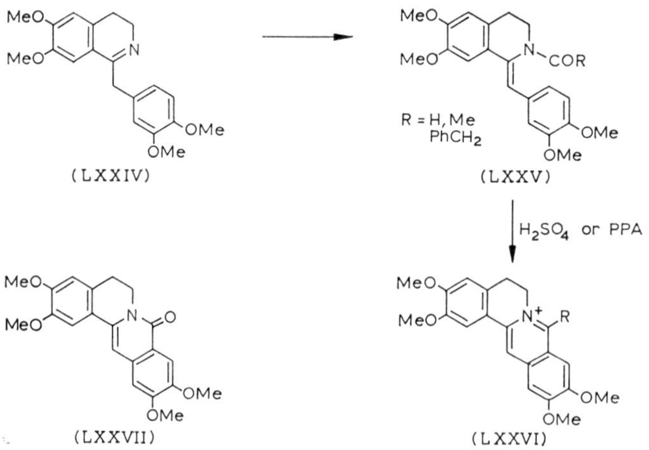
MeO
MeO
N
OMe
OMe
(LXXIV)
MeO
MeO
N
COR
R = H, Me
PhCH2
OMe
OMe
(LXXV)
H2SO4 or PPA
MeO
MeO
N
O
OMe
OMe
(LXXVII)
MeO
MeO
N+
R
OMe
OMe
(LXXVI)

When papaverine is subjected to the Vilsmeier reaction, the product is the fully aromatic cation LXXVIII (*N. T. LeQ. Thuan* and *J. Gardent*, Bull. Soc. chim. Fr., 1966, 2401). Similar conditions may also be used with a 1-benzyl-3,4-dihydroisoquinoline (*Kametani et al.*, Yakugaku Zasshi, 1974, **94**, 478; C.A., 1974, **81**, 37690). Once again the 2,3,10,11-tetraoxygenation pattern is obtained in these reactions.

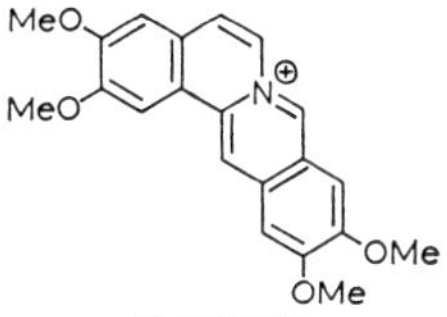

(LXXVIII)

The first synthesis of berberine can be classified as a Type 1 approach (*Perkin, J. N. Ray* and *Robinson*, J. chem. Soc., 1925, **127**, 740); it is summarised in Scheme 81. Treatment of LXXIX with zinc in acetic acid led to oxyberberine by reduction of the double bond, opening of the lactone ring,

SCHEME 81

THE FIRST SYNTHESIS OF BERBERINE

(LXXIX)

Oxyberberine

Berberine

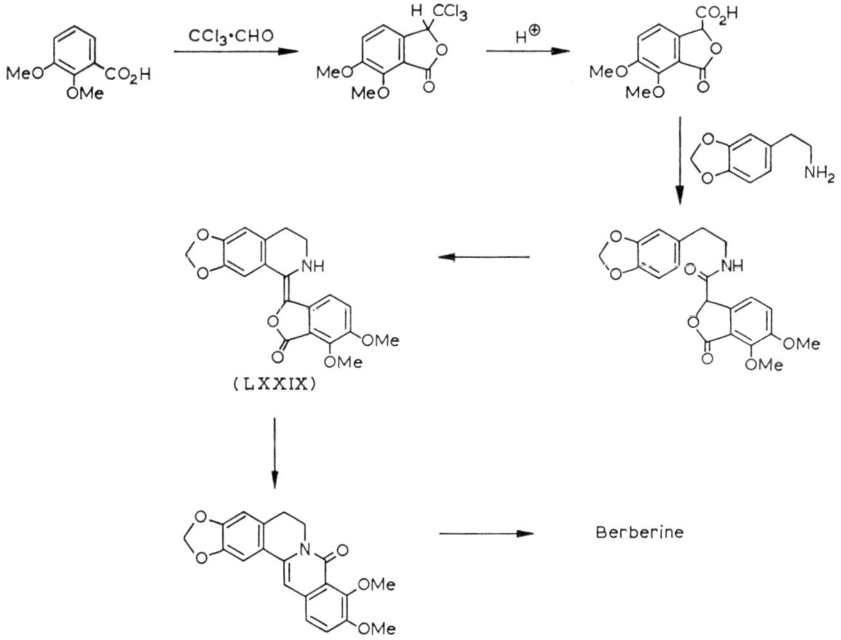

ring closure and dehydration. The phthalideisoquinolines (section 8, p. 152) are easily converted into protoberberine derivatives, for example hydrastine → dihydroberberine metho salt (*R. Mirza* and *Robinson*, Nature, 1950, **166,** 271; *T. R. Govindachari et al.*, J. chem. Soc., 1957, 2943):

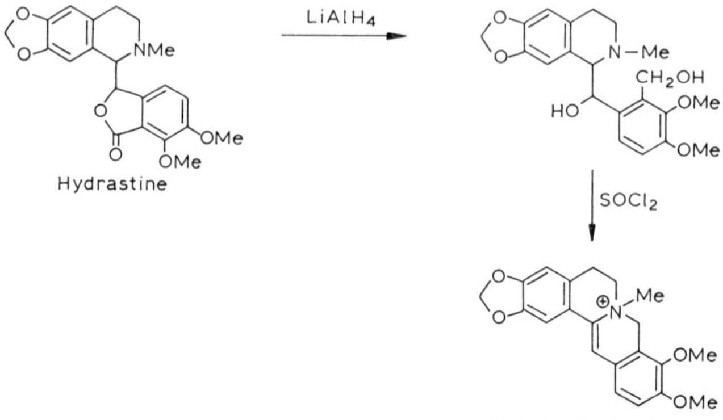

Hydrastine

Dihydroberberine metho ion

A further method of synthesis of the berbine skeleton involves the partial hydrolysis of homophthalimides of the type LXXX, followed by conversion to the ester amide LXXXI and cyclisation to LXXXII (*Haworth, Perkin* and *H. S. Pink, ibid.*, 1925, **127,** 1709). Presumably cyclisation to a 3,4-dihydroisoquinoline occurs first, so that the method may be classified as a Type 1 and not a Type 3 synthesis. It is possible for homophthalimides to be opened hydrolytically to give amido acids of type LXXXIII rather than the

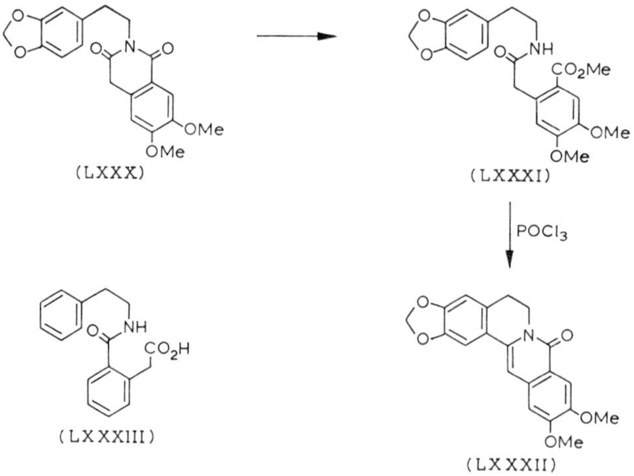

(LXXX)

(LXXXI)

(LXXXIII)

(LXXXII)

required one (*J. W. Huffman* and *E. G. Miller*, J. org. Chem., 1960, **25**, 90). Although some such examples are known, it seems to have been reported far less frequently than might be expected. The major disadvantage of this synthetic approach, however, is the relative inaccessibility of the required dialkoxyhomophthalic acids.

SCHEME 82

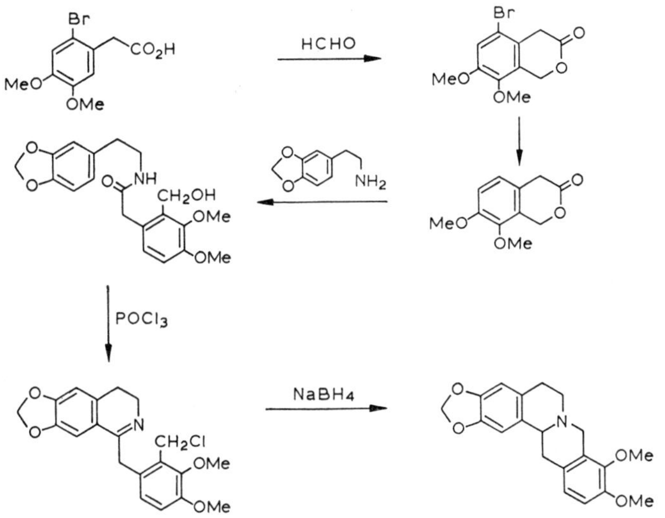

A further modification of the Type 1 route utilises isochroman-3-ones (*T. S. Stevens*, J. chem. Soc., 1935, 663; *Battersby et al., ibid.*, 1966, 1052). The method is illustrated in Scheme 82.

A recent development is due to *Kametani*, who used benzocyclobutenes as intermediates, LXXXIV → LXXXVI. Presumably the triene LXXXV is an intermediate (*Kametani, K. Ogasawara* and *T. Takahashi*, Tetrahedron, 1973, **29**, 73). Similar syntheses of LXXXVII and LXXXVIII have been reported (*Kametani et al.*, J. heterocycl. Chem., 1974, **11**, 179). In a modification to this method, the benzocyclobutene LXXXIX was heated with 6,7-dimethoxy-3,4-dihydroisoquinoline to yield the tetrahydroprotoberberine XC directly (*idem*, J. org. Chem., 1974, **39**, 447). Further examples have been described (*idem*, Tetrahedron, 1974, **30**, 1043; Chem. and Pharm. Bull., Japan, 1973, **21**, 907):

(LXXXIV) → Δ → (LXXXV) → (LXXXVI)

(LXXXVII): R¹ = H; R² = R³ = Me
(LXXXVIII): R¹ = Me; R² = R³ = H

(LXXXIX) + → Δ 150° → (XC)

R = H or Me

(ii) Type 2 syntheses

Since *N*-benzylisoquinolinium salts with the required oxygenation pattern, for example XCI, are readily available, attempts have been made to utilise them in berbine syntheses. Reduction of XCI to the *N*-benzyltetrahydroiso-quinoline (XCII) is easily achieved, but the products do not react with methylal or acetal (*Haworth* and *Perkin*, J. chem. Soc., 1925, **127**, 1434, 1444). The isoquinoline-1-amide (XCIII) was quaternised with 2,3-di-methoxybenzyl bromide to XCIV and this was reduced to XCV, but the latter compound could not be cyclised or hydrolysed to the carboxylic acid *(idem, loc. cit.)*.

Isoquinoline Reissert compounds, for example XCVI, form cyclic perchlorates (XCVII), which can be reduced catalytically, or with sodium tetrahydridoborate to an amide XCVIII (*J. W. Elliot* and *J. O. Leflore*, J. org. Chem., 1963, **28**, 3181). The latter cyclises with polyphosphoric acid giving XCIX (*S. F. Dyke et al.*, Tetrahedron, 1969, **25**, 1881). However, the isomeric amide C was extremely difficult to cyclise so this particular modification of the Type 2 route seems to hold little promise.

The most successful synthesis of protoberberines, utilising the Type 2 method is due to *C. K. Bradsher* (Scheme 83) (*Bradsher* and *N. L. Dutta*, Nature, 1959, **184,** 1943; J. Amer. chem. Soc., 1960, **82,** 1145; J. org. Chem., 1961, **26,** 2231; *Bradsher* and *J. H. Jones, ibid.,* 1958, **23,** 430; *H. F. Andrew* and *Bradsher*, Tetrahedron Letters, 1966, 3069; *Dutta* and *Bradsher*, J. org. Chem., 1962, **27,** 2213; *S. A. Telang* and *Bradsher, ibid.,* 1965, **30,** 752; *W. Augstein* and *Bradsher, ibid.,* 1969, **34,** 1349).

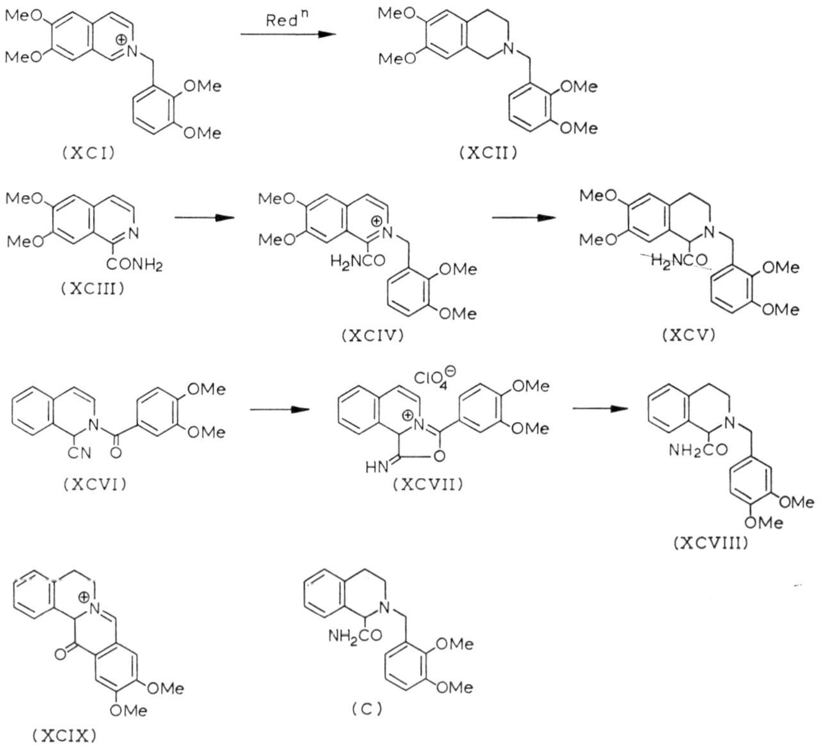

The poorest step is the first one, the oxidation of CI to CII (typically about 30%). Better yields in the quaternisation and cyclisation were realised if the oxime or acetal of CII were used instead of the free aldehyde. The reaction has been extended to prepare 13-methyltetrahydroprotoberberines, *e.g.* Scheme 84 (*A. A. Bindra, M. S. Wadia* and *Dutta*, Indian J. chem., 1969, **7,** 744).

SCHEME 83
BRADSHER'S SYNTHESIS OF PROTOBERBERINES

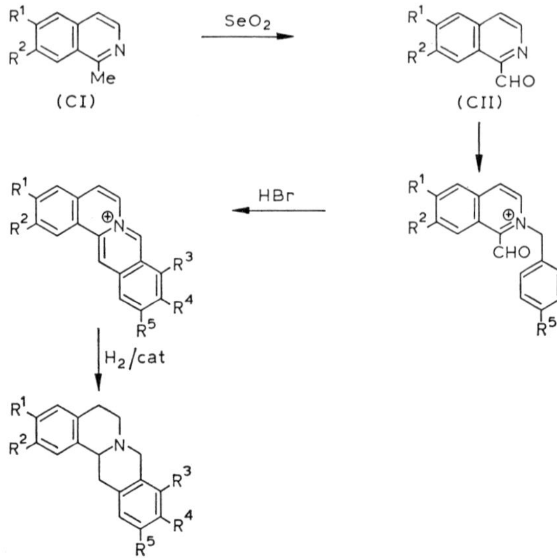

Typical compounds made:

R^1	R^2	R^3	R^4	R^5
OMe	OMe	H	OMe	OMe
OMe	OMe	OMe	OMe	H
OMe	OMe	OH	OMe	H
OMe	OMe	OMe	OMe	OH
O–CH$_2$–O		OMe	OH	H

SCHEME 84

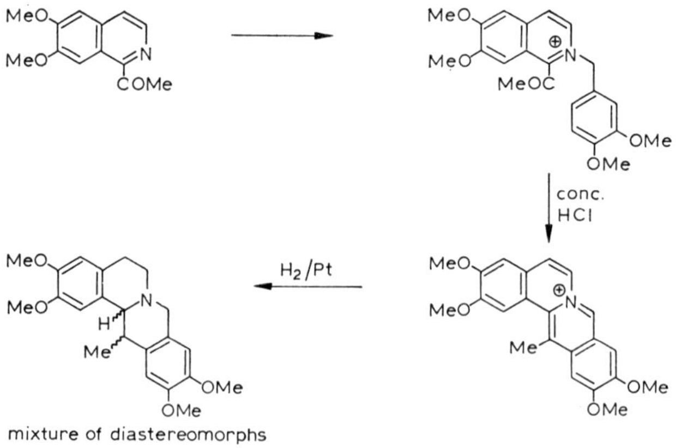

mixture of diastereomorphs

(iii) Type 3 syntheses

The most successful method classified under this heading was first used
by *Battersby* for the synthesis of **coreximine** and **norcoralydine** and utilises
the enamine character of 1,2-dihydroisoquinolines (Scheme 85) (*Battersby
et al.*, Tetrahedron, 1961, **14**, 46). Since isoquinolinium salts undergo disproportionation to a 1,2-dihydroisoquinoline and an isocarbostyril in the
presence of strong base, a simplification of the above method was achieved

SCHEME 85

1,2-DIHYDROISOQUINOLINES IN THE SYNTHESIS OF
TETRAHYDROPROTOBERBERINES

Norcoralydine   R = Me
Coreximine   R = H

by treating the quaternary salt of type CIII successively with strong base and
strong acid. The 1,2-dihydroisoquinoline such as CIV formed in the disproportionation, underwent cyclisation to the berbine derivative as shown in
Scheme 85. Unexpectedly, the isocarbostyril CV was also cyclised by acid
to give the 8-oxoberbine CVI (*D. W. Brown* and *S. F. Dyke*, Tetrahedron,
1966, **22**, 2429). To prepare 2,3,9,10-tetraoxytetrahydroprotoberberines by
this method requires a 7,8-dioxyisoquinoline. Such compounds are now

SCHEME 86

H$_2$/Pt or NaBH$_4$

6 N HCl room temp.

N. B. S.

HCl

available (Scheme 86) (*Dyke et al., ibid.*, 1969, **25**, 1881). New syntheses of tetrahydroberberine and tetrahydropalmitine have been achieved *via* the appropriate 1,2-dihydroisoquinolines *(idem, loc. cit.)*. Since the cyclisation of the 1,2-dihydroisoquinolines of type CVIII require acid conditions similar to those used to cyclise benzylaminoacetals an attempt was made to prepare a berbine derivative by the double cyclisation of acetals such as CVII, but the product proved to be CX, formed from the intermediate 4-hydroxytetrahydroisoquinoline CIX *(idem, loc. cit.)*.

(CVII)

(CVIII)

(CIX)

(CX)

(iv) Type 4 syntheses

This method of approach requires a pre-formed 3-arylisoquinoline derivative. When CXI was treated successively with glycidol, sodium periodate and mineral acid, the 5-hydroxytetrahydroprotoberberine CXII was obtained in good yield (Scheme 87) (*idem*, Tetrahedron, 1971, **27**, 3495). The first synthesis of berberastine (CXIII) was achieved along very similar lines (*Dyke* and *E. P. Tiley, ibid.*, 1975, **31**, 561).

An alternative approach involving 3-arylisocoumarin derivatives is shown in Scheme 88 (*Dyke et al.*, J. chem. Soc., C, 1971, 3219).

(e) Biosynthesis

It was suggested some years ago (*E. Winterstein* and *G. Trier*, "Alkaloids", Bornträger, Berlin, 1910, p. 307; *R. Robinson*, "Structural Relations of Natural Products", Oxford University Press, 1955, p. 86) that protoberberines and tetrahydroprotoberberines are derived in the plant from 1-benzyl-1,2,3,4-tetrahydroisoquinolines. The biosynthesis of the latter has been discussed (see p. 24). In the last 15 years or so it has been possible to study

(±)Reticuline   Berberine (CXIV)

these early proposals by using ^{14}C, ^{3}H or ^{2}H tracers (*A. R. Battersby*, Abh. Deut. Akad. Wiss. Berlin Kl. Chem. Geol. Biol., 1966, **3**, 295). It was suggested by *D. H. R. Barton* (Hugo Miller Lecture, Proc. chem. Soc., 1963, 293) and by *Battersby* (Tilden Lecture, *ibid.*, p. 189) that the "berberine bridge" carbon ($C_{(8)}$ on the protoberberine numbering) is derived *in vivo* from the *N*-methyl group of a 1-benzyl-1,2,3,4-tetrahydroisoquinoline precursor. When (±)reticuline ^{14}C-labelled at the $C_{(6)}$-OMe and *N*-methyl groups was fed to *Hydrastis canadensis* plants, the berberine isolated contained activity only at the methylenedioxy group and at $C_{(8)}$ (CXIV) (*Barton, R. H. Hesse* and *G. W. Kirby*, J. chem. Soc., 1965, 6379). The iminium ion CXV is probably an intermediate, which can arise either from the *N*-oxide or the carbinolamine. Cyclisation to the tetrahydroproto-

SCHEME 87
THE SYNTHESIS OF 5-HYDROXYBERBINES

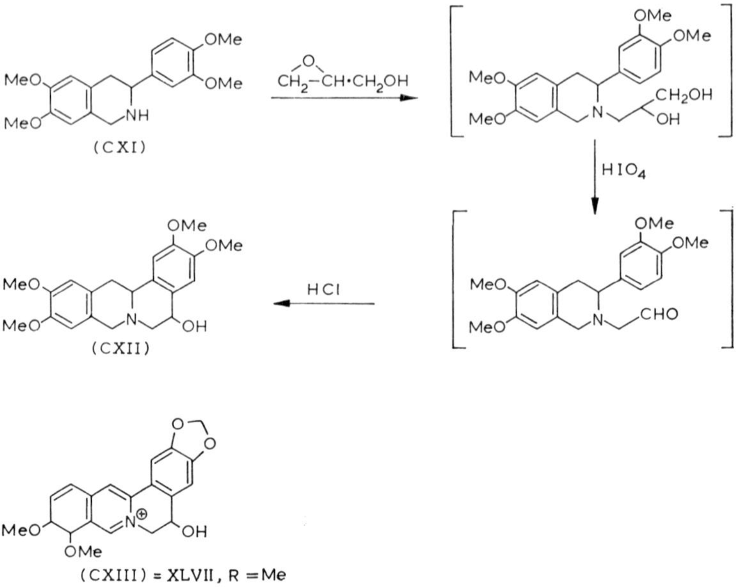

(CXI)

(CXII)

(CXIII) = XLVII, R = Me

SCHEME 88
BERBINES FROM ISOCOUMARINS

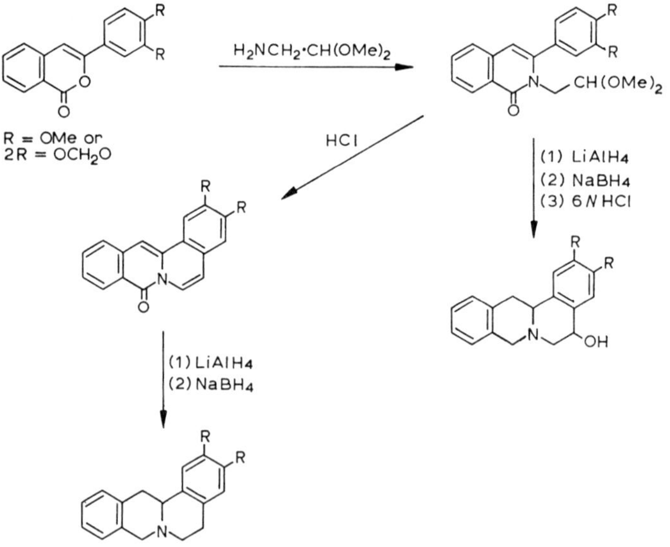

R = OMe or
2R = OCH$_2$O

berberine is then facilitated by the *ortho*-phenolic grouping on the potential ring D (Scheme 89). This experiment also demonstrated that the methylenedioxy group arises from an 2-methoxyphenol. Further experiments showed that $(+)(S)$-reticuline is incorporated very efficiently into berberine, and is far superior to $(-)$reticuline. In *Chelidonium majus* it has been established that $(+)(S)$-reticuline is converted successively into $(-)$ **scoulerine** and $(-)(S)$-**stylopine** (*Battersby et al.*, Chem. Comm., 1965, 89;

SCHEME 89

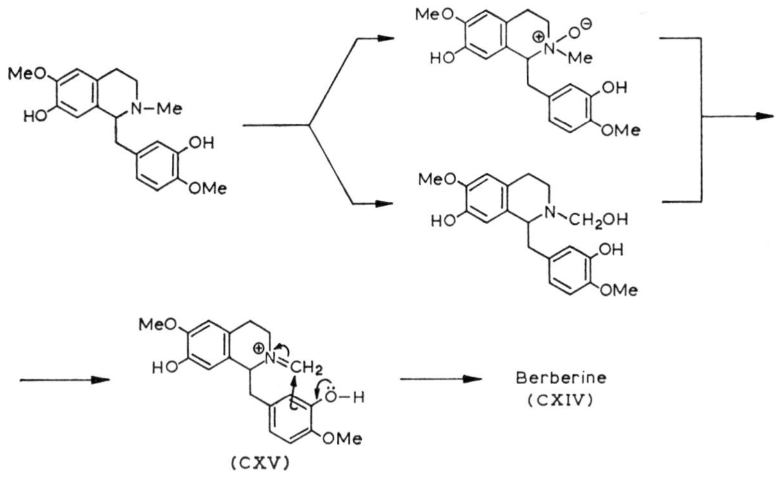

Berberine
(CXIV)

(CXV)

J. chem. Soc., Perkin I, 1975, 1140). Subsequently it was established that $(+)$reticuline is the precursor of the majority of the 2,3,9,10-tetraoxyprotoberberines. **Coreximine** also is derived from reticuline since it has been reported that $(\pm)$reticuline labelled at $C_{(3)}$ is incorporated into it in opium poppies (*E. Brockmann-Hanssen, C.-C. Fu* and *G. Zanati*, J. pharm. Sci., 1971, **60**, 873). Labelled reticuline is also incorporated specifically into **corydaline** in *Corydalis cava* (*G. Blaschke*, Arch. Pharm., 1968, **301**, 439). It is possible that the $C_{(13)}$-Me group arises by cyclisation of the precursor CXVI, but it is more likely that *C*-alkylation of the enamine CXVII occurs (*idem, loc. cit.*, and Arch. Pharm., 1970, **303**, 358; *I. D. Spenser et al.*, Canad. J. Chem., 1974, **52**, 2818):

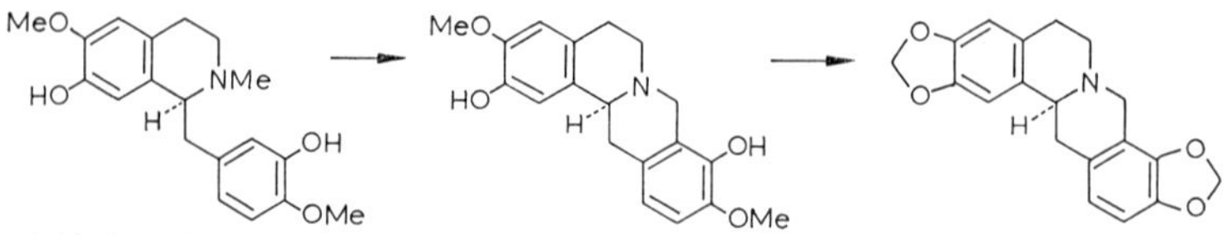

(+)(*S*)-Reticuline → (−)Scoulerine → (−)(*S*)-Stylopine

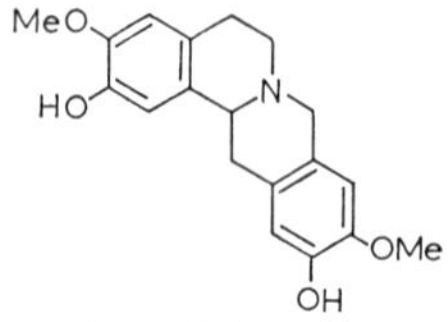

Coreximine

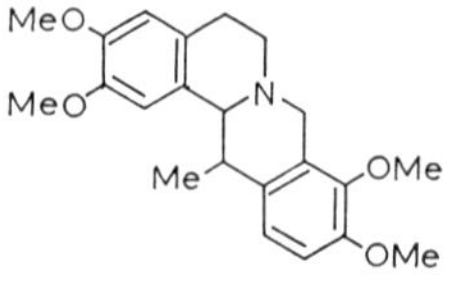

Corydaline

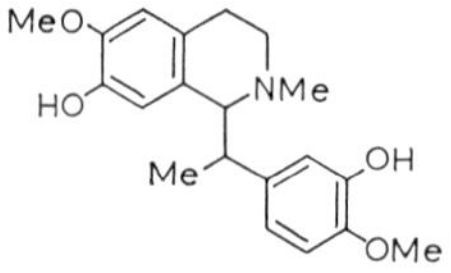

(CXVI)

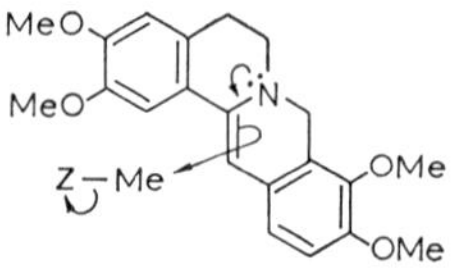

(CXVII)

It is also likely that **ophiocarpine** arises from an enamine reaction with a dihydroprotoberberine similar to CXVII, but is also possible for condensation of **dopamine** with CXVIII (a metabolite of noradrenaline) to yield CXIX, followed by cyclisation to ophiocarpine.

It has been established that berberastine (CXIII, Scheme 87) is biosynthesised from noradrenaline (*I. Monkovic* and *Spenser*, Canad. J. Chem.,

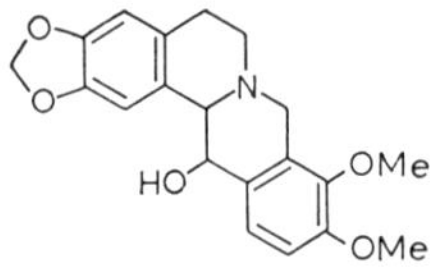

Ophiocarpine

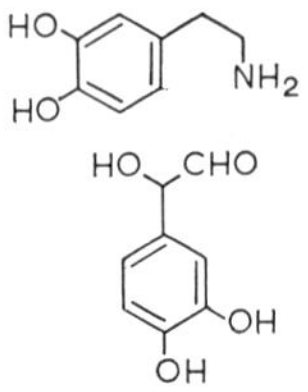

(CXVIII)

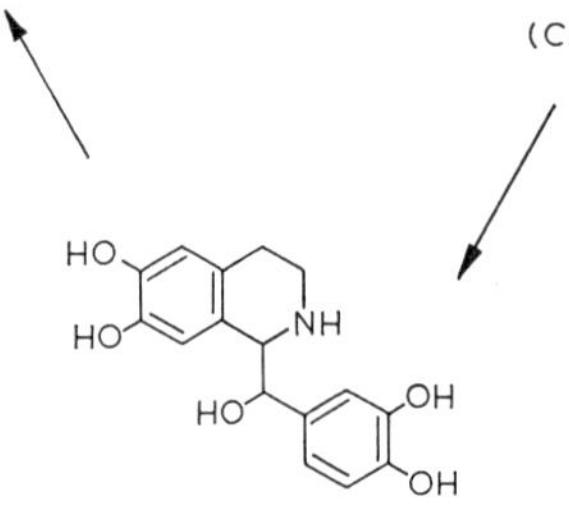

(CXIX)

1965, **43**, 2017). Additionally, it is possible that 4-hydroxynorlaudanosoline (CXX) is a key precursor, like norlaudanosoline itself, of a range of alkaloids of the pavinane, isopavinane, aporphine, erythrina and perhaps the benzo-[*c*]phenanthridine groups (*Dyke et al.*, Tetrahedron Letters, 1969, 1515).

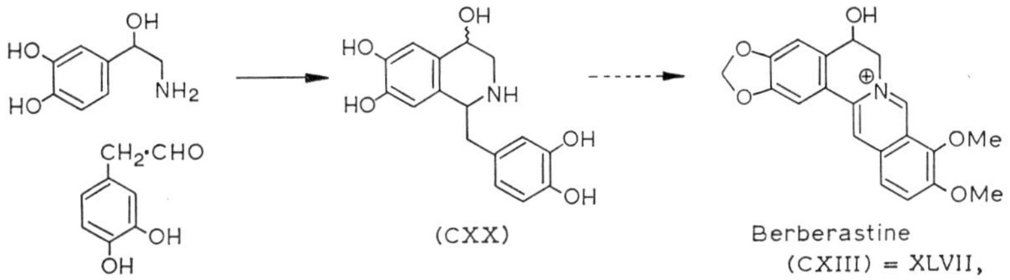

(CXX)

Berberastine
(CXIII) = XLVII,
R = Me

7. The protopine alkaloids

Although the protopine-type alkaloids do not possess an isoquinoline ring they are derived both *in vivo* and *in vitro* from 1-benzylisoquinoline derivatives. They occur in Berberidaceae, Fumariaceae, Papaveraceae, Ranunculaceae and Rutaceae families, and are characterised by the existence of a ten-membered ring that contains an *N*-methyl group and a ketone carbonyl function at $C_{(14)}$ (I). All of the alkaloids of this group contain at least four

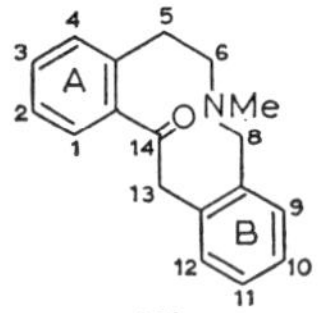

(I)

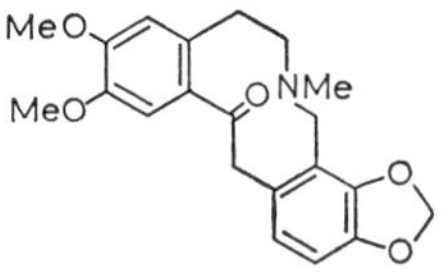

Cryptopine

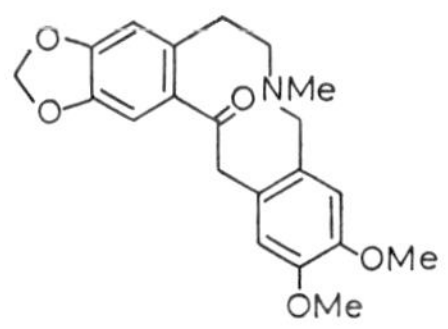

Fagarine II

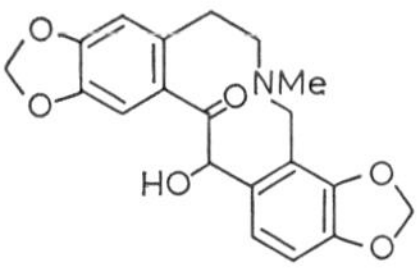

Ochrobirine

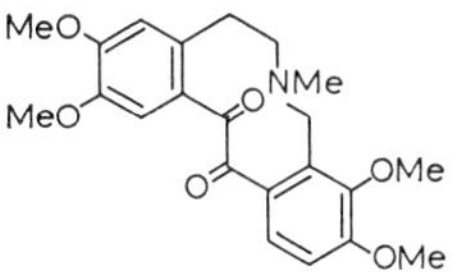

13-Oxomuramine

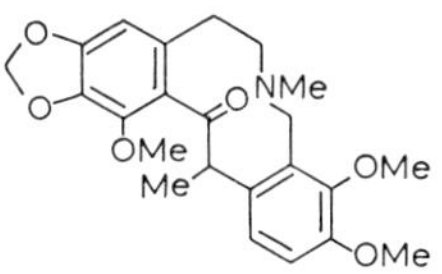

Corycavidine

oxygen functions, two in each of the aromatic rings. The ring A substituents are located at $C_{(2)} + C_{(3)}$ and the ring B oxygen functions are more commonly found at $C_{(9)} + C_{(10)}$, *e.g.* **cryptopine,** although some examples are known with the $C_{(10)} + C_{(11)}$ pattern, *e.g.* **fagarine-II.** Additional hydroxyl or methoxyl groups may be present at $C_{(1)}$ or $C_{(13)}$; the latter position may alternatively carry carbonyl oxygen, *e.g.* **ochrobirine** (note that a spiro-benzylisoquinoline alkaloid of the same name is known), and 13-**oxo-muramine.** Some derivatives are known that possess a $C_{(13)}$-methyl group, *e.g.* **corycavidine.** The protopine alkaloids have been reviewed by *M. Shamma*, "The Isoquinoline Alkaloids; Chemistry and Pharmacology", Academic Press, New York, 1972, p. 345, and by *F. Santavy*, in "The Alkaloids", 1970, **12,** 390.

SCHEME 90

THE STRUCTURE OF CRYPTOPINE

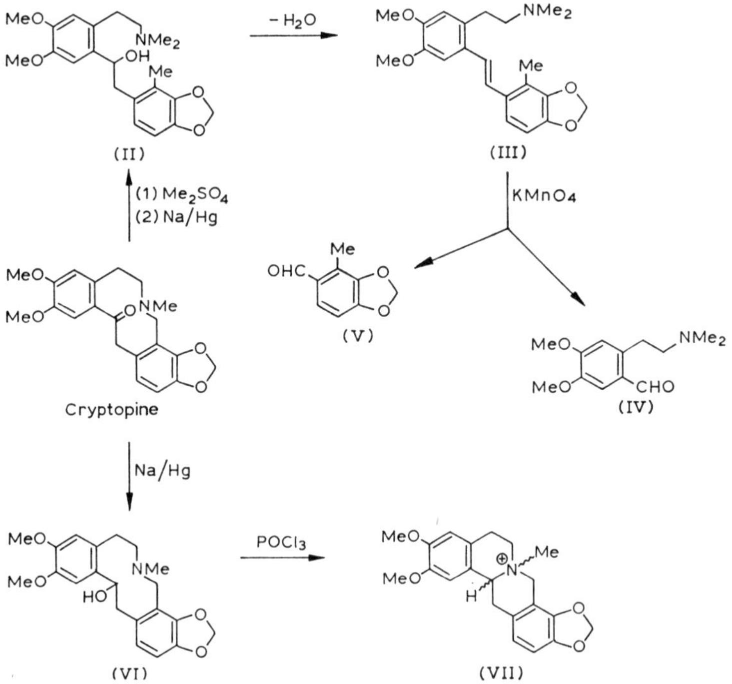

(a) Structure determination

The first alkaloid of this group to be examined was cryptopine, $C_{21}H_{23}NO_5$, a minor constituent of opium. The structure was elucidated by *W. H. Perkin Jr.*, who described his work in a monumental paper published in 1916 (J. chem. Soc., 1916, **109,** 815). In all over fifty transformation products were described, but in Scheme 90 only the essential data required to elucidate the structure of cryptopine, are summarised. Such degradative sequences became the routine approach for studying other members of the group. The product II, from the Emde degradation of cryptopine methosulphate was dehydrated to the stilbene III and cleaved to a mixture of the two aldehydes IV and V. On this evidence the ketone carbonyl group of cryptopine could be located at $C_{(13)}$ or $C_{(14)}$. However, when cryptopine itself was reduced with sodium amalgam, the hydroxyamine VI was easily cyclised to a diastereomeric mixture of tetrahydroprotoberberine metho salts VII. Hence, the carbonyl group of cryptopine must be at $C_{(14)}$. Cryptopine itself may be cyclised *(Perkin, loc. cit.)* to the quaternary salt isocryptine chloride, which can be demethylated and reduced to (±)**sinactine** (Scheme 91) (*K. Goto* and *Z. Kitasato*, J. chem. Soc., 1930, 1234; *E. Spath* and *E. Mosettig*, Ber., 1931, **64,** 2048):

SCHEME 91

Cryptopine

Isocryptopine chloride

(±)Sinactine

Cryptopine has also been cyclised photochemically in high yield to **epi-berberine** (*X. A. Dominguez et al.*, Tetrahedron Letters, 1967, 2493).

Cryptopine (and protopine) does not exhibit the expected ketonic properties to any degree, and in the i.r. spectrum of the base the carbonyl frequency is at 1675 cm^{-1}, suggesting an amide type of carbonyl group

(*F. A. L. Anet, A. S. Bailey* and *R. Robinson,* Chem. Ind., 1953, 944; *E. H. Mottus, H. Schwartz* and *L. Marion,* Canad. J. Chem., 1953, **31,** 1144; *N. J. Leonard, T. W. Milligan* and *T. L. Brown,* J. Amer. chem. Soc., 1960, **82,** 4075). This is due to an interaction between the carbonyl group and the adjacent nitrogen lone pair. In cryptopine perchlorate the carbonyl band is entirely lacking indicating that the salt has the structure VIII. As expected, the carbonyl frequency in cryptopine methiodide (IX), is higher (1686 cm^{-1}) than in cryptopine itself.

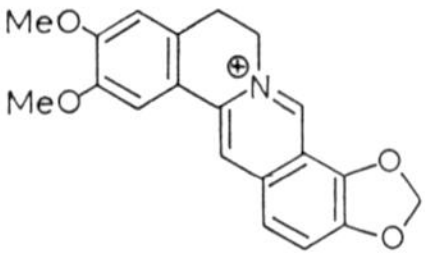

Epiberberine 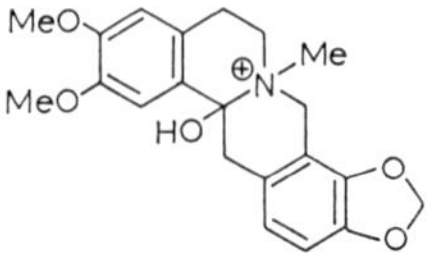   (VIII) 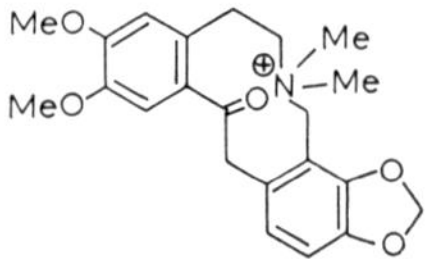   (IX)

The proton magnetic resonance spectrum of **cryptopine** (*Shamma* and *L. R. Bennet,* unpublished results, 1971, quoted by *Shamma* in his book; *S. F. Dyke* and *D. W. Brown,* unpublished results — see *Brown,* Ph.D. Thesis, University of Bath, 1966) is summarised in X and that of **protopine** in XI (Varian, n.m.r. Spectral Catalogue, Spectrum No. 339); both spectra were measured at room temperature. An interesting change occurred when the spectrum of protopine was redetermined at $-45°$ (*Anet* and *M. A. Brown,*

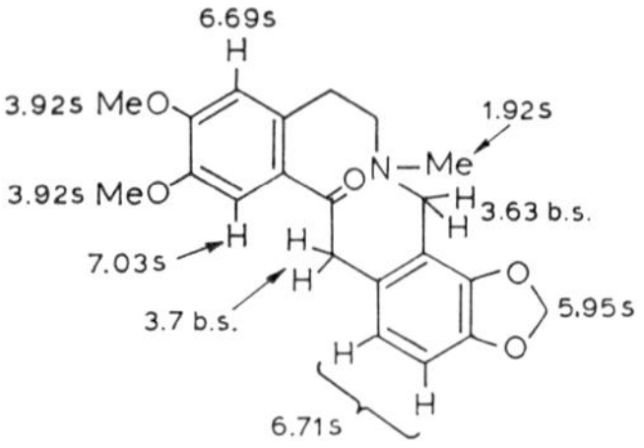

(X) 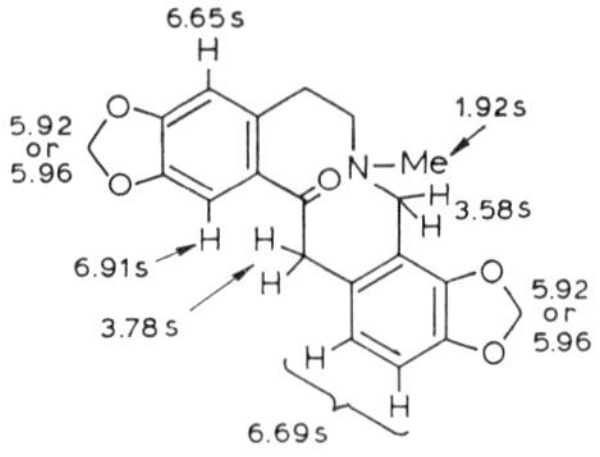   (XI)

Tetrahedron Letters, 1967, 4881). The broad singlets at 3.58 and 3.78 δ in the room temperature spectrum became AB quartets, partially superimposed upon each other. This was interpreted to mean that the inversion of the ten-membered ring had been slowed down at the lower temperature.

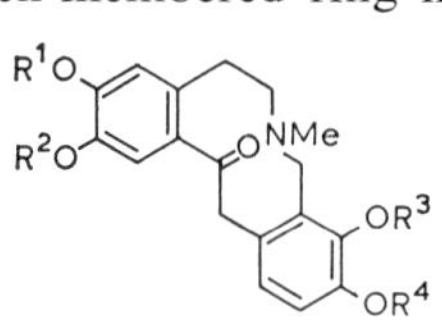

(XII):   R^1 + R^2 = CH$_2$; R^3 = H; R^4 = Me
(XIII):   R^1 = R^2 = R^3 = Me; R^4 = H

The ^{13}C-nuclear magnetic resonance spectra of allocryptopine, cryptopine, **hunnemannine** (XII), **muramine** (XIII) and protopine have been analysed (*T. T. Nakashima* and *G. E. Maciel*, Org. Magnetic Resonance, 1973, **5**, 9). Evidence for transannular amino-carbonyl interaction was found. The base peak in the mass spectrum of cryptopine (M^+ *m/e* 369) is at *m/e* 148 and results from the fragmentation shown in XIV → XV (*L. Dolejs, V. Hanus* and *J. Slavik*, Coll. Czech. chem. Comm., 1964, **29**, 2479):

Among the more interesting of the many derivatives of cryptopine described by Perkin is **anhydrocryptopine,** which is formed when isocryptopine chloride is treated with base. Anhydrocryptopine undergoes a variety of transformations when treated with acids, and *Perkin (loc. cit.)* allotted structures to most of these products. However, some of this work has been corrected (*Dyke* and *Brown*, Tetrahedron, 1968, **24**, 1455; 1969, **25**, 5375). Only the corrected structures are shown in Scheme 92.

(b) Synthesis

(*i*) *Chemical synthesis*

The main method used relies upon the preparation of the appropriately substituted tetrahydroprotoberberine and its ring-opening and oxidation. The synthesis of **cryptopine** (*R. D. Haworth* and *Perkin*, J. chem. Soc., 1926, 1769) illustrates the method (Scheme 93). Considerable speculation exists on the mechanism of the transformation of XVI into cryptopine. One proposal is summarised in XVI → XVII → cryptopine (*P. B. Russell*, J. Amer. chem. Soc., 1956, **78**, 3115).

When epiberberine *N*-oxide is treated with potassium chromate it yields the carbinolamine which can be *N*-methylated and ring-opened to cryptopine (*K. W. Bentley* and *A. W. Murray*, J. chem. Soc., 1963, 2491).

13-Oxoprotopines can be obtained by oxidation of the protopine derivative; 13-oxomuramine has been prepared in this way (*Leonard* and *R. R. Sauers*, J. org. Chem., 1957, **22**, 63).

SCHEME 92
SOME REACTIONS OF ANHYDROCRYPTOPINE

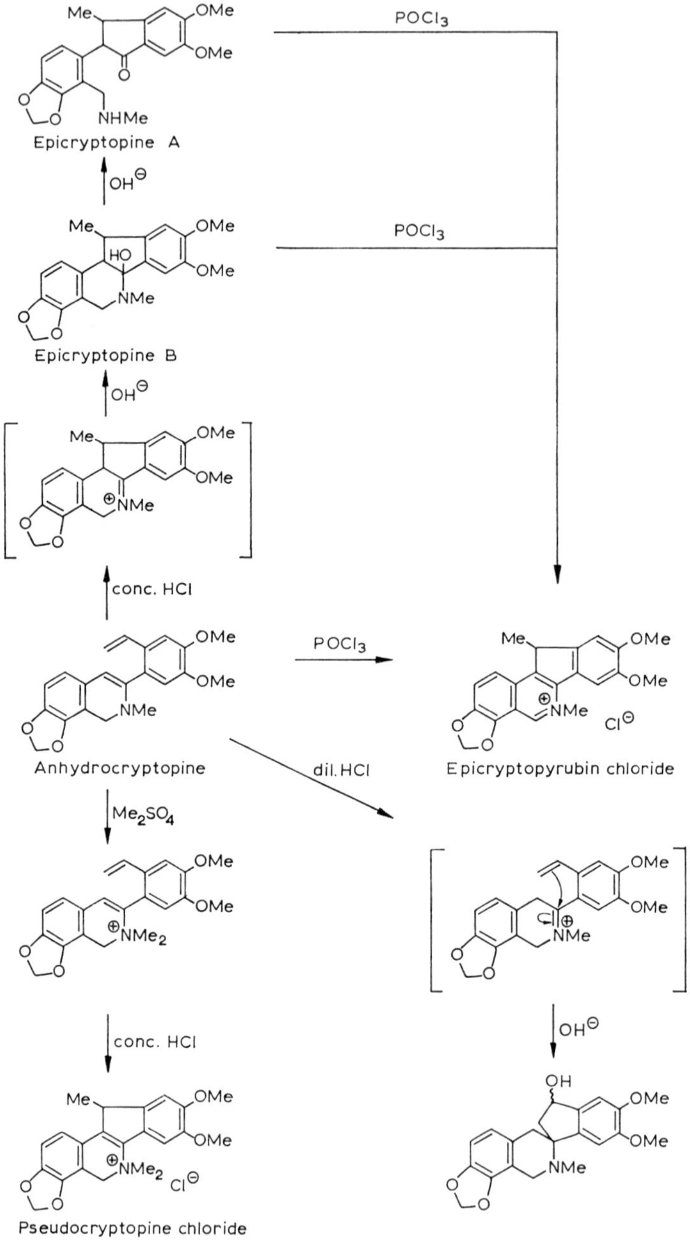

SCHEME 93

THE SYNTHESIS OF CRYPTOPINE

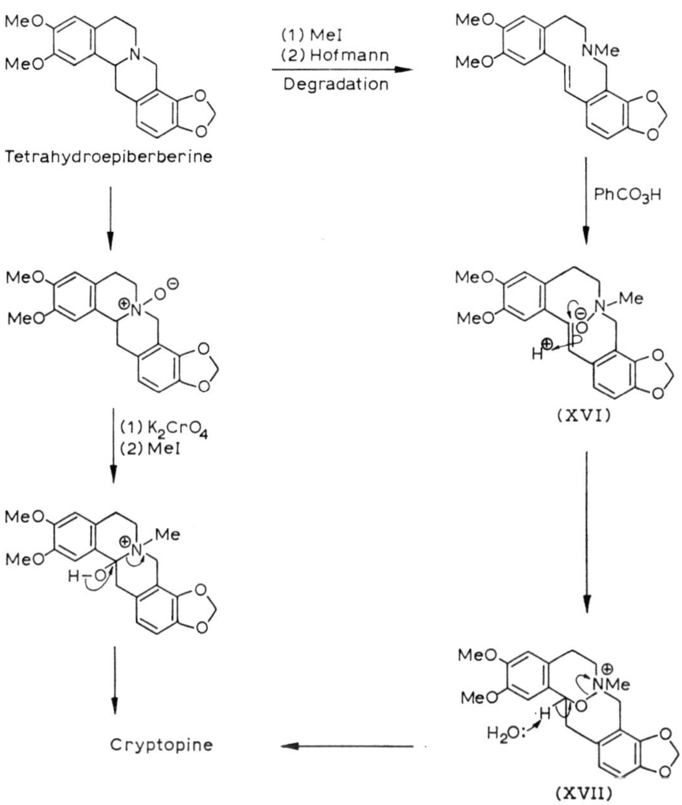
MeO
MeO
N
Tetrahydroepiberberine
(1) MeI
(2) Hofmann
Degradation
NMe
$PhCO_3H$
Me
(XVI)
(1) K_2CrO_4
(2) MeI
Me
H_2O:
NMe
(XVII)
Cryptopine

(*ii*) *Biosynthesis*

It was suggested some time ago that protopines originate from tetrahydro-protoberberines and this has been substantiated by feeding experiments with *Dicentra spectabilis* (*D. H. R. Barton, R. H. Hesse* and *G. W. Kirby*, Proc. chem. Soc., 1963, 267) and especially with *Chelidonium majus* (*A. R. Battersby et al.*, J. chem. Soc., Perkin I, 1975, 1147). The route that has been established is summarised in Scheme 94. However, the steps between stylopine metho salt and protopine have not yet been elucidated.

SCHEME 94

THE BIOSYNTHESIS OF PROTOPINE

(+)(*S*)-Reticuline   (−)(*S*)-Scoulerine

(−)(*S*)-Stylopine

(−)α-Narcotine

8. The phthalideisoquinoline alkaloids

Alkaloids of this group, based upon the nucleus I, are small in number but are important because (−)α-**narcotine** is one of the major alkaloids of opium. Phthalideisoquinolines have also been found in the Berberidaceae, Fumariaceae and Ranunculaceae. All possess *N*-methyl groups and oxygen functions at $C_{(6)}$, $C_{(7)}$, $C_{(4')}$ and $C_{(3')}$. Narcotine is unique in having an additional methoxyl group at $C_{(8)}$. The stereochemistry at $C_{(1)}$ and $C_{(9)}$ may also vary from one member to another. The structural elucidation and

(I)

(−)α-Narcotine

synthesis of narcotine and hydrastine were reported in the first two decades of this century, although the absolute stereochemistry has been established only in the last 15 years. The subject has been reviewed on several occasions (*A. R. Pinder*, Chem. Reviews, 1964, **64**, 551; *J. Stanek*, in "The Alkaloids", 1967, **9**, 117; *F. Santavy, ibid.*, 1970, **12**, 397).

(a) *Structure determination*

(i) *(−)α-Narcotine*

Classical methods showed that this tertiary base, $C_{22}H_{23}NO_7$, contains three methoxy groups, one *N*-methyl and one methylenedioxy group. It was also apparent that a lactone ring is present, thus accounting for all of the oxygen atoms. Narcotine is readily cleaved hydrolytically, oxidatively and reductively (Scheme 95). Identification of the fragments **cotarnine** and **meconine,** together with the recognition of opianic acid, **hydrocotarnine** and 2,3-dimethoxyphthalic anhydride led to the overall structure of narcotine, with the stereochemistry as yet undefined. If the metho salt of narcotine is heated with base, **narceine** (III) is produced by way of II (*W. Roser*, Ann., 1888, **247**, 167; *C. R. Addinall* and *R. T. Major*, J. Amer. chem. Soc., 1933, **35**, 1202, 2153). Narceine occurs in opium and the analogous compounds

SCHEME 95

CLEAVAGE REACTIONS OF NARCOTINE

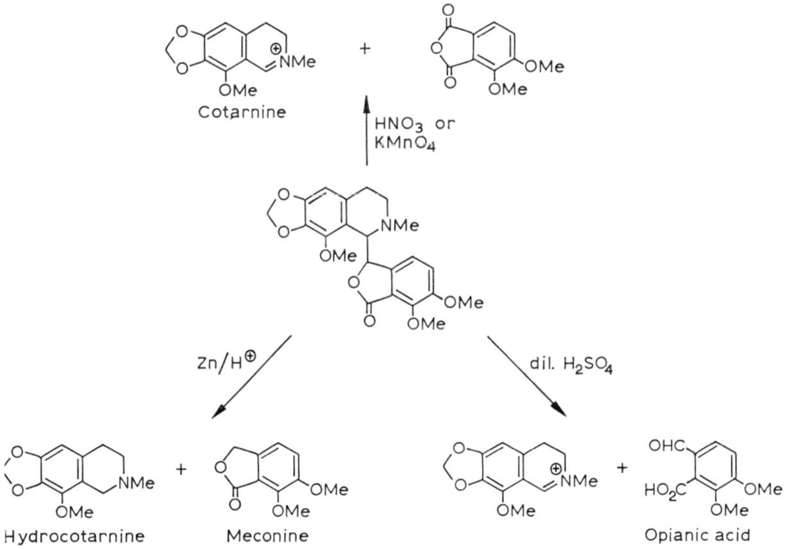

adlumidiceine (IV) and **adlumiceine** (V) have been isolated from *P. rhoes* (*Santavy et al.*, Phytochemistry, 1973, **12**, 2513). If *N*-benzylnarcotine is subjected to Hofmann degradation, followed by debenzylation, the product is **nornarceine** (*W. Kloetzer, S. Teitel* and *A. Brossi*, Monatsh., 1972, **103**, 1210).

Narcotine methochloride $\xrightarrow[\text{Base}]{\Delta\ +}$ (II) $\longrightarrow$

(III) $R^1 + R^2 = CH_2$; $R^3 = OMe$; $R^4 = R^5 = Me$

(IV) $R^1 + R^2 = R^4 + R^5 = CH_2$; $R^3 = H$

(V) $R^1 = R^2 = Me$; $R^3 = H$; $R^4 + R^5 = CH_2$

Nornarceine

(ii) (−)β-Hydrastine

This is narcotine without the $C_{(8)}$-methoxyl group and its chemistry is very similar to that of narcotine, except that hydrastine does not undergo the hydrolytic or reductive cleavage that is so characteristic of narcotine. The reason for this difference is obscure. Oxidation of hydrastine yields opianic acid and hydrastinine, which can be reduced to the tetrahydroisoquinoline. The same substance may be obtained by reductive cleavage of hydrocotarnine (*F. L. Pyman* and *F. G. P. Remfry*, J. chem. Soc., 1912, **101**, 1595).

(b) Absolute configuration

(−)α-Narcotine can be reduced with lithium tetrahydridoaluminate to the diol VI, the di-*O*-acetate of which can be hydrogenolysed to the substituted 1-benzyltetrahydroisoquinoline (VII) (Scheme 96). This compound exhibits a positive Cotton Effect near 295 nm, hence the $C_{(1)}$ hydrogen is alpha — *i.e.* it belongs to the (*R*)-series (*M. Ohta et al.*, Chem. and Pharm. Bull., Japan, 1964, **12**, 1080). When (−)α-narcotine is reacted with hot, methanolic potassium hydroxide solution, an equilibrium is set up between it and a diastereomorphic form, (−)β-narcotine. This material, when subjected to

SCHEME 96

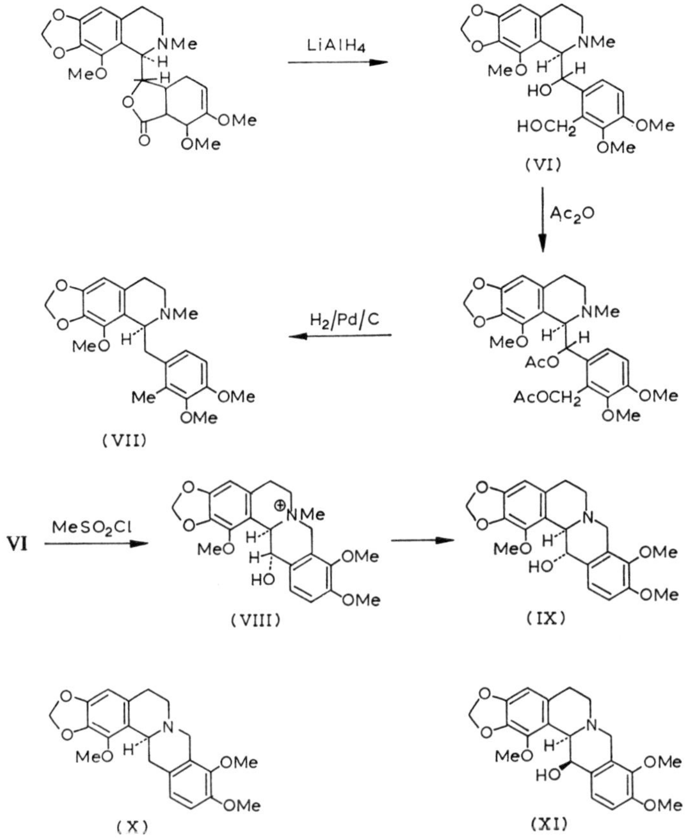

the sequence illustrated in Scheme 96 gives the *same* compound VII. Hence the α- and β-narcotines differ in the configuration at $C_{(9)}$.

When the diol VI from α-narcotine is reacted with methanesulphonyl chloride the quaternary salt VIII is formed. This can be *N*-demethylated to IX, then hydrogenolysed to the tetrahydroprotoberberine X. This latter compound exhibits a strong negative rotation, indicating that the $C_{(14)}$-H atom is α, as shown. The same sequence of reactions applied to β-narcotine gave XI which, upon hydrogenolysis gave the same tetrahydroproto-berberine X, thus confirming the alpha configuration for the $C_{(14)}$-H atom (*A. R. Battersby* and *H. Spenser*, J. chem. Soc., 1965, 1087). If VI is reacted with *p*-toluenesulphonyl chloride, the product is XII and this, with acetic

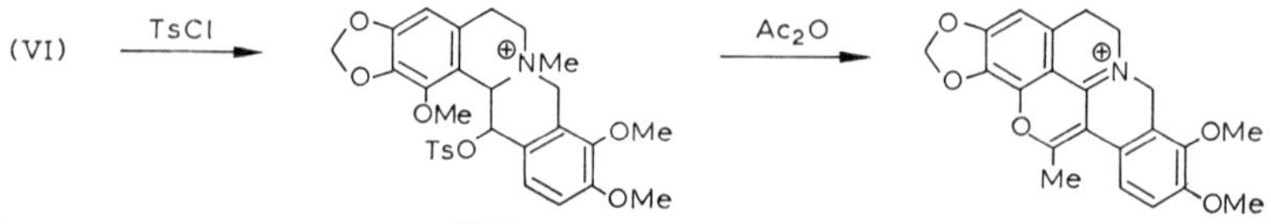

(VI) $\xrightarrow{\text{TsCl}}$ (XII) $\xrightarrow{\text{Ac}_2\text{O}}$ (XIII)

anhydride is converted into XIII (*V. Simanek* and *A. Klasek*, Tetrahedron Letters, 1971, 4133).

The coupling constant $J_{13,\,14}$ in IX is 9.0 Hz, corresponding to a dihedral angle of about 160°, thus indicating a *trans* orientation of these two hydrogen atoms. In XI, the value of $J_{13,\,14}$ is only 1.5 Hz, in agreement with a *cis* orientation and a dihedral angle of 60°. It follows that the configuration of the two centres in $(-)\alpha$-narcotine are *erythro* (1R,9S) as shown on p. 152.

The relative stereochemistry and absolute configurations of $(-)\beta$-hydrastine (the naturally occurring isomer), and of the epimerised $(-)\alpha$-hydrastine, have been elucidated by a sequence of reactions similar to those

SCHEME 97

THE CONFIGURATION OF THE HYDRASTINES

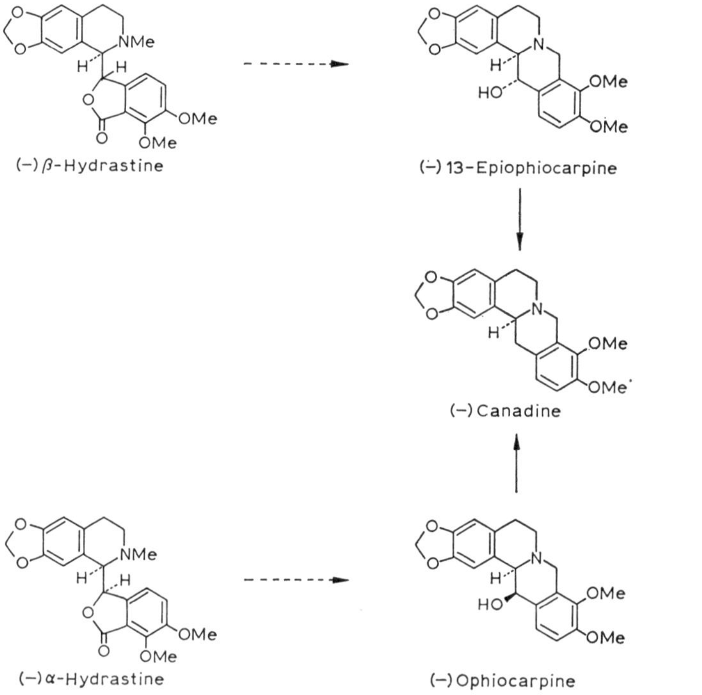

$(-)\beta$-Hydrastine

$(-)$ 13-Epiophiocarpine

$(-)$ Canadine

$(-)\alpha$-Hydrastine

$(-)$ Ophiocarpine

SCHEME 98

INTERCONVERSIONS OF PHTHALIDEISOQUINOLINES

(−)β-Hydrastine — pyridine / hydrochloride →

BBr$_3$ / CH$_2$Cl$_2$

CH$_2$Cl$_2$/DMSO / NaOH →

(−)Bicuculline

(−)β-Hydrastine — BCl$_3$ → — CH$_2$N$_2$ →

(−)Cordrastine -II

used for the narcotines (*Ohta, H. Tani* and *S. Morozumi*, Chem. and Pharm. Bull., Japan, 1964, **12**, 1072) (Scheme 97).

The absolute configurations of the phthalideisoquinolines may also be deduced from o.r.d. measurements (*Ohta et al., ibid.*, 1964, **12**, 1080; *G. Snatzke et al.*, Tetrahedron, 1969, **25**, 5059) and also from c.d. spectra (*idem, loc. cit.*; *T. Kikuchi* and *T. Nishinaga*, Tetrahedron Letters, 1969, 2519). In o.r.d. spectra three Cotton effects are apparent; the first between 320–335 nm, reflects the stereochemistry at $C_{(9)}$, as does the third one at 235–255 nm. Positive Cotton effects mean a (9R)-configuration. The second Cotton effect, between 280–300 nm depends upon the configuration at $C_{(1)}$ and is positive for the (R)-configuration.

Although mass spectra have been recorded they have not been of much value in structural elucidation. The molecular ion is not recorded because of the very ready cleavage of the $C_{(9)}$–$C_{(1)}$ bond (*M. Ohashi et al.*, J. Amer. chem. Soc., 1963, **85**, 2807).

Interconversions within the phthalideisoquinoline group are now easily possible. Epimerisation at $C_{(1)}$ can be brought about by (*1*) successive treatment with cyanogen bromide and mineral acid (*G. Gaa'l, P. Kerekes* and *B. Bognar*, J. pr. Chem., 1971, **313**, 935) and (*2*) treatment with triphenylmethylammonium butyrate (*Gaa'l et al.*, Pharmazie, 1971, **26**, 431). Selective *O*-demethylations and re-alkylations may also be achieved (Scheme 98) (*Simanek* and *Klasek, loc. cit.*; *Teitel, J. O'Brien* and *Brossi*, J. org. Chem., 1972, **37**, 3368).

(c) Synthesis

(i) Chemical synthesis

The basic molecular framework of the phthalideisoquinoline alkaloids was first constructed (*W. H. Perkin* and *R. Robinson*, J. chem. Soc., 1911, **99**, 775) by heating an ethanolic solution containing meconine and cotarnine. Presumably the anion was formed from the former, which then added to $C_{(1)}$ in the imminium ion of cotarnine. $(\pm)\alpha$-Narcotine was formed in low yield and was resolved. None of the other diastereomorphic pair was formed. Yields of condensation products were improved substantially by replacing meconine with nitromeconine (*E. Hope* and *Robinson, ibid.*, 1914, **105**, 2085). The nitro group was subsequently removed from XIV by standard methods to give $(\pm)\beta$-narcotine. When nitromeconine was replaced by iodomeconine, the product of the condensation, after reductive removal of the iodo substituent, proved to be $(\pm)\alpha$-narcotine once more.

When nitromeconine was added to hydrastinine, a mixture of diastereomorphs was obtained and reductive removal of the nitro group yielded a diastereomorphic mixture of hydrastines (*idem*, Proc. chem. Soc., 1912, **28**, 17; *Hope et al.*, J. chem. Soc., 1931, 236). A better synthesis of hydrastine is summarised in Scheme 99 (*R. D. Haworth* and *Pinder, ibid.*, 1950, 1776; *Haworth, Pinder* and *Robinson*, Nature, 1950, **165**, 529).

SCHEME 99

A SYNTHESIS OF HYDRASTINE

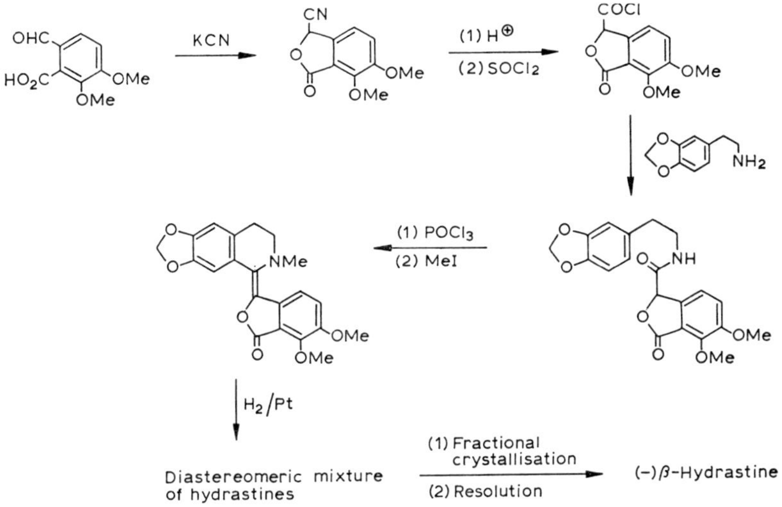

In a novel approach to the synthesis of phthalideisoquinoline alkaloids, an aziridinium intermediate XVII was prepared from XV and XVI. The aziridinium ring present in XVII could undergo ring-opening to give either a rhoeadine-type skeleton XVIII (path a) or a phthalideisoquinoline system (route b) (*T. Kametani et al.*, J. chem. Soc., Perkin I, 1976, 1221):

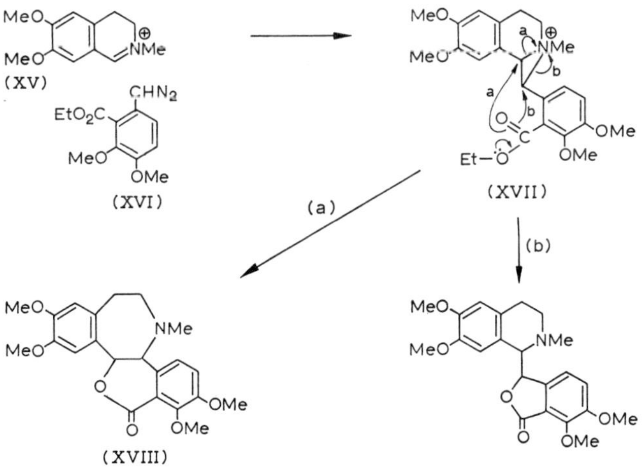

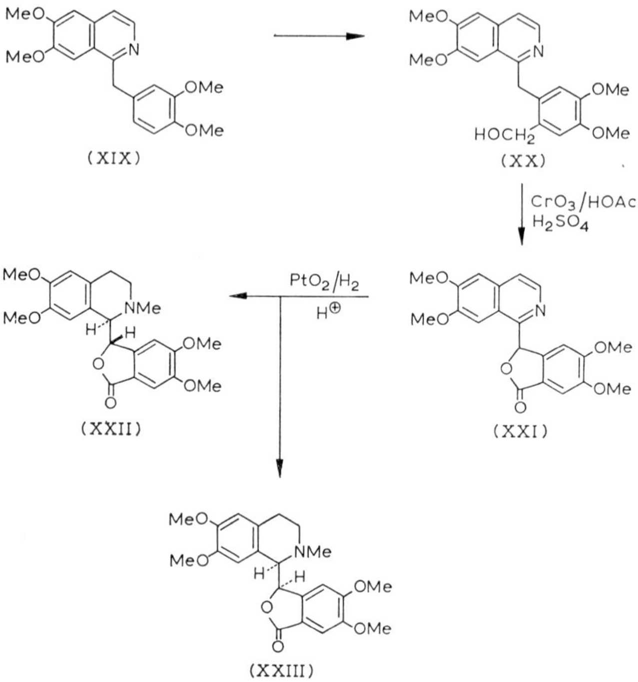

In a simple synthesis of analogues of the alkaloids, papaverine (XIX) was converted to XX, then oxidised to XXI, which was reduced to a mixture of XXII and XXIII (*M. Shamma* and *V. St. Georgiev*, Tetrahedron Letters, 1974, 2339).

Spirobenzylisoquinolines have provided the starting materials for another method of synthesis of phthalideisoquinolines (Scheme 100) (*D. B. MacLean et al.*, Canad. J. Chem., 1973, **51**, 3287). A further method invented by *Kametani et al.* (J. chem. Soc., Perkin I, 1974, 2509; Heterocycles, 1975, **3**, 405) is shown in Scheme 101.

(ii) Biosynthesis

It has been established that tyrosine is the precursor of hydrastine in *Hydrastis canadensis* L. (*I. D. Spenser* and *J. R. Gear*, J. Amer. chem. Soc., 1962, **84**, 1059; *G. Kleinschmidt* and *K. Mothes*, Z. Naturforsch., 1959,

SCHEME 100

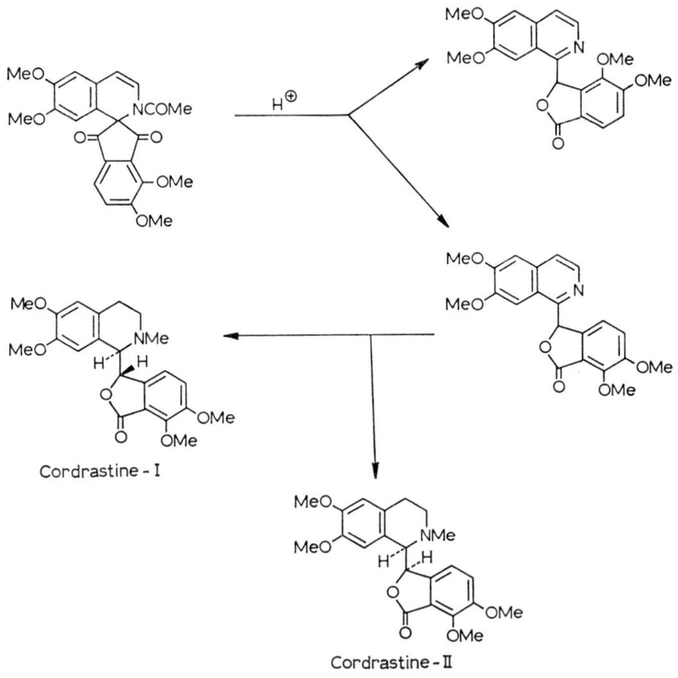
H⊕
Cordrastine - I
Cordrastine - II

SCHEME 101

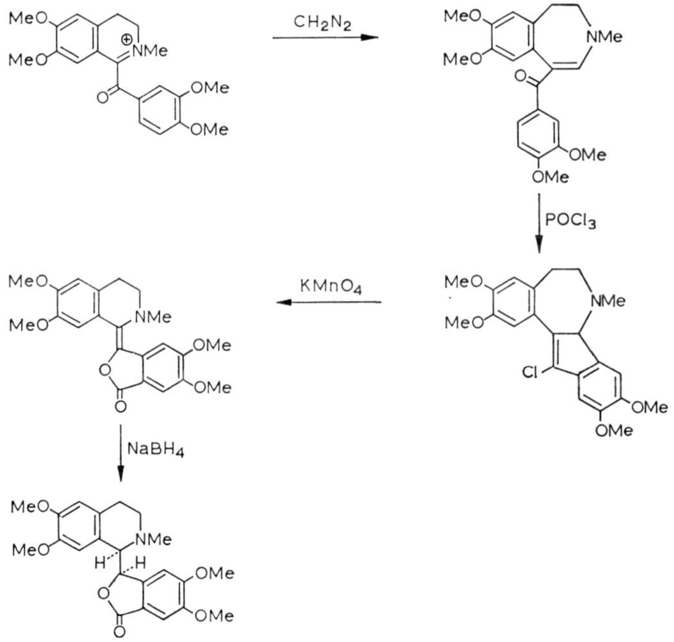
CH2N2
POCl3
KMnO4
NaBH4

B14, 52) and of narcotine in *Papaver somniferum* (*Battersby et al.*, J. chem. Soc., C, 1968, 2163; *Kleinschmidt* and *Mothes, loc. cit.*; *Gear* and *Spenser*, Canad. J. Chem., 1963, **41,** 78). As a result of these and other studies involving the feeding of appropriate precursors to *H. canadensis* or *P. somniferum* (*R. N. Gupta* and *Spenser*, Biochem. Biophys. Res. Commun., 1963, **13,** 115; *Battersby et al., loc. cit.*; *Battersby* and *M. Hirst*, Tetrahedron Letters, 1965, 669; *Battersby et al.*, Chem. Comm., 1967, 602; J. chem. Soc., Perkin I, 1975, 1140, 1147) it is apparent that norlaudanosoline is first formed from tyrosine, then converted into the reticulines. Incorporation of (+)reticuline into narcotine is more efficient than that of (−)reticuline; it was evident that epimerisation was occurring to give the correct reticuline for further elaboration. The biosynthetic pathway to narcotine can be summarised in Scheme 102.

SCHEME 102

BIOSYNTHESIS OF NARCOTINE

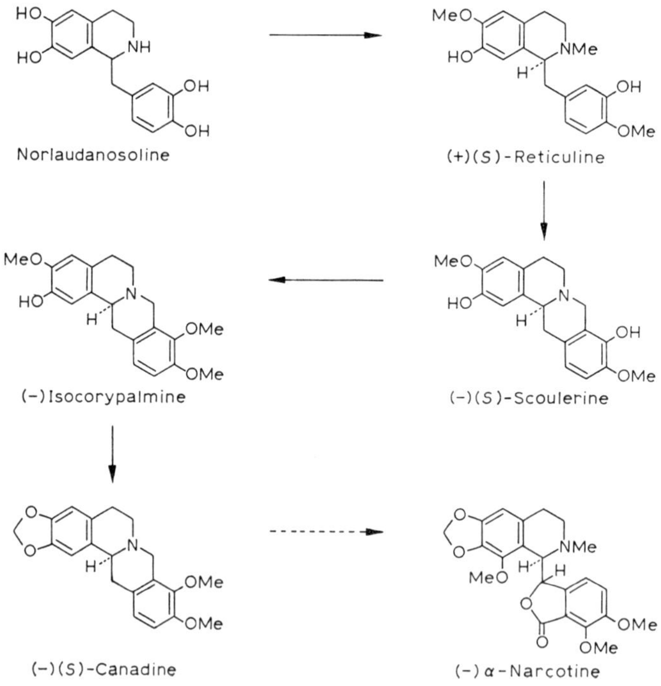

Norlaudanosoline

(+)(*S*)-Reticuline

(−)Isocorypalmine

(−)(*S*)-Scoulerine

(−)(*S*)-Canadine

(−)α-Narcotine

9. The rhoeadines and papaverrubines

About 30 alkaloids are known based upon the ring system I; some are acetals (R = Me) and some are hemiacetals (R = H). All possess oxygen functions at $C_{(7)}$, $C_{(8)}$, $C_{(12)}$ and $C_{(13)}$; the B/D ring junction is *cis* in some examples, but *trans* in others. A further variation concerns the stereochemistry at $C_{(14)}$. The papaverrubines are des-*N*-methyl rhoeadines. The subject has been reviewed by *F. Santavy* ("The Alkaloids", 1970, **12**, 398) and by *S. Pfeifer* (Pharmazie, 1971, **26**, 328).

R = H or Me
(I)

(a) Structure determination

(+)**Rhoeadine**, $C_{21}H_{21}NO_6$, was the first member of this group to be investigated thoroughly. The usual chemical tests indicated the presence of two methylenedioxy groups, one *N*-methyl and one *O*-methyl group (as part of an acetal function) (*Santavy et al.*, Coll. Czech. chem. Comm., 1965, **30**, 335; *H. Hrbek, Santavy* and *L. Dolejs, ibid.*, 1970, **35**, 3712). When treated with dilute hydrochloric acid, the acetal function of rhoeadine is lost to give the hemiacetal **rhoeagenine** (which is also a natural product) (*O. Hesse*, Ann. Chem. Pharm., 1969, **149**, 35; *Santavy et al.*, Coll. Czech. chem. Comm., 1959, **24**, 3493; 1960, **25**, 1901). Oxidation of this hemiacetal with potassium permanganate yielded hydrastic acid and 3,4-methylenedioxyphthalic acid as the main products, thus giving some information about the relative positions of the substituents. Treatment of rhoeagenine with nitric acid yielded hydrastinine as the major product (*W. Awe* and *W. Winkler*, Arch. Pharm., 1957, **290**, 367) whereas distillation with zinc dust gave, amongst other products, some isoquinoline. Whereas rhoeadine is stable to reduction with lithium tetrahydridoaluminate or catalytic hydrogenation, the hemiacetal rhoeagenine is reduced to rhoeageninediol, which can be oxidised by manganese dioxide to the lactone oxyrhoeagenine (i.r. band at 1725 cm^{-1}). The same lactone is also obtained by the treatment of rhoeagenine with this reagent. Oxyrhoeagenine may be reconverted into rhoeagenine with lithium tetrahydridoaluminate (Scheme 103). Although two successive Hofmann degradations were carried out on rhoeadine, neither the methine nor the bismethine were fully characterised, and so were of limited value in structural elucidation. Subsequently it was found that in

SCHEME 103

Rhoeadine   LiAlH$_4$ Rhoeagenine

MnO$_2$

Rhoeageninediol   MnO$_2$ Oxyrhoeagenine

addition to the expected methine II, two further products, III and IV, were formed, resulting from rearrangement and oxidation (*M. D. Rozwadowska* and *J. W. ApSimon*, Tetrahedron, 1972, **28**, 4125). The structure of rhoeadine was established by some further chemistry with rhoeageninediol, and by mass spectral studies (*Santavy et al.*, Coll. Czech. chem. Comm., 1965, **30**, 335, 3479; 1970, **35**, 3712; *Dolejs* and *V. Hanus*, Tetrahedron, 1967, **23**, 2997). The mass spectrum, which is characterised by three intense peaks (Scheme 104), is insensitive to the geometry at the B/D ring junction; isorhoeadine exhibits essentially the same spectrum. However, the hemiacetal alkaloid rhoeagenine undergoes an entirely different fragmentation pathway (Scheme 104).

An alternative degradation procedure has been applied to **alpinigenine,** in which the derived oxime is converted into the cyano compound, which is then subjected to the Hofmann degradation. The product, VI, can be cleaved by periodate to yield a mixture of VII and VIII (*H. Ronsch*, Tetrahedron Letters, 1969, 5124).

(i) *Relative configuration*

When **glaudine** is reacted with very dilute mineral acid, isomerisation occurs to give **epiglaudine.** This latter base may be isomerised further to

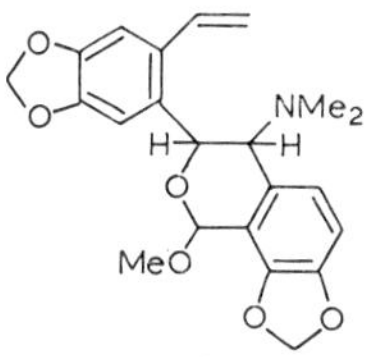

(II 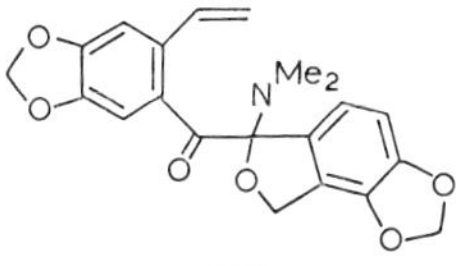)   (III 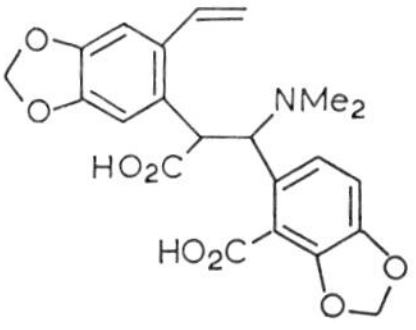)   (IV)

oreodine, which cannot be reconverted to glaudine. The B/D ring junction in glaudine and epiglaudine is *trans* because the $J_{1,2}$ value of 9 Hz was found on examining the n.m.r. spectra, whereas in oreodine, $J_{1,2}$ was only 2 Hz

SCHEME 104

MASS SPECTRAL FRAGMENTATION PATTERNS

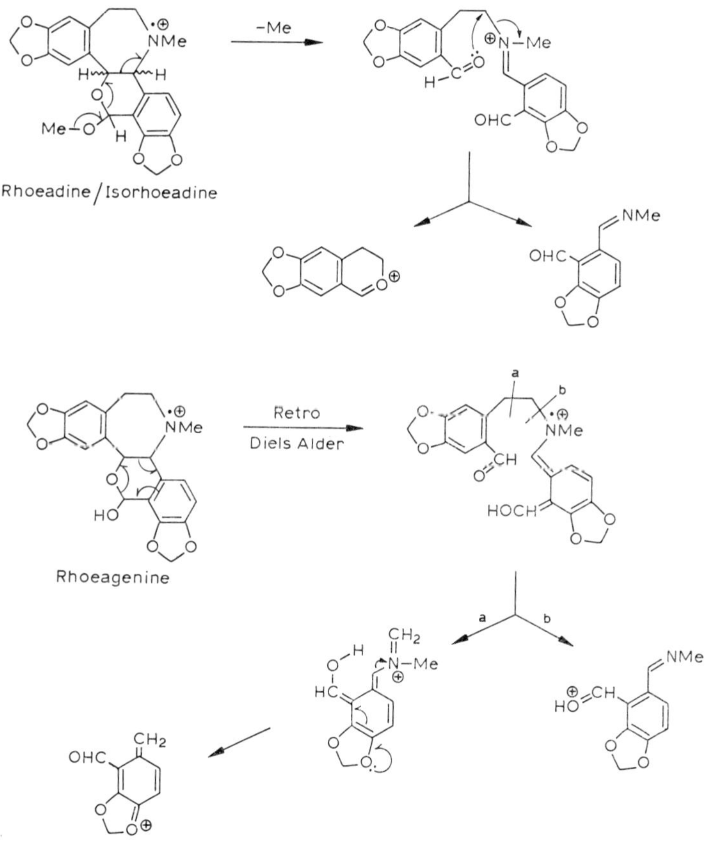

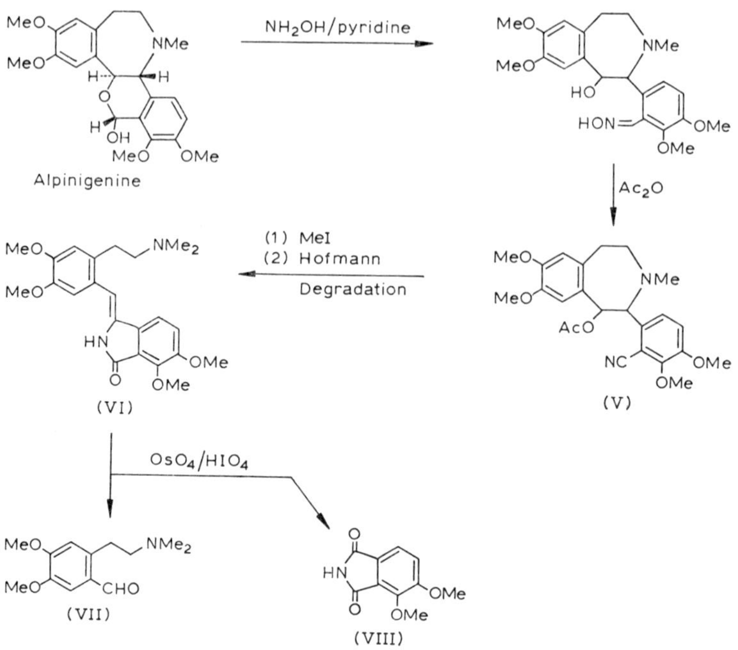

NH$_2$OH/pyridine

Alpinigenine

Ac$_2$O

(1) MeI
(2) Hofmann

Degradation

(VI)

(V)

OsO$_4$/HIO$_4$

(VII)

(VIII)

(*Santavy et al.*, Coll. Czech. chem. Comm., 1965, **30**, 335, 3479; 1970, **35**, 3712; *A. D. Cross, I. Mann* and *S. Pfeifer*, Pharmazie, 1966, **21**, 181; *A. Guggisberg et al.*, Helv., 1967, **50**, 621; *M. Maturovo et al.*, Coll. Czech. chem. Comm., 1967, **32**, 419; *M. Shamma et al.*, Chem. Comm., 1968, 212). By a combination of n.m.r. data and relative rates of quaternisation the relative stereochemistries were deduced to be as shown with the B/D ring junction of oreodine *cis* and the $C_{(14)}$-methoxyl group on the same side as the bridgehead hydrogen atoms. Furthermore it was established that in glaudine the methoxyl group at $C_{(14)}$ is beta, and alpha in epiglaudine — relative to the $C_{(1)}$-hydrogen as alpha-orientated.

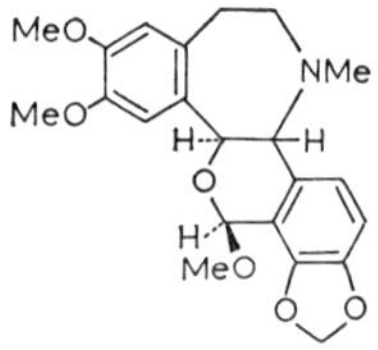

Glaudine
$J_{1,2}$ = 9 Hz

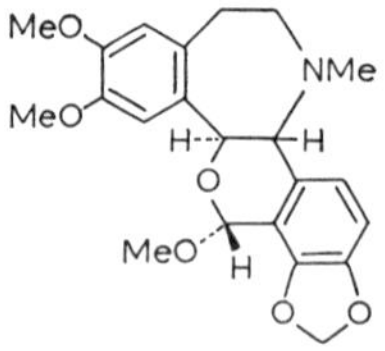

Epiglaudine
$J_{1,2}$ = 9 Hz

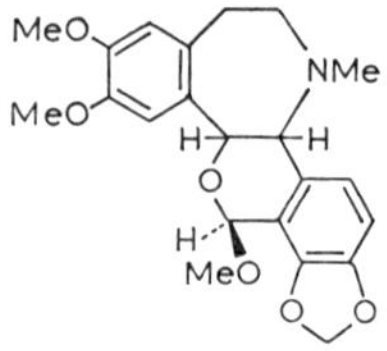

Oreodine
$J_{1,2}$ = 2 Hz

(ii) Absolute configuration

A study of o.r.d. and c.d. curves for a number of rhoeadine alkaloids led to ambiguity in the assignment of absolute configuration (*E. Brockmann-Hanssen et al.*, J. pharm. Sci., 1968, **57**, 30; *Santavy, Hrbek* and *K. Blaha*, Coll. Czech. chem. Comm., 1967, **32**, 4452). However, a Davydov splitting was observed in the c.d. spectrum of (+)glaudine, so by application of the

SCHEME 105

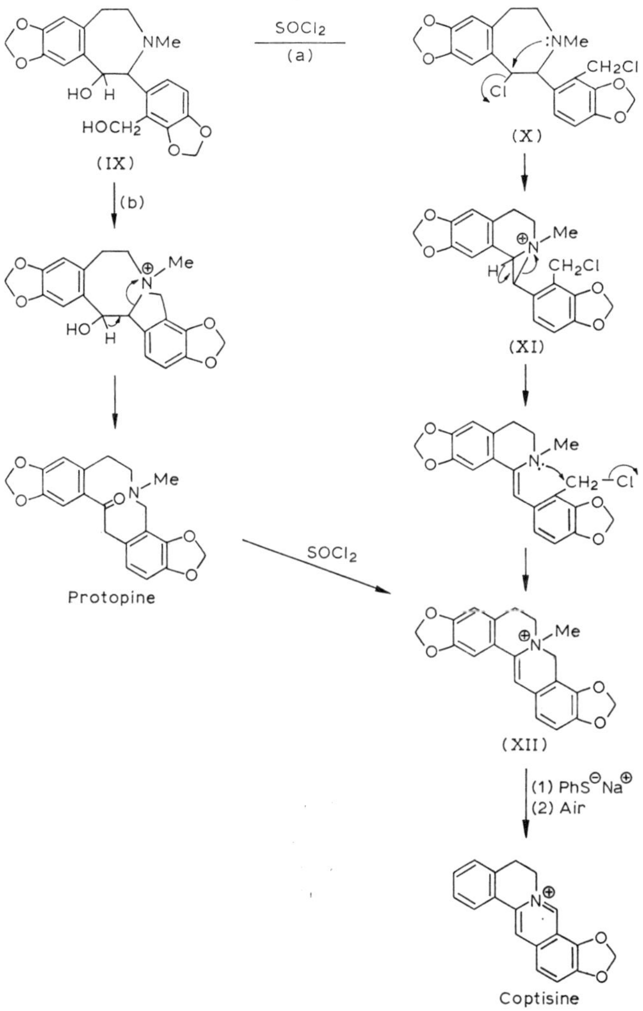

aromatic chirality method, the absolute stereochemistry was assigned (*Shamma et al.*, Tetrahedron Letters, 1971, 4207). Since all of the known rhoeadines and papaverrubines are strongly dextrarotatory, and since their relative configurations are known, the absolute configurations follow. An X-ray analysis of rhoeagenine methiodide has confirmed the assignment of configuration as 1*R*, 2*R*, 14*S* (*C. S. Huber*, Acta Cryst., 1972, **B28**, 982).

(b) Some rearrangement reactions

(*1*) When rhoeageninediol (IX) is boiled with thionyl chloride, it is converted into the quaternary salt XII which, upon catalytic hydrogenation, yields the metho salt of tetrahydrocoptisine. Alternatively the salt XII can be demethylated and oxidised to **coptisine** (Scheme 105) (*A. Klasek, V. Simanek* and *Santavy*, Tetrahedron Letters, 1968, 4549). A mechanism, involving the dichloride X and the aziridinium salt XI has been suggested (*Shamma*, "The Isoquinoline Alkaloids", Academic Press, New York, 1972, p. 407). An alternative mechanism is also shown in Scheme 105 involving initial ring-opening of IX to **protopine,** followed by recyclisation to XII.

(*2*) When **papaverrubine A** is treated with dilute hydrochloric acid, a deep red colour is produced, due to the formation of the quaternary salt XIII. A similar colour reaction is given by all of the papaverrubines (*J. Slavik*, Chem. Listy, 1958, **52**, 1957; Coll. Czech. chem. Comm., 1959, **24**, 2506; *D. Walterova* and *Santavy, ibid.*, 1968, **33**, 1623; *Pfeifer* and *I. Mann*, Pharmazie, 1965, **20**, 643).

(c) Synthesis

(i) Chemical synthesis

Partial syntheses from phthalideisoquinolines and from spirobenzylisoquinolines have been achieved, and since these two groups of compounds have themselves been synthesised, these conversions represent formal total syntheses of the rhoeadine alkaloids.

(—)α-**Narcotine** has been transformed into XIV, an analogue of glaudine by the method outlined in Scheme 106 (*A. Brossi et al.*, J. Amer. chem. Soc.,

SCHEME 106

RHOEADINES FROM PHTHALIDEISOQUINOLINES

(XIV)

1971, **93**, 4321; Canad. J. Chem., 1972, **50**, 2022). Essentially the same route has been followed using (−)**bicuculline,** thus providing the first synthesis of (+)**rhoeadine** (*W. Klotzer, S. Teitel* and *Brossi,* Helv., 1971, **54**, 2057; 1972, **55**, 2228):

(−)Bicuculline   (+)Rhoeadine

SCHEME 107

A SYNTHESIS OF RHOEAGENINEDIOL

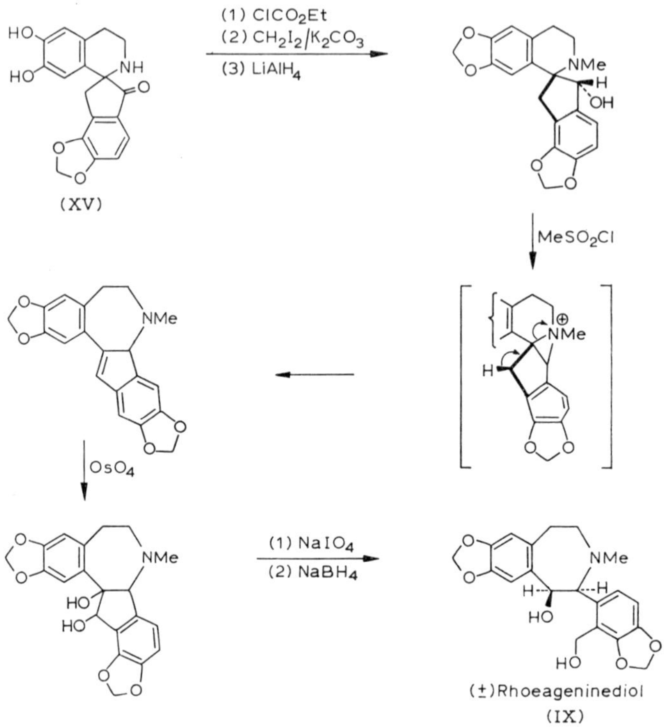

(±)**Rhoeageninediol** (IX) has been synthesised from the synthetic spiro-benzylisoquinoline XV (Scheme 107) (*H. Irie, S. Tani* and *H. Yamane,* Chem. Comm., 1970, 1713; J. chem. Soc., Perkin I, 1972, 2986). The product, IX, had previously been transformed into rhoeadine during structural studies. In an analogous manner (±)alpinigenine has been prepared from XVI *(ibid., idem)*:

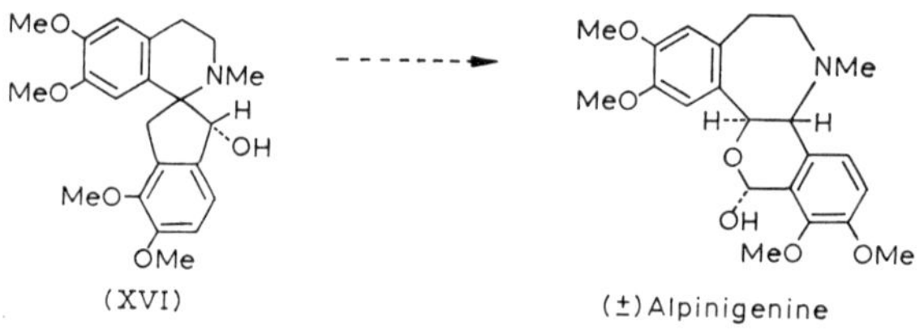

SCHEME 108

HCHO / HCl
NaCN
(1) / (2) Na/Hg
(1) HCl / (2) KOH
(1) SOCl2 / (2) AlCl3
KOH
Triton B / O2
(XVII)
Rose Bengal / air
(XVIII)
(1) NaBH4 / (2) HCl
AlH(Bu^i)2
(±)cis-Alpinigenine

SCHEME 109

80 %

98% | excess NaOH + PhCOCl

KOBu^t/DMSO
56%

SCHEME 110

A different approach, possibly along biosynthetic lines, has been used for the synthesis of (±)*cis*-alpinigenine (Scheme 108). A key step was the photosensitized oxidation of the enaminoketone XVII to XVIII (*K. Orito, R. H. Manske* and *R. Rodrigo*, J. Amer. chem. Soc., 1974, **96**, 1944).

A simple preparation of the ring skeleton of the thermodynamically less stable *trans*-B/D rhoeadines, which is of potential importance for the alkaloids themselves, is due to *Shamma* and *L. Toke* (Scheme 109) (Chem. Comm., 1973, 740).

(*ii*) *Biosynthesis*

The biosynthesis of the rhoeadines has not yet been elucidated, although it is known that two molecules of tyrosine are involved (*H. Bohm* and *Ronsch*, Z. Naturforsch., 1968, **B23,** 1552; *Ronsch*, Tetrahedron Letters, 1969, 5121). A postulated route (*Santavy et al.*, Coll. Czech. chem. Comm., 1965, **30,** 3479) is shown in Scheme 110.

10. The benzo[c]phenanthridines

The benzo[c]phenanthridine ring-system (or 1,2-benzphenanthridine, 5-azachrysene, α-naphthophenanthrene) is numbered as shown in I, although *M. Shamma* recommends that shown in II, to emphasise the proven biosynthetic relationship to the protoberberine ring system III (*M. Shamma*, "The Isoquinoline Alkaloids", Academic Press, New York, 1972, p. 315):

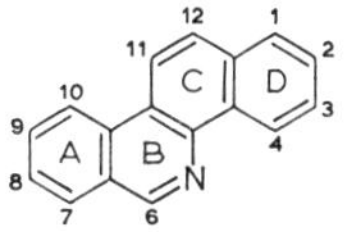

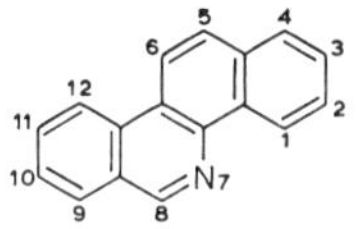

The parent compound was first prepared in 1900, as shown in Scheme 111 (*C. Graebe* and *F. Honigsberger*, Ann., 1900, **311,** 257; *Graebe* and *R. Guehm, ibid.*, 1905, **335,** 113; *Graebe, ibid.*, 1904, **335,** 122). Very little additional work was reported on the ring system until the 1930's when the structures of some alkaloids, isolated much earlier, were investigated. About two dozen benzo[c]phenanthridine alkaloids are now known (Tables 5 to 8).

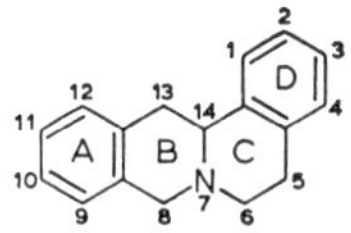

Sanguinarine was first isolated in 1829 (*J. F. Dana*, Mag. Pharm., 1829, **23,** 125) and until 1959 only sanguinarine, chelerythrine, chelidonine and homochelidonine were known. *H. R. Arthur, W. H. Hui* and *Y. L. Ng* (J. chem. Soc., 1959, 1840, 4007) isolated **avicine** and **nitidine** and determined the structures by classical methods. Since then *J. Slavik* and his co-workers (*Slavik, L. Slavikova* and *K. Haisova*, Coll. Czech. chem. Comm., 1967, **32,** 4420) have characterised **chelirubine, chelilutine, sanguirubine, sanguilutine** and **marcarpine** by the use of n.m.r. and mass spectral methods. **Toddaline** has been shown (*T. R. Govindachari* and *B. S. Thyagarajan*, J. chem. Soc., 1956, 769) to be identical with chelerythrine. **Bocconine,** which was originally reported (*M. Onda et al.*, Chem. and Pharm. Bull., Japan, 1970, **18,** 1435) to be represented by structure IV has now (*Slavik* and

SCHEME 111

THE SYNTHESIS OF BENZO[c]PHENANTHRIDINE

Chrysene

$K_2Cr_2O_7$

CO_2H
CO_2H

NH_3

$CONH_2$
CO_2H

CO_2H
$CONH_2$

Br_2/KOH

Br_2/KOH

Santavy, Coll. Czech. chem. Comm., 1972, **37**, 2804) been shown to be identical with **chelirubine.**

The structure of chelirubine was first reported as V (*Slavik et al., ibid.*, 1967, **32**, 4420; 1968, **33**, 1619) on the basis of n.m.r. and mass spectral data. However, this has now been modified to VI on the basis of some oxidation studies (*H. Ishii et al.*, Tetrahedron Letters, 1975, 319). It is possible that all of the 11-hydroxy structures in Table 6 will need to be modified.

Three alkaloids have been isolated which are without *N*-methyl groups *viz.* **norchelerythrine** (*Govindachari* and *N. Viswanathan*, Indian J. Chem., 1967, **5**, 280), **norchelidonine** (*Slavik*, Coll. Czech. chem. Comm., 1959, **24**, 3141) and **norsanguinarine** (*P. Balderstone* and *S. F. Dyke*, J. Chromato-

TABLE 5

THE BENZO[c]PHENANTHRIDINE ALKALOIDS

A: B:

Name	Formula	R^1	R^2	R^3	R^4	R^5	Ref.
Sanguinarine	A	O–CH₂–O		H	O–CH₂–O		1,2
Chelerythrine	A	OMe	OMe	H	O–CH₂–O		1,4
Avicine	A	H	O–CH₂–O		O–CH₂–O		5,6
Nitidine	A	H	OMe	OMe	O–CH₂–O		5,7
Fagaronine	A	H	OMe	OMe	OH	OMe	8,9
Chelidonine	B	O–CH₂–O		H	O–CH₂–O		1,3
Homochelidonine	B	OMe	OMe	H	O–CH₂–O		1

References

1 *R. H. F. Manske*, in "The Alkaloids", 1954, **4**, 253.
2 *S. F. Dyke, B. J. Moon* and *M. Sainsbury*, Tetrahedron Letters, 1968, 3933; J. chem. Soc., C, 1970, 1797.
3 *W. Oppolzer* and *K. Keller*, J. Amer. chem. Soc., 1971, **93**, 3836.
4 *A. S. Bailey* and *R. Robinson*, J. chem. Soc., 1950, 1375.
5 *H. R. Arthur, W. H. Hui* and *Y. L. Ng, ibid.*, 1959, 1840, 4007.
6 *K. W. Gopinath, T. R. Govindachari* and *N. Viswanathan*, Tetrahedron, 1961, **14**, 322.
7 *Arthur* and *Ng*, J. chem. Soc., 1959, 4010.
8 *W. M. Messmer et al.*, J. pharm. Sci., 1972, **61**, 1858.
9 *J. P. Gillespie, L. G. Amoros* and *F. R. Stermitz*, J. org. Chem., 1974, **39**, 3239.

TABLE 6[a]

Name	R^1	R^2	R^3	R^4	R^5	R^6	R^7	Ref.
Chelirubine	O–CH₂–O		H	O–CH₂–O		OMe	H	1,2
Chelilutine	OMe	OMe	H	O–CH₂–O		OMe	H	1,2
Sanguirubine	O–CH₂–O		H	OMe	OMe	OMe	H	1,2
Sanguilutine	OMe	OMe	H	OMe	OMe	OMe	H	1,2
Macarpine	O–CH₂–O		H	O–CH₂–O		OMe	OMe	2

[a] See text p. 173, concerning these structures.

References

1 *J. Slavik et al.*, Coll. Czech. chem. Comm., 1968, **33**, 1619.
2 *Slavik, L. Slavikova* and *K. Haisova, ibid.*, 1967, **32**, 4420.

TABLE 7

Corynoline[1,2,3]

Corynoloxine [4]

Corynolamine [5]

Bocconoline [5]

12-β-Hydroxycorynoline[6]

11-Epicorynoline [6]

14-Epicorynoline[7]

References

1 *N. Takao*, Chem. and Pharm. Bull., Japan, 1963, **11,** 1306, 1312.
2 *S. Naruto, S. Arakawa* and *H. Kaneko*, Tetrahedron Letters, 1968, 1705.
3 *T. Kametani, ibid.*, 1972, 3729; J. chem. Soc., Perkin II, 1973, 1605.
4 *Takao*, Chem. and Pharm. Bull., Japan, 1971, **19,** 247.
5 *H. Ishi, K. Hosya* and *Takao*, Tetrahedron Letters, 1971, 2429.
6 *G. Nonaka* and *I. Nishioka*, Chem. and Pharm. Bull., Japan, 1975, **23,** 521.
7 *Takao et al.*, Tetrahedron Letters, 1974, 805.

graphy, 1977, **132,** 359). **Methoxychelidonine** is believed to be VII, but final proof is still lacking. Two isomers of **corynoline** have been isolated recently in which the ring B/C junction is *trans* rather than the more usual *cis* (Table 7). **Chelidimerine** (VIII) is the acetonyl dimer of sanguinarine, and the acetonyl dimer of chelerythrine is also known (*D. B. MacLean et al.*, Canad. J. Chem., 1969, **47,** 1951; *N. R. Farnsworth et al.*, Lloydia, 1970, **33,** 267; 1972, **35,** 87).

TABLE 8

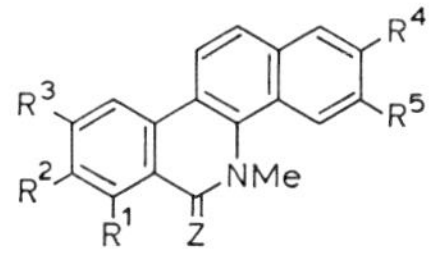

Name	R^1	R^2	R^3	R^4	R^5	Z	Ref.
Dihydrochelerythrine	OMe	OMe	H	$O–CH_2–O$		2H	1
Oxychelerythrine	OMe	OMe	H	$O–CH_2–O$		O	1
Dihydrosanguinarine	$O–CH_2–O$		H	$O–CH_2–O$		2H	2
Oxysanguinarine	$O–CH_2–O$		H	$O–CH_2–O$		O	3
Dihydroavicine	H	$O–CH_2–O$		$O–CH_2–O$		2H	4
Oxyavicine	H	$O–CH_2–O$		$O–CH_2–O$		O	4
Dihydronitidine	H	OMe	OMe	$O–CH_2–O$		2G	4
Oxynitidine	H	OMe	OMe	$O–CH_2–O$		O	4

References

1 *P. J. Schener, M. Y. Chang* and *C. E. Swanholm,* J. org. Chem., 1962, **27**, 1472.
2 *L. Slavikova* and *J. Slavik,* Coll. Czech. chem. Comm., 1956, **21**, 211.
3 *E. Spath et al.,* Ber., 1937, **70**, 1677.
4 *H. R. Arthur, W. H. Hui* and *Y. L. Ng,* J. chem. Soc., 1959, 1840, 4007.

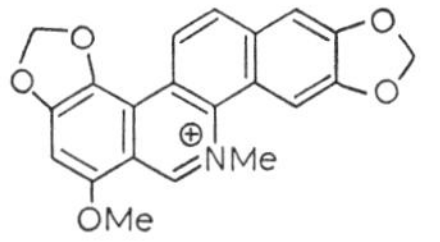

(IV)

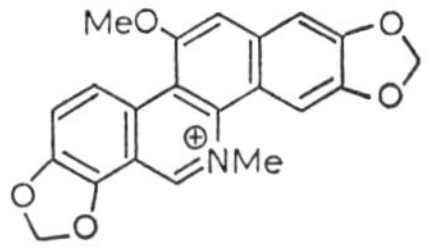

(V)

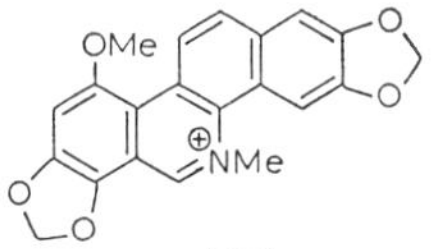

(VI)

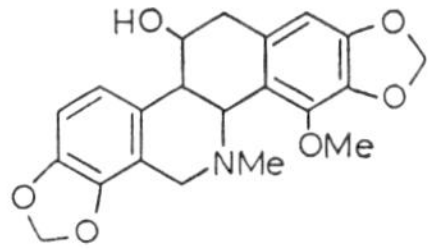

(VII)

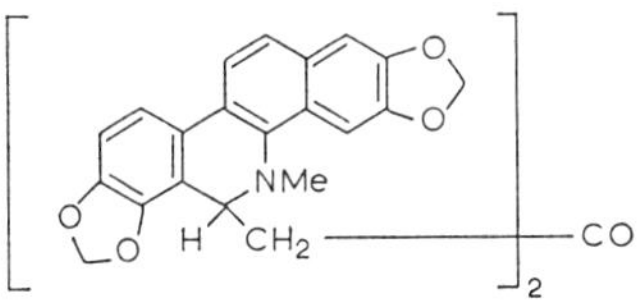

(VIII)

It is well known that in alkaline solution quaternary isoquinolinium salts undergo a disproportionation reaction IX → X + XI, so that the compounds listed in Table 8 may be artefacts.

The benzo[c]phenanthridine alkaloids are found mainly in the two botanical orders Rhoeadales and Rutales (*J. Hutchinson*, "The Families of Flowering Plants", Oxford University Press, 1959, Vol. I; *R. F. Raffauf*, "A Handbook of Alkaloids and Alkaloid-Containing Plants", Wiley-Interscience, New York, 1970). Within these orders, the families Papaveraceae and Fumaraceae are the most common ones containing benzo[c]phenanthridines, and the alkaloids are most abundant in *Chelidonium majus* L. and *Sanguinaria canadensis* L.

(a) Structure and stereochemistry

The elucidation of the structures of chelidonine, homochelidonine, sanguinarine and chelerythrine and their interrelationships were achieved by *Gadamer*, by *Bruckhausen* and *Bersch* and by *Spath* and *Kuffner*. This early work has been reviewed (*T. A. Henry*, "The Plant Alkaloids", Churchill, London, 1949, p. 277; *J. V. Crawford*, in "The Chemistry of Heterocyclic Compounds: Six-membered Heterocyclic Nitrogen Compounds with Four Condensed Rings", ed. *C. F. H. Allen*, Interscience, New York, 1951, p. 160; *L. P. Walls*, in "Heterocyclic Compounds", ed. *R. C. Elderfield*, Wiley, New York, 1952, **4**, 613; *R. H. F. Manske*, in "The Alkaloids", 1954, **4**, 253; 1960, **7**, 430; 1968, **10**, 485) and will be summarised briefly here.

Chelerythrine (XII), upon distillation with zinc dust gives benzo[c]-phenanthridine itself. When oxidised under mild conditions with potassium permanganate, the product is XIII, whereas more energetic conditions of oxidation yields 4,5-methylenedioxyphthalic acid. Mild reduction yields dihydrochelerythrine which after cleavage of the methylenedioxy group, followed by *O*-methylation with diazomethane affords the tetramethoxy compound XIV. The same substance has been obtained from **sanguinarine** (XV) (see below) and the structure has been confirmed by synthesis (see p. 189):

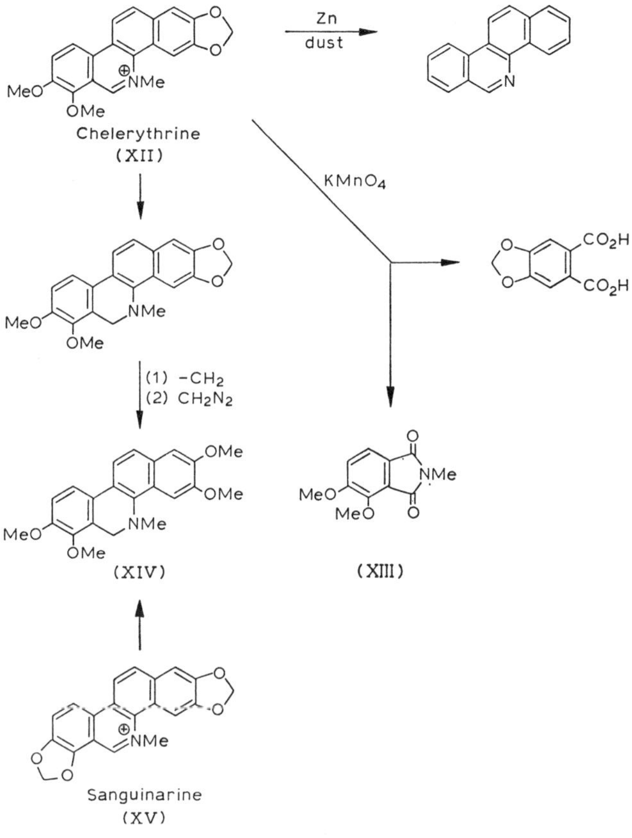

Hence the structure of chelerythrine must be XII and this has been confirmed by synthesis. The structure of sanguinarine (XV) can be deduced in a similar fashion, although it was investigated after the structure of chelidonine had been formulated.

The structures of **avicine** (XVI, 2R = OCH$_2$O) and **nitidine** (VI, R = OMe) were established (*Arthur, Hui* and *Ng, loc. cit.*) in a similar manner. Both were converted to the known tetramethoxy compound XVII.

The structure of (+)**chelidonine** was established in 1930 and was the first member of this group of alkaloids to be examined. The molecular formula was eventually found to be $C_{20}H_{19}NO_5H_2O$, and the molecule was shown, by standard methods, to possess two methylenedioxy groups, one N-methyl function and one hydroxyl group. Zinc dust distillation gave benzo[c]phenanthridine. Oxidation of the alkaloid with potassium permanganate gave XVIII and 3,4-methylenedioxyphthalic acid. Hofmann degradation of chelidonine gave a methine, $C_{21}H_{21}NO_5$, which was oxidised with potassium permanganate to a mixture of 3,4-methylenedioxyphthalic acid and XIX. Hence part-structures for the methine XX and for chelidonine (XXI) can be written. The remaining problem is the location of the hydroxyl group. Since chelidonine is not a phenol, the hydroxyl group cannot be located on rings A or D, and $C_{(6)}$ and $C_{(14)}$ are ruled out because the alkaloid does not possess the properties of a pseudobase (carbinolamine). This leaves $C_{(11)}$, $C_{(12)}$ or $C_{(13)}$ for the location of the OH group, but because this function is retained in the methine, it cannot be at $C_{(13)}$. With a hydroxyl group at either $C_{(11)}$ or $C_{(12)}$, dehydration should be facile, and this is found to be so. Treatment of chelidonine with thionyl chloride gives deoxychelidonine (XXII). When chelidonine is treated with acetic anhydride at 140°, an optically inactive N-acetate XXIII is produced which is oxidised by nitric acid to 1,2,4-benzenetricarboxylic acid. However, when chelidonine reacts with acetic anhydride at room temperature, an optically active O-acetate is produced which is oxidised by mercuric acetate to sanguinarine (XV).

The methine base XX, when treated with methyl iodide and base gave a compound, the structure of which was expressed as XXIV without rigorous proof. In the light of subsequent evidence, this is now known to be correct.

Oxidation of chelidonine under Oppenauer conditions gave hydroxydihydrosanguinarine (XXV) rather than the expected ketone, and when mercuric acetate was used, a substance called didehydrochelidonine was among the products. It is only recently that the structure XXVI of this substance has been elucidated (*M. H. Benn* and *R. E. Mitchell*, Canad. J. Chem., 1969, **47**, 3701). The structure of chelidonine was expressed as XXVII but the position of the hydroxyl function remained in doubt. The ready formation of XXVI is very strong evidence in favour of the accepted structure, and more recently n.m.r. spectral evidence supports this assign-

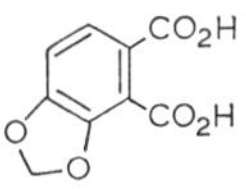

(XVIII)

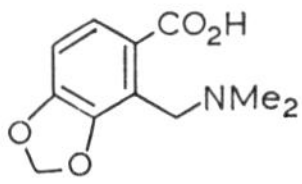

(XIX)

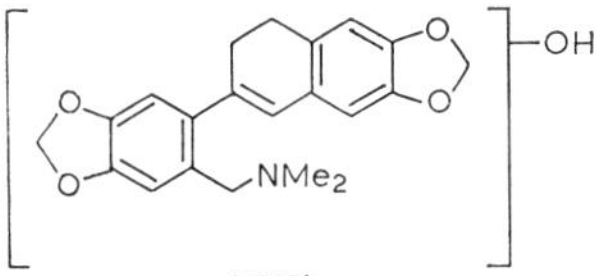

(XX)

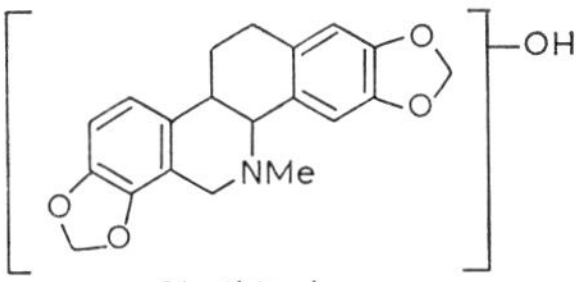

Chelidonine
(XXI)

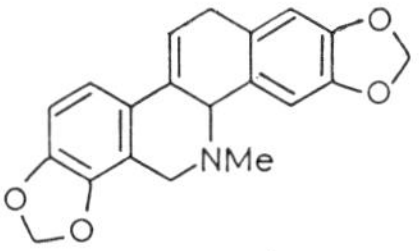

(XXII)

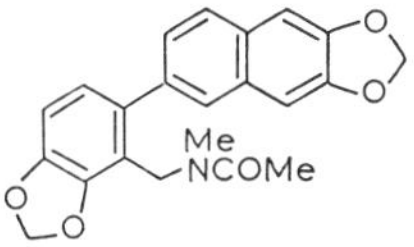

(XXIII)

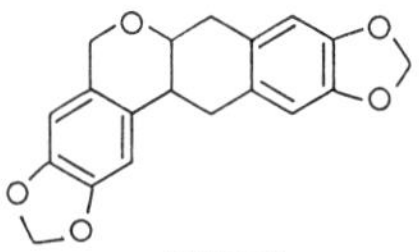

(XXIV)

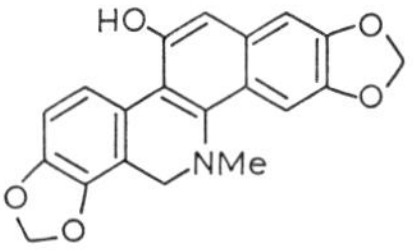

(XXV)

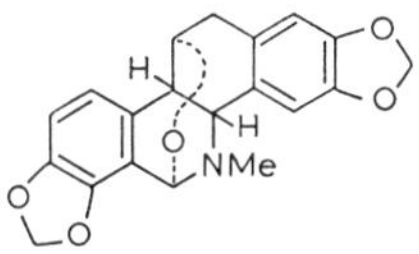

(XXVI)

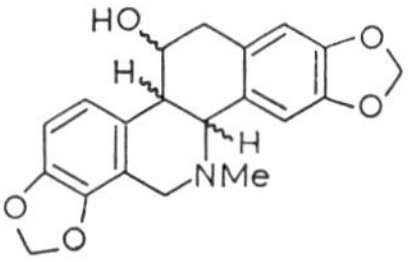

(XXVII)

ment. The final proof of structure was provided in 1972 when the racemate was synthesised for the first time (see p. 194).

It was *H. W. Bersch* (Arch. Pharm., 1958, **291**, 491) who first considered the relative stereochemistry of chelidonine. He noted that the infrared spectrum of the base exhibited a strongly hydrogen-bonded OH absorption, and consequently pointed out that the hydroxyl group must be located at $C_{(11)}$. The strongest hydrogen-bonding is possible if the OH group is axial and if the rings B/C are *cis*-fused. These conclusions have been amply confirmed by detailed n.m.r. studies (*E. Seoane*, Anales Real Soc. Espan. Fis. Quim., Madrid, Ser. B, 1965, **61**, 755; *C.-Y. Chen* and *Maclean*, Canad. J. Chem., 1967, **45**, 3001; *S. Naruto, S. Arakawa* and *H. Kaneko*, Tetrahedron Letters, 1968, 1705). These showed that rings B and C are *cis*-fused in the half-chair conformation and the $C_{(11)}$-OH is axial XXVIII. The absolute

(XXVIII)

(XXIX)

(XXX)

(XXXI)   11(*R*), 13(*S*), 14(*R*)

stereochemistry of chelidonine has been determined by two methods (*G. Snatzke et al.*, Tetrahedron, 1970, **26**, 5013). The first method employs the correlation, developed by *Horeau* of absolute configuration of secondary alcohols with the kinetic partial resolution during esterification with α-phenylbutyric anhydride. In this method an excess of (±)α-phenylbutyric anhydride is added to the optically active secondary alcohol. The unreacted anhydride is recovered and hydrolysed and the optical rotation of the acid is measured. It has been shown empirically that when the configuration of the alcohol is written as XXIX, the recovered α-phenylbutyric acid has the (*S*)(+)isomer in excess. On this basis, chelidonine can be represented as XXX, which is equivalent to XXXI since the relative configurations are known. It was shown *(idem, loc. cit.)* that by an analysis of the c.d. spectrum of (+)chelidonine employing a nonempirical method, the same conclusions emerge.

The structures of **corynoline** and **corynoloxine,** two $C_{(13)}$-methyl benzo[*c*]-phenanthridine derivatives were determined (*N. Takao*, Chem. and Pharm. Bull., Japan, 1963, **11**, 1306, 1312; 1971, **19**, 247; *N. Takao, Bersch* and *S. Takao, ibid.*, 1971, **19**, 259) by chemical methods (Scheme 112). Coryno-line, (XXXII), $C_{21}H_{21}NO_5$, contains one alcoholic hydroxyl group, one *N*-methyl group, two methylenedioxy functions and a *C*-methyl group. It is readily dehydrated to deoxycorynoline, which can be hydrogenated to dihydrodeoxycorynoline (XXXIII). Two successive Emde degradations yielded XXXIV, which, with selenium at 300° was degraded to XXXV. Since corynoline is transformed into the benzo[*c*]phenanthridine XXXVI on heating, the isolation of XXXV fixes the position of the *C*-methyl group in a benzo[*c*]phenanthridine skeleton. The hydroxyl group was placed at $C_{(11)}$ by analogy with chelidonine. When corynoline is oxidised with aqueous potassium permanganate at room temperature, one of the products is

SCHEME 112

THE STRUCTURES OF CORYNOLINE AND CORYNOLOXINE

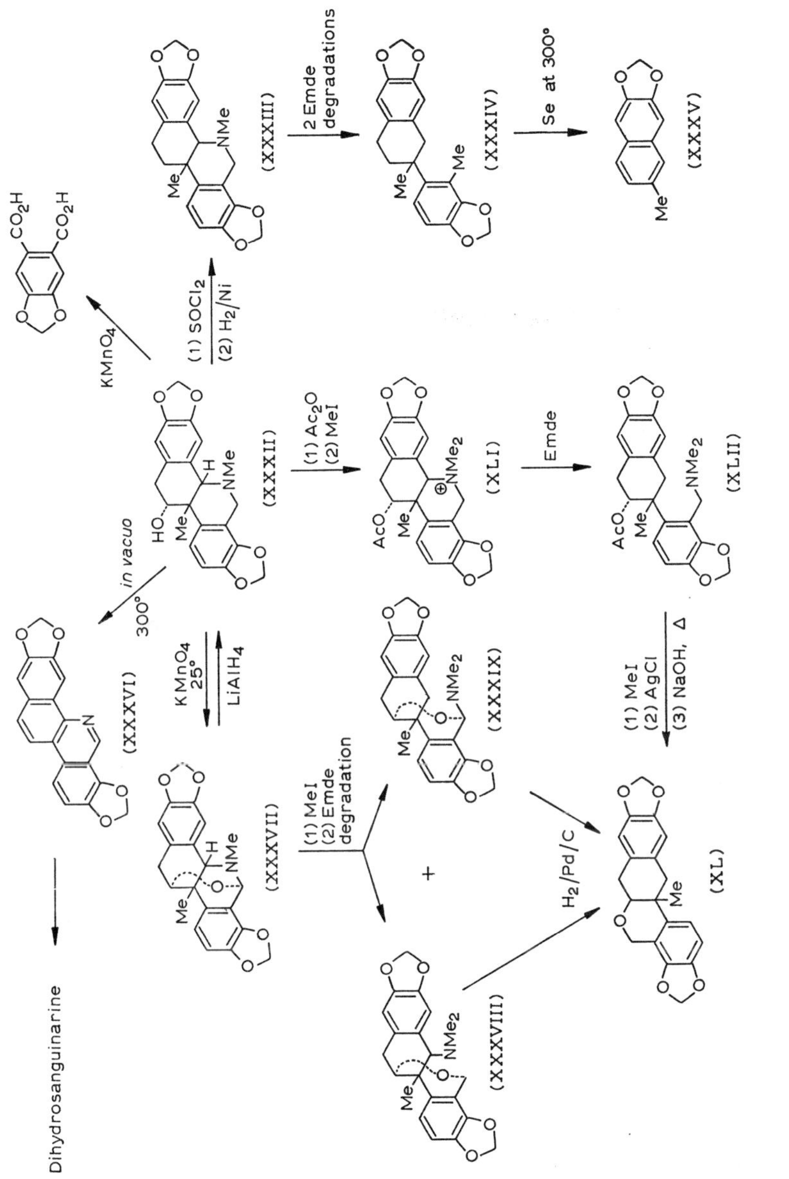

(XLIII)

(XLIV): $2R = OCH_2O$
(XLV): $R = OMe$

corynoloxine (XXXVII), which is also isolated from the plant with coryno-line. The formation of XXXVII, together with an intramolecular hydrogen bond absorption in the i.r. of corynoline, demonstrates that the rings B/C are *cis*-fused in both alkaloids. Emde degradation of corynoloxine metho-chloride gave both of the possible products XXXVIII and XXXIX. Further degradation of either gave the polycyclic ether XL which was also obtained from corynoline (XXXII) *via* XLI and XLII. The proposals concerning the stereochemistry and conformations (*Naruto, Arakawa* and *Kaneko, loc. cit.*) of corynoline have been substantiated by an X-ray analysis of its *p*-bromobenzoate (*T. Kametani et al.*, Tetrahedron Letters, 1972, 3729).

(b) Spectral characteristics

The ultraviolet spectra of benzo[*c*]phenanthridine alkaloids are quite characteristic and are a useful aid in structural elucidation since a 2,3,8,9-tetraoxy compound can be distinguished from an isomer with the 2,3,7,8-tetraoxygenation pattern (*J Holubek* and *O. Strouf*, "Spectral Data and Physical Constants of Alkaloids", Czech. Acad. Sci., Prague, Vol. I onwards, 1965, onwards).

Mass spectral studies are best carried out on the pseudocyanide XLIII (R = CN), acetonyl derivative XLIII (R = $CH_2 \cdot COMe$) or the dihydro compound XLIII (R = H), rather than on the involatile quaternary salts. The pseudocyanides exhibit strong $M^{\oplus}$ ion peaks (which are also the base peaks) and a second strong peak corresponds to $(M–CN)^{\oplus}$. Further fragmentations from the latter involve loss of CH_3, CH_2O, CO and CHO (*Slavik et al.*, Coll. Czech. chem. Comm., 1968, **33**, 1619; *MacLean et al.*, Canad. J. Chem., 1969, **47**, 1951). The *N*-desmethyl compounds are relatively easily prepared, and these tertiary bases provide more useful spectra. Thus, breakdown of **noravicine** (XLIV) is summarised in Scheme 113 (*Dyke, B. J. Moon* and *M. Sainsbury*, unpublished; *Moon*, Ph.D. Thesis, University of Bath, 1969). These fragmentations have been substantiated by the appearance of the appropriate metastable ions and doubly charged species. After loss of HCN, no recognisable fragmentation pattern could be observed. The mass spectrum of norsanguinarine is similar to that of noravicine, but those of **nornitidine** (XLV) and of 2,3,8,9-tetramethoxybenzo[*c*]phenanthri-

SCHEME 113

MASS SPECTRAL FRAGMENTATIONS OF NORAVICINE

m/e 317   M$^{\oplus}$

$\xrightarrow{-\text{HCHO}}$

m/e 287

$\downarrow{-\text{CO}}$

m/e 259

$\xleftarrow{-\text{HCHO}}$

m/e 229

$\downarrow{-\text{CO}}$

m/e 201

$\xrightarrow{-\text{HCN}}$ *m/e* 174

dine are more complex, presumably due to the methoxyl groups, each of which may fragment in more than one way.

The n.m.r. spectra of benzo[*c*]phenanthridines have attracted some attention, and are now of considerable importance in structural studies. The spectra of noravicine, nornitidine and norsanguinarine in trifluoracetic acid solution are highly characteristic (*Dyke, Sainsbury* and *Moon*, Tetrahedron, 1968, **24**, 1467; J. chem. Soc., C, 1970, 1797). The methylenedioxy group attached to ring D always appears at higher field than a similar group attached to ring A — whether at $C_{(7)} + C_{(8)}$ or $C_{(8)} + C_{(9)}$. The hydrogen atoms at $C_{(11)}$ and $C_{(12)}$ appear as a quartet ($J = 9$ Hz) in all compounds studied. The quartet due to the *ortho* hydrogens at $C_{(9)} + C_{(10)}$ in norsanguinarine have $J = 10$ Hz. The aromatic proton at highest field is always that at $C_{(1)}$ and the $C_{(6)}$ hydrogen is always at lowest field.

The quaternary alkaloids have been studied either as the dihydro derivative or as the pseudocyanide (*Slavik, L. Dolejs* and *A. D. Cross*, Coll. Czech. chem. Comm., 1968, **33**, 1619). The data for sanguinarine pseudocyanide in dimethyl sulphoxide solution are summarised in XLVI:

(XLVI)

(c) Synthesis

(i) Chemical synthesis

It is convenient to consider the fully aromatic derivatives, such as chelery-thrine separately from the chelidonine type of oxidation level.

The first successful synthesis of a tetra-oxybenzo[c]phenanthridine (*T. Richardson, R. Robinson* and *E. Seijo*, J. chem. Soc., 1937, 835) utilised the Bischler–Napieralski reaction and is summarised in Scheme 114. The

SCHEME 114

ROBINSON'S BENZO[c]PHENANTHRIDINE SYNTHESIS

essential feature is the preparation of the required 2-aryl-1-naphthylamine derivative. This route was used to synthesise **dihydronitidine** (*Arthur* and *Ng*, *ibid.*, 1959, 4010) and **oxyavicine** (*K. W. Gopinath, Govindachari* and *N. Visnawathan*, Tetrahedron, 1961, **14**, 322):

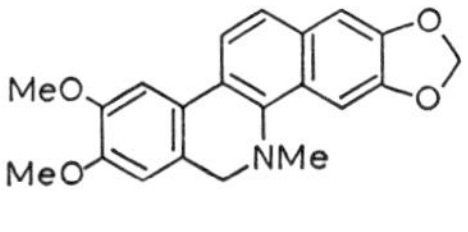

Dihydronitidine

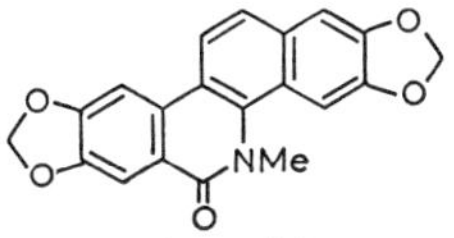

Oxyavicine

In a slight modification of Robinson's procedure (*Govindachari et al.*, J. chem. Soc., 1957, 4760), the Bischler–Napieralski ring-closure was carried out on the aromatic naphthylamine *e.g.* XLVII, rather than on the tetra-hydronaphthalene derivative. A series of benzo[*c*]phenanthrine derivatives were thus prepared. The yields in the nitidine synthesis have been improved (*K.-Y. Zee-Cheng* and *C. C. Cheng*, J. heterocycl. Chem., 1973, **10**, 85) by a closer study of the conditions of each step. The required tetralone XLIX has also been prepared (*Kametani et al., ibid.*, p. 31) by arylation of the α-tetralone XLVIII in the presence of potassamide:

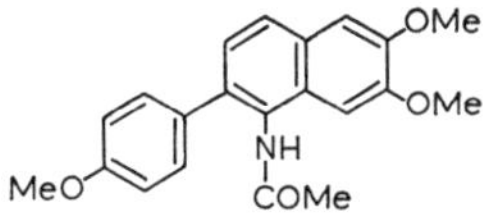

(XLVII)

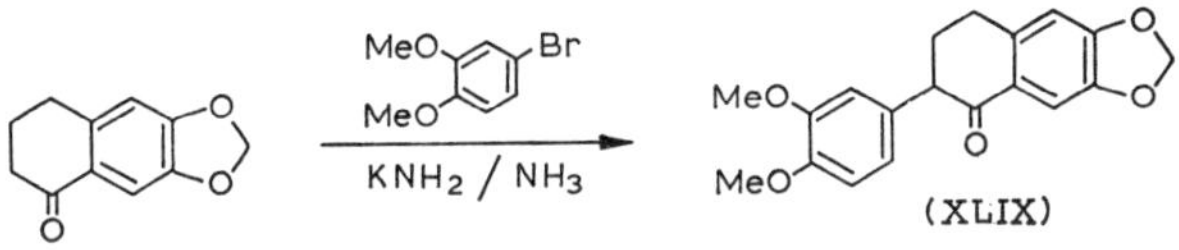

(XLVIII)

KNH$_2$ / NH$_3$

(XLIX)

In an attempt to obtain the 2,3,7,8-tetraoxygenation pattern of chelerythrine and sanguinarine (*A. S. Bailey* and *Robinson*, J. chem. Soc., 1950, 1375) the bromo derivative L was subjected to the Bischler–Napieralski reaction, but without success:

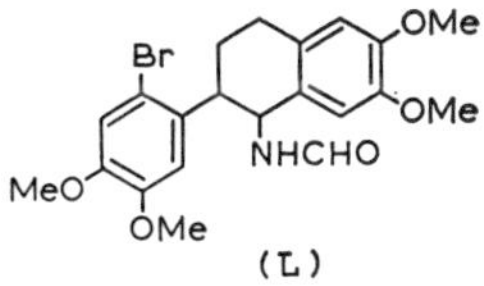

(L)

In order to achieve the required orientation of substituents, the original route (Scheme 114) was followed using opianic acid in place of veratral-dehyde (*Bailey* and *Robinson, loc. cit.*; *Bailey, Robinson* and *R. S. Staunton*,

SCHEME 115

THE FIRST SYNTHESIS OF CHELERYTHRINE

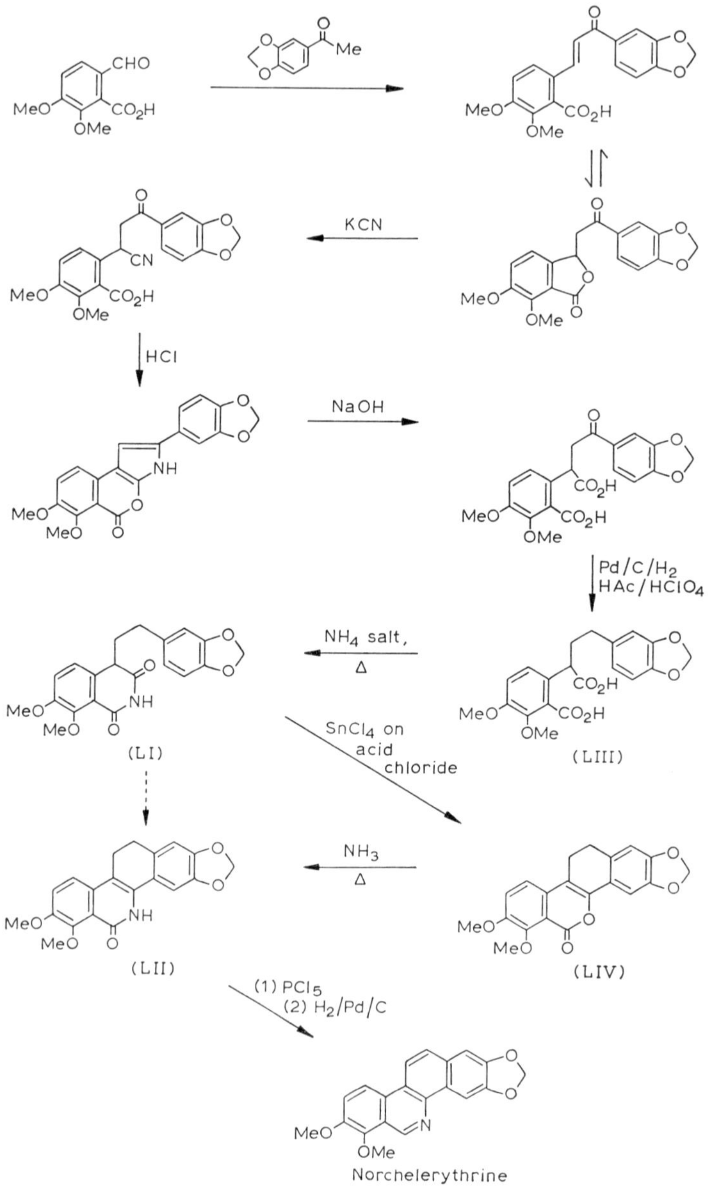

(LV)

(LVI)

J. chem. Soc., 1950, 2277; *Bailey* and *C. R. Worthing*, *ibid.*, 1956, 4535).
In Scheme 115 the synthesis of **chelerythrine** is presented. Some modification
had to be made when it was found that the homophthalimide LI could not
be cyclised to LII. By converting the acid LIII to the isocoumarin LIV,
then reacting with ammonia, this difficulty was surmounted.

The application of the Pschorr reaction to the synthesis of benzo[*c*]-
phenanthridine has attracted considerable attention over the years. However, cyclisations of aniline derivatives of the type LV have not been successful. It was reported in 1963 (*R. A. Abramovitch* and *G. Tertzakian*, Canad.
J. Chem., 1963, **41**, 2265) that Pschorr cyclisation to a benzo[*c*]phenanthri-
dine does occur in LVI. When a method was discovered (*Dyke, Sainsbury*
and *Moon, loc. cit.*) for the preparation of appropriately substituted iso-
quinoline-4-acetic acids, a versatile synthesis of the alkaloids of this group
was evolved. In Scheme 116, the synthesis of **sanguinarine** *(idem, ibid.)*

SCHEME 116

THE SYNTHESIS OF SANGUINARINE

illustrates the method, which has also been used to synthesise avicine and nitidine.

S. V. Kessar et al. (*Kessar, R. Gopal* and *M. Singh*, Tetrahedron, 1973, **29,** 167; *Kessar, D. Pal* and *Singh, ibid.*, 1973, **29,** 177) have found that treatment of *o*-chloroaniline derivatives of the type LVII with potassium amide causes cyclisation, *via* an aryne to a benzo[*c*]phenanthridine. The method has been applied to the synthesis of chelerythrine from LVIII (*Kessar, Singh* and *P. Balakrishnan*, Indian J. Chem., 1974, **12,** 323) and of **fagaronine** (*J. P. Gillespie, L. G. Amoros* and *F. R. Stermitz*, J. org. Chem., 1974, **39,** 3239).

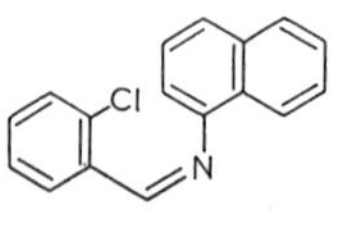

(LVII) 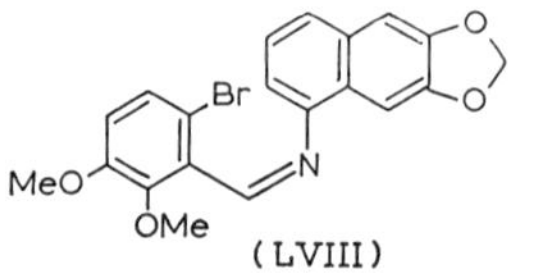 (LVIII) 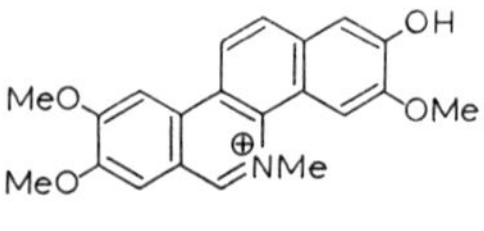 Fagaronine

Photochemical methods have been applied with some success to the synthesis of alkaloids of the benzo[*c*]phenanthridine group. The synthesis of 2,3,8,9-tetramethoxybenzo[*c*]phenanthridine (*Dyke* and *Sainsbury*, Tetrahedron, 1967, **23,** 3161) illustrates one approach (Scheme 117).

SCHEME 117

PHOTOCHEMICAL SYNTHESIS OF BENZO[c]PHENANTHRIDINES

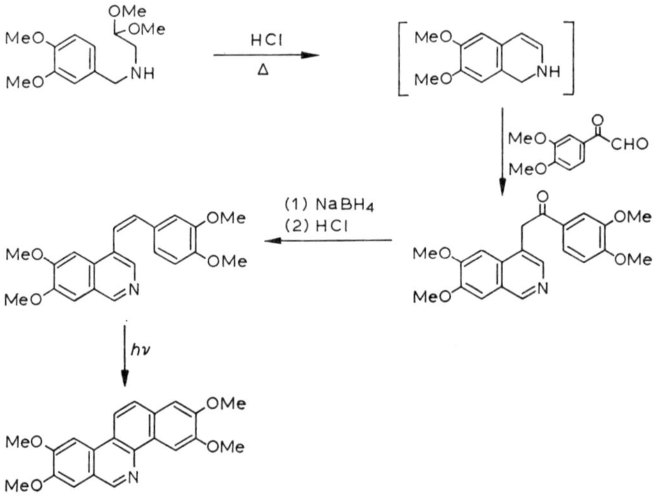

Successful photochemical cyclisations of amides LXIX have recently been described (*Kessar, Singh* and *Balakrishnan*, Tetrahedron Letters, 1974, 2269; *I. Ninomiya et al.*, J. chem. Soc., Perkin I, 1975, 762) leading to effective syntheses of nitidine and avicine:

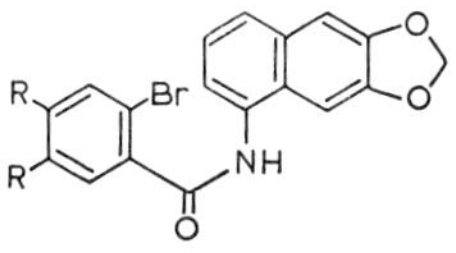

(LXIX)

R = OMe or 2R = OCH₂O

SCHEME 118
CONVERSION OF PROTOPINE INTO SANGUINARINE

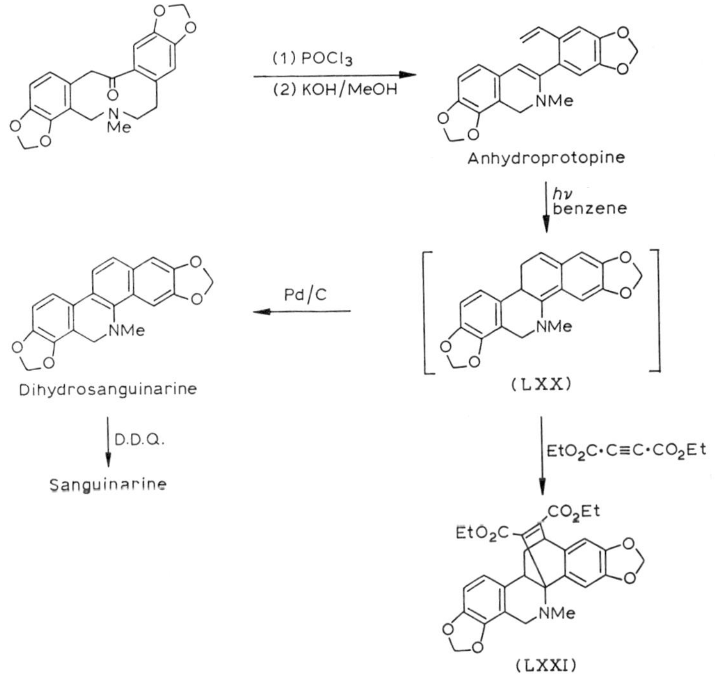

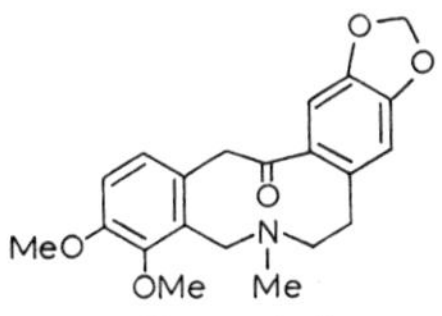

Anhydroprotopine

Dihydrosanguinarine

Sanguinarine

(LXX)

(LXXI)

Allocryptopine

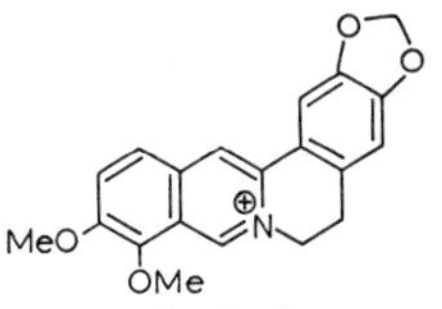

Berberine

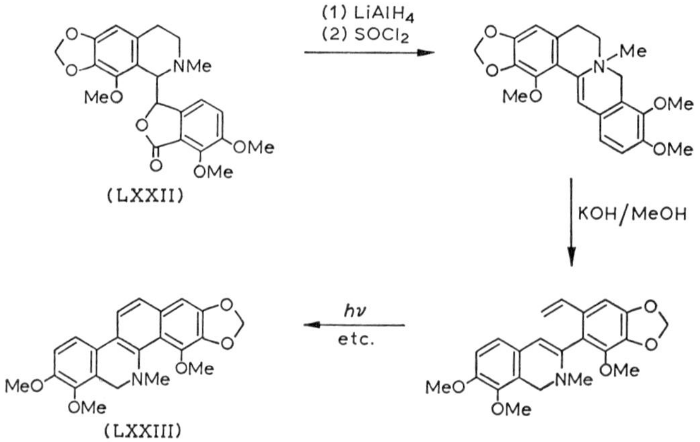

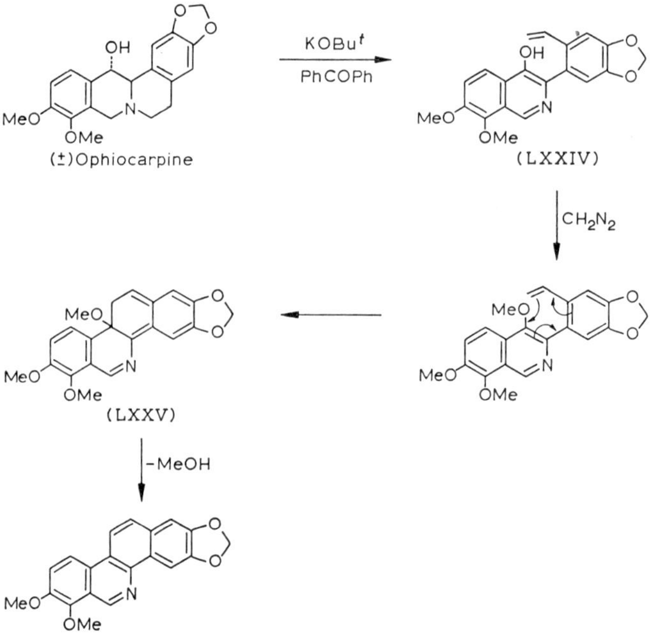

An extremely interesting synthesis of sanguinarine has been described (*M. Onda, K. Yonezawa* and *K. Abe*, Chem. and Pharm. Bull., Japan, 1969, **17**, 404; 1971, **19**, 31) from protopine by a combination of chemical and photochemical methods (Scheme 118). Anhydroprotopine was irradiated in benzene solution at room temperature with a high pressure mercury lamp. The photoproduct LXX was immediately dehydrogenated to dihydrosanguinarine in 50% yield. The structure of LXX was established (*Onda et al., ibid.,* 1971, **19,** 317) by trapping with diethyl acetylenedicarboxylate to give LXXI. Chelerythrine was synthesised similarly; the required diene for photolysis was prepared either from allocryptopine or berberine. In a logical extension to this work (*Onda* and *K. Kawakami, ibid.,* 1972, **20,** 1484) 1-α-narcotine (LXXII) has been transformed into a benzo[c]phenanthridine derivative LXXIII. The shortest synthesis of chelerythrine yet reported (*V. Smula, R. H. F. Manske* and *R. Rodrigo,* Canad. J. Chem., 1972, **50,** 1544) starts from (±)ophiocarpine. Modified Oppenauer oxidation yields LXXIV, which after *O*-methylation is photocyclised to *N*-norchelerythrine *via* LXXV. Essentially the same method has been used to convert berberine *via* LXXVI into a mixture of LXXVII and LXXVIII (*Onda et al.,* Chem. and Pharm. Bull., Japan, 1973, **21,** 1333):

SCHEME 119

Many attempts have been made to synthesise the chelidonine type of benzo[*c*]phenanthrine alkaloid. A photochemical route has been studied by *Ninomiya et al.* (*Ninomiya, T. Naito* and *T. Mori*, Tetrahedron Letters, 1969, 3643; J. chem. Soc., Perkin I, 1973, 505, 1696) (Scheme 119). Irradiation of LXXIX gave the B/C *trans* compound LXXX, which was isomerised to LXXXI with selenium. The first synthesis of corynoline has been achieved by the photocyclisation and further manipulations of XXXII (*Ninomiya, O. Yamamoto* and *Naito*, Chem. Comm., 1976, 437):

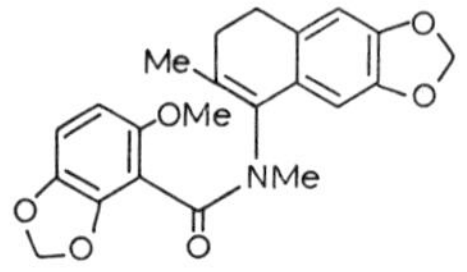

(LXXXII)

The major achievement in recent years in this group of alkaloids has been the total synthesis of ($\pm$)chelidonine (*W. Oppolzer* and *K. Keller*, J. Amer. chem. Soc., 1971, **93**, 3836). It was found that when the model compound LXXXIII (R = H) was thermalised, a 90% yield of the *trans* B/C benzo[*c*]-phenanthridine LXXXIV was produced, but when the methyl carbamate LXXXIII (R = CO$_2$Me) was heated, the required B/C *cis* compound was

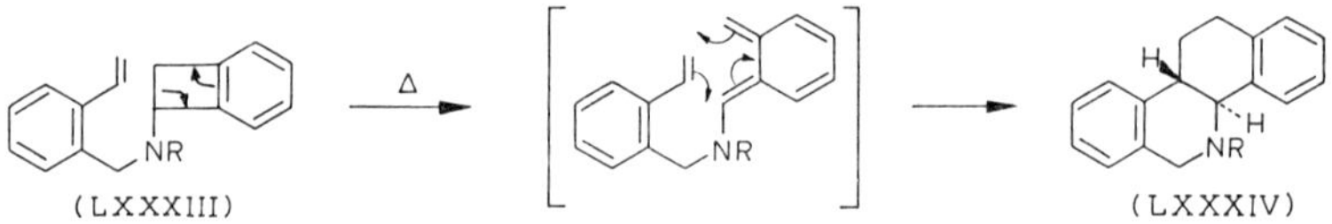

(LXXXIII) $\xrightarrow{\Delta}$ (LXXXIV)

formed in 78% yield. Based upon this experience, the synthesis of chelidonine was achieved as shown in Scheme 120. The required styrene LXXXV was obtained by degradation of 7,8-methylenedioxy-1,2,3,4-tetrahydroisoquinoline and the known nitrile LXXXVI was converted by standard methods into the benzyl carbamate LXXXVII. Condensation of LXXXV and LXXXVII gave LXXXVIII, which was brominated and dehydrobrominated to LXXXIX. Thermal reorganisation of the latter gave XC which was hydroborated and oxidised to a mixture of the required *cis* and the *trans* B/C fused systems (XCI). The *cis*-product was separated and, since the hydroxyl group had the axial orientation, was oxidised to the ketone XCII and stereoselectively reduced; the *N*-protecting group was removed to yield ($\pm$)*N*-**norchelidonine.**

(*ii*) *Biosynthesis*

It was suggested some time ago (*F. Bruchhausen* and *Bersch*, Ber., 1930, **63**, 2520) that the biosynthesis of the benzo[*c*]phenanthridine skeleton

SCHEME 120
THE SYNTHESIS OF (±)CHELIDONINE

(LXXXV)   (LXXXVI)   (LXXXVII)

(LXXXVIII)   (LXXXIX)

(XCI)   (XC)

B_2H_6 etc.

(1) Separation
(2) Oxidation

(XCII)   *N*-Norchelidonine

(1) $NaBH_4$
(2) H_2

involved oxidative cleavage of the tetrahydroprotoberberine system, followed by recyclisation (Scheme 121). A series of experiments involving the feeding of labelled precursors to *Chelidonium majus* plants has confirmed this overall scheme and established some of the detail. Thus, when (±)[2-^{14}C]tyrosine was fed to the plant, the chelidonine isolated was shown to be labelled specifically, although unequally, at $C_{(11)}$ and $C_{(14)}$. Radioactive sanguinarine was also isolated (*E. Leete*, J. Amer. chem. Soc., 1963, **85**, 473). When [1-^{14}C]dopamine was fed to the plant (*Leete* and *J. B. Murrill*, Tetrahedron Letters, 1964, 147) the resultant chelidonine was found to be labelled only at $C_{(11)}$, but since radioactive stylopine was isolated as well, these authors (Phytochemistry, 1967, **6**, 231) fed (±)[6-^{14}C]stylopine to *C. majus*; it was incorporated into chelidonine. Further work established that the key intermediate is (+)reticuline, and by using a series of ^{14}C and ^{3}H-labelled reticulines, the overall process was established (*A. R. Battersby et al.*, Chem. Comm., 1965, 89; 1967, 602; J. Chem. Soc., 1965, 3323; *ibid.*, Perkin I,

SCHEME 121

1975, 1147). Thus, when (*S*)-reticuline, labelled as shown, was used as a precursor, the chelidonine isolated, although ^{14}C-labelled at $C_{(6)}$ and $C_{(11)}$, possessed no tritium. This was explained by postulating that the reticuline is first converted into scoulerine, then into (*S*)-stylopine (T at $C_{(14)}$), which is subsequently oxidised to the 1,2-dihydroisoquinoline derivative XCIII, which then cyclises to chelidonine. In a series of experiments scoulerine (XCIV) with ^{14}C- and ^{3}H-labels were separately fed to *C. majus*. The resulting chelidonine was analysed for tritium. The results are summarised below:

Stylopine

(+)Reticuline

(XCIII)

Scoulerine
(XCIV)

When [6-^{14}C,5-^{3}H] (XCIV) was fed, $C_{(5)}$-T retained in product
When [6-^{14}C,14-^{3}H] (XCIV) was fed, $C_{(14)}$-T completely lost
When [6-^{14}C,6-^{3}H] (XCIV) was fed, $C_{(6)}$-T 50% lost
When [6-^{14}C,13-^{3}H] (XCIV) was fed, $C_{(13)}$-T some lost

These results correspond to a stereospecific oxidation at $C_{(6)}$ in the proto-berberine precursor and a stereospecific generation of the $C_{(14)}$-N double bond, which then isomerises to the 1,2-dihydroisoquinoline in a non-specific manner so that loss of hydrogen (or tritium) from $C_{(13)}$ involves an isotope effect. There are several problems remaining to be solved in the above scheme. The precise steps between stylopine and chelidonine are not known, although it is tempting to suggest that the bond cleavage is initiated by α-hydroxylation at $C_{(6)}$ and $C_{(14)}$ to give XCV, which then generates the aldehyde XCVI as shown. *N*-Methylation may occur before or after the hydroxylation.

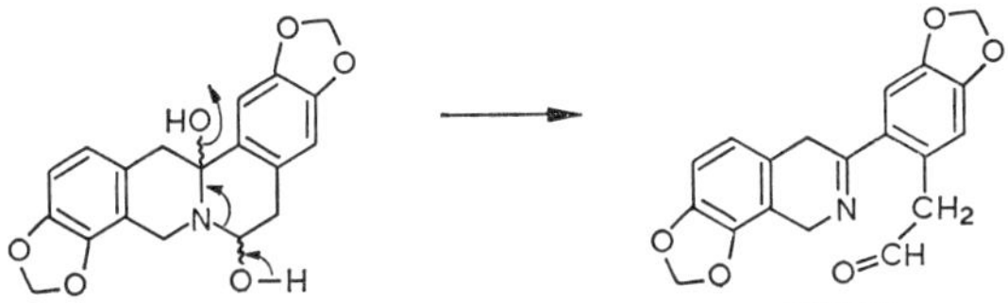

The biosynthesis of sanguinarine is not completely known. It may arise from chelidonine by dehydration and dehydrogenation, but it may equally be formed in a separate branch from a much earlier precursor.

Presumably avicine and nitidine are derived from coreximine in a way analogous to the chelidonine biosynthesis. It is possible that corynoline is derived from tetrahydrocorysamine (XCVII) and this seems reasonable since both corynoline and XCVII occur in the same plant, *Corydalis incisa* (*C. Tani et al.*, Yakugaku Zasshi, 1962, **82**, 748). It is known that reticuline is incorporated into corydaline when it is fed to *Corydalis cava* (*G. Blaschke*, Arch. Pharm., 1968, **301**, 439).

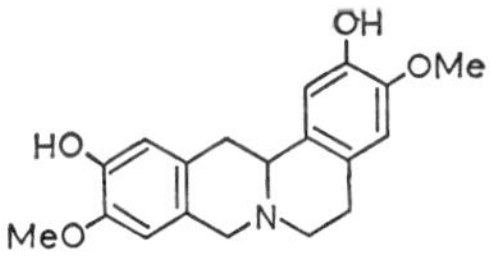

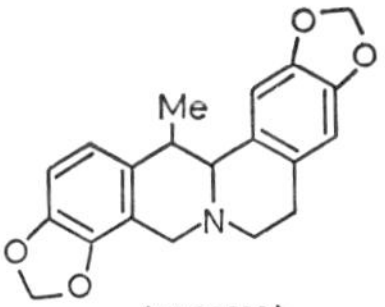

11. The spirobenzylisoquinolines

The spirobenzylisoquinoline alkaloids each possess the ochotensane ring system I, and since it is highly probable that they are derived in nature from tetrahydroprotoberberines (II), the numbering system shown in I has been adopted.

All of the alkaloids isolated so far (about 20) have been found in Fumaria-ceae plants. All have oxygen functions (OH, OMe or OCH_2O groups) at $C_{(2)}$, $C_{(3)}$, $C_{(9)}$ and $C_{(10)}$ and an *N*-methyl group. Some, such as (+)**ochotensimine,** possess an exocyclic methylene group at $C_{(13)}$. A second sub-division contains those alkaloids, such as (−)**fumaricine,** that carry a hydroxyl group at $C_{(8)}$, whereas others, such as (+)**fumariline,** are in a higher oxidation state with a $C_{(8)}$-carbonyl oxygen atom. In a higher oxidation state still are those derivatives, for example, (+)**ochrobirine** with hydroxyl groups at $C_{(8)}$ and $C_{(13)}$. Obvious further variations are exemplified by **sibiricine** and **fumarofine.** If the hydroxyl group at $C_{(8)}$ or $C_{(13)}$ lies on the *N*-side of the system (see Ib), it is classified as *syn,* whereas if the hydroxyl group at either of these sites is projecting on the side of ring A it is classified as an *anti* substituent.

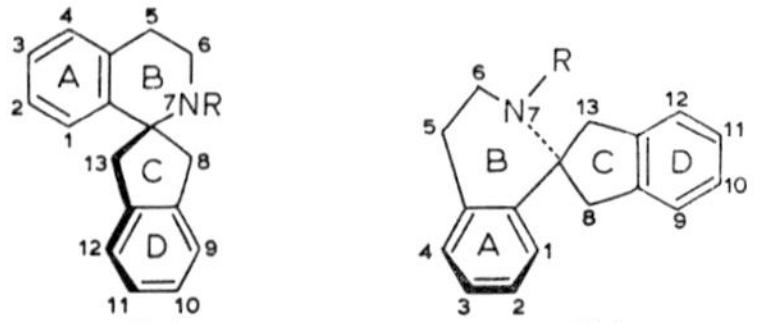

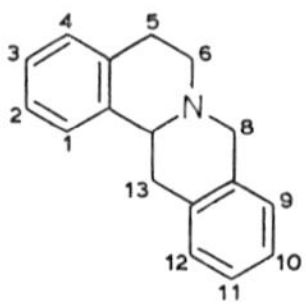

Several reviews of the spirobenzylisoquinoline alkaloids have appeared recently (*M. Shamma*, "The Isoquinoline Alkaloids", Academic Press, New York, 1972, Chap. 20; in "The Alkaloids", ed. *R. H. F. Manske*, Vol. 13, 1971, Chap. 2; *S. McLean* and *J. Whelan*, in "Alkaloids", ed. *K. Wiesner*, MTP International Review of Science, Organic Chemistry, Series One, Vol. 9, Butterworths, London, 1973, p. 161).

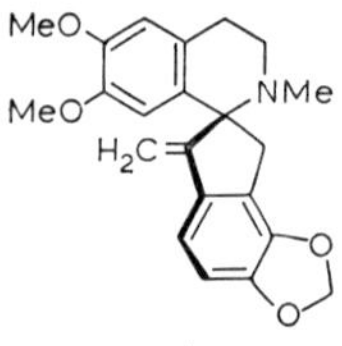

(+)Ochotensimine

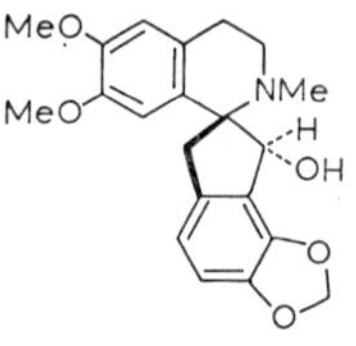

(−)Fumaricine

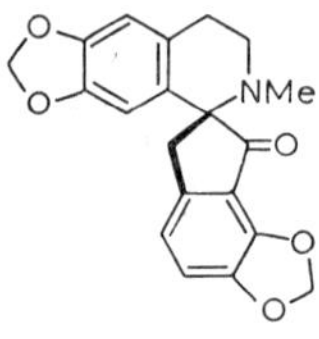

(−)Fumariline

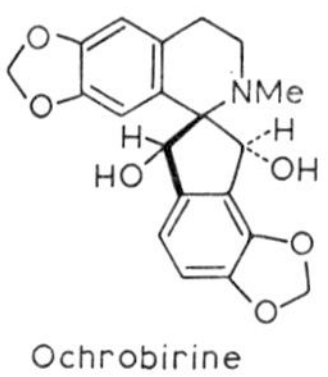

Ochrobirine

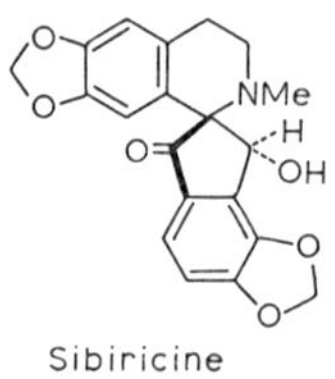

Sibiricine

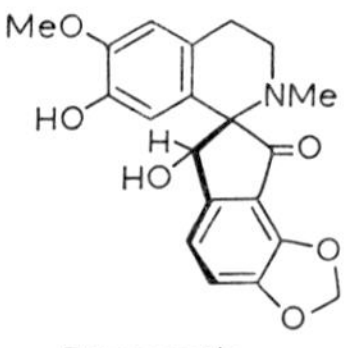

Fumarofine

(a) Structure determination

(i) Ochotensine and related alkaloids

Although ochotensine was first isolated in 1936 (*Manske*, Canad. J. Res., 1936, **14B, 354**), and ochotensimine a little later (*idem, ibid.*, 1940, **18B**, 75), it was 1964 before the structures were elucidated, largely as a result of the interpretation of n.m.r. spectra (*S. McLean, M.-S. Lin* and *Manske*, Canad. J. Chem., 1966, **44**, 2449); these alkaloids proved to be the first examples of a novel variant of the 1-benzylisoquinoline structure.

Ochotensine, $C_{21}H_{21}NO_4$, is an optically active, phenolic base containing one methoxyl group. *O*-Methylation with diazomethane gave *ochotensimine*. Hofmann degradation of the latter did not give characterisable products, but Emde degradation yielded a crystalline dihydrobase. The signals in the proton n.m.r. spectrum of ochotensimine, obtained first at 60 MHz, and later at 100 MHz, are summarised in Table 9, together with the interpretations. The one-hydrogen singlets at 5.63 and 4.90 δ suggested an exocyclic methylene group, and this was confirmed by the fact that these two signals were replaced by bands characteristic of the CH_3–$CH\!<$ } grouping in dihydroochotensimine. The AB quartet at 3.45 and 2.95 δ has a very large coupling constant ($J = 18$ Hz), associated with geminal coupling. Since

TABLE 9

N.M.R. SPECTRUM OF OCHOTENSIMINE

Chemical shift δ	Integral	Multiplicity	Assignment
7.12	1	d $J = 8$	
6.80	1	d $J = 8$	
6.50	1	s	
6.30	1	s	
5.97	2	s	
5.63	1	s	$CH_2 = C<^C_C$
4.90	1	s	
3.83	3	s	OMe
3.62	3	s	OMe
3.45	1	d $J = 18$	$CH_2<^C_C$
2.95	1	d $J = 18$	
2.90	4	m	$ArCH_2CH_2\text{-}N<$
2.13	3	s	NMe

SCHEME 122

none of these lines is further split, the $-CH_2-$ group has no α-hydrogen atoms attached to it.

From these data, some possible structural fragments can be deduced (Scheme 122). Some degradations of ochotensimine were carried out and are summarised in Scheme 123.

Since Hofmann degradation of VI yields a methine III without olefinic protons in its n.m.r. spectrum, the part-structure VII can be considered. The second stage Hofmann degradation III $\rightarrow$ IV $\rightarrow$ V confirmed the

SCHEME 123

DEGRADATION OF OCHOTENSIMINE

presence of the $ArCH_2 \cdot CH_2N\!\!<$ moiety since the n.m.r. spectrum of V clearly showed signals associated with $ArCH_2 \cdot CH_3$.

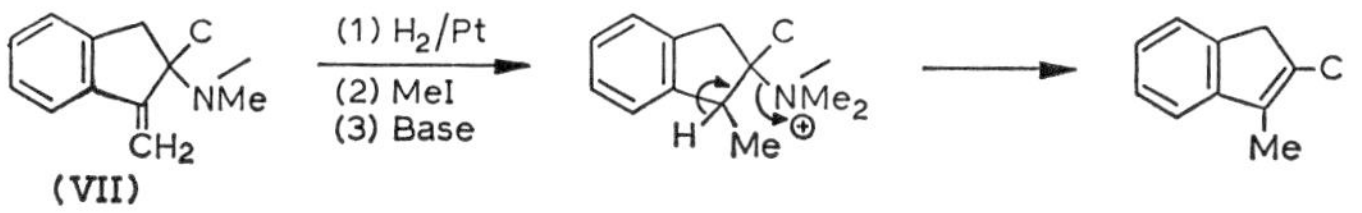

On the basis of this evidence the structure proposed for ochotensimine *(idem, loc. cit.)* was VIII (R = Me) although it was recognised that the placing of the oxygen functions on the two aromatic rings (both nature and position) was in doubt. The spectral data are summarised in IX.

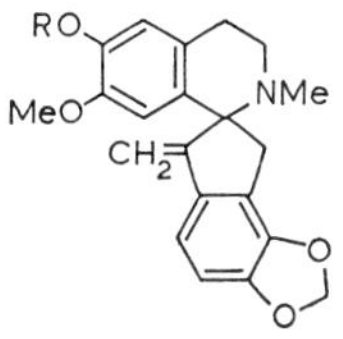

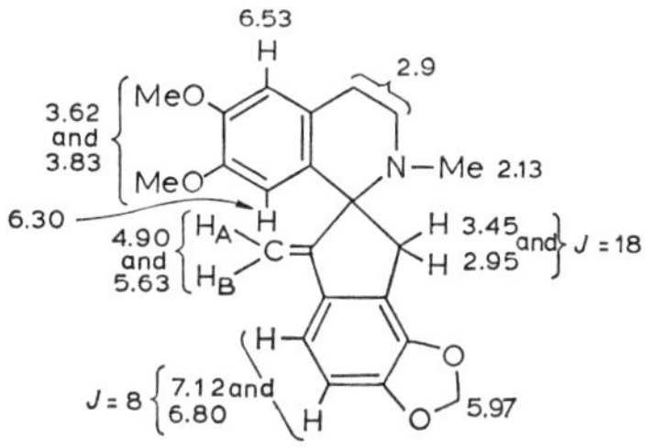

The structure of ochotensine was proved to be VIII (R = H) by X-ray analysis of the methiodide (*McLean et al.*, Tetrahedron Letters, 1966, 185; *A. C. Macdonald* and *J. Trotter*, J. chem. Soc., B, 1966, 929). The structure VIII (R = Me) for ochotensimine followed from this.

It is interesting to note that all uncertainties about structure could have been removed by a study of Nuclear Overhauser Effects (NOE) in the n.m.r. spectrum — a technique that was not developed until later (*R. A. Bell* and *J. K. Saunders*, "Topics in Stereochemistry", 1971, **7**, 2). Thus, saturation of $II_{(13)}$ (see IX) produced a positive NOE of 24% at the proton on $C_{(12)}$, and a negative effect (7% decrease in area) at the proton on $C_{(11)}$. Irradiation of the $C_{(5)}$ multiplet caused a NOE at the singlet at 6.53 δ — hence this is at $C_{(4)}$. Similarly, irradiation of the methoxyl resonances in turn caused NOE enhancements at $C_{(1)}$-H (6.30 δ) and $C_{(4)}$-H. These experiments remove all doubts about the nature and orientation of the oxygen functions attached to rings A and D.

Fumaricine (X), fumaritine (XI) and fumariline (XII) are three closely related alkaloids, the structures of which were established almost entirely by spectroscopic methods (*D. E. MacLean et al.*, Canad. J. Chem., 1968, **46**, 2837; 1969, **47**, 3593). Fumariline (XII), $C_{20}H_{17}NO_5$, contains two OCH_2O groups, and one NMe group. The i.r. spectrum of fumariline exhibits a strong band at 1709 cm^{-1}, indicative of a conjugated five-membered ring ketone. The spectral characteristics of XII were found to be very similar to those of the synthetic ketone XIII, which was an intermediate in the synthesis of ochotensimine (see later). The protons of the methylene group ($C_{(13)}$) in the alkaloid showed the expected AB quartet with $J = 18$ Hz, but each signal showed further fine splitting, attributed to long range coupling to an aromatic proton. This was not observed with XIII, where it is known that the $C_{(8)}$ protons are adjacent to an oxygen function (at $C_{(9)}$). Decoupling of the protons in XII at $C_{(13)}$ and $C_{(12)}$ confirmed the interpretation. Since the $C_{(12)}$-H is a doublet ($J = 8$), it follows that a methylenedioxy group is present at ($C_{(9)} + C_{(10)}$).

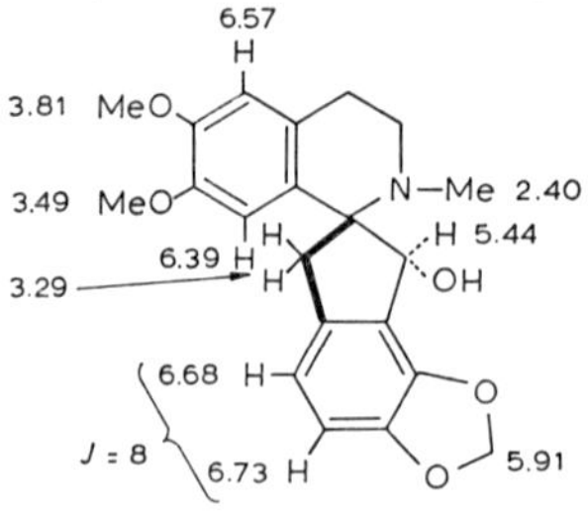

Fumaricine
(X)

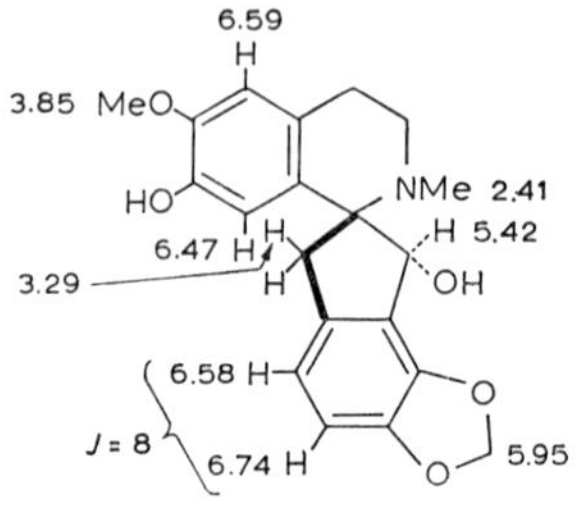

Fumaritine
(XI)

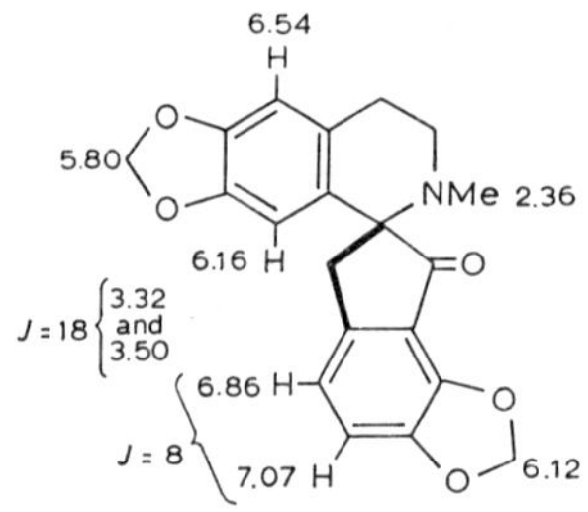

Fumariline
(XII)

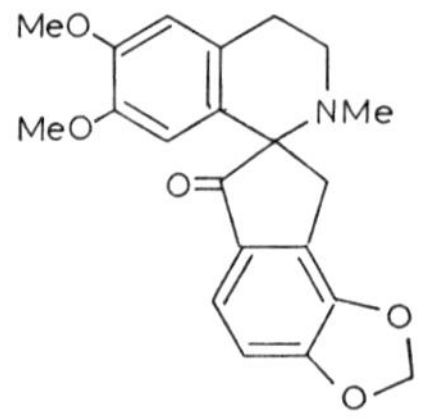

(XIII)

Fumaricine (X) was quickly shown to be the *O*-methyl ether of fumaritine (XI). A band at 3560 cm^{-1} in the i.r. spectrum (carbon disulphide solution) does not shift on dilution, thus indicating the presence of an intramolecular hydrogen bond in the alkaloid. From this it was concluded that an OH group at $C_{(8)}$ was *cis* to the *N*-Me group. However, an observable NOE between the *N*-methyl and the $C_{(8)}$-H required the opposite conclusion, and it was then postulated that the OH group must lie over aromatic ring D to account for its spectral properties. A NOE was also found between the $C_{(13)}$-hydrogen atoms and the aromatic one proton doublet at 6.68 δ ($C_{(12)}$-H).

Hence the oxygenation of ring D must be $C_{(9)} + C_{(10)}$. The one proton singlet at 6.39 δ ($C_{(1)}$-H) was also enhanced in intensity when the $C_{(13)}$ hydrogens were irradiated.

Finally, the position of the phenolic hydroxyl of fumaritine (XI) was established when it was found that irradiation of the methoxyl group at 3.85 δ caused a 24% enhancement of the signal at 6.59 assigned to the $C_{(4)}$-H.

(ii) Other alkaloids

The structures of other members of the spirobenzylisoquinoline group have been elucidated by a combination of i.r. and proton n.m.r. spectroscopy. The use of NOE has been especially valuable.

Now that a number of alkaloids and synthetic spirobenzylisoquinolines is available, it has been found possible to correlate u.v. spectral characteristics; especially useful is the distinction between a $C_{(8)}$ and a $C_{(13)}$ ketone function (*MacLean et al.*, Canad. J. Chem., 1971, **49**, 3020; *F. Santavy et al.*, Coll. Czech. chem. Comm., 1970, **35**, 2418). However, it would seem that the correlation of mass spectral fragmentation patterns with structure should provide a powerful aid to future structural studies (*C. K. Yu* and *MacLean*, Canad. J. Chem., 1971, **49**, 3025). The fragmentation pattern is critically dependent upon the number and character of oxygen functions in the five-membered ring (ring C). In those alkaloids that lack an oxygen function in ring C (*e.g.* ochotensimine) the parent ion is the base peak in the spectrum. However, with one hydroxyl group in ring C (*e.g.* fumaricine) cleavage occurs between the carbon carrying the OH group and the *N*-methyl function

SCHEME 124

MASS SPECTRAL FRAGMENTATIONS OF
RING *C* SUBSTITUTED OCHOTANSANES

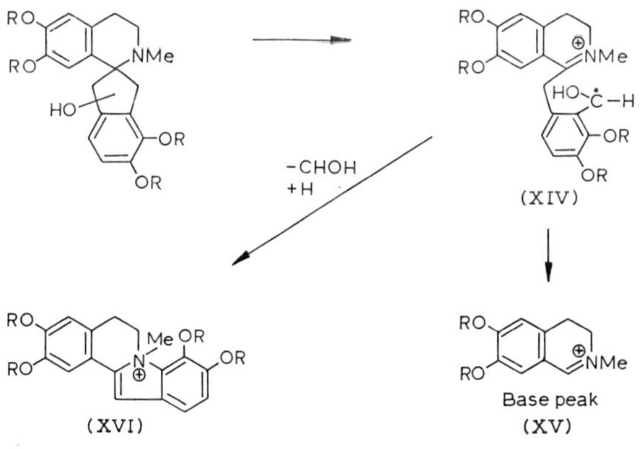

(Scheme 124) to give the ion XIV which either loses the 1-benzyl group to yield XV as the base peak, or loses 31 units to give XVI. This pattern depends upon the H of the OH group and is changed in the *O*-acetate. In the ring C ketones the spectrum is dominated by an $(M - 29)^{\oplus}$ peak. In those alkaloids that possess both a hydroxyl group *and* a carbonyl group in ring C, the parent ion is the base peak and the rest of the spectrum is a complex pattern that may be accounted for as the superposition of patterns due to the fragmentation of a hydroxy compound and of a ketone. When ring C carries *two* hydroxyl groups the fragmentation is again different and distinctive.

SCHEME 125

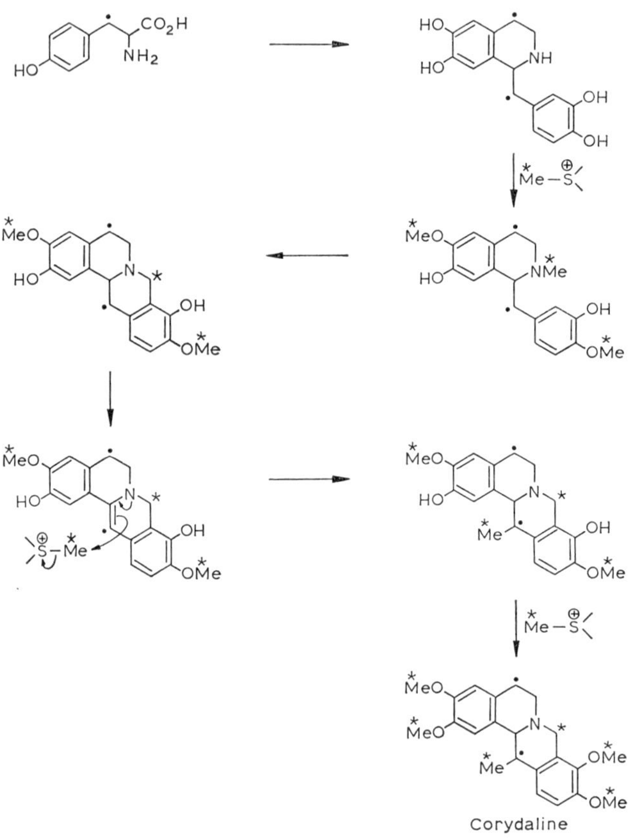

Corydaline

(b) Synthesis

(i) Biosynthesis

The biosynthesis of **corydaline** and of **ochotensimine** in corydalis plants have been examined (*I. D. Spenser et al.*, Canad. J. Chem., 1974, **52**, 2818). The route to the former is shown in Scheme 125. It was established that ochotensimine arises in the plant from corydaline. The results of the labelling experiments are summarised in Scheme 126.

SCHEME 126

(ii) Chemical synthesis

The first synthesis of a spirobenzylisoquinoline alkaloid was reported in 1968 (*S. McLean, M.-S. Lin* and *Whelan*, Tetrahedron Letters, 1968, 2425; Canad. J. Chem., 1970, **48**, 948) and this is summarised in Scheme 127. Similar syntheses of ochotensimine, fumaricine and analogues have been achieved independently (*B. A. Beckett* and *R. B. Kelly*, J. heterocycl. Chem., 1968, **5**, 685; Canad. J. Chem., 1969, **47**, 2501; *H. Irie, T. Kishimoto* and *S. Uyeo*, J. chem. Soc., C, 1968, 3051; 1969, 1645, 2600; 1971, 1644).

A synthesis of (±)**ochrobirine** (*T. Kametani, S. Hibino* and *S. Takano*, Chem. Comm., 1971, 925) was achieved, utilising the appropriately substituted ninhydrin in the Pictet–Spengler reaction (Scheme 128). In a slight modification the bromo derivative **XVII** has been obtained; it was converted into the hydroxyketone **XVIII**, which thus allows a distinction to be made between the two oxygen functions of the ring C. This method then leads to a stereo-selective synthesis of (±)ochrobirine (*McLean* and *Whelan*, Canad. J. Chem., 1973, **51**, 2457) (see p. 207).

SCHEME 127

THE SYNTHESIS OF (±)OCHOTENSIMINE

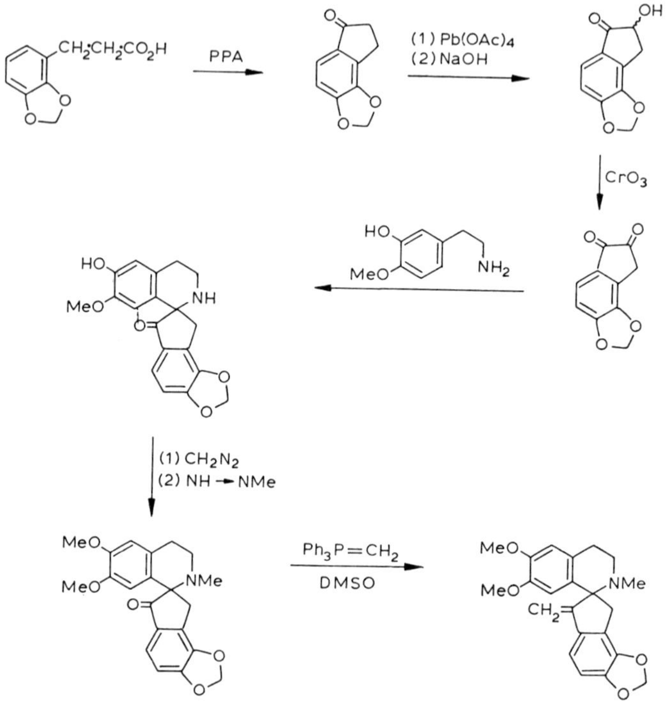

SCHEME 128

A SYNTHESIS OF (±)OCHROBIRINE

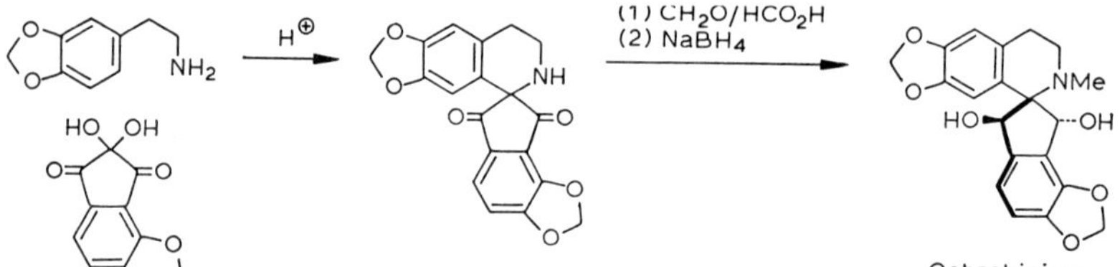

Ochrobirine

The indanone XIX ($R^1 = R^2 = $ Me) has been used in the synthesis of *cis*-alpinigenine (see p. 171) (*K. Orito, Manske* and *R. Rodrigo*, J. Amer. chem. Soc., 1974, **96**, 1944). It, together with the analogue XIX ($R^1 + R^2 = $ CH$_2$) has been converted also into the spirobenzylisoquinolines XX ($R^1 = $ R$^2 = $ Me) and XX ($R^1 + R^2 = $ CH$_2$), respectively (*Manske et al.,* Tetra-

(XVII): R^1 = Br; R^2 = H
(XVIII): R^1 = OH; R^2 = H

hedron Letters, 1974, 3243). This latter compound had previously been transformed into ochotensimine. The indanone XIX has also been converted *via* XXI into XXII, a useful precursor of alkaloids such as **yenhusomine** XXIII (*Kametani et al.*, J. chem. Soc., Perkin I, 1976, 63).

(1) HCl
(2) Br$_2$/HAc
(3) Et$_3$N

(XIX)

(XX)

(1) KOH/EtOH
(2) Triton B

(XXI)

(XXII)

(XXIII)
Yenhusomine

A slightly different approach, in which the indanone ring is closed at a much later stage in the synthesis, has yielded an analogue with a $C_{(10)}, C_{(11)}$-methylenedioxy group, instead of the $C_{(9)}, C_{(10)}$-oxygenation pattern (*idem*, J. heterocycl. Chem., 1969, **6**, 49) (Scheme 129).

SCHEME 129

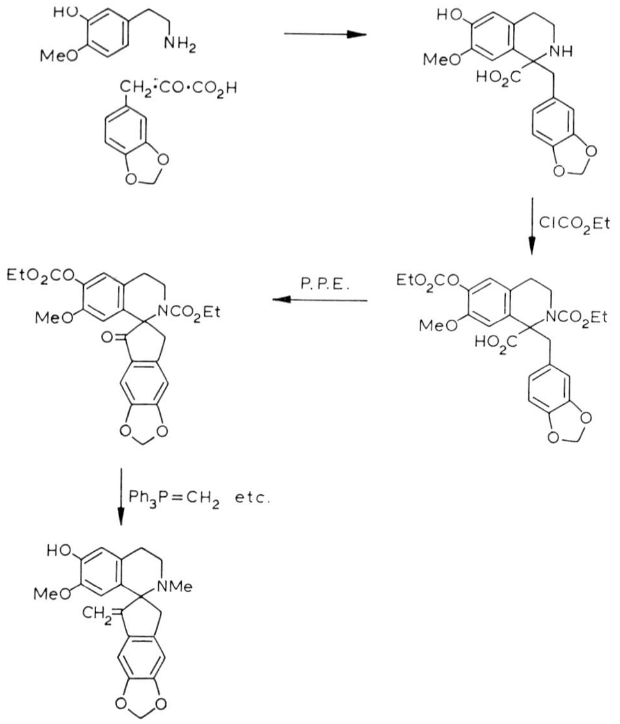

SCHEME 130

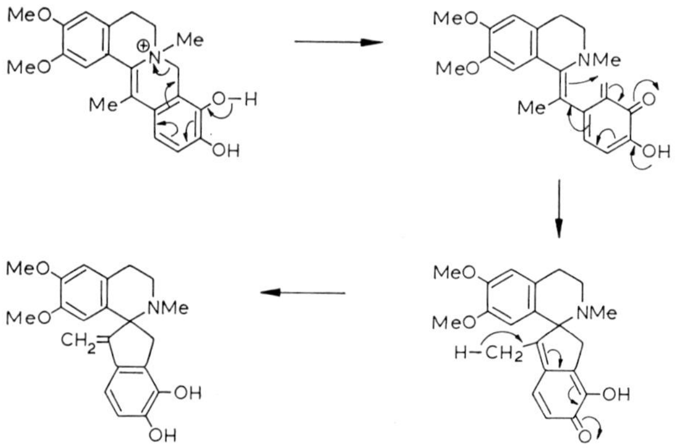

A series of base-catalysed transformations of phenolic protoberberine derivatives has been carried out, inspired by a proposal for the biosynthesis of spirobenzylisoquinolines. In the initial work (*M. Shamma* and *C. D. Jones*, J. Amer. chem. Soc., 1969, **91**, 4009; 1970, **92**, 4943) it was considered that ochotensimine might arise as shown in Scheme 130. This proposal was tested with the synthetically simpler, but otherwise equivalent, protoberberine derivative XXIV as shown in Scheme 131.

SCHEME 131

A SYNTHESIS OF SPIROBENZYLISOQUINOLINES
FROM PROTOBERBERINES

NaOH/EtOH/H₂O

DMSO

CHCl₃

(XXIV)

(XXVI)

(XXV)

When XXIV was heated under reflux with alkali, the quinone methide XXV was formed. This was converted into XXVI when DMSO was added, and the re-conversion of XXVI to XXV was achieved upon dissolution in alcohol or chloroform.

In a later paper (*Shamma* and *J. F. Nugent*, Tetrahedron Letters, 1970, 2625) a refinement of the biogenetic proposal (Scheme 132) led to the conversion of XXVIII → XXVIII (Scheme 133).

Then it was found that the monophenolic dihydroprotoberberine XXIX could be converted by base into the spirobenzylisoquinoline XXX (*Idem*, Chem. Comm., 1971, 1647) (Scheme 134), and the isomeric phenol XXXI upon prolonged base treatment was isomerised to XXXII (see p. 212).

SCHEME 132

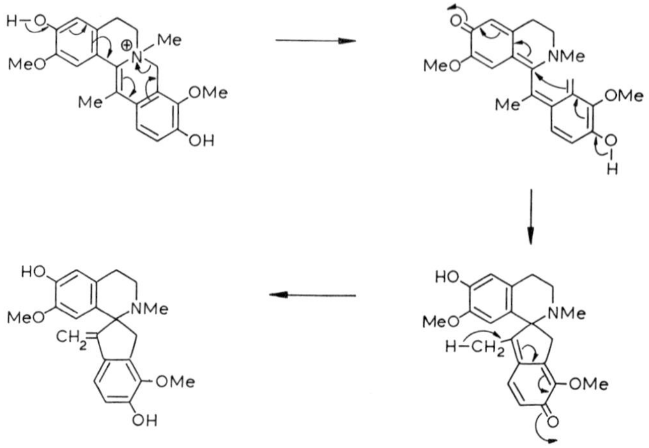

SCHEME 133

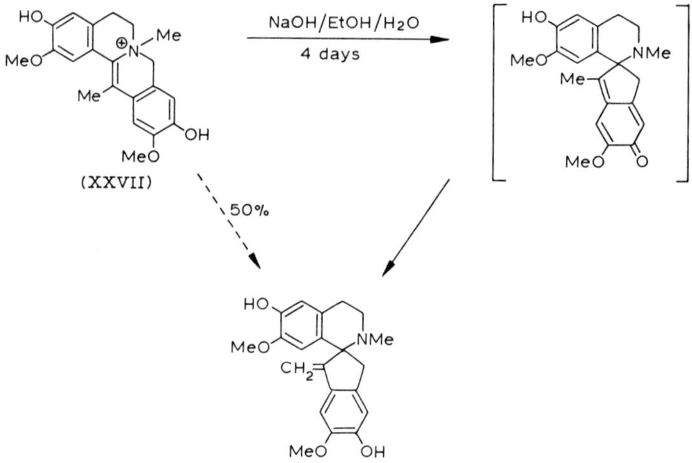

NaOH/EtOH/H₂O
4 days

(XXVII)

50%

(XXVIII)

However, when the phenolic dihydroprotoberberine XXXIII was reacted with base the reaction took an entirely different course (Scheme 135) to give XXXIV.

In a slightly different approach berberine phenol betaine (XXXV) (p. 126) has been converted into the oxospirobenzylisoquinoline XXXVI by the method outlined in Scheme 136 (see p. 212) (*D. B. MacLean et al.*, Tetrahedron Letters, 1973, 2795).

SCHEME 134

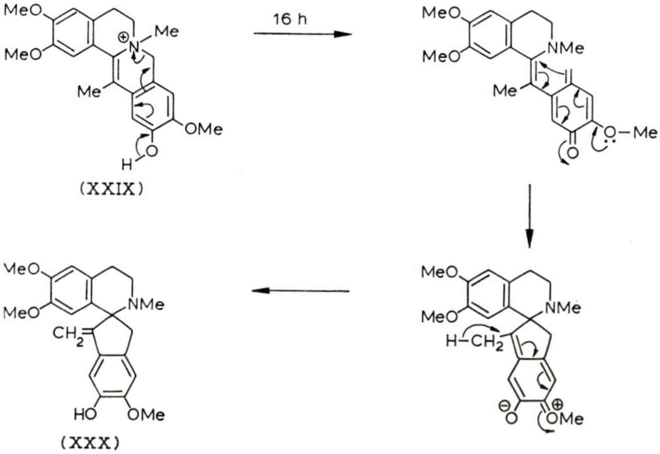

16 h

(XXIX)

(XXX)

SCHEME 135

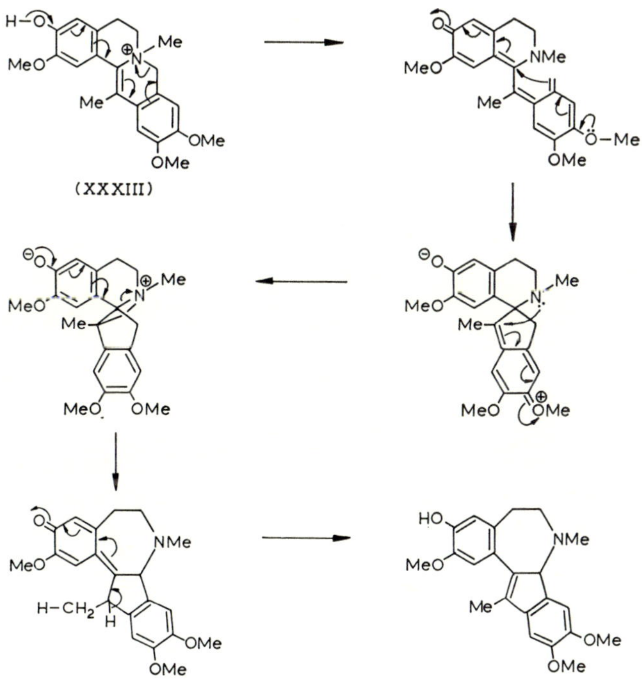

(XXXIII)

(XXXIV)

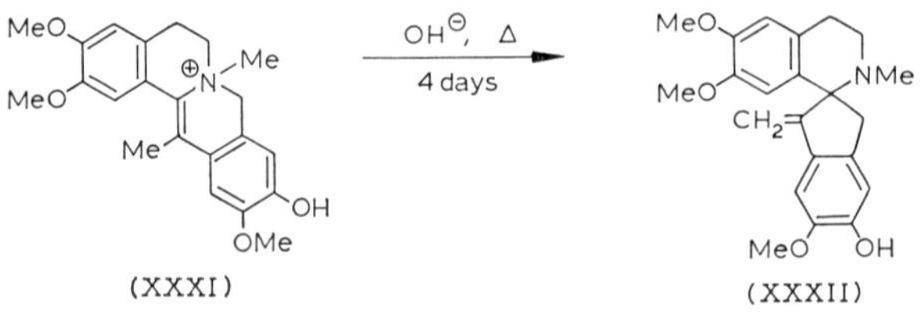

SCHEME 136

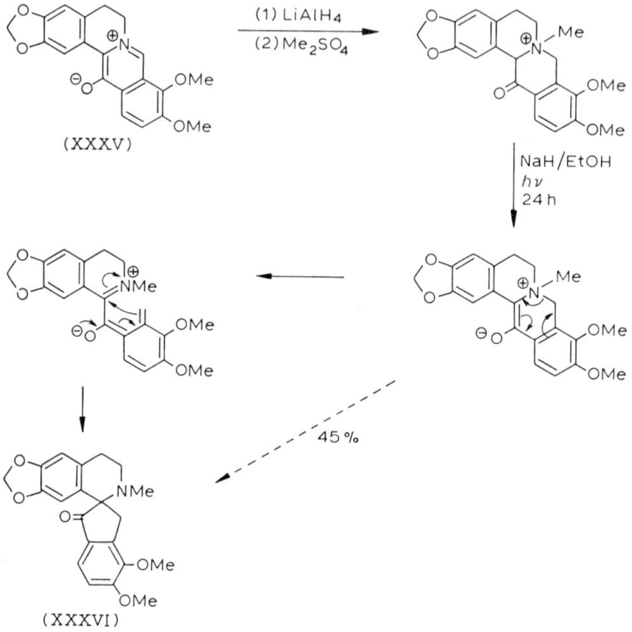

A new approach to spirobenzylisoquinolines is illustrated in Scheme 137 for (±)ochroberine (*N. E. Cundasawmy* and *MacLean*, Canad. J. Chem., 1972, **50,** 3028). *Kametani* has further developed the use of benzcyclobutanes into a synthesis of spirobenzylisoquinolines (Scheme 138) (*Kametani, T. Takahashi* and *K. Ogasawara*, Tetrahedron Letters, 1972, 4847; J. chem. Soc., Perkin I, 1973, 1464). The hydrochloride of XXXVII (R = H) is stable at room temperature in organic solvents, but the free base in chloroform is oxidised in air to XXXVII (R = OH), which then undergoes rearrangement to XXXVIII (*Kametani et al.*, J. chem. Soc., Perkin I, 1974, 2141; Heterocycles, 1974, **2,** 339).

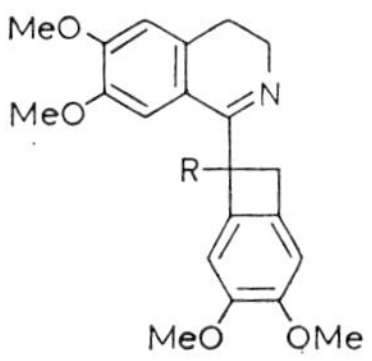

MeO
MeO
N
R
MeO
OMe

(XXXVII)

MeO
MeO
NH
O
MeO
OMe

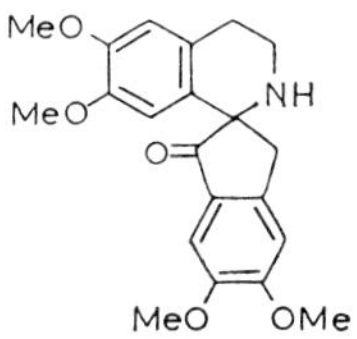

(XXXVIII)

SCHEME 137

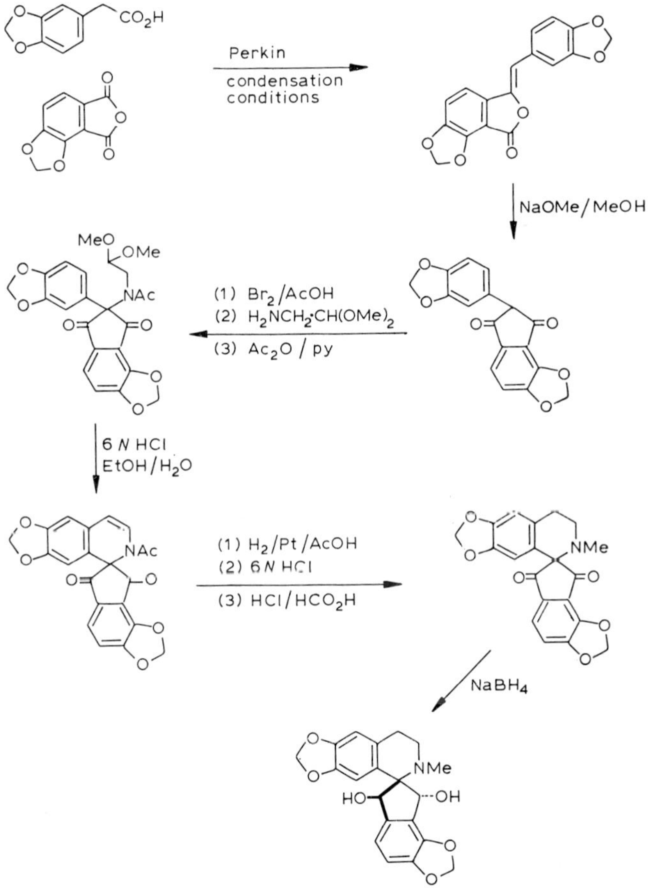

CO2H
Perkin
condensation
conditions
NaOMe/MeOH
MeO
OMe
NAc
(1) Br2/AcOH
(2) H2NCH2·CH(OMe)2
(3) Ac2O / py
6 N HCl
EtOH/H2O
NAc
(1) H2/Pt/AcOH
(2) 6 N HCl
(3) HCl/HCO2H
NMe
NaBH4
NMe
HO
OH

(±)Ochroberine

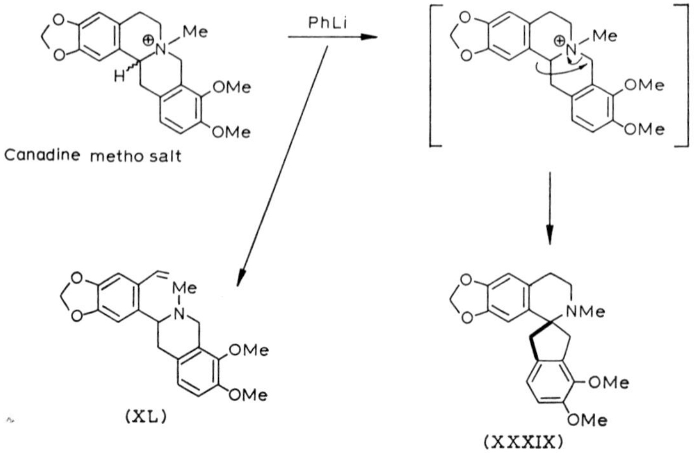

Canadine metho salt

(XL)

(XXXIX)

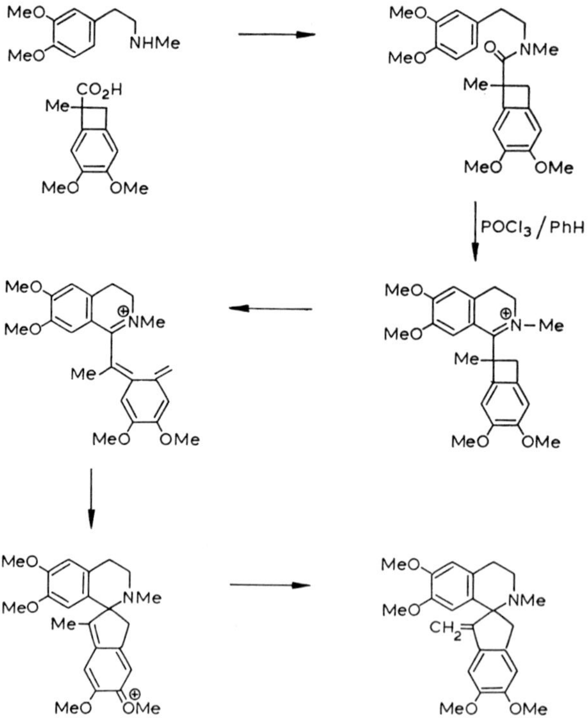

SCHEME 138

Optically active ochotensanes can be obtained by treating either $(+)$ or $(-)$ canadine with phenyl-lithium or other strong bases. The Stevens product, XXXIX, is accompanied by the expected Hofmann degradation product XL (*J. Imai, Y. Kondo* and *T. Kakemoto*, Tetrahedron, 1976, **32,** 1973).

12. The Ipecacuanha alkaloids

Ipecacuanha is the native name for a small shrub-like plant found in the tropical rain forests, especially of Brazil. Ipecac root has been used for more than 300 years as an emetic and, in recent times, as a specific remedy for amoebic dysentery. Commercial ipecac is the root bark of two species

SCHEME 139

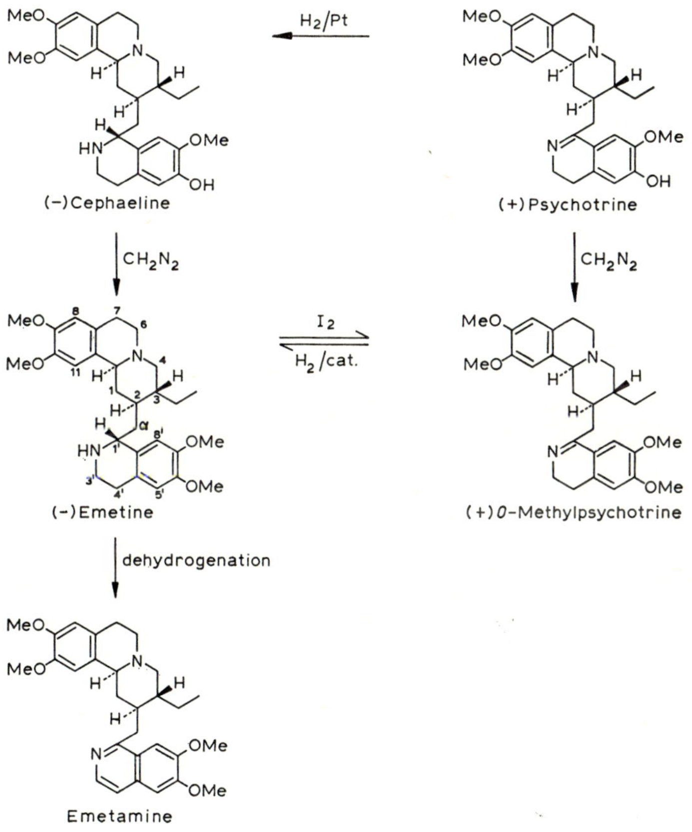

in the Rubiaceae family, *Psychotria ipecacuanha* Stokes (*Cephaelis ipecacuanha* Rich) and *Psychotria granadensis* Benth (*Cephaelis acuminata* Karsten). Other, related species of Rubiaceae contain smaller amounts of the alkaloids. The principal alkaloid is **emetine** (and is the pharmacologically active constituent), but the related bases **emetamine, cephaeline, psychotrine** and *O*-**methylpsychotrine** are also present. The inter-relationship of these alkaloids is summarised in Scheme 139. Accounts of structural elucidation and synthesis of these bases have been published, which also contain all the early literature (*H. T. Openshaw*, in "Chemistry of the Alkaloids", ed. *S. W. Pelletier*, Van Nostrand–Reinhold, 1970, p. 85; *M. Shamma*, "The Isoquinoline Alkaloids", 1972, p. 426; *A. Brossi, G. V. Parry* and *S. Teitel*, "The Alkaloids", 1971, Vol. 13, p. 189; *R. H. F. Manske, ibid.*, Vol. 7, p. 419). A brief summary of the structural elucidation of emetine is given here. (An expanded discussion is to be found in *K. W. Bentley*, "The Isoquinoline Alkaloids", Pergamon Press, Oxford, 1965, p. 217 — but no references are included). Emetine, $C_{29}H_{40}N_2O_4$, contains four methoxyl groups, one *C*-methyl and one secondary amino group; no *N*-methyl groups are present. The second nitrogen is, however, tertiary. Oxidation under a variety of conditions led to the fragments I–IV. Since emetine contains no reducible groups, the formation of I must have involved a dehydrogenation as well as an oxidation. It was decided from the yield of II that there must be *two* 6,7-dimethoxyisoquinoline moieties present in emetine.

$$ (I) \qquad (II) \qquad (III) \qquad (IV) $$

A number of investigations had been made on the course of Hofmann degradation of emetine, and the results are summarised in Scheme 140. From this work it was concluded that the secondary nitrogen must be present in a ring, whereas the tertiary nitrogen is at a bridgehead of two rings. The structure deduced for emetine from these, and much other, data had also been arrived at by *R. Robinson* (Nature, 1948, **162**, 524) from biogenetic considerations. The biosynthetic route postulated, however, has since been shown to be erroneous *(vide infra)*.

The structures of the remainder of the ipecac alkaloids follow from this work (Scheme 139), except for the position of the phenolic hydroxyl group in cephaeline. This problem was solved by carrying out Hofmann degradations on the *O*-ethyl ether (Scheme 141).

SCHEME 140
DEGRADATION OF EMETINE
(1) MeI
(2) Hofmann degradation
(1) H2/Pt
(2) MeI
(3) Hofmann
(1) H2/Pt
(2) MeI
(3) Hofmann
(1) MeI
(2) Hofmann
(1) H2/Pt
(2) MeI
(3) Hofmann
Ba(MnO4)2
oxidation
O3
reduction
oxidation
proved by synthesis

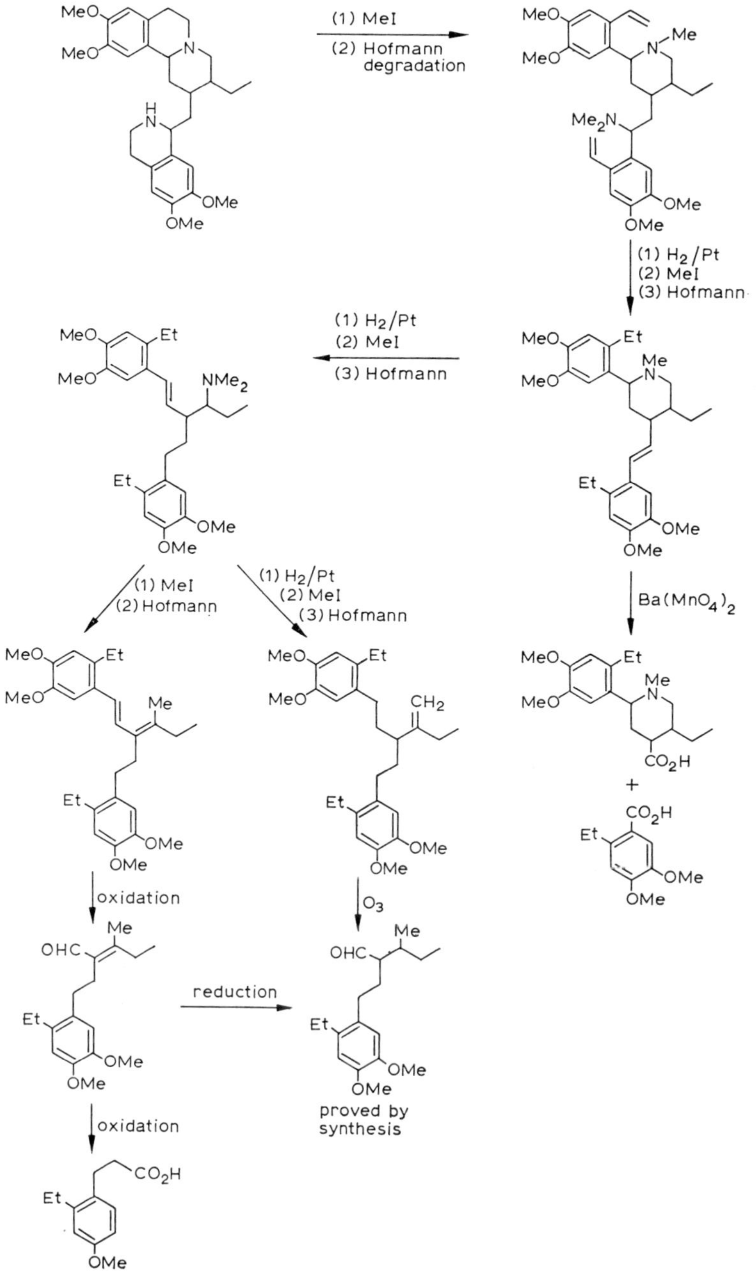

SCHEME 141

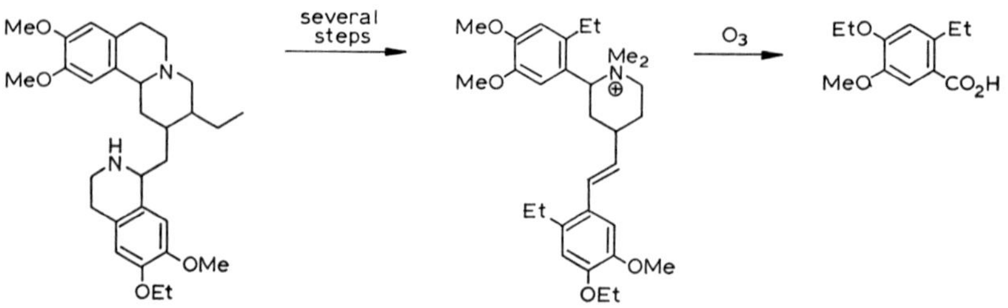

SCHEME 142

THE STRUCTURE OF PROTOEMETINE

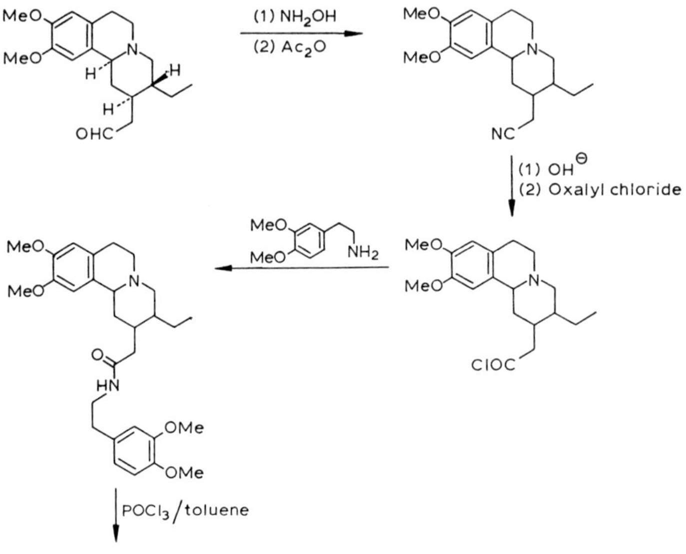

(+)O-Methylpsyochotrine

Protoemetine, $C_{19}H_{27}NO_3$, was isolated from ipecac roots in 1957 (*Battersby* and *B. J. T. Harper*, J. chem. Soc., 1959, 1748). The structure was established when a conversion to O-methylpsychotrine was achieved (Scheme 142). The stereochemistry shown was established for emetine by the methods discussed below.

When psychotrine is reduced, it affords a mixture of cephaeline and iso-cephaeline, and these can be O-methylated to emetine and isoemetine, respectively. Hence, emetine differs from isoemetine (not known as a natural product) in the stereochemistry at $C_{(1)}$.

(a) Configuration of emetine and related compounds

(i) Emetine

The alkaloid contains four different asymmetric centres (at $C_{(2)}$, $C_{(3)}$, $C_{(11b)}$ and $C_{(1')}$). When *N*-acetylemetine was subjected to successive Hofmann degradations (with reduction of the double bonds introduced after the first stage only), the diene V was produced. Ozonolysis of V led to the (+)acid VI, which had been correlated previously with VII, of known absolute configuration:

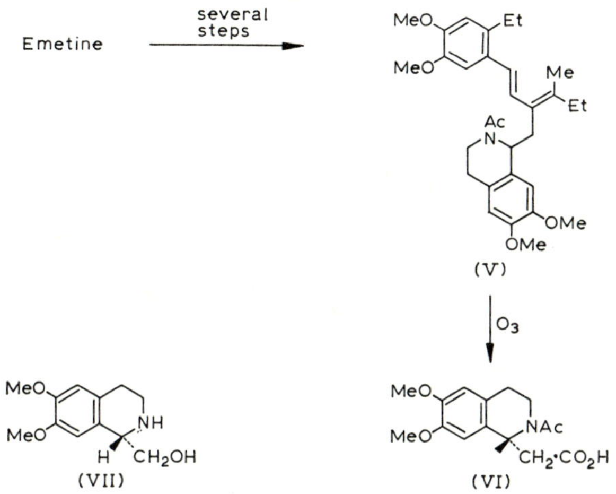

The o.r.d. curves of emetine and its salts will be governed by the asymmetry at positions 11b and 1', since these are in close proximity to the chromophoric groups (*E. E. van Tamelen, P. E. Aldrich* and *J. B. Hester*, J. Amer. chem. Soc., 1957, **79**, 4817; 1959, **81**, 507, 6214). Emetine hydrobromide exhibits almost constant specific rotation in the wavelength range 350–700 nm. Isoemetine hydrobromide, however, shows an increase in positive rotation as the wavelength decreases. This indicates that in emetine the centres 11b and 1' have opposite configurations, separate dispersion effects of which cancel each other. Emetine exhibits Bohlmann bands at about 2750 cm^{-1} in the i.r., characteristic of a *trans*-quinolizidine structure. Hence part structure VIII may be written for emetine. Reduction of IX under a wide variety of conditions always results in only one isomer of X. This implies an axial orientation of the $C_{(11b)}$-H in emetine. The thermodynamically less stable $C_{(11b)}$-epimer of X can be obtained, together with X by reduction of IX with tin and hydrochloric acid.

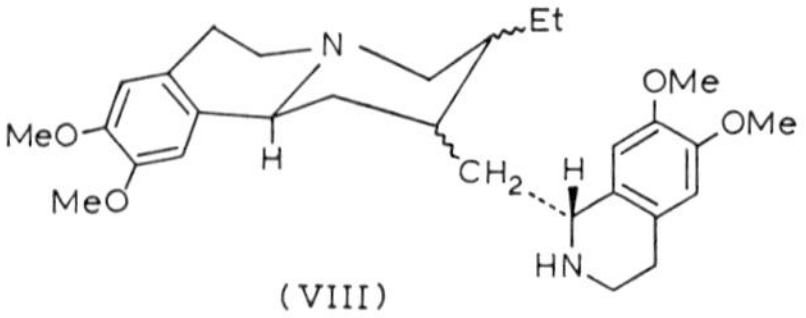

(VIII)

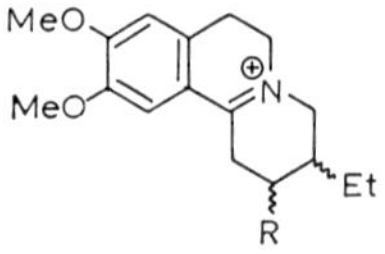

(IX 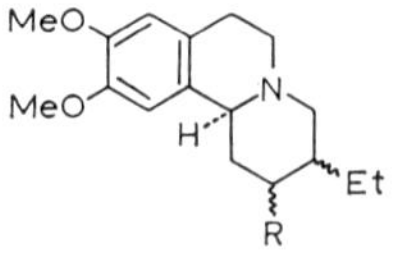)   (X)

***O*-Methylpsychotrine** can be converted into the bisbenzo salt XI and oxidation of this yields the betaine XII. Hydrogenolysis then gives rise to the acid, the methyl ester of which was recovered unchanged from heating with methanolic sodium methoxide. Thus, epimerisation had not occurred so that the carboxyl group of XIII must occupy the more stable equatorial conformation. The structure of emetine can be expanded to XIV.

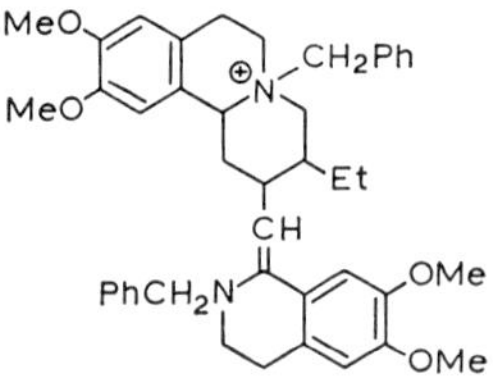

oxidation

(XI)

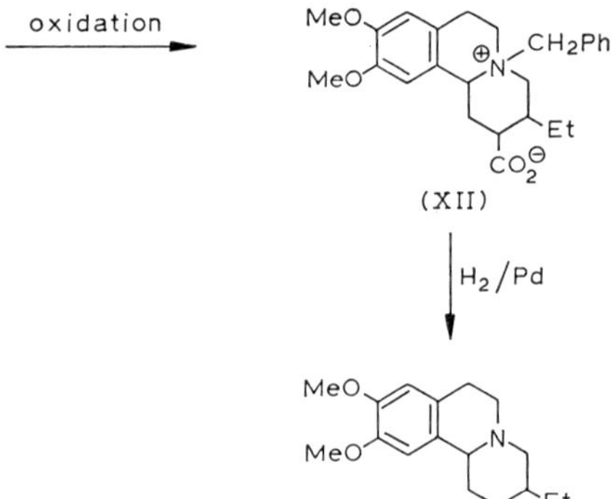

(XII)

H_2/Pd

(XIII)

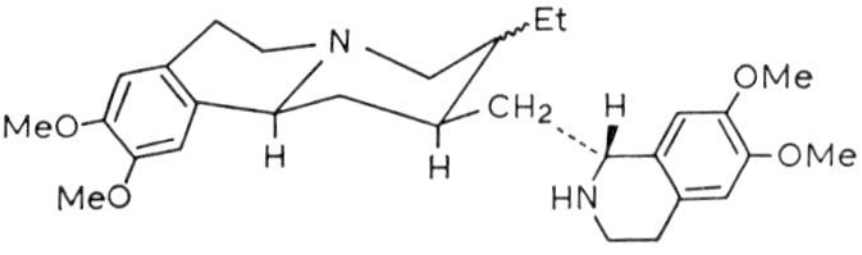

Emetine
(XIV)

It was established that no stereochemical inversion had occurred during the preparation of XIII by re-forming *O*-methylpsychotrine by a series of reactions not involving the asymmetric centres. Proemetine, which possesses the same configuration at the relevant centres (2, 3, 11b) as emetine (since

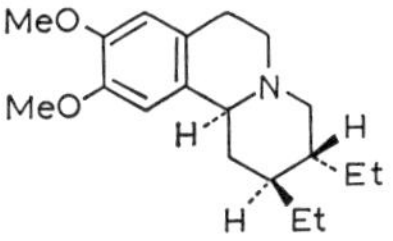

(XV)

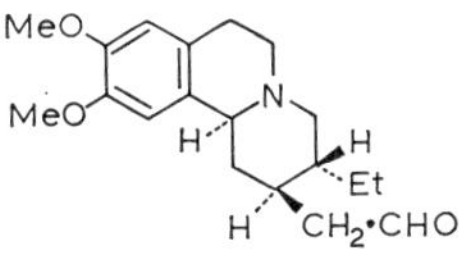

Protoemetine
(XVIII)

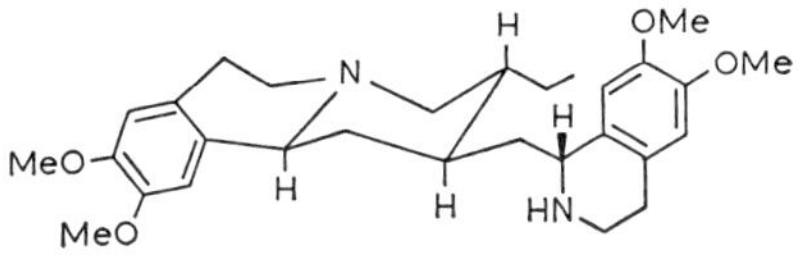

Emetine
(XIX)

it had been converted into *O*-methylpsychotrine) can be reduced by the
Wolff–Kishner method into the diethyl compound XV. That the two ethyl
groups are *trans* to each other was proved by synthesis from *trans*-3,4-
diethylcyclopentanone (*Van Tamelen* and *J. B. Hester*, J. Amer. chem. Soc.,
1959, **81**, 507; *Van Tamelen, P. E. Aldrich* and *Hester, ibid.*, p. 6214) (Scheme
143). The intermediate XVI has also been correlated with XVII which
has been converted into emetine (see p. 227).

SCHEME 143

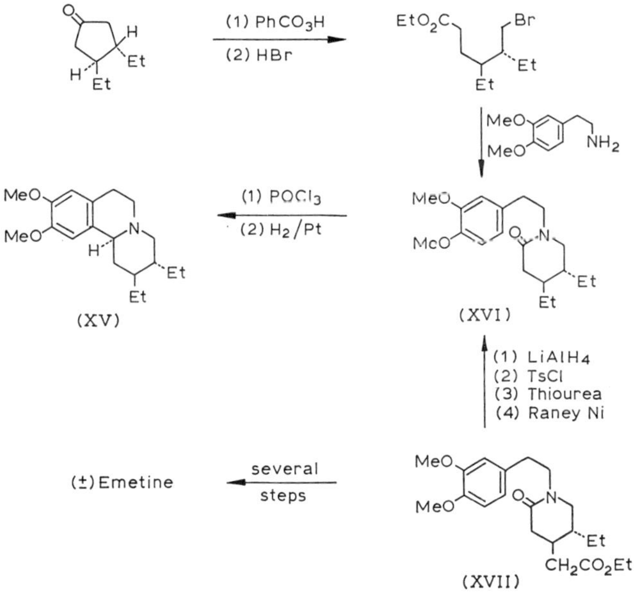

(1) PhCO₃H
(2) HBr

(XV)

(1) POCl₃
(2) H₂/Pt

(XVI)

(1) LiAlH₄
(2) TsCl
(3) Thiourea
(4) Raney Ni

(±)Emetine

several
steps

(XVII)

The absolute configuration of protoemetine was shown to be that shown in XVIII by a comparative study of molecular rotation changes in a series of derivatives (*M. Terashima*, Chem. and Pharm. Bull., Japan, 1960, **8**, 517).

Thus, the configurations of the centres in emetine correspond to a *trans*-quinolizidine with all the substituents equatorial XIX.

SCHEME 144

THE STRUCTURE OF IPECOSIDE

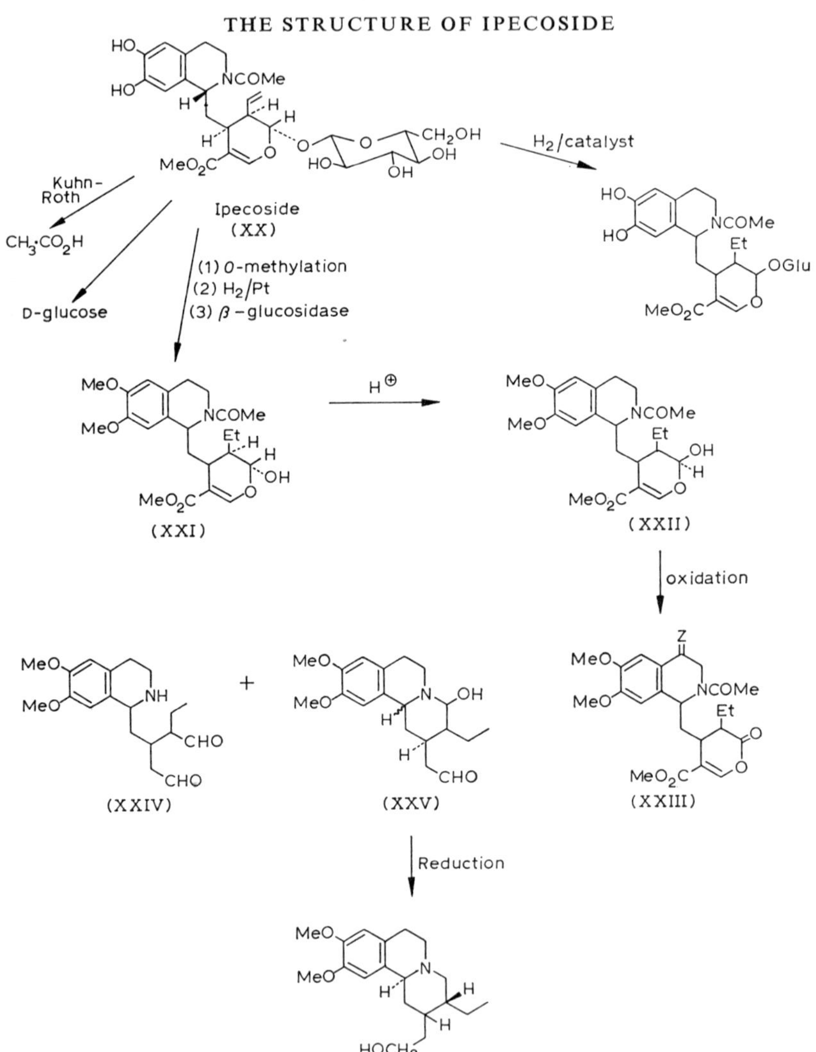

Ipecoside, (XX), $C_{27}H_{35}NO_{12}$, was isolated from ipecac root in 1952 and the elucidation of its structure involved spectroscopic methods (*A. R. Battersby et al.*, Chem. Comm., 1967, 219). Mild acid hydrolysis gave D-glucose, and since the enzyme β-glucosidase also hydrolyses the alkaloid, a β-glucosidic linkage can be assigned to the anomeric centre. Ipecoside exhibits an u.v. spectrum in accord with a tetrahydroisoquinoline chromophore, with λ_{max} 285 nm. This undergoes a bathochromic shift in alkali, indicating phenolic functionality. A second band at λ_{max} 238 nm was assigned to a $MeO_2C-C=C-CO-$ chromophore, also deduced from the i.r. spectrum ν_{max} at 1690 cm^{-1}). An amide carbonyl group is also present (1630 cm^{-1}). In dihydroipecoside, obtained upon catalytic hydrogenation, the u.v. and i.r. spectral characteristics in the carbonyl region are unchanged; only an isolated double bond has been reduced. β-Glucosidase also hydrolyses O,O-dimethyldihydroipecoside to give an aglycone, formulated as XXI (Scheme 144). This is isomerised by acid to a more stable compound, XXII, which can be oxidised to the enol lactones XXIII (Z = 2H) and XXIII (Z = O). Vigorous acid hydrolysis of O,O-dimethyldihydroipecoside gave a mixture of XXIV and XXV. Reduction of the latter gave (−)dihydroprotoemetine, identical with an authentic sample.

Alangiside (XXVI) has been obtained from *Alangium lamarckii*. This monoterpenoid lactam analysed for $C_{25}H_{31}NO_{10}$. Spectral data suggested a close resemblance with desacetylipecoside. In fact when desacetylipecoside (XXVII) hydrochloride was treated with a weak base, lactamisation occurred to give XXVIII, *O*-methylation of which provided *O*-methylalangiside (XXIX) (*Battersby et al.*, Chem. Comm., 1971, 904).

(*ii*) *Emetine analogues*

Early studies on the Indian plant *Alangium lamarckii* Thwaites enabled emetine, cephaeline and psychotrine to be isolated from the seed kernels and root bark (*S. C. Pakrashi* and *P. P. Ghosh-Dastidar*, Indian J. Chem., 1964, **2**, 379; *H. Budzikiewicz*, *Pakrashi* and *H. Vorbrüggen*, Tetrahedron, 1964, **20**, 399). Further examination of the stem and root bark led to two new alkaloids, **alangicine,** identified as 11-hydroxypsychotrine (XXX) and a desmethylpsychotrine (XXXI) or (XXXII) (*Pakrashi* and *E. Ali*, Tetrahedron Letters, 1967, 2143). A third alkaloid **tubulosine** (XXXIII) was

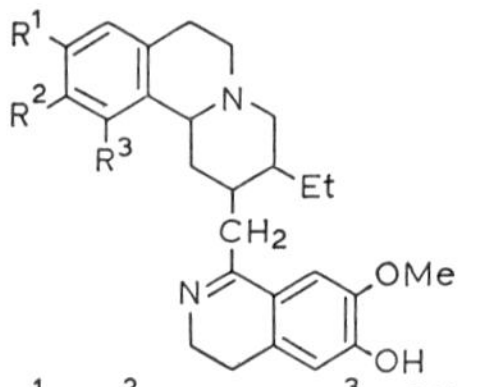

(XXX) : $R^1 = R^2 = OMe$; $R^3 = OH$
(XXXI) : $R^1 = OH$; $R^2 = OMe$; $R^3 = H$
(XXXII): $R^1 = OMe$; $R^2 = OH$; $R^3 = H$

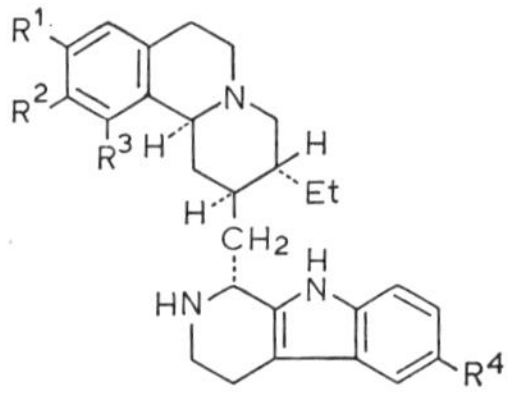

	R^1	R^2	R^3	R^4
(XXXIII) :	OMe	OMe	H	OH
(XXXIV) :	OMe	OH	H	OH
(XXXV) :	OH	OMe	H	OH
(XXXVI) :	OMe	OMe	H	H
(XXXVII):	OMe	OMe	OH	OH

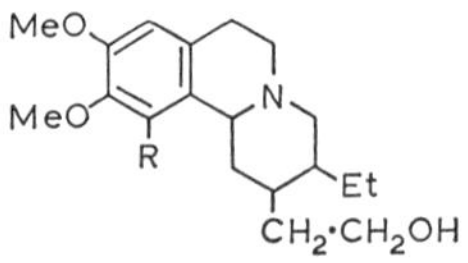

(XXXVIII): R = H
(XXXIX) : R = OH

already known. Subsequently two more new alkaloids were isolated from the bark and shown to be **isotubulosine** (the $C_{(1')}$-epimer of XXXIII, and a demethyltubulosine (XXXIV) or (XXXV) (*A. Popelak, E. Haack* and *H. Spingler, ibid.*, 1966, 1081, 5077). Two further indole analogues of emetine, *viz.* **deoxytubulosine** (XXXVI) and **alangimarckine,** originally believed to be the 11-hydroxy derivative (XXXVII) of XXXVI, were isolated from the leaves of this plant, together with **dihydroprotoemetine** (XXXVIII) and **ankorine** believed to be XXXIX (*Battersby et al., ibid.*, 1966, 4965).

The structures of these alkaloids were determined largely by mass spectrometry, together with n.m.r. spectroscopy.

The mass spectral fragmentations of the ipecac alkaloids usually gives sets of peaks which are characteristic of the benzo[*a*]quinolizine and isoquinoline moieties (*Pakrashi* and *Ghosh-Dastidar, loc. cit.*; *Budzikiewicz, Pakrashi* and *Vorbrüggen, loc. cit.*). The precise mode of fragmentation seems to depend upon the degree of saturation of ring D. The two most

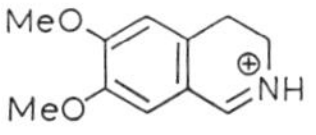

m/e 178
(XL)

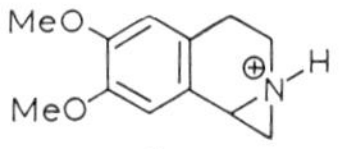

m/e 192
(XLI)

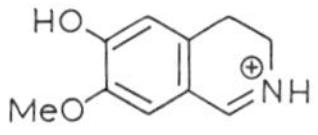

m/e 164
(XLII)

SCHEME 145

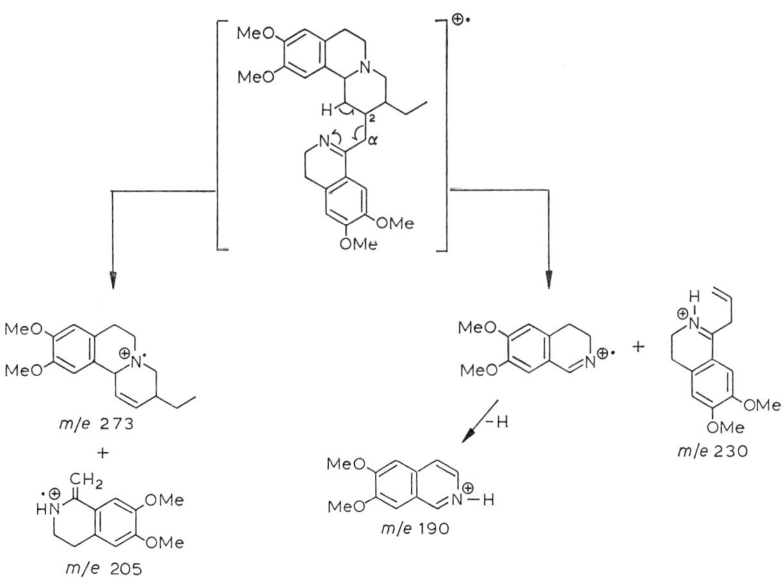

m/e 273

+

m/e 205

m/e 190

m/e 230

intense peaks, at m/e 192 and 178, in the spectrum of cephaeline are thought to be due to the ions XL–XLII. *O*-Methylpsychotrine, however, has an imine function in ring D and fragments in two main ways (Scheme 145). In the first mode, cleavage occurs at the $C_{(2)}$–C_α bond to give the ions at m/e 273 and m/e 205. Alternatively a retro Diels–Alder reaction occurs in ring C to form the stable ions at m/e 190 and m/e 230 *(idem, loc. cit.)*. Comparison of mass spectra between members of the same class of structures can be instructive. For example, in the mass spectrum of alangimarckine, the peaks corresponding to the benzoquinolizine fragment are all m/e 16 units higher than those in **deoxytubulosine** (XXXVI), whilst the signals due to the β-carboline moiety in each are the same. Hence the phenolic hydroxyl group present in alangimarckine must be in the benzoquinolizine fragment.

It is very difficult to make generalisations concerning the ^{1}H-n.m.r. spectra of the emetine bases (see *Teitel* and *Brossi*, J. Amer. chem. Soc., 1966, **88**, 4068). However, the one aromatic proton of the benzoquinolizine part of the structure of alangimarckine is at high field, and was assigned to $C_{(8)}$, thus placing the phenolic hydroxyl group at $C_{(11)}$. Similar data were recorded for ankorine (*Battersby et al.*, Tetrahedron Letters, 1966, 4965). Again, the position of the single aromatic proton (at 6.24 δ) in the benzo-quinolizine part led to the assignment of the 11-hydroxydihydroproto-emetine structure. Oxidation of *O*-ethylankorine with potassium ferricyanide gave 3-ethoxy-4,5-dimethoxyphthalic acid.

However, all four possible racemic forms of XXXIX have been synthesised, and none of them is identical with ankorine. The structure of the alkaloid was, accordingly, modified to XLIII, and the relative stereochemistry shown was assigned from further mass spectral studies (*C. Szantay, E. Szentirmay* and *L. Szabo, ibid.*, 1974, 3725). This structure has been confirmed, and the absolute stereochemistry assigned as indicated in XLIII by a synthesis of (−)ankorine (*S. Yoshifuji* and *T. Fujii, ibid.*, 1975, 1965; see also *Fujii, Yoshifuji* and *S. Kamada, ibid.*, 1975, 1527).

It is believed that the structures of alangicine and alangimarckine also require modification with the location of the phenolic hydroxyl groups at $C_{(8)}$ instead of at $C_{(11)}$. The relative and absolute stereochemistry of each at $C_{(1b)}$, $C_{(2)}$ and $C_{(3)}$ correspond to those in emetine.

(iii) Structure of rubremetine

When emetine is oxidised under mild conditions (ferric chloride, iodine or mercuric acetate), a red, optically active compound, **rubremetine** is produced. The structure assigned to this cation is shown in Scheme 146 (*Battersby, Openshaw* and *H. C. S. Wood*, Experientia, 1949, **5**, 114; *Openshaw* and *Wood*, J. chem. Soc., 1952, 391; *Battersby et al.*, Experientia, 1966, **22**, 134).

SCHEME 146

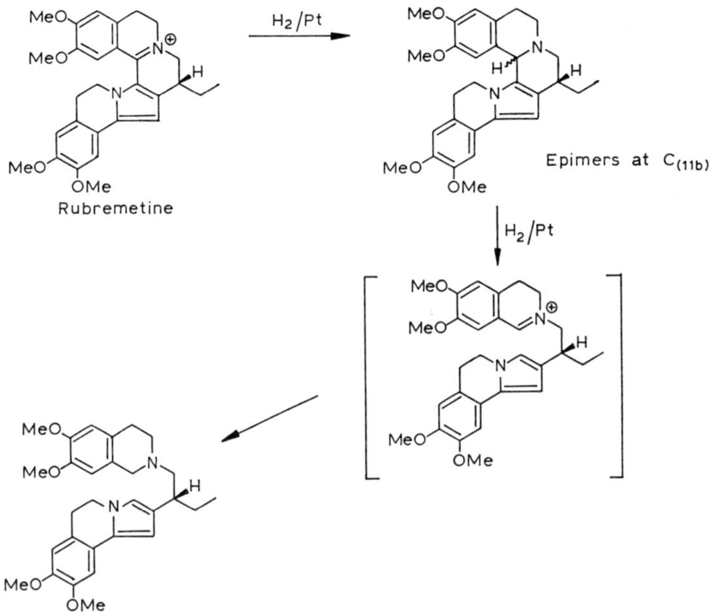

(b) Synthesis of emetine

(i) Chemical synthesis

More than a dozen syntheses of emetine have been described; some have been used on a commercial scale.

The first synthesis was described by *N. A. Preobrazhenskii et al.* in 1950 (Doklady Akad. Nauk, S.S.S.R., 1950, **75**, 539; Tetrahedron, 1958, **4**, 223). The structures of some of the intermediates later required revision (*H. Rapoport et al.*, 1965, **87**, 4221; *A. W. Frahm*, Arch. Pharm., 1968, **301**, 621), and these have been incorporated into Scheme 147. There was no stereochemical control in the synthesis; diastereomers were separated at various

SCHEME 147

THE FIRST SYNTHESIS OF EMETINE

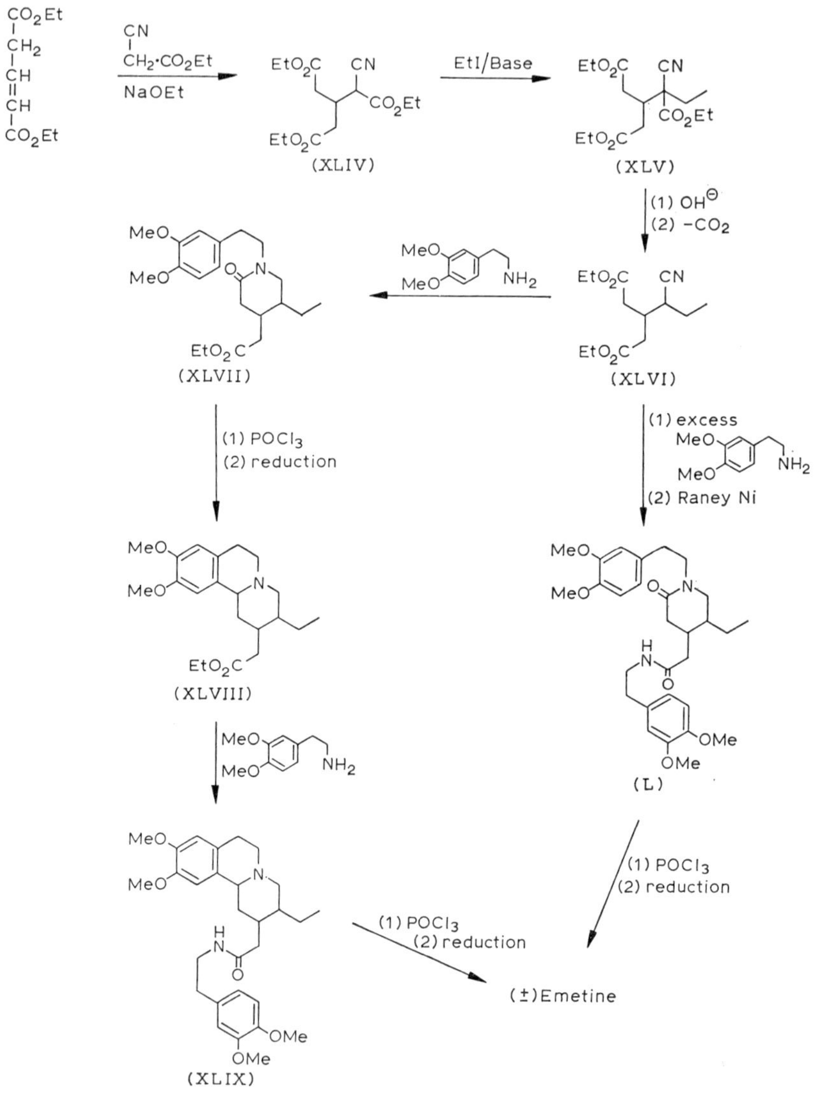

points. Michael addition of ethyl cyanoacetate to diethyl glutaconate gave XLIV, which was ethylated to give XLV. Selective hydrolysis of this, followed by thermal decarboxylation gave XLVI which was reacted with homoveratrylamine to yield the tetrahydropyridone XLVII. Bischler–Napieralski cyclisation followed by reduction yielded the key intermediate XLVIII. This could then be reacted with more homoveratrylamine to yield XLIX cyclisation, followed by reduction of which gave ($\pm$)emetine. In a modification, the cyanoester XLVI could be reacted with an excess of homoveratrylamine to yield the bisamide L, double cyclisation of which gave emetine.

SCHEME 148

A STEREOSPECIFIC SYNTHESIS OF EMETINE

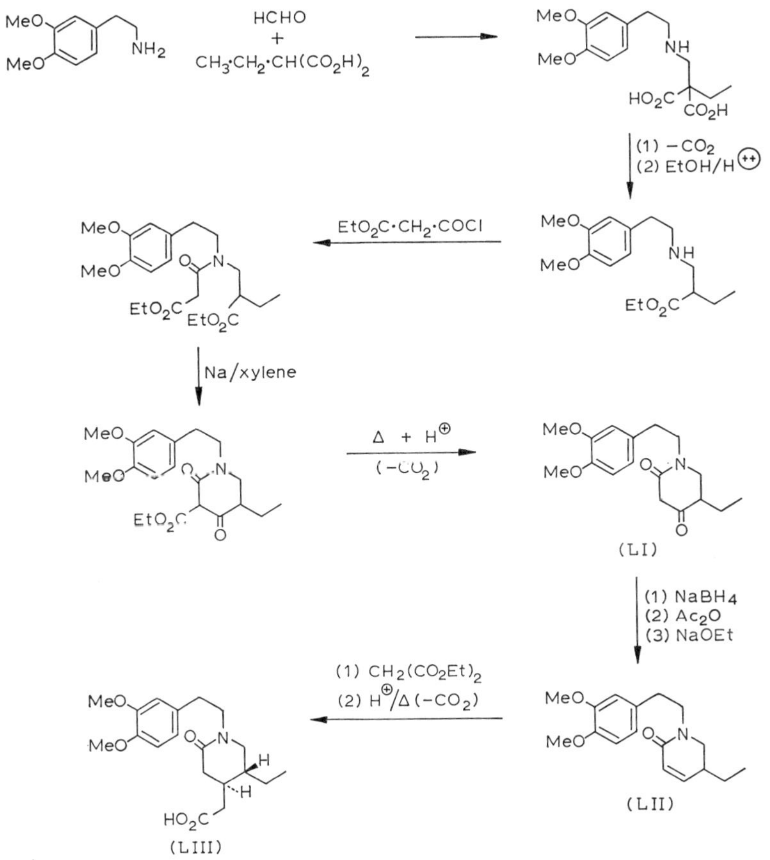

A stereospecific synthesis was reported by *Battersby* and *J. C. Turner* (J. chem. Soc., 1960, 717). The important intermediate LI had previously been described (*Y. Ban*, Pharm. Bull., 1955, **3**, 53), but its preparation is included in Scheme 148. A Michael condensation with the unsaturated lactam LII gave the thermodynamically preferred *trans* compound LIII, identical with material derived from proemetine (apart from optical activity). Emetine had previously been obtained from proemetine.

A unique route to optically active emetine, which has been used on a commercial scale, is summarised in Scheme 149 (*Openshaw* and *N. Whittaker*, J. chem. Soc., 1963, 1449, 1461; *Whittaker, ibid.*, C, 1969, 85). The interaction of the imine LIV with the Mannich base LV gave an almost quantitative yield of LVI. When such oxobenzoquinolizines are treated with acid, ring-opening can occur to give the intermediate quaternary ion LVII. Hence, when the resolution of the latter was carried out with ($-$)camphor-10-sulphonic acid, the desired laevorotatory salt crystallised out. Moreover, the dextro isomer of LVI was racemised *via* LVII under these acidic conditions, so that the laevo salt was continuously separated. The conversion of

SCHEME 149

THE BURROUGHS–WELLCOME SYNTHESIS OF EMETINE

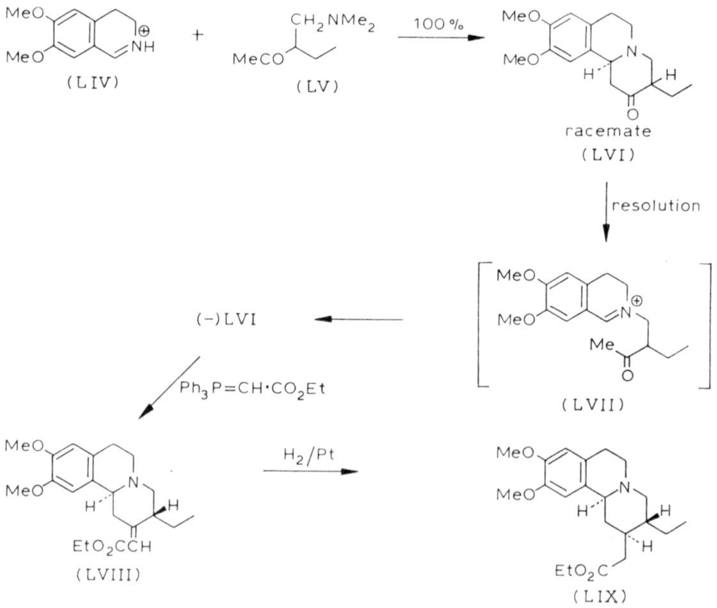

$(-)$LVI to LVIII and thence to LIX followed standard lines. The remaining steps to optically active emetine are standard.

Of the other syntheses of emetine described, mention should be made of the method due to *A. W. Burgstahler* and *Z. J. Bithos* (J. Amer. chem. Soc., 1960, **82,** 5466); see also *Van Tamelen et al., ibid.,* 1969, **91,** 7359; *D. E. Clark et al.,* J. chem. Soc., 1962, 2490; *Brossi,* Pure appl. Chem., 1969, **19,** 171; *A. Grüssner et al.,* Helv., 1959, **42,** 2431; *M. Barasch, J. M. Osbond* and *J. C. Wickens,* J. chem. Soc., 1959, 3530) (Scheme 150). The key intermediate $(-)$LXI from which various ipecac alkaloids have been synthesised, has been prepared recently from the ethyl ester of $(+)$**cincholoipin** (LX), an oxidative degradation product of dihydrocinchonine (*Fujii* and *Yoshifuji,* Tetrahedron Letters, 1975, 731).

SCHEME 150

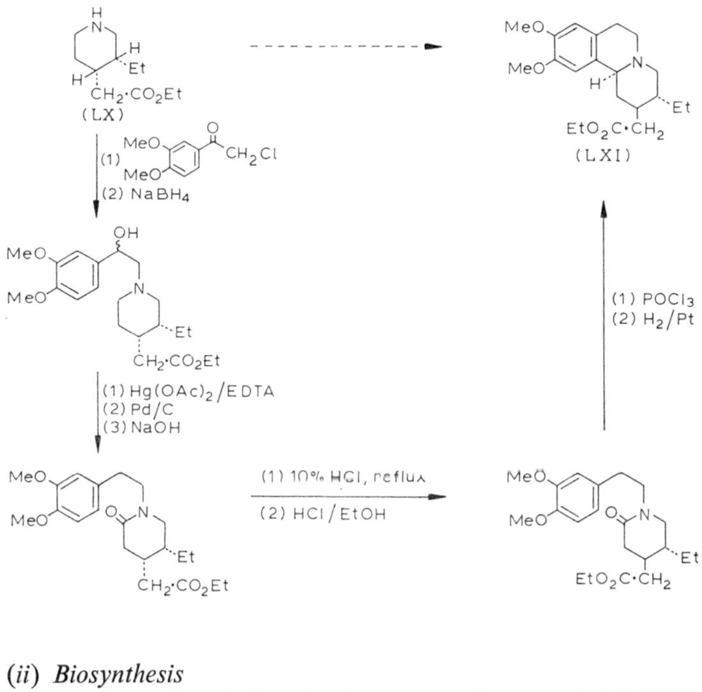

(ii) Biosynthesis

It has been found that, in *C. ipecacuanha,* geraniol (LXII) is converted into **loganin,** which is then a precursor of ipecoside and cephaeline (*Battersby* and *B. Gregory,* Chem. Comm., 1968, 134; *Battersby, R. S. Kapil* and *R. Southgate, ibid.,* 1968, 131; *D. Arigoni et al., ibid.,* 1968, 136 and 137).

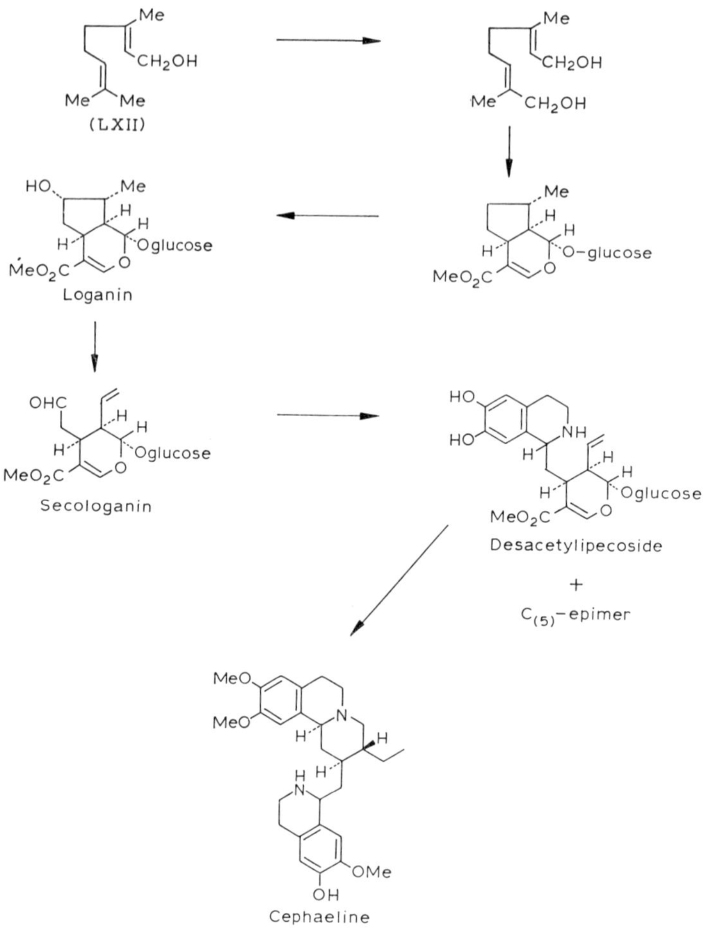

(LXII)

Loganin

Secologanin

Desacetylipecoside

+

$C_{(5)}$-epimer

Cephaeline

Loganin is converted into **secologanin,** which is then incorporated into desacetylipecoside and desacetylisoipecoside; the former, but not the latter, may then be transformed into ipecoside, cephaeline and emetine. A stereochemical inversion ($C_{(5)} \rightarrow C_{(11b)}$) occurs (*Battersby* and *R. J. Parry*, Chem. Comm., 1971, 901).

13. Colchicine

Colchicine is the main active principle of *Colchicum autumnale* L., which has been in use, probably for thousands of years, in the treatment of gout. About 40 years ago the possibility of the use of colchicine in cancer treatment attracted attention. The biological activity of colchicine has been reviewed (*O. J. Eigsti* and *P. Dustin*, "Colchicine in Agriculture, Medicine, Biology and Chemistry", Iowa State College Press, Ames, Iowa, 1955).

Structural work on colchicine began about 1820 and spanned approximately 100 years. The structure I is based upon a vast store of information, an X-ray analysis and several total syntheses. The chemistry of colchicine and similar alkaloids has been reviewed (*W. C. Wildman*, in "Chemistry of the Alkaloids", ed. *S. W. Pelletier*, Van Nostrand–Reinhold, New York, 1970, p. 199; *Wildman*, in "The Alkaloids", 1960, Vol. 6, p. 247; *J. W. Cook* and *J. D. Loudon*, in "The Alkaloids", 1952, Vol. 2, p. 261; *Wildman* and *B. A. Pursey*, in "The Alkaloids", 1968, Vol. 11, p. 407).

(*a*) *Structure determination*

Early work led to II as a possible structure, but the many rearrangements suffered by the molecule of colchicine caused difficulties in the interpretation of the data. The correct structure was deduced very largely from a re-interpretation of earlier data. In this account a very abbreviated summary of the evidence is presented and the incorrect structures will receive very little further attention.

Colchicine analysed for $C_{22}H_{25}NO_6$, and is not an alkaloid in the narrow meaning of the term because it is neutral, not basic. Hydrolysis of colchicine by acids yields methanol and **colchiceine**, $C_{21}H_{23}NO_6$. The initial interpretation was that a methyl ester was undergoing hydrolysis, but ultimately it was realised that colchicine contains an enol methyl ether function. When colchiceine was reacted with diazomethane, the expected colchicine was accompanied by **isocolchicine.** The latter can be hydrolysed by acids to colchiceine and **allocolchicine.** Further acidic hydrolysis of colchiceine yielded acetic acid and a primary amine, trimethylcolchinic acid, $C_{19}H_{21}NO_5$. This amino acid contains three methoxyl groups which may be successively hydrolysed by acids. The sixth oxygen function of the original colchicine was eventually shown to be part of a particularly unreactive carbonyl group. These data can be summarised in the expression III. Colchicine therefore contains eleven double-bond equivalents. Catalytic hydrogenation of the alkaloid yields hexahydrocolchicine which still contains one double bond (since it can be epoxidised), but is probably tetra-substituted, since it is resistant to reduction. In this reduction (to the hexahydro derivative) a new hydroxyl group is formed; it must have come from the carbonyl system. Since colchicine can be oxidised to 3,4,5-trimethoxyphthalic acid, it must contain an aromatic ring. From the number of double-bond equivalents and the four double bonds found, together with the amide carbonyl group, colchicine must be tricyclic.

N-Acetylcolchinol methyl ether, IV, can be prepared from colchiceine by the action of alkaline hydrogen peroxide and subsequent methylation of the reaction product with diazomethane. Hydrolysis of IV yields colchinol methyl ether V. The latter can be oxidised to 4-methoxyphthalic acid. It was not appreciated until much later that a ring contraction has occurred in going from colchiceine to IV. A ring contraction is also involved in the formation of allocolchicine (VI), when colchicine is treated with sodium methoxide in methanol.

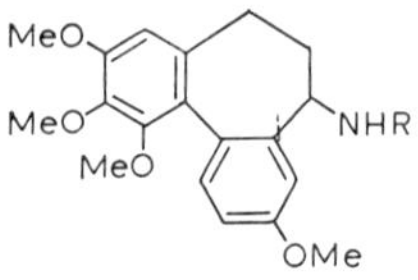

MeO MeO MeO NHR OMe (IV) R = Ac (V) R = H

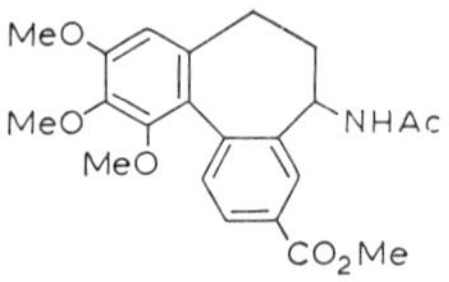

MeO MeO MeO NHAc CO₂Me (VI)

When IV is treated with phosphorus pentoxide, the acetamide group is lost to yield deaminocolchinol methyl ether (VII) and its double-bond isomer VIII. Both yield the same dihydro derivative on catalytic hydrogenation. Hofmann degradation of IV gives VII only, and oxidation of this material with potassium dichromate leads to IX as the major product, with some X as well. These structures were established by synthesis, and compound IX showed for the first time the orientation of the three methoxyl groups in ring A of colchicine. The production of X suggested that ring B is seven-membered. This feature was also deduced from the fact that when VII was treated with osmium tetroxide, followed by lead tetra-acetate, cleavage and re-cyclisation occurred to give the aldehyde XII *via* XI. Similar treatment

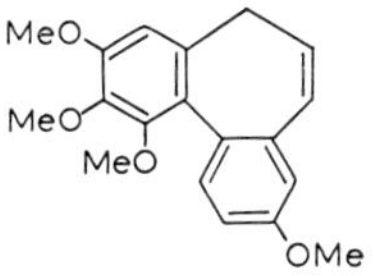

(VII)

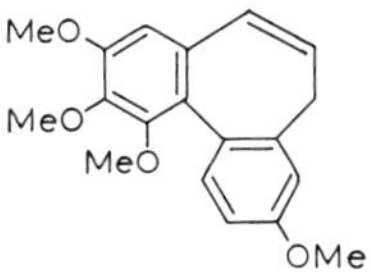

(VIII)

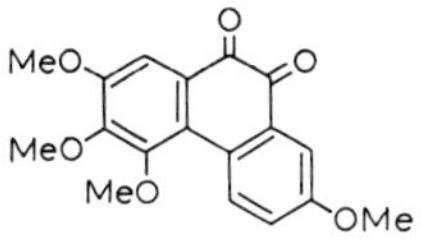

(IX)

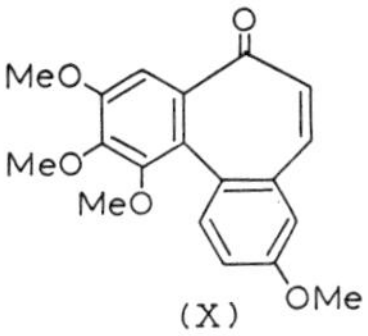

(X)

of VIII led to XIII. The structures of these phenanthrene aldehydes were proved by synthesis.

The synthesis of X was accomplished as shown in XIV → XVI → X.

The structures of *N*-acetylcolchinol methyl ether (IV) and of V, and hence of the seven-membered ring B, were confirmed by synthesis XVII → XXI → V → IV.

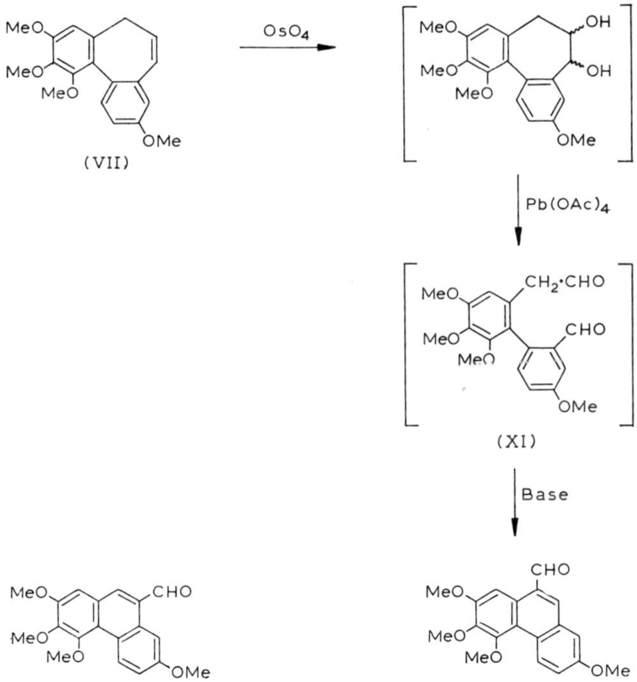

(VII)

OsO_4

$Pb(OAc)_4$

(XI)

Base

(XIII)

(XII)

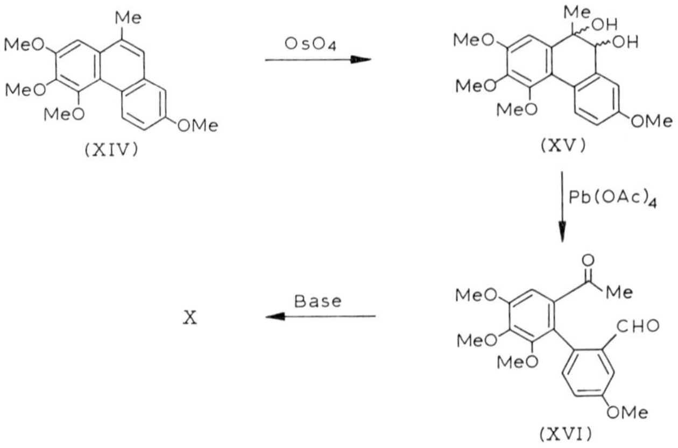
Me
MeO
MeO
MeO
OMe
(XIV)
OsO4
Me
OH
OH
MeO
MeO
MeO
OMe
(XV)
Pb(OAc)4
O
Me
MeO
MeO
MeO
CHO
OMe
(XVI)
Base
X

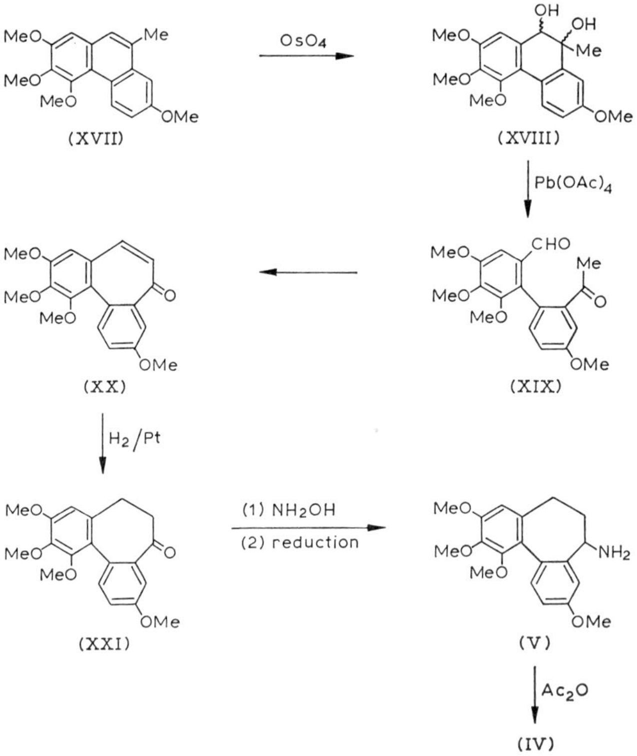
MeO
MeO
MeO
Me
OMe
(XVII)
OsO4
OH
OH
Me
MeO
MeO
MeO
OMe
(XVIII)
Pb(OAc)4
CHO
Me
O
MeO
MeO
MeO
OMe
(XIX)
MeO
MeO
MeO
O
OMe
(XX)
H2/Pt
MeO
MeO
MeO
O
OMe
(XXI)
(1) NH2OH
(2) reduction
MeO
MeO
MeO
NH2
OMe
(V)
Ac2O
(IV)

N-Benzoyltrimethylcolchinic acid can be oxidised to the anhydride XXII which, with phosphorus pentoxide is transformed into deaminocolchinic acid anhydride. The latter structure has been confirmed by synthesis.

Although the above evidence seemed to suggest that ring C of colchicine is six-membered, it was *Dewar* who, after reviewing the relevant data, first proposed that it is a tropolone ring (*M. J. S. Dewar*, Nature, 1945, **155,**

141; see also *P. L. Pauson*, Chem. Reviews, 1955, **55,** 9; *T. Nozae*, Fortschr. Chem. org. Naturstoffe, 1956, **13,** 247; *Cook* and *Loudon*, Quart. Reviews, 1951, **5,** 99). Two structures for the alkaloid then required consideration, I and XXIII; one of these represents colchicine, the other isocolchicine. Considerable difficulty was experienced in deciding between these structures. The X-ray analysis did show that rings B and C are seven-membered but did not resolve the disposition of carbonyl and methoxyl groups in ring C. The problem was solved by the preparation of a series of hydrogenated colchicines that retained the carbonyl oxygen atom in the same position

as it occupied in colchicine itself (*H. Rapoport et al.*, J. Amer. chem. Soc., 1954, **76,** 3693; *Rapoport* and *J. B. Lavigne*, ibid., 1955, **77,** 667; *Rapoport, J. E. Campion* and *J. E. Gordon*, ibid., p. 2389). Reduction of colchicine leads to the production of XXIV, which can also be obtained from certain derivatives, *e.g.* XXV, as summarised in Scheme 151. The position of the

SCHEME 151

REDUCTION PRODUCTS OF COLCHICINE

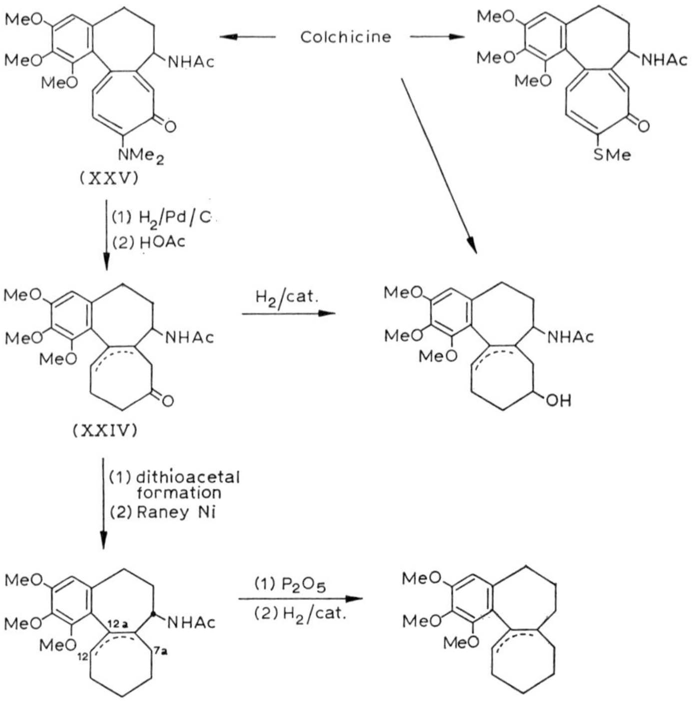

Colchicine

(XXV)

(1) H$_2$/Pd/C
(2) HOAc

(XXIV)

H$_2$/cat.

(1) dithioacetal
   formation
(2) Raney Ni

(XXVII)

(1) P$_2$O$_5$
(2) H$_2$/cat.

remaining double bond was still not certain until synthesis of XXVI was achieved by the route outlined in Scheme 152 (*H. J. E. Loewenthal*, J. chem. Soc., 1961, 1429). The ethylene acetal of XXVI was found to be identical with that produced from colchicine.

The synthesis of XXVII has been achieved as part of a total synthesis of colchicine (see p. 244). The compound XXVIII with a $\Delta^{7a,12a}$ double bond was produced unambiguously, and it differs from the substance formed as

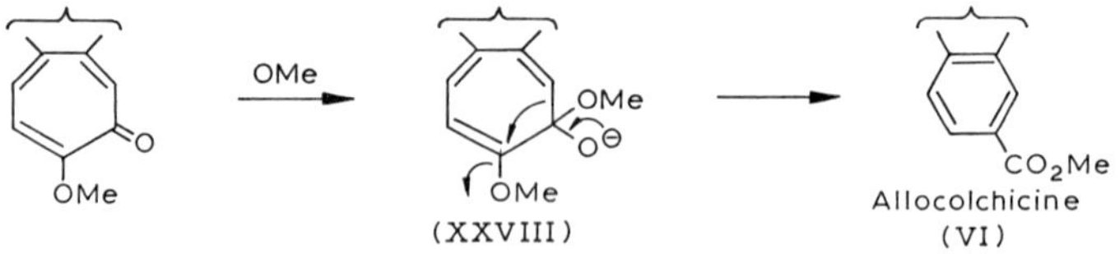

(XXVIII)

Allocolchicine
(VI)

shown in Scheme 151. However, treatment with boron trifluoride caused isomerisation of the double bond to the $\Delta^{12,12a}$ position and this proved to be identical with the product obtained from colchicine (G. *Sunagawa*, T. *Nakamura* and J. *Nakazawa*, Chem. and Pharm. Bull., Japan, 1961, **9**, 81).

The transformation of colchicine (I) into allocolchicine (VI) can be seen as an example of the benzylic acid rearrangement, *via* XXVIII.

SCHEME 152

SYNTHESIS OF TETRAHYDRODEMETHOXYCOLCHICINE

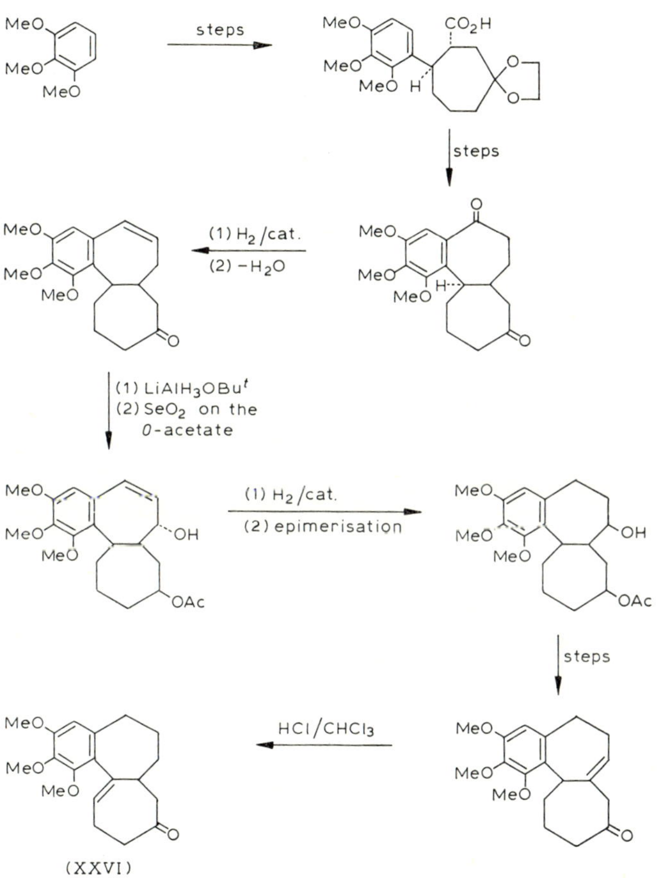

(XXVI)

(XXIX)

The absolute configuration of colchicine was established by oxidation to *N*-acetyl-L-glutamic acid (XXIX) (*H. Corrodi* and *E. Hardeggar*, Helv., 1955, **38**, 2030).

Colchiceine is believed to be stabilised in the iso series (*R. M. Horowitz* and *G. E. Ullyot*, J. Amer. chem. Soc., 1952, **74**, 587):

(V)

HNO$_2$

Carbinol I
C$_{19}$H$_{22}$O$_5$

Carbinol II

−H$_2$O

Isodeaminocolchinol
methyl ether

(VII)

(XXX)

(XXXI)

An interesting and puzzling situation arose when colchinol methyl ether (V) was treated with nitrous acid (*Cook, J. Jack* and *Loudon*, J. chem. Soc., 1952, 607). Two carbinols were produced, "carbinol I", $C_{19}H_{22}O_5$ and an optically inactive "carbinol II". Dehydration of carbinol I yielded a mixture of deaminocolchinol methyl ether (VII) and isodeaminocolchinol methyl ether. The alcohol XXX was synthesised and shown not to be either carbinol I or II. The structure of carbinol II was shown by synthesis to be XXXI, and it was then realised that carbinol I is the laevo isomer of XXXI. Thus, treatment of V with nitrous acid has involved a Demjanov ring B contraction reaction to form carbinols I and II, and conversion of carbinol I into VII has involved a ring-expansion reaction.

(b) Optical properties

(i) Rotation

A number of factors seem to affect the specific rotation of colchicine and some of its derivatives, and solvent variation is one of the most important (*P. Bellet* and *P. Regnier*, Ann. pharm. franç., 1952, **10**, 340). Some derivatives of isocolchicine exhibit mutarotation (*Rapoport* and *Lavigne*, J. Amer. chem. Soc., 1956, **78**, 2455; *R. F. Raffauj, E. F. Bumbier* and *Ullyot*, ibid., 1954, **76**, 1707; *L. Velluz* and *G. Muller*, Bull. Soc. chim. Fr., 1955, 198). An explanation has been made involving the non-coplanarity of rings A and C (*Rapoport, R. H. Allen* and *M. E. Cisney*, J. Amer. chem. Soc., 1955, **77**, 670). Usually isocolchicine and its ring C derivatives have considerably higher negative rotation values than corresponding derivatives of colchicine (*M. Sorkin*, Helv., 1946, **29**, 246; *F. Santavy*, Chem. Listy, 1952, **46**, 280; *Horowitz* and *Ullyot, loc. cit.*).

MeO
MeO
MeO
H
NH₂
OH
Colchinol

Both colchicine and isocolchicine exhibit double negative Cotton effects (330 and 280 nm). The effect at 330 nm is probably related to the absorption of the tropolone ring, since it is absent from colchinol. The band at 280 nm corresponds to the K band of the biaryl system, and the sign agrees with that predicted from the X-ray analysis (*Santavy et al.*, Coll. Czech. chem. Comm., 1964, **29**, 2822).

(ii) Nuclear magnetic resonance

In their review ("The Alkaloids", 1968, **11**, 420) *Wildman* and *Pursey* summarised the data collected by *V. Delaroff* and *P. Rathle* (Bull. Soc. chim. Fr., 1965, 1621). The aromatic and tropolone methoxyls absorb as singlets at about 3.9 δ, and the $C_{(4)}$-H at 6.55 δ. Neither of these types of protons are influenced by changes in ring C or at $C_{(7)}$. The proton at $C_{(8)}$ varies in the normal and the iso series, and is also affected by changes at $C_{(7)}$. The effect due to these changes at $C_{(7)}$ are more marked in the iso-series. The $C_{(11)} + C_{(12)}$ protons constitute an AB quartet which shifts slightly in δ-value with changes in ring C-substituents and also with changes at $C_{(7)}$.

(c) Colchicine-like alkaloids

Several alkaloids are known, the structures of which are variations on the colchicine skeleton, for example **cornigerine** (*A. D. Cross et al.*, Coll. Czech. chem. Comm., 1964, **29**, 1187) and "**substance CC-12**" (*idem, ibid.*, 1966, **31**, 374). However, one of the most important developments, because of its bearing upon the biosynthesis of colchicine (p. 252) was the isolation of **androcymbine**, $C_{21}H_{25}NO_5$ from *A. melanthioides*. The structure (XXXII) was deduced mainly from physical, but also from some chemical evidence

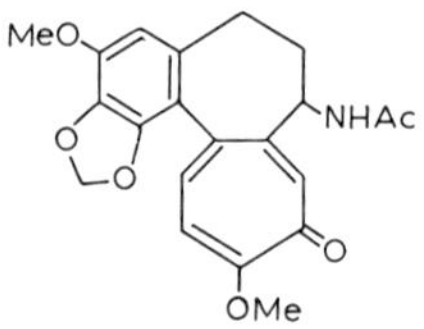

Cornigerine 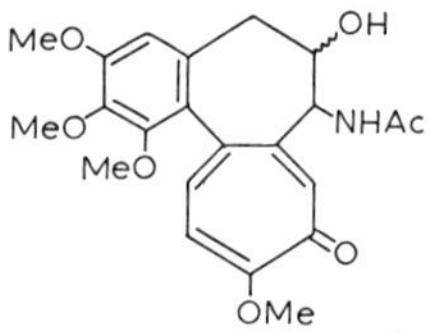   "Substance CC-12" 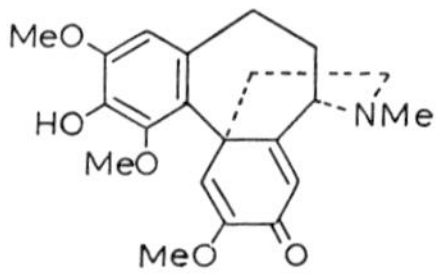   Androcymbine (XXXII)

(*A. R. Battersby* and *R. B. Herbert*, Chem. Comm., 1965, 228; J. chem. Soc., Perkin I, 1972, 1730, 1736, 1741). The u.v. and i.r. spectra (especially bands at 1615, 1635 and 1665 cm^{-1}) are distinctive of a cross-conjugated cyclohexadienone system. The u.v. spectrum also suggested the presence of an additional isolated aromatic chromophore. The presence of *N*-methyl and phenolic hydroxyl groups were deduced by chemical methods, and confirmed by n.m.r. data. In addition to signals for two OMe groups and one aromatic proton, the spectrum exhibited two singlet olefinic proton absorptions at 6.27 and 6.83 δ. Reduction of *O*-methylanhydrocymbine with sodium in liquid ammonia gave the 1-phenethyltetrahydroisoquinoline (XXXIII), the structure of which was established by synthesis. The stereochemistry explicit in XXXII was allocated on the basis of a positive Cotton effect in the 265–278 nm region, together with the observation that anhydrocymbine and salutaridine (XXXIV), of known configuration, have opposite Cotton effects at 270 nm. Thus, anhydrocymbine may be classified as a homomorphinanodienone alkaloid (see C.C.C., Vol. IV G, p. 319).

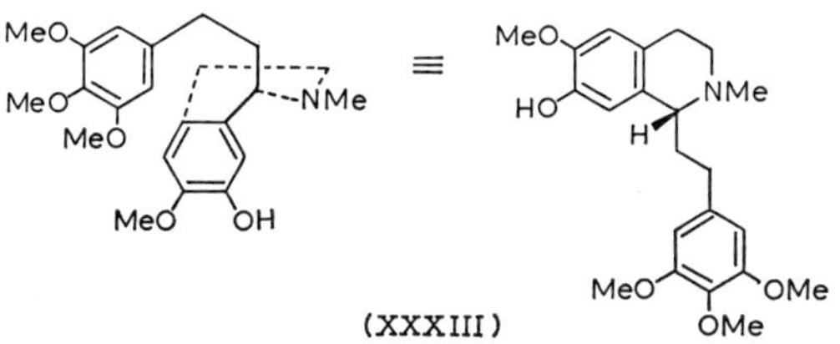

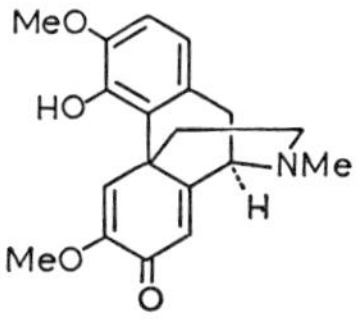

Three methods have been developed for the synthesis of anhydrocymbine-type compounds; phenol oxidation (*T. Kametani et al.*, J. chem. Soc., C, 1969, 1295), Pschorr reaction (*idem, ibid.*, 1968, 3084), and the photolysis of diazonium salts (*Kametani, M. Koizumi* and *K. Fukumoto, ibid.*, 1971, 1792; J. org. Chem., 1971, **36**, 3729). The last method is illustrated in Scheme 153.

SCHEME 153

A SYNTHESIS OF ANHYDROCYMBINE

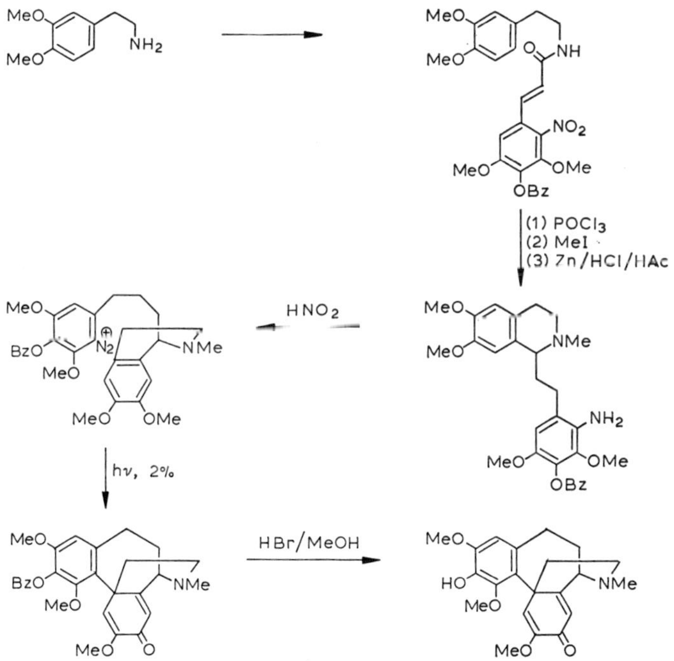

(d) *The synthesis of colchicine*

(i) *Chemical syntheses*

A number of syntheses of colchicine have been invented during the last few years. They may be classified, in the main as A → AB → ABC or A → AC → ABC types depending upon the order of construction of the rings. Most of the routes end at (±)colchiceine or (±)desacetylcolchiceine, since both of these substances had previously been converted into colchicine.

One example of the A → AB → ABC approach (Scheme 154) is due to *Van Tamelen et al.* (J. Amer. chem. Soc., 1959, **81,** 6341; Tetrahedron, 1961, **14,** 8).

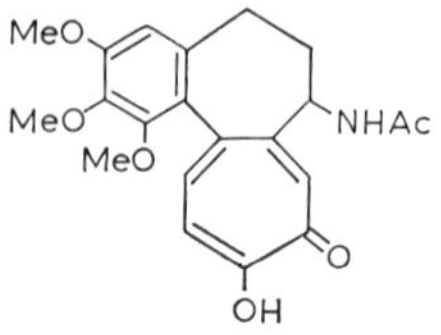

(±)Colchiceine

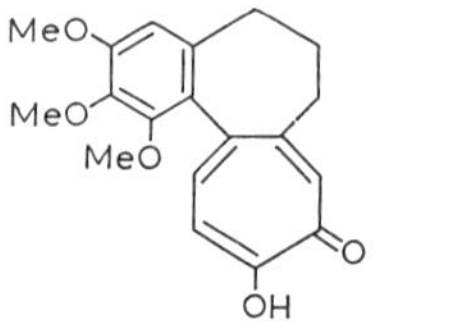

(±)Desacetylcolchiceine

In Eschenmoser's synthesis the starting material was purpurogallin tri-methyl ether (Scheme 156, p. 248) (*A. Eschenmoser et al.*, Helv., 1961, **44,** 540). Note that the free hydroxyl group is important in the reaction of XXXV with methyl propiolate, since it proceeds as shown in XXXV → XXXVII → XXXIX → XXXVI:

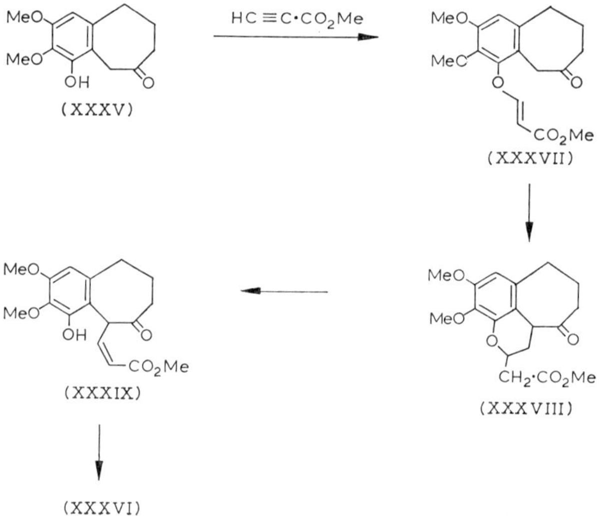

$HC \equiv C \cdot CO_2Me$

(XXXV)

(XXXVII)

(XXXVIII)

(XXXIX)

(XXXVI)

SCHEME 154

VAN TAMELEN'S SYNTHESIS OF COLCHICINE

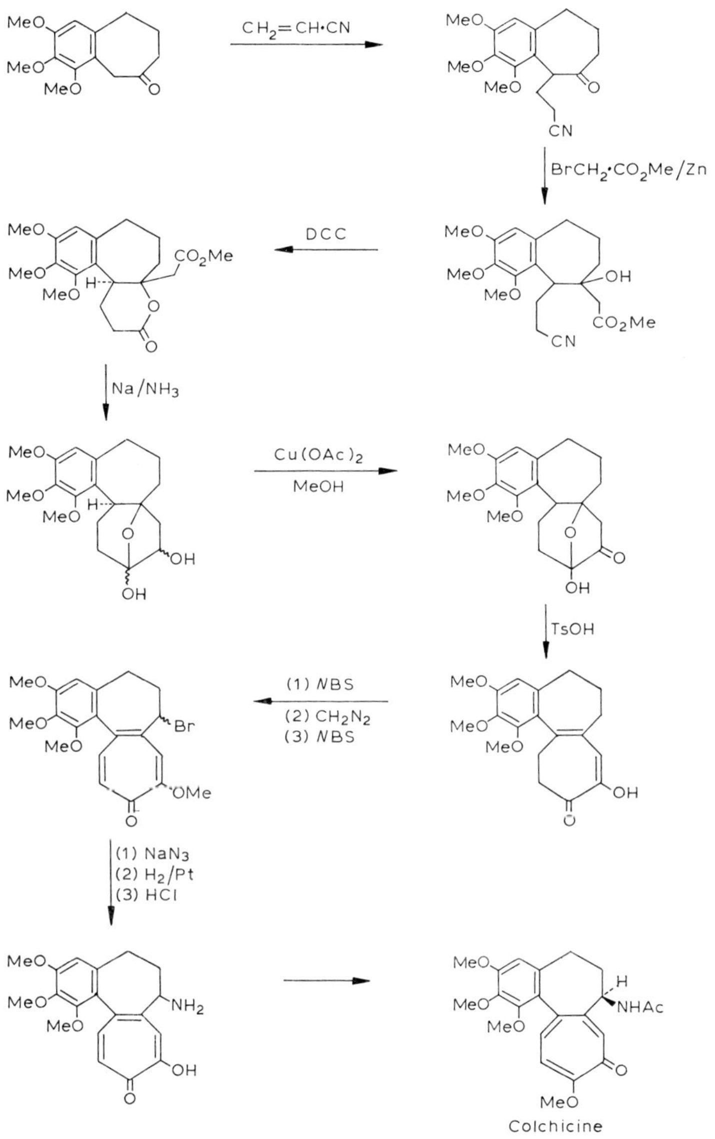

SCHEME 155

WOODWARD'S SYNTHESIS OF COLCHICINE

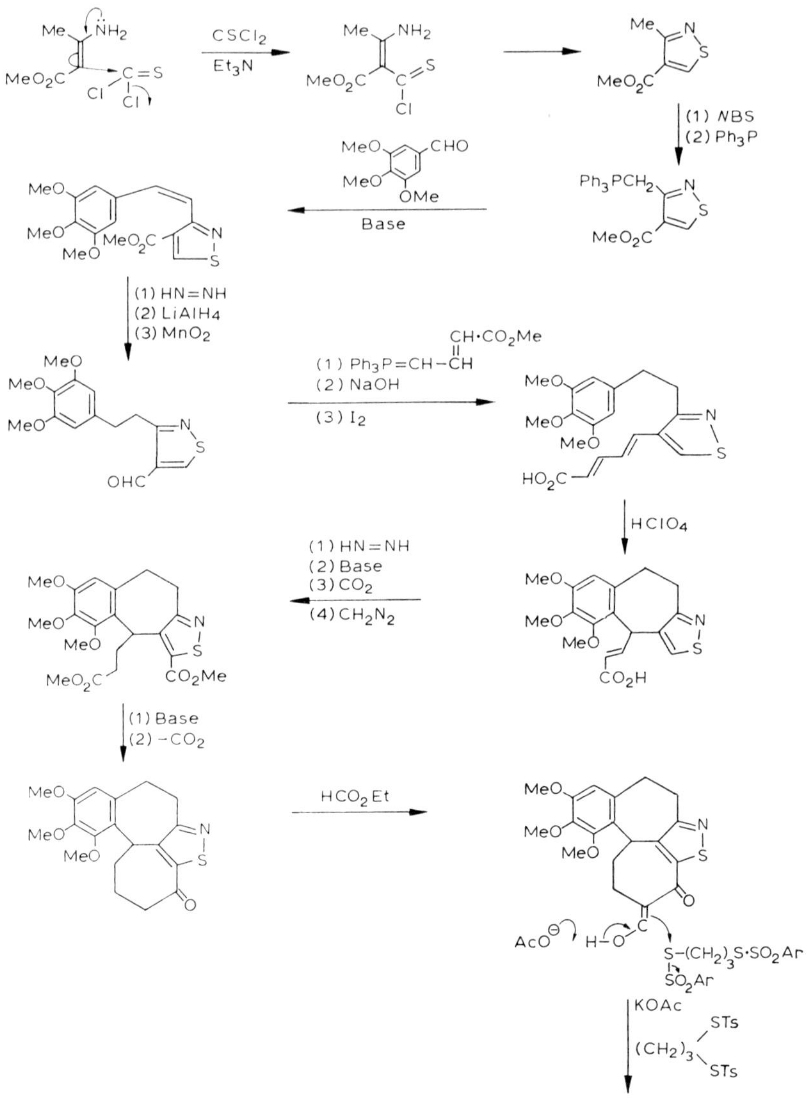

SCHEME 155 (II)

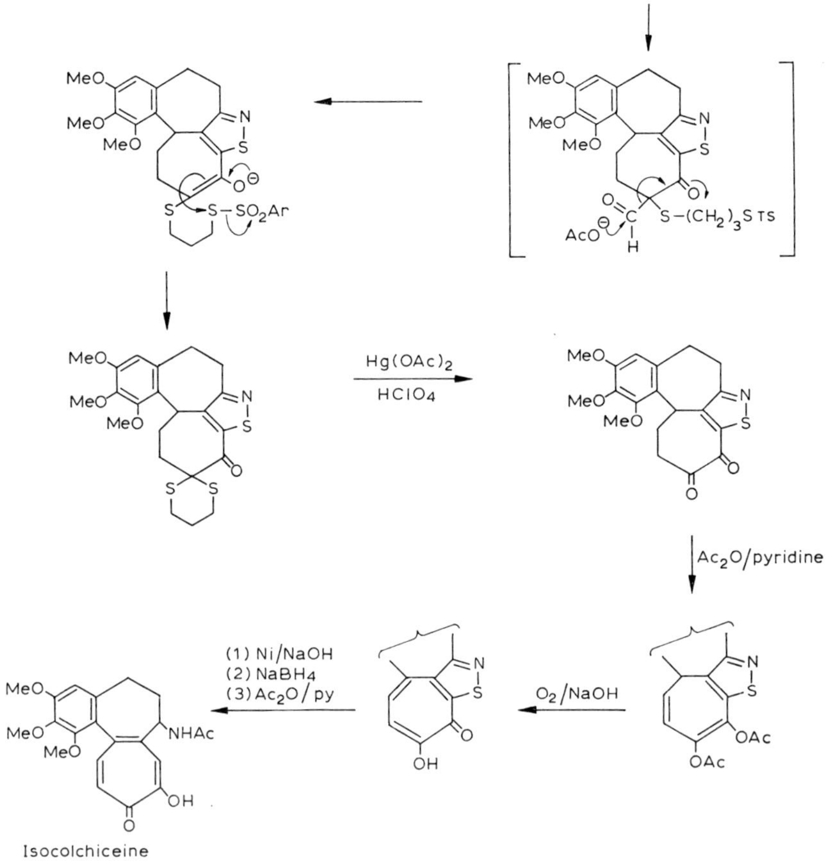
MeO
MeO
MeO
SO2Ar
S-(CH2)3STs
AcO
Hg(OAc)2
HClO4
Ac2O/pyridine
O2/NaOH
(1) Ni/NaOH
(2) NaBH4
(3) Ac2O/py
NHAc
OH
OAc
Isocolchiceine

Woodward's synthesis is quite different from the other two since it utilises isoxazole chemistry (*R. B. Woodward*, The Harvey Lectures, 1963, **59**, 31) (Scheme 155).

A. I. Scott's synthesis is based upon phenolic coupling (*Scott et al.*, J. Amer. chem. Soc., 1963, **83**, 3040) (Scheme 157).

A recent Japanese synthesis illustrates another way to build up ring C (*E. Kotani*, *F. Miyazaki* and *S. Tobinaga*, Chem. Comm., 1974, 300) (Scheme 158, p. 250).

SCHEME 156

ESCHENMOSER'S SYNTHESIS OF COLCHICINE

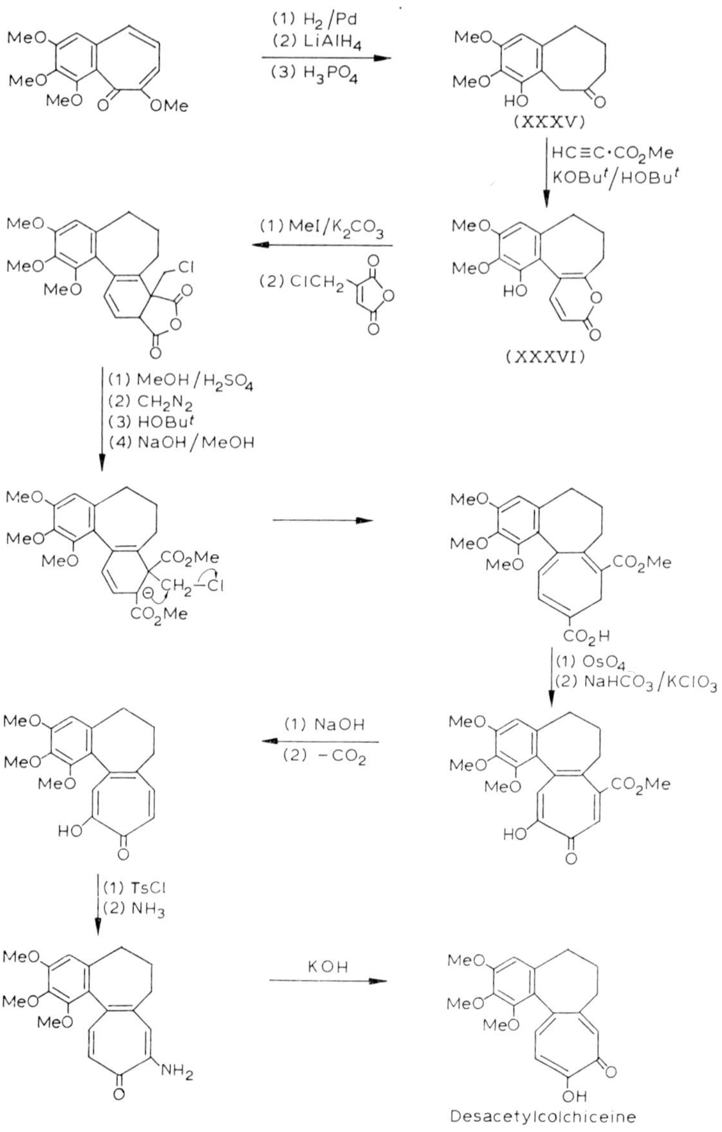

SCHEME 157

SCOTT'S SYNTHESIS OF COLCHICINE

(ii) Biosynthesis

The biosynthesis of colchicine has attracted considerable interest, especially the origin of the tropolone ring C. A number of hypotheses have been advanced (*Wildman* and *Pursey*, in "The Alkaloids", 1968, Vol. 11, p. 448) and a considerable amount of feeding experiments have been performed, especially using *C. autumnale* and *C. byzantium*. The results of these studies, and the degradation schemes used to establish the positions of the ^{14}C-labels have been summarised by *Wildman* and *Pursey, loc. cit.*). The essential results appear in Scheme 159 (p. 251), and from these it became apparent that phenylalanine is not converted into tyrosine in *Colchicum*, that phenylalanine is first transformed into cinnamic acid before incorporation into colchicine, and that the tropolone ring is derived by ring-expansion of the aromatic ring of tyrosine, incorporating the benzylic carbon atom. The biosynthetic route to colchicine has been established by the work of *E. Leete* (*Leete* and *P. E. Nemeth*, J. Amer. chem. Soc., 1960, **82,** 6055; 1961, **83,** 2193; *Leete, ibid.*, 1963, **85,** 3666; Tetrahedron Letters, 1965, 333),

SCHEME 158

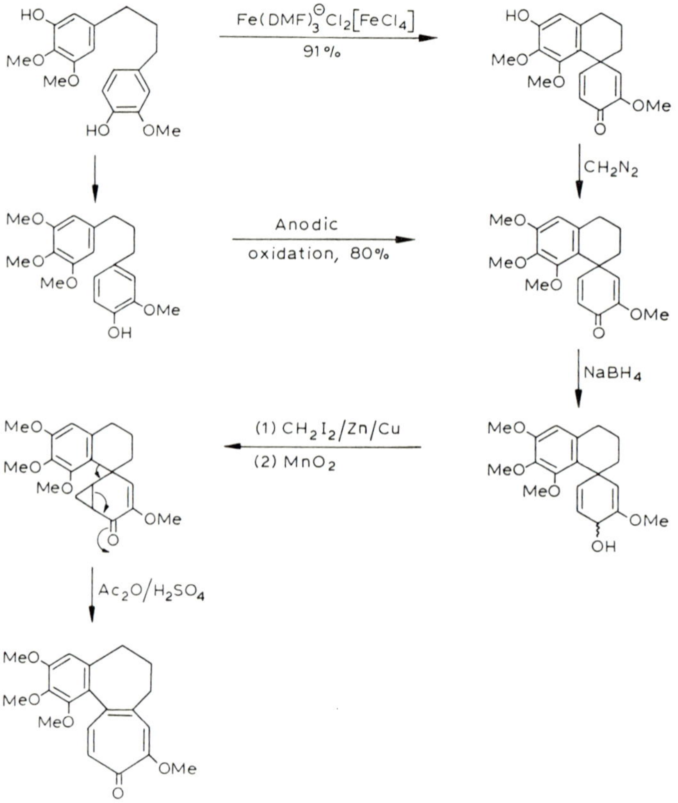

Fe(DMF)$_3^{\ominus}$Cl$_2$[FeCl$_4$]

91%

CH$_2$N$_2$

Anodic oxidation, 80%

NaBH$_4$

(1) CH$_2$I$_2$/Zn/Cu

(2) MnO$_2$

Ac$_2$O/H$_2$SO$_4$

Desacetylamidoisocolchicine

Scott (*Scott, H. Guilford* and *E. Lee*, J. Amer. chem. Soc., 1971, **93**, 3534 and refs. therein; *Scott* and *K. J. Wiesner*, Chem. Comm., 1972, 1075), but principally of *Battersby* and his co-workers (Proc. chem. Soc., 1960, 346; 1964, 86, 260; J. chem. Soc., 1964, 4257; Chem. Comm., 1965, 228; 1966, 603; 1967, 390; Tetrahedron Letters, 1974, 3315; J. chem. Soc., Perkin I, 1972, 1730, 1736, 1741, 2355; Pure and appl. Chem., 1967, **14**, 117).

These results were initially interpreted in terms of Scheme 160 for the biosynthesis of colchicine. The initial precursor XL was thought to undergo phenol coupling to give XLI, which then can be ring-expanded, *via* XLII to the alkaloid. However, tritiated XL was not incorporated into colchicine. An analogy did exist for the ring-expansion reaction (*O. Chapman* and *P. Fitton*, J. Amer. chem. Soc., 1963, **85**, 41) (XLIII → XLIV → XLV).

SCHEME 159

$[3-^{14}C]$ Phenylalanine $\longrightarrow$ $C_{(5)}$

$[2-^{14}C]$ Phenylalanine $\longrightarrow$ $C_{(6)}$

$[1-^{14}C]$ Phenylalanine $\longrightarrow$ $C_{(7)}$

$[2-^{14}C]$ Cinnamic acid $\longrightarrow$ $C_{(6)}$

$[3-^{14}C]$ Cinnamic acid $\longrightarrow$ $C_{(5)}$

$[1-^{14}C]$ Sodium acetate $\longrightarrow$ $N-CO \cdot CH_3$

$[3-^{14}C]$ Tyrosine $\longrightarrow$ $C_{(12)}$ (80%)

$\left.\begin{array}{l} [1-^{14}C] \text{ Tyrosine} \\ [2-^{14}C] \text{ Tyrosine} \end{array}\right\}$ no specific incorporation

$\longrightarrow$ $C_{(9)}$

SCHEME 160

(XL)

(XLI)

(XLII)

Colchicine

SCHEME 161

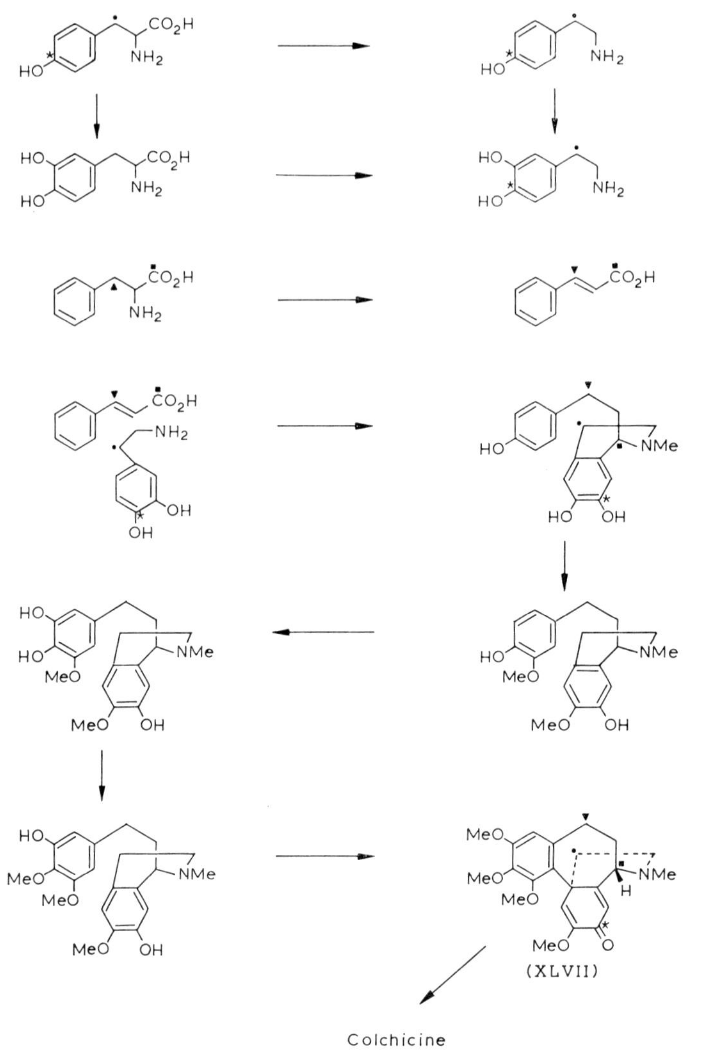

At this point a new alkaloid, **androcymbine,** was isolated from *C. Autumnale* and its structure was quickly established as XXXII (see p. 242), and its biogenesis from a 1-phenethylisoquinoline derivative, **autumnaline** (XLVI), seemed plausible. When tritiated *O*-methylandrocymbine was fed to *C. autumnale* plants, very high and specific incorporation into colchicine

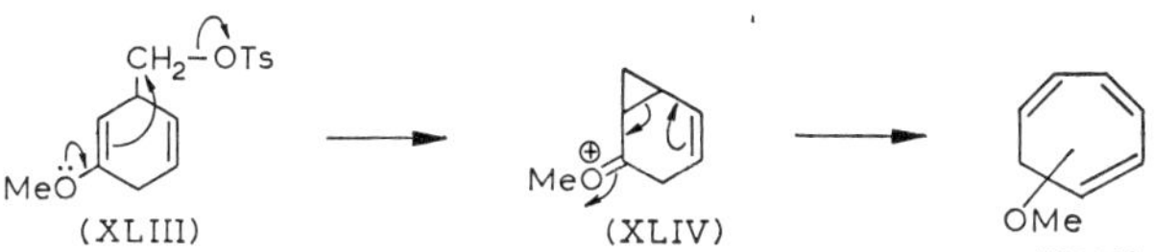

(XLIII) (XLIV) (XLV)

was observed. Another important piece of evidence was that 15-*N*-tyrosine was incorporated intact into the alkaloid. [9-^{14}C]Autumnaline was efficiently and specifically incorporated into colchicine, hence establishing that colchicine is a modified 1-phenethylisoquinoline alkaloid. Furthermore the *S*-isomer (as XLVI) is incorporated far more efficiently than its enantiomer.

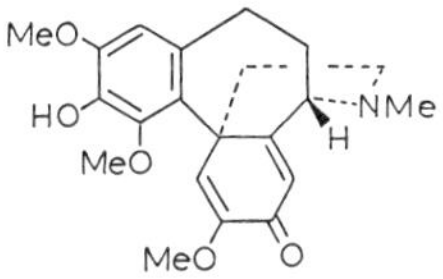

Androcymbine
(XXXII)

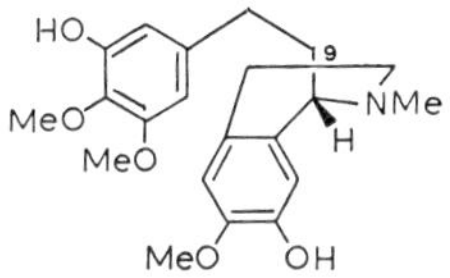

Autumnaline
(XLVI)

The sequence of the biosynthesis of colchicine can now be summarised as Scheme 161. The order of events of hydroxylation and methylation have been examined by the administration of 1-phenethylisoquinolines variously substituted with hydroxyl and methoxyl groups. Variously substituted

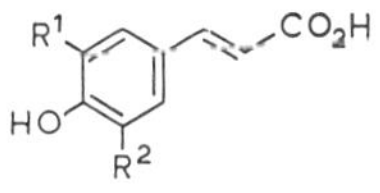

(XLVIII)

R^1	R^2
H	H
OMe	H
OH	H
OH	OMe

cinnamic acids (XLVIII) were also studied, but none was incorporated. It is not known whether reduction of the cinnamic acid double bond occurs before or after construction of the isoquinoline derivative.

The ring-expansion of XLVII may occur *via* XLIX, L and LI, where OX is a good leaving group, possibly phosphate.

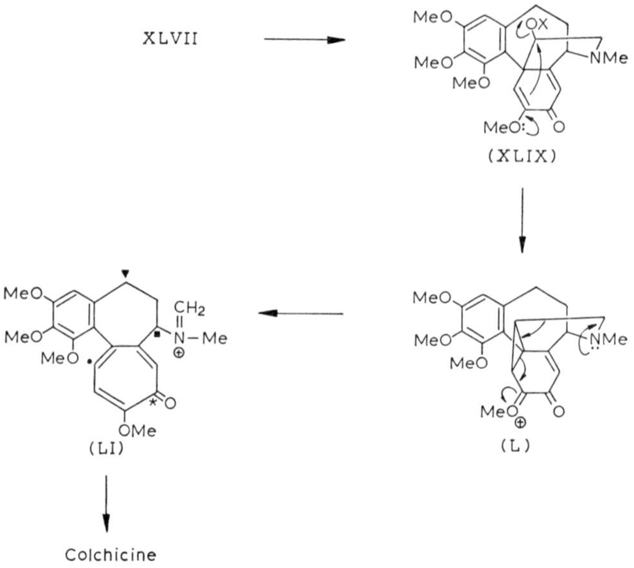

XLVII ⟶ (XLIX)

(LI) (L)

Colchicine

14. 1-Phenethylisoquinolines

The biosynthesis of colchicine was found eventually to involve a 1-phenethyltetrahydroisoquinoline as precursor (p. 252), and an interesting development has been the realisation that a number of alkaloids exist which are based upon the 1-phenethylisoquinoline building block in parallel to some of the alkaloids derived from 1-benzylisoquinolines (for a review see *T. Kametani* and *M. Koizumi*, "The Alkaloids", 1973, Vol. 14, p. 265). The types of structures so far recognised are simple 1-phenethyltetrahydroisoquinoline, bisphenethylisoquinoline, homoaporphine, homoproaporphine, homomorphinandienones and homoerythrina, but more types of structure are almost certain to come to light. The latter two types will not be considered further in this account.

(a) Simple 1-phenethyltetrahydroisoquinolines

The only one reported so far is **autumnaline,** isolated from *Colchicum cornigerum* (*F. Santavy et al.*, Coll. Czech. chem. Comm., 1969, **34**, 3540; *A. R. Battersby et al.*, Chem. Comm., 1969, 1066; *Kametani et al.*, J. chem. Soc., C, 1968, 271). The structure was established by comparison with an

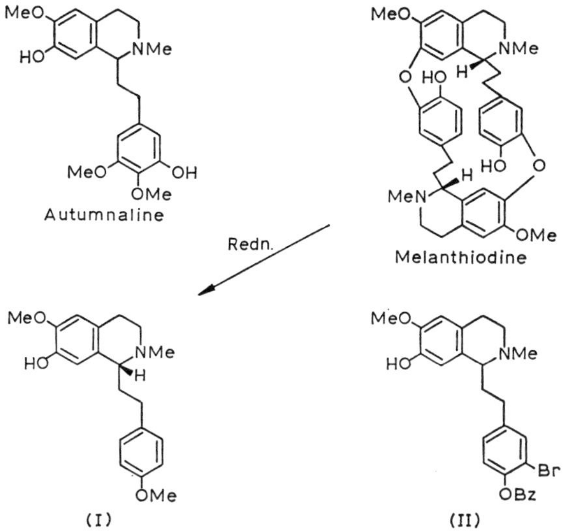

Autumnaline

Redn.

Melanthiodine

(I)

(II)

authentic specimen. The biosynthesis of the alkaloid has not yet been firmly established, although it is presumed to arise from dopamine and a cinnamic acid derivative.

(b) Bisphenethyltetrahydroisoquinolines

So far only one has been fully characterised, **melanthioidine** (*Battersby, R. B. Herbert* and *Santavy*, Chem. Comm., 1965, 415; *Battersby et al.*, J. chem. Soc., C, 1967, 1739). The structure followed from the observation that reduction of the alkaloid with sodium in liquid ammonia gave the 1-phenethylisoquinoline I almost exclusively; the (*R*)-configuration followed from the o.r.d. spectrum. The alkaloid has been synthesised by carrying out a double Ullmann reaction with II (*idem, loc. cit.; Kametani et al.*, Chem. and Pharm. Bull., Japan, 1968, **16**, 663).

(c) The homoproaporphines and homoaporphines

Three homoaporphine alkaloids are known, and the first of these to be recognised was **kreysigine**, $C_{22}H_{27}NO_5$, which has been found both in the racemic (*Battersby et al.*, Chem. Comm., 1967, 450) and (−)-forms (*Santavy et al.*, Coll. Czech. chem. Comm., 1969, **34**, 3540). The structural assignment was due to *Battersby et al. (loc. cit.)*. The u.v. spectrum was thought to be similar to the spectra of aporphines and the chemical shift positions of three of the four methoxyl groups present are very similar to those observed in aporphines. The fourth methoxyl, with δ 3.59, absorbed at a substantially higher field. The structure III was then considered for kreysigine, and this was confirmed by a synthesis involving oxidation of the diphenolic tetrahydroisoquinoline IV (R = H) to the dienone V, followed by rearrangement to VI. *O*-Methylation gave a mixture of fully methylated material (VII) and kreysigine (III) (*Idem*, Chem. Comm., 1967, 450; *A. Brossi, J. O'Brien* and *S. Teitel*, Helv., 1969, **52,** 678; *Kametani et al.*, J. org. Chem., 1968, **33,** 690). In an alternative synthesis the monophenolic tetrahydroisoquinoline IV (R = Me) was treated with lead tetra-acetate when ($\pm$)*O*-acetylkreysigine was obtained in 18% yield (*O. Hoshino, T. Toshioka* and *B. Umezawa*, Chem. Comm., 1972, 740).

The diphenolic product VI was subsequently shown to be identical with a second homoaporphine alkaloid **multifloramine** which occurs with kreysigine and a third alkaloid of this type, **floramultine** has been isolated from

Kreysigine
(III)

CH_2N_2

(IV)

49% $K_3Fe(CN)_6/KOH$

$H^{\oplus}$

Multifloramine
(VI)

(V)

CH_2N_2

(VII)

Kreysigia multiflora. Floramultine, $C_{21}H_{25}NO_5$, is also diphenolic; the other three oxygen atoms are accounted for in methoxyl groups. Partial *O*-methylation of floramultine gave kreysigine, thus establishing the homoaporphine skeleton and the presence of a phenolic hydroxyl group at $C_{(1)}$. Floramultine exhibits a three-hydrogen singlet at 3.55 δ in the ^{1}H-n.m.r. indicating that a methoxyl group is at $C_{(12)}$. From colour tests it was concluded that the alkaloid does not contain a catechol system, so the second phenolic group must be at $C_{(10)}$. The absolute configuration was established by a synthesis of (−)multifloramine (VI) from the (−)isomer of IV, the configuration of which is known (*Brossi, O'Brien* and *Teitel, loc. cit.*). A study of o.r.d. spectra for homoaporphines and homoproaporphines led to the same conclusions (*A. F. Beecham et al.,* Austral. J. Chem., 1968, **21,** 2829).

Kreysiginone (IX), was the first homoproaporphine to be isolated and examined (*Battersby et al.,* Chem. Comm., 1967, 934). The structure was established when it was shown that one of the two isomeric products of oxidation of the diphenolic tetrahydroisoquinoline VIII is identical with

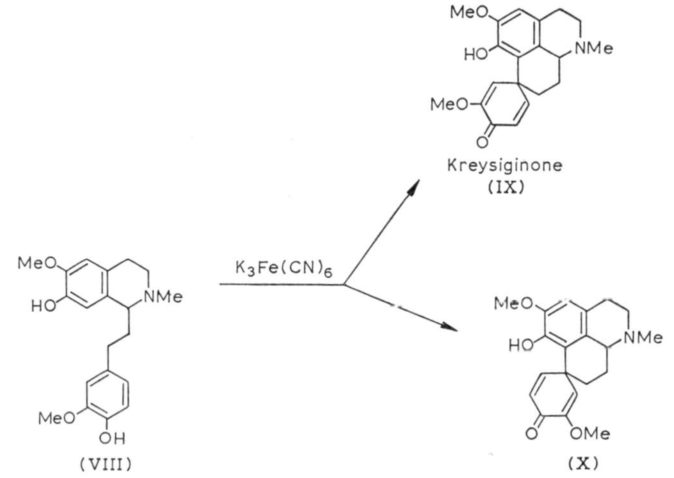

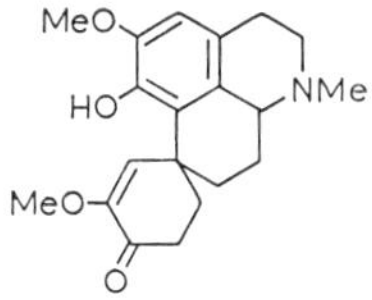

Kreysiginone
(IX)

$K_3Fe(CN)_6$

(VIII)

(X)

(XI)

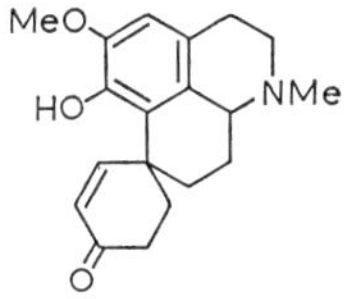

Bulbocodine
(XII)

the alkaloid. Dihydrokreysiginone (XI) was also isolated from the same plant. A further dihydrohomoproaporphine, **bulbocodine** (XII) is known (*Santavy et al.*, Planta Med., 1968, **16**, 357; Helv., 1971, **54**, 1084).

The synthesis of kreysiginone by phenolic oxidation acted as a spur for a number of other oxidative coupling reactions that gave rise to either the homopropaporphine or to the homoaporphine structure. Yields are usually very much higher than in the formation of proaporphines and very little, if any, of the anhydrocymbine type of structure is formed (*Battersby et al.*, Chem. Comm., 1967, 934; *Kametani et al.*, *ibid.*, 1967, 878, 1103; J. chem. Soc., C, 1968, 271, 1003; 1970, 382; J. org. Chem., 1968, **33**, 690; Chem. and pharm. Bull., Japan, 1969, **17**, 814). Photolysis of the diazonium salt XIII yielded the deaminated compound IV (R = Me) as the major product, together with some of kreysigine (III) and a small amount of *O*-methyl-anhydrocymbine (XIV) (*Kametani et al.*, J. chem. Soc., C, 1971, 1923).

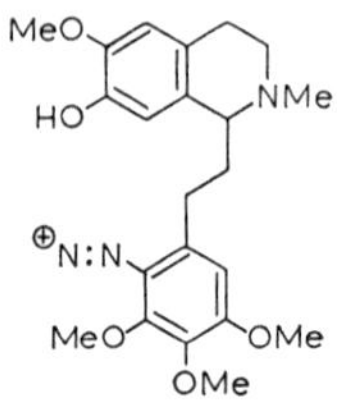

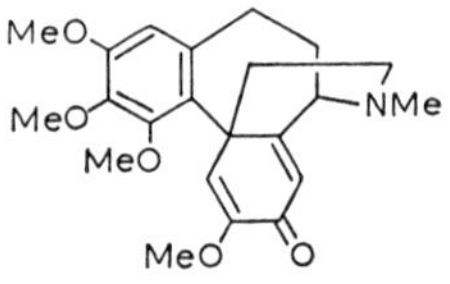

By feeding ^{14}C-labelled autumnaline to *K. multiflora*, it has been established that kreysigine arises in the plant by direct coupling, rather than by initial conversion to the homoproaporphine, followed by rearrangement (*Battersby et al.*, Chem. Comm., 1969, 1066).

Chapter 37

Fused Heterocyclic Systems having a Nitrogen Atom in Common to Two or More Rings

NEIL CAMPBELL

Pyrrole and pyridine may be fused to another ring in such a manner that the annealing atoms are a nitrogen and an adjacent carbon atom. This type of fusion is seen in pyrrolizine (C.C.C., Vol. IV A, p. 80), indolizine (I) and quinolizine (II) in which a pyridine ring is fused respectively to a pyrrole and to a pyridine ring through nitrogen and carbon.

(I)   (II)

1. Indolizines*

(a) Indolizine and its derivatives

A general method for the preparation of 2-substituted indolizines is that of *Tschitschibabin* from 2-picoline and bromoketones such as bromoacetone to yield quaternary salts (*e.g.* III), which undergo ring-closure using sodium hydrogen carbonate to give, for instance, 2-methylindolizine (IV):

(III)   (IV)   (V)   (VI)

---

* Formerly known as pyrrocolines.
W. L. Mosby, "Heterocyclic Systems with Bridgehead Nitrogen Atoms", Part 1, p. 233 (Interscience, New York and London, 1961); *M. H. Palmer*, "The Structure and Reactions of Heterocyclic Compounds", Edward Arnold, London, 1967, pp. 304, 335.

The parent compound can be prepared by the action of acetic anhydride on 2-methylindolizine to give, by a complex mechanism, 1,3-diacetylindolizine (V), which is readily deacetylated by acid to give a 10% yield of indolizine (*M. Scholtz*, Ber., 1912, **45**, 734). Indolizine is also prepared in 50% yield by heating 3-(2-pyridyl)-1-propanol (VI) with palladised charcoal, ring-closure being accompanied by dehydrogenation and loss of water (*V. Boekelheide* and *R. J. Windgassen*, J. Amer. chem. Soc., 1959, **81**, 1456).

The indolizines possess planar molecules encircled by ten π-electrons and sustain a ring current in an applied magnetic field (*W. W. Paudler* and *H. L. Blewitt*, J. heterocycl. Chem., 1966, **3**, 33). The aromatic character thus indicated is attributed to resonance in which the unshared nitrogen electrons participate (Ia $\leftrightarrow$ Ib $\leftrightarrow$ Ic):

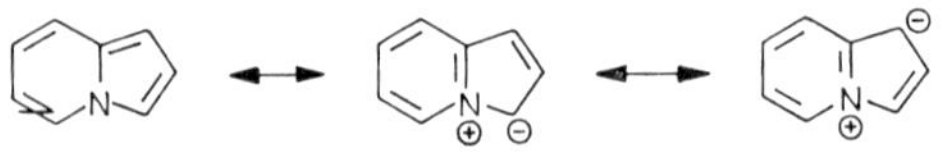

(Ia)   (Ib)   (Ic)

^{13}C Chemical shifts indicate that the bridgehead nitrogen does not disturb π-electron delocalisation in the indolizine molecule and confirm the high degree of aromaticity of indolizine (*R. J. Pugmire et al.*, J. Amer. chem. Soc., 1971, **93**, 1887).

It is not unexpected therefore that indolizines readily undergo electrophilic substitution. Nitrosation (*E. T. Borrows et al.*, J. chem. Soc., 1946, 1075), Friedel–Crafts acylation (*D. E. Ames, T. F. Grey* and *W. A. Jones, ibid.*, 1959, 620) and methylation and formylation (*E. D. Rossiter* and *J. E. Saxton, ibid.*, 1953, 3654) occur at $C_{(3)}$, but nitration mainly at position 1,2-methylindolizine thus yielding 1-nitro-2-methylindolizine (62%) and 3-nitro-2-methylindolizine (1.5%) (*Borrows et al., ibid.*, 1946, 1077). Protonisation likewise occurs at these positions rather than at the nitrogen atom, and again mainly at the 3-position (*D. H. Reid et al., ibid.*, 1962, 3288; B, 1966, 44), but sometimes at $C_{(1)}$ depending on the substituent present (*W. L. F. Armarego, ibid.*, 1966, 191). For example, protonated 1,2,3-trimethylindolizine (the indolizine perchlorate in trifluoroacetic acid) contains a proton τ 4.71 as a quartet. This can only mean that protonisation has occurred at a methyl-bearing carbon atom (*e.g.* VII), in the five-membered ring. That $C_{(3)}$ is the site of protonisation is shown by the value of τ which indicates the expected deshielding by the neighbouring positively charged nitrogen.

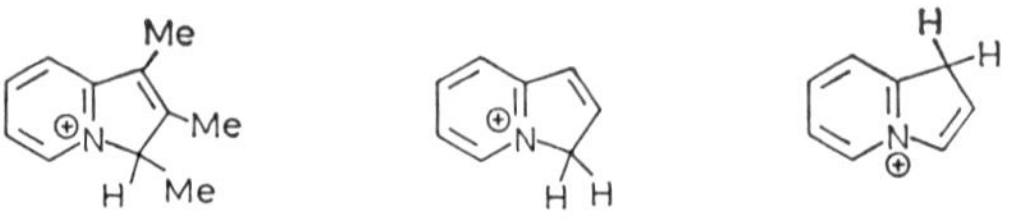

(VII)   (VIII)   (IX)

It is known that methylene protons in VIII absorb at about τ 4.4 while the 1-methylene protons (IX) absorb at about τ 5.8. Certain substituted indolizines such as 3-methylindolizine protonate preferably at $C_{(1)}$, but in dilute hydrochloric acid exist as a mixture of $C_{(1)}$ and $C_{(3)}$ protonated species.

It is of interest that indolizine absorbs in the u.v. region at 237, 282 and 346 nm thus resembling naphthalene which absorbs in this region (*J. D. Bower*, J. chem. Soc., 1957, 4510; *Armarego, ibid.*, 1964, 4227).

Indolizines exhibit many of the properties of pyrrole and indole and give a positive pine splint test (Baeyer test for pyrrole), a violet-red colour when fused with oxalic acid (Angeli test for indoles), and (when positions 1 and 3 are unsubstituted) a blue or violet Ehrlich test.

Indolizine and its derivatives undergo ring-fission when oxidised with perhydrol and this method is very useful in structural determinations (*O. Diels* and *R. Meyer*, Ann., 1934, **513**, 129). For example, oxidation of 2-*p*-acetylphenylindolizine with perhydrol gives picolinic acid *N*-oxide and 4-acetylbenzoic acid *(Burrows et al., loc. cit.)*.

Indolizine, C_8H_7N, is a colourless crystalline solid, m.p. 75°, b.p. 205°, with an odour of naphthalene. It forms a red *picrate*, m.p. 101°, and with benzenediazonium chloride yields 3-benzeneazoindolizine. For the reaction of indolizine with dimethyl acetylenedicarboxylate see p. 272. In the presence of hydrogen bromide and palladium indolizine is hydrogenated to give 1,2-*dihydro*-3H-*indolizinium bromide* (X), m.p. 176–179°, but in the absence of acid the pyridine nucleus is first attacked (*O. G. Lowe* and *L. C. King*, J. org. Chem., 1959, **24**, 1200):

(X)   (XI)   (XII)

Further hydrogenation yields **octahydroindolizine, indolizidine,** b.p. 75°/3 mm, *picrate*, m.p. 228–229°, *hydrobromide*, m.p. 196°, which is identical with the optically inactive base δ-**coniceine**, obtained from (±)coniine (*K. Löffler et al.*, Ber., 1909, **42**, 3420). It is also obtained by the reduction of 3-oxoindolizidine (XI) with sodium and ethanol (*G. R. Clemo* and *G. R. Ramage*, J. chem. Soc., 1932, 2969) or by the mild hydrogenation of β-2-pyridylpropionitrile (XII) with a platinum catalyst (*Boekelheide et al.*, J. Amer. chem. Soc., 1953, **75**, 3243). In the latter reaction the pyridine ring is first reduced, then undergoes intra-molecular alkylation, and finally loses ammonia to yield indolizidine. For another synthesis see *K. Arh-Lipovac* and *R. Seiwerth*, Monatsh., 1953, **84**, 992.

Selective cleavage of the central carbon–nitrogen bond in indolizidine quaternary salts can be effected by various reagents and is exemplified by the Emde reaction on the 9-phenyl derivative (XIII) to give 1-methyl-5-phenylazacyclononane (XIV) when treated with lithium and ammonia (*M. G. Reinecke* and *R. F. Francis*, J. org. Chem., 1972, **37**, 3494):

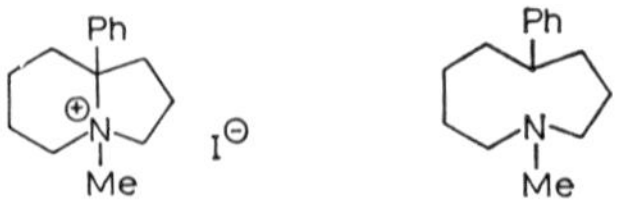

(XIII)   (XIV)

Non-reductive cleavage leaving the pyrrole moiety intact is sometimes encountered, the indoloindolizine derivative XV with potassium cyanide yielding the indole derivative XVI (*G. H. Foster* and *J. Harley-Mason*, Chem. Comm., 1968, 1440):

(XV)   (XVI)

Indolizine in many of its chemical reactions resembles azulene (*M. Fraser* and *Reid*, J. chem. Soc., 1963, 1421). Both substances are polar with the five- and seven-membered rings bearing negative and positive charges respectively. Both possess pronounced electron density at $C_{(1)}$ and $C_{(3)}$ and undergo protonation and electrophilic substitution at these positions (p. 260). In both, methyl groups at $C_{(1)}$ lose hydride ions, while methyl groups at $C_{(4)}$ and $C_{(6)}$ in azulene and $C_{(5)}$ in indolizine are acidic. Finally, like azulene-1-carboxylic acid indolizine-1-carboxylic acid is easily decarboxylated (*D. R. Bragg* and *D. G. Wibberley*, ibid., 1963, 3277).

(b) Benzoindolizines

These substances are described in the literature under a variety of names. The benzoindolizine nomenclature will be used in the text, but in Chemical Abstracts they are designated as derivatives of pyrrole or pyridine, *e.g.*:

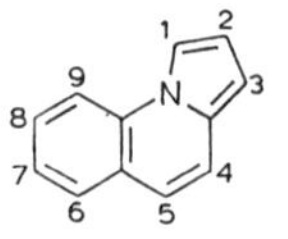

Benzo [*e*] indolizine
Pyrrolo [1,2 -*a*] quinoline

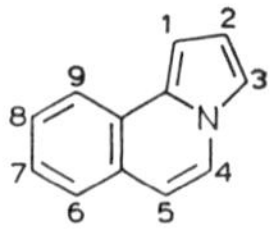

Benzo [*g*] indolizine
Pyrrolo [2,1 *a*] isoquinoline

Benzo[*b*]indolizine (XVIII) is indexed in Chemical Abstracts as pyrido[1,2-*a*]indole and benzo[*a*]indolizine (XXI) as pyrido[1,2-*a*]isoindole.

Two methods of preparation of the benzoindolizines may be mentioned. 2-Benzylpyridine (XVII) when heated with copper gives benzo[*b*]indolizine

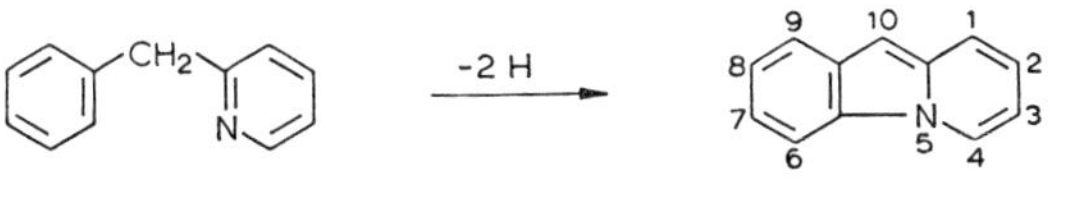

(XVII)   (XVIII)

(XVIII) in 40% yield (*J. von Braun*, Ber., 1937, **70**, 1767), while irradiation of 2-bromo-*N*-benzylpyridinium bromide (XIX) with u.v. light yields 6[*H*]-benzo[*a*]indolizinium bromide (XX), which with sodium carbonate gives **benzo[*a*]indolizine** (XXI), yellow crystals, m.p. 225° (decomp.) (*A. Fozard* and *C. K. Bradsher*, J. org. Chem., 1967, **32**, 2966):

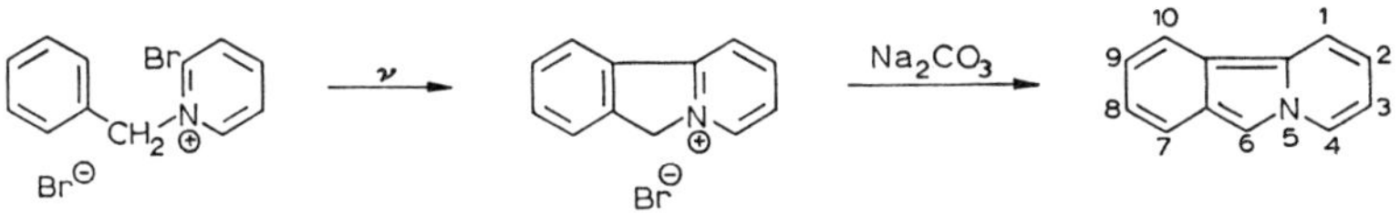

(XIX)   (XX)   (XXI)

The benzoindolizines are weak bases, soluble in concentrated hydrochloric acid but insoluble in dilute acids. Their conversion into indolizinium salts presents points of interest, addition of alkyl halides to benzo[*b*]indolizines, for example, involving addition at $C_{(10)}$. This is shown by 6,9,10-trimethylbenzo[*b*]indolizine (XXII) with ethyl iodide giving the same salt XXIII as 10-ethyl-6,9-dimethylbenzo[*b*]indolizine (XXIV) gives with methyl iodide (*R. Robinson* and *J. E. Saxton*, J. chem. Soc., 1952, 976):

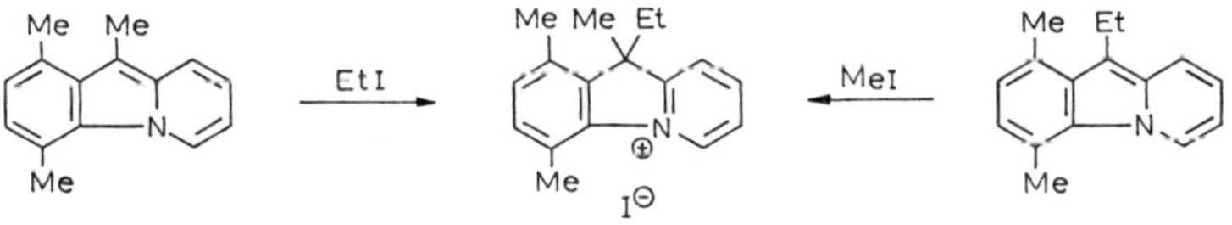

(XXII)   (XXIII)   (XXIV)

Benzo[*b*]indolizine (XVIII), m.p. 175–176°, *picrate*, m.p. 138°; *methiodide*, m.p. 231° (*Y. Arata, T. Ohashi* and *K. Uwai*, J. pharm. Soc., Japan, 1955, **75**, 265; C.A., 1955, **49**, 10279). **Benzo[*e*]indolizine**, m.p. 115–116°, exhibits a light blue fluorescence and gives a blue colour with Ehrlich's reagent (*E. M. Roberts, M. Gates* and *Boekelheide*, J. org. Chem., 1955, **20**, 1442). **Benzo[*g*]indolizine**, m.p. 83.5–84° (*Boekelheide* and *J. C. Godfrey*, J. Amer. chem. Soc., 1953, **75**, 3679). Other methods of synthesising benzo[*e*]indolizines include the action of methyl propiolate on quinoline in methanol to give *methyl benzo*[e]-*indolizine-3-carboxylate*, m.p. 135–136.5° (*R. M. Acheson* and *M. S. Verlander*, J. chem. Soc., C, 1969, 2311. See also *W. J. Irwin* and *Wibberley*, *ibid.*, Perkin I, 1974, 250).

2. Quinolizines and their derivatives*

(a) Quinolizines

Quinolizine, or pyridocoline as it was formerly called, has not been obtained in the pure state. The product obtained from tetraethyl quinolizine-1,2,3,4-tetracarboxylate was shown to be a mixture of quinolizine and indolizine (*O. Diels* and *K. Alder*, Ann., 1933, **505**, 103), while a substance considered to be quinolizine was shown to be 3-methylindolizine (*Diels* and *H. Schrum*, *ibid.*, 1937, **530**, 68).

Quinolizine can be represented by three formulae (I, II and III) and is clearly not aromatic:

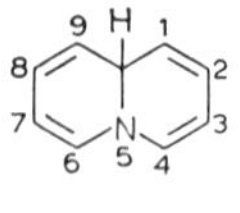

(I) 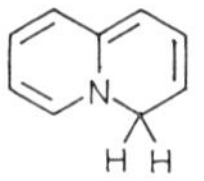   (II) 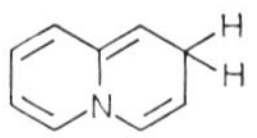   (III)

The 9a*H*-forms (*e.g.* I) exist, but are unstable and readily tautomerise to the stable 4*H*-quinolizines (*e.g.* II), the greater conjugation of which is reflected in their u.v. spectra. Tetramethyl quinolizine-1,2,3,4-tetracarboxylate (IV), for example, isomerises to the 4*H*-quinolizine (V), and the course of the tautomerism can be followed by p.m.r. spectroscopy (*R. M. Acheson, R. S. Feinberg* and *J. M. F. Gagan*, J. chem. Soc., 1965, 948):

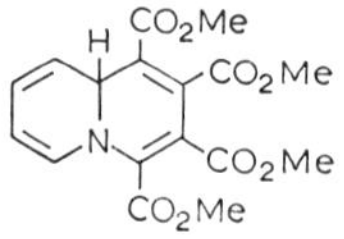

(IV) 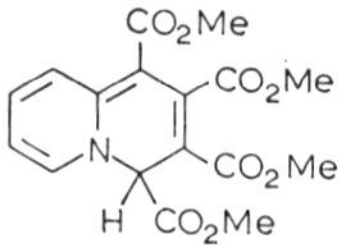   (V) 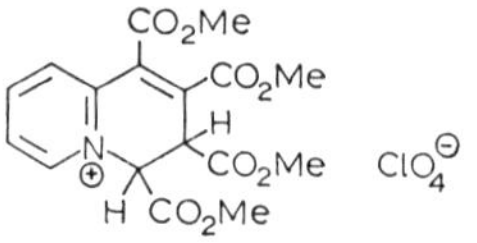   (VI)

The quinolizines form salts with strong acids, protonation occurring at $C_{(3)}$ (*e.g.* VI), the stable derivative V for instance, forming the perchlorate VI (*Acheson* and *G. A. Taylor*, *ibid.*, 1960, 1691). Somewhat unexpectedly in certain substituted quinolizines such as tetramethyl 9-methyl-4*H*-quinolizine-1,2,3,4-tetracarboxylate protonisation occurs at $C_{(1)}$ to give the salt VII:

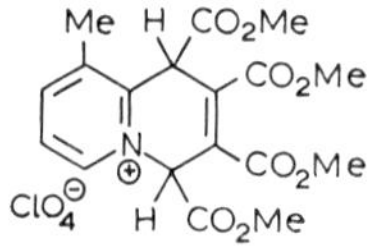

(VII) 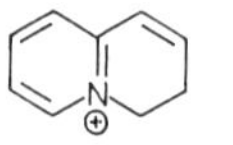   (VIII) 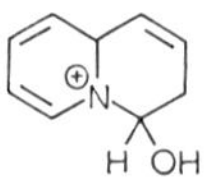   (IX) 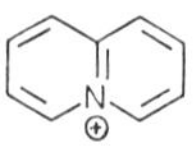   (X)

---

* *M. H. Palmer*, "The Structure and Reactions of Heterocyclic Compounds", p. 165, Arnold, London, 1967; *R. M. Acheson*, Advances in heterocyclic Chem., 1963, **1**, 125.

Although quinolizine does not possess an aromatic structure the quinolizinium salts*, obtained by oxidising the dihydro-cations and dehydrating the products (VIII → IX → X) are isoelectronic with naphthalene and are stable. The quinolizinium ion (X) in the form of the *iodide*, m.p. 220–230°, *picrate*, m.p. 180–181°, and *bromide*, m.p. 260–261°, has been synthesised (*Boekelheide* and *W. G. Gall*, J. Amer. chem. Soc., 1954, **76**, 1832; *cf. E. E. Glover* and *G. Jones*, J. chem. Soc., 1959, 1686). The 2- and 4-methyl groups in methylquinolizinium salts readily condense with aldehydes to form styryl derivatives, thus resembling the 2- and 4-picolinium salts (C.C.C. Vol. IV F, p. 152) (*H. V. Hansen* and *E. M. Amstutz*, J. org. Chem., 1963, **28**, 393).

Formulae II and III for quinolizine suggest that the corresponding oxo-compounds should be stable and these have in fact been prepared. Treatment of 2-hydroxyquinolizinium bromide with aqueous sodium carbonate yields 2*H*-**quinolizin-2-one** (XI), m.p. 128–129.5°, v_{max} 1634 cm^{-1}, which is reduced by hydrogen and an Adams catalyst to 6,7,8,9-*tetrahydro*-2H-*quinolizin-2-one*, m.p. 133–135° (*A. Fozard* and *Jones*, J. chem. Soc., 1964, 2761):

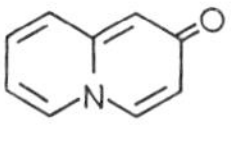

(XI) 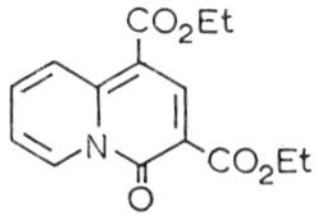   (XII) 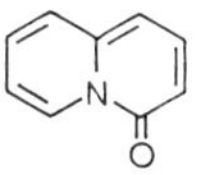   (XIII)

Ethyl 2-pyridylacetate condenses with diethyl ethoxymethylenemalonate to give diethyl 4-quinolizinone-1,3-dicarboxylate (XII), which when hydrolysed and decarboxylated yields 4*H*-**quinolizin-4-one** (XIII), yellow crystals, m.p. 71–72°, v_{max} 1666 cm^{-1}, soluble in water and many organic solvents giving a blue fluorescence, *picrate*, m.p. 137° (*Boekelheide* and *J. P. Lodge*, J. Amer. chem. Soc., 1951, **73**, 3681; *S. I. Goldberg* and *A. H. Lipkin*, J. org. Chem., 1970, **35**, 242). It is identical with the product obtained by dehydrogenation of quinolizidin-4-one (*E. Späth* and *F. Galinovsky*, Ber., 1936, **69**, 761).

A derivative of cyclopenta[*c*]quinolizine (XV) is obtained by boiling the enamine XIV with dimethyl acetylenedicarboxylate in toluene (*W. K. Gibson* and *D. Leaver*, J. chem. Soc., C, 1966, 324). This unique conversion of an indolizine into a quinolizine ring system involves a remarkable rearrangement with loss of dimethylamine.

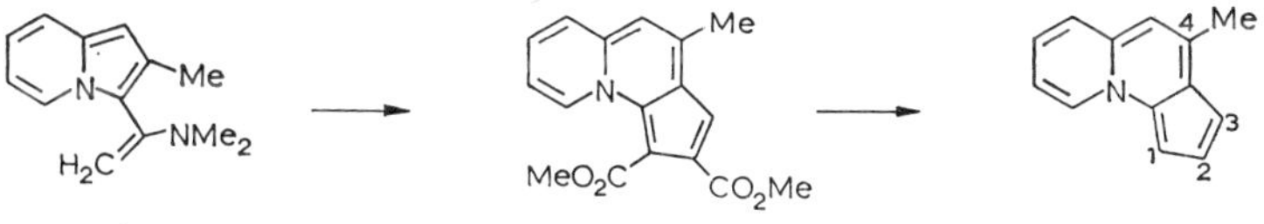

(XIV)   (XV)   (XVI)

---

* Aromatic Quinolizines, *B. S. Thyagarajan*, Advances in Heterocyclic Chemistry, Vol. 5, p. 291, Academic Press, New York, 1965.

Heating the dimethyl ester XV with hydrochloric acid gives the parent compound, 4-*methylcyclopenta*[c]*quinolizine* (XVI), deep red crystals, m.p. 77–78°. As in indolizine, electrophilic substitution occurs at $C_{(1)}$ and $C_{(3)}$, the Vilsmeier reaction yielding 1,3-*diformyl-*, m.p. 229–231°, and acetic anhydride giving 1-*acetyl-*, m.p. 110°, and 1,3-*di-acetyl-4-methylcyclopenta*[c]*quinolizine*, m.p. 217–219°. The reverse process whereby a quinolizine is converted into an indolizine derivative is illustrated by the conversion of tetramethyl 7,9-dimethyl-4*H*-quinolizinetetracarboxylate (XVII) into *trimethyl 6,8-dimethylindolizinetricarboxylate* (XVIII), m.p. 159°, by treatment with dilute nitric acid (*Acheson* and *Taylor, loc. cit.*) or by photolysis of the 9a*H*-isomer (*cf.* IV) in dilute benzene solution with 1-bromonaphthalene as photosensitiser (*A. O. Plunkett*, Chem. Comm., 1969, 1044):

(XVII) ⟶ (XVIII)

Calculations suggest that the order of stability of quinolizines is $4H > 2H > 9aH$ (*Acheson* and *D. M. Goodall*, J. chem. Soc., 1964, 3225), but no 2*H*-quinolizine derivatives were known until 1974 when *Plunkett* (Tetrahedron Letters, 1974, 4181) irradiated tetramethyl 4a*H*-benzo[c]quinolizine-1,2,3,4-tetracarboxylate (XIX) as above and obtained a mixture containing *tetramethyl 3H-benzo*[c]*quinolizine-1,2,3,4-tetracarboxylate* (XX), pale yellow crystals, m.p. 146°:

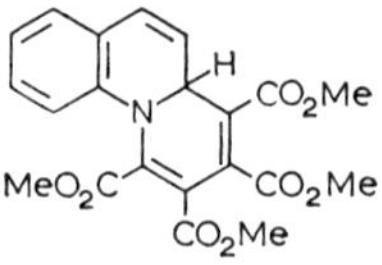

(XIX) 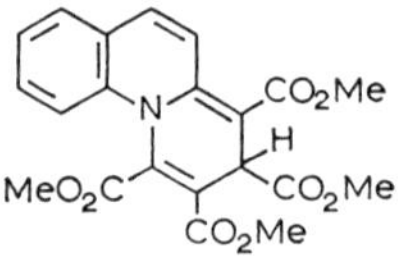   (XX)

Tetramethyl 9a*H*-quinolizine-1,2,3,4-tetracarboxylate (IV) when treated with a molecular equivalent of sodium hydroxide is preferentially hydrolysed to give the salt XXI, which is decarboxylated on acidification to give a mixture of *trimethyl* 4H-*quinolizine-1,2,3-tricarboxylate* (XXII), yellow needles, m.p. 165° (decomp.) with two protons at τ 5.38, and the 2H-*isomer* (XXIII), red needles, m.p. 172–173°, with singlets at τ 2.58 and 5.04. The constitution of XXIII is confirmed by u.v. absorption at 465 nm, indicative of a highly conjugated chromophore (*Acheson, S. J. Hodgson* and *R. G. McR. Wright*, J. chem. Soc., Perkin I, 1976, 1911):

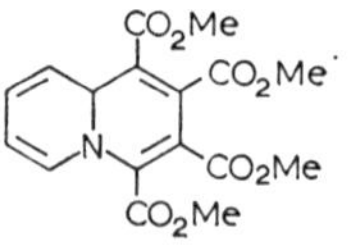

(IV) 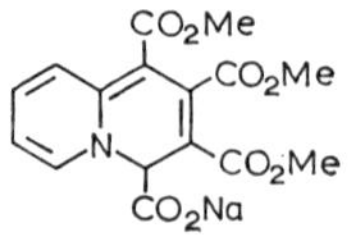   (XXI) 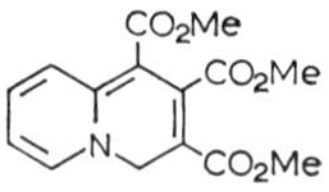   (XXII) 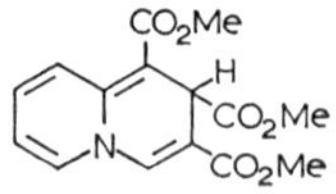   (XXIII)

(b) *Quinolizidine* and its derivatives*

(i) *Quinolizidine and the quinolizidones*

Quinolizidine (XXV), otherwise known as norlupinane or octahydro-quinolizine, is the central component in the molecule of the lupine alkaloids (see Chap. 38, p. 285) and can be prepared either by the decarboxylation of lupinic acid by heating with soda-lime or by the Curtius degradation of lupinic acid hydrazide (*G. R. Clemo* and *G. R. Ramage*, J. chem. Soc., 1931, 437, 3190). It has been synthesised from 2-(ω-bromobutyl)piperidine (XXIV) by loss of hydrogen bromide (*Clemo et al., ibid.*, 1932, 2959), or by the thermal cyclisation of ethyl γ-(2-piperidyl)butyrate (XXVI) to give 4-oxoquinolizidine (XXVII), followed by electrolytic or catalytic (Pt) reduction (*F. Galinovsky* and *E. Stern*, Ber., 1943, **76**, 1034).

The stereochemistry of quinolizidine has been investigated and a group of strong i.r. bands in the 2700–2800 cm^{-1} range is used as evidence for the *trans*-fusion ($C_{(9a)}$ hydrogen and nitrogen lone pair of electrons) of the two rings (*F. Bohlmann*, Chem. Ber., 1958, **91**, 2157; *T. Masamune* and *T. Takasugi*, Chem. Comm., 1967, 625). This is attributed to two axial CH bonds at $C_{(4)}$ and $C_{(6)}$ which are *trans* to the nitrogen lone pair of electrons and which occur in the *trans*- but not in the *cis*-compounds (see also *T. A. Crabb, R. F. Newton* and *D. Jackson*, Chem. Reviews, 1971, **71**, 109). The *trans*-conformation has been confirmed by p.m.r. spectroscopy (*H. P. Hamlow, S. Okuda* and *N. Nakagawa*, Tetrahedron Letters, 1964, 2553).

Quinolizidine, $C_9H_{17}N$, b.p. 84°/21 mm, is found in the bulbs of *Leontuce Smirnovii* (C.A., 1971, **75**, 137, 536) and forms a *picrate*, m.p. 199–201°, *methiodide*, m.p. 322° (333–334° also reported) (*Y. Arata et al.*, Yakugaku Zasshi, 1969, **89**, 389). The mass spectrum shows a base peak at *m/e* 138 (loss of hydrogen from the molecular ion *m/e* 139) and indicates fission at the $C_{(9a)}$–$C_{(1)}$ bond (*M. Hussain, J. S. Robertson* and *T. R. Watson*, Austral. J. Chem., 1970, **23**, 772).

Bicyclic α-aminoketones undergo rearrangement on Clemmensen reduction and examples of this are found in the reduction of the quinolizidones. Reduction of 1-quinolizidone (XXVIII, R = H) by the Wolff–Kishner method yields quinolizidine, but Clemmensen reduction yields 1-azabicyclo-

---

* *N. J. Leonard*, in "The Alkaloids", eds. *Manske* and *Holmes*, Vol. 3, p. 137 and Vol. 7, p. 263, Academic Press, New York, 1957, 1960; *M. L. Mosby*, in "The Chemistry of Heterocyclic Compounds", ed. *A. Weissberger*, "Heterocyclic Systems with Bridgehead Nitrogen Atoms", 1961, Part 2, p. 1019.

[5.3.0]decane (XXIX, R = H). It is the ketonic ring which undergoes contraction since 2-methyl-1-quinolizidone (XXVIII, R = Me) under the same conditions yields 8-methyl-1-azabicyclo[5.3.0]decane (XXIX, R = Me) (*N. J. Leonard* and *W. C. Wildman*, J. Amer. chem. Soc., 1949, **71**, 3089; *Clemo, R. Raper* and *V. J. Vipond*, J. chem. Soc., 1949, 2095. For the mechanism of the reaction see *J. H. Brewster*, J. Amer. chem. Soc., 1954, **76**, 6364).

(XXVIII)   (XXIX)   (XXX)   (XXXI)

2-Quinolizidone by Clemmensen reduction yields quinolizidine, but 3-quinolizidone gives mainly the racemates of 3-methylindolizidine (XXX) (*Leonard* and *S. H. Pines, ibid.*, 1950, **72**, 4931). Skeletal rearrangements were also observed in the Clemmensen reduction of 1-azabicyclo[5.3.0]decan-6-one (XXXI) to quinolizidine (*Leonard et al., ibid.*, 1949, **71**, 3100; 1951, **73**, 5210) and of 1-azabicyclo[5.4.0]hendecan-3-one (XXXII) to racemic 4-methylquinolizidine (XXXIII).

(XXXII)   (XXXIII)   (XXXIV)

The four quinolizidones have been prepared and are characterised by the melting-points of their picrates except for the 4-oxo compound which does not form a stable picrate:

	b.p. (°C/mm)	*picrate* m.p. (°C)
1-*Quinolizidone*	104/12	167–168
2-*Quinolizidone*	64– 66/0.5	211
3-*Quinolizidone*	62– 63/0.65	180–182
4-*Quinolizidone*	100–103/2	*perchlorate* m.p. (°C)
		170–171

They participate in the Fischer indole reaction to give indoloquinolizidines (*W. A. Reckhow* and *D. S. Tarbell, ibid.*, 1952, **74**, 4960), 1-quinolizidone thus yielding the reduced carboline derivative XXXIV. This synthesis is of service in the study of the yohimbine alkaloids.

Lupinane, 1-*methylquinolizidine*, $C_{10}H_{19}N$, is obtained in two epimeric forms which can be separated by fractional crystallisation. β-Lupinane (XXXV) is obtained by the reduction of lupinine tosylate by lithium tetrahydridoaluminate and the α-compound XXXVI by a similar reduction of epilupinine (*Galinovsky* and *H. Nesvadba*, Monatsh., 1954, **85**, 1300; *Bohlmann et al.*, Ber., 1961, **94**, 3151):

(XXXV)   (XXXVI)   (XXXVII)

They both absorb in the 2780–2810 cm^{-1} range, thus indicating that in both compounds the $C_{(9a)}$ hydrogen atom is *trans* to the nitrogen lone pair of electrons (see p. 267). The β-compound has been synthesised and found to be identical with the lupinane obtained from lupinine (*K. Winterfield* and *F. W. Holschneider*, Ber., 1933, **66**, 1751). The structure was confirmed by degradation with cyanogen bromide to give 2-δ-bromobutyl-1-cyano-3-methylpiperidine (XXXVII).

The two asymmetric centres at carbon atoms 1 and 9a account for isomerism of the lupinanes and a detailed study by *T. M. Moyneham et al.* (J. chem. Soc., 1962, 2637) showed that the *β-compound*, b.p. 72–73°/15 mm, *picrate*, m.p. 163.5–164°, is the rac.-*trans*-compound (XXXV) and the α-compound is the rac.-*cis*-compound (XXXVI), b.p. 73–74°/16 mm, *picrate*, m.p. 186.5–187.5°.

(ii) Benzoquinolizinium ions*

One of the first, if not the first, quinolizinium ions to be described is the dibenzoquinolizinium ion **Coralyn,** an alkaloid derivative (*W. Schneider* and *K. Schroeter*, Ber., 1920, **53**, 1459):

Coralyn

Benzo[*b*]quinolizinium ion
Acridizinium ion

The monobenzoquinolizinium ions have been synthesised, sometimes with surprising ease (*A. Richards* and *T. S. Stevens*, J. chem. Soc., 1958, 3067; *Bradsher* and *L. E. Beavers*, J. Amer. chem. Soc., 1959, **77**, 4812; *Bradsher* and *J. P. Sherer*, J. org. Chem., 1967, **32**, 733). For example, 2-formyl-pyridine or preferably its oxime with benzyl bromide yields the salt XXXVIII

---

* *C. K. Bradsher*, Accounts of Chemical Research, 1969, **2**, 181.

which with hydrogen bromide is smoothly converted into the acridizinium ion (*Bradsher, T. W. G. Solomons* and *F. R. Vaughan, ibid.*, 1960, **25**, 757):

(XXXVIII)   (a)   (XXXIX)   (b)

The **acridizinium ion** has been more thoroughly investigated than the other two isomeric ions and undergoes some interesting reactions. It is stable to acids, but with bases gives rise to a product which p.m.r. and i.r. spectra show to be a mixture of the pseudo-base (XXXIXa) and the open-chain aldehyde (XXXIXb). Formation of an oxime is additional evidence for the aldehyde structure.

Other examples of nucleophilic attack at $C_{(6)}$ are known, benzyl cyanide and alkali, for instance, yielding the product XL (*Bradsher* and *J. H. Jones, J. Amer. chem. Soc.*, 1959, **81**, 1938):

(XL)   (XLI)

The acridizinium ion also undergoes transannular addition of the Diels–Alder type, maleic anhydride thus yielding the product XLI (*Bradsher* and *Solomons, ibid.*, 1958, **80**, 933). Contrary to many Diels–Alder reactions certain "inverse" reactions occur in which the diene reacts as an electron-deficient component, while the dienophile is electron-rich (*J. Sauer* and *H. Wiest*, Angew. Chem. intern. Edn., 1962, **1**, 269). The acridizinium ion is clearly an electron-deficient diene.

3. Azepine derivatives containing a bridge nitrogen atom

5*H*-**Pyrrolo**[1,2-*a*]azepine (I), whose octahydro derivative has already been mentioned (p. 268), has not been prepared. It would be devoid of aromatic properties, but some of its derivatives might exhibit aromatic stability. The 5-oxo compound, 3*a*-**azaazulene-4-one,** m.p. 49–51°, can be represented by

(I)   (II)   (III)

the canonical formulae II and III and has been prepared by the cyclisation of methyl ω-(2-pyrrolyl)butadienecarboxylate (IV) by sodium hydride in boiling toluene (*W. Flitsch, B. Müter* and *U. Wolf*, Chem. Ber., 1973, **106,** 1993). The participation of the polar carbonyl group in the conjugated double bond system of both rings is shown by i.r. absorption at 1652 cm^{-1} (*cf.* 2-quinolinone, 1656 cm^{-1}), some 40 cm^{-1} lower than that (1700 cm^{-1}) of the ketone V. P.m.r. spectra also indicate aromatic character of the compound as the consequence of charge separation and resulting delocalisation of the cyclic π-electrons.

The contribution of III is shown by the deshielding of the $C_{(3)}$ proton by the flanking quaternary nitrogen atom, τ 1.95 compared with τ 2.80 for the corresponding proton in V.

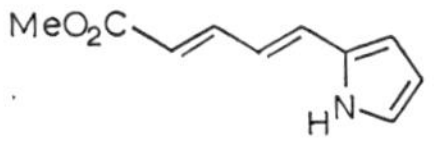

(IV) 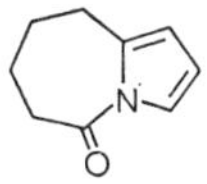   (V)

2,3,5,6,7,8,9,9a-**Octahydro-1***H*-**pyrrolo[1,2-*a*]azepine,** 1-azabicyclo[5.3.0]decane, $C_9H_{17}N$, (VI), b.p. 79–80°/17 mm, *picrate*, m.p. 213–214°, *methiodide*, m.p. 282°, is prepared by methods already mentioned (pp. 267, 268). Its structure follows from its synthesis by the ring-closure of the dibromoamine VII (*V. Prelog* and *R. Seiwerth*, Ber., 1939, **72,** 1638):

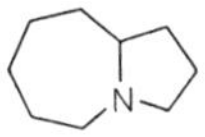

(VI) 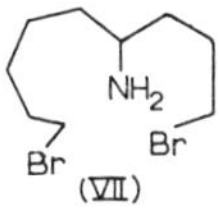   (VII)    (VIII)   (IX)

The partially reduced compound, **cyclohepta[*a*]pyrrole,** $C_9H_{13}N$, (VIII), b.p. 89–95°/10 mm, m.p. 36°, has been prepared and when heated to 650° is rearranged to **cyclohepta[*b*]-pyrrole** (IX), m.p. 104°, ν_{max} 3400–3450 cm^{-1} (pyrrole NH) (*J. M. Patterson* and *S. Soedigdo*, J. org. Chem., 1967, **32,** 2969).

The pyridine analogue, **pyrido[1,2-*a*]azepine** (X) has not been prepared and a fully aromatic form cannot be obtained, although a betaine structure (XI) might give rise to stabilisation.

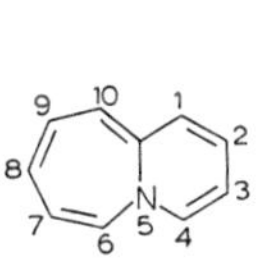

(X) 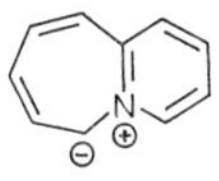   (XI) 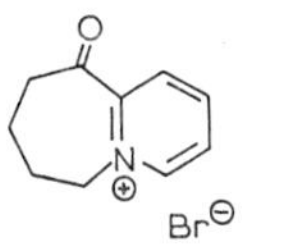   (XII) 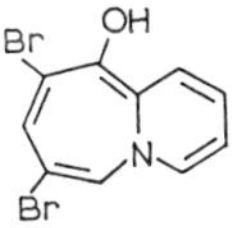   (XIII)

10-**Oxo**-7,8,9,10-**tetrahydro-6***H*-**pyrido[1,2-*a*]azepininium bromide** (XII), m.p. 154–156°, ν 1700 cm^{-1}, has been prepared and converted into 7,9-**dibromo-10-hydroxypyrido[1,2-*a*]-**

azepine (XIII), which decomposes above 100° (*A. Fozard* and *G. Jones*, J. org. Chem., 1965, **30**, 1523).

4. Tricyclic compounds with nitrogen common to all three rings

(*a*) *Cyclazines*

A number of tricyclic compounds and their derivatives are known which contain a nitrogen atom common to all three rings. An example, cycl-[3.2.2]azine (II), is described below and other cyclazines are included in the sequel.

(*i*) *Cycl*[3.2.2]*azine, pyrrolo*[2,1,5-cd]*indolizine* (II)

This interesting tricyclic system was first prepared and investigated by *V. Boekelheide* and his collaborators. It is prepared from 5-methyl-indolizine by forming the lithium compound (I, R = Li) which with dimethylformamide yields the aldehyde (I, R = CHO). Ring-closure of the aldehyde yields the pyrroloindolizine II (*R. J. Windgassen Jr.*, *W. H. Saunders Jr.* and *Boekelheide*, J. Amer. chem. Soc., 1959, **81**, 1459):

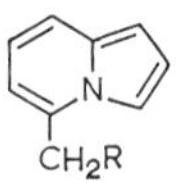

(I)    (II)   (IIa)

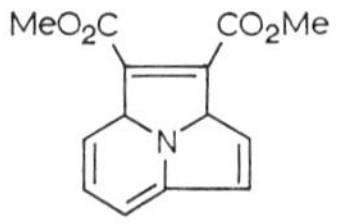

(III) 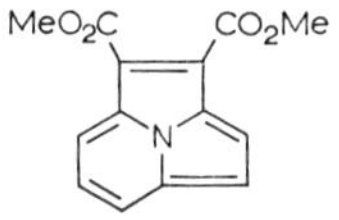   (IV) 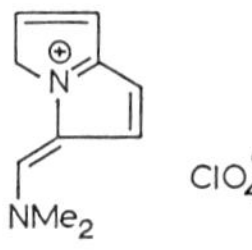   (V) 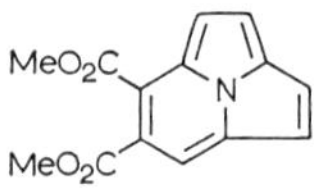   (VI)

It can be obtained by the interaction of indolizine and dimethyl acetylene-dicarboxylate in the presence of a dehydrogenating agent (palladised charcoal) to give the 1,2-dicarboxylic ester III (10–15%) and hence the dehydrogenated product IV (65–75%). Hydrolysis of the ester followed by decarboxylation by heating with copper in quinoline yielded the parent compound II (*A. Galbraith et al., ibid.*, 1961, **83**, 453). The 6,7-diester, VI, is obtained by treating the quaternary salt V with dimethyl acetylene-dicarboxylate and sodium hydride in dimethylformamide and can be converted by standard procedures into the parent compound II (*M. A. Jessep* and *D. Leaver*, Chem. Comm., 1970, 790).

Pyrrolo[2,1,5-*cd*]indolizine (II) is a crystalline, fluorescent compound, yellow needles, m.p. 64–65°, *trinitrobenzene* derivative, m.p. 222–226°, and is stable to light, air and heat (*cf.* the indolizines). It can be represented by the canonical forms II and IIa and the molecule is obviously symmetrical with positions 1 and 4 (also 2 and 3) equivalent. This is confirmed by the p.m.r. spectrum which is composed of an A_2B multiplet for the three protons of the six-membered ring and two identical quartets for the protons of the five-membered rings (*L. M. Jackman, Q. N. Porter* and *G. R. Underwood*, Austral. J. Chem., 1965, **18**, 1221; *Boekelheide et al.*, Helv., 1963, **46**, 1951). The aromaticity of the molecule is also established by the p.m.r. spectrum, a resonance energy of 52 kcal/mol, as well as by electrophilic substitution at position 1. 1-Bromo-, 1-nitro- and 1-acetyl-pyrrolo-indolizine are obtained with bromine, cupric nitrate and acetic anhydride, and acetyl chloride and stannic chloride, respectively (*Boekelheide* and *T. Small*, J. Amer. chem. Soc., 1961, **83**, 462). The acetyl compound can be converted into the corresponding carboxylic acid, the methyl ester of which is identical with the ester obtained by the addition of methyl propiolate to indolizine. Protonation also occurs at $C_{(1)}$ (*F. Gerson et al.*, Helv., 1963, **46**, 1940).

The mass spectrum of the 1-methyl compound shows a peak at 330 and another at 329, probably the result of loss of hydrogen and ring-enlargement to the ion VII (*R. M. Acheson* and *D. A. Robinson*, J. chem. Soc., C, 1968, 1633):

(VII)

An interesting related compound is **cyclopenta[*h*]cycl[4.2.2]azine** (IX), m.p. 193–194°, green plates, formed by the condensation of 2*H*-pyrrolizine and the fulvene VIII by means of sodium hydride (*Jessep* and *Leaver*, *loc. cit.*). Its aromaticity (14 π-electrons and τ 1.24–2.56) is reflected in the similarity of its u.v. spectrum to that of azulene and the occurrence of electrophilic substitution at positions 6 and 8:

(VIII)   2*H*-Pyrrolizine   (IX)

(*ii*) Cycl[3.3.2]azine

Although the parent compound **cycl[3.3.2]azine, pyrrolo-[2,1,5-*de*] quinolizine** is not known, derivatives have been prepared of which the first was *tetramethyl 1,2-dihydro-2-phenylcycl[3.3.2]azine-2a,3,4,5-tetracarboxylate* (X, R = CO_2Me), orange-brown crystals, m.p. 195–196°, obtained by the action of dimethyl acetylenedicarboxylate on 2-styrylpyridine (*Acheson* and *R. S. Feinberg*, J. chem. Soc., C, 1968, 351):

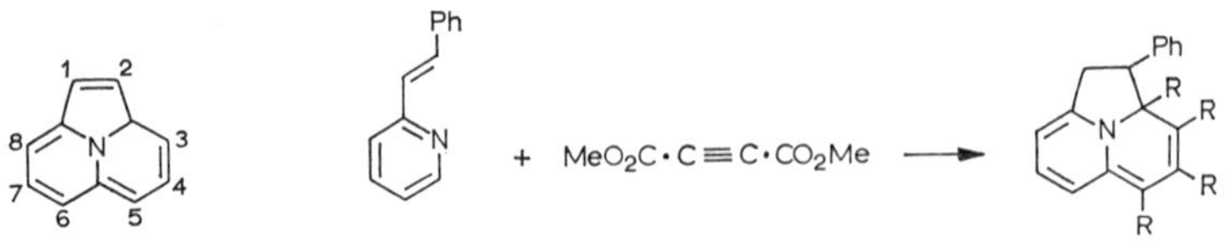

Cycl[3.3.2]azine   2-Styryl-pyridine $+ \ MeO_2C \cdot C \equiv C \cdot CO_2Me \longrightarrow$ (X)

Other derivatives have been prepared (*W. K. Gibson et al.*, Chem. Comm., 1967, 214) and the *Orchidaceae* alkaloid **crepidine** is a substituted deca-hydrocycl[3.3.2]azine (*P. Kierkegaard, A. M. Pilotti* and *K. Leander*, Acta Chem. Scand., 1970, **24**, 3757).

(*iii*) *Cycl[3.3.3]azine*

Cycl[3.3.3]azine, pyrido[2,1,6-*de*]quinolizine (XI), is prepared by heating *tert*-butyl 4*H*-quinolizine-4-ylideneacetate (XII) and *tert*-butyl propiolate in nitrobenzene (*D. Farquhar, T. T. Gough* and *Leaver*, J. chem. Soc., Perkin I, 1976, 341). The reaction involves dehydrogenation of the initial product accompanied by loss of isobutene and carbon dioxide:

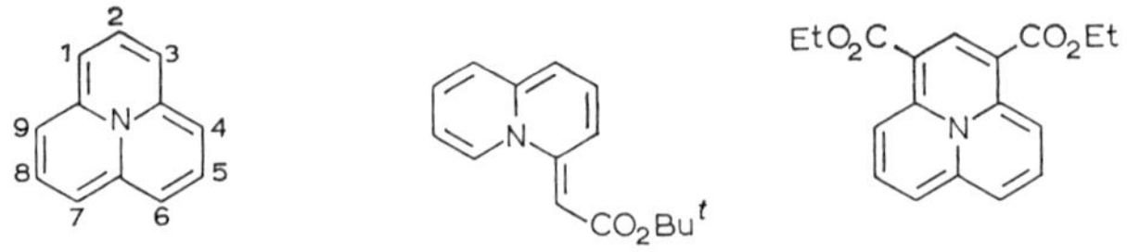

(XI)   (XII)   (XIII)

The pyridoquinolizine (XI), purplish-brown needles, m.p. 140–142° (sealed tube under nitrogen), gives yellow solutions in ethers or hydrocarbons, but decomposes rapidly when exposed to air or when dissolved in chloroform or hydroxylic solvents.

The structure of cycl[3.3.3]azine was confirmed by the p.m.r. spectrum with a triplet at τ 7.37 ($H_{(2)}$, $H_{(5)}$ and $H_{(8)}$) and a doublet at τ 7.93 (remaining 6 protons). The striking high degree of shielding suggests a conjugated non-aromatic system of 12 π-electrons in contrast to the aromatic 10 π-electrons in the aromatic pyrroloindolizine (τ 2.1–2.8) and confirms the prediction of *M. J. S. Dewar* and *N. Trinajstic* (J. chem. Soc., A, 1969, 1754) that the compound would be anti-aromatic with alternating single and double bonds. Lack of aromatic character is found in the reaction of diethyl cycl[3.3.3]azine-1,3-di-carboxylate (XIII) with dimethyl acetylenedicarboxylate to give the Diels–Alder adduct XIV, red prisms, m.p. 255–256°:

The instability of the parent compound results in attempts at substitution proving abortive, but the chemical properties of the accessible diethyl ester XIII are interesting. Hydrogenation with an Adams catalyst gives the tetrahydro compound, *diethyl* 3a,4,5,6-*tetrahydropyrido*[2,1,6-de]*quinolizine*-1,3-*dicarboxylate* (XV), red needles, m.p. 104–105°. The ester also undergoes electrophilic substitution, formylation with dimethylformamide and phosphoryl chloride yielding the 4-*formyl* (XVI, R^1 = CHO; R^2 = H), m.p. 186–190° and the 6-*formyl* derivative, (XVI, R^1 = H; R^2 = CHO), m.p. 181–182°, while nitration with copper nitrate and acetic anhydride gives the 4-*nitro* (XVI, R^1 = NO_2; R^2 = H), dark blue needles, m.p. 193–194°, and 6-*nitro* derivative (XVI, R^1 = H, R^2 = NO_2), dark blue plates, m.p. 211–212°. These reactions may be typical enamine reactions.

Cycl[3.3.3]azine is readily oxidised with chlorine, bromine or antimony pentachloride to give the blue radical ion XVII or the dication XVIII. The e.s.r. spectrum of the radical ion is closely related to that of the phenalenyl radical.

A naturally occurring cycl[3.3.3]azine derivative is the ladybird alkaloid, **coccinellin,** needles, m.p. ~235° (decomp.), an *N*-oxide obtained from *Coccinella septempunctata* (*R. Karlsson* and *D. Losman*, Chem. Comm., 1972, 626). The oxygen-free base is known as **precoccinelin:**

Precoccinelin

Dimethyl acetylenedicarboxylate reacts with cyclopenta[*c*]quinolizines (*e.g.* XIX) to give **cyclopenta[*i,j*]pyrido[2,1,6-*de*]quinolizines** (also termed **cyclopenta[*cd*]cycl[3.3.3]azines**), *e.g.* XX (*R. P. Cunningham et al.*, J. chem. Soc., C, 1969, 239):

Hydrolysis and decarboxylation yield the parent compounds, *e.g.* XX with CO_2Me replaced by H. These are crystalline, highly coloured compounds which are stable in air. The compounds are aromatic in character with ^1H chemical shifts of τ 2.5–3.7 and methyl group protons strongly deshielded (τ 7.60) in CS_2 solution relative to that observed (τ 8.28) in C_6D_6. The 1,2 bond is comparable with the 1,2 bond in azulene and is part of the aromatic system. In addition to two non-polar forms, dipolar forms can also contribute to the resonance hybrid. XX also undergoes electrophilic substitution, protonation and benzoylation with benzoyl chloride and sodium hydrogen carbonate occurring at $C_{(1)}$.

Another 12-π electron peripheral compound is **cycl[4.3.2]azine** (XXI), a derivative of which has been prepared (*W. Flitsch* and *B. Müter*, Angew. Chem. intern. Edn., 1973, **12**, 501) and shown by the high field p.m.r. signals to be paratropic.

*(iv) Julolidine and lilolidine**

Julolidine, 2,3,6,7-tetrahydro-1*H*,5*H*-benzo[*ij*]quinolizine, a derivative of which was first prepared by *A. Reissert* in 1891 (see *D. J. Cook et al.*, J. Amer. chem. Soc., 1950, **72**, 4989), was shown by synthesis to have the structure XXIII. It can be synthesised in various ways including heating 1,2,3,4-tetrahydroquinoline (XXII) with 1,3-dibromopropane (*D. B. Glass* and *A. Weissberger*, Org. Synth., 1946, **26**, 40), heating di-(3-hydroxypropyl)aniline (XXIV) with phosphorus pentoxide in boiling xylene (*R. E. Rindfusz* and *V. L. Harnack*, J. Amer. chem. Soc., 1920, **42**, 1720), and by the interaction of propargyl aldehyde and the enamine XXV (*Bohlmann* and *O. Schmidt*, Chem. Ber., 1964, **97**, 1354).

---

* *W. L. Mosby*, "Heterocyclic Systems with Bridgehead Nitrogen Atoms", 1961, part 2, p. 1069.

Julolidine, $C_{12}H_{15}N$, m.p. 39–40°, b.p. 155–156°/17 mm, *picrate*, m.p. 174°, *methiodide*, m.p. 186°, may be compared with 2,6-dimethyl-*N*,*N*-dimethylaniline and this suggests that it will undergo electrophilic substitution at $C_{(9)}$. A high electron density at this position is indicated by the ease with which julolidine is oxidised to 9,9′-bijulolidyl. This is also the reactive position for bromination and attack in the Vilsmeier–Haack reaction, *N*-methylformanilide and *N*-methylbenzanilide giving the 9-*aldehyde*, m.p. 83°, p-*nitrophenylhydrazone*, m.p. 254° and 9-*benzoyl* derivative, m.p. 108–109°, *oxime*, m.p. 202–203°, respectively (*P. A. S. Smith* and *Tung-Yin Yu*, J. org. Chem., 1952, **17**, 1281).

Julolidine with hydrogen and a platinum oxide catalyst gives two hexahydrojulolidines (*M. Protiva* and *V. Prelog*, Helv., 1949, **32**, 621; *Boekelheide* and *G. P. Quinn*, J. Amer. chem. Soc., 1948, **70**, 2830), one of which, b.p. 124°/14 mm, *picrate*, m.p. 186–187°, *methiodide*, 299–300°, is identical with the product obtained by the hydrogenation of the oxime of diethyl cyclohexanone-2,6-di-β-propionate with a copper chromite catalyst at 265°/350 at. (*N. J. Leonard* and *W. J. Middleton*, *ibid.*, 1952, **74**, 5144). It is a racemate

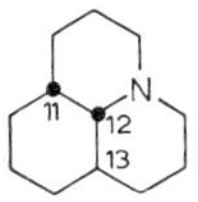

(XXVI) 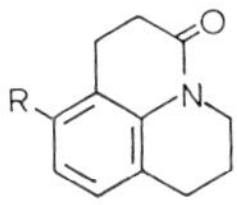   (XXVII)

which is partially resolved chromatographically on a D-lactose column. Three isomeric hexahydrojulolidines are theoretically possible, depending on the relative positions of the hydrogen atoms at carbon atoms 11, 12 and 13, namely *cis–cis*, *cis–trans* and *trans–trans*. Only the *cis–trans* isomer can be racemic, while the others are *meso* compounds. The hexahydrojulolidine under discussion must therefore be the *cis–trans* isomer XXVI.

Ring-fission is observed when julolidine methochloride is reduced with sodium amalgam to give the *base* XXVIII, b.p. 144–148°, *picrate*, m.p. 189°, *methiodide*, m.p. 178° (*von Braun et al.*, Ber., 1918, **51**, 1215). The structure of this base is supported by oxidation to isophthalic acid:

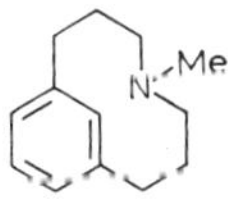

(XXVIII)

Julolid-3-one-10-carboxylic acid (XXVII, R = CO_2H), m.p. 286°, is a degradation product of the alkaloid **annotinine** (*K. Wiesner et al.*, Chem. and Ind., 1956, 1019) and is decarboxylated to *julolid-3-one* (XXVII, R = H), m.p. 102–103°.

Structurally related to julolidine is **piperidino[3,4,5-*i*,*j*]quinolizidine,** $C_{11}H_{20}N_2$, (XXIX), m.p. 54–57°, *dipicrate*, m.p. 265° (decomp.), which when heated with palladised asbestos is dehydrogenated to give 1,2,3,5,6,7-*hexahydropyrido*[3,4,5-i,j]-*quinolizine*, $C_{11}H_{14}N_2$, (XXX), m.p. 62–66°, *picrate*, m.p. 222–223°, a degradation of the lupine alkaloid **matrine** and named *nordehydro-α-matrinidine*. This compound has been synthesised (*K. Tsuda et al.*, J. org. Chem., 1956, **21**, 598, 1481).

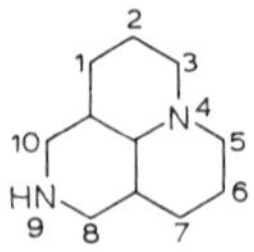

(XXIX) 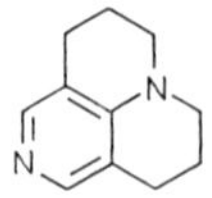 (XXX)

Lilolidine, an analogue of julolidine, has the structure XXXIII. The parent compound, 4*H*-pyrrolo[3,2,1-*i,j*]quinoline (XXXI), has not been prepared and is known only in the form of its hydrogenated derivatives such as lilolidine.

Lilolidine can be prepared by reacting indolinine (XXXII) with 1,3-bromochloropropane or preferably with β-chloropropionyl chloride which yields the indoline derivative (XXXIV). This undergoes ring-closure with aluminium trichloride to give the ketone XXXV, which is reduced by lithium tetrahydridoaluminate to lilolidine (XXXIII) (*G. Hallas* and *D. C. Taylor*, J. chem. Soc., 1964, 1518). It is a low melting solid, b.p. 112–113.5°/5 mm, m.p. 17–18°, *picrate*, m.p. 170–171°.

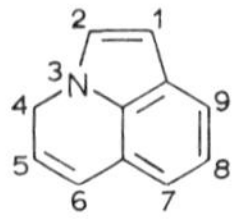

(XXXI) (XXXII) Br(CH₂)₃Cl (XXXIII)

(XXXIV) (XXXV)

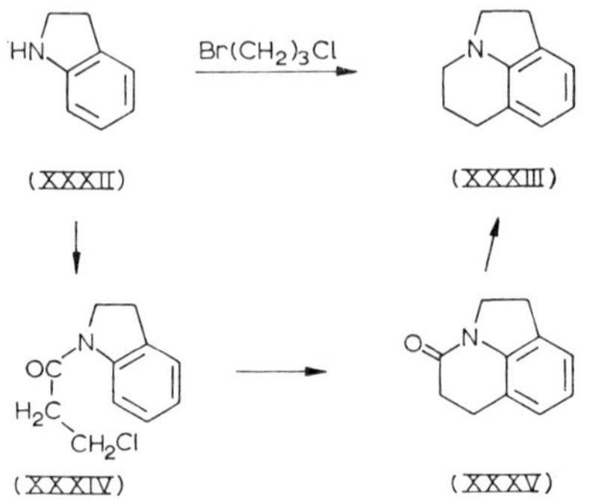

5. Bridged ring compounds*

(a) *Compounds with nitrogen common to two rings*

Bicyclic piperidine and related compounds annealed in the 1,3- and 1,4-positions have been prepared. These bridged ring compounds are characterised by the ease with which they are formed and by their great chemical stability towards reagents such as boiling hydriodic acid or acid potassium permanganate. This furnishes good evidence as was first noted by *J. Meisenheimer*

---

* For the systematic nomenclature of these bicyclic systems see C.C.C., Vol. II C, p. 2 *et seq.*

(Ann., 1920, **420,** 190) that the molecules in compounds such as quinuclidine are strainless and consequently the valencies of the nitrogen atom must have a pyramidal configuration.

Bicyclic piperidine derivatives bridged at the 1,3- and 1,4-positions can be prepared by the intramolecular alkylation of 3- and 4-bromoalkylpiperidines. Thus 3-β-bromoethylpiperidine (I) undergoes ring-closure to yield 1-**azabicyclo**[3.2.1]**octane** (II), a water-soluble solid, m.p. 85°, b.p. 150°, *picrate*, m.p. 294–295° (*V. Prelog* and *K. Balenovic*, Ber., 1941, **74,** 1510):

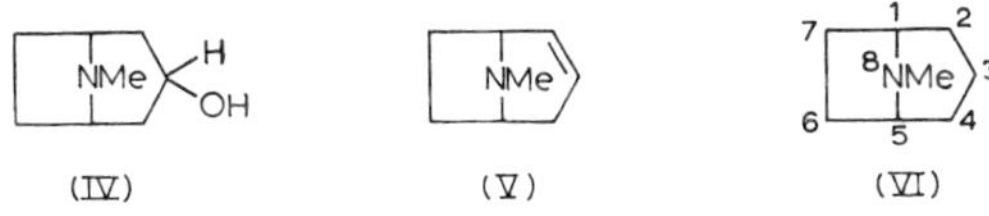

The symmetrical compound 1-**azabicyclo**[3,3,1]**nonane** (III), m.p. 114°, b.p. 175°, *picrate*, m.p. 283°, has also been prepared (*Prelog et al.*, Ber., 1939, **72,** 1319).

(b) Bicyclic systems having a nitrogen bridge

Bicyclic compounds bridged by a nitrogen atom are known and are exemplified by tropane (VI) and *N*-methylgranatanine (XII). These compounds are the parent bodies of alkaloids such as tropine (IV) and ψ-pelletierine (VIII) and interest in them is derived not only from this relationship but also from their stereochemistry which has been intensively investigated.

Tropine (IV) when heated with sulphuric acid and acetic acid is converted into **tropidine,** $C_8H_{13}N$, (V), b.p. 163°, *picrate*, m.p. 289–295°. This in turn can be reduced to **tropane,** 8-*methyl*-8-*azabicyclo*[3.2.1]*octane*, $C_8H_{15}N$ (VI), b.p. 167.5–168.5°, *picrate*, m.p. 284–285°, *picrolonate*, m.p. 207° (*W. L. Archer, C. J. Cavallito* and *A. P. Gray*, J. Amer. chem. Soc., 1956, **78,** 1227), which has the unusual property of dissolving in cold but not in hot water.

Nortropane, $C_7H_{13}N$, (VI, NMe replaced by NH), m.p. ∼60°, b.p. 161°, 61°/11 mm, *hydrochloride*, m.p. 285°(decomp.), N-*benzoyl* deriv., m.p. 94–95°, is obtained by heating tropidine or nortropine with phosphonium iodide and hydriodic acid. It is a strong base which absorbs carbon dioxide from the atmosphere.

Chemical evidence that in the tropane compounds the piperidine ring occurs in the chair form, *e.g.* as in VII, has been confirmed by X-ray investigations (*e.g. E. J. Gabe* and *W. H. Barnes*, Acta Cryst., 1963, **16**, 796):

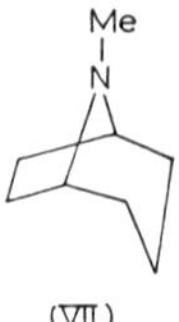

(VII)

ψ-**Pelletierine**, *9-methyl-9-azabicyclo*[3.3.1]*non-3-one* (VIII), $C_9H_{15}NO$, m.p. 64–65°, *hemihydrate*, m.p. 48–49°, the alkaloid from pomegranate root bark, is reduced by lithium tetrahydridoaluminate or Raney nickel to *3-α-granatanol, 9-methylgranatanin-3α-ol* (IX), m.p. 69°, *picrate*, m.p. 275–276°, and by sodium in ethanol or *n*-butanol to *3-β-granatanol*, a conformer of the *β*-isomer (see below), m.p. 99–100°, *picrate*, m.p. 264–265° (*C. L. Zirkle et al.*, J. org. Chem., 1961, **26**, 395):

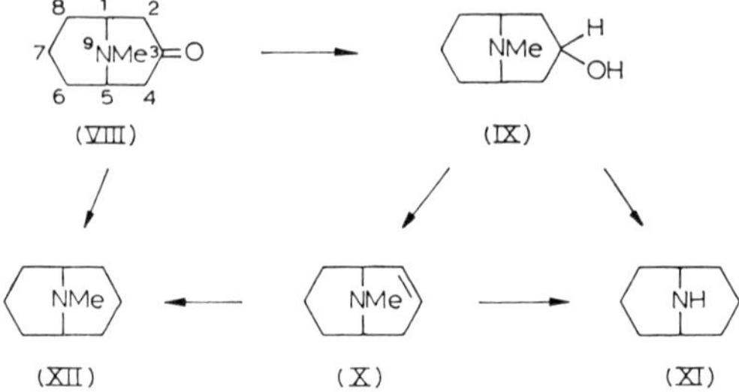

(VIII)   (IX)

(XII)   (X)   (XI)

Dehydration of 3-α-granatanol with sulphuric acid and acetic acid yields the unsaturated N-*methylgranatenine, 9-methyl-9-azabicyclo*[3.3.1]*non-2-ene*, $C_9H_{15}N$ (X), m.p. 17°, b.p. 71–72°/17 mm, *picrate*, m.p. 286° (decomp.), *methiodide*, m.p. > 315°, which with hydrogen and a platinum catalyst yields N-*methylgranatanine, 9-methyl-9-azabicyclo*-[3.3.1]*nonane* (XII), $C_9H_{17}N$, *perchlorate*, m.p. 270° (decomp.), *picrate*, m.p. 297–298° (*S. Wawzonek* and *P. J. Thelen*, J. Amer. chem. Soc., 1950, **72**, 2118; *N. J. Leonard* and *D. F. Morrow*, *ibid.*, 1958, **80**, 371). This compound is also prepared by Wolff–Kishner reduction of *ψ*-pelletierine (VIII).

Granatanine, *9-azabicyclo*[3.3.1]*nonane* (XI), $C_8H_{15}N$, m.p. 50–60°, N-*nitroso* deriv., m.p. 148°, N-*benzoyl* deriv., m.p. 111°, is obtained by heating 3-α-granatanol or *N*-methylgranatenine with phosphorus and hydriodic acid.

P.m.r. evidence points to a chair–chair conformation XIII for *ψ*-pelletierine (*C.-Y. Chen* and *R. J. W. LeFevre*, J. chem. Soc., B, 1966, 539; *cf. Leonard, Morrow* and *M. T. Rogers*, J. Amer. chem. Soc., 1957, **79**, 5476):

(XIII)   (XIV)

3-β-Granatanol exists predominately in a similar conformation (XIII, CO replaced by CHOH), while 3-α-granatanol has a chair–boat conformation (XIV).

(c) *Compounds with nitrogen common to three rings*

(i) *Quinuclidine**

Quinuclidine, 1-*azabicyclo*[2.2.2]*octane*, is of interest both because of its rigid structure and the consequent stereochemical implications and because its nucleus forms part of the molecule of a number of alkaloids including quinine and cinchonine. It can be synthesised in a number of ways, frequently from piperidine derivatives (*Leonard* and *S. Elkin*, J. org. Chem., 1962, **27,** 4635). 4-β-Bromoethylpiperidine (XV), for example, with dilute alkali yields (68 %) quinuclidine (*R. Lukeš, O. Strouf* and *M. Ferles*, Chem. Listy, 1956, **50,** 1624):

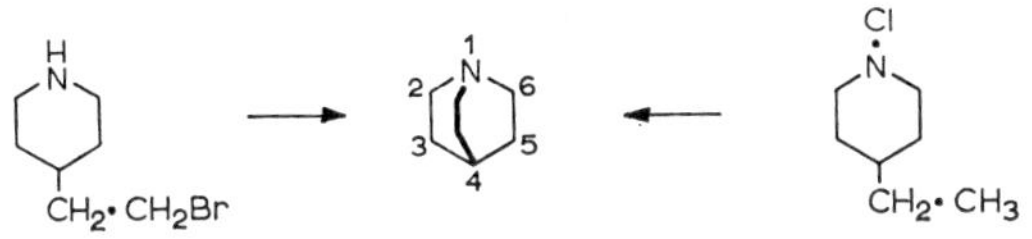

(XV)   Quinuclidine   (XVI)

Quinuclidine is also obtained when *N*-chloro-4-ethylpiperidine (XVI) is irradiated in sulphuric acid and then treated with alkali, 7-methyl-1-azabicyclo[2.2.1]heptane also being formed (*S. Wawzonek* and *T. C. Wilkinson*, J. org. Chem., 1966, **31,** 1732).

γ-Pyrone and pyran derivatives are converted into the bridged piperidine system (*Prelog et al.*, Ann., 1937, **532,** 69; 1938, **535,** 37). γ-Pyrone, for example, by a series of reactions yields the pyran-alcohol XVII, which with hydrobromic acid gives tris(β-bromoethyl)methane (XVIII). This is converted by ammonia into quinuclidine:

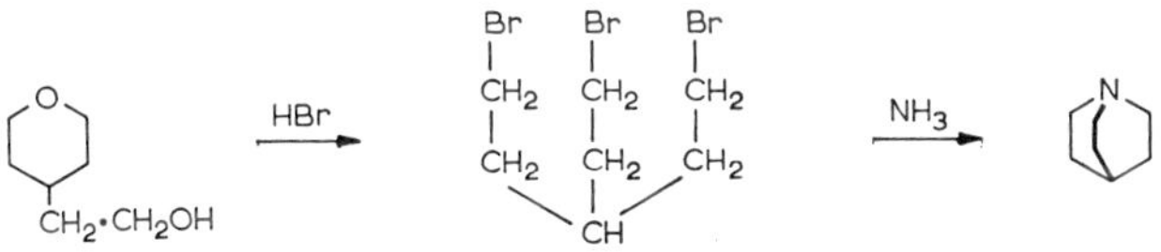

(XVII)   (XVIII)

* *L. N. Yakhontov,* Advances in heterocycl. Chem., 1970, **11,** 473; Russian Chem. Reviews, 1969, **38,** 470; *W. L. Mosby*, "Heterocyclic Systems with Bridgehead Nitrogen Atoms", p. 1331, Interscience, New York, 1961; *H. R. Ing*, "Heterocyclic Compounds", Vol. 3, p. 374, Wiley, New York, 1952.

Reduction of 2- and 3-oxoquinuclidine by the Clemmensen method yields quinuclidine and it is interesting that no rearrangement occurs similar to that undergone by other cyclic α-aminoketones (p. 267) (*Leonard et al.*, J. Amer. chem. Soc., 1952, **74**, 1704; 1953, **75**, 6249).

Substituted quinuclidines are prepared from compounds containing the necessary substituents. 4-Cyanoquinuclidine (XXI), for instance, can be prepared by ring-closing the 4-cyano-*N*-methylpiperidine derivative XIX to give the methochloride XX, pyrolysis of which removes methyl chloride to yield XXI (*C. Grob et al.*, Helv., 1954, **37**, 1672, 1681):

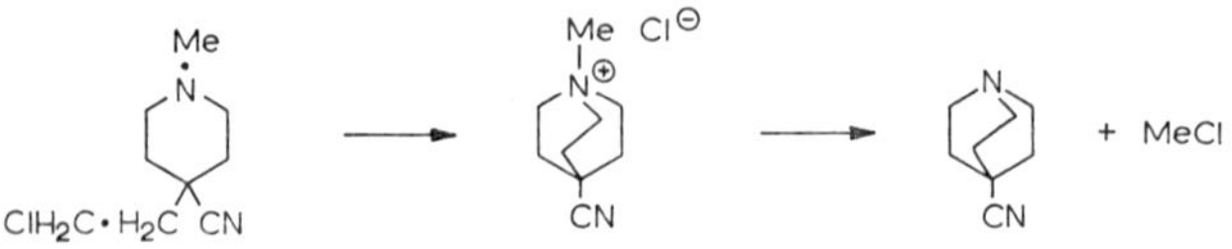

$$\text{(XIX)} \longrightarrow \text{(XX)} \longrightarrow \text{(XXI)} + \text{MeCl}$$

Quinuclidine, $C_7H_{13}N$, is a volatile, crystalline compound, m.p. 158° (sealed tube), *picrate*, m.p. 275–276° (decomp.), *ethiodide*, m.p. 170–171°, which sublimes readily and is soluble in water and organic solvents. It is a strong base, pK_a 10.58, comparable to the trialkyl-amines and is characterised by its great chemical stability, remaining unchanged when heated with sulphuric acid, nitric acid or aqueous potassium permanganate.

Quinuclidine undergoes ring-fission smoothly when heated with palladised charcoal to give 4-ethylpiperidine, and Hofmann degradation of the methiodide gives N-*methyl-4-vinylpiperidine*, *picrate*, m.p. 145–146° (*R. Lukeš, O. Strouf* and *M. Ferles*, Coll. Czech. chem. Comm., 1957, **22**, 1173). Quinuclidine and *N*-methyl-4-(2′-hydroxyethyl)piperidine are also obtained. It is cleaved by phenyl chloroformate to phenyl 4-(2-chloroethyl)-piperidine-*N*-carboxylate, the constitution of which follows from reduction with lithium tetrahydridoaluminate to give *N*-methyl-4-ethylpiperidine (*J. D. Hobson* and *J. G. McCluskey*, J. chem. Soc., C, 1967, 2015):

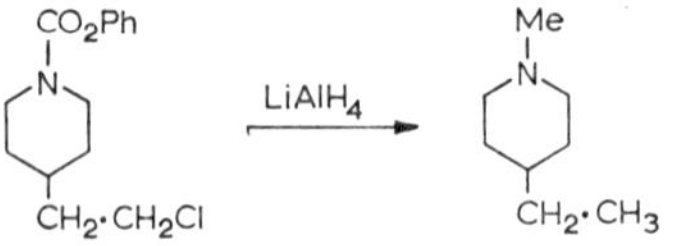

$$\xrightarrow{\text{LiAlH}_4}$$

X-Ray examination of quinuclidinyl benzilate hydrobromide, a psychotomimetic drug shows that quinuclidine contains a boat-shaped piperidine ring as postulated on the basis of classical stereochemical concepts (*A. Meyerhöffer* and *D. Carlström*, Acta Cryst., 1969, ₘ**25**, 1119).

(±)-**3-Ethylquinuclidine**, $C_9H_{17}N$, b.p. 190–192°/720 mm, *picrate*, m.p. 153–154.5°, *methiodide*, m.p. ∼55°, has been synthesised and is structurally closely related to the 3-vinylquinuclidine moiety of the cinchona alkaloid molecule.

3-Oxoquinuclidine, $C_7H_{11}NO$ (XXIII), m.p. 138°, *picrate*, m.p. 210°, is prepared by the internal condensation of the piperidine derivative XXII by means of potassium in toluene (*Grob, A. Kaiser* and *E. Renk*, Helv., 1957, **40**, 2170). With hydrazoic acid it undergoes ring-fission to give the *N*-substituted 1,2,3,6-tetrahydropyridine derivative XXIV, while the oxime of the ketone undergoes the Beckmann transformation with PPA to give the ring-enlarged product XXV (*E. E. Mikhlina et al.*, Zhur. organ. Khim., 1965, **1**, 1336; C.A., 1965, **63**, 1325). The oxime tosylate with potassium hydroxide undergoes fragmentation to give 4-cyanopiperidine (XXVI) and its *N*-tosylate (*Grob et al.*, Helv., 1963, **46**, 1190).

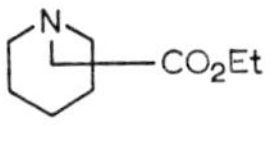

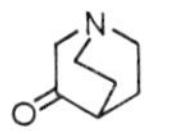

The quinuclidine ring in the cinchona alkaloids undergoes a type of bond rupture known as *hydramine fission*, when heated with acetic acid. The alkaloid XXVII is thus isomerised to a toxin XXVIII (*P. Rabe* and *W. Schneider*, Ann., 1909, **365**, 377):

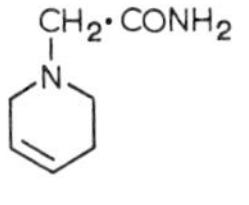

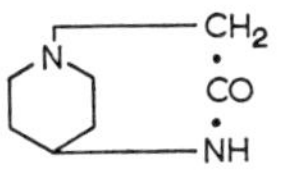

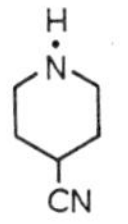

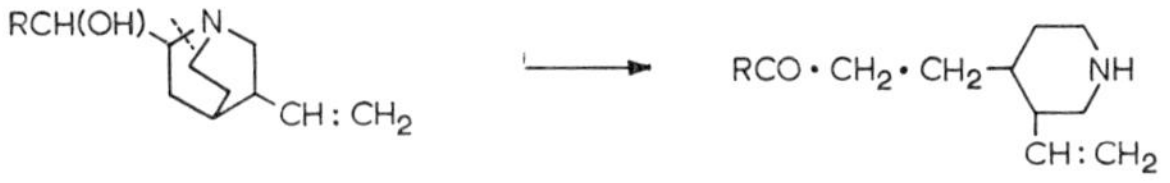

R = 1-Quinolinyl radical

Hydramine fission, first observed by *L. Pasteur* in 1853 occurs when β-hydroxyamines are heated with acetic acid:

$$RCH(OH) \cdot CN \rightarrow RCO \cdot CH + NH$$

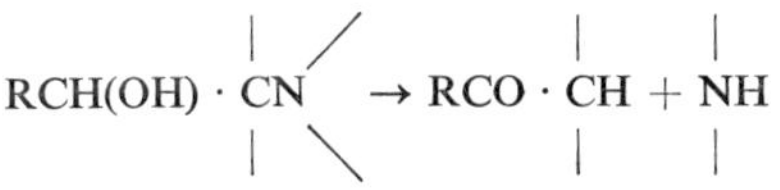

With carbinols containing a neighbouring methylene group the reaction takes a different course, α-phenyl-β-piperidino-ethanol (XXIX), for instance, when heated with phosphoric acid (acetic acid is ineffective) yielding phenylacetaldehyde and piperidine instead of the expected acetophenone and piperidine (*F. Kröhnke* and *A. Schulze*, Ber., 1942, **75**, 1154; *K. Bodendorf* and *K. Dettke*, Chem. Ber., 1956, **89**, 114):

$$PhCH(OH) \cdot CH_2NC_5H_{10} \rightarrow PhCH_2 \cdot CHO + NHC_5H_{10}$$
$$(XXIX)$$

(XXX)　　　　　(XXXI)

The lower homologue of quinuclidine, **1-azabicyclo-**[2.2.1]**heptane** (XXXI, R = H), m.p. 78–79°, *picrate*, m.p. 185°, is obtained by heating the tribromide **XXX** with ammonia (*V. Prelog* and *E. Chernikov*, Ann., 1936, **525,** 192). It forms a *methiodide*, m.p. 320°, with almost explosive violence. The 7-*methyl* derivative (XXXI, R = Me) is converted by the Hofmann degradation to N-*methyl* 4-*vinylpiperidine*, b.p. 146°/743 mm, *picrate*, m.p. 146° (*Lukeš, Strouf* and *Ferles*, Coll. Czech. chem. Comm., 1958, **23,** 326).

Chapter 38

Lupinane and Quinolizidine Alkaloids*

H. C. S. WOOD AND R. WRIGGLESWORTH

Several biogenetically distinct groups of alkaloids containing the quinolizidine nucleus are known.

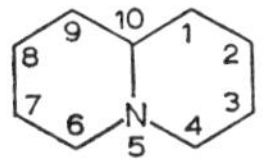

These groups are the Lupinane, Cryptopleurine, Nuphar, Ormosia and Lythraceae alkaloids. Many new alkaloids have been isolated from natural sources in the last fifteen years and the structural determination of these has relied to an ever increasing extent on spectroscopic methods.

The presence of characteristic absorption in the i.r. spectra of quinolizidines at 2700–2800 cm^{-1} has been attributed by *F. Bohlmann* (Chem. Ber., 1958, **91**, 2157) to C–H deformation in the *trans*-fused quinolizidine ring system and this has proved extremely useful in determination of configuration (*M. Wiewiorowski, O. E. Edwards* and *M. D. Bratek-Wiewiorowski*, Canad. J. Chem., 1967, **45**, 1447).

Protons adjacent to nitrogen in the quinolizidine ring system appear away from the main aliphatic hydrogen envelope in the n.m.r. spectrum. They have been studied by deuterium-labelling experiments (*Bohlmann, D. Schumann* and *H. Schulz*, Tetrahedron Letters, 1965, 173) which show that α-protons oriented *cis* to the electron pair in nitrogen undergo a de-shielding effect whereas "*trans*-hydrogens" are shielded. The analysis of the ^{13}C-n.m.r. spectra of several lupin alkaloids has been reported recently (*F. Bohlmann* and *R. Zeisberg*, Chem. Ber., 1975, **108**, 1043). Mass spectroscopy (*N. Neuner-Jehle, Schumann* and *G. Spiteller*, Monatsh., 1968, **99**, 390; 1967, **98**, 836) and optical rotatory dispersion (*S. Iskandarov, R. A. Shaimardanov* and *S. Yu. Yunosov*, Khim. prir. Soedin., 1971, **5**, 636; *J. P. Ferris et al.*, J. Amer. chem. Soc., 1971, **93**, 2963) have also proved useful in the determination of structure and configuration in the quinolizidines.

---

* This review covers the literature to June 1976.

1. Lupinane alkaloids

The Lupinane alkaloids are found mainly in the plant families *Leguminosae* (broom, gorse, laburnum, lupin) and Papilionaceae but also to a lesser extent in the Chenopodiaceae, Berberidaceae, Papaveraceae and Solanaceae. The main structural types of lupin alkaloids have been characterised for some time and more recent work in this area has been concerned with variations within the structural types, with the determination of relative and absolute stereochemistry and with synthesis.

(a) Bicyclic lupinane alkaloids

Lupinine, $C_{10}H_{19}NO$, is a laevorotatory alkaloid found in yellow lupin seeds (*G. Baumert*, Ber., 1881, **14,** 1321, 1880, 1882; 1882, **15,** 631, 634). It is a saturated base containing a tertiary nitrogen atom but no *N*-methyl group. It is also a primary alcohol which on oxidation gives lupininic acid, $C_{10}H_{17}NO_2$. This on decarboxylation gives norlupinane.

Three stages of Hofmann degradation were required to give a nitrogen-free product (*R. Willstätter* and *E. Fourneau, ibid.*, 1902, **35,** 1910) implying that lupinine contains a bicyclic system with the nitrogen atom common to both rings. Introduction of a hydrogenation stage after each Hofmann degradation gave a saturated alcohol, $C_{10}H_{22}O$, which was shown to be 4-hydroxymethylnonane (I) by dehydration and subsequent ozonolysis (*P. Karrer et al.*, Helv., 1928, **11,** 1062):

CH_2OH

Me   Me

(I)

CH_2OH

8   9   10   1   2   3   N   7   6   5   4

Lupinine

(II)

On the basis of these observations the structure 1-hydroxymethylnorlupinine (II) was proposed for lupinine. Consideration of this structure shows the presence of two asymmetric carbon atoms ($C_{(1)}$ and $C_{(10)}$) and thus implies the existence of 2 racemic pairs. (−)Lupinine was degraded to (−)4-methyl-nonane, the absolute stereochemistry of which was subsequently determined as R by an independent synthesis (*P. A. Levene* and *R. E. Marker*, J. biol. Chem., 1931, **91,** 761) thus establishing the stereochemistry of (−)lupinine at $C_{(1)}$ as R. The *cis* relationship of the hydrogens at $C_{(1)}$ and $C_{(10)}$ was established by synthetic work (*V. Boekelheide* and *J. P. Lodge Jr.*, J. Amer. chem. Soc., 1951, **73,** 3681). From this information *R. C. Cookson* (Chem.

and Ind., 1953, 339) deduced the absolute configuration of (−)lupinine as
(IIa) $1R:10R$:

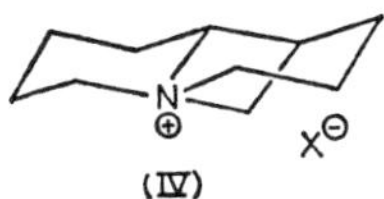

Lupinine with the axial hydroxymethyl group at $C_{(1)}$ is metastable with
respect to the diastereomeric epilupinine (III).

\ For example, lupinine is converted by refluxing in benzene containing sodium
to epilupinine (*K. Winterfeld* and *F. W. Holschneider*, Ber., 1931, **64**, 137;
C. Schöpf et al., ibid., p. 683). This conversion may also be carried out
photolytically using acetophenone as sensitiser (*S. Paszyc* and *H. Wroblewska*,
Bull. Acad. Polon. Sci. Ser. Sci. Chim., 1970, **18**, 15; C.A., 1970, **72**, 100950q).

Several other independent lines of evidence support the above stereochemical assignments. The i.r. spectrum of lupinine (but not epilupinine)
shows the presence of intramolecular hydrogen bonding in carbon disulphide
solution (*L. Marion, D. A. Ramsey* and *R. N. Jones*, J. Amer. chem. Soc.,
1951, **73**, 305; *A. F. Thomas, H. J. Vipond* and *Marion*, Canad. J. Chem.,
1955, **33**, 1290). This is consistent with the axial hydroxymethyl group in
lupinine. Vectorial addition of the group dipole moments in lupinine and
epilupinine as represented above are in agreement with the observed dipole
moments of 3.07 D and 2.30 D, respectively (*J. Ratuský* and *F. Šorm*, Coll.
Czech. chem. Comm., 1954, **19**, 340). Finally the tosylate of lupinine (but
not that of epilupinine) quaternises on heating to the endomethylene
quinolizidinium salt (IV) (*F. Galinovsky* and *H. Nesvadba*, Monatsh., 1954,
85, 1300; *H. Podkowinska* and *Wiewiorowski*, Bull. Acad. Polon. Sci. Ser.
Sci. Chim., 1967, **15**, 467; C.A., 1968, **68**, 78460y) the structure of which
was proved by X-ray diffraction (*C. S. Huber*, Acta Cryst. Sect. B, 1969,
25, 1140):

(+)**Epilupinine** (III) has been isolated from *Lupinus pilosus* and the "tetra-lupine" isolated by *J. F. Couch* (J. Amer. chem. Soc., 1934, **56**, 2434) from *L. palmeri*, has been shown to be (±)epilupinine (*Thomas, Vipond* and *Marion, loc. cit.*).

An early synthesis by *G. R. Clemo et al.* (J. chem. Soc., 1937, 965) confirmed the gross structure which had been elucidated for lupinine. Since that time many syntheses of the lupinine skeleton have been devised. *E. E. van Tamelen* and *R. Foltz* (J. Amer. chem. Soc., 1960, **82**, 502, 2400; 1969, **91**, 7372) have developed a synthesis of (±)epilupinine as shown in Scheme I. This was modelled on the biogenetic scheme proposed by *R. Robinson* and *Schöpf* for lupinine and involved reactions under mild conditions only:

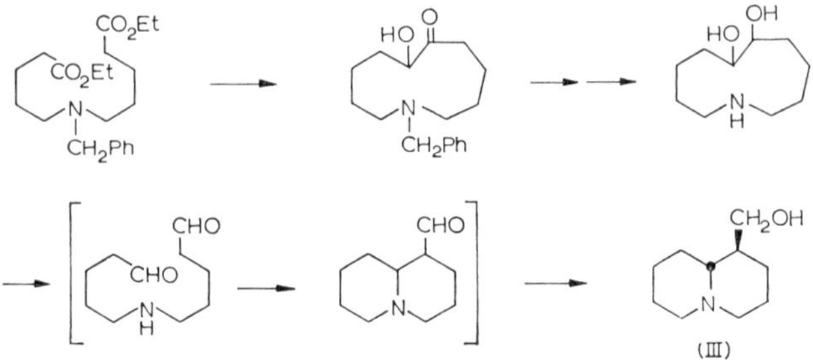

Scheme I

Reduction of the unstable lupinaldehyde with lithium tetrahydridoaluminate gave only the more stable epilupinine (III). Another synthetic route by these authors involved decarboxylation of the malonic acid derivative V. Subsequent esterification of the epimeric monocarboxylic acids VI and reduction with tetrahydridoaluminate gave a separable mixture of 20% (±)lupinine and 80% (±)epilupinine:

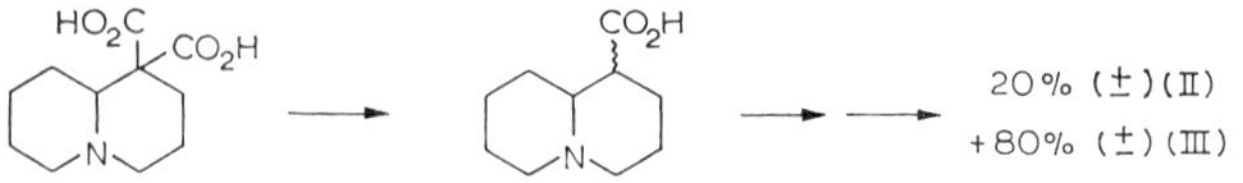

An asymmetric synthesis by *S. I. Goldberg* and *I. Ragade* (J. org. Chem., 1967, **32**, 1046) achieved a 10% optical yield of (−)lupinine (IIa) from the optically active menthoxycarbonyl ester VII by reduction first with tetrahydridoborate and subsequently with lithium tetrahydridoaluminate:

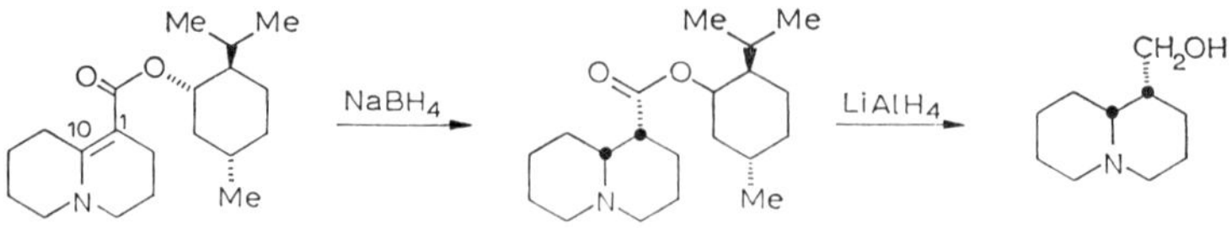

Deuteration experiments showed the site of hydride attack to be at $C_{(10)}$, subsequent protonation occurring at $C_{(1)}$. The fact that no epilupinine is formed in this synthesis shows that formation of the second asymmetric carbon atom (at $C_{(1)}$ by protonation) is 100% stereospecific but that the hydride addition to $C_{(10)}$ directed by the optically active menthoxycarbonyl moiety is only 10% stereoselective.

Application of the general alkaloid synthesis (based on the formation of fused piperidine ring systems) developed by *E. Wenkert* (Acc. Chem. Res., 1968, **1**, 78) has provided another route to the lupinine skeleton (*Wenkert, K. G. Dave* and *R. V. Stevens*, J. Amer. chem. Soc., 1968, **90**, 6177). Other syntheses have been described by *Winterfeld* and *H. von Cosel* (Arch. Pharm., 1940, **278**, 70); *Boekelheide et al.* (J. Amer. chem. Soc., 1953, **75**, 3243); *Bohlmann* and *O. Schmidt* (Chem. Ber., 1964, **97**, 1354); *T. Kappe* (Monatsh., 1967, **98**, 1852) and *Schöpf, H. Arm* and *H. Koop* (Ann., 1968, **712**, 168; *G. C. Gerrans, A. S. Howard* and *B. S. Orlek* (Tetrahedron Letters, 1975, 4171).

(−)*Lupinine*, m.p. 70–71°; b.p. 255–257°; $[a]_D$ −19° to −26° (ethanol); *hydrochloride*, m.p. 212–213°, *methiodide*, m.p. 295–296°, *picrate*, m.p. 136–137°, O-*benzoate*, m.p. 49–50°, *phenylcarbamate*, m.p. 94–95°, *picrolonate*, m.p. 192°. (+)*Lupinine*, m.p. 68°, $[a]_D$ +19.9° (ethanol). (+)*Epilupinine*, m.p. 76–78°, $[a]_D$ +20.3 (water). (±)*Lupinine*, m.p. 59°, *methiodide*, m.p. 303° (decomp.), *picrate*, m.p. 127°. (±)*Epilupinine*, m.p. 81°, *methiodide*, m.p. 248°, *picrate*, m.p. 139°. (−)*Lupininic acid*, m.p. 255°, *hydrochloride*, m.p. 275°, $[a]_D$ −13.1° (methanol).

Lamprolobine, $C_{15}H_{24}N_2O_2$, isolated from *Lamprolobium fruticosum*, has the following structure as shown by its hydrolysis to glutaric acid and (+)1-aminomethylquinolizidine (*N. K. Hart, S. R. Johns* and *J. A. Lamberton*, Austral. J. Chem., 1968, **21**, 1619):

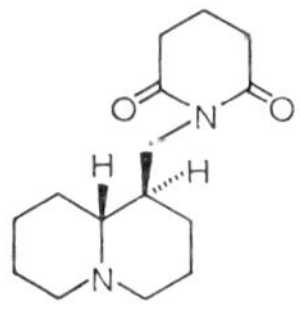

Lamprolobine

The absolute stereochemistry follows from the interrelationship of (+)1-aminomethylquinolizidine and (+)epilupinine and is as shown above. The synthesis of (±)lamprolobine described by *Wenkert* and *R. A. Jeffcoat* (J. org. Chem., 1970, **35**, 515) is another example of the fused piperidine ring-system synthesis (see lupinine). Other syntheses have been described by *Goldberg* and *A. H. Lipkin* (*ibid.*, 1970, **35**, 242) and *Y. Yamada, K. Hatano* and *M. Matsui* (Agr. biol. Chem., 1970, **34**, 1356).

Lamprolobine is an oil, [α]$_D$ +29°, (ethanol) although there seems to be some confusion in the literature as to the sign of the optical rotation, *picrate*, m.p. 153–154°.

Lusitanine, $C_{12}H_{20}N_2O$, m.p. 186, $[\alpha]_D^{27}$ + 6°, from *Genista lusitanica*, on catalytic hydrogenation gives the epimeric 1-acetylaminolupinines. Hydrolysis and subsequent reduction with sodium tetrahydridoborate gives a mixture of lupinine and epilupinine. These facts and spectroscopic evidence support the formulation shown for lusitanine (*E. Steinegger* and *K. Wicky,* Pharm. Acta Helv., 1965, **40**, 610, 658):

Lusitanine

Mamanine, $C_{15}H_{22}N_2O_2$, m.p. 171–172°. $[\alpha]_D^{24.3}$ + 31.7° (ethanol) was isolated from *Sophora chrysophylla* along with **pohakuline,** $C_{15}H_{26}N_2O_2$, m.p. 170–171°, $[\alpha]_D^{25}$ − 17.2° (ethanol) (*M. Kadooka et al.,* Tetrahedron, 1976, **32**, 919).

(−)*trans*-(4′-**Hydroxycinnamoyl) lupinine** was isolated from *Lupinus luteus* seedlings by *I. Murakoshi et al.,* (Phytochemistry, 1975, **14**, 2714):

Mamanine

Pohakuline

(-)*trans*-(4′- Hydroxycinnamoyl) lupinine

(b) Phenanthronorlupinane alkaloids

Cryptopleurine, $C_{24}H_{27}NO_3$, is not a lupin alkaloid but is considered here because it is structurally related to quinolizidine. It was originally isolated from *Cryptocarya pleurosperma* by *I. S. de la Lande* (Austral. J. exp. biol. med. Sci., 1948, **26**, 181) and more recently it has been obtained from *Boehmeria cylindrica* (*N. R. Farnsworth et al.,* Austral. J. Chem., 1969, **22**, 1805). Cryptopleurine was shown to be a saturated tertiary base containing three methoxyl groups and to have u.v. absorption characteristic of a phenanthrene system by *E. Gellert* and *N. V. Riggs* (*ibid.*, 1954, **7**, 113). The structure was determined shortly afterwards by X-ray analysis (*J. Fridrichsons* and *A. McL. Mathieson* (Nature, 1954, **173**, 732; Acta cryst., 1955, **8**, 761). Two early syntheses of cryptopleurine have been described (*C. K. Bradsher* and *H. Berger*, J. Amer. chem. Soc., 1957, **79**, 3287; *P. Marchini* and *B. Belleau*, Canad. J. Chem., 1958, **36**, 581). More recently a synthesis based on bio-genetic considerations has been described and is outlined below (*J. M. Paton*, *P. L. Pauson* and *T. S. Stevens*, J. chem. Soc., C, 1969, 1309; see also *E. Kotani*, *M. Kitizawa* and *S. Tobinaga*, Tetrahedron, 1974, **30**, 3027):

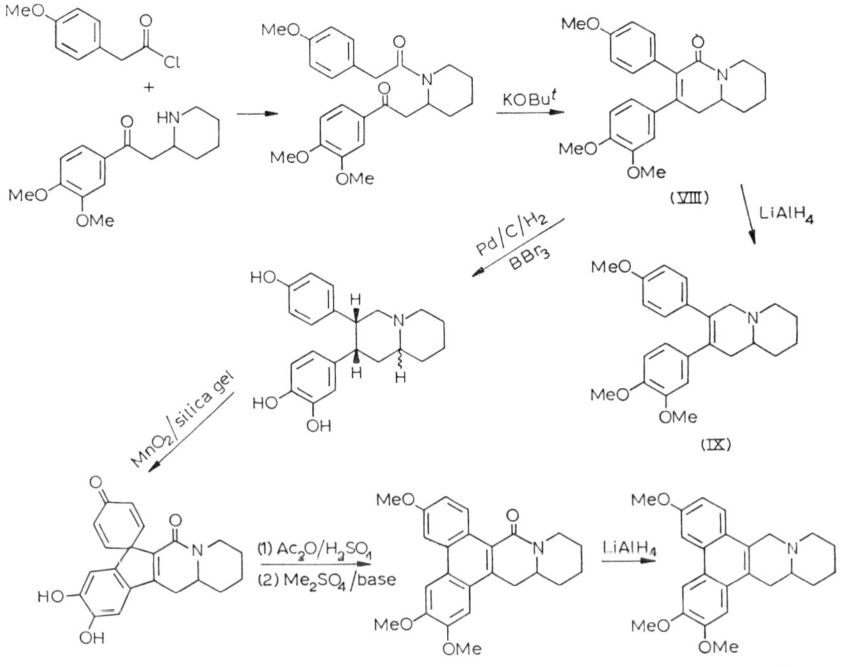

Cryptopleurine

Support for the biogenetic aspects of this synthesis comes from the isolation of cryptopleurine and a phenanthroquinolizidine alkaloid IX from *Boehmeria platyphylla* (*Hart*, *Johns* and *Lamberton*, Austral. J. Chem., 1968, **21**, 2579). The latter alkaloid can also be formed from the synthetic intermediate VIII by reduction with lithium tetrahydridoaluminate.

Cryptopleurine has m.p. 197–198°, $[\alpha]_D$ −106° (chloroform); *hydrochloride*, m.p. 262° (decomp.), *picrate*, m.p. 221–222° (decomp.) and *methiodide*, m.p. 215–217°.

Cryptopleuridine, $C_{23}H_{23}NO_4$, m.p. 196–197°, $[\alpha]_D$ + 90° (chloroform), has recently been isolated from *Cryptocarya pleurosperma* (*Johns et al.*, *ibid.*, 1970, **23**, 353). The following structure has been postulated:

Cryptopleuridine

(c) Tricyclic lupin alkaloids

Cytisine, $C_{11}H_{14}N_2O$, occurs widely in the plant family Papilionaceae. Details of its isolation have been described by numerous investigators (*inter alia, A. Partheil*, Chem. Ber., 1891, **24**, 634; *H. R. Ing.* J. chem. Soc., 1931, 2200; 1933, 504; *L. H. Briggs* and *J. Ricketts, ibid.*, 1937, 1795; *Briggs* and *W. E. Russell, ibid.*, 1942, 507, 555). Cytisine is a monoacidic optically active base. One nitrogen atom is secondary and is more basic than the other tertiary nitrogen. The former is contained in a ring as two stages of Hofmann degradation are required to eliminate it as trimethyl-amine (*Partheil*, Arch. Pharm., 1892, **230,** 448). The alkaloid exhibits certain aromatic properties (*e.g.* nitration, halogen substitution reactions). U.v. and i.r. data support the original suggestion of *Ing* (J. chem. Soc., 1931, 2195) of the presence of an α-pyridone ring in cytisine. This is further supported by the production of pyridine on zinc dust distillation of the alkaloid.

On electrolytic reduction cytisine gives "tetrahydrodeoxycytisine", $C_{11}H_{20}N_2$, by saturation of the α-pyridone ring. Hofmann degradation of this compound (*E. Späth* and *F. Galinovsky*, Ber., 1932, **65,** 1526; *Galinovsky* and *E. Stern, ibid.*, 1944, **77,** 132) showed that the tertiary nitrogen was common to two rings. Further Hofmann degradation on "*N*-acetyltetra-hydrodeoxycytisine" eliminated the tertiary nitrogen. Dehydrogenation and oxidation of this product gave 5-methylnicotinic acid (X) thus establishing the presence of the secondary nitrogen in a piperidine ring. From the above data formula XI was proposed for cytisine and structure XII thus follows for "tetrahydrodeoxycytisine":

(X)

(XI)
Cytisine

(XII)
Tetrahydrodeoxycytisine

Cytisine on heating with phosphorus and hydriodic acid gives cytisoline which proved to be 2-hydroxy-6,8-dimethylquinoline by synthesis (*Späth*, Monatsh., 1919, **40,** 15, 93). The mechanism of this process is thought to involve elimination of the secondary nitrogen as ammonia, ring-opening and alternative cyclisation at the 5-position of the α-pyridone ring:

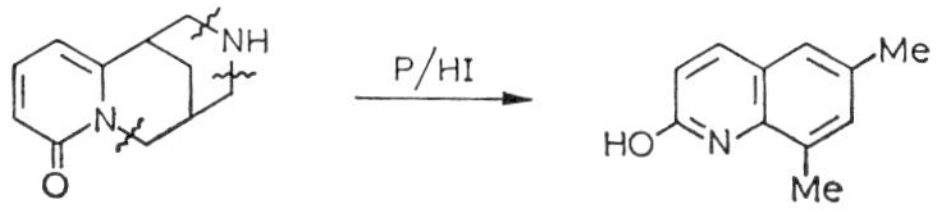

(XI) Cytisoline

Hofmann degradation studies on cytisine itself support the above formulation (*Späth* and *Galinovsky*, *loc. cit.*).

Three syntheses of cytisine have been described. *E. E. van Tamelen* and *J. S. Baran* (J. Amer. chem. Soc., 1955, **77,** 4944) condensed 2-(α-pyridyl) allylmalonic acid (XIII) with benzylamine and formaldehyde to give the carboxylic acid XIV. Reduction of the ethyl ester of this acid with lithium tetrahydridoaluminate furnished the alcohol XV, which with hydrobromic acid underwent intramolecular quaternisation to give the tricyclic pyridinium bromide XVI. Oxidation with alkaline ferricyanide and debenzylation with hydriodic acid give (±)cytisine which was resolved using (+)camphorsulphonic acid to give the laevorotatory isomer identical with the natural alkaloid:

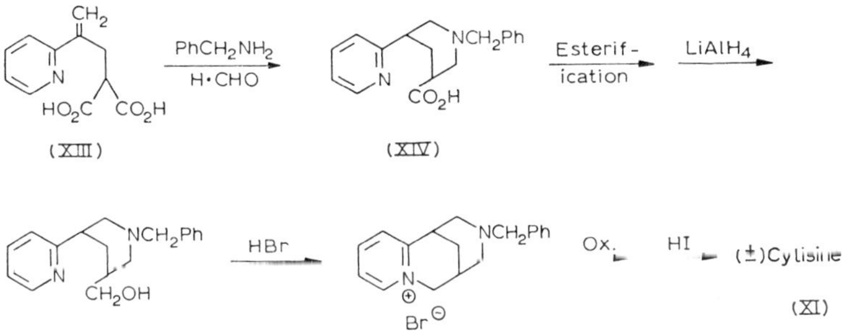

(XIII) (XIV) (XV) (XVI) (XI)

In a synthesis described by *F. Bohlmann et al.* (Angew. Chem., 1955, **67,** 708; Chem. Ber., 1956, **89,** 792) tetrahydrocytisine was formed *via* the quinolizidine XVIII, which in turn was made from the Michael addition of diethyl malonate to the vinylpyridine derivative XVII:

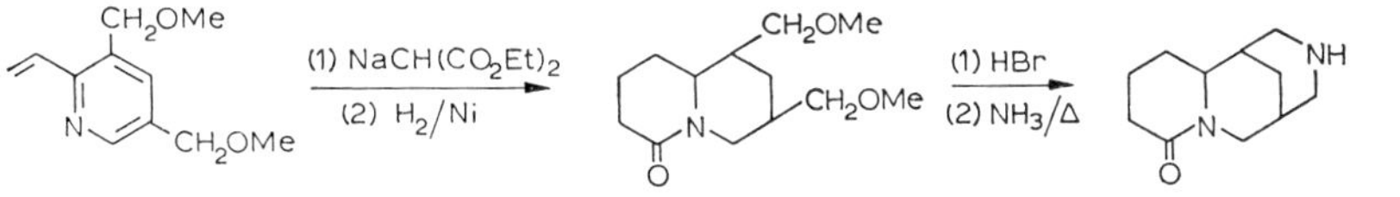

(XVII) (XVIII) (±)Tetrahydrocytisine

The product was converted to ($\pm$)cytisine by the method involving nitrogen protection, dehydrogenation and subsequent deprotection, developed earlier by *Galinovsky, O. Vogl* and *W. Moroz* (Monatsh., 1954, **85**, 1137).

T. R. Govindachari et al. (J. chem. Soc., 1957, 3839) condensed ethyl 5-cyano-2-methylnicotinate (XIX) with ethoxymethylenemalonate to give diethyl 7-cyano-4-quinolizone-3,9-dicarboxylate, which was converted in several steps to 7-aminomethyl-9-hydroxymethyl-4-quinolizone. Selective reduction of ring B and cyclisation to form ring C completed the synthesis giving ($\pm$)cytisine:

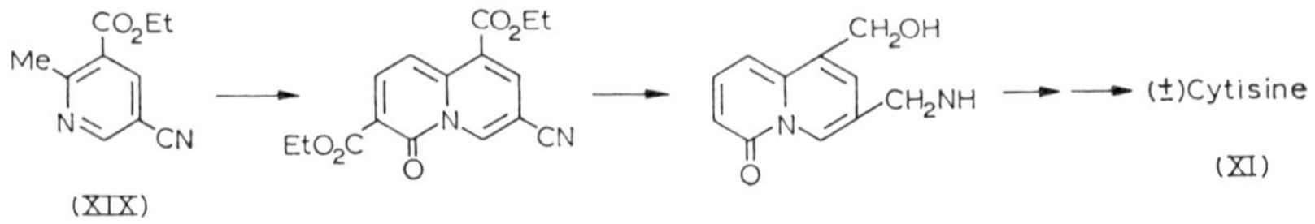

(i) Stereochemistry

Cytisine has two asymmetric centres at $C_{(7)}$ and $C_{(9)}$ but as these are fixed in a *cis*-relationship there is only one racemic form which is resolvable.

More recently the absolute stereochemistry has been established by correlation of ($-$)cytisine with ($-$)anagyrine of known absolute configuration (see below). Thus *S. Okuda, K. Tsuda* and *H. Kataoka* (Chem. and Ind., 1961, 1751) converted ($-$)cytisine by hydrogenation and tosylation to the N-tosyl derivative XX. This product was also obtained from ($-$)($7R:9R:11R$)anagyrine as outlined below:

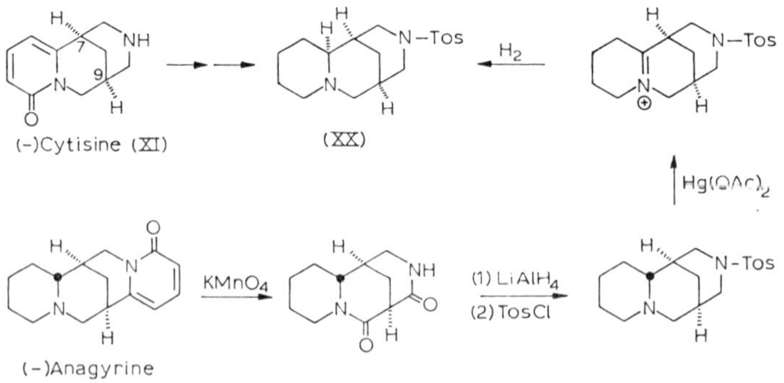

The absolute stereochemistry of ($-$)cytisine (XI) follows as $7R:9S$.

Argentine, $C_{23}H_{26}N_4O_3$, from *Ammodendron argenteum* has been assigned the following dicytisine structure from spectroscopic data (*Kh. A. Aslanov et al.*, Biochem. physiol. Alk., 4th Int. Symp., 1969, 463):

Caulophylline, $C_{12}H_{16}N_2O$, was first isolated from *Caulophyllum thalictroides* by *F. B. Power* and *A. H. Solway* (J. chem. Soc., 1913, **103**, 191) but it occurs also in many *Cytisus* spp. On *N*-demethylation it gives cytisine and, conversely, methylation of the latter gives caulophylline, which therefore must be *N*-methylcytisine. The absolute stereochemistry follows from that of cytisine (see above) as $7R:9S$.

Dimethamine isolated from *Thermopsis alterniflora* has been assigned the following structure by *I. Iskandarov et al.* (Khim. prir. Soedin., 1972, **2**, 176):

Angustifoline, $C_{14}H_{22}N_2O$, has been isolated from *Lupinus angustifolius* and from *L. perennis* by *M. Wiewiorowski, Galinovsky* and *M. D. Bratek* (Monatsh., 1957, **88**, 663) and from *L. polyphyllus* by *Bohlmann* and *E. Winterfeldt* (Chem. Ber., 1960, **93**, 1956). It has also been shown to be identical with **jamaicensine** isolated from *Ormosia jamaicensis* and from *O. panamensis* (*H. A. Lloyd*, J. org. Chem., 1961, **26**, 2143). Angustifoline forms an *N*-acetyl derivative and has a terminal double bond as oxidation gives formaldehyde. This evidence and i.r. and n.m.r. data led *L. Marion, Wiewiorowski* and *Bratek* (Tetrahedron Letters, 1960, **19**, 1) and *Bohlmann* and *Winterfeldt (loc. cit.)* independently and concurrently to propose structure XXI for angustifoline:

HCHO

Angustifoline
(XXI)

β-Epihydroxylupanine

(XXII)

Confirmation of this structure was provided by the observation that angustifoline combined with formaldehyde to give 13-epihydroxylupanine. This reaction is stereospecific and is thought to occur *via* the cationic intermediate XXII, which is then stereospecifically attacked by the incoming nucleophile because of steric reasons. Evidence in support of this mechanism is the isolation (as the perchlorate) of the cationic intermediate of the analogous deoxyangustifoline (*G. I. Birnbaum et al.*, J. chem. Soc., B, 1967, 1368).

The reverse of the above cyclisation reaction also occurs readily. The tosylate of 13-epihydroxylupanine undergoes ring-opening when treated with tri-ethylamine in acetonitrile to give a mixture of angustifoline and tetrahydro-rhombifoline (see below) (*Bohlmann* and *D. Schumann*, Chem. Ber., 1965, **98**, 3133). Dehydrogenation of angustifoline with N-bromosuccinimide gave a compound XXIII, which was identical with the alkaloid "W-102" isolated earlier from *L. albus* (*Bratek* and *Wiewiorowski*, Rocz. Chem., 1959, **33**, 1187):

(XXIII)

Angustifoline (XXI) has been synthesised from cytisine (XI) by *Bohlmann et al.* (Chem. Ber., 1962, **95**, 944) as shown below:

Cytisine (XI)

Angustifoline (XXI)

The Grignard alkylation step is stereospecific; the approach of the reagent from the underside of the molecule is sterically hindered.

[Although XI above represents the (+)enantiomer of cytisine, no implication is to be made from this concerning the *absolute* stereochemistry of natural (−)angustifoline, as no optical rotation data were reported in the above paper. The representation of the (+)cystisine (XI) is for convenience only.]

Tinctorine, $C_{15}H_{20}N_2O$, an alkaloid isolated from *Genista tinctoria* by *D. Knoefel* and *H. R. Schuette* (J. pr. Chem., 1971, **312**, 887) is identical with **alteramine** isolated from *Thermopsis alterniflora* by *R. A. Shaimardanov et al.* (Khim. prir. Soedin., 1970, **6**, 276). The structure was determined from the hydrogenation products, which were related to angustifoline, as *N*-methyl-11-(prop-2-enyl)cytisine:

Tinctorine

Dehydroalbine, $C_{14}H_{18}N_2O$, is an alkaloid which has been isolated from *Lupinus albus*. The structure was determined from spectroscopic data and from the hydrogenation product XXIV, which is the $C_{(11)}$ epimer of di-hydrodeoxyangustifoline (*Wiewiorowski* and *J. Wolinska-Mocydlarz*, Bull. Acad. Polon. Sci. Ser. Sci. Chim., 1964, **12**, 217):

Dehydroalbine   (XXIV)

N-**Methylalbine,** $C_{15}H_{22}N_2O$, also isolated from *L. albus* (originally called alkaloid "u 4/5") by the above authors (*ibid.*, 1961, **9,** 709; 1964, **12,** 213) gave *N*-methyldihydrodeoxyangustifoline on hydrogenation:

N-Methylalbine

Rhombifoline, $C_{15}H_{20}N_2O_2$, isolated from *Thermopsis rhombifolia* (*R. H. F. Manske* and *Marion*, Canad. J. Research, 1942, **20B,** 265; 1943, **21B,** 144) is a ditertiary monoacidic base which on cleavage with hydriodic acid gives cytisine. The reverse of this process can be achieved by alkylation with but-3-enyl bromide. Rhombifoline is therefore *N*-but-3-enylcytisine (*W. F. Cockburn* and *Marion*, Canad. J. Chem., 1952, **30,** 92). The absolute stereochemistry follows from that of cytisine as $(-)7R:9S$ as shown:

Rhombifoline

Tetrahydrorhombifoline, $C_{15}H_{24}N_2O$, has been isolated from *L. angustifolius* (*Bohlmann et al.*, Chem. Ber., 1963, **96,** 2254). 11-*Oxotetrahydrorhombifoline*, $C_{15}H_{22}N_2O_2$, (XXV), isolated from *Ormosia coutinhoi* by *S. McLean, A. G. Harrison* and *D. G. Murray* (Canad. J. Chem., 1967, **45,** 751) had been synthesised previously from 13-epihydroxylupanine tosylate by *Bohlmann* and *Schumann (loc. cit.)*:

Tetrahydrorhombifoline   (XXV)

Physical data on the tricyclic lupin alkaloids are recorded in Table 1.

TABLE 1*

TRICYCLIC LUPIN ALKALOIDS

Alkaloid	m.p. (°C)	b.p. (°C/mm)	$[\alpha]_D°$
Angustifoline	79–80		−7.5, a
Caulophylline	138		−221, d
Cytisine	153	218/2	−119.6, d
Dehydroalbine	50		+103
N-Methylalbine	67.5		−559
Rhombifoline	—	120/0.2	−232.4, a
Tetrahydrorhombifoline	—	160/0.001	+81, a

* Solvents for $[\alpha]_D$ measurements: a, Ethanol; b, Methanol; c, Chloroform; d, Water; e, Acetone.

(d) Tetracyclic lupin alkaloids

(i) Anagyrine, lupanine and sparteine alkaloids

In this series of alkaloids it is convenient to define three basic structural types according to oxidation level:

A   B   C

"Anagyrine-like"   "Lupanine-like"   "Sparteine-like"

A description of the elucidation of the stereochemistry of these alkaloids, which has been achieved by interrelation between the three structural types, is a necessary prelude to any discussion of the individual compounds within the structural types.

There are different stereochemical restraints on the three groups defined above, thus:

A has three asymmetric carbon atoms (at $C_{(7)}$, $C_{(9)}$ and $C_{(11)}$) but steric requirements make the methylene bridge between $C_{(7)}$ and $C_{(9)}$, cis, and the asymmetries at $C_{(7)}$ and $C_{(9)}$ are consequently interdependent. This makes possible the existence of 4 stereoisomers (two racemic pairs).

Scheme 2

A

(+) Anagyrine
(— -trans)

(+) Thermopsine
(— -cis)

H_2/Pt

H_2/Pt

B

(+)Lupanine
(cis-trans)

(trans-cis)

(trans-trans)

(+)α-Isolupanine
(cis-cis)

$H_2/Pt/HCl$

$H_2/Pt/HCl$

$H_2/Pt/HCl$

C

(−)Sparteine
(cis-trans)

(+)β-Isosparteine
(trans-trans)

(−)α-Isosparteine
(cis-cis)

B has four asymmetric carbon atoms (at $C_{(6)}$, $C_{(7)}$, $C_{(9)}$ and $C_{(11)}$). As the same restraint applies to $C_{(7)}$ and $C_{(9)}$ as in *A*, there are eight possible stereoisomers (4 racemic pairs).

C has four asymmetric carbon atoms as in *B*, with the same $C_{(7)}$–$C_{(9)}$ bridge, but in addition the molecule is symmetrical; this reduces the total number of possible stereoisomers to six (three racemic pairs).

All of these possible structures are shown in Scheme 2, which also shows the methods of chemical interrelation which have been developed within the three structural types (*N. J. Leonard* and *L. Marion*, Canad. J. Chem., 1951, **29**, 355).

The "vertical" hydrogenation reactions (in Scheme 2) have been supplemented by "horizontal" interconversions (*i.e.* change in stereochemistry but not in "oxidation state"). The latter have been achieved by a sequence of stereospecific mercuric acetate dehydrogenation and catalytic rehydrogenation (*idem, loc. cit.*; *B. P. Moore* and *Marion*, Canad. J. Chem., 1953, **31**, 187; *D. Kettelhack, M. Rink* and *K. Winterfeld*, Arch. Pharm., 1954, **287**, 1; *E. E. van Tamelen* and *J. S. Baran*, J. Amer. chem. Soc., 1955, **77**, 437), For example, (−)sparteine, $C_{15}H_{26}N_2$, on dehydrogenation with mercuric acetate produces, first, dehydrosparteine, $C_{15}H_{24}N_2$, and then didehydrosparteine, $C_{15}H_{22}N_2$; the former product is rehydrogenated back to (−)sparteine, but the latter gives (−)α-isosparteine:

(-) Sparteine

Hg(OAc)₂

cat. H

Hg(OAc)₂

Hg(OAc)₂

cat. H₂

(-)α-Isosparteine

The two unnamed lupanine isomers in Scheme 2 are not found naturally.

The correctness of the above assignments has been shown by X-ray diffraction studies on $(-)\alpha$-isosparteine monohydrate by *M. Przybylska* and *W. H. Barnes* (Acta Cryst., 1953, **6**, 377).

That the structures shown above in Scheme 2 represent the correct absolute stereochemistry of the compounds has been shown by *S. Okuda, K. Tsuda* and *H. Kataoka* (Chem. and Ind., 1961, 1115, 1751) who related $(-)$anagyrine (the enantiomer of the structure shown above) to $(+)$epilupinine of known absolute configuration, as shown below:

KMnO₄

O₃

hydrolysis

(-)Anagyrine

Δ, $-CO_2$

LiAlH₄

NH₃/EtOH

TosCl/pyr.

(+)Epilupinine

From the interrelations shown above, the whole series of alkaloids discussed in Scheme 2 have the following absolute stereochemistry:

$(+)$Anagyrine	$7S:9S:11S$
$(+)$Thermopsine	$7S:9S:11R$
$(+)$Lupanine	$6R:7S:9S:11S$
$(+)\alpha$-Isolupanine	$6R:7S:9S:11R$
$(-)$Sparteine	$6R:7S:9S:11S$
$(+)\beta$-Isosparteine	$6S:7S:9S:11S$
$(-)\alpha$-Isosparteine	$6R:7S:9S:11R$

(1) Anagyrine-like alkaloids. **Anagyrine,** $C_{15}H_{20}N_2O$, occurs in several species of *Lupinus* and *Genista* and has been isolated from gorse *(Ulex europaeus)*. It resembles cytisine in many of its properties although both nitrogens are tertiary. The aromatic properties and i.r. and u.v. data were indicative of an α-pyridone ring system being present. Three stages of Hofmann degradation were required to remove the more basic nitrogen as trimethylamine (*H. R. Ing*, J. chem. Soc., 1933, 504). The other product of the degradation was hexahydroanagyryline, $C_{15}H_{23}NO$, ozonolysis of which gave a lactam, $C_{11}H_{21}NO$. Hydrolysis and oxidation of the latter gave nonane-2,4-dicarboxylic acid (*H. N. Rydon, ibid.*, 1936, 1444). The structures of hexahydroanagyryline and the lactam follow as XXVI and XXVII, respectively:

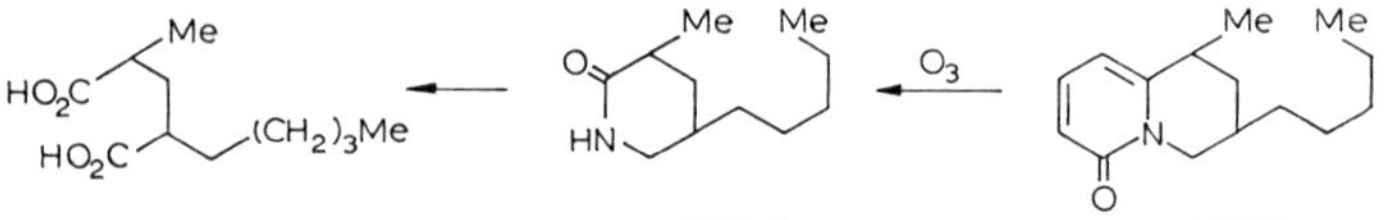

(XXVII)   (XXVI)

On catalytic hydrogenation in acetic acid anagyrine is converted into lupanine (*G. R. Clemo et al., ibid.*, 1931, 429; *Ing, loc. cit.*) and on electrolytic reduction in dilute sulphuric acid, sparteine is produced (*F. Galinovsky* and *E. Stern*, Ber., 1944, **77**, 132).

The last named authors oxidised anagyrine with barium permanganate to 17-oxyanagyrine (XXVIII) which on catalytic hydrogenation gave 17-oxysparteine (XXIX) already synthesised by *Clemo et al.* (J. chem. Soc., 1936, 1025).

On the basis of the above observations the following structure was proposed for anagyrine:

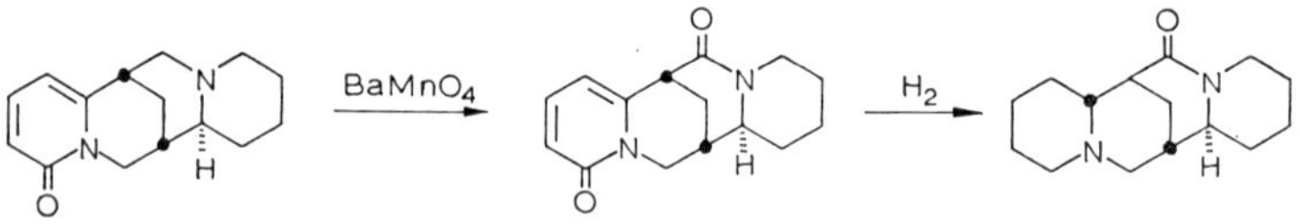

Anagyrine   (XXVIII)   (XXIX)

Anagyrine has been synthesised from α-picoline by *Van Tamelen* and *Baran* (J. Amer. chem. Soc., 1956, **78**, 2913). Condensation of α-picoline with formaldehyde and acetic anhydride gave 2-(α-acetoxymethylvinyl)pyridine, condensation of which with diethyl malonate produced the vinylpyridine derivative XXX. Condensation of the latter with 2,3,4,5-tetrahydropyridine gave the carboxylic acid XXXI, which on esterification under equilibrating conditions gave the single isomer XXXII. This material underwent reduction, bromination and ring closure to the quaternary salt XXXIII, which on oxidation with ferricyanide produced anagyrine:

An improved route to the intermediate XXXII from α-picoline has recently been described by *S. I. Goldberg* and *A. H. Lipkin* (J. org. Chem., 1972, **37**, 1823).

Baptifoline, $C_{15}H_{20}N_2O_2$, has been isolated from *Baptisia* species (*Marion* and *W. F. Cockburn*, J. Amer. chem. Soc., 1948, **70**, 3472; *Marion* and *F. Turcotte, ibid.*, 1948, **70**, 3253) and more recently from *Caulophyllum thalictroides* (*M. S. Flom, R. W. Doskotch* and *J. L. Beal*, J. pharm. Sci., 1967, **56**, 1515). The presence of a hydroxyl group was detected by i.r. spectroscopy which also supported other evidence indicating an α-pyridone ring system. Hydrogenation of baptifoline using platinum in acetic acid resulted in absorption of two mols. of hydrogen and formation of $(-)13$-hydroxylupanine from which the structure of baptifoline follows (*M. Martin-Smith* and *Marion*, Canad. J. Chem., 1957, **35**, 37):

Solvolysis of baptifoline gives the more stable equatorial $C_{(13)}$ epimer **epibaptifoline,** which occurs naturally in *Retama* species (*M. D. Vazquez et al.,* An. Real. Soc. Espan. Fis. Quim. Ser. B, 1966, **62,** 837; C.A., 1967, **66,** 76216h). A synthesis from cytisine by a route analogous to the synthesis of angustifoline gave epibaptifoline (*F. Bohlmann et al.,* Chem. Ber., 1962, **95,** 944).

Camoensine, $C_{14}H_{18}N_2O$, has been isolated from *Camoensia maxima* and characterised by *J. Santamaria* and *F. Khuong Huu Laine* (Phytochemistry, 1975, **14,** 2501):

Camoensine

Leontidine, $C_{14}H_{18}N_2O$, has been isolated from *Leontice alberti* and on the basis of the mass spectrum has been assigned the structure shown below, which has been confirmed by synthesis from cytisine (*S. Iskandarov et al.,* Khim. prir. Soedin., 1971, **5,** 631):

Leontidine

The absolute stereochemistry, $7R:9R:11S$, as shown above has been determined from o.r.d. studies.

Thermopsine, $C_{15}H_{20}N_2O$, has been found as the $(+)$ form in *Lupinus* species and as the $(-)$ form in *Thermopsis* species. I.r. evidence suggested the presence of an α-pyridone ring and this fact with the observation that electrolytic reduction gave α-isosparteine hydrate, $C_{15}H_{28}N_2O$, led to the proposal by *Cockburn* and *Marion* (Canad. J. Chem., 1951, **29,** 13, 22) that thermopsine was the $C_{(11)}$ epimer of anagyrine. Anagyrine on dehydrogenation with mercuric acetate and subsequent reduction produced thermopsine (*Van Tamelen* and *Baran, loc. cit.*).

Thermopsine has also been synthesised by an intriguing route from 13-hydroxylupanine as shown below:

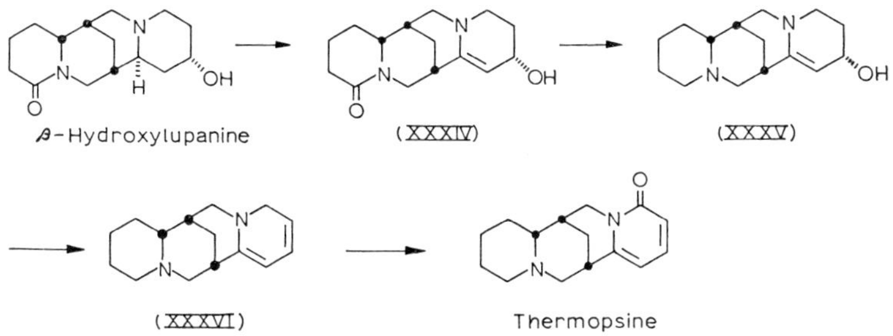

β-Hydroxylupanine   (XXXIV)   (XXXV)

(XXXVI)   Thermopsine

Dehydrogenation of 13-hydroxylupanine gave the enamine XXXIV, reduction of which with lithium tetrahydridoaluminate removed the lactam group to give XXXV. Dehydration with phosphorus pentoxide then produced the diene XXXVI, which was oxidised by ferricyanide to thermopsine *(Bohlmann et al., loc. cit.)*. Physical data on these anagyrine-like alkaloids are listed in Table 2.

TABLE 2

ANAGYRINE-LIKE ALKALOIDS

Alkaloid	m.p. (°C)	b.p. (°C/mm)	[α]$_D$°
Anagyrine	—	210–215/4	−165.3
Baptifoline	210	—	−147.7 a
Leontidine	118–119	—	−192
Thermopsine	205–205.5	—	−159.6 a

(2) Lupanine-like alkaloids. **Lupanine,** $C_{15}H_{24}N_2O$, is found as (+), (−) and (±) forms in blue lupin seeds. It is a monoacidic tertiary base having no *N*-methyl group. The oxygen atom is part of an amide group, thus accounting for the lack of basicity of the second nitrogen. On reduction with phosphorus and hydriodic acid, lupanine gives sparteine *(Clemo et al., J. chem. Soc., 1928, 1811; 1931, 429; Galinovsky and Stern, loc. cit.)*, and as mentioned above, lupanine is the product of catalytic hydrogenation of anagyrine. From these relationships and the above data *Clemo* and *R. Raper* (J. chem. Soc., 1933, 644) proposed the structure for lupanine shown in the following relay synthesis for racemic lupanine *(Clemo et al., ibid., 1956, 3390)*:

(XXXVII)

(XXXVIII)   (XXXIX)

(±)Lupanine

Thus, ethyl β-hydroxy-α-6-methyl-2-pyridyl)acrylate (XXXVII) (*idem, ibid.*, 1954, 2693), on condensation with ethyl 2-pyridylacetate gave the quinolizone which on oxidation, esterification, catalytic hydrogenation and cyclisation gave the tetracyclic intermediate XXXVIII. By successive reduction, chlorination and Hofmann degradation, the latter was converted into 2-methyl-Δ^2-dehydrosparteine (XXXIX) which had previously been prepared by the action of methylmagnesium iodide on lupanine (*K. Winterfeld* and *E. Hofmann*, Arch. Pharm., 1937, **275**, 5). This material was converted in several steps to (±)lupanine. The synthesis of (±)anagyrine by *Van Tamelen* and *Baran* also constitutes a synthesis of (±)lupanine, as the former gives the latter on catalytic hydrogenation.

Trilupine, isolated by *J. F. Couch* (J. Amer. chem. Soc., 1937, **59**, 1469) from *Lupinus laxus*, originally formulated as the bis-*N*-oxide of lupanine, has been shown by *Marion* (Canad. J. Chem., 1952, **30**, 386) to be (+)lupanine, isolated as the hydrochloride dihydrate. More recently lupanine has been obtained from *Leontice leontopelatum* along with **Leontiformine** (XL; R = −CHO) and **Leontiformidine** (XL: R = −H) (*P. P. Panov, L. N. Panova* and *N. M. Mollov*, Doklady Bolg. Akad. Nauk, 1972, **25**, 55):

(XL)

Camoensidine, $C_{14}H_{22}N_2O$, has been isolated along with camoensine *(vide supra)* from *Camoensia maxima (Santamaria* and *Khuong Huu Laine, loc. cit.)*:

Camoensidine

α-Isolupanine, $C_{15}H_{24}N_2O$, found in several *Lupinus* species, is converted by lithium tetrahydridoaluminate into α-isosparteine, and has an i.r. spectrum identical with that of tetrahydrothermopsine, hence it is assigned the following structure, which is diastereoisomeric with lupanine:

Isolupanine

Dehydrogenation with mercuric acetate and subsequent catalytic hydrogenation convert lupanine into α-isolupanine *(Marion* and *N. J. Leonard,* Canad. J. Chem., 1951, **29,** 355). This transformation has also been carried out by reduction of the iminium cation XLI, formed from lupanine $N_{(16)}$-oxide (XLII), with tetrahydridoborate *(P. Baranowski, M. Wiewiorowski* and *L. Lompa-Krzymien,* Rocz. Chem., 1966, **40,** 73):

(XLII)   (XLI)

Lupanine   Isolupanine

13-Hydroxylupanine, $C_{15}H_{24}N_2O$, has been isolated from several *Lupinus* species. On reduction with phosphorus and hydriodic acid it gives lupanine (*A. Beckel*, Arch. Pharm., 1910, **248**, 451). Catalytic reduction and subsequent Oppenauer oxidation gives an oxosparteine, thus establishing the secondary nature of the hydroxyl group (*F. Galinovsky* and *M. Pöhm*, Monatsh., 1949, **80**, 864). The position of the hydroxyl group as $C_{(13)}$ was established by oxidation with chromic acid and the structure of 13-hydroxylupanine follows as XLIII:

(XLIII)

The configuration of the hydroxyl group at $C_{(13)}$ was determined by an investigation of the rate of hydrolysis of $C_{(13)}$ epimeric esters against suitable model compounds (*Bohlmann et al.*, Chem. Ber., 1961, **94**, 1767). Thus the widely occurring alkaloid is the 13-axial isomer (*i.e.* OH is α in XLIII). This is confirmed by the observation that reaction of angustifoline with formaldehyde gives the equatorial 13-epihydroxylupanine (OH is β in XLIII). The latter is found as **Jamaidine** in *Ormosia* species but is much less common than the axial epimer (*H. A. Lloyd*, J. org. Chem., 1961, **26**, 2143). Both of the epimeric tosylates of 13-hydroxylupanine are converted to angustifoline and tetrahydrorhombifoline on treatment with base (see above) but the equatorial isomer reacts much faster. Similarly solvolysis of the tosylates with methanol gives 13-**epimethoxylupanine** identical with the alkaloid from *Lupinus angustifolius* (*M. D. Bratek* and *Wiewiorowski*, Bull. Acad. Polon. Sci. Ser. Sci. Chim., 1961, **9**, 705):

13-Epimethoxylupanine

Recently many esters of 13α-hydroxylupanine (XLIII) have been found in nature. The list in Table 3 gives the names, structures and sources of these alkaloids.

TABLE 3

ACYLOXYLUPANINES

Alkaloid	R	Source	Ref.
Benzoate ester	—C(=O)—Ph	*L. angustifolius* *L. polyphyllus*	1
Calpurnine (Oroboidine)	—C(=O)-pyrrol-2-yl	*Calpurnia lasiogyne* *Virgilia capensis*	2
Cinnamate esters (*cis*- and *trans*-)	—C(=O)—CH=CHPh	*L. angustifolius*	1
Cinegalleine	—C(=O)-(3-OH-4-OMe-5-OMe-phenyl)	*Genista cinerea*	3
Cinegalline	—C(=O)-(3-OH-4-OMe-5-OH-phenyl)	*Genista cinerea*	4
Cineverine	—C(=O)-(3-OMe-4-OMe-phenyl)	*Genista cinerea*	5
Formylcinegalleine	—C(=O)-(3-OCHO-4-OMe-5-OMe-phenyl)	*Genista cinerea*	6
Sarodesmine	—C(=O)-(3-OMe-4-OMe-5-OMe-phenyl)	*Sarothamnus catalaunicus*	7
Tiglate ester	—C(=O)—C(Me)=CH(Me)	*L. angustifolius*	1

References
1   *M. D. Bratek-Wiewiorowski, M. Wiewiorowski* and *I. Reifer*, Bull. Acad. Polon. Sci. Ser. Sci. Chim., 1963, **11**, 629.
2   *G. C. Gerrans* and *J. Harley-Mason*, J. chem. Soc., 1964, 2202; *A. Goosen, ibid.*, 1963, 3067.
3   *G. Faugeras* and *R. R. Paris*, Compt. rend., 1970, **270D**, 203.
4   *Idem, ibid.*, 1968, **267D**, 538.
5   *Idem, ibid.*, 1966, **263D**, 436.
6   *Idem, ibid.*, 1970, **271D**, 1219.
7   *Idem, ibid.*, 1970, **271D**, 611.

Lupanoline, $C_{15}H_{24}N_2O_2$, isolated from *L. sericeus* (*Marion, Leonard* and *B. P. Moore*, Canad. J. Chem., 1953, **31**, 181, 187) has hydroxyl group absorption in the i.r. spectrum and is reduced by lithium tetrahydridoaluminate to $(-)\beta$-isosparteine. Acetic anhydride dehydrates the alkaloid to anhydrolupanoline, $C_{15}H_{22}N_2O$, which on hydrogenation over platinum gives 17-oxo-β-isosparteine. This evidence establishes the oxo function at $C_{(17)}$ and because reduction with tetrahydridoaluminate removes the hydroxyl this must be adjacent to nitrogen (methanolamine), and the structure of lupanoline must therefore be as shown:

Lupanoline → 17-Oxo-β-isosparteine

Nuttalline, $C_{15}H_{24}N_2O_2$, extracted from *Lupinus nuttallii*, on reduction with sodium tetrahydridoborate gives $(-)4\alpha$-hydroxysparteine, on Oppenauer oxidation 2,4-dioxosparteine and on dehydration with phosphorus pentoxide 2-oxo-3,4-dehydrosparteine. From this evidence nutalline has been assigned the structure 4α-hydroxy-2-oxosparteine (*S. I. Goldberg* and *V. M. Balthis*, J. chem. Soc., D, 1969, 660):

Nuttalline

17-Oxolupanine, $C_{15}H_{22}N_2O_2$, isolated from *Lupinus polyphyllus* (F. *Bohlmann* and *E. Winterfeldt*, Chem. Ber., 1960, **93**, 1956) and from *L. angustifolius* (*Bratek* and *Wiewiorowski*, Rocz. Chem., 1959, **33**, 1187), had been synthesised earlier by oxidation of lupanine with potassium permanganate or potassium ferricyanide (*O. E. Edwards, F. H. Clarke* and *B. Douglas*, Canad. J. Chem., 1954, **32**, 235). Confirmation of $C_{(17)}$ as the location of the oxo group in 17-oxolupanine came from reduction with lithium tetrahydridoaluminate, which gave sparteine:

17-Oxolupanine

Physical data are recorded in Table 4.

TABLE 4

LUPANINE-LIKE ALKALOIDS

Alkaloid	m.p. (°C)	b.p. (°C/mm)	$[\alpha]_D°$
Calpurnine	154		+59
Cineverine	149–151		+65 a
13-Epihydroxylupanine	194.5–195		+63.7 a
13-Hydroxylupanine	172–174		+61.1 d
13-Hydroxylupanine			
benzoate ester	205		+35
cinnamate ester			
(cis)	145		+26.1
(trans)	166		+42
tiglate ester	perchlorate m.p. 240		+28.1
α-Isolupanine	75–76		+65.9 a
Lupanine	44	185–186/0.08	+61.4 e
Lupanoline	174–176		+31 a
Nuttalline	108–109		+25.3 a
17-Oxolupanine	154		+139

(3) Sparteine-like alkaloids. **Sparteine (pachycarpine)**, $C_{15}H_{26}N_2$, is very common as $(+)$, $(-)$ and $(\pm)$ forms in many species of *Baptisia*, *Lupinus*, *Cytisus* and *Genista*. It is a saturated diacidic tertiary base which on methylation gives two stereoisomeric monomethiodides, thus suggesting a symmetrical structure. Six stages of Hofmann degradation completely remove both nitrogens giving an unsaturated hydrocarbon **sparteilene**, $C_{15}H_{20}$. Inclusion of a hydrogenation stage after each Hofmann degradation (*P. Karrer et al.*, Helv., 1930, **13**, 1292) gave a saturated hydrocarbon, $C_{15}H_{32}$, which was subsequently shown by synthesis (*M. Schirm* and *H. Besendorf*, Arch. Pharm., 1942, **280**, 64) to be 6,8-dimethyltridecane.

As discussed above, anagyrine on electrolytic reduction, and lupanine on reduction with phosphorus and hydriodic acid both give sparteine. These relationships led to the proposal of the following structure for sparteine (*Clemo* and *Raper*, *loc. cit.*):

Sparteine

Support for this formulation was provided by the product of oxidation of sparteine with potassium ferricyanide. This was shown to be 17-oxosparteine by comparison with an authentic, synthetic sample (see below).

The reverse of this oxidation reaction provided one of the early syntheses of ($\pm$)sparteine. 17-Oxosparteine gives ($\pm$)sparteine on reduction with lithium tetrahydridoaluminate (*Clemo et al.*, J. chem. Soc., 1949, 663), or by electrolytic reduction (*F. Galinovsky et al.*, Monatsh., 1948, **79,** 322; 1949, **80,** 112). Because of their relationship the syntheses of lupanine described above are also syntheses of sparteine. A synthesis of sparteine following the biogenetic hypothesis of *C. Schöpf* has been described by *E. E. van Tamelen* and *R. L. Foltz* (J. Amer. chem. Soc., 1960, **82,** 2400). Condensation of acetone, formaldehyde and piperidine under mild conditions gives a Mannich base, XLIV, which on dehydrogenation with mercuric acetate spontaneously cyclises to 8-oxosparteine. The latter affords ($\pm$)sparteine on Wolff–Kishner reduction:

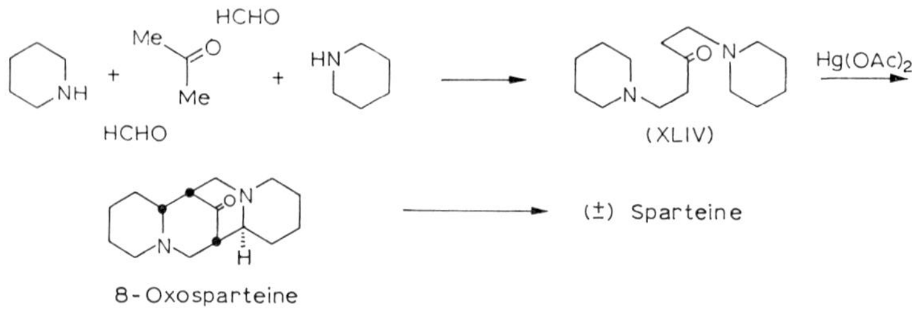

8-Oxosparteine

Another synthesis has been described from epilupinine. 1-Piperidino-quinolizidine (XLV) formed *via* the bromolupinine from the enantiomer of III (p. 287), cyclises on dehydrogenation with mercuric acetate to the intermediate XLVI, which on reduction with sodium tetrahydridoborate gives ($\pm$)sparteine. An alternative mode of cyclisation results ultimately in allomatridine (*F. Bohlmann, Winterfeldt* and *U. Friese*, Chem. Ber., 1963, **96,** 2251):

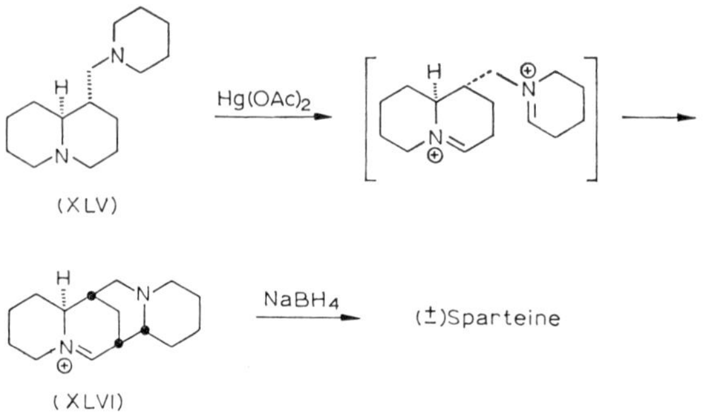

(XLV)

(XLVI)

Other syntheses have been described by *F. Šorm* and *B. Keil* (Coll. Czech. Chem. Comm., 1947, **12,** 655; 1948, **13,** 544) and by *N. J. Leonard* and

R. E. Beyler (J. Amer. chem. Soc., 1949, **71**, 757; 1950, **72**, 1316).

On oxidation with sodium hypobromite sparteine gives two isomeric products of molecular formula $C_{30}H_{48}N_2$, called α- and β-diplospartyrine, respectively. The elegant rationalisation of this reaction by *Schöpf* and *K. Keller* (Naturwiss., 1956, **43**, 325) was confirmed by these authors by an alternative synthesis of β-diplospartyrine. By analogy with the oxidation of sparteine under other reaction conditions, it was suggested that two products, XLVII and XLVIII, were formed initially on oxidation with sodium hypobromite and that under the acid conditions these products condensed together to give α- and β-diplospartyrines, which were stereoisomeric at $C_{(17')}$ as shown below:

Sparteine

NaOBr

(XLVII)

$H^{\oplus}$

(XLVIII)

NaOBr

$-H^{\oplus}$

Diplospartyrine

Conformational studies of sparteine in solution by n.m.r. spectroscopy show that ring C exists in the boat form shown on p. 311 (*Bohlmann, D. Schumann* and *C. Arndt*, Tetrahedron Letters, 1965, 2705). An X-ray study of sparteine-$N_{(16)}$-oxide, however, showed that in this case all four piperidine rings were in the chair form (*P. J. Krueger* and *J. Skolik*, Tetrahedron, 1967, **23**, 1799; *S. N. Srivastava* and *M. Przybylska*, Tetrahedron Letters, 1968, 2697). Circular dichroism spectra of seventeen compounds of the sparteine series have been described (*W. Klyne et al.*, J. chem. Soc., Perkin II, 1974, 2565).

$(-)\alpha$-**Isosparteine,** $C_{15}H_{26}N_2$, isolated from *Lupinus caudatus* and *Cytisus scoparius* is also formed from $(-)$sparteine by dehydrogenation with mercuric acetate and subsequent catalytic reduction (*Winterfeld et al.,* Arch. Pharm., 1935, **273**, 521). This process has also been carried out by heating with aluminium chloride (*Galinovsky* and *P. Knoth*, Naturwiss., 1954, **41**, 454) and by heating with platinum at 103° (*Wiewiorowski et al.,* Bull. Acad. Polon. Sci. Ser. Sci. Chim., 1965, **13**, 757). Catalytic hydrogenation of thermopsine (p. 304) gives α-isosparteine (*L. Marion et al.*, J. Amer. chem. Soc., 1951, **73**, 1769), the structure and stereochemistry of which was placed on a firm basis by X-ray diffraction studies on the monohydrate (*Przybylska* and *W. H. Barnes*, Acta cryst., 1953, **6**, 377).

A synthesis of (±)α-isosparteine from 1,3-di(2-piperidyl)-2-propanone has been described recently by *Schöpf et al.* (Ann., 1972, **755**, 86). Condensation of the latter compound with formaldehyde generated the acetal XLIX, which on heating with acetic anhydride gave a 17% yield of 8-oxo-α-isosparteine (L), converted into α-isosparteine by reduction of the semi-carbazone:

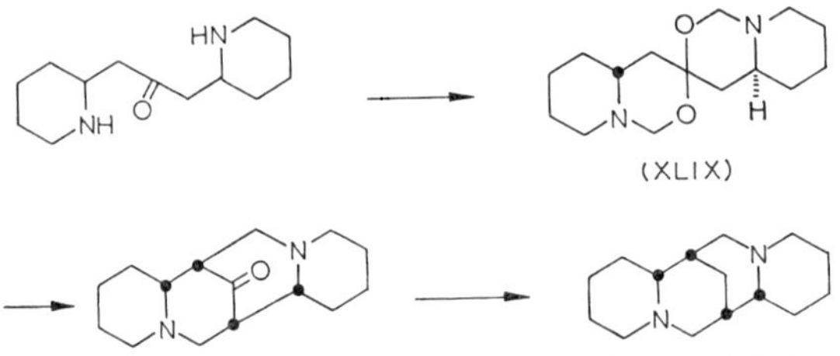

(XLIX)

(L)   α-Isosparteine

(−)*β*-**Isosparteine**, $C_{15}H_{26}N_2$, has been isolated from several *Lupinus* species and as spartalupine has been obtained from *L. sericeus* (*M. Carmack et al.*, J. Amer. chem. Soc., 1955, **77**, 4435). *R. Greenhalg* and *Marion* (Canad. J. Chem., 1956, **34**, 456) have shown that pusilline, originally assigned a piperidylquinolizidine structure, is in fact (−)*β*-isosparteine. Dehydrogenation with mercuric acetate and subsequent catalytic hydrogenation of *β*-isosparteine gives sparteine from which the structure follows:

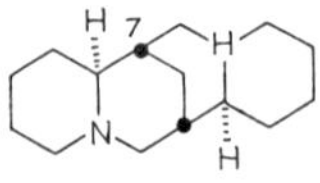

β-Isosparteine

(−)7-**Hydroxy-*β*-isosparteine** has recently been isolated from *L. sericeus* by *Carmack, S. I. Goldberg* and *E. W. Martin* (J. org. Chem., 1967, **32**, 3045). The structure of the alkaloid was confirmed by an X-ray structure determination on the perchlorate (*J. M. H. Pinkerton* and *L. K. Steinrauf, ibid.,* 1967,

32, 1828) but the racemic material had been synthesised earlier by *Bohlmann et al.* (Chem. Ber., 1962, **95,** 2365).

(+)17-**Oxosparteine,** $C_{15}H_{24}N_2O$, isolated from *Genista monosperma* (*E. P. White*, New Zealand J. Sci. Technol., 1957, **B38,** 712) was identical with the alkaline ferricyanide oxidation product of (+)sparteine (*Clemo* and *G. C. Leitch*, J. chem. Soc., 1928, 1811). Racemic 17-oxosparteine was synthesised from ethyl 2-pyridylacetate by *Clemo et al.* (*ibid.*, 1936, 1025):

17-Oxosparteine

Aphylline, $C_{15}H_{24}N_2O$, found in *Anabasis aphylla* is a monoacidic tertiary base containing a lactam system which on catalytic hydrogenation gives sparteine. *A. P. Orekhov* (J. gen. Chem. U.S.S.R., 1937, **7,** 2048) has proposed the structure 10-oxosparteine, diastereoisomeric with 17-oxosparteine, to explain the difference in the reactivity of the lactam groupings between the former and the latter (see also *Galinovsky* and *E. Stern*, Ber., 1944, **77,** 132). On acid hydrolysis aphylline undergoes ring-opening to the amino acid aphyllic acid (R = H), the methyl ester of which (R = Me), has been found in *Anabasis aphylla* by *Kh. A. Aslanov, S. Z. Mukhamedzhanov* and *A. S. Sadykov* (Nauch. Tr. Tashkent Gos. Univ., 1966, **286,** 71):

Aphylline (LI)   Aphyllic acid

A synthesis of (±)aphylline has been described by *Bohlmann et al.* (Chem. Ber., 1957, **90,** 653). A similar method was adopted by *Van Tamelen* and *J. S. Baran* (J. Amer. chem. Soc., 1956, **78,** 2913) who modified their synthesis of anagyrine by catalytically reducing the intermediate XXXII, followed by heating to cause cyclisation with formation of (±)aphylline (LI):

(±)Aphylline

(XXXII)

Aphyllidine, $C_{15}H_{22}N_2O$, also found in *Anabasis aphylla*, on careful hydrogenation gives aphylline. Placement of the double bond at $C_{(5)}$–$C_{(6)}$ follows from Hofmann degradation studies (*Orekhov et al.*, Chem. Ber., 1932, **65,**

234; 1934, **67,** 1974) and from oxidation with chromic acid (*A. S. Sadykhov* and *R. N. Nurridinov*, Doklady Akad. Nauk S.S.S.R., 1955, **102,** 755) which generates 3-aminobutanoic acid and quinolizidine-1,3-dicarboxylic acid. The structure is thus represented as LII:

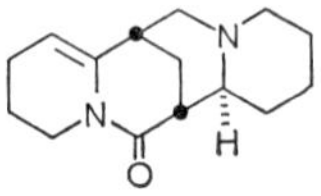

Aphyllidine (LⅡ)   Argyrolobine (LⅢ)

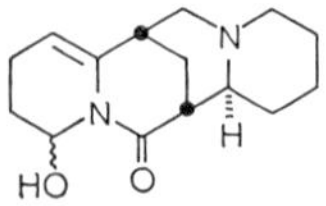

Aphyllidine is also found in *Argyrolobium megarhizum* along with **argyro-lobine,** $C_{15}H_{22}N_2O_2$ (*Y. Tsuda* and *Marion,* Canad. J. Chem., 1964, **42,** 764). [As the natural alkaloids, (+)aphylline and (+)aphyllidine generate (+)sparteine on reduction, in contrast to (+)argyrolobine which gives (−)sparteine, the absolute configurations of the former are represented correctly as enantiomeric with LI and LII, whereas LIII correctly represents the absolute stereochemistry of (+)argyrolobine]. 17-**Oxoaphyllidine** has also been isolated from *Anabasis aphylla* by *Sadykov* and *Nurridinov* (Zhur. obshcheĭ Khim., 1960, **30,** 1736, 1739). (−)**Lindenianine,** $C_{15}H_{24}N_2O_2$, isolated from *Lupinus lindenianus* was shown to be (−)13β-hydroxy-10-oxo-sparteine by chemical and spectroscopic methods. The preferred conformation was established from the c.d. spectrum (*T. Nakano, A. C. Spinelli* and *A. M. Méndez,* J. org. Chem., 1974, **39,** 3584):

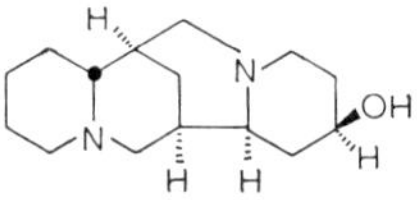

Lindenianine

Monspessulanine, $C_{15}H_{22}N_2O$, isolated from *Cytisus monspessulanus* by *White* (New Zealand J. Sci. Technol., 1951, **B33,** 54), was assigned the structure 5,6-dehydro-10-oxo-α-isosparteine from spectral data and by its conversion to the $C_{(11)}$ epimer of aphylline (*idem,* J. chem. Soc., 1964, 4613):

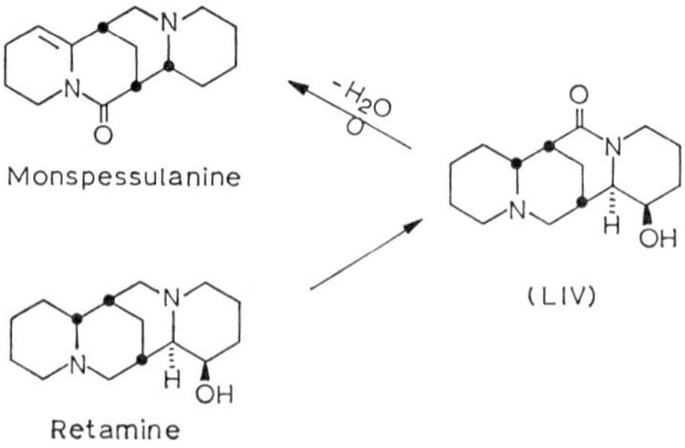

Monspessulanine

Retamine

(LIV)

The structure has been confirmed by synthesis, by oxidation of retamine (see below) with ferricyanide to 17-oxoretamine (LIV), which on dehydration gave monspessulanine (*Bohlmann* and *Schumann*, Tetrahedron Letters, 1965, 2433).

(+)**Retamine,** $C_{15}H_{26}N_2O$, isolated from *Retama sphaerocarpa* and from several species of *Genista* is a ditertiary di-acidic base which on reduction with phosphorus and hydriodic acid gives sparteine. An early suggestion by *I. Ribas et al.* (Anales fis. y quim. Madrid, 1946, **42**, 516) that the base contained a tertiary hydroxyl group led to several wrong structures but this was corrected later by the same author (Tetrahedron Letters, 1960, **11**, 77) who proved the presence of a secondary hydroxyl group. *Bohlmann et al.* (Chem. Ber., 1962, **95**, 2365; 1965, **98**, 653, 659) considered the structure of retamine to be either a 12- or 14-hydroxysparteine derivative on the earlier evidence. Comparison with the authentic, synthetic epimers of 14-hydroxysparteine eliminated the latter alternative. Spectral data and solvolysis experiments indicated the presence of an axial hydroxyl group and thus the structure of retamine follows as 12-hydroxysparteine. Further support for this structure was provided by a synthesis from 3-pyridinol. Alternatively retamine has been produced by reduction with tetrahydridoborate and subsequent oxidation of various dehydrosparteines (*Ribas, J. L. Castedo* and *A. Garcia,* Tetrahedron Letters, 1965, 3181; *K. H. Shim, L. Fonzes* and *Marion,* Canad. J. Chem., 1965, **43**, 2012). [As the natural (+)alkaloid gives (+)sparteine on hydrogenation the absolute stereochemistry is enantiomeric with that shown above.]

Thermopsamine, $C_{15}H_{26}N_2O$, reported recently from *Thermopsis lanceolata* is a 13-hydroxysparteine of unknown stereochemistry (*S. Yu. Yunusov et al.,* Khim. prir. Soedin., 1971, **7**, 463).

(−)**Virgiline,** $C_{15}H_{24}N_2O_2$, isolated from *Virgilia oroboides (V. capensis)* was originally given the wrong molecular formula, $C_{16}H_{26}N_2O_2$ (*White,* New Zealand J. Sci. Technol., 1946, **27B**, 478). This was corrected by *G. C. Gerrans* and *J. Harley-Mason* (Chem. and Ind., 1963, 1433; J. chem. Soc., 1964, 2202) who also deduced the structure. Evidence for the presence of a hydroxyl group was obtained from the i.r. spectrum and from acetylation, and catalytic hydrogenation gave 12-hydroxysparteine. The structure thus follows as (−)13-hydroxyaphylline (R = H):

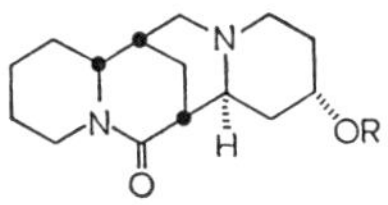

Thermopsamine

(R = H)

O-(2-**Pyrrolylcarbonyl**)**virgiline** (above formula, R =

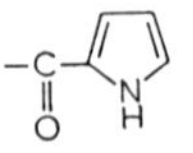

has also been obtained from *V. oroboides* (*White*, J. chem. Soc., 1964, 5243) and more recently the 2,3-dehydro derivative (LV) has been isolated from *Readea membranaceae*. The double-bond position was assigned from the n.m.r. spectrum which lacked the $C_{(2)}$ methylene group and showed evidence for the presence of –CH=CH–N–C– (*A. H. Manchanda, J. Nabney*

and *D. W. Young*, J. chem. Soc., C, 1968, 615):

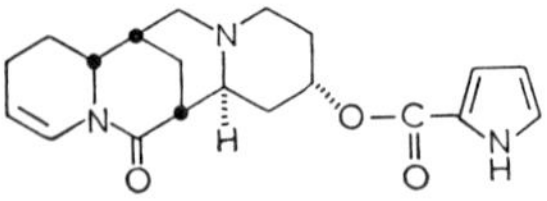

(LV)

Multiflorine, $C_{15}H_{22}N_2O$, isolated from *Lupinus varius* as Alkaloid LVI, from *L. albus* as Alkaloid "b-109" and from *L. multiflorus*, on hydrogenation takes up three moles of hydrogen to give (−)sparteine. The structure as 2,3-dehydro-4-oxosparteine was assigned on the basis of spectral data (*W. D. Crow*, Austral. J. Chem., 1959, **12**, 474; see also *Goldberg* and *R. F. Moates*, J. org. Chem., 1967, **32**, 1832). 13-**Hydroxymultiflorine** and 5,6-**dehydro**-13-**hydroxymultiflorine** have been isolated from *L. albus* also (*M. Wiewiorowski, J. Bartz* and *W. Wysocka*, Bull. Acad. Polon. Sci. Ser. Sci. Chim., 1961, **9**, 715):

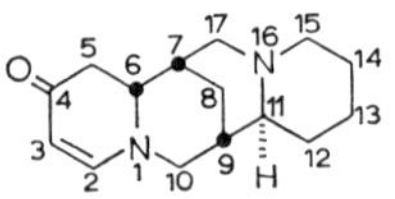

Multiflorine

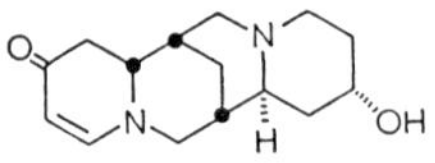

13-Hydroxymultiflorine

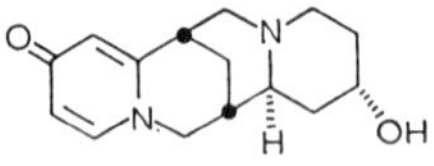

5,6-Dehydro-13-hydroxymultiflorine

Physical data on these sparteine-like alkaloids are recorded in Table 5.

TABLE 5

SPARTEINE-LIKE ALKALOIDS

Alkaloid	m.p. (°C)	b.p. (°C/mm)	[α]$_D$°
Aphyllidine	112–112.5		+5.57 b
Aphylline	57		+10.3 b
α-Isosparteine	108–110		−51.3 a
β-Isosparteine	32.2–32.4	100–110/0.2	−15.4 a
Lindenianine	212–213		−11 c
Monspessulanine	101		+117 a
Multiflorine	109		−299
17-Oxoaphyllidine	184		−21 b
17-Oxosparteine	82–84		+20 a
O-2-Pyrrolylcarbonylvirgiline	271		−23 a
Retamine	168		+43.15 a
Sparteine	—	188/18	−17 a
Thermopsamine	154–155		+26.4 a
Virgiline	248		−46 a

(ii) Matrine-like alkaloids

This series of tetracyclic lupin alkaloids is structurally related to the groups discussed above in section (i) and may be represented by the skeletal (matridine) formula shown below:

The numbering system shown above is the one at present in use by Chemical Abstracts, although others have been used in the past (R. H. F. Manske, "The Alkaloids", Vol. VII, Academic Press, London and New York, 1960, p. 296; A. M. Patterson, L. T. Capell and D. F. Walker, "The Ring Index", 2nd Edition, 1960, RRI 5048, p. 684).

Matrine, $C_{15}H_{24}N_2O$, isolated from several species of *Sophora* and from *L. angustifolius* as "lupanidine" (M. Rink, Arch. Pharm., 1952, **285**, 187) is a saturated monoacidic tertiary base. Alkaline hydrolysis gives the salt of an amino acid, matrinic acid, $C_{15}H_{26}N_2O_2$ (H. Kondo, ibid., 1928, **266**, 1) suggesting the presence of a lactam group. This is also supported by reduction with lithium tetrahydridoaluminate, which gives matridine, $C_{15}H_{26}N_2$. Matrinic acid on distillation with zinc dust gives 1-methylquinolizidine (K. Winterfeld and A. Kneuer, Ber., 1931, **64**, 150) and on distillation with soda-lime gives, among other products, a base named α-matrinidine,

$C_{12}H_{20}N_2$ (*Kondo et al.*, Ber., 1935, **68**, 570). This material on distillation with zinc dust gives 1-methylquinolizidine and on dehydrogenation loses two molecules of hydrogen to give dehydro-α-matrinidine, $C_{12}H_{16}N_2$. The reactivity of the methyl group in this molecule and u.v. studies (*S. Okuda*, Pharm. Bull., Tokyo, 1956, **4**, 257) support the structure LVI for the latter, which was originally proposed by *Kondo et al.* (Ber., 1935, **68**, 1899). This structure has been confirmed by the synthesis shown below (*K. Tsuda et al.*, J. org. Chem., 1956, **21**, 598, 1481):

Dehydro-α-matrinidine
(LVI)

(LVII)

Dehydrogenation of matrine itself with palladised asbestos gives octadehydro-matrine, $C_{15}H_{16}N_2O$, which exhibits colour reactions characteristic of an α-pyridone, thus supporting the presence of a lactam. The other dehydro-genation product was a base, $C_{14}H_{20}N_2$, which was synthesised from dehydro-α-matrinidine (LVI) by treatment with butyl-lithium and ethyl bromide (*Tsuda*, Ber., 1936, **69**, 429) and must therefore be LVII.

From the accumulated evidence discussed matrine and matrinic acid were formulated as:

Matrine

Matrinic acid

The above structure for matrine has four asymmetric carbon atoms (at $C_{(5)}$, $C_{(6)}$, $C_{(7)}$ and $C_{(11)}$) and obviously matrine is therefore one of several stereochemical possibilities. This is exemplified by the observation that on heating with platinum and hydrogen, matrine is isomerised to **allomatrine,** a compound which exhibits similar properties to the former and gives rise to a whole series of allo derivatives which parallel the normal matrine compounds (*E. Ochiai, S. Okuda* and *H. Minato,* J. pharm. Soc., Japan, 1952, **72,** 781).

The assignment of the relative stereochemistry of matrine and allomatrine by *Tsuda* and *H. Mishima* (Chem. and Pharm. Bull., Japan, 1957, **5,** 285) using conformational analysis, was made from the following assumptions and observations:

(*1*) The $C_{(11)}$–$C_{(12)}$ bond was equatorial to ring C in both series. This assumption decreases the number of possible racemates from eight to four and is based on the observation that matrinic and allomatrinic acids are not isomerised on boiling with alcoholic base.

(*2*) The relative inaccessibility of $N_{(1)}$ in matrine relative to allomatrine.

(*3*) The relative rates of dehydrogenation of matrine and allomatrine. The greater rate observed in the former reflects the greater number of *cis* hydrogens.

(*4*) Conversion from the "normal" to the allo-series represents formation of the thermodynamically more stable conformer.

The structures of matrine and allomatrine were thus assigned as LVIIIa and LIX:

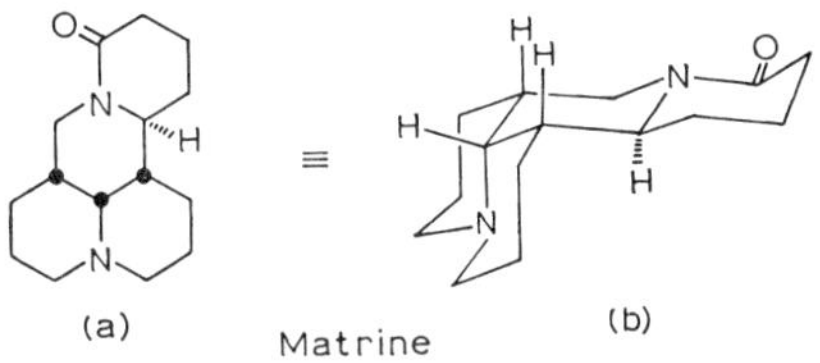

(a)   (b)

Matrine
(LVIII)

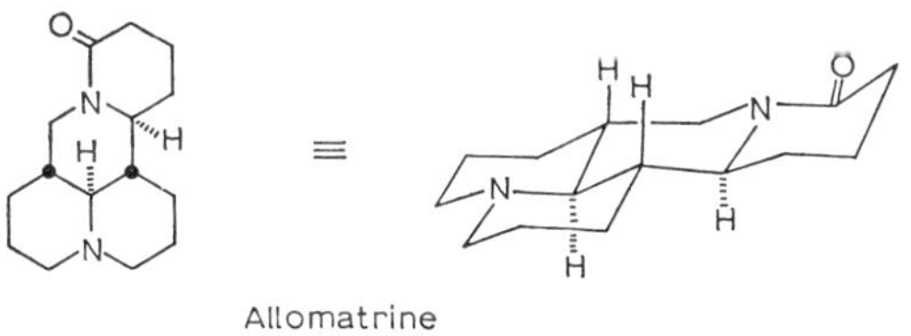

Allomatrine
(LIX)

Support for these assignments has been obtained from dehydrogenation experiments with mercuric acetate (*F. Bohlmann, W. Weise* and *D. Rahtz,* Angew. Chem., 1957, **69,** 641, 642), dipole moment measurements (*B. Eda, Tsuda* and *M. Kubo,* J. Amer. chem. Soc., 1958, **80,** 2426) and from further isomerisation experiments (*Kh. A. Aslanov et al.,* Khim. prir. Soedin., 1969, **2,** 93, 96). Infrared (*F. Bohlmann et al.,* Chem. Ber., 1958, **91,** 2176) and n.m.r. data (*Bohlmann* and *D. Schumann,* Tetrahedron Letters, 1965, 2435) also support the above assignments.

The absolute configuration of (+)matrine has been determined by *Okuda et al.* (Chem. and Pharm. Bull., Japan, 1966, **14**, 314). The two epimeric 14-chloromatrines (LXa and LXb) prepared from (+)matrine (*idem*, Chem. and Ind., London, 1962, 1326) were separated by chromatography and assigned the structures shown by n.m.r. (from the size of the coupling constants between $C_{(14)}$–hydrogen and the adjacent hydrogens on $C_{(13)}$). This fixed the stereochemistry of the new asymmetric centre (at $C_{(14)}$) relative to the other centres of asymmetry in the molecule.

(LXa) R^1 = H, R^2 = Cl
(LXb) R^1 = Cl, R^2 = H
(LVIIIb) R^1 = R^2 = H

The α-chloromatrine (LXa) on degradation with nitric acid gave (+)chlorosuccinic acid of known absolute stereochemistry, thus fixing the absolute stereochemistry at $C_{(14)}$ and by extension also at $C_{(5)}$, $C_{(6)}$, $C_{(7)}$ and $C_{(11)}$. The absolute configuration of (+)matrine is thus given by LVIIIb and LVIIIa represents (−)matrine. The assignment was supported by the results of an asymmetric induction experiment based on the method of *A. Horeau* (Tetrahedron Letters, 1962, 965) and by a combination of o.r.d. and n.m.r. work (*S. Iskandarov et al.*, Khim. prir. Soedin., 1971, **7**, 174) although the opposite configuration for (+)matrine has been reported by *O. Cervinka* (Z. Chem., 1967, **7**, 190) based on rotational shift analogies with the sparteine series.

A large body of synthetic work has further confirmed the structural assignments in the matrine series.

(±)Allomatridine was synthesised by *Tsuda* and *Mishima (loc. cit.)* from dehydro-α-matrinidine (LVI) by a base-catalysed condensation with diethyl ethoxymethylenemalonate. Subsequent saponification and decarboxylation gave LXI identical with octadehydromatrine. This was converted to (±)allomatridine by high-pressure hydrogenation:

(LVI)   (LXI)

Allomatridine   (LXII)

An alternative synthesis described by *C. Schöpf* and *W. Schweter* (Naturwiss., 1953, **40**, 165) converted the piperideine trimer LXII to allomatridine by catalytic hydrogenation and heating, presumably by ring opening and alternative recyclisation. Another synthesis described by *Bohlmann, E. Winterfeldt* and *U. Friese* (Chem. Ber., 1963, **96**, 2251), involved mercuric acetate dehydrogenation of epilupinine piperidide and subsequent reduction (see sparteine, p. 311). Other syntheses have been described by *T. K. Kasymov et al.* (Khim. prir. Soedin., 1969, **5**, 458) and by *G. Kobayashi et al.* (Chem. and Pharm. Bull., Japan, 1970, **18**, 124). ($\pm$)Matridine (LXIII) the thermodynamically less stable isomer of allomatridine has been synthesised by the method outlined below, which has also been modified to give ($\pm$)matrine (LVIIIa, b) (*L. Mandell* and *K. P. Singh*, J. Amer. chem. Soc., 1961, **83**, 1766; *Mandell et al., ibid.*, 1963, **85**, 2682; 1965, **87**, 5234):

Matridine (LXIII) R = H_2
Matrine (LVIII a,b) R = O

Matrine (LVIIIa, b) has also been synthesised from aminolupinane by condensation with ethyl 4-formylbutanoate. Subsequent dehydrogenation and cyclisation gave a mixture of matrine (LVIIIa, b), allomatrine (LIXa, b) and lupanine (p. 305) (*Bohlmann et al.*, Chem. Ber., 1966, **99**, 3358).

Lupanine

Albertine, $C_{15}H_{22}N_2O_2$, provisionally assigned as 13-hydroxy-7,11-dehydro-matrine, **albertamine,** $C_{15}H_{24}N_2O_2$, and **albertidine,** $C_{15}H_{24}N_2O$, of unknown structure have been isolated from *Leontice albertii* (*S. Yu. Yunusov et al.*, Khim. prir. Soedin., 1967, **3**, 26; 1969, **5**, 409; 1974, **10**, 377). **Darvasolin,** $C_{15}H_{24}N_2O_2$, from *Leontice darvasica* was converted into matridine by lithium tetrahydridoaluminate and into sophoramine by phosphorus pentoxide at high temperature. The structure, determined by spectroscopic methods, is as shown below (*S. Yu. Yunusov et al., ibid.*, p. 115):

Darvasolin

Goebeline, $C_{30}H_{44}N_4O_2$, isolated from *Sophora pachycarpa* (*Y. I. Pakanaev* and *A. S. Sadykov*, Zhur. obshcheĭ Khim., 1961, **31**, 2428) was originally assigned several monomeric matrine-like structures but more recently has been shown to have the dimeric structure from the mass spectra of the alkaloid and its derivatives (*S. Iskandarov et al.*, Khim. prir. Soedin., 1972, 347):

Albertine (?) Albertamine Goebeline

(+)**Isometrine,** $C_{15}H_{24}N_2O$, from *Sophora flavescens* was characterised as [5R,6R,7S,11R]17-oxomatridine by X-ray crystallography on the hydrobromide (*A. Ueno et al.*, Chem. and Pharm. Bull., Japan, 1975, **23**, 2560) and **lehmannine,** $C_{15}H_{22}N_2O$, from *Ammothamnus lehmanii*, which gave matrine on hydrogenation, was shown by spectroscopic methods to have the structure shown below (*Y. K. Kushmuradov, K. A. Aslanov* and *S. Kuchkosov*, Khim. prir. Soedin, 1975, **11**, 377):

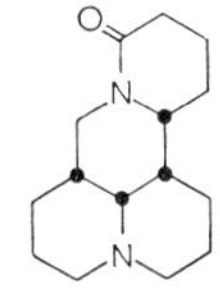

(+)Isomatrine   Lehmannine

Leontine, $C_{15}H_{24}N_2O$ (also known as isoleontine), was characterised by *T. F. Platonova, A. D. Kuzovkov* and *P. S. Massagetov* (Zhur. obshcheĭ Khim., 1953, **23**, 880) from *L. eversmanni* (see also *Yunusov* and *L. G. Sorokina, ibid.*, 1949, **19**, 1955). It is the (−)enantiomer of allomatrine and thus has the structure and absolute stereochemistry shown in LIX (p. 321).

Leontalbine, $C_{15}H_{22}N_2O$, and the isomeric **leontalbinine** were isolated from *Leontice albertii* (*Yunusov et al.*, Doklady Akad. Nauk S.S.S.R., 1964, **21**, 32). The structure of the former as the enantiomer of 5,17-dehydromatrine was assigned on the basis of hydrogenation experiments and n.m.r. data, and was confirmed by synthesis (*idem*, Khim. prir. Soedin., 1966, **2**, 66). The following structure has been proposed for the latter alkaloid:

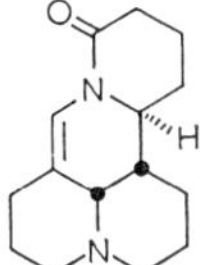

Leontalbine

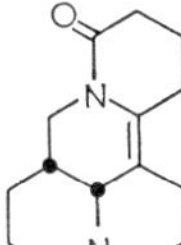

Leontalbinine

Oxymatrine, $C_{15}H_{24}N_2O_2$, isolated from *Sophora flavescens* (*Bohlmann, D. Rahtz* and *C. Arndt*, Chem. Ber., 1958, **91**, 2189) is the *N*-oxide of (+) matrine and is converted to the latter by sulphur dioxide or by phosphorus and hydriodic acid.

Sophoramine, $C_{15}H_{20}N_2O$, **isosophoramine,** $C_{15}H_{20}N_2O$ and **sophocarpine,** $C_{15}H_{22}N_2O$, are related alkaloids which have been isolated from *Sophora pachycarpa* (*W. F. Proskurnina* and *Kuzovkov*, Doklady Akad. Nauk S.S.S.R., 1953, **91**, 1145; *Sadykov, Y. K. Kushmuradov* and *K. A. Aslanov, ibid.*, 1962, **145**, 829). All of the alkaloids have been synthesised from matrine as shown below (*S. Okuda et al.*, Chem. and Ind., London, 1962, 1326; *Okuda, H. Kamata* and *K. Tsuda*, Chem. and Pharm. Bull., Japan, 1963, **11**, 1349):

Matrine gives the dichloro derivative **LXIV** on treatment with thionyl chloride and sulphuryl chloride. Catalytic hydrogenation removes one chlorine to give the monochloromatrine (**LXV**) used in the determination of the absolute configuration of matrine. Elimination of hydrogen chloride gives the product **LXVI**, identical with sophocarpine. On dehydrogenation the α-pyridone **LXVII** is obtained, identical with sophoramine. On treatment of sophoramine or the dichloromatrine (**LXIV**) with pyridine at high temperature epimerisation occurs at $C_{(6)}$ (presumably *via* a ring-opened intermediate) to give isosophoramine. A seven-step stereoselective total synthesis of (+)isophoramine from nicotinonitrile has been described (*E. Wenkert et al.*, J. Amer. chem. Soc., 1973, **95**, 8427).

Sophoranol, $C_{15}H_{24}N_2O_2$, isolated from *Sophora flavescens* (*Okuda et al.*, Chem. and Pharm. Bull., Japan, 1965, **13**, 482) is 5-hydroxymatrine:

Sophoranol

Sophoridine, $C_{15}H_{24}N_2O$, isolated from *L. albertii* (*D. Kamalitdinov, Iskandarov* and *Yunusov*, Khim. prir. Soedin., 1967, **3**, 352) was assigned the structure and relative stereochemistry shown in LXVIII (stereoisomeric with matrine) on the basis of i.r. data and isomerisation experiments (*A. I. Begisheva et al.*, Doklady Akad. Nauk S.S.S.R., 1967, **24**, 25; Khim. prir. Soedin., 1969, **5**, 455):

Sophoridine
(LXVIII)·

Physical data on the matrine-like alkaloids are recorded in Table 6.

TABLE 6

MATRINE-LIKE ALKALOIDS

Alkaloid	m.p. (°C)	b.p. (°C)	$[a]_D^{\circ}$
Albertine	161		−101.5 a
Goebeline	231		−12.9
Isosophoramine	149		+53.3
Leontalbine	oil		−167 a
Leontalbinine	107–108		−135.5 a
Leontine	103–104		
Matrine	77, 84, 87 (polymorphous)		+40.9 d
Oxymatrine	208		+47.7 a
Sophocarpine	53–54 (monohydrate)		−29.4 a
Sophoramine	164–165		−90.8 a
Sophoranol	171		+66 d
Sophoridine	108–109		+59.3 d

2. Nuphar alkaloids

Several of the Nuphar alkaloids from the plant family Nymphaeceae (water-lilies) have quinolizidine structures, although they are biogenetically unrelated to the lupin alkaloids, being of sesquiterpenoid origin. The alkaloids were first detected by *G. Dragendorff* (Pharm. Z. Russland, 1879, **18,** 514) but only recently have the majority of the alkaloids been isolated and their structures determined.

Nupharidine, $C_{15}H_{23}NO_2$, and **deoxynupharidine,** $C_{15}H_{23}NO$, were first isolated from *Nuphar japonicum* (*J. Arima* and *T. Takahashi*, Bull. chem. Soc., Japan, 1931, **52,** 815; *M. Kotake et al.*, Proc. Imp. Acad., Tokyo, 1943, **19,** 490) and have also been found in *Nuphar luteum* (yellow water lily) (*O. Achmatowicz* and *M. Mollowna*, Rocz. Chem., 1939, **19,** 493). On reduction with sulphur dioxide or with phosphorus and hydriodic acid, nupharidine gives deoxynupharidine, suggesting that the former is an *N*-oxide of the latter. This was confirmed by reversing the process using hydrogen peroxide. It therefore only remained to solve the structure of deoxynupharidine (see Scheme 3). On catalytic hydrogenation deoxynupharidine gave a tetrahydro derivative and a hexahydro derivative. The former product on Hofmann degradation (one stage) and subsequent ozonolysis and oxidation gave tetrahydrofuran-3-carboxylate (I). Under similar degradation deoxynupharidine gave furan-3-carboxylic acid (II) (*Y. Arata* and *T. Ohashi*, J. pharm. Soc., Japan, 1957, **77,** 229). The i.r. spectrum of deoxynupharidine also indicated the presence of a furan grouping (*Arata, ibid.*, p. 321). Oxidation of deoxynupharidine and distillation of the product with soda-lime gave 1,7-dimethylquinolizidine (III). These two pieces of information established the basic skeleton of the molecule, and it remained to confirm and position the various substituents. This was achieved by a combination of Hofmann degradation and ozonolysis reactions. Thus "hexahydrodeoxynupharidine" underwent two Hofmann degradation steps (with one catalytic hydrogenation step interposed) to give the nitrogen-free product $C_{15}H_{28}O$. Ozonolysis of this material gave a mixture of four products: formic acid, 3-methylbutanoic acid, δ-(3′-tetrahydrofuryl)-α-methylpentanoic acid, $(C_9H_{17}O)CO_2H$ (IV) and methyl δ-methyl ω-(3′-tetrahydrofuryl)heptyl ketone, $(C_{12}H_{23}O)CO \cdot CH_3$, (V), establishing the above nitrogen-free product as a mixture of two isomeric olefins VI and VII. Hydrogenation of this mixture gave a single product, $C_{15}H_{30}O$, (VIII), which was degraded to 4,8-dimethylpelargonic acid (IX) (*Arata, M. Koseki* and *K. Sakai, ibid.*, 1957, **77,** 232). These facts established the presence and position of the two methyl groups attached to the skeleton and also the position of the 3-furyl substituent and thus led to the proposal of structure X for deoxynupharidine and **XI** for nupharidine (see also *Kotake, S. Kusumoto* and *T. Ohara*, Ann., 1957, **606**, 148):

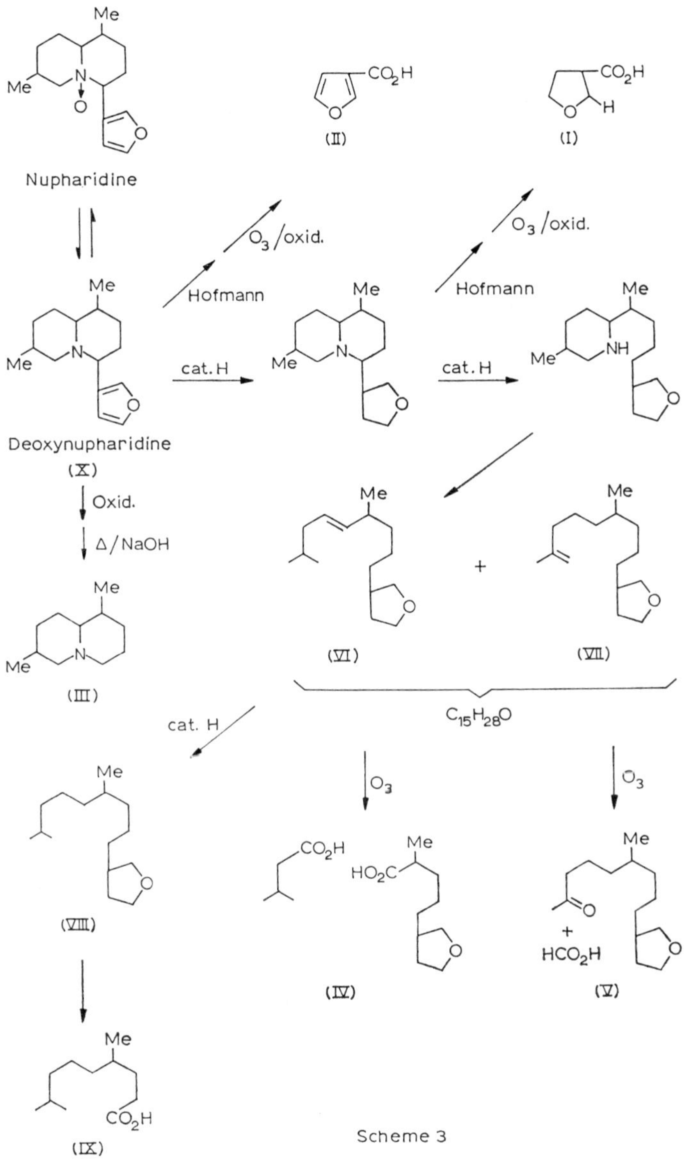

Scheme 3

The above assignment of structure to deoxynupharidine is in accord with the mass spectral fragmentation pattern (*Achmatowicz et al.*, Tetrahedron Letters, 1964, 927).

Following on from the gross-structure determination the configuration of the groups attached to the quinolizidine nucleus of ($\pm$)deoxynupharidine was determined by *F. Bohlmann et al.* (Chem. Ber., 1961, **94**, 3151), who synthesised the racemate from 1,7-dimethyl-4-oxoquinolizidine (XII) of known configuration by a modification of a method of *I. Murakoshi* (J. pharm. Soc., Japan, 1958, **78**, 594) as shown below:

The configuration of the quinolizidone XII was determined by i.r. comparisons of the reduction product (the quinolizidine) using lithium tetrahydridoaluminate with authentic 1,7-dimethylquinolizidines obtained from hydroxymethyl compounds of known configuration. This established the relative stereochemistry of ($\pm$)deoxynupharidine as Xa:

The absolute configuration of naturally occurring ($-$)deoxynupharidine was determined separately by *Kotake et al.* (Bull. chem. Soc., Japan, 1962, **35**, 1335) and by *Arata, N. Hazama* and *Y. Kojima* (J. pharm. Soc., Japan, 1962, **82**, 326). Degradation of ($-$)deoxynupharidine to ($-$)α-methyladipic acid was achieved as shown:

Unfortunately the $(-)\alpha$-methyladipic acid was assigned the wrong absolute configuration and this led to the wrong assignment for $(-)$deoxynupharidine. This was corrected when the configuration of $(-)\alpha$-methyladipic acid was redetermined as R (*I. Kawasaki* and *T. Kaneko*, Bull. chem. Soc. Japan, 1968, **41**, 1482). The correct absolute stereochemistry of $(-)$deoxynupharidine then follows as shown in Xa (*C. F. Wong, E. Auer* and *R. T. LaLonde*, J. org. Chem., 1970, **35**, 517). This has been confirmed more recently by a similar degradation of Xa to $R(+)$methylsuccinic acid (*Kawasaki, I. Kusumoto* and *Kaneko*, Bull. pharm. Soc. Japan, 1968, **41**, 1264), and by an X-ray crystal structure determination (*K. Oda* and *H. Koyama*, J. chem. Soc., B, 1970, 1450). The ^{13}C-n.m.r. spectra of nine Nuphar alkaloids have been reported (*R. T. LaLonde, T. N. Donvito* and *A. I. M. Tsai*, Canad. J. Chem., 1975, **53**, 1714).

Several syntheses of deoxynupharidine have been described. The first by *Kotake et al.* (Ann., 1960, **636**, 158) started from 2,5-dimethylpyridine, which on condensation with formaldehyde gave the diol XIV; on dehydration, acetylation and condensation with diethyl malonate this gave XV which was catalytically reduced to the piperidine derivative XVI. The steps from this intermediate to deoxynupharidine are described in the above discussion of stereochemistry (*Murakoshi* method).

The product consisted of only two diastereoisomeric racemates (out of a total possible of eight). The hydrogenation of the pyridine ring and the reduction of the dehydrodeoxynupharidine (XIII) are stereospecific. Chromatographic separation of the two racemates and resolution using (−)tartaric acid gave (−)deoxynupharidine identical with the natural material. Some slight modifications of this method (besides the synthesis by *Bohlmann* described above) have been developed (*Arata, T. Nakanishi* and *Y. Asaoka*, Chem. and Pharm. Bull., Japan, 1962, **10**, 675; *I. Jezo, M. Karwas* and *K. Tihlarik*, Chem. Zvesti, 1961, **15**, 283). An alternative synthesis of the intermediate piperidine derivative (ethyl ester) XVI has been described (*J. T. Wróbel* and *Z. Dabrowski*, Rocz. Chem., 1965, **39**, 1239) as shown:

Dehydrodeoxynupharidine, $C_{15}H_{21}NO$, (XIII) has been isolated from *N. luteum* (*Y. Arata*, Chem. and Pharm. Bull., Japan, 1965, **13**, 392). The structure was assigned from spectral data and from the relationship to deoxynupharidine which was produced on reduction with tetrahydridoborate.

Castoramine, $C_{15}H_{23}NO_2$, is structurally related to the above alkaloids but is found in the scent gland of the beaver, *Castor canadensis*, from which it has been isolated (*Z. Valenta* and *A. Khaleque*, Tetrahedron Letters, 1959, 1). The structure and stereochemistry have been confirmed by synthesis (*F. Bohlmann et al.*, Tetrahedron, 1963, **19**, 195):

Castoramine

Nupharolidine, $C_{15}H_{23}NO_2$, was isolated from *Nuphar luteum*. The structure was determined from spectral data (*J. T. Wróbel* and *A. Iwanow*, Rocz. Chem., 1969, **43**, 997):

Nupharolidine

Several sulphur-containing alkaloids have been isolated from *Nuphar luteum*. **Thiobinupharidine (thionuphlutine A)**, $C_{30}H_{42}N_2O_2S$, isolated by *O. Achmatowicz* and *Z. Bellen* (Tetrahedron Letters, 1962, 1121), **neothiobinupharidine**, isolated by *Achmatowicz* and *J. T. Wróbel* (*ibid.*, 1964, 129) and **thiophlutine B** (*LaLonde, C. F. Wong* and *W. P. Cullen, ibid.*, 1970, 4477) are stereoisomers having the gross structure XVII (R = R′ = H):

Neothiobinupharidine

(XVII)

The relative and absolute configurations of these molecules have recently been defined by a combination of X-ray crystallography, c.d. spectra, chemical transformations and relation to model compounds (*LaLonde, Wong* and *K. C. Das*, J. Amer. chem. Soc., 1973, **95**, 6342; J. org. Chem., 1974, **39**, 2892; *Wróbel et al.*, Canad. J. Chem., 1973, **51**, 2810; *Wong* and *LaLonde*, J. org. Chem., 1973, **38**, 3225).

Several hydroxylated derivatives of the above alkaloids also have been found in *N. luteum*; these are listed in Table 7.

TABLE 7

NUPHAR ALKALOIDS

Alkaloid	m.p. (°C)	b.p. (°C/mm)	$[\alpha]_D{}^\circ$
Castoramine	65–66		hydrochloride m.p. 215° decomp.
Dehydrodeoxynupharidine	—	110–120/3	−130.1 c
Deoxynupharidine	21–22	112–115/8	−112.5
Nupharidine	222	—	+14.5
Nupharolidine	110	—	
Neothiobinupharidine	159–160		−93.09
Thionuphlutine A	129–130		+49.8 d

6,6′-Dihydroxythionuphlutine B, 6,6′-dihydroxythiobinupharidine (**nuph-leine;** (XVII, R = R′ = OH) (*LaLonde, Wong* and *Cullen, loc. cit.*; *M. E. Perelson, T. N. Ilinskaya* and *O. N. Tolkachev*, Khim. prir. Soedin, 1975, **11,** 768).

6-Hydroxythionuphlutine B, 6-hydroxythiobinupharidine (thionupharoline; (XVII, R = OH, R′ = H) (*LaLonde, Wong* and *Das, loc. cit.*; *T. I. Martin et al.,* Canad. J. Chem., 1974, **52,** 2705).

6-Hydroxythiobinupharidine (XVII, R = H, R′ = OH) (*LaLonde, Wong* and *Das,* J. org. Chem., 1974, **39,** 2892).

3. Ormosia alkaloids

Alkaloids were detected in *Ormosia* species (plant family Leguminosae) in 1919 (*K. Hess* and *F. Merck*, Ber., 1919, **52,** 1976) but structural elucidation has required X-ray and spectroscopic techniques. It is convenient to divide the alkaloids into two groups, the pentacyclic series and the hexacyclic series. The structures have been obtained by inter-relationships between the alkaloids based on an unequivocal structure obtained from X-ray crystallographic studies. Thus **jamine,** $C_{21}H_{35}N_3$(I), isolated from *O. jamaicensis* and *O. panamensis* (*P. Naegeli, W. C. Wildman* and *H. A. Lloyd*, Tetrahedron Letters, 1963, 2069) has the following structure determined by X-ray methods (*I. L. Karle* and *J. Karle, ibid.,* 1963, 2065). The structures of the other *Ormosia* alkaloids followed from this:

Jamine (I)

Jamine occurs naturally as the racemate (as do several of the other *Ormosia* alkaloids) and therefore the above structure gives no information about absolute stereochemistry.

Ormosanine, $C_{20}H_{35}N_3$, isolated from *O. panamensis* (*Lloyd* and *E. C. Horning*, J. Amer. chem. Soc., 1958, **80,** 1506), from *O. jamaicensis* as "Alkaloid A" (*C. H. Hassall* and *E. M. Wilson*, J. chem. Soc., 1964, 2656; *Z. Valenta et al.,* Tetrahedron Letters, 1963, 1559) and from *Pipthanthus nanus* as piptamine (*R. A. Konowalova, B. S. Diskina* and *M. S. Rabinovich,* Zhur. obshcheĭ Khim., 1951, **21,** 733) condenses with formaldehyde to give jamine (I) and the structure and stereochemistry follow from this as II in the list given below.

| | Hydrogen at | | | |
	$C_{(6)}$	$C_{(11)}$	$C_{(16)}$	$C_{(18)}$
Ormosanine (II)	α	β	α	α
Piptanthine (III)	β	β	α	α
18-Epiormosanine (IV)	α	β	α	β
16-Epiormosanine (IVa)	α	β	β	α
Dasycarpine (V)	α	α	α	β
Isotetrahydro-ormojanine (VI)	β	α	α	β
(−)Templetine (VIa)	α	β	β	β
Tetrahydro-ormojanine (VII)	β	α	β	β

Piptanthine, $C_{20}H_{35}N_3$, isolated from *Piptanthus nanus* (*U. Eisner* and *F. Šorm,* Coll. Czech. chem. Comm., 1959, **24,** 2348) is also formed from ormosanine by a catalytic epimerisation reaction (at $C_{(6)}$). The structure III has been assigned on the basis of the above relationship and of spectroscopic data (*P. Deslongchamps, Valenta* and *J. S. Wilson,* Canad. J. Chem., 1966, **44,** 2539).

18-**Epiormosanine,** $C_{20}H_{35}N_3$, has recently been isolated from *O. semicastrata* by *S. McLean et al.* (ibid., 1971, **49,** 1976; 1972, **50,** 1639) and has structure IV as its name implies. The isomeric ($\pm$)16-**epiormosanine** isolated from *Hevea linearis* has the structure IVa, determined by X-ray crystallography (*J. A. Lamberton et al.,* Tetrahedron Letters, 1975, 3875).

Dasycarpine, $C_{20}H_{35}N_3$, from *O. dasycarpa* (*R. T. Clarke* and *M. F. Grundon,* J. chem. Soc., 1960, 41) is yet another stereoisomer of ormosanine having structure V. An analogous epimerisation to the above ormosanine–piptanthine conversion leads from dasycarpine to isotetrahydro-ormojanine (VI), which is one of the two stereoisomeric hydrogenation products of ormojanine (the other being tetrahydro-ormojanine, VII). The structures of the latter, synthetic materials were determined by *Deslongchamps, Valenta* and *Wilson (loc. cit.)* and the structure of dasycarpine followed from this.

(−)**Templetine,** $C_{20}H_{35}N_3$, m.p. 120.5–122°, $[\alpha]_D$ — 52° (ethanol), from *Templetonia retusa,* has the structure and absolute configuration represented by VIa, established by a combination of spectroscopic methods and X-ray crystallography (*J. R. Cannon et al.,* Tetrahedron Letters, 1974, 1683).

Podopetaline, $C_{20}H_{33}N_3$, from *Podopetalum ormondii* (*N. K. Hart et al.,* ibid., 1972, 5333) is a dehydro derivative of ormosanine. **Ormocastrine** from *O. semicastrata,* originally assigned the podopetaline structure *(McLean et al., loc. cit.)* is the hydrochloride of the latter (*McLean* and *R. Misra,* Canad. J. Chem., 1974, **52,** 1907). The position of the double bond at Δ^{16-17}

followed from n.m.r. data and the identity of the hydrogenation products, ormosanine and 16-epiormosanine. The absolute configuration has been established by X-ray crystallography; the formula below represents the (−)enantiomer (*M. F. Mackay, L. Satzke* and *A. M. Mathieson*, Tetrahedron, 1975, **31**, 1295):

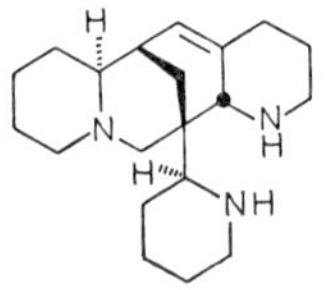

Ormocastrine

The second group of *Ormosia* alkaloids is exemplified by **panamine,** $C_{20}H_{33}N_3$, isolated from *O. panamensis* (*Lloyd* and *Horning, loc. cit.*).

On reduction with sodium tetrahydridoborate panamine is converted into ormosanine (II) and on catalytic reduction to a mixture of ormosanine and piptanthine (III). From these results the structure VIII was postulated (*E. M. Wilson*, Chem. and Ind., 1965, 472; *Deslongchamps, Valenta* and *Wilson, loc. cit.*) and this has been confirmed by X-ray crystallographic data (*Karle* and *Karle*, Tetrahedron Letters, 1966, 1659).

Ormosajine, $C_{20}H_{33}N_3$, isolated by *Hassall* and *Wilson (loc. cit.)* from *O. jamaicensis* was converted by reduction to dasycarpine (V) and thus has the structure and stereochemistry shown in IX:

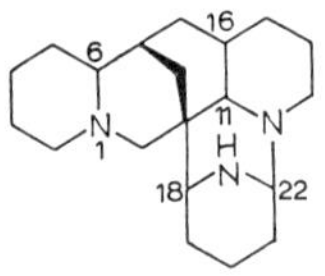

| | Hydrogens at | | | | |
	$C_{(6)}$	$C_{(11)}$	$C_{(16)}$	$C_{(18)}$	$C_{(22)}$
Panamine (VIII)	α	β	α	α	α
Ormosajine (IX)	α	α	α	β	β

Ormosinine, from *O. panamensis* originally assigned as the $C_{(11)}$ epimer of panamine (*Naegeli, Wildman* and *Lloyd, loc. cit.*), has a dimeric structure $(C_{40}H_{66}N_6)$ based on panamine, as shown by high vacuum sublimation (giving panamine) and by osmometric molecular weight measurement. Structure X has been proposed by *Deslongchamps, Valenta* and *Wilson (loc. cit.)*.

Ormojine, found in *O. jamaicensis* is also formed by acid treatment of ormosajine (IX). Mass spectral data suggested a molecular formulae of $C_{40}H_{66}N_6$ and a dimeric structure XI has been proposed by *A. P. Davies* and *Hassall* (Tetrahedron Letters, 1966, 6291) on spectroscopic evidence and on the similarity of this acid-catalysed dimerisation to the formation of the diplospartyrines from sparteine (*C. Schöpf* and *E. K. Keller*, Naturwiss., 1956, **43**, 325):

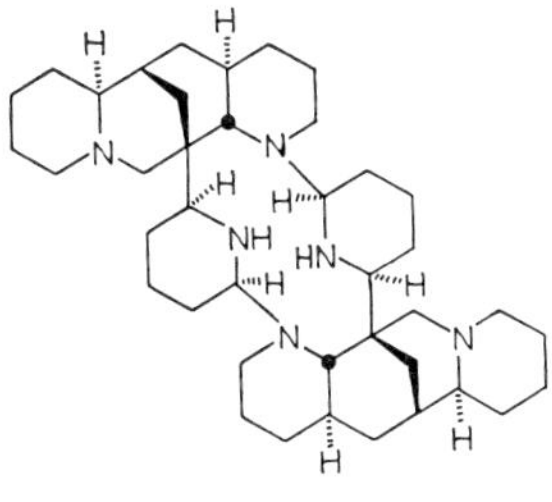

Ormosinine (X)

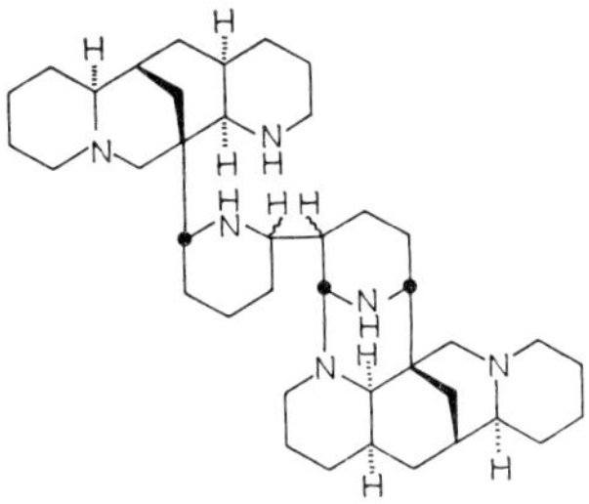

Ormojine (XI)

Ormojanine, $C_{20}H_{31}N_3$, from *O. jamaicensis* (*Hassall* and *Wilson, loc. cit.*; *Valenta et al.*, Tetrahedron Letters, 1963, 1559) was assigned the structure XII from a study of the hydrogenation products and from the catalytic dehydrogenation product (*idem*, Canad. J. Chem., 1966, **44**, 2525, 2539).

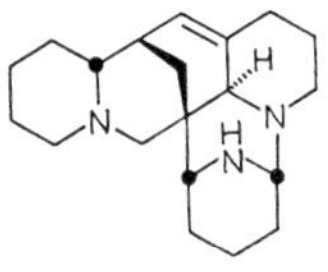

Ormojanine (XII)

Recently a total synthesis of ormosanine has been described (subsequent $C_{(6)}$ epimerisation to piptanthine and oxidation to panamine provided synthetic routes to these alkaloids also) (*Valenta, H. J. Liu* and *T. T. J. Yu*,

TABLE 8

ORMOSIA ALKALOIDS

Alkaloid	m.p. (°C)	b.p. (°C/mm)	$[\alpha]_D^0$
Dasycarpine	oil	160–165/0.5	+12.5 a
18-Epiormosanine	267 decomp.		−10
Jamine	154		0
Ormocastrine	263 decomp.		−29 b
Ormojanine	126		−144 a
Ormojine	154–155		+24 a
Ormosajine	oil		+23 a
			picrate
			m.p. 138–139
Ormosanine	168		0
Ormosinine	220		+9.0
Panamine	36–40		−11
Piptanthine	143		24.3 a

J. chem. Soc., D, 1970, 1116; *H. J. Liu et al.*, Canad. J. Chem., 1976, **54,** 97).
The synthesis proceeded *via* the key intermediate dilactam XIII which was
made by the route outlined below:

reduction

Ormosanine (II)

N-chloro-succinimide

epimerisation

(XIII)   (VIII)   (III)

4. *Lythraceae* alkaloids

Alkaloids from *Lythraceae* species of the Leguminosae were first isolated
in 1962 and subsequently several were shown to contain the quinolizidine
nucleus. Structural elucidation has relied on X-ray studies and spectroscopic
methods as many of the alkaloids have the same basic structure, different
only in the degree and position of substitution and in their stereochemistry.

Lythrine, $C_{26}H_{29}NO_5$, isolated from *Heimia salicifolia* (*B. Douglas et al.*,
Lloydia, 1964, **27,** 25; *R. N. Blomster, A. E. Schwarting* and *J. M. Bobbitt,*
ibid., 1964, **27,** 15; *Hg. Appel, A. Rother* and *Schwarting, ibid.*, 1965, **28,** 84),
along with several other novel *Lythraceae* alkaloids was shown to have the
structure and absolute stereochemistry shown in I by X-ray studies (*B.
Douglas et al.*, Experientia, 1965, **21,** 247), on the *O*-methyl hydrobromide.
The alkaloids from the above source and seven new alkaloids isolated from
Decodon verticillatus by *J. P. Ferris et al.* (J. org. Chem., 1962, **27,** 2985)
were all shown to be related in gross structure to lythrine by the latter authors
(*idem, ibid.*, Tetrahedron Letters, 1966, 3641) who were also able to classify
them into 4 groups, A, B, C and D, differing only in stereochemistry and

substitution pattern. The alkaloids **decinine,** $C_{26}H_{31}NO_5$ and **lyfoline,** $C_{25}H_{27}NO_5$, were related to lythrine as "Group A" alkaloids and have the structure II and III, respectively (*idem*, J. Amer. chem. Soc., 1971, **93**, 2942). Another "Group A" alkaloid **lythridine,** $C_{26}H_{31}NO_6$, was originally assigned a wrong structure from mass spectral data (*Appel* and *H. Achenbach*, Tetrahedron Letters, 1966, 5789), which was corrected to IV by X-ray studies (*Douglas et al.*, Chem. and Ind., London, 1966, 1795):

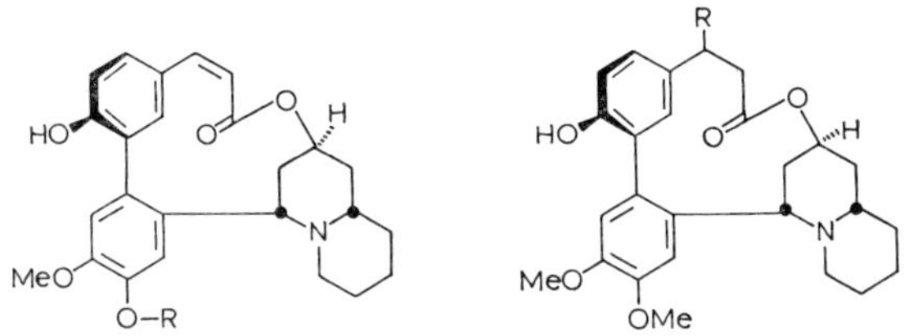

Lythrine (I) R = CH₃
Lyfoline (III) R = H

Decinine (II) R = H
Lythridine (IV) R = H

'Group A'

The "Group B" alkaloid, **vertine,** $C_{26}H_{29}NO_5$, was recognised as a diastereoisomer of lythrine. As the latter has four asymmetric centres (*i.e.* 8 possible racemates) and a quinolizidine ring system capable of existence in one *trans* and two *cis* forms, vertine is one of a total of twenty-four racemic possibilities. O.r.d. experiments showed the biphenyl chromophore to have the same chirality in the A and B series therefore the difference between vertine and lythrine must reside within the quinolizidine moiety. Evidence from i.r. (Bohlmann bands), n.m.r. and rates of quaternisation suggested the A series contained a *trans*-fused quinolizidine whereas the B group contained a *cis*-fused system. The difference between lythrine and vertine has to be in their stereochemistry at $C_{(10)}$, this was confirmed by degradative experiments. Vertine is thus represented as V. Vertine on catalytic hydrogenation absorbed one molecule of hydrogen and gave a product identical with **decamine,** $C_{26}H_{31}NO_5$, which was therefore assigned as VI:

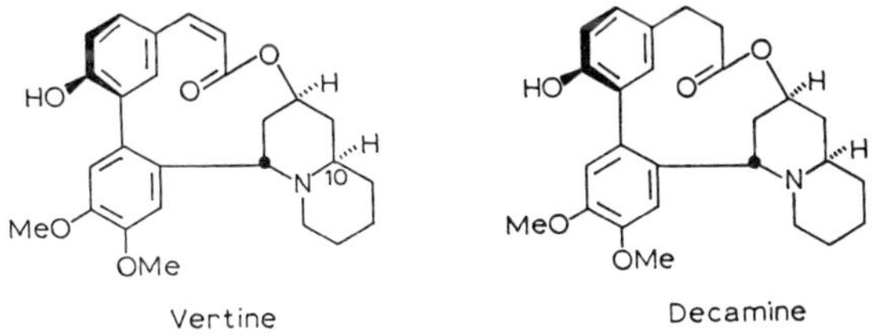

Vertine
(V)

Decamine
(VI)

Series B

Alkaloids of Groups C and D, typified by **nesodine,** $C_{26}H_{29}NO_5$, and **verticillatine,** $C_{25}H_{27}NO_5$, respectively, have different u.v. absorption spectra from groups A and B, which suggests a different substitution pattern in the aromatic rings. This was confirmed by n.m.r. The configurations of the quinolizidine ring differed in groups C and D as in A and B above and thus the structures for nesodine and verticillatine follow as VII and VIII, respectively. The group C alkaloid **decodine,** $C_{25}H_{29}NO_5$, is represented as IX and the group D alkaloids **lagerstroemine,** $C_{26}H_{31}NO_5$, and **dihydroverticillatine,** $C_{25}H_{29}NO_5$, isolated from *Lagerstroemia indica* by *Ferris et al.* (J. Amer. chem. Soc., 1971, **93**, 2958) are represented as X and XI, respectively:

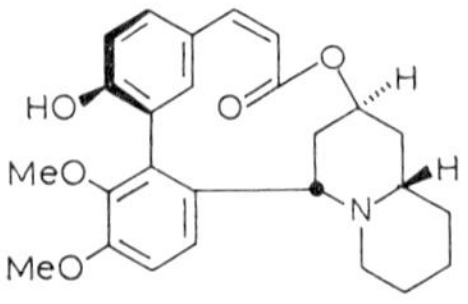

Nesodine (VII)

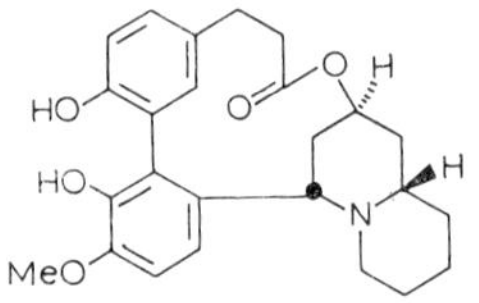

Decodine (IX)

Group C

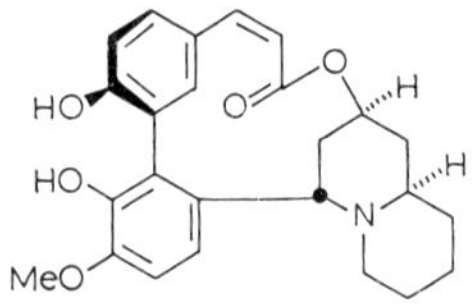

Verticillatine
(VIII)

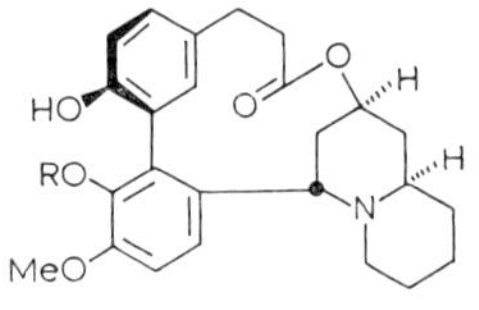

Lagerstroemine (X; R = Me)
Dihydroverticillatine (XI; R = H)

Group D

Vertaline and **decaline,** $C_{26}H_{31}NO_5$, do not belong to the four groups described above. The presence of a biphenyl ether link was suspected from the spectroscopic and analytical data and this was confirmed by the X-ray crystallographic study of vertaline which was shown to be XII (*J. A. Hamilton* and *L. K. Steinrauf, ibid.,* 1971, **93**, 2939). The structure of the diastereoisomeric decaline was shown to be the $C_{(10)}$ epimer XIII by investigations similar to those described above (*Ferris et al., ibid.,* p. 2954). The des-methyl derivatives of vertaline and decaline, found in *D. verticillatus*, have been tentatively assigned as the 4″-hydroxy derivatives XIV and XV on biogenetic grounds:

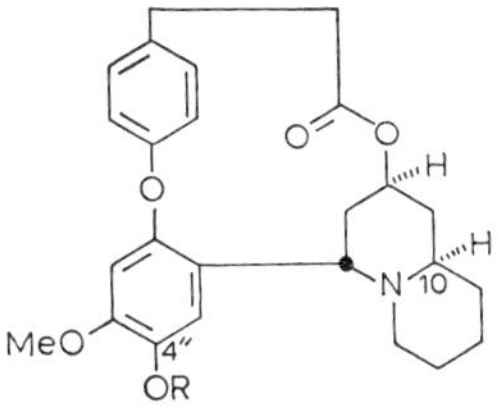

Vertaline (XII ; R = Me)
(XIV ; R = H)

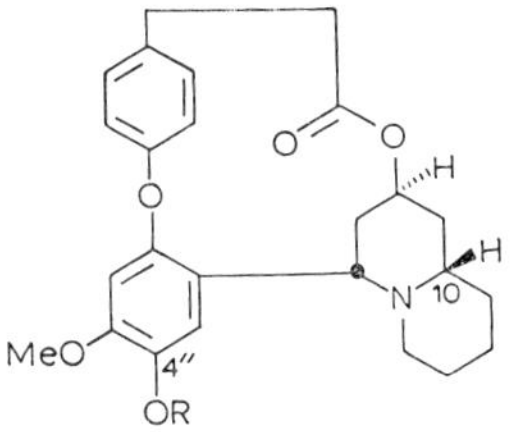

Decaline (XIII ; R = Me)
(XV ; R = H)

Lagerine, $C_{25}H_{29}NO_5$, isolated from *L. indica* (*idem.* J. Amer. chem. Soc., 1971, **93,** 2958) was assigned a structure which was shown subsequently to be incorrect by comparison with synthetic material of this structure (*Y. Arata et al.*, Chem. and Pharm. Bull., Japan, 1974, **22,** 1684). The structure was corrected by spectroscopy and synthesis of methyllagerine (*M. Hanacka et al.*, Tetrahedron Letters, 1974, 2533). More recently a similar synthesis of ($\pm$)lagerine has been described (*Hanacka, M. Kamei* and *Arata*, Chem. and Pharm. Bull., Japan, 1975, **23,** 2191):

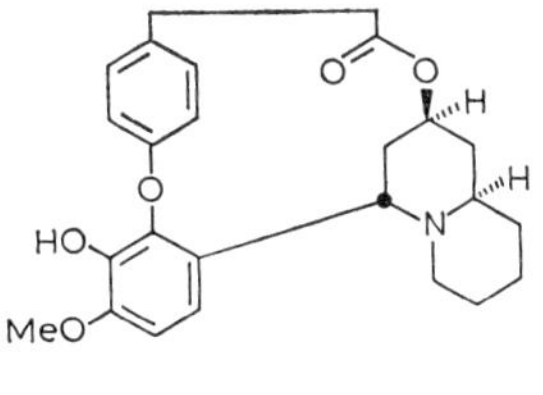

Lagerine

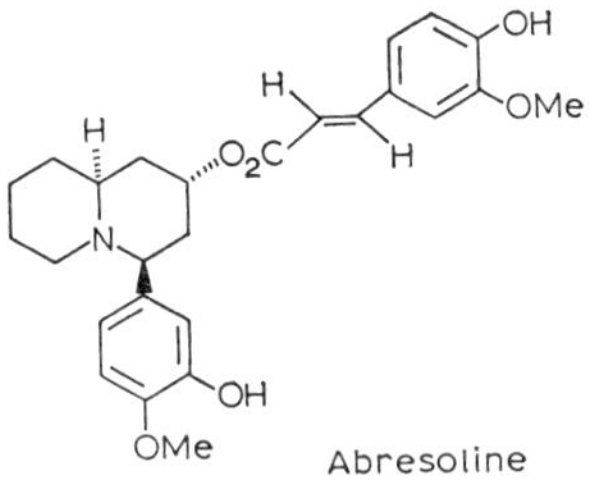

Abresoline

Abresoline, $C_{26}H_{30}NO_6$, a minor alkaloid isolated recently from *Heimiu salicifolis*, has the biogenetically significant structure, based on spectral data and synthesis of the dihydro compound, shown above (*R. B. Norhammer, A. E. Schwarting* and *J. M. Edwards*, J. org. Chem., 1975, **40,** 656).

The absolute configuration of the *Lythrum* alkaloids has been determined by *Ferris et al.* (*ibid.*, p. 2963) by a comparison of the o.r.d. spectra with model compounds of known absolute configuration and is correctly represented in the above structures.

Recently *E. Fujita* and *Y. Saeki* (J. chem. Soc., D, 1971, 368; J. chem. Soc., Perkin I, 1972, 2141) have described the isolation and characterisation of seven related alkaloids from *Lythrum anceps*. X-Ray crystallography of **lythrancine II** and **IV** has confirmed these structural assignments and established the absolute stereochemistry of the alkaloids (*M. J. Barrow,*

P. D. Cradwick and *G. A. Sim*, J. chem. Soc., Perkin II, 1974, 1812). The structures of these alkaloids are given below:

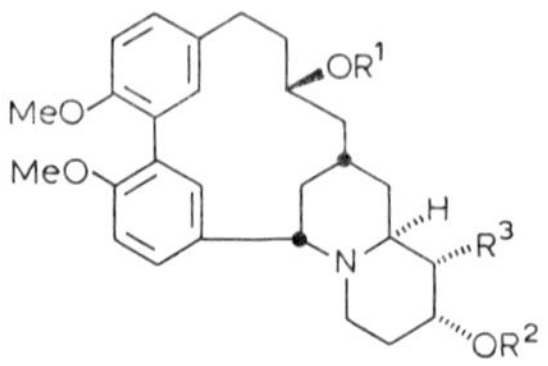

Lythrancine	I	$R^1 = R^2$	$= H; R^3 = OH$
	II	$R^1 = Ac; R^2 = H; R^3 = OH$	
	III	$R^1 = R^2$	$= Ac; R^3 = OH$
	IV	$R^1 = R^2$	$= Ac; R^3 = OAc$
Lythrancepine	I	$R^1 = R^2$	$= R^3 = H$
	II	$R^1 = Ac; R^2 = R^3$	$= H$
	III	$R^1 = R^2$	$= Ac; R^3 = H$

TABLE 9

LYTHRACEAE ALKALOIDS

Alkaloid	m.p. (°C)	$[\alpha]_D°$
Decaline	80–81	−136 c
Decamine	222	−142 c
Decinine	222	−145 c
Decodine	193	−97 c
Dihydroverticillatine	260–263 decomp.	—
Lagerstroemine	230	−137 c
Lyfoline	224	—
Lythridine	217–219	−174
Lythrine	243	+32.5
Nesodine	190	—
Vertaline	194	−170 c
Verticillatine	312	+119 b
Vertine	245	+39 c

Following the development of synthetic approaches to the Lythraceae alkaloids (*J. Rosazza, Bobbitt* and *Schwarting*, J. org. Chem., 1970, **35,** 2564; *J. T. Wróbel* and *M. W. Golebiewski*, Rocz. Chem., 1971, **45,** 705; *M. Hanaoka et al.*, Chem. and Pharm. Bull., Japan, 1975, **23,** 1573), several alkaloids and analogues, *e.g.* (±)decaline (*Hanaoka, N. Ogawa* and *Y. Arata, ibid.*, 1975, **23,** 2140), (±)demethyldecaline (*idem, ibid.*, 1974, **22,** 1945), (±)decinine (*I. Lantos* and *B. Loev*, Tetrahedron Letters, 1975, 2011), (±)methyldecinine (*Hanaoka, H. Sassa* and *C. Shimezawa*, Chem. and Pharm. Bull., Japan, 1975, **23,** 2478; *Hanaoka et al., ibid.*, 1974, **22,** 1216; *Loev, Lantos* and *H. van Hoeven*, Tetrahedron Letters, 1974, 1101), and (±)vertaline (*Hanaoka, Ogawa* and *Arata, ibid.*, 1976, **24,** 1045; 1974, **22,** 973; *E. J. Corey, K. C. Nikolaou* and *L. S. Melvin Jr.*, J. Amer. chem. Soc., 1975, **97,** 654), have now been synthesised.

Chapter 39

Compounds Containing Two Fused Five- and Six-Membered Heterocyclic Rings each with One Hetero Atom

NEIL CAMPBELL

Compounds with two fused heterocyclic nuclei fall into four main classes: (*1*) those in which both hetero rings are fused to a homocyclic nucleus; (*2*) those in which two hetero rings are fused to one another through carbon atoms; (*3*) those in which a hetero atom, nitrogen, is common to two or three fused ring, and (*4*) bridged ring compounds. The inclusion of certain tricyclic compounds such as the cyclazines has been made for reasons of completeness and continuity.

The compounds described in this chapter have for the most part either one oxygen-containing ring and one nitrogen-containing ring or two nitrogen-containing rings.

Compounds with two fused furan rings have been described in Vol. IV A, p. 193 *et seq.* and those with two fused pyrrole rings, the pyrrolizidines, in Vol. IV A, p. 480 *et seq.* Compounds with two oxygen-containing rings fused to a benzene ring are described in Vol. IV E, and include the furocoumarins (p. 126 *et seq.*), furochromones (p. 155 *et seq.*), pyranocoumarins (p. 134 *et seq.*) and pyranoxanthones (p. 329).

1. Compounds containing two hetero rings fused to an aromatic system

(*a*) Bz-*Pyrroloquinolines*

The *Bz*-pyrroloquinolines contain both a pyridine and a pyrrole nucleus fused to benzene. They exhibit a blue fluorescence in solution and their ultraviolet spectra differ considerably from those of the carbolines.

3*H*-**Pyrrolo**[2,3-*c*]**quinoline,** m.p. 228–230°, is prepared by heating ethyl pyruvate 3-quinolylhydrazone with zinc chloride (*T. R. Govindachari, S. Rajappa* and *V. Sudarsanam,* Tetrahedron, 1961, **16**, 1):

3 *H*-Pyrrolo[2,3-*c*]quinoline

The N-*methyl ether*, m.p. 115–116°, *picrate*, m.p. 253°, is obtained by the oxidation of calycanthine and by heating echitamyrine with selenium.

Vomipyrine, a degradation product of the alkaloid vomicine, is a derivative of 2H-*pyrrolo*[2,3-h]*quinoline*, m.p. 216–217°, N-*methyl ether*, m.p. 112°, which has been synthesised by the cyclodehydration of 7-formamido-8-methylquinoline by means of potassium *tert*-butoxide (*R. Robinson* and *A. M. Stephen*, Nature, 1948, **162,** 177). Among other isomers which have been synthesised are 1H-*pyrrolo*[2,3-f]*quinoline*, m.p. 236–238°, and 1H-*pyrrolo*[3,2-h]*quinoline*, m.p. 94–96° (*H. Wieland* and *L. Horner*, Ann., 1938, **536,** 89; *Horner*, Ann., 1939, **540,** 73):

3 *H*-Pyrrolo[2,3-*h*]-
quinoline

1 *H*-Pyrrolo[2,3-*f*]-
quinoline

1 *H*-Pyrrolo[2,3-*h*]-
quinoline

(b) Phenanthrolines*

The phenanthrolines are the most important representatives of these tricyclic compounds and contain two pyridine nuclei fused to a benzene ring. They are in fact diazaphenanthrenes containing a nitrogen atom in each of the outer rings, in contrast to the benzonaphthyridines which contain nitrogen atoms in adjacent rings. All ten phenanthrolines are known. Three are prepared by the double Skraup reaction on *o*-, *m*- and *p*-phenylenediamine, which thus yield *o*-(1,10-)-, *m*-(1,7-)- and *p*-(4,7-)-phenanthroline (I, II and III), respectively:

(I)

(II)

(III)

The structure I for *o*-phenanthroline follows unequivocally from its synthesis, but from the other two phenylenediamines linear products might result. As in other Skraup syntheses the products, however, are the angular compounds. Their structure is established by the oxidation with permanganate

---

* *W. O. Kermack* and *J. E. McKail*, "Heterocyclic Compounds", Vol. 7, p. 344, Wiley, New York, 1961; *B. Graham*, "Chemistry of Heterocyclic Compounds", Vol. 12, p. 386 (1958).

of *m*- and *p*-phenanthroline, to dibasic acids (*e.g.* IV) which on decarboxylation yield 2,3'-dipyridyl (V) and 3,3'-dipyridyl, respectively:

(II)   (IV)   (V)

A more general method for preparing all the phenanthrolines is the photo-cyclisation of 1,2-dipyridylethenes, the intermediate dihydro-products undergoing dehydrogenation by air (*H. H. Perkampus* and *G. Kassebeer*, Ann., 1966, **696,** 1). This is exemplified by the preparation of 4,7-phenanthroline (III) from 1,2-di(β-pyridyl)ethene (VI) *via* the dihydro intermediate VII:

(VI)   (VII)   (III)

The ultraviolet spectra of the phenanthrolines show some similarity to that of phenanthrene (*G. M. Badger, R. S. Pearce* and *R. Pettit*, J. chem. Soc., 1951, 3199; *A. Sucharda-Sobczyk et al.*, Roczniki Chem., 1974, **48,** 1265). The p.m.r. spectra are informative, the symmetrical compounds giving rise to four signals and the unsymmetrical to eight (*Perkampus* and *Kassebeer, loc. cit.*; *J. Mlochowski*, Roczniki Chem., 1974, **48,** 2145). Mass spectra of the phenanthrolines indicate that under electron bombardment all give identical final decomposition products (*H. Budzikiewicz, K. Kramer* and *Perkampus*, Z. Naturforsch., B, 1970, **25,** 178).

The phenanthrolines with two basic nitrogen atoms undergo mono- and di-protonation in strongly acid media (*A. A. Schilt* and *W. E. Dunbar*, Tetrahedron, 1974, **30,** 401). 1,10-Phenanthroline was for long regarded as a mono-acid base, but in fact diprotonated 10-phenanthrolium salts can be isolated (*L. H. Berka* and *S. W. Press*, Inorg. Nucl. Chem. Lett., 1970, **6,** 439). The reaction of the phenanthrolines with methyl iodide has not been fully investigated. Mono- and di-methiodides may be formed, 1,10-phenanthroline, for example, forming the monomethiodide, while 4,7-phenanthroline yields the mono- and di-methiodide.

With hydrogen peroxide in acetic acid at 80°, the phenanthrolines give mono- and di-*N*-oxides, but 1,10-*phenanthroline* yields only the 1-*oxide*, m.p. 182–183° (*Mlochowski* and *K. Kloc*, Roczniki Chem., 1973, **47,** 727). 1,7-*Phenanthroline* 7-*oxide*, m.p. 188°, 1,7-*dioxide*, m.p. 223–224°. 1,8-*Phenanthroline* 8-*oxide*, m.p. 232–233°, 1,8-*dioxide*, m.p. 270–271°, 4,7-*Phenanthroline* 4-*oxide*, m.p. 239°, 4,7-*dioxide*, m.p. 334°.

The phenanthrolines not unexpectedly undergo electrophilic substitution in the middle (benzenoid) ring. Nitration with nitric and sulphuric acids at 120–180° yields the following products. 6-**Nitro**-1,7-**phenanthroline,** m.p. 170°. 5-*Nitro*-, m.p. 207° and 6-*nitro*-1,8-*phenanthroline*, m.p. 178–179°. 5-*Nitro*-4,7-*phenanthroline*, m.p. 230–231°. 5-*Nitro*-1,10-*phenanthroline*, m.p. 197–198° (*Mlochowski* and *Z. Skrowaczewska, ibid.*, 1973, **47,** 2255).

Bromination with bromine in oleum at 120° follows a similar pattern although attack in the outer rings may occur to a minor degree (*Mlochowski, ibid.*, 1974, **48,** 2145). The main products are: 6-**bromo-,** m.p. 113° and 5,6-**dibromo-1,7-phenanthroline,** m.p. 154°; 5- and 6-*bromo-* and 5,6-*dibromo-1,8-phenanthroline,* m.p. 215°; 5-*bromo-,* m.p. 148° and 5,6-*dibromo-4,7-phenanthroline,* m.p. 262°; 5-*bromo-,* m.p. 118° and 5,6-*dibromo-1,10-phenanthroline,* m.p. 223°.

TABLE 1

PHENANTHROLINES

	m.p.(°C)	*Ref.*
1,7	78– 78.5	2,3
1,8	100–104	1,4
1,9	90	1
1,10	117–117.5	1
2,7	143	1
2,8	109–110	1
2,9	142.5	1
3,7	115–118	1
3,8	139–141	1
4,7	168	1

References

1   *H.-H. Perkampus* and *G. Kassebeer,* Ann., 1966, **696,** 1.
2   *G. F. Smith* and *C. A. Getz,* Chem. Reviews, 1935, **16,** 113.
3   *A. R. Surrey* and *R. A. Cutler,* J. Amer. chem. Soc., 1954, **76,** 1109.
4   *R. F. Homer,* J. chem. Soc., 1958, 1574.

1,10-**Phenanthroline** (I), o-*phenanthroline,* is prepared by the Skraup reaction from o-nitraniline or from o-phenylenediamine (*G. F. Smith* and *C. A. Getz,* Chem. Reviews, 1935, **16,** 113). It crystallises with 1 H_2O and forms complex compounds with ferrous and other metallic ions (*W. W. Brandt, F. P. Dwyer* and *E. C. Gyarfas, ibid.,* 1954, **54,** 959). It gives a red colour with ferrous salts in solution and the complex with ferrous chloride is an excellent high potential oxidation–reduction indicator. The phenanthroline in solution absorbs at 265 nm and mono- and di-protonation are accompanied by spectral changes to 270 and 275 nm, respectively (*W. A. E. McBryde,* Canad. J. Chem., 1965, **43,** 3472).

1,7-**Phenanthroline** (II), m-*phenanthroline,* crystallises with 2 H_2O and is prepared from m-phenylenediamine or from 5-aminoquinoline (*R. Lukeš* and *J. Pliml,* Chem. Listy, 1955, **49,** 836).

4,7-**Phenanthroline** (III), p-*phenanthroline,* is prepared from p-phenylenediamine and from 6-aminoquinoline. It forms a *picrate,* m.p. 255–256°, *monomethiodide,* m.p. 268–269°, and *dimethiodide,* m.p. 271°.

Isomeric with the phenanthrolines are the linear compounds, 1,5- (VIII), m.p. 240–242° (monohydrate) and 1,8-**anthrazoline (diazaanthracene)** (IX), m.p. 164.5–165°, *dipicrate*, m.p. 262° (decomp.), which have been prepared by indirect methods (*P. Ruggli et al.*, Helv., 1938, **21**, 1066; 1939; **22**, 478):

(VIII) (IX)

The ease with which phenanthrene is oxidised to phenanthraquinone is not paralleled by the phenanthrolines, which are inert to many oxidising agents and undergo ring-fission with others. The quinones, however, can be obtained by oxidation of 5- or 6-hydroxy-phenanthrols with a mixture of nitric and sulphuric acids (*J. Druey* and *P. Schmidt*, Helv., 1950, **33**, 1080; 1957, **40**, 350). The quinones form oximes, etc., yield phenazines with *o*-phenylenediamine, and with alkali undergo ring-contraction to yield diazofluorenones. 1,7-**Phenanthroline-5,6-quinone** (X), *1,7-phenanthroline-5,6-dione*, pale yellow, m.p. 255°, *oxime*, m.p. 225°, thus yields 1,5-*diazafluorenone* (XI), m.p. 159°, reduced by the Wolff–Kishner method to 1,5-*diazafluorene*, m.p. 108°:

(X) (XI)

4,7-**Phenanthroline-5,6-quinone**, m.p. 295°, *oxime*, m.p. 250°, *dioxime*, m.p. *ca.* 300°, *semicarbazone*, m.p. 195° (decomp.). 1,8-*Diazafluorenone*, pale yellow crystals, m.p. 205°. 1,8-*Diazafluorene*, m.p. 170°. 1,10-**Phenanthroline-5,6-quinone**, orange crystals, m.p. 250°, *dioxime*, m.p. 189°. 4,5-*Diazafluorenone*, pale yellow crystals, m.p. 214–215°. 4,5-*Diaza-fluorene*, m.p. 172° (*J. E. Dickeson* and *L. A. Summers*, Austral. J. Chem., 1970, **23**, 1023). The quinone has toxic properties (*R. D. Gillard, R. E. E. Hill* and *R. Maskill*, J. chem. Soc., A, 1970, 1447).

2. Compounds containing two hetero rings fused through adjacent carbon atoms

(a) Furopyridines

The furo[2,3-*b*]pyridine nucleus is found in a number of alkaloids (p. 350); but it is only recently that the parent compound was prepared (*H. Sliwa*, Bull. Soc. chim. Fr., 1970, 646). Treatment of the 3-substituted pyridinone (I) with silver hydroxide in methanol yields (80%) 7-*azacoumaran-3-one*, $C_7H_5NO_2$ (II), m.p. 95°, *dinitrophenylhydrazone*, m.p. 175°, with a strong i.r. band at 1725 cm^{-1} comparable with the 1720 cm^{-1} band in coumaran-3-

one. Reduction with lithium tetrahydridoaluminate gives *7-azacoumaran-3-ol* $C_7H_7NO_2$, m.p. 90°, the acetate of which, when pyrolysed, gives **furo[2,3-*b*]pyridine**, C_7H_5NO (III), b.p. 95°/22 mm in 94% yield:

The aromaticity of the molecule is indicated by its n.m.r. spectrum (*J. W. McFarland et al.*, J. heterocycl. Chem., 1971, **8**, 735). Reduction with hydrogen and a platinum oxide catalyst gives *2,3-dihydrofuro*[2,3-b]*pyridine*, *7-azacoumaran*, b.p. 90°/15 mm, with strong i.r. bands at 985 and 945 cm⁻¹ characteristic of coumaran. *4-Methylfuro*[2,3-b]-*pyridine*, b.p. 118°/20 mm, *picrate*, m.p. 151°, has a smell of pyridine (*R. Robinson* and *S. S. Watt*, J. chem. Soc., 1934, 1537).

Heating the 3-(bromoacetyl)-4-ethoxypyridinium bromide (IV) in a sealed tube at 150° gives a 25% yield of the hydrobromide of 5-azacoumaran-3-one (V), which is converted in three stages into **furo[3,2-*c*]pyridine** (VI), b.p. 85°/15 mm, *picrate*, m.p. 165–180° (*G. Lhommet, H. Sliwa* and *P. Maitte*, Bull. Soc. chim. Fr., 1972, 1442). It has also been synthesised from furfuraldehyde (*F. Eloy* and *A. Deryckere*, J. heterocycl. Chem., 1971, **8**, 57) and from ethyl 2-formylfuran-3-carboxylate (*J. D. Bourzat* and *E. Bisagni, ibid.*, 1971, 1727). *2-Methylfuro*[3,2-c]*pyridine*, b.p. 106°/16 mm, *picrate*, m.p. 183–187° (*Eloy* and *Deryckere, loc. cit.*).

(*b*) *Furoquinolines**

Furo[2,3-*b*]quinoline is the parent compound of a number of alkaloids in which a quinoline ring is annealed to a furan ring as in dictamnine, obtained from the roots of *Dictamnus albus* L., a member of the Rutaceae (*R. Storer* and *D. W. Young*, Tetrahedron, 1973, **29**, 1217). The structure of these compounds has been elucidated by oxidative degradation to derivatives of 4-hydroxy-2-quinolinones (*e.g.* VII, R = H) and by hydrogenolysis to 3-ethyl-4-hydroxy-2-quinolinone derivatives (*e.g.* VII, R = Et). Structural confirmation has been achieved by synthesis.

A useful synthesis involves ring-closure of 2-quinolinonecarbaldehydes typified by VIII, which with polyphosphoric acid yields dictamnine (*J. F. Collins et al.*, J. chem. Soc., Perkin I, 1973, 94).

---

* *J. R. Price*, Fortschr. der Chemie org. Naturstoffe, 1956, **13**, 302; *H. T. Openshaw*, "The Alkaloids", Vol. VII, p. 233, Academic Press, New York, 1960.

Dictamnine   (VII)   (VIII)

In a related synthesis the starting compounds are *N*-methyl-2-quinolinone-carbaldehydes, the aldehyde IX, for example, with polyphosphoric acid giving **isodictamnine** (X), m.p. 185–188° (47%), along with the angular *isomer* XI (R = Me), m.p. 129–130°. These isofuroquinolines are seldom found in nature and are characterised by their tendency to assume a pink colour:

(IX)   (X)   (XI)

The furo[2,3-*b*]quinoline alkaloids are mostly linear three-ringed compounds derived from dictamnine, containing one or more methoxyl groups and occasionally a substituted butoxy group attached to the benzene nucleus. They are readily distinguished from their angular analogues by u.v., i.r. or p.m.r. spectra (*E. A. Clarke* and *M. F. Grundon*, J. chem. Soc., 1964, 4190). It is noteworthy that the angular compounds are thermodynamically more stable than the linear isomers and isodictamnine when heated with polyphosphoric acid is partly converted into the *compound* XI (R = Me), m.p. 130°. Dictamnine similarly yields the angular *analogue* XI (R = H), m.p. 239–241° (*Grundon et al., ibid.,* 1955, 4284; Perkin 1, 1973, 94). The *compound*, m.p. 225°, claimed by *Y. Asahina* and *M. Inubuse* (Ber., 1932, **65**, 63) to have formula XI (R = Me), must have some other structure.

The 4-methoxyfuroquinolines undergo a reaction characteristic of α- and γ-alkoxyquinolines, namely, isomerisation when heated in a sealed tube with methyl iodide. Thus dictamnine yields isodictamnine (X). The reaction sometimes occurs under milder conditions (see, *e.g.* acronycidine, p. 352).

The structures of the furoquinolines have been determined partly by chemical methods of which examples are given in the sequel and partly by spectral methods, particularly p.m.r. measurements (see, *e.g. A. V. Robertson,* Austral. J. Chem., 1963, **16**, 451; *R. H. Prager et al., ibid.,* 1962, **15**, 301). Dictamnine, for example, exhibits two doublets at τ 2.47 and 3.55 (J 3.0 Hz) corresponding to H_2 and H_3, respectively, while the three protons on the methoxyl group absorb at τ 4.22. Infrared spectra are useful (*L. H. Briggs* and *L. D. Colebrook*, J. chem. Soc., 1960, 2458; *N. J. McCorkindale*, Tetrahedron, 1960, **14**, 223). The mass spectral fragmentation patterns have been rationalised (*D. M. Clugston* and *D. B. MacLean*, Canad. J. Chem., 1965, **43**, 2516).

Dictamnine, 4-*methoxyfuro*[2,3-b]*quinoline*, $C_{12}H_9NO_2$, m.p. 132–133°, *picrate*, m.p. 164–165°, occurs in *Dictamnis albus* and *Skimmis repens*, Nakai. It is oxidised by potassium permanganate to *dictamnal*, 4-*methoxy-2-quinolinone-3-carbaldehyde* (VII, R = CHO), m.p. 259–260°, *phenylhydrazone*, m.p. 228°, and to the corresponding *dictamnic acid* (VII, R = CO_2H), m.p. 260° (decomp.), which is readily demethylated and decarboxylated to 4-hydroxy-2-quinolinone. Dictamnine when treated with hydrogen and a platinum catalyst undergoes hydrogenolysis to yield 3-*ethyl-4-methoxy-2-quinolinone* (VII, R = Et), m.p. 186–187°. Dictamnine and dictamnic acid have both been synthesised (*R. G. Cooke* and *H. F. Haynes*, Austral. J. Chem., 1958, **11**, 224; *Grundon* and *McCorkindale*, J. chem. Soc., 1957, 2177; *J. F. Collins et al., loc. cit.*).

Pteleine, 6-*methoxydictamnine*, $C_{13}H_{11}NO_3$, m.p. 136–138°, *picrate*, m.p. 197–198°, is obtained from *Medicosma cunninghamii* Hook, and when heated with methyl iodide yields 6-*methoxyisodictamnine (isopteleine)*, m.p. 212–213° (*E. Bianchi, C. C. J. Culvenor* and *J. A. Lamberton*, Austral. J. Chem., 1968, **21**, 2357). Its structure has been established (*F. Werny* and *P. J. Scheuer*, Tetrahedron, 1963, **19**, 1293) and confirmed by synthesis (*Y. Kuwayama et al.*, Yakagaku Zasshi, 1968, **88**, 1050; C.A., 1969, **70**, 29125).

(±)-**Platydesmine**, $C_{15}H_{17}NO_3$, m.p. 137–138°, *picrate*, m.p. 107–108°, is a 2,3-dihydrodictamnine with an $Me_2C(OH)$-group at position 2 and is found in *Zanthoxylum parviflorum* Benth. (*J. A. Diment, E. Ritchie* and *W. C. Taylor*, Austral. J. Chem., 1967, **20**, 565; 1969, **22**, 1797; *R. M. Bowman* and *Grundon*, J. chem. Soc., C, 1966, 1504). It is oxidised by lead tetra-acetate and iodine to dictamnine. In nature it occurs as the (+)-isomer.

Evolitrine, $C_{13}H_{11}NO_3$, m.p. 114–115°, *picrate*, m.p. 201–202°, is 7-methoxydictamnine and is obtained from *Evodia littoralis* Ende. It has been synthesised (*Cooke* and *Haynes*, *loc. cit.*).

Robustine, 8-*hydroxydictamnine*, yellow crystals, m.p. 149°, is isolated from the roots of *Haplophyllum robustum* and has been synthesised (*V. N. Ramachandran et al.*, Indian J. Chem., 1973, **11**, 1088). With diazomethane it yields γ-**fagarine**, 8-*methoxydictamnine*, m.p. 142°, *picrate*, m.p. 177°, a constituent of *Fagara coca* and *Aegle marmelos*. γ-Fagarine is oxidised by potassium permanganate to γ-*fagarine aldehyde*, 4-*hydroxy-8-methoxy-quinolinone-3-carbaldehyde*, m.p. 185°, *phenylhydrazone*, m.p. 207° (decomp.) and γ-*fagaric acid*, 4,8-*dimethoxyquinolin-2-one-3-carboxylic acid*, m.p. 215° (*V. Deulofeu et al.*, J. Amer. chem. Soc., 1942, **64**, 2326). It has been synthesised (*Grundon* and *McCorkindale*, *loc. cit.; Collins et al., loc. cit.*).

Two furoquinoline alkaloids, haploperine and haplopine, have been isolated from the seeds of *Haplophyllum perforatum* (*G. P. Sidyakin* and *S. Yu. Yunusov*, C.A., 1962, **57**, 15170). Haploperine is identical with evoxine (p. 351) and when fused with potassium hydroxide gives **haplopine**, 7-*hydroxy-8-methoxydictamnine*, m.p. 205°. With diazomethane haplopine yields skimmianine and it has been synthesised *(Ramachandran et al., loc. cit.)*

Kokusaginine, 6,7-*dimethoxydictamnine*, $C_{14}H_{13}NO_4$, m.p. 171°, *picrate*, m.p. 218–219°, is obtained from the bark of *Flindersia collinus* Bail. along with flindersiamine. It is oxidised by potassium permanganate to 3-*formyl-4,6,7-trimethoxyquinolin-2-one*, m.p. 248–249° and the corresponding *carboxylic acid*, m.p. > 350° (*F. A. L. Anet et al.*, Austral. J. Sci. Res., 1952, **5**, 412). It has been synthesised (*Y. Kuwayama*, Yakagaku Zasshi, 1962, **82**, 702; C.A., 1963, **58**, 5741). **Maculine**, 6,7-*methylenedioxydictamnine*, $C_{13}H_9NO_4$, m.p. 205–210° (decomp.), *picrate*, m.p. 198° (decomp.), occurs in *Flindersia maculosa* (*R. C. F. Brown et al.*, Austral. J. Chem., 1954, **7**, 181) and yields on hydrogenolysis 3-*ethyl-6,7-methylenedioxy-4-hydroxyquinolin-2-one*, m.p. 300–303° (*R. J. Gell et al.*, ibid., 1955, **8**, 422). It has been synthesised *(Kuwayama, loc. cit.).*

Skimmianine, 7,8-*dimethoxydictamnine*, $C_{14}H_{13}NO_4$, m.p. 176°, *picrate*, m.p. 195–197°, is obtained from *Skimmia japonica* and has been synthesised (*Asahina* and *S. Nakanishi*, Ber., 1930, **63**, 2057; *Collins et al., loc. cit.*). It is oxidised by potassium permanganate to *skimmianal*, 4,7,8-*trimethoxy-3-formylquinolin-2-one* (XII, R = CHO), m.p. 238° and *skimmianic acid* (XII, R = CO$_2$H), m.p. 248°, *methyl ester*, m.p. 233°, which has been synthesised (*Brown et al.*, Austral. J. Chem., 1955, **8**, 121). The acid is partially demethylated and decarboxylated with hydrochloric acid to give 4-*hydroxy-7,8-dimethoxyquinolin-2-one*, m.p. 250°. Hydrogenolysis of skimmianine gives 3-*ethyl-4,7,8-trimethoxyquinolin-2-one*, identical with a synthetic specimen (*T. Ohta* and *Y. Mori*, Pharm. Bull. Japan, 1955, **3**, 396; C.A., 1956, **50**, 13055).

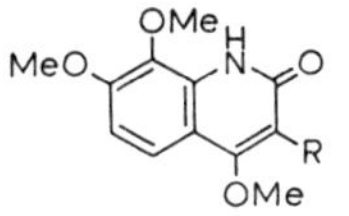

(XII) 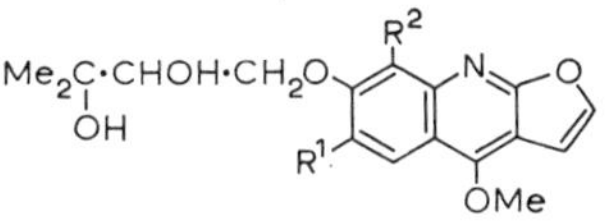   (XIII)

Maculosidine, 6,8-*dimethoxydictamnine*, m.p. 184°, and **maculosine**, 4-(2,3-*dihydroxy-3-methylbutoxy*)-6,7-*methylenedioxyfuro*[2,3-b]*quinoline*, m.p. 229–230°, $[\alpha]_D^{25} + 36°$ (pyridine), are obtained from Australasian *Flindersia* species (*R. H. Prager, E. Ritchie* and *W. C. Taylor*, Austral. J. Chem., 1960, **13**, 380).

Flindersiamine, 8-*methoxy-6,7-methylenedioxydictamnine*, m.p. 211–212°, *picrate*, m.p. 200°, is oxidised to 4,8-*dimethoxy-6,7-methylenedioxyquinollnon-3-carboxylic acid*, m.p. 249–250°, which is decarboxylated by hydrochloric acid into 4-*hydroxy-8-methoxy-6,7-methylenedioxyquinolinone*, m.p. 295° *(Anet et al., loc. cit.).* It has been synthesised (*T. R. Govindachari* and *S. Prabhakar*, Indian J. Chem., 1963, **1**, 348).

Evoxine, 7-(2,3-*dihydroxy-3-methyl-n-butoxy-8-methoxydictamnine* (XIII, R^1 = H, R^2 = OMe), m.p. 154–155°, is isolated from *Evodia xanthoxyloides* F. Muell. (*F. W. Eastwood et al.*, Austral. J. Chem., 1954, **7**, 87). It is identical with **haploperine** (p. 350), which is obtained from the seeds of *Haplophyllum perforatum* (*Sidyakin* and *Yunusov, loc. cit.*). Its structure was indicated by fusion with potassium hydroxide to give a phenol which with ethyl iodide yielded 2,4-dihydroxy-7-ethoxy-8-methoxy-3-ethylquinoline.

Evolatine, 7-(2,3-*dihydroxy-3-methyl-n-butoxy*)-6-*methoxydictamnine* (XIII, R^1 = OMe, R^2 = H), m.p. 201–202°, is isomeric with evoxine and is obtained from *Evodia alata* (*R. J. Gell et al.*, Austral. J. Chem., 1955, **8**, 114). When fused with potassium hydroxide it yields a phenol which when methylated gives kokusaginine and which on ethylation yields the 7-ethyl ether, converted by hydrogenolysis into 7-*ethoxy*-3-*ethyl*-2,4-*dihydroxy*-6-*methoxyquinoline*, m.p. 243–244°.

Acronycidine, 5,7,8-*trimethoxydictamnine*, m.p. 136.5–137.5°, *picrate*, m.p. 181.5–182.5°, is isolated from the bark of *Acronychian baueri* and *Melicope fareana* (*T. J. Batterham* and *Lamberton, ibid.*, 1965, **18**, 859). It is oxidised to 3-*formyl*-4,5,7,8-*tetramethoxyquinolin*-2-*one*, m.p. 219.5–220.5°, and the corresponding carboxylic acid, 4,5,7,8-*tetramethoxy*-*quinolinone*-3-*carboxylic acid*, m.p. 211° (decomp.), which when heated with hydrochloric acid gives 4-*hydroxy*-5,7,8-*trimethoxyquinolin*-2-*one*, m.p. 231–232°. Under mild conditions with methyl iodide it forms isoacronycidine (XIV) (*cf.* isodictamnine, p. 349) (*J. R. Price, ibid.*, 1959, **12**, 458).

Isomaculosidine, m.p. 170–172°, obtained from the root of *Dictamnus albus* L., is an example of a naturally occurring isofuroquinoline and is 6,8-*dimethoxy*-N-*methyl-iso-dictamnine* (XV) (*Storer* and *Young, loc. cit.*). It is obtained from maculosidine by treatment with methyl iodide and on hydrogenolysis yields 3-*ethyl*-4-*hydroxy*-6,8-*dimethoxy*-N-*methylquinolin*-2-*one*, m.p. 212–216°:

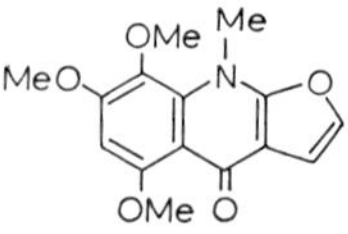

(XIV) 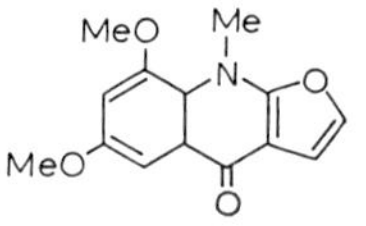   (XV)

(c) Pyranopyridines

The pyranopyridine nucleus is found in certain alkaloids (see below), but the simpler derivatives have not been intensively investigated. Some representative examples of this class of compound are discussed in the sequel, their formulation and nomenclature being based on that used in Chemical Abstracts, *i.e.*

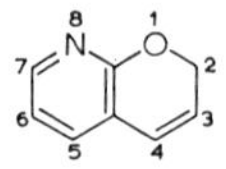

2H-Pyrano[2,3-b]pyridine   1H-Pyrano[4,3-b]pyridine

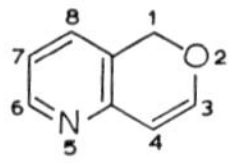

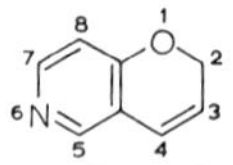

2H-Pyrano[3,2-c]pyridine   1H-Pyrano[3,4-c]pyridine

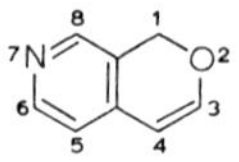

(i) 2H-*Pyrano*[2,3-b]*pyridine and derivatives*

The 2,3-disubstituted pyridine derivative XVI when heated with hydrobromic acid at 100° yields 2H-*pyrano*[2,3-b]*pyridinium bromide* (XVII), m.p. 149°, treatment of which with silver hydroxide gives 2*H*-**pyrano**[2,3-*b*]-**pyridine**, C_8H_7NO (XVIII), b.p. 64°/0.07 mm, m.p. 10.5°:

The formula is confirmed by the u.v. spectrum with a strong band at λ_{max} 305 nm (log ε 3.89) and by comparing its n.m.r. spectrum with that of the protonated intermediate XVII (*cf. I. C. Smith* and *W. G. Schneider*, Canad. J. Chem., 1961, **39**, 1158), a low-field displacement being observed particularly for the β- and γ-protons when the compound is protonated. Hydrogenation with a palladium catalyst gives 8-**azachroman**, C_8H_9NO, m.p. 30°, b.p. 56°/1 mm (*H. Sliwa*, Bull. Soc. chim. Fr., 1970, 631).

The sulphur analogue 2*H*-**thiapyrano**[2,3-*b*]**pyridine**, C_8H_7NS (XX), b.p. 91–92°/0.08 mm, is obtained by ring closure of 2-ethylthio-3-γ-hydroxy-propionylpyridine (XIX) with hydrobromic acid, followed by treatment of the product with silver hydroxide:

The p.m.r. spectrum is similar to that of the pyranopyridine XVIII. Note-worthy is the τ value of 6.37 for the CH_2 protons compared with that of 5.00 for the corresponding protons in XVIII, an observation attributed to the smaller electronegativity of the sulphur atom compared with that of the oxygen atom.

Pyridino-γ-pyrones were unknown until 1967 when *C. Bonsall* and *J. Hill* (J. chem. Soc., C, 1967, 1836) converted certain 3-substituted pyridinones into 5,7-*dimethyl-2-phenyl-8-azachromone* (XXI, R = Ph), 5,7-*dimethyl-8-azaflavone*, m.p. 184°, and 2,5,7-*trimethyl-8-azachromone* (XXI, R = Me), m.p. 116°. 8-**Azaflavone**, yellow crystals, m.p. 96°, is obtained by the ring-closure of the 2,3-disubstituted pyridine derivative XXII by means of hydro-bromic acid. Its i.r. and p.m.r. spectra resemble those of flavone *(Sliwa, loc. cit.)*:

(*ii*) 1H-*Pyrano*[4,3-b]*pyridines*

A derivative of 1*H*-pyrano[4,3-*b*]pyridine is known in the form of 3,4-*dihydro*-cis-1,3-*dimethyl-1H-pyrano*[4,3-b]*pyridine* (XXIII), b.p. 58°/0.1 mm (*C. Eskenazi, Sliwa* and *P. Maitte*, Bull. Soc. chim. Fr., 1971, 2951):

(XXIII) (XXIV) (XXV)

(iii) 2H-*Pyrano*[3,2-c]*pyridines*

2*H*-**Pyrano**[3,2-*c*]**pyridine**, C_8H_7NO (XXV), b.p. 69°/0.1 mm, is obtained in 26% yield by heating the ethoxypyridine derivative XXIV in a sealed tube at 160° (*G. Lhommet, Sliwa* and *Maitte, ibid.*, 1972, 1435). The p.m.r. spectrum shows two deshielded protons α to the nitrogen atom and τ 5.1 for the methylene protons (*cf.* τ 5.25 in 2*H*-benzopyran) as well as "zig-zag" coupling between H_4 and H_8, similar to that observed in 2*H*-benzopyran.

6-**Azachromone**, $C_8H_5NO_2$ (XXVI, R = H), m.p. 150° and 6-**azaflavone** (XXVI, R = Ph), yellow crystals, m.p. 167°, have a strongly deshielded proton at $C_{(5)}$ as the result of the influence of the nitrogen atom and neighbouring carbonyl group. Other spectral data include v_{max} 1660 cm^{-1} (conjugated CO), 1623 cm^{-1} (conjugated C:C), and λ_{max} 305 nm (log ε 4.96) (*I. Belsky*, Tetrahedron Letters, 1970, 4597).

(XXVI) (XXVII)

(iv) 1H-*Pyrano*[3,4-c]*pyridines*

The 1*H*-pyrano[3,4-*c*]pyridine nucleus is found in alkaloids isolated from various *Gentiana* species. The dihydro compound (XXVII, R = H), $C_9H_{11}NO$, m.p. 159–160° and **oliveridine**, $C_{10}H_{13}NO_2$ (XXVII, R = OMe), m.p. 260° are obtained from *Gentiana olivieri* (*T. U. Rakhmatullaev et al.*, C.A., 1970, **73**, 84624).

Other structurally related alkaloids are pyridinal lactones. The parent compound of this group is *gentianadine*, $C_8H_7NO_2$ (XXVIII, R = H), m.p. 77–78°, *hydrochloride*, m.p. 196–197°, *picrate*, m.p. 156–158° from *G. turkestanorum* (*A. Samatov et al.*, C.A., 1967, **67**, 117,007). It is oxidised by potassium permanganate to pyridine-3,4-dicarboxylic acid and has been synthesised (*J. Dolby et al.*, Acta Chem. Scand., 1971, **25**, 735).

Gentianidine, $C_9H_9NO_2$ (XXVIII, R = Me), m.p. 131–132°, is 6-methylgentianadine and is obtained from *G. Macrophylla*. It is oxidised to berberonic acid (pyridine-2,4,5-tricarboxylic acid) and has been synthesised by heating 4,6-dimethylnicotinic acid with formaldehyde and hydrochloric acid at 100° (*Liang Xiao-Tien et al.*, Scientia Sinica, 1965, **14**, 869; C.A., 1965, **62**, 5309), a synthetic method of general application for this type of compound.

Gentianine, $C_{10}H_9NO_2$ (XXIX), m.p. 82–83°, *hydrochloride*, m.p. 171–172°, *methiodide*, m.p. 190–191°, has been isolated from various species such as *Gentiana, Swertia* and *Enicostemina*, sometimes as an artefact. Its structure follows from oxidation to pyridine-

(XXVIII)   (XXIX)   Gl = β-D-glucose
(XXX)

3,4,5-tricarboxylic acid and from synthesis by heating 4-methyl-5-vinylnicotinic acid with formaldehyde and sodium hydrogen carbonate (*T. R. Govindachari, K. Nagarajan* and *S. Rajappa*, J. chem. Soc., 1957, 2725. The structure has been confirmed by p.m.r. spectroscopy (*D. Lavie* and *R. Taylor-Smith*, Chem. and Ind., 1963, 781), and is reminiscent of that of gentiopicroside (formerly known as gentiopicrin) (XXX) (*L. Canonica et al.*, Tetrahedron, 1961, **16**, 192).

Gentioflavine, $C_{10}H_{11}NO_3$, m.p. 218–222° (decomp.), is found in various *Gentiana* species and has formula XXXI. The formyl group is shown by a positive Tollens' reaction and the formation of an oxime. The alkaloid with concentrated nitric acid yields pyridine-2,3,4,5-tetracarboxylic acid which under the conditions of oxidation is decarboxylated to pyridine-3,4,5-tricarboxylic acid (*N. L. Marekov* and *S. S. Popov*, Tetrahedron, 1968, **24**, 1323). It can be converted into gentianidine.

(XXXI)   (XXXII)   (XXXIII)

A reduced pyranopyridine derivative XXXIII is obtained by heating methyl 2-dehydro-quinuclidine-3-carboxylate methiodide (XXXII), methyl iodide being lost in the process.

(d) Pyrrolopyridines or azaindoles

These substances are described in the literature under a variety of names which include in addition to those given above, diazaindenes, *Bz*-azaindoles and pyrindoles. The name azaindole is frequently found in the chemical journals and will be used here, but the *Chemical Abstracts* nomenclature invokes both the pyridine and the pyrrole nucleus present in the molecule, *i.e.*

4-Azaindole	1*H*-Pyrrolo[3,2-*b*]pyridine
5-Azaindole	1*H*-Pyrrolo[3,2-*c*]pyridine
6-Azaindole	1*H*-Pyrrolo[2,3-*c*]pyridine
7-Azaindole	1*H*-Pyrrolo[2,3-*b*]pyridine

---

* See "Monoazaindoles", *R. E. Willette*, Advances in Heterocycl. Chem., 1968, **9**, 27; *L. N. Yakhontov*, Russian Chem. Reviews, 1968, **37**, 551.

Azaindoles are prepared synthetically but 7-azaindole (XXXIV) has been isolated from coal tar (*O. Kruber*, Ber., 1943, **76**, 128) and 7-methyl-6-azaindole is a degradation product of the alkaloids harmine and harmaline.

(XXXIV)   (XXXV)   (XXXVI)

As would be expected the azaindoles are prepared by methods closely related to those used to synthesise indole and its derivatives. The Madelung cyclisation of formamidopicolines gives azaindoles, 2-formamido-3-picoline (XXXV), for instance, with sodium anilide yielding 7-azaindole (XXXIV) (*M. M. Robison* and *B. L. Robison*, J. Amer. chem. Soc., 1955, **77**, 6554). This method is often unsatisfactory and the use of formamidines such as XXXVI, thought to be intermediates in the Madelung reaction, is sometimes preferable. Thus the formamidine XXXVI when heated in boiling *N*-methylaniline with sodium *N*-methylanilide gives 7-azaindole (XXXIV) in 80% yield (*R. R. Lorenz et al.*, J. org. Chem., 1965, **30**, 2531; *cf. R. Herbert* and *D. G. Wibberley*, J. chem. Soc., C, 1969, 1505).

Naphthyridine derivatives can be converted into azaindole-3-carboxylic acids by the Süs ring-contraction of *o*-quinone diazides such as XXXVII, which when irradiated yields 6-azaindole-3-carboxylic acid (XXXVIII) (*O. Süs* and *K. Möller*, Ann., 1956, **599**, 233; 1958, **612**, 153):

(XXXVII)   (XXXVIII)   (XXXIX)

All four azaindole-3-carboxylic acids can thus be prepared and readily undergo decarboxylation to give the parent azaindole *e.g.* 6-azaindole (XXXIX), except in the case of the 4-aza acid. Related to this method is the photochemically induced ring-contraction of diazotised aminohydroxy-naphthyridines (*S. Okuda* and *M. M. Robison*, J. org. Chem., 1959, **24**, 1008). In this way diazotized 3-amino-4-hydroxy-1,6-naphthyridine (XL) yields 5-azaindole-3-carboxylic acid (XLI):

(XL)   (XLI)

Phenylhydrazones of arenecarbaldehydes and ketones are smoothly converted into indoles (see C.C.C. 2nd Edn., Vol. IV A, p. 398 *et seq.*) and some pyridylhydrazones similarly undergo conversion to azaindoles although harsher conditions are necessary (*A. H. Kelly* and *J. Parrick*, Canad. J. Chem., 1966, **44**, 2455; see also *Herbert* and *Wibberley, loc. cit.*). Acetophenone 2-pyridylhydrazone is thus converted in 63% yield into *2-phenyl-7-azaindole*, m.p. 201–202°.

The azaindoles are crystalline compounds which exhibit a more intense fluorescence than the indoles. They give no positive pine-splint test nor do they react with Ehrlich's reagent, but they give a strong bluish-green colour with sodium nitroprusside and sodium hydroxide. They are sensitive to oxidising agents, but the oxidation of the benzoyl derivatives can

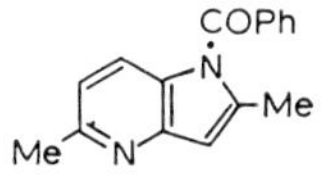

(XLII) 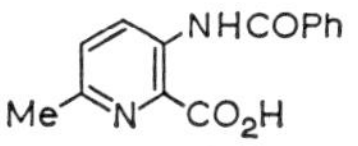   (XLIII)

give results of value for structural determination. 1-Benzoyl-2,5-dimethyl-4-azaindole (XLII), for example, is oxidised by potassium permanganate to 3-benzamido-6-methyl-pyridine-2-carboxylic acid (XLIII). Oxidation of the azaindoles with hydrogen peroxide in acetic acid yields aminonicotinic acids, 5-azaindole (XLIV), for example, giving 4-aminonicotinic acid (XLV) (*B. A. J. Clark* and *Parrick*, Tetrahedron, 1974, **30**, 475).

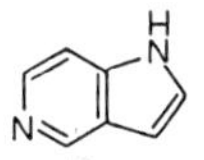

(XLIV) 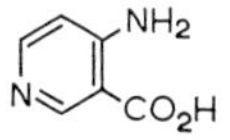   (XLV)

There is evidence that the azaindoles contain the imino group in the five-membered ring, 7-azaindole, for instance possessing the stable form XLVIa rather than the tautomeric form XLVIb:

(a)   (XLVI)   (b)

This is not unexpected since XLVIa possesses an aromatic pyridine nucleus whereas in XLVIb a quinonoid structure obtains. Structure XLVIa fits in with methylation (methyl iodide and sodamide) and acylation at $N_{(1)}$, and quaternary salt formation and protonation at the pyridine nitrogen atom (see, however, p. 359).

As required by theory, electrophilic substitution of 7-azaindole occurs most readily at $C_{(3)}$ (for examples, see p. 358), whereas nucleophilic attack occurs at $C_{(4)}$ and $C_{(6)}$ in the pyridine ring. 6-Chloro-4-methyl-7-azaindole, for example, is attacked at $C_{(6)}$ by sodium methoxide at 300° to the extent of 30% in contrast to the complete replacement of chlorine in 2-chloropyridine by the same reagent at 70°. This sluggishness is also shown by the failure of 4-methyl-7-azaindole to condense with aldehydes.

It is difficult to predict with certainty the nature of the products obtained by the reduction of the azaindoles. 5-Azaindole is not reduced by hydrogen and a platinum catalyst in the presence of acid, whereas 7-azaindole yields 2,3,3a,4,5,6-hexahydro-1*H*-pyrrolo-[2,3-*b*]pyridine (XLVII), a product also of the high pressure hydrogenation of 7-azaindole in acid solution. In neutral solution at 200° the product is the 2,3-dihydro compound, **7-azaindolenine,** m.p. 83–85°, while at higher temperatures 2-amino-3-ethylpyridine results. The indolenines are useful in preparative work, 7-azaindolenine, for example, yielding with nitric acid the *N*-nitro compound 1-nitro-7-azaindolenine (XLVIII), which with sulphuric acid gives 5-*nitro-7-azaindolenine* (XLIX), m.p. 260.5–261.5°. Dehydrogenation with palladium gives 5-*nitro-7-azaindole* (L), m.p. 275–280° (*M. M. Robison, B. L. Robison* and *F. P. Butler*, J. Amer. chem. Soc., 1059, **81**, 743).

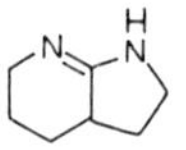

(XLVII) 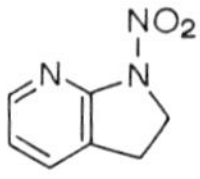   (XLVIII) 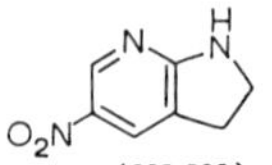   (XLIX) 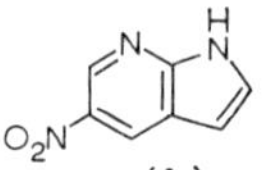   (L)

4-**Azaindole,** m.p. 127–128°, *picrate*, m.p. 242–244°, *oxalate*, m.p. 201–202°, is prepared by the cyclisation of 3-formamido-2-picoline by sodium anilide (*T. K. Alder* and *A. Albert*, J. chem. Soc., 1960, 1795; *Lorenz et al.. loc. cit.*). 3-Acetamido-2-picoline similarly yields 2-*methyl-4-azaindole*, m.p. 193–194°.

5-**Azaindole,** m.p. 111.5–112.5°, *oxalate*, m.p. 192–193° (*Okuda* and *Robison*, J. org. Chem., 1959, **24**, 1008; *cf. W. Herz* and *D. R. K. Murty, ibid.*, 1961, **26**, 122). Protonation is accompanied by deshielding of the pyrrol moiety protons and this is interpreted to mean that the proton adds at $N_{(6)}$ while $N_{(1)}$ assumes a positive charge (*G. G. Dvoryantzeva et al.*, Tetrahedron Letters, 1972, 2857):

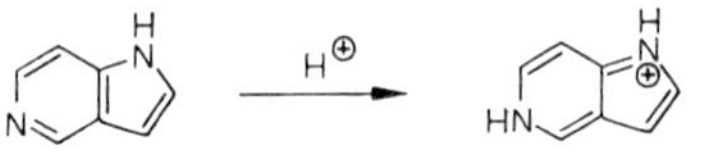

6-**Azaindole** (LII, R = H), m.p. 133–135°, *picrate*, m.p. 212–213°, is obtained by the reductive ring-closure of the keto ester LI, followed by hydrolysis and subsequent decarboxylation of the resulting ethyl 6-azaindole-2-carboxylate (LII, R = CO₂Et) (*M. H. Fisher* and *A. R. Matzuk*, J. heterocycl. Chem., 1969, **6**, 775).7-*Methyl-6-azaindole*, **apoharmine,** m.p. 185°, is prepared by cyclisation of 2-acetylpyrrole aminoacetal by a

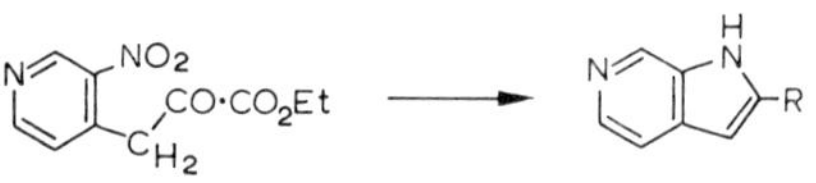

(LI)   (LII) 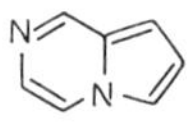   (LIII)

modified Pomeranz–Fritsch synthesis (C.C.C. Vol. IV F, p. 376). Pyrrole-2-carbaldehyde similarly yields 6-azaindole. Pyrrolopyrazines such as LIII are also formed in this synthesis.

7-**Azaindole** (XLVIa), m.p. 106–107°, b.p. 100–107°/0.5 mm, *picrate*, m.p. 232–233°, *acetyl* deriv., m.p. 67°, *hydrochloride*, m.p. 163°, with bromine in chloroform yields 3-*bromo-7-azaindole*, m.p. 186–187° (*Herbert* and *Wibberley, loc. cit.*) and with fuming nitric acid gives 3-*nitro-7-azaindole*, m.p. > 300° (*Robison, Robison* and *Butler, loc. cit.*).

7-Azaindole is a tautomeric substance (XLVIa ⇄ XLVIb) (*cf.* p. 357). With methyl iodide or methyl *p*-toluenesulphonate it yields the 7-methyl quaternary salt LIV, which with alkali gives the yellow, hygroscopic *7-methyl-7H-pyrrolo*[2,3-*b*]*pyridine* (LV), m.p. 48.5–50.5° (sealed tube) (*Robison* and *Robison*, J. Amer. chem. Soc., 1955, **77**, 6554), the structure of which follows from hydrogenation of the compound with a platinum catalyst to give the 3-substituted *N*-methylpiperidine derivative LVI. On the other hand, 7-azaindole, when methylated with methyl iodide in the presence of sodamide gives the colourless *1-methyl-7-azaindole* (LVII), b.p. 112–116°/21 mm. Insight into the tautomerism of 7-azaindole is afforded by its u.v. spectrum and those of the methyl derivatives LV and LVII

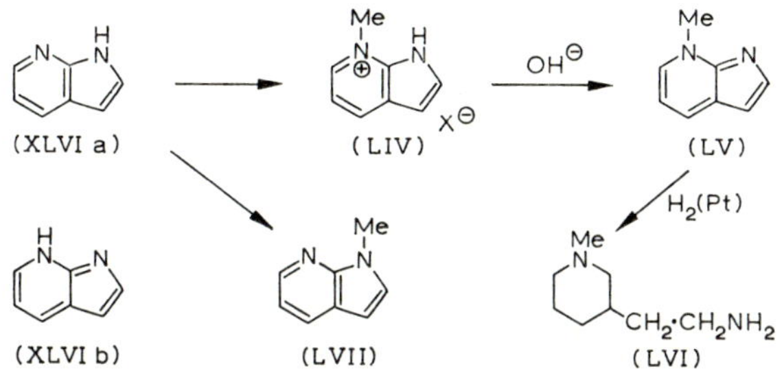

in cyclohexane. The spectrum of LVII is quite different from that of LV, but very similar to that of 7-azaindole, thereby supporting formula XLVIa. There is evidence, that in non-hydrogen bonding solvents 7-azaindole exists as an equilibrium mixture of the two tautomers XLVIa and XLVIb (*K. C. Ingham* and *M. A. El-Bayoumi, ibid.*, 1974, **96**, 1674). Such solutions exhibit two fluorescence bands, one due to XLVIa and the other to XLVIb, and the bands have been correlated with the fluorescence bands of the *N*-methyl derivatives LV and LVII.

(*e*) Carbolines or pyridindoles*

The linear tricyclic compounds known as carbolines or pyridoindoles contain an indole ring fused to a pyridine ring as shown in the formulae below. Various systems of nomenclature have been used to designate these sub-

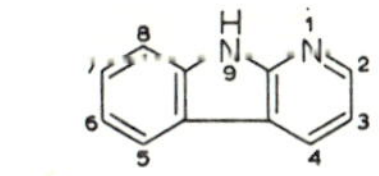

α-Carboline
9[*H*]-Pyrido[2,3-*b*]indole

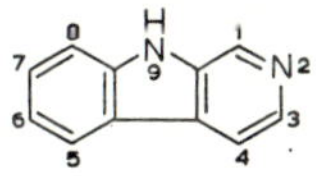

β-Carboline
9[*H*]-Pyrido[3,4-*b*]indole

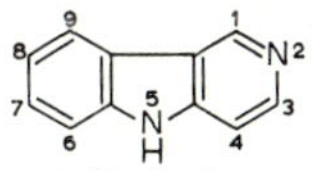

γ-Carboline
5[*H*]-Pyrido[4,3-*b*]indole

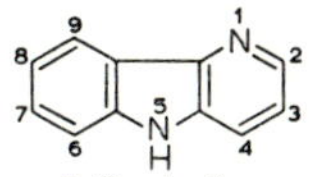

δ-Carboline
5[*H*]-Pyrido[3,2-*b*]indole

---

* See "The Carbolines", by *R. A. Abramovitch* and *I. D. Spenser*, Advances in heterocycl. Chem., 1964, **3**, 79; "The Carbolines", by *W. O. Kermack* and *J. E. McKail*, in Heterocyclic Compounds, ed. Elderfield, Vol. 7, p. 237, Wiley, New York, 1961.

stances and the *Chemical Abstracts* nomenclature, given under the formulae below, is widely used, but the simpler carboline nomenclature is also encountered in the literature and will be used in this chapter.

Clearly the carbolines are azacarbazoles and this structural relationship is shown by their mode of preparation by the pyrolysis or photolysis of the appropriate triazoles (*cf.* the Graebe–Ullmann synthesis of carbazole). The four carbolines are prepared in this way, 1-β-pyridylbenzotriazole (LVIII), for example, when heated with zinc chloride losing nitrogen to give a mixture of β- and δ-carbolines:

(LVIII)   β-Carboline   δ-Carboline

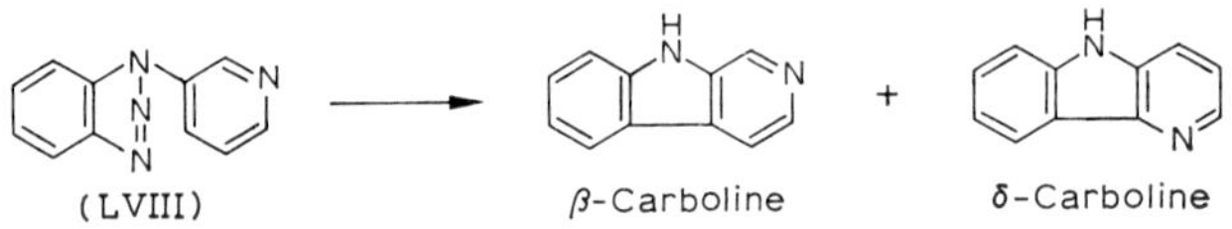

The Fischer indole synthesis has been applied to the preparation of tetrahydrocarbazoles and is used in the preparation of tetrahydrocarbolines. 1-Benzyl-4-piperidone phenylhydrazone is thus converted into tetrahydro-γ-carboline and hence by dehydrogenation and debenzylation into γ-carboline itself in 30% overall yield (*N. P. Buu-Hoï et al.*, J. chem. Soc., 1964, 208).

In a reaction analogous to the conversion of diphenylamine into carbazole, anilinopyridines when irradiated in various solvents yield carbolines, 2-anilinopyridine in tetrahydrofuran, thus yielding α-carboline in 80% yield (*V. M. Clark, A. Cox* and *E. J. Herbert, ibid.*, C, 1968, 831):

α-Carboline

3-Anilinopyridine similarly yields a mixture of β- and δ-carbolines in the ratio 1:1.7.

Ring-closure of nitrenes can yield carbolines, the reduction of 6-chloro-2-methyl-3-nitro-4-phenylpyridine, for example, with triethyl phosphite yielding a nitrene which cyclises to 3-chloro-1-methyl-β-carboline. This substance loses chlorine when treated with lithium tetrahydridoaluminate to give 1-methyl-β-carboline (harman) (*T. Kametani, K. Ogasawara* and *T. Yamanaka, ibid.*, C, 1968, 1006).

The pyrrole ring of the carbolines possesses no basicity, but as expected, the pyridine nitrogen undergoes protonation and forms quaternary ammonium salts with alkyl halides. These give anhydro salts with alkalis, δ-carboline giving a methiodide LIX, which with alkali yields the orange anhydro salt LXa:

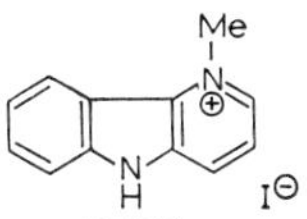

In this compound the pyrrole nitrogen is clearly a site for nucleophilic attack and with methyl iodide yields a methiodide LXI, identical with that obtained from $N_{(5)}$-methyl-δ-carboline and methyl iodide (*R. A. Abramovitch, K. A. H. Adams* and *A. D. Notation*, Canad. J. Chem., 1960, **38**, 2152). *J. W. Armit* and *R. Robinson* in a classical paper (J. chem. Soc., 1925, **127**, 1604) forecast the theory of resonance by stressing that such anhydro-salts can be written in either a non-polar *o*-quinonoid form (LXb) or a polar form (LXa) in which a tendency towards aromatic stability in terms of Kekulé formulae is balanced against neutralisation of charges. The authors' statement that "the formulae represent extreme conditions and the true structure doubtless lies between the two" is an example of remarkable scientific prescience.

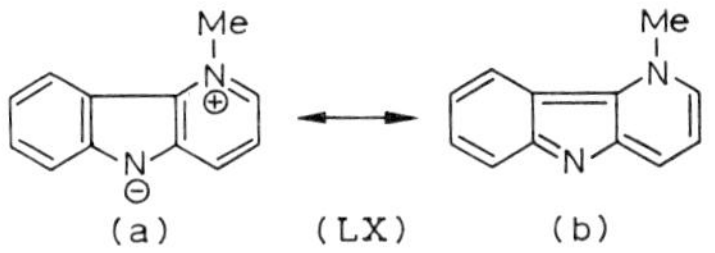

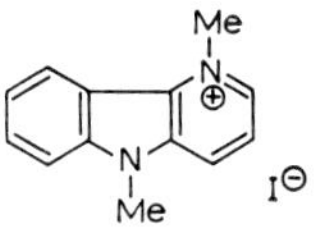

The carbolines behave as heteroaromatic compounds as indicated by their p.m.r. spectra, the nuclear protons possessing τ values between 1.6 and 2.8 (*P. J. Black* and *M. L. Heffernan*, Austral. J. Chem., 1965, **18**, 353; *Abramovitch* and *I. D. Spenser*, Canad. J. Chem., 1964, **42**, 954). They react with electrophilic reagents, not unexpectedly at the carbon atoms of the benzene ring *ortho* and *para* to the NH group (see *e.g. L. Paoloni*, Gazz., 1960, **90**, 1530). 1-Methyl-β-carboline, for example, with concentrated nitric acid yields 1-*methyl-6-nitro-β-carboline* (57% yield), m.p. 299–300° and (probably) 1-*methyl-8-nitro-β-carboline*, m.p. 209–210° (*H. R. Snyder et al.*, J. Amer. chem. Soc., 1948, **70**, 222). α-Carboline gives 6-*nitro-α-carboline*, m.p. >315° (*R. R. Burtner*, U.S.P. 2,690,441/1955; C.A., 1955, **49**, 13297). δ-Carboline gives mainly the 8-*nitro* (56%), m.p. >360° and 6-*nitro-δ-carboline* (17%), m.p. 232–233°, the relatively low m.p. of the second product suggesting hydrogen bonding.

The carbolines undergo nucleophilic attack in the pyridine ring, β-carboline, for instance, giving 1-*amino-β-carboline*, m.p. 199.5–200.5°, with amide anion (*Snyder, H. G. Walker* and *F. X. Werber*, J. Amer. chem. Soc., 1949, **71**, 527).

α-**Carboline**, $C_{11}H_8N_2$, m.p. 215–216°, *picrate*, m.p. 272–274°, is formed by heating 1-α-pyridylbenzotriazole with polyphosphoric acid (*B. Witkop, ibid.*, 1953, **75**, 3361).

β-**Carboline, norharman,** m.p. 199–201°, *methiodide*, m.p. 243–244°, is the parent substance of the harmala alkaloids (see C.C.C., Vol. IV B, p. 75). **Harman,** $C_{12}H_{10}N_2$, m.p. 228–230°, *picrate*, m.p. 262° (decomp.), is 1-methyl-β-carboline and is identical with the alkaloids **aribine,** obtained from *Ariba rubra*, and **loturnine,** from *Symplocos racemosa*. The methyl group condenses with benzaldehyde in the presence of alkali to give a benzylidene derivative, oxidation of which gives β-*carboline-1-carboxylic acid*, m.p. 272°

(decomp.), which is found in the stem bark of *Pleiocarpa mutica* Benth. as the *methyl ester*, m.p. 168° (*H. Achenbach* and *K. Biemann*, J. Amer. chem. Soc., 1965, **87**, 4177; *C. Kump, J. Seibl* and *H. Schmid*, Helv., 1963, **46**, 498).

γ-**Carboline**, m.p. 225° (228–229° also reported), *picrate*, m.p.'s of 242°, 250° and 262° are reported. It is synthesised by the Fischer method (p. 260) and from 3-formylindole (*L. K. Dalton et al.*, Austral. J. Chem., 1969, **22**, 185).

δ-**Carboline**, m.p. 214–215°, *picrate*, m.p. 262–266° (277–279° also reported).

(f) Dihydro-furoquinolines and -pyranoquinolines

It is convenient to group both types of compounds together, particularly as both are found in the *Lunasia* alkaloids (*S. Goodwin et al.*, J. Amer. chem. Soc., 1959, **81**, 6209). The alkaloids are derivatives of the *linear* quinolin-4-ones LXII and LXIII and the *angular* quinolin-2-ones LXIV and LXV:

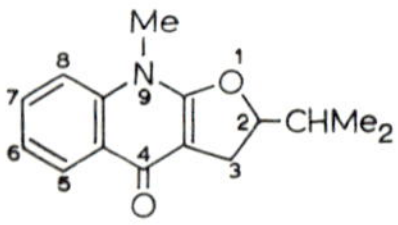

(LXII) 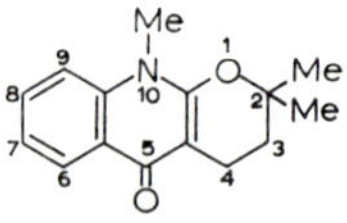   (LXIII)

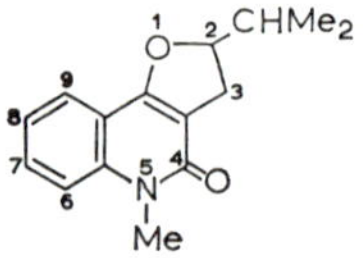

(LXIV) 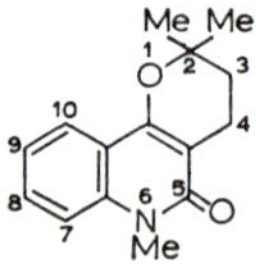   (LXV)

These differences in structure are reflected in the chemical and physical properties of the compounds, the quinolin-2-ones, for example being so weakly basic that their u.v. spectra in neutral and acidic solvents are identical.

The structures of the dihydro-furoquinolines and -pyranoquinolines have been established partly by chemical means and partly by spectroscopic evidence. I.r. spectroscopy is frequently useful (*L. H. Briggs et al.*, Anal. Chem., 1957, **29**, 904), but p.m.r. spectroscopy has proved to be invaluable. The protons at $C_{(5)}$ in the *linear* compound LXII and at $C_{(6)}$ in LXIII, for instance, are strongly deshielded ($\tau \sim 1.9$) by the neighbouring carbonyl group, whereas in the *angular* compounds LXIV and LXV, deshielding is less marked (*A. V. Robertson*, Austral. J. Chem., 1963, **16**, 451). The dihydrofuran and dihydropyran structures (*e.g.* LXIV and LXV) are also easily differentiated, the *C*-methyl protons in the former showing resonance whereas in the pyrans singlets are observed. Integration values of the oxygen-ring protons (Me not considered) are also decisive, *e.g.* the furo-compounds LXIV and LXV, showing relative shift intensities of 2:1:1 and the pyrans 2:2

(see also *Goodwin, J. N. Shoolery* and *L. F. Johnson*, J. Amer. chem. Soc., 1959, **81**, 3065).

Linear dihydrofuro- and -pyrano-quinolines result in high yield when epoxides such as LXVI are treated with a base (*R. M. Bowman* and *M. F. Grundon*, J. chem. Soc., 1966, 1504). The products LXVII and LXVIII when boiled with methyl iodide undergo isomerisation to yield the alkaloids balfourodine and isobalfourodine:

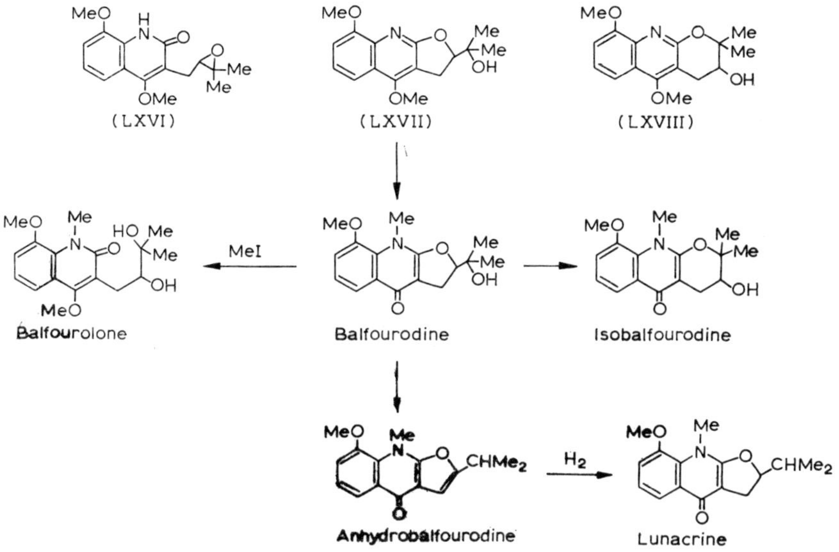

(LXVI)　　　　(LXVII)　　　　(LXVIII)

Balfourolone　　　Balfourodine　　　Isobalfourodine

Anhydrobalfourodine　　　Lunacrine

The dihydrofuroquinolines undergo a number of interesting intramolecular transformations including isomerisation to pyranoquinolines and angular furoquinolines. These changes are outlined below with balfourodine as an example.

(+)-**Balfourodine**, $C_{16}H_{19}NO_4$, $[\alpha]_D + 49°$, m.p. 191–192°, is obtained from the Brazilian and Argentinian tree *Balfourodendron riedelianum* Engl. Its structure (*H. Rapoport* and *K. G. Holden*, J. Amer. chem. Soc., 1960, **82**, 4395) has been confirmed by synthesis of the (±)-compound (*E. A. Clarke* and *Grundon*, J. chem. Soc., 1964, 438) and the enantiomers by asymmetric epoxidation of dimethylallylquinolones (see LXVI above) (*Bowman, J. F. Collins* and *Grundon, ibid.*, Perkin I, 1973, 626). With methyl iodide and mild alkali it undergoes a typical 4-quinolone reaction to give the dimethoxy product *balfourolone*, $C_{17}H_{23}NO_5$ $[\alpha]$ −36°, m.p. 100°, which is isolated as an artefact in the extraction of lunacrine from natural sources.

The absolute configuration of (+)(R)-balfourodine, (−)(S)-isobalfourodine and (−)(R)-balfourolone have been established (*idem. loc. cit.*).

Balfourodine with acetic anhydride and pyridine gives the acetate of isobalfourodine, and both the parent compound and the acetate with alkali are isomerised to ψ-isobalfourodine. A second product of the action of alkali on balfourodine is ψ-balfourodine, a pyrano[3,2-c]quinoline derivative (*Rapoport* and *Holden, loc. cit.*; *Grundon* and *K. J. James*, Chem. Comm., 1970, 337):

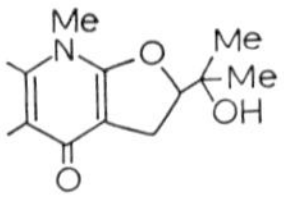

Balfourodine

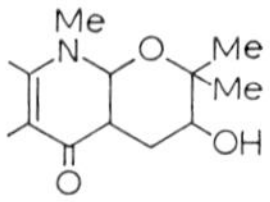

Isobalfourodine

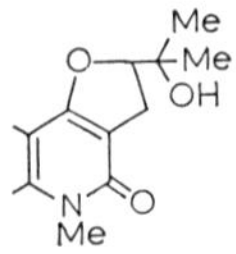

ψ-Balfourodine

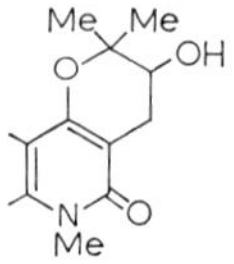

ψ-Isobalfourodine

Balfourodine is dehydrated by sulphuric acid to *anhydrobalfourodine*, m.p. 126–126.5°, the u.v. spectrum of which is identical with that of ψ-isofagarine. Reduction of the anhydro-compound gives lunacrine.

The alkaloid (+)**isobalfourodine** $C_{16}H_{19}NO_4$, has m.p. 204–205°, $[\alpha]_D + 15°$, *racemic form*, m.p. 184–186°. **Lunasia 11** is the *laevo* enanthiomorph (*H. C. Beyermann* and *R. W. Rooda*, Proc. Kon. Ned. Akad. Wetenschap., 1960, **63**ₘ, 154; *Rapoport* and *Holden, loc. cit.*).

(—)**Lunacrine**, $C_{16}H_{19}NO_3$, m.p. 117–119°, and (+)**lunacridine**, $C_{17}H_{23}NO_4$, m.p. 86–87°, are extracted from *Lunasia amara* Blanco, the second compound possibly as an artefact (*Goodwin* and *E. C. Horning*, J. Amer. chem. Soc., 1959, **81**, 1908, 3065). Both compounds have been synthesised (*Clarke* and *Grundon*, J. chem. Soc., 1964, 438).

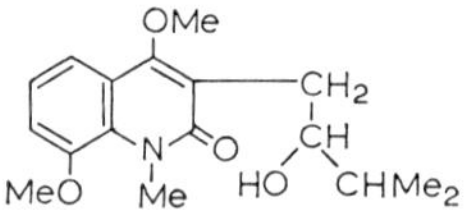

Lunacridine

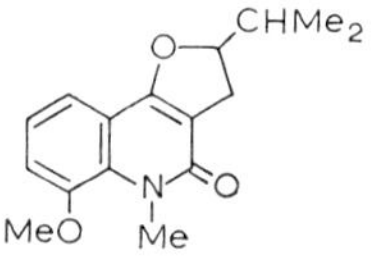

(LXIX)

Lunacrine is converted into lunacridine by heating with potassium hydroxide followed by methylation of the product with diazomethane. This and other evidence shows that lunacridine has the above formula and this is confirmed by the cyclisation of lunacridine to give some lunacrine, but mainly the *angular 2,3,4,5-tetrahydro-2-isopropyl-6-methoxy-5-methyl-4-oxofuro-[3,2-c]quinoline* (LXIX), m.p. 76–78°, which is the structure assigned to the alkaloid (—)**lunasia VI** (*Beyermann* and *Rooda, loc. cit.*; *J. R. Price*, Austral. J. Chem., 1959, **12**, 458).

(±)-*Lunacrine*, m.p. 145–147°, *picrate*, m.p. 213–214°; (±)*lunacridine*, m.p. 73–74°. For the asymmetric synthesis of lunacrine see *Bowman, G. A. Gray* and *Grundon*, J. chem. Soc., Perkin I, 1973, 1051.

A major constituent of *Lunasia quercifolia* is (—)**lunasine,** a quaternary ammonium salt, which can be obtained as the *iodide,* m.p. 129.5°, by the action of methyl iodide on lunacrine. (—)**Lunine,** $C_{16}H_{17}NO_4$, m.p. 228–229°, $[a]_D$ —39° (CHCl₃), is converted by methyl iodide into O-*methylluninium iodide* (LXX) (*N. K. Hart* and *Price, ibid.,* 1966, **19,**

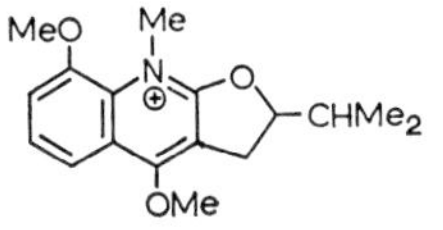

Lunasine

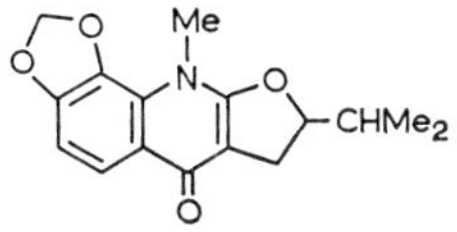

Lunine

2185). The presence of the methylenedioxy group is shown by the Labat test and the marked deshielding of the $C_{(5)}$ proton in the p.m.r. spectrum shows it to be unsubstituted. The coupling constant with the remaining proton in the ring is 8.5 Hz clearly indicating *ortho*-coupling. The oxymethylene group must therefore occupy the 7 and 8 positions.

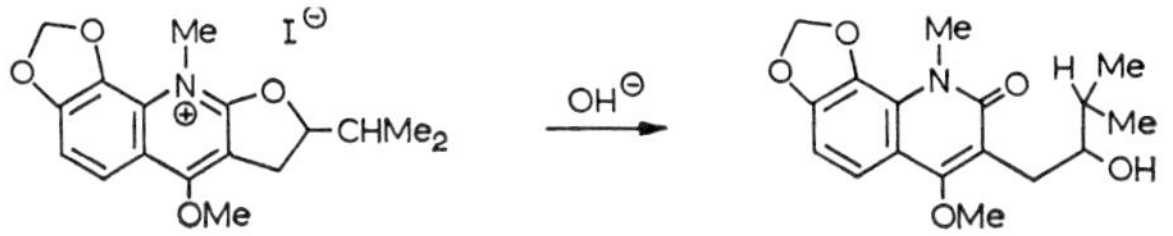

(LXX)

Lunidine

O-Methylluninium iodide (LXX) with alkali yields the alkaloid **lunidine,** which is obtained from *Lunasia amara* Blanco. It is isolated as the (+)*isomer,* m.p. 65–66.5°, $[a]_D^{24}$ +28.6° (EtOH). The constitution of lunidine has been established by p.m.r. evidence (*A. Rüegger* and *D. Stauffacher,* Helv., 1963, **46,** 2329).

(+)**Platydesmin,** m.p. 138–139°, $[a]_D$ + 47° (CHCl₃), and its *acetate,* m.p. 126–127°, $[a]_D$ +23° (CHCl₃) are found in various *Rutaceae* species (*S. R. Johns* and *J. A. Lamberton,* Austral. J. Chem., 1966, **19,** 1991). **Khaplofoline,** m.p. 272–274°, N-*methyl* deriv., m.p. 120–121°, is obtained from *Haplophyllum foliosum* (*Bowman* and *Grundon,* J. chem. Soc., C, 1966, 1084):

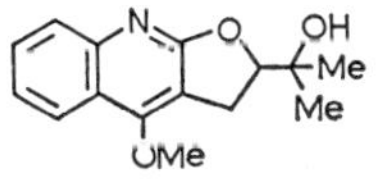

Platydesmin

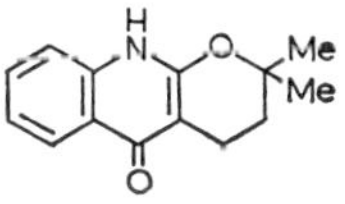

Khaplofoline

Angular pyranoquinolines are obtained in high yield by treatment of epoxides such as LXXI with a base (*Bowman, Grundon* and *James,* J. chem. Soc., Perkin I, 1973, 1055):

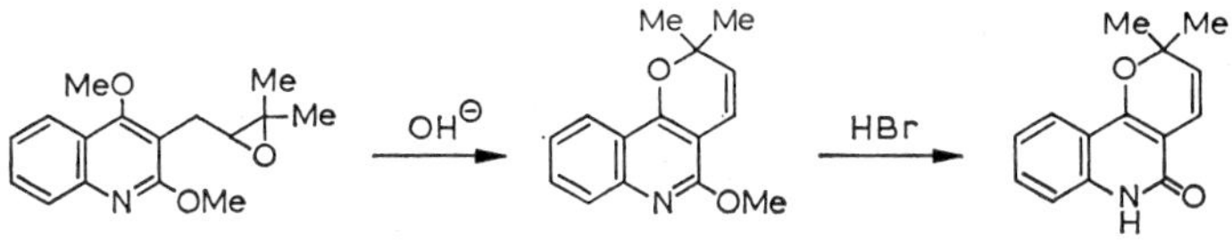

(LXXI)

Flindersine

In this instance the product is 2,2-*dimethyl-5-methoxy*-2H-*pyrano*[3,2-c]*quinoline,* m.p. 47–48°, which with hydrobromic acid yields the alkaloid **flindersine,** 5,6-*dihydro*-2,2-*dimethyl*-2H-*pyrano*[3,2-c]*quinol-5-one,* m.p. 192–195° (depending on the rate of heating), *picrate*, m.p. 178°, *bromo* deriv., m.p. 218°. It is obtained from the Australian tree *Flindersia australis* (*R. F. C. Brown et al.*, Austral. J. Chem., 1954, **7**, 348; 1956, **9**, 277). Its structure follows from its degradation with 30% alkali to 4-hydroxyquinol-2-one, oxidation to 4-hydroxyquinol-2-one-3-carboxylic acid, synthesis and its p.m.r. spectrum. Its N-*methyl* derivative, m.p. 83–85°, occurs in *Spathelia sorbifolia* L. (*C. D. Adams, D. R. Taylor* and *J. M. Warner*, Phytochemistry, 1973, **12**, 1359).

Naturally occurring substances are known in which a pyran ring is fused to the benzene ring of quinoline. **Acronycine,** 3,12-*dihydro-6-methoxy-3,3,12-trimethyl-7H-pyrano*[2,3-c]-*acridin-7-one*, $C_{20}H_{19}NO_3$, m.p. 175–176°, is obtained from the Australian tree *Acronychia baueri* Schott:

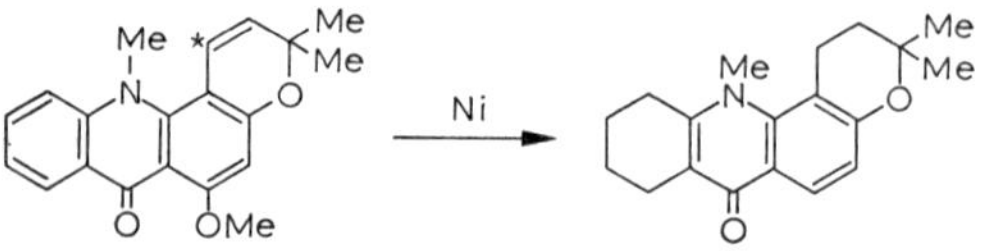

Acronycine   (LXXII 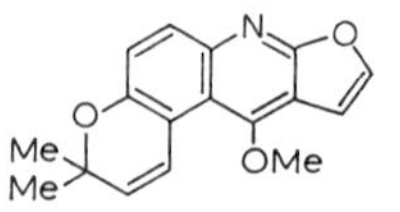)   Medicosimine

Evidence for this structure was provided by degradation to 1,3-dimethoxy-4-methoxy-carbonyl-*N*-methylacridone (*P. L. Macdonald* and *A. V. Robertson*, Austral. J. Chem., 1966, **19**, 275) and by treatment of the 1-*O*-tosylate with Raney nickel to yield the hydrogenated compound LXXII (*T. R. Govindachari et al.*, Tetrahedron, 1966, **22**, 3245). The structure was confirmed by synthesis (*J. Hlubucek et al.*, Austral. J. Chem., 1970, **23**, 1881; *J. R. Beck et al.*, J. Amer. chem. Soc., 1968, **90**, 4706). Useful evidence was supplied by the nuclear Overhauser effect which demonstrated the proximity of the NMe group and the hydrogen atom marked with an asterisk.

Medicosimine, $C_{17}H_{15}NO_3$, m.p. 138.5–139.5°, *picrate*, m.p. 190–191° (*J. A. Lamberton* and *Price*, Austral. J. Chem., 1953, **6**, 173) is clearly a quinoline derivative linked both to a furan and to a pyran ring.

(g) *Naphthyridines or diazanaphthalenes**

Naphthyridine is the name given to the ring system containing two fused pyridine rings with two carbon atoms in common. The six theoretically possible naphthyridines have been prepared, generally from suitably substituted pyridines, by methods similar to those used in the synthesis of quinolines and isoquinolines. They are similar in their chemical behaviour to pyridine, are soluble in water, form salts with acids, and react with picric acid, etc. to form π-complexes. They form mono- and di-*N*-oxides and form quaternary ammonium salts with methyl iodide (see below).

---

* See *W. W. Paudler* and *T. J. Kress*, Advances in heterocycl. Chem., 1970, **11**, 123; *M. J. Weiss* and *C. R. Hauser*, "Heterocyclic Compounds", Vol. 7, p. 198 (Wiley, New York, 1961); *C. F. H. Allen*, Chemical Reviews, 1950, **47**, 275.

The naphthyridines are weaker bases than quinoline and show the following pK_a values (in water at 20°): 1,5-, 2.91; 1,6-, 3.78; 1,7-, 3.63; 1,8-, 3.39; quinoline 4.94 (*W. W. Paudler* and *T. J. Kress*, J. heterocycl. Chem., 1968, **5**, 561). These authors also show that protonation and *N*-methylation occur at the same sites, 1,7-naphthyridine thus yielding the cation LXXIII (R = H or Me). This conclusion is based on the observed p.m.r. cross-nitrogen coupling which occurs when the nitrogen lone pair has been removed by protonation or alkylation.

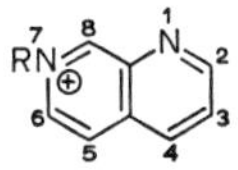

(LXXIII)

The naphthyridines form monomethiodides, which behave in expected fashion. The *monomethiodide* of *1,5-naphthyridine*, m.p. 254–255°, for example, with alkali gives the pseudobase LXXIV and there is no evidence that this exists in an open-chain form to any great extent (*J. W. Bunting* and *W. G. Meathrel*, Canad. J. Chem., 1972, **50**, 917; *Bunting*, J. chem. Soc., Perkin I, 1974, 1833):

(LXXIV)

Oxidation of the pseudo-base by alkaline potassium ferricyanide yields *1-methyl-1,5-naphthyridin-2-one*, m.p. 104–105°, which after successive treatment with methyl iodide and potassium ferricyanide gives *1,5-dimethyl-1,5-naphthyridine-2,6-dione*, m.p. 220–222°.

1,5-Naphthyridine with dimethyl sulphate gives *1,5-dimethyl-1,5-naphthyridinium dimethosulphate* (LXXV), m.p. 200°, which with alkali gives the mono-pseudo-base LXXVI (*J. E. Dickson et al., ibid.*, 1973, 2885). This seems to apply to the other naphthyridines, for the *N,N'*-dimethylnaphthyridinium salts obtained by methylation with "magic methyl", methyl fluorosulphonate (*D. J. Pokorny* and *Paudler*, Canad. J. Chem., 1973, **51**, 476) yield mono- and not di-pseudo-bases:

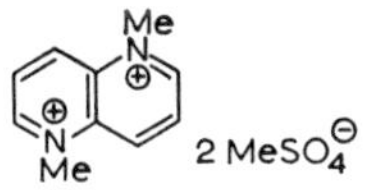

(LXXV)

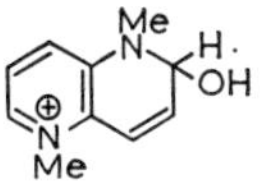

(LXXVI)

It is noteworthy that with bases such as 1,6-naphthyridine, pseudo-base formation occurs preferentially at the quinolinoid site rather than at the isoquinolinoid site, the 1,6-dication LXXVII thus yielding the pseudo-base

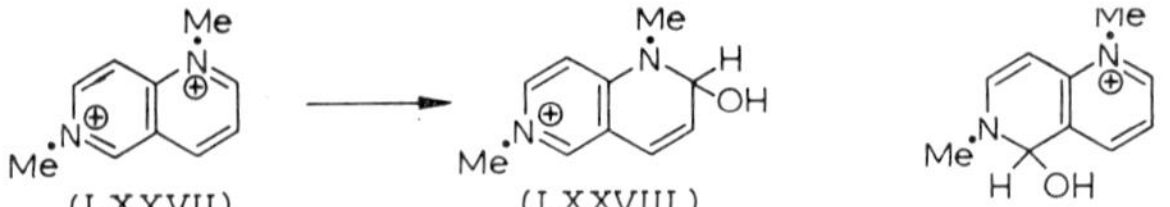

(LXXVII) → (LXXVIII) (LXXIX)

LXXVIII and not LXXIX. In spite of the fact that oxidation of the pseudo-bases with alkaline potassium ferricyanide yields diones as shown above, only small quantities of dipseudo-bases are present in aqueous solution (*Bunting, loc. cit.; Bunting* and *Meathrel,* Canad. J. Chem., 1974, **52,** 962).

The naphthyridines form Reissert compounds (*cf.* Vol. IV F, p. 263) (*Y. Kobayashi, I. Kumadaki* and *H. Sato,* Chem. Phar. Bull. Japan, 1969, **17,** 2614).

The naphthyridines are planar with the exception of the 1,8-isomer, which, as the result of repulsion of the nitrogen lone pairs of electrons, is non-planar (*A. Clearfield et al.,* Acta Cryst., 1972, **B28,** 350; *M. Brufani, G. Duranti* and *G. Giacomello,* Gazz., 1959, **89,** 2328). The molecule assumes planarity when it is chelated to a metal atom. The bond lengths in the naphthyridines are indicative not only of aromatic character, but also of some degree of bond fixation (*cf.* naphthalene and quinoline) and the resonance energy of 31.09 kcal/mol is close to that of quinoline (32.95 kcal/mol) (*M. J. S. Dewar, A. J. Harget* and *N. Trinajstic,* J. Amer. chem. Soc., 1969, **91,** 6321):

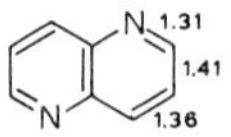

1,5-Naphthyridine
Bond lengths (Å)

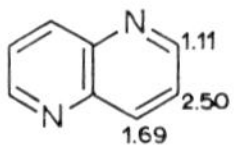

1,5-Naphthyridine
p.m.r. spectrum (CDCl$_3$)

The aromatic character of the naphthyridines is confirmed by their p.m.r. spectra. In the spectrum of 1,5-naphthyridine, for example, the deshielding characteristic of aromatic protons is obvious (see figure) (*W. L. F. Armarego* and *T. J. Batterham,* J. chem. Soc., B, 1966, 750, who also give the p.m.r. spectra of other naphthyridines). The three u.v. absorption bands of naphthalene, quinoline and isoquinoline are observed also in the naphthyridine spectra.

Electrophilic attack occurs at positions *meta* to the nitrogen atoms (see Table 2), whereas nucleophilic substitution by means of potassium amide and liquid ammonia occurs at carbon atoms *ortho* and *para* to the nitrogen atoms (*Paudler* and *Kress,* J. org. Chem., 1968, **33,** 1384). Another example of nucleophilic action is given by 1,3,6,8-tetrachloro-2,7-naphthyridine (p. 373). For a review of nucleophilic substitution of the naphthyridines see *R. G. Shepherd* and *J. L. Fedrick,* Advances in heterocyclic Chem., 1965, **4,** 377.

The naphthyridines with perbenzoic or *m*-chloroperbenzoic acid yield mixtures of mono- and di-*N*-oxides (*Paudler, Pokorny* and *S. J. Cornrich,*

J. heterocycl. Chem., 1970, **7**, 291). This is summarised in Table 2 which also gives the products obtained by bromination with bromine in carbon tetrachloride or by decomposition of the bromine–naphthyridine complexes (*Paudler* and *Kress*, *loc. cit.*). The action of bromine on pyridine at 500° differs sharply from electrophilic bromination at lower temperatures and similar results are encountered in the naphthyridine series, 1,5-naphthyridine, for example, yielding 2-bromo- ($\sim$ 10%), 2,6-dibromo- ($\sim$ 30–40%) and smaller quantities of 2,3,6- and 2,4,6-tribromo-1,5-naphthyridine (*J. Pomorski* and *H. J. den Hertog*, Roczniki Chem., 1973, **47**, 1973). The site of substitution patently differs from that noticed above.

Reduction of the naphthyridines with sodium and ethanol yields decahydronaphthyridines (*Armarego*, J. chem. Soc., C, 1967, 377). Hydrogenation with a palladium–carbon catalyst frequently yields 1,2,3,4-tetrahydroproducts, but sometimes both rings are attacked. 1,7-Naphthyridine, for instance, yields 1,2,3,4- (57%) and 5,6,7,8-tetrahydro-1,7-naphthyridine (43%).

TABLE 2

NAPHTHYRIDINES

Naphthyridine	N-Oxidation	Bromination	Amination
1,5-	1 (67%)	3 (27%) 3,7 (10%)	2 (33%)
1,6-	1 (12%) 6 (38%)	3 (18%) 8 (23%) 3,8 (11%)	2 (45%)
1,7-	1 (3%) 7 (45%)	5 (25%) 3,5 (?%)	8 (56%)
1,8-	1 (47%)	3 (5%) 3,6 (0.5%)	2 (30%)
2,7-	2 (72%)		

The replacement of 2- and 4-hydroxyl groups in the pyridine nucleus by halogen is observed in the naphthyridines, 2,4-dihydroxy-1,8-naphthyridine, for example, with phosphorus pentachloride and phosphoryl chloride yielding 2,4-dichloro-1,8-naphthyridine (*G. Koller*, Ber., 1927, **60**, 1572, 1918). The ease of decarboxylation of pyridine- and quinoline-carboxylic acids (C.C.C., Vol. 4 F, pp. 186, 334 *et seq.*) is paralleled by the behaviour of 4-hydroxy-1,8-naphthyridine-2,3-dicarboxylic acid which yields the 3-

carboxylic acid when heated at 170° or when boiled with 20% hydrochloric acid. Further decarboxylation to 4-hydroxy-1,8-naphthyridine is achieved by heating at 300°:

The structures of naphthyridines and substituted naphthyridines have been established by chemical methods and synthesis with the aid of spectroscopy. U.v. spectra are sometimes useful and can be used to distinguish between 1,5- and 1,7-naphthyridine (*Rapoport* and *Batcho*, J. org. Chem., 1963, **28**, 1753). Infrared absorption bands in the 1000–670 cm^{-1} region are caused by C–H bending vibrations which depend on the number of adjacent free hydrogen atoms in an aromatic nucleus. The use of this method is illustrated by the following examples *(idem, loc. cit.)*. All four compounds contain three adjacent hydrogen

TABLE 3

INFRARED C–H BENDING VIBRATIONS

	3H	2H	1H
1,5-	816	—	—
(Me)	823	—	890
1,7-	818	838	943
(Cl)	801	841	—

atoms in at least one ring and show bands in the 800–820 cm^{-1} range. Naphthyridines with isolated hydrogen atoms absorb in the 890–940 cm^{-1}, while those with two such hydrogen atoms show bands near 840 cm^{-1}. Three adjacent hydrogen atoms are indicated by still lower wave numbers. These data must, however, be used with some caution (*Armarego, G. B. Barlin* and *E. Spinner*, Spect. Acta, 1966, **22**, 117).

P.m.r. spectra have proved invaluable in the naphthyridine field (*Paudler* and *Kress*, Advances in heterocyclic Chem., Vol. 11, p. 123; *Armarego* and *Battenham, loc. cit.*). The simple spectrum of the symmetrical 1,5-naphthyridine, for example, contrasts with the more complex spectrum of the 1,6-isomer (*Rapoport* and *Batcho, loc. cit.*). Other examples are provided by 4-methyl-2,6-naphthyridine which exhibits only one pair of protons showing measurable spin–spin coupling, and by 2-methyl-1,8-naphthyridine in which the methyl protons give rise to singlets whereas in the 4-methyl isomer the methyl protons are coupled to the $C_{(3)}$-proton (*cf.* methylquinolines).

Mass spectra of all naphthyridines show the same general fragmentation pattern (*Paudler* and *Kress*, *loc. cit.*), the most abundant ions having *m/e* values of 103 (by loss of HCN), 104 (loss of C_2H_2), or 76 by the further loss of HCN from the ion, *m/e* 103.

1,5-Naphthyridine, yellow needles, m.p. 75°, b.p. 150–154°/54 mm, *picrate*, m.p. 198–199°, *methiodide*, m.p. 254–255°, is obtained from 3-aminopyridine by the Skraup reaction (*A. Albert*, J. chem. Soc., 1960, 1790; *Rapoport* and *Batcho*, *loc. cit.*). No 1,7-isomer is detected in the product. 1,5-Naphthyridine forms salts which contain acid and base in the ratio of 1:1, but the hydrochloride has the formula $C_8H_6N_2$, 2HCl (*cf. F. G. Mann* and *J. Watson*, J. org. Chem., 1948, **13**, 502). It forms an N-*oxide*, m.p. 127–128° and 1,5-*di*-N-*oxide*, m.p. 299–301° (*E. P. Hart*, J. chem. Soc., 1954, 1879). 2-*Amino*-1,5-*naphthyridine*, m.p. 196–198°. 3-*Bromo*-, m.p. 106–107° and 3,7-*dibromo*-1,5-*naphthyridine*, m.p. 239–240°.

1,6-Naphthyridine, m.p. 31.5°, *picrate*, m.p. 219–220°, is prepared in 40% yield by heating 4-aminopyridine with "sulpho-mix" (fuming sulphuric acid, nitrobenzenesulphonic acid and glycerol) (*Kress* and *Paudler*, Chem. Comm., 1967, 3). For another method see *N. Campbell* and *R. Ferrier*, Proc. roy. Soc. Edinb., 1959–1960, **65A**, 231). 1-N-*Oxide*, m.p. 158–160°, 6-N-*oxide*, m.p. 158°. The *di*-N-*oxide*, m.p. 278°, undergoes the Meisenheimer reaction with phosphoryl chloride to give 3,5- (42.1%), 2,5- (33.8%), 4,5-dichloro-1,6-naphthyridine (15%) and a little 5-chloro-1,6-naphthyridine (*Paudler* and *Pokorny*, J. heterocycl. Chem., 1972, **9**, 1151). Protonation occurs at position 6 and methyl iodide yields 6-*methylnaphthyridinium iodide*, m.p. 153–155°, which is oxidised by alkaline potassium ferricyanide to 6-*methyl*-5-*oxo*-5,6-*dihydro*-1,6-*naphthyridine*, m.p. 97–98° (*Paudler* and *Kress*, *ibid.*, 1968, **5**, 561). 2-*Amino*-1,6-*naphthyridine*, m.p. 238–240°. 3-*Bromo*-, m.p. 125–126°, 8-*bromo*-, m.p. 84–86° and 3,8-*dibromo*-1,6-*naphthyridine*, m.p. 187–189°.

1,7-Naphthyridine, m.p. 64°, *picrate*, m.p. 196.5–197.5°, is obtained from 2-cyano-3-pyridylacetonitrile by treatment in ether with hydrogen bromide to give 6-*amino*-8-*bromo*-1,7-*naphthyridine*. By standard procedures this yields 1,7-naphthyridine (*R. Tan* and *A. Taurins*, Tetrahedron Letters, 1966, 1233; *cf. N. Ikekawa*, Chem. Pharm. Bull., Tokyo, 1958, **6**, 401; C.A., 1959, **53**, 3226):

As noted above 3-aminopyridine undergoes the Skraup reaction to yield 1,5-naphthyridine, but 2-substituted 3-aminopyridines yield 1,7-naphthyridine derivatives, 3-amino-2-hydroxypyridine thus yielding 8-hydroxy-1,7-naphthyridine (*Albert* and *A. Hampton*, J. chem. Soc., 1952, 4985). 3-Aminopyridine-*N*-oxide likewise undergoes ring-closure to yield 1,7-naphthyridine derivatives, giving, for instance with ethoxymethylenemalonic ester *ethyl* 4-*hydroxy*-1,7-*naphthyridine*-3-*carboxylate* (*J. G. Murray* and *C. R. Hauser*, J. org. Chem., 1954, **19**, 2008):

1,7-Naphthyridine 1-N-*oxide*, m.p. 190–191° and 7-N-*oxide*, m.p. 150–151°. *7-Methyl-1,7-naphthyridinium iodide*, m.p. 226–228°, is oxidised with alkaline potassium ferrocyanide to *7-methyl-8-oxo-7,8-dihydro-1,7-naphthyridine*, an oil. *8-Amino-1,7-naphthyridine*, m.p. 165–166°. *5-Bromo-*, m.p. 69–71° and *3,5-dibromo-1,7-naphthyridine*, m.p. 149–151°.

1,8-Naphthyridine, m.p. 98–99°, sublimes at 80°/1 mm, *picrate*, m.p. 207–208°, *methiodide*, m.p. 180–181° (sealed tube), is prepared from 2-aminopyridine by the Skraup reaction using "sulpho-mix" (*Paudler* and *Kress*, J. org. Chem., 1967, **32**, 832). It has a simple p.m.r. spectrum (*E. M. Hawes* and *D. G. Wibberley*, J. chem. Soc., C, 1967, 1564) and p.m.r. spectra may profitably be used to establish the site of substitution in its derivatives (*Paudler et al.*, J. heterocycl. Chem., 1970, **7**, 291). It forms an N-*oxide*, m.p. 132–133°. *2-Amino-1,8-naphthyridine*, m.p. 141–142°. *3-Bromo*, m.p. 155–156° and *3,8-dibromo-1,8-naphthyridine*, m.p. 300°. 4-Chloro derivatives when heated with hydrazines undergo ring-fission to yield 1,3-disubstituted pyrazoles and in this way 4-chloro-7-methyl-1,8-naphthyridine when heated with phenylhydrazine yields the *N*-phenyl derivative of *2-amino-6-methyl-3-(pyrazol-3-yl)pyridine*, m.p. 136–137° (*R. A. Bowie* and *B. Wright*, J. chem. Soc., Perkin I, 1972, 1109):

A useful derivative is *2-hydroxy-1,8-naphthyridine*, m.p. 198–199°, which is prepared from 2-amino-6-hydroxypyridine by means of the Skraup reaction (*W. Roszkiewicz* and *M. Wozniak*, Synthesis, 1976, 691). It is converted by phosphoryl bromide into 2-bromo-1,8-naphthyridine, which with ammonia yields 2-amino-1,8-naphthyridine. Nitration of the hydroxy compound yields 3-nitro-2-hydroxy-1,8-naphthyridine and hence *2-chloro-3-nitro-1,8-naphthyridine*, m.p. 267–268°. The chloro compound by means of *p*-tosylhydrazine gives *3-nitro-1,8-naphthyridine*, m.p. 247–248°, which is readily reduced to the *3-amino-1,8-naphthyridine*, m.p. 141–142°.

2,6-Naphthyridine, m.p. 118–119°, *monopicrate*, m.p. 222–224°, *dipicrate*, m.p. 207.5–209.5°, is obtained by a series of reactions from 4-cyano-3-pyridylacetate (*Tan* and *Taurnis*, Canad. J. Chem., 1974, **52**, 843, *cf. G. Giacomello et al.*, Tetrahedron Letters, 1965, 1117). *4-Methyl-2:6-naphthyridine*, C,H$_8$N$_2$, m.p. 78°, is isolated from *Antirrhinum majus* (*K. J. Harkiss* and *D. Swift*, Tetrahedron Letters, 1970, 4773). The p.m.r. spectrum shows two coupled doublets at τ 2.22 (C$_{(4)}$, C$_{(8)}$) and τ 1.15 (C$_{(3)}$, C$_{(7)}$) and a singlet at τ 0.71 (C$_{(1)}$, C$_{(5)}$). A dibenzo derivative is **calycanine**, 6,12-*diazochrysene*, C$_{16}$H$_{10}$N$_2$, m.p. 310°, *dihydrochloride*, m.p. 350°, a degradation product of the alkaloid calycanthine (*R. B. Woodward et al.*, Proc. chem. Soc., 1960, 76):

Calycanine

It has been synthesised (*T. Kobayashi* and *R. Kikumoto*, Tetrahedron, 1962, **18**, 813).

2,7-**Naphthyridine, copyrine,** m.p. 92–94°, *picrate*, m.p. 240° (*Ikekawa*, C.A., 1959, **53**, 372), N-*oxide*, m.p. 228–229°. 1,3,6,8-Tetrachloro-2,7-naphthyridine (LXXX) has been synthesised and with methanol and potassium carbonate yields 3,6-*dichloro*-1,8-*dimethoxy*-2,7-*naphthyridine* (LXXXI, R = Cl), m.p. 155–157°, and with hydrogen, palladium chloride and potassium carbonate in methanol gives 1,8-*dimethoxy*-2,7-*naphthyridine* (LXXXI, R = H), m.p. 108–110° (*Ferrier* and *Campbell*, J. chem. Soc., 1960, 3513):

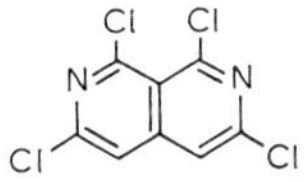

(LXXX) 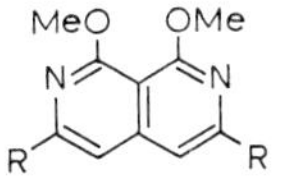   (LXXXI)

Chapter 40

Compounds containing a Six-membered Ring having two Hetero-atoms from Group VIB of the Periodic Table: Dioxanes, Oxathianes and Dithianes

MALCOLM SAINSBURY

1. Dioxanes and related compounds

The most familiar six-membered heterocycle containing two oxygen atoms is 1,4-dioxane which is frequently employed as a solvent in preparative organic chemistry. 1,3-Dioxanes are methylene ethers of 1,3-diols and 1,2-dioxanes are cyclic peroxides.

Nomenclature. The three possible dioxanes are shown below (formulae I–III):

(I)	(II)	(III)
1,2-Dioxane	1,3-Dioxane	1,4-Dioxane
(*o*-dioxane)	(*m*-dioxane)	(*p*-dioxane)

The term dioxin is reserved for the least hydrogenated forms, thus I and III lead to 1,2-dioxin (IV) and 1,4-dioxin (VI), respectively, but the only possible dehydro analogue of II must contain two more hydrogen atoms than either IV or VI, thus 1,3-dioxin (V) is more correctly named $2H,4H$-1,3-dioxin or 2,4-dihydro-1,3-dioxin.

(IV)	(V)	(VI)
1,2-Dioxin	$2H,4H$-1,3-Dioxin	1,4-Dioxin
(*o*-dioxin)	(*m*-dioxin)	(*p*-dioxin)

Dihydro derivatives of IV and VI are, of course, isomeric with $2H,4H$-1,3-dioxin; some examples will be encountered in the text, and are referred to as 3,4- or 3,6-dihydro-1,2-dioxins or 2,3-dihydro-1,4-dioxins, respectively.

(a) *1,2-Dioxanes, 1,2-dioxins and related structures*

1,2-Dioxane would be the cyclic peroxide corresponding with 1,4-tetramethylene glycol, but such simple peroxides are rarely encountered. *T. H. Bergemann* and co-workers (J. Agr. Food Chem., 1968, **16,** 679) report that a six-membered cyclic peroxide is formed by the auto-oxidation, at 37°, of methyl linoleate. It is accompanied by other peroxides. Succinoyl peroxide (C.C.C., Vol. I D, p. 307) was once supposed to be 1,2-dioxane-3,6-dione, but this is more probably a non-cyclic polymer (*L. Vanino* and *E. Thiele*, Ber., 1896, **29,** 1724). Many examples are known in which oxygen adds across the terminal carbon atoms of 1,3-dienes (see, for example, *K. Gollnick* and *G. O. Schenck*, in "1,4-Cycloadditions", ed. *J. Hamer*, Academic Press, London and New York, 1967, pp. 255–344 and refs. cited therein). Normally, the reaction is achieved by irradiation (u.v.) in the presence of a sensitizer. Thus 1,3-cyclohexadiene and oxygen give *norascaridole* (2-*cyclohexene*-1,4-endo-*peroxide*) (VII) (*Schenck* and *K. Ziegler*, Naturwiss., 1944, **32,** 157; *Schenck, H. G. Kinkel* and *H. J. Mertens*, Ann., 1953, **584,** 125; see also C.C.C. Suppl. Vol. II AB, p. 161), an unstable solid, m.p. 89°, which explodes at 100°:

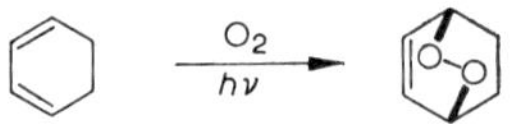

(VII)

Similarly 1,1′-bicyclohexenyl yields the *cis* peroxide VIII (*K.-H. Schulte-Elte*, Diss. Univ. Göttingen, 1961) and 1,2,3,4-tetraphenylbutadiene yields IX (*G. Rio* and *J. Berthelot*, Bull. Soc. chim. Fr., 1971, 2938):

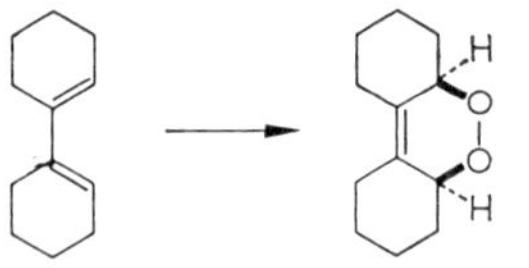

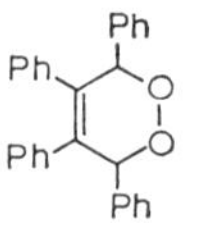

(VIII)   (IX)

(For other syntheses of 3,6-dihydro-1,2-dioxins by this approach see *K. Kondo* and *M. Matsumoto*, Chem. Comm., 1972, 1332).

Mesityl oxide (4-methylpent-3-en-2-one) undergoes photo-oxidation in methanol containing metal salts to afford products of type X (*T. Sato et al.*, J. chem. Soc., Perkin I, 1976, 779):

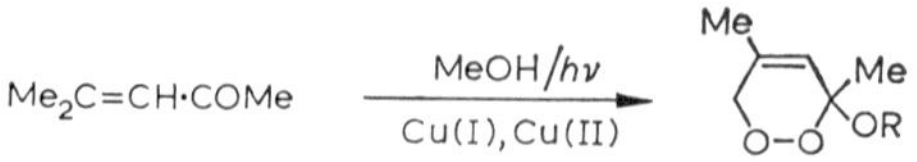

(X)   R = H or Me

Phthaloyl peroxide (XI) can be regarded as a benzo derivative of 3,6-dihydro-1,2-dioxin-3,6-dione (*M. R. De Camp* and *M. Jones*, J. org. Chem., 1972, **37**, 3942). In methanol, at room temperature, it affords the monomethyl ester of monoperphthalic acid, which undergoes deoxygenation on warming, thus forming monomethyl phthalate:

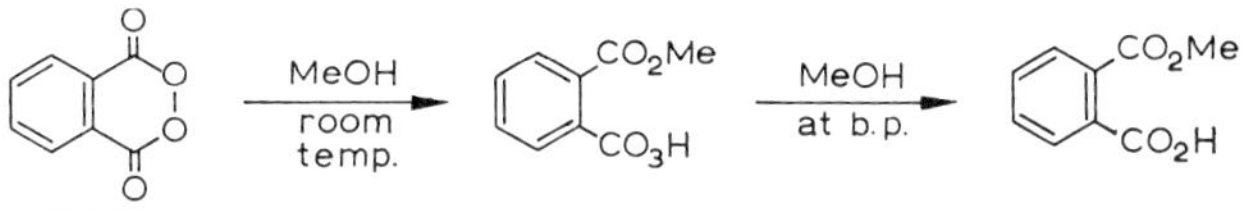

At lower temperature, and in argon, XI may be photolysed to give benzo-2-propenolide and 1-carbonyl-3,5-cyclohexadien-2-one (XII) (*O. L. Chapman et al.*, J. Amer. chem. Soc., 1973, **95**, 4061).

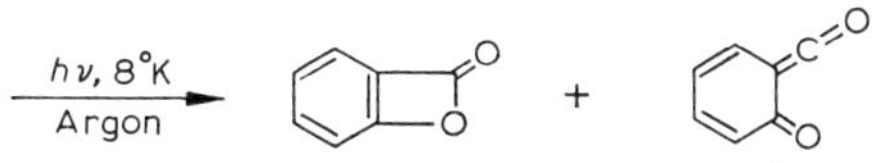

When 1,1-diphenyl-2-methoxyethene is photo-oxygenated in benzene with dinaphthalenethiophene, as sensitizer, one equivalent of oxygen is utilized; the main product is the γ-lactone 2-phenylbenzo-3-butenolide (XV), but smaller amounts of the ester, methyl (2-hydroxydiphenyl)acetate (XIII) and the 3,4-dihydro-1,2-benzodioxin (XIV) are formed. Both XIII and XIV yield XV on heating or upon treatment with acid (*C. S. Foote et al.*, *ibid.*, 1973, **95**, 586):

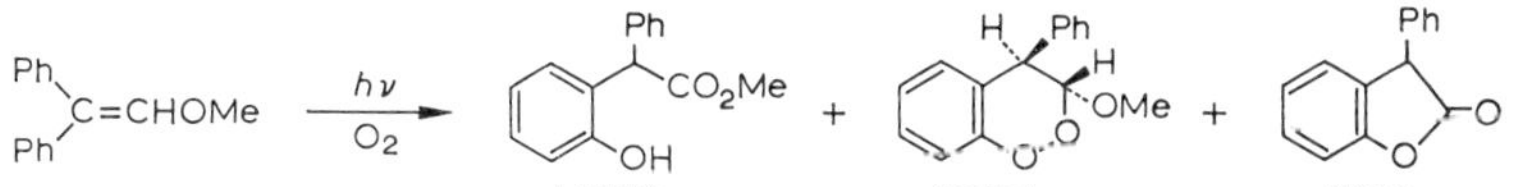

(b) 1,3-Dioxanes and related compounds

(i) 1,3-Dioxanes

1,3-Dioxanes are cyclic acetals, usually made by the reaction of 1,3-glycols with aldehydes or acetals (*L. Henry*, Bull. Acad. roy. Belge, 1902, 460; *J. W. Hill* and *W. H. Carothers*, J. Amer. chem. Soc., 1935, **57**, 925; see also *D. Rhodes* and *R. Newton*, Ger. Offen. 2,454,281/1975) or with ketones (*E. J. Salmi*, Ber., 1938, **71**, 1803) in the presence of acid catalysts. The aldehyde or glycol may be generated from an unsaturated compound in the reaction mixture; *e.g.* styrene with formaldehyde gives 4-phenyl-1,3-dioxane (*H. J. Prins*, Chem. Weekblad, 1919, **16**, 1510; *E. Fourneau*, Bull.

Soc. chim Fr., 1930, **47**, 860) and acetylene with glycols and mercuric salts gives cyclic acetals (*H. S. Hills et al.*, J. Amer. chem. Soc., 1923, **45**, 3108; 1928, **50**, 2242).

Hydrolysis. Like the hydrolysis of acetals, that of 1,3-dioxanes is a second order uni-molecular reaction since the rate is determined by the rate at which the cation XVI is formed (see *F. Aftalion et al.*, Bull. Soc. chim. Fr., 1965, 1512; 1950, 1958; *P. A. Laurent, A. G. da Silva Pinto* and *J. L. C. Priera, ibid.*, 1960, 926; *M. L. Bender* and *M. S. Silver*, J. Amer. chem. Soc., 1963, **85**, 3006):

$$\text{(reaction scheme, cation XVI)}$$

4,4-Dialkyl-1,3-dioxanes on hydrolysis may afford conjugated 1,3-dienes directly (*V. I. Isagulyants* and *M. G. Safarov*, Zhur. org. Khim., 1965, **1**, 974; *L. A. Mikeska* and *E. A. Arundale*, U.S.P. 2,350,517/1944; C.A., 1944, **38**, 4965; *H. O. Mottern*, U.S.P. 2,335,691/ 1943; C.A., 1944, **38**, 2965). Thus 4,4-dimethyl 1,3-dioxane (XVII) gives 2-methylbuta-1,3-diene (XVIII), whereas 4,4,5-trimethyl-1,3-dioxane with dilute mineral acid yields 2,3-dimethylbuta-1,3-diene.

Ion-exchange resins are also recommended for the hydrolysis of 1,3-dioxanes (*D. E. Conrad et al.*, J. org. Chem., 1961, **26**, 3571; U.S.P. 3,000,905/1959; C.A., 1962, **56**, 1460).

Alcoholysis. Under certain conditions the alcoholysis of 4,4-dialkyl-1,3-dioxanes may lead to the formation of ethers as well as dienes (*M. Hellin et al.*, Bull. Soc. chim. Fr., 1963, 800; *W. E. Grigsby*, U.S.P. 2,426,015/1947; C.A., 1948, **42**, 588; *M. I. Farberov* and *N. K. Shemyakina*, Zhur. obshcheĭ Khim., 1956, **26**, 2749). Here, the formation of the ether results from attack of the alcohol upon an intermediate cation formulated as XIX (*A. A. Korotkov, L. F. Roguleva* and *V. A. Tsitokhtsev, ibid.*, 1960, **30**, 2298):

$$\text{(reaction schemes, XVII, XVIII, XIX)}$$

Hydrogenolysis. 1,3-Dioxanes may form either the corresponding diols or their mono-ethers during hydrogenation in alcoholic media depending upon the reaction conditions employed. Thus Raney nickel and hydrogen at 180 atm. afford 1,3-alkoxyalkanols (*R. Ströbele*, G.P. 877,601/1953; C.A., 1958, **52**, 10154a). Similar results are obtained if lithium tetrahydroaluminate or boron trifluoride/ether are used (*E. Eliel et al.*, J. Amer. chem. Soc., 1959, **81**, 6087; 1962, **84**, 2371; *Abdul-Raham Abdun-Nur* and *C. H. Issidorides*, J. org. Chem., 1962, **27**, 67; *B. E. Legelter, U. E. Diner* and *R. K. Brown*, Canad. J. Chem., 1964, **65**, 2113).

The use of palladium normally affords good yields of 1,3-diols, but on some occasions using platinum as catalyst no reaction occurs (*P. E. Verkade et al.*, Rec. Trav. chim., 1941, **60**, 112).

Conformation. An X-ray study of 2-(4'-chlorophenyl)-1,3-dioxane establishes the molecular parameters shown in formula XX (*A. J. de Kok* and *C. Romers, ibid.*, 1970, **89**, 313; see also *G. M. Kellie, P. Lust* and *F. G. Riddell*, J. chem. Soc., Perkin II, 1972, 2384):

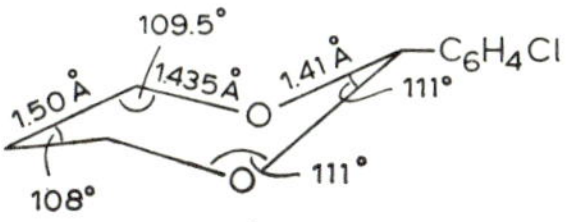

Torsional angles τ

CO–CO   63°
CO–Cl   59°
OC–CC   55°

(XX)

These results had already been anticipated through analyses of models (*Eliel* and *M. C. Knoebel*, J. Amer. chem. Soc., 1968, **90**, 3444) and 'H n.m.r. spectroscopy (*H. R. Buys*, Rec. Trav. chim., 1969, **88**, 1003). For a review of the stereochemistry of 1,3-dioxanes see *Eliel* (Accounts of Chemical Research, 1970, **3**, 1) and *E. Bernert, M. Anteunis* and *D. Tavernier* (Bull. Soc. chim. Belg., 1974, **83**, 457, and refs. cited therein).

1,3-**Dioxane**, m.p. —42°, b.p. 106°, forms a *complex* (decomp., 126°) with mercuric chloride and an unstable *picrate*, m.p. 57° (*H. T. Clarke*, J. chem. Soc., 1912, **101**, 1788). 2-*Methyl-1,3-dioxane*, b.p. 110–112° *(Hill et al., loc. cit.)*; 2,2-*dimethyl-1,3-dioxane*, b.p. 123–125°, has a camphor-like smell, as have some 5,5-dimethyl derivatives.

5-**Hydroxy**-1,3-**dioxane**, b.p. 193°, is obtained, together with 4-hydroxymethyl-1,3-dioxolane, by the condensation of formaldehyde with glycerol; the two may be separated by recrystallisation of their *benzoates*, m.p. 74° and 26°, respectively. 6-Hydroxy-1,3-dioxane derivatives are formed in certain polymerisations of aldehydes, *e.g.* 6-hydroxy-2,4-dimethyl-1,3-dioxane from acetaldehyde and aldol (*E. Späth et al.*, Ber., 1943, **76**, 57).

1,3-Dioxan-4-ones variously substituted in positions 2, 5 and 6 are obtained by the acid-catalysed condensation of β-hydroxy acids and carbonyl compounds (*M. Farines et al.*, Bull. Soc. chim., Fr., 1976, 825):

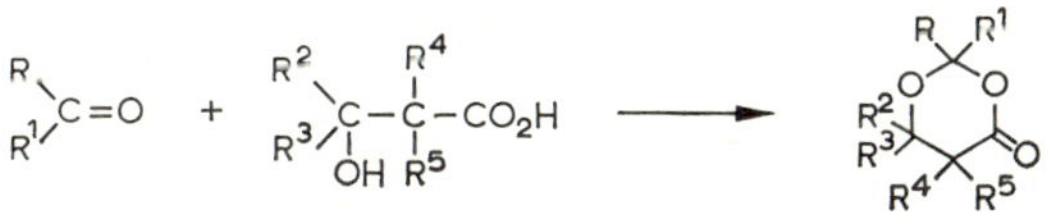

The ring structure of certain 4-oxo-1,3-dioxanes usually attain the half-chair conformation although specific substitution may force the ring into a boat or twisted boat form (*P. Äyras*, Org. Mag. Resonance, 1975, **7**, 177 and refs. cited therein).

Meldrum's acid is **isopropylidene malonate** (2,2-*dimethyl-1,3-dioxane-4,6-dione*) (XXI), m.p. 97°; it is obtained from malonic acid and acetone in acetic anhydride containing sulphuric acid (*A. N. Meldrum*, J. chem. Soc., 1908, **93**, 598; *D. Davidson* and *S. A. Bernhardt*, J. Amer. chem. Soc., 1948, **70**, 3426):

It condenses again with acetone in the presence of pyridine and molecular sieves to give isopropylidene isopropylidenemalonate (XXII) (*G. Swoboda, J. Swoboda* and *F. Wessely*,

Monatsh., 1961, **95,** 1283) which reacts as a dienophile and with 1,3-pentadiene, for example, at 130°/12 h affords the adduct XXIII (*W. G. Dauben, A. P. Kozikowski* and *W. T. Zimmerman*, Tetrahedron Letters, 1975, 515):

3-Hydroxypropyl 2,4,6-trinitrophenyl ether cyclises in tertiary butanol solution containing potassium *tert*-butoxide to the Meisenheimer complex XXIV (*V. N. Drozd et al.*, Zhur. org. Khim., 1975, **11,** 1135; C.A., 1975, **83,** 79167e):

(ii) 1,3-Dioxins

Few dehydrogenated derivatives of 1,3-dioxane have been described.

2,4-Dihydro-1,3-dioxin has b.p. 78° (*C. H. Snyder* and *A. R. Soto*, J. org. Chem., 1964, **29,** 742; *R. Camerlynck* and *Anteunis*, Tetrahedron, 1975, **31,** 1837); it exists as the 1,2-diplanar (sofa) C_2 form. Recently *E. J. Corey* and *J. W. Suggs* (Tetrahedron Letters, 1975, 3775) have prepared a number of derivatives of 1,3-dioxin as intermediates in the synthesis of ketones.

2-Phenyl-2,4-dihydro-1,3-**dioxin,** b.p. 72–75°/0.1 mm, has been prepared by distilling the *p*-toluenesulphonate of 5-hydroxy-2-phenyl-1,3-dioxane with potassium hydroxide (*H. O. L Fischer et al.*, Ber., 1931, **64,** 611).

(iii) 2,4-Dihydro-1,3-benzodioxin and related compounds

2,4-Dihydro-1,3-benzodioxin (XXV) is the methylene ether of saligenin (2-hydroxybenzyl alcohol). It has been prepared by condensing saligenin with methylene sulphate in aqueous alkaline medium (*W. Baker*, J. chem. Soc., 1931, 1770):

The heterocycle has considerable stability and the compound reacts as a disubstituted benzene towards electrophilic reagents; nitration for example, gives the 6-nitro and 6,8-dinitro derivatives (*F. Calvert* and *M. C. Carnero*, Anal. Fis. Quim., 1932, **30**, 445).

2,4-Dihydro-1,3-benzodioxins are also formed when a phenol with the *para*-position hindered and a free *ortho*-position reacts with a suitable aldehyde (2 mol. equiv.):

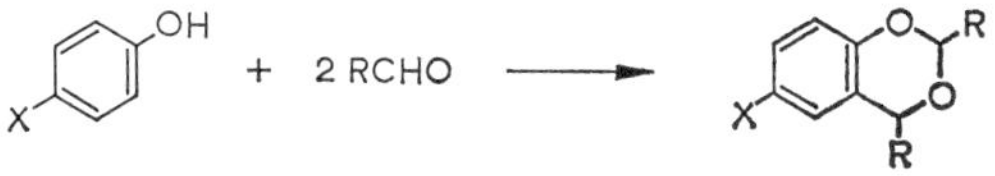

Thus 4-nitrophenol with formaldehyde in the presence of dilute sulphuric acid gives 6-**nitro-2,4-dihydro**-1,3-**benzodioxin** (*W. Borsche* and *A. D. Berkhout*, Ann., 1904, **330**, 82). Similarly, *E. D. Chattaway* and his co-workers combined 4-hydroxybenzoic acid, 4-aminophenol and phenol-4-sulphonic acid individually with chloral (J. chem. Soc., 1927, 685, 2013; 1928, 1088) and 4-hydroxybenzoic acid and dichloroacetaldehyde (*ibid.*, 1931, 1737) to obtain the corresponding 2,4-dihydro-1,3-benzodioxins.

A. J. Brink (Monatsh., 1973, **104**, 619) has examined the ^{1}H n.m.r. spectra of a number of 3-substituted 2,4-dihydro-1,3-benzodioxins and comments upon the magnetic non-equivalence of the methylene protons in the heterocycle.

Under electron impact two "retro" Diels–Alder fragmentations are observed in the mass spectra of 2,4-disubstituted 2,4-dihydro-1,3-benzodioxins; one direct from the molecular ion, the second involving hydrogen transfer (*J. F. Grunstein et al.*, Org. Mass Spectrom., 1974, **9**, 1166). Nitro derivatives exhibit "retro" Diels–Alder reactions, occurring from negative ions (*J. H. Bowie* and *A. C. Ho*, J. chem. Soc., Perkin II, 1975, 724).

Tetrachlorobenzyne and acetone in the presence of butane-2,3-dione give XXVI (R = R′ = Me), if diethyl ketone or ethyl methyl ketone replace acetone, XXVI (R = R′ = Et) or XXVI (R = Me, R′ = Et) are formed respectively (*H. Heaney* and *C. T. McCarty*, Chem. Comm., 1970, 123). With sulphuric and acetic acids these products afford XXVII (R = OAc) and with hydrogen bromide in acetic acid XXVII (R = Br):

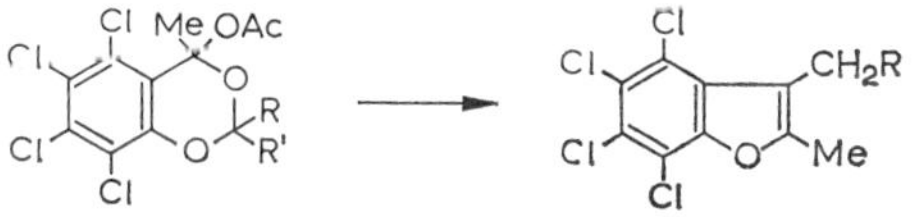

'XXVI)   (XXVII)

The salt XXVIII is obtained from 2-acetoxybenzoyl chloride by the action of antimony pentachloride in dichloromethane at $-10°$ (*H. Brinkmann* and *C. Ruechardt*, Tetrahedron Letters, 1972, 5221):

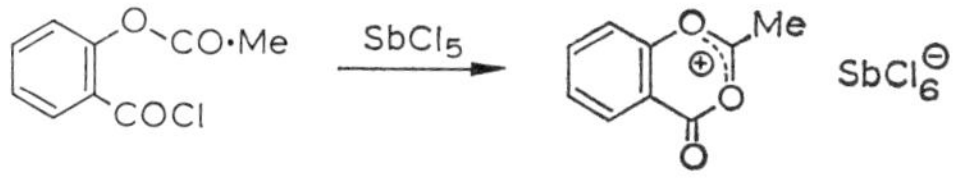

(XXVIII)

Similarly, Friedel–Crafts reactions between *O*-acylsalicyloyl chlorides and aromatic hydrocarbon derivatives (ArH) having high anionoid activity, give 2-alkyl-2-aryl-2,4-dihydro-1,3-benzodioxin-4-ones (*W. Lonsky* and *W. Mayer*, Chem. Ber., 1975, **108,** 1593); again the reaction requires the participation of a carbonium species similar to XXVIII:

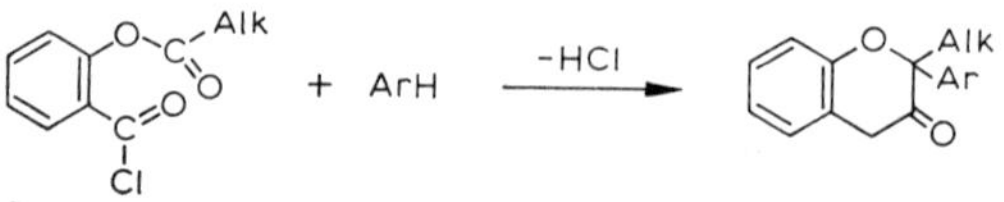

2,4-**Dihydro**-1,3-**benzodioxin** is a highly refracting oil, b.p. 208–209°/749 mm, m.p. 12.5°. 6-*Nitro*-2,4-*dihydro*-1,3-*benzodioxin*, m.p. 148°, is oxidised by chromic acid to 6-*nitro*-4-*oxo*-2H-1,3-*benzodioxin*, m.p. 110°, hydrolysed by alkali to 5-nitrosalicylate. It can be halogenated to 8-*chloro*-, m.p. 152–153°, and 8-*bromo*-6-*nitro*-2,4-*dihydro*-1,3-*benzodioxin*, m.p. 165–166°. With excess of formaldehyde 6-nitro-2,4-dihydro-1,3-benzodioxin forms the 8,8′-*methylenebis* derivative, m.p. 225°, which can be oxidised by chromic acid to the 8,8′-*dione*. This product chars without melting above 285° (*Chattaway* and *R. M. Goepp*, J. chem. Soc., 1933, 699). 6-*Amino*-2,4-*dihydro*-1,3-*benzodioxin* has m.p. 68°, *acetyl* derivative, m.p. 128–130°.

(*c*) *1,4-Dioxanes and 1,4-dioxins*

(*i*) *1,4-Dioxanes*

The fully saturated 1,4-dioxanes are much better known and more important than their unsaturated derivatives and will be discussed first. They are cyclic ethers and, as such, resemble the aliphatic ethers in methods of preparation and in properties. The synthetic routes are exemplified by those applicable to 1,4-dioxane itself.

(*1*) Ethane-1,2-diol is etherified with 1,2-dibromoethane (*A. V. Lourenço*, Ann. chim., 1803, [iii], **67,** 284).

(*2*) Ethane-1,2-diol is self-etherified by treatment with sulphuric acid or other catalysts (*A. Favorski*, J. Soc. phys. chem. Russe, 1906, **38,** 741; *J. van Alphen*, Rec. Trav. chim., 1930, **49,** 1040).

(*3*) Ethylene oxide is passed over a heated acid catalyst such as sodium hydrogen sulphate (*I. G.*, B.P. 346,550/1930), propylene oxide forms 2,5-dimethyl-1,4-dioxane.

(*4*) A di-β-halogenoalkyl ether is treated with alkali. Other methods of forming the 1,4-dioxane ring appear under the descriptions of individual compounds.

1,4-**Dioxane,** which is manufactured on a large scale, is a colourless liquid at ordinary temperatures, b.p. 101°, m.p. 11.8° (12.5–13.0°), $d_{4°}^{15°}$ 1.0392, n_D^{20} 1.42241. It is very hygroscopic, miscible with water (constant boiling mixture, b.p. 86.8°/742 mm, with

20% of water) and with many organic solvents and is of importance as a solvent for organic substances and inorganic salts. It forms numerous addition compounds, *e.g.* a *dibromide*, $C_4H_4O_2Br_2$, m.p. 66°, an *iodide*, m.p. 85° and 1,4-dioxanium salts, *e.g. sulphate*, m.p. 101°, *picrate*, m.p. 66° and a mercurichloride. The bromide is a convenient brominating agent and the sulphur trioxide adduct, (1-dioxanio)sulphate, a sulphonating agent (*C. M. Suter et al.*, J. Amer. chem. Soc., 1938, **60**, 538). Dioxane forms complexes with boron trichloride, mol. ratios 1:1 and 2:3, and with boron tribromide, 1:1 (*M. J. Frazer et al.*, Chem. and Ind., 1958, 1263).

In chemical properties 1,4-dioxane resembles the aliphatic ethers (C.C.C., 2nd Edn., Vol. I B, p. 46). Peroxide formation occurs when it is exposed to light and air, but this is conditioned by the presence of impurities (*E. Eigenberger*, J. pr. Chem., 1931, [ii], **130**, 75). Ring-fission with acyl halides, acyl anhydrides or hydrogen bromide may involve one or both ether links. Dioxane may be used as a solvent in catalytic reductions but it reacts violently with hydrogen and Raney nickel above 210°.

Although dioxane is commonly used as a solvent in photochemical experiments, it has been observed (*P. H. Mazzocchi* and *M. W. Bowen*, J. org. Chem., 1975, **40**, 2689) to form the radical species XXIX, which may dimerize to 2,2'-bi-(1,4-dioxanyl) (as a pair of diastereoisomers) or afford acetaldehyde. The latter reacts further with the radical XXIX to give the diastereomeric alcohols XXX and XXXI:

A number of chloro-1,4-dioxanes are known (see Table 1), the conformations and in some cases the configurations of these compounds have been established by ^{1}H-n.m.r. spectroscopy (*R. E. Ardrey* and *L. A. Cort*, J. chem. Soc., Perkin II, 1975, 959) and by other techniques (see reference 6 below Table 1).

Mono- and di-chlorodioxanes are reactive compounds resembling *α*-chloro-ethers but further chlorination causes a progressive reduction in reactivity. By reaction of the mono- and di-chloro compounds with alkyl- or aryl-magnesium halides, alkyl and aryl derivatives of dioxane are obtained, *e.g. 2-chloro-1,4-dioxane*, b.p. 57–65°/25 mm (by addition of hydrogen chloride to 1,4-dioxin) gives *2-phenyl-1,4-dioxane*, m.p. 46°; *2,3-diphenyl-1,4-dioxane* has m.p. 49–50°. Alkyl-1,4-dioxanes are also obtained by etherification of 1,2-glycols and by the action of alkali on *ββ'*-dihalogeno ethers. A report that *2,5-diphenyl-*

TABLE 1

CHLORO-1,4-DIOXANES

1,4-Dioxane	m.p. (°C)	b.p. (°C/mm)	Ref.
cis-2,3-dichloro-	52–53		1
trans-2,3-dichloro-	30–31		2
2,2-dichloro-		45–50/4	3
trans-2,5-dichloro-	118–120		4
2,2,3-trichloro-		59–61/0.7	3
cis-trans-2,3,5,6-tetrachloro-	70		5,6
cis-anti-cis-2,3,5,6-tetrachloro-	142		5,6
trans-syn-trans-2,3,5,6-tetrachloro-	101		6,7
trans-anti-trans-2,3,5,6-tetrachloro-	55		6

References
1 E. Caspi et al., J. org. Chem., 1962, **27**, 3183.
2 R. K. Summerbell and H. E. Lunk, J. Amer. chem. Soc., 1957, **79**, 4802.
3 Idem, J. org. Chem., 1958, **23**, 499.
4 L. N. Brian et al., J. Amer. chem. Soc., 1950, **72**, 2207.
5 R. Christ and Summerbell, ibid., 1933, **55**, 4547.
6 C. Romers et al., Topics Stereochem., 1969, **4**, 60.
7 M. Lüdicke and W. Stumpf, Naturwiss., 1953, **40**, 363.

1,4-*dioxane* was formed by the action of diazomethane on benzoyl bromide could not be confirmed (see *W. Bradley* and *R. Robinson*, J. chem. Soc., 1928, 1312) but *Brian et al. (loc. cit.)* claim to have prepared the compound from 2,5-dichloro-1,4-dioxane (see above) in two stereoisomeric forms, m.p. 121–122° and 173°.

By evaporating an aqueous solution containing glyoxal and ethane-1,2-diol, a substance was obtained having no definite melting point but crystallising from acetone, which is probably 2,3-dihydroxy-1,4-dioxane since it is converted by phosphorus pentachloride into *trans*-2,3-dichloro-1,4-dioxane, from which it can again be obtained by hydrolysis (*F. S. H. Head*, J. chem. Soc., 1955, 1036).

The dichloro- compound reacts with more ethane-1,2-diol to give two isomeric products, $C_6H_{10}O_4$, m.p. 109–112° and 133–136° (*W. Baker* and *F. B. Field*, ibid., 1932, 88). The higher melting isomer, identical with a compound formed from the interaction of glyoxal sulphate and ethane-1,2-diol is cis-*hexahydro*-1,4-*dioxino*-[2,3-b]-1,4-*dioxin* (XXXII, R = H), whereas the lower melting component is 2,2'-*bi*(1,3-*dioxolan-2-yl*) (XXXIII, R = H):

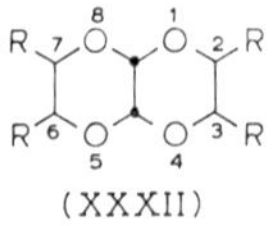

(XXXII) 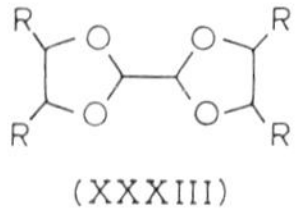　　　　(XXXIII) 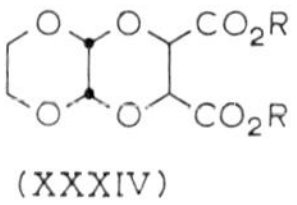　　　　(XXXIV)

Similarly, *trans*-2,3-dichloro-1,4-dioxane reacts with concentrated sulphuric acid at room temperature to give at least two products, one of which, $C_6H_6Cl_4O_4$, m.p. 137–139°, was originally considered to be either 2,3,6,7-tetrachlorohexahydro-1,4-dioxino-[2,3-*b*]-1,4-dioxin (XXXII, R = Cl) or 4,4',5,5'-*tetrachloro-bi*(1,3-*dioxolan-2-yl*) (XXXIII, R = Cl)

(*Cort*, J. chem. Soc., 1960, 3167; *H. H. Huang*, *ibid.*, B, 1971, 1024), but recent mass, [1]H-n.m.r., i.r. and Raman spectroscopic studies (*Cort et al.*, J. chem. Soc., Perkin I, 1972, 177) substantiate the latter as the correct representation.

With (+)dimethyl tartarate the same substrate affords 2,3-*methoxycarbonylhexahydro*-1,4-*dioxino*-[2,3-b]-1,4-*dioxin* (XXXIV, R = Me), m.p. 73°, $[a]_D$ −65°, in which the ring-fusion is *cis* (*Cort et al.*, J. chem. Soc., C, 1968, 1213). The *diethyl* analogue (XXXIV, R = Et) has m.p. 94° (*J. Böeseken et al.*, J. Amer. chem. Soc., 1933, **55**, 1284).

2,3-*Diethoxy*-1,4-*dioxane*, b.p. 96–97°/15 mm, is obtained from the dichloro compound with sodium ethoxide (*Böeseken et al.*, *loc. cit.*). The dimeric forms of glycollic and other a-hydroxy aldehydes (C.C.C., 2nd Edn., Vol. I D, p. 44 *et seq.*) are 2,5-di-hydroxy-1,4-dioxanes.

Dioxo-derivatives of dioxane are represented by the dimeric lactides of a-hydroxy acids, *e.g.* glycollide and lactide, which are 2,5-dioxo compounds, and diglycollic anhydride is 2,6-dioxo-1,4-dioxane (C.C.C., 2nd Edn., Vol. I D, p. 92). 2,3-**Dioxo**-1,4-**dioxane**, m.p. 144°, is the ethylene ester of oxalic acid (*C. A. Bischoff* and *P. Walden*, Ber., 1894, **27**, 2939). 2-Oxo-1,4-dioxane, has m.p. 31°, b.p. 214°, and is obtained from 1,2-ethandiol and ethyl chloroacetate (*E. Hollo, ibid.*, 1928, **61**, 902) or from 2,3-dichloro-1,4-dioxane and formic acid at 170° (*R. K. Summerbell* and *H. E. Lunk*, J. Amer. chem. Soc., 1958, **60**, 604); it reacts with arylmagnesium halides giving diaryl glycol ethers which may be cyclised to 2,2-diaryl-1,4-dioxanes (*Summerbell et al.*, J. org. Chem., 1958, **23**, 932).

(*ii*) *1,4-Dioxins*

2,3-**Dihydro**-1,4-**dioxin**, b.p. 94°, is obtained by dehalogenation of 2,3-dichloro-1,4-dioxane, *e.g.* by the action of magnesium and magnesium iodide. By the same method 1,4-**dioxin**, b.p. 75°, is formed from 2,3,5,6-tetrachloro-1,4-dioxane (*Summerbell* and *R. R. Umhoefer*, J. Amer. chem. Soc., 1939, **61**, 3016, 3020). Molecular orbital energies, electron distribution and π-bond orders of 1,4-dioxin have been calculated (*C. Decoret* and *B. Tinland*, Spectrosc. Letters, 1970, **3**, 175) and its infra-red and Raman spectra in liquid and gas phases have been studied (*J. E. Connett et al.*, Spectrochim. Acta, 1960, **22**, 1859).

(*iii*) *1,4-Benzodioxin and its derivatives*

2,3-**Dihydro**-1,4-**benzodioxin** (XXXV, R = H), b.p. 212–214°/757 mm, is frequently, but incorrectly named 1,4-benzodioxane; it and related compounds are obtained from 1,2-dihydric phenols and 1,2-dihalogenoethanes (*cf. D. Voländer*, Ann., 1894, **280**, 205; *B. N. Ghosh*, J. chem. Soc., 1915, **107**, 1591; *H. Fijita* and *M. Yamashita*, Nippon Kagaku Kaishi, 1974, 1148). The parent compound may thus be prepared from catechol and 1,2-dibromoethane by heating with sodium carbonate and copper bronze in glycerol at 190–200°.

(XXXV)

Similarly, pyrogallol and 1,2-dichloroethane afford 2,3-dihydro-5-hydroxy-1,4-benzodioxin (XXXV, R = OH) (*Fugita* and *Yamashita, loc. cit.*).

2-(2-Hydroxyphenoxy)ethanol may be cyclodehydrated by treatment with mercuric oxide and iodine to 2,3-dihydrobenzo-1,4-dioxin (*A. Gossen* and *C. D. W. McCleland*, Chem. Comm., 1975, 655) the yield is, however, only 14%. 6-Halogeno derivatives of the heterocycle may be prepared similarly. Other methods of synthesis are reviewed by *Dauksas* ("Advances in the chemistry of 1,4-benzodioxanes", Khim. geterot. Soedin., 1975, 1155).

The [1]H-n.m.r. spectrum of 2,3-dihydro-1,4-benzodioxin has been analysed (*G. Misiunas* and *H. Jionantis*, C.A., 1971, **75**, 13111k). It is established that the proton signals of the heterocycle undergo more pronounced chemical shifts on change of solvent than do the signals due to the benzenoid hydrogen atoms. This is interpreted as the solvent dependence of the position of equilibrium between associated and unassociated forms in solution. For the mass spectrum of 2,3-dihydro-1,4-benzodioxin see *P. Vouros* and *K. Biemann* (Org. Mass Spectrom., 1970, **3**, 1317), and for the infra-red spectrum see *V. Salna* and co-workers (C.A., 1970, **73**, 87143c).

Certain 2-methylamino derivatives of 2,3-dihydro-1,4-benzodioxin show interesting pharmacological properties including sedative and tranquillising actions (*R. Jonas* and *H. Mueller-Calgan*, Ger. Offen., 1,959,562/1971; C.A., 1971, **75**, 76806n).

C. Moureu (Compt. rend., 1899, **128**, 559; Bull. Soc. chim. Fr., 1899, **21**, 295) was the first to describe 1,4-**benzodioxin** (XXXVIII), see also *B. Becker* and *Barthell* (Monatsh., 1947, **77**, 80), *A. R. Katritsky et al.* (Tetrahedron, 1966, **22**, 931) and *W. Schroth et al.* (Z. Chem., 1967, **4**, 152):

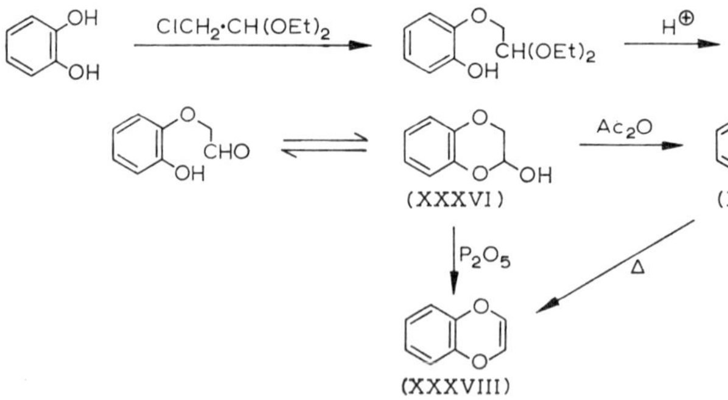

1,4-**Benzodioxin** is obtained in 15% yield by heating 2,3-dihydro-2-hydroxy-benzo-1,4-dioxin (XXXVI) with phosphorus pentoxide in quinoline; this yield may be increased to 80% if XXXVI is first acetylated and then the 2-acetoxy derivative XXXVII is heated at 450°.

The high resolution [1]H-n.m.r. spectrum of 1,4-benzodioxin has been studied (*K. K. Deb, J. E. Bloor* and *T. C. Cole*, Org. Magn. Resonance, 1970, **2**, 431) and also its mass spectrum (*I. C. Calder, R. B. John* and *J. M. Desmarchelier*, Org. Mass Spectrom., 1970, **4**, 121).

(*iv*) *Lignan derivatives*

Some natural products containing a 2,3-dihydrobenzo-1,4-dioxin unit are known, thus (—)**eusiderin** occurs in certain species of the *Lauraceae* (*J. J. Hobbs* and *F. E. King*, J. chem. Soc., 1960, 4732; *O. R. Gottlieb*, Rivista Latinoamer. Quim., 1974, **5**, 1):

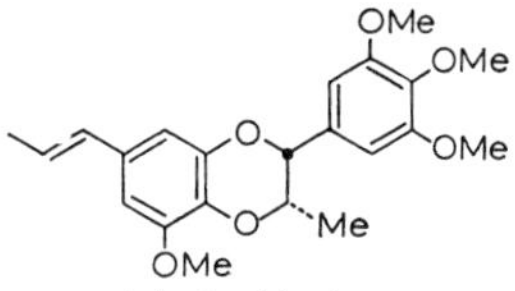

(−)-Eusiderin 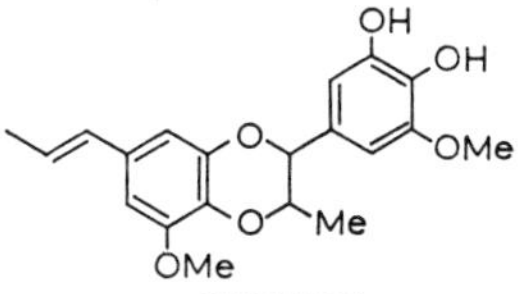   (XXXIX)

The structure shown, is not totally secure, but rests on the fact that 1-(4,5-dimethoxy-3-hydroxyphenyl)-2-propene on oxidation with silver oxide gives the dihydric phenol XXXIX, which on methylation affords (±)-eusiderin (*L. Merlini* and *A. Zanarotti*, Tetrahedron Letters, 1975, 3621).

Silybin, the first flavanolignan (*A. Pelter* and *R. Hänsel*, ibid., 1968, 2911; *D. J. Abraham et al., ibid.,* 1970, 2670), has the structure shown (*Hänsel et al.,* Chem. Ber., 1975, **108**, 1482):

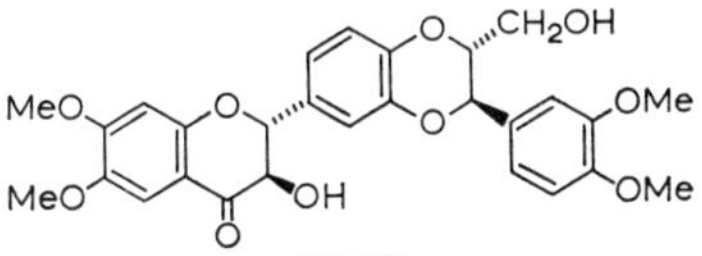

Silybin

Dibenzo-1,4-dioxin *(diphenylene dioxide)* (XL), m.p. 119°, is obtained by heating sodium or potassium 2-chlorophenoxide. For other methods see *N. M. Cullinane et al.,* J. chem. Soc., 1934, 717. 2-Aminodibenzo-1,4-dioxin undergoes ready condensation with ω-bromomethylketones to give 2-aroylmethylaminodibenzo-1,4 dioxins (*G. Suint-Ruf et al.,* J. heterocycl. Chem., 1975, 1069).

(XL)

(*v*) *Polychlorinated dibenzo-1,4-dioxins*

Certain polychlorodibenzo-1,4-dioxins, formed during the manufacture of polychlorophenols, are **extremely toxic** to mammals and should be handled with the care normally reserved for highly radioactive material. These compounds are stable in the ecosystem and thus their presence has caused concern (for a review see "Chlorodioxins, Origin and Fate", *E. H. Blair,*

Ed., Advances in Chemistry Series, No. 120, Amer. chem. Soc., Washington D.C., 1973).

During the course of a general synthetic study of these products, *A. P. Gray, S. P. Ceta* and *J. F. Cantrell* (Tetrahedron Letters, 1975, 2873) found that self-condensations of certain potassium phenolates may proceed *via* Smiles rearrangements. Thus potassium 2-bromo-3,4,5-trichlorophenate affords the anion XLI, which rapidly equilibrates through XLII to the rearranged anion XLIII. The anions XLI and XLIII then form the expected product 1,2,3,6,7,8-**hexachlorodibenzo**-1,4-**dioxin** (XLIV), m.p. 285–286°, and the unexpected compound 1,2,3,7,8,9-**hexachlorodibenzo**-1,4-**dioxin** (XLV), m.p. 243–244°, respectively:

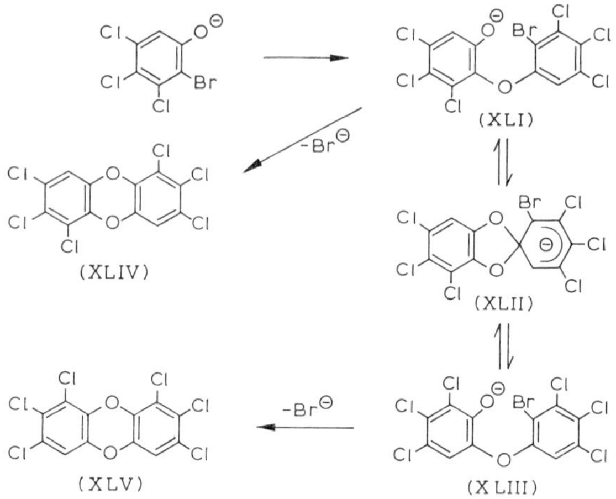

In a parallel study *A. S. Kende* and *M. R. de Camp* (*ibid.*, 1975, 2877) condensed catechol derivatives with polychloronitrobenzenes to form the corresponding diphenyl ethers. In the example cited below, demethylation of the resultant ether with lithium bromide in dimethyl sulphoxide gave XLIV and XLV in the same proportions as found by *Gray* and his co-workers:

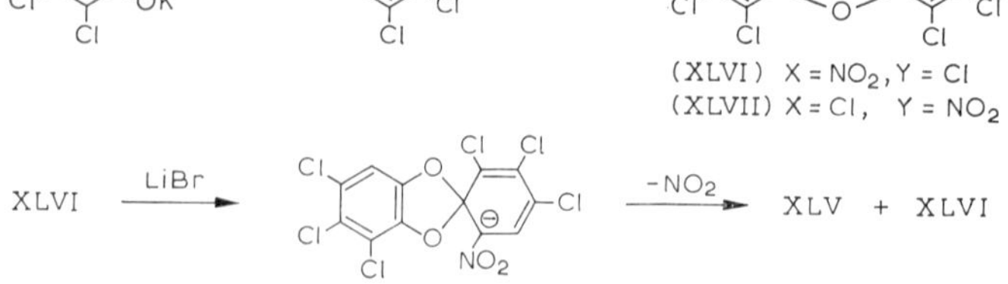

The results of this work explain how 1,2,3,7,8,9-hexachlorodibenzo-1,4-dioxin arises as an environmental contaminant, a fact that hitherto had been rather puzzling (*cf. J. C. Wootton* and *W. L. Courchene*, J. Agr. Food Chem., 1964, **12**, 94).

2. Dithanes, dithiins and their benzo derivatives

Structures equivalent to the dioxanes, in which both oxygen atoms are replaced by sulphur, are named dithanes, whereas structures analogous to the dioxins are known as dithiins, thus nomenclature and numbering systems follow those already given (p. 375).

(*a*) *1,2-Dithanes, 1,2-dithiins and their benzo derivatives*

(*i*) *1,2-Dithianes*

1,2-**Dithane,** b.p. 60°/5 mm, m.p. 32–33°, is conveniently prepared in a yield of 68 % by the oxidation of 1,4-butanedithiol with dimethyl sulphoxide (*T. J. Wallace*, U.S.P. 3,376,313/1968):

$$\text{(HS–CH}_2\text{CH}_2\text{CH}_2\text{CH}_2\text{–SH)} \xrightarrow{\text{Me}_2\text{SO}} \text{1,2-dithiane}$$

Alternatively 1,4-butanedithiol may be treated with toluene-4-sulphonyl chloride in the presence of base (*L. Field* and *R. B. Barbee*, J. org. Chem., 1969, **34**, 36):

$$\xrightarrow[\text{2 OH}^{\ominus}]{\text{ArSO}_2\text{Cl}} \left[\text{ArSO}_2\text{S(CH}_2)_4\text{S}^{\ominus}\right] \xrightarrow{-\text{ArSO}_2^{\ominus}}$$

Lead 1,4-butanedithiolate, on heating with sulphur, also affords 1,2-dithiane. For other methods see *J. G. Affleck* and *G. Dougherty, ibid.*, 1950, **15**, 865; *J. A. Barltrop, P. M. Hayes* and *M. Calvin*, J. Amer. chem. Soc., 1954, **76**, 4348; *A. Schöberl* and *H. Gräffe*, Ann., 1958, **614**, 66; *Schöberl*, G.P. 963,557/1957; C.A., 1959, **53**, 20908 and *S. F. Birch, T. V. Cullum* and *R. A. Dean*, J. Inst. Petroleum, 1953, **39**, 206.

Some representative 1,2-dithianes are listed in Table 2.

TABLE 2

SOME 1,2-DITHIANES

Substituent	m.p. (°C)	b.p. (°C/mm)	Ref.
3,6-Me$_2$	—	66–69/10	1
3,3,6,6-Me$_4$	—	oil	2
3-CO$_2$H	75–76	—	3
4-CO$_2$H	115–116	—	4
cis-3,6-(CO$_2$H)$_2$	199	—	5
trans-3,6-(CO$_2$H)$_2$	275 (decomp.)	—	5
4,4-(CO$_2$H)$_2$	152 (decomp.)	—	3
cis-4,5-(OH)$_2$	128–129	—	6
trans-4,5-(OH)$_2$	129–130	—	6

References

1   *T. L. Cairns et al.*, J. Amer. chem. Soc., 1952, **74**, 3982.
2   *G. Claeson, G. M. Androes* and *M. Calvin*, ibid., 1961, **83**, 4357.
3   *Claeson*, Acta Chem. Scand., 1959, **13**, 1709.
4   *Claeson* and *A. Lansjoen*, ibid., 1959, **13**, 840.
5   *A. Fredga*, Ber., 1938, **71B**, 289.
6   *A. Lüttringhaus et al.*, Z. Naturforsch., 1961, **16b**, 761.

When oxidised with potassium metaperiodate 1,2-dithiane gives the 1,1-*dioxide*, (I), m.p. 53–55°, in 67% yield (*Field* and *Barbee, loc. cit.*). Hydrogen peroxide in acetic acid at room temperature may also be used. *E. J. Goethals et al.* (Bull. Soc. chim. Belg., 1969, **78**, 191), give the m.p. as 61°.

Using an eight molar excess of hydrogen peroxide in dioxane containing acetic and sulphuric acids, the 1,1,2,2-*tetra-oxide*, (II), m.p. 239–242°, is formed:

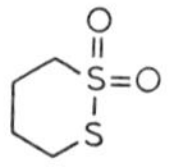

(I) 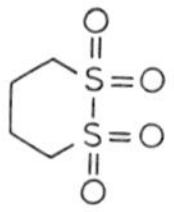   (II) 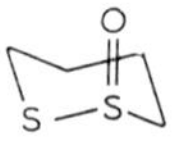   (III)

1,2-*Dithiane*-1-*oxide*, (III), m.p. 67–74°, is prepared by treating 1,4-butanedithiol with hydrogen peroxide in acetic acid at 10° (*D. N. Harpp* and *J. G. Gleason*, J. org. Chem., 1971, **36**, 1314). In this compound the oxygen atom is considered to adopt an axial orientation as shown in formula III.

(ii) 1,2-Dithiin and derivatives

1,2-**Dithiin** (*o-dithiin*), an orange red compound, b.p. 45°/4 mm, λ_{max} 451 (log ε 2.88) nm, is obtained by the oxidation of *cis,cis*-1,3-butadiene-1,4-dithiol with ferric chloride in methanol (*W. Schroth, F. Billig* and *H. Langguth*, Z. Chem., 1965, **5**, 353).

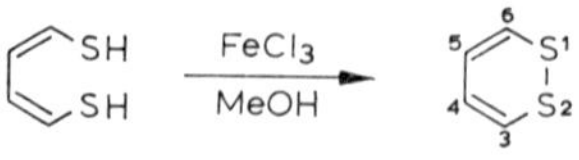

The same molecule and several other derivatives (see Table 3) are also prepared by reacting phenylmethanethiol with *cis-cis*-1,4-disubstituted buta-1,3-dienes, debenzylating the products, *cis,cis*-1,4-bis(benzylthio)buta-1,3-dienes (IV), with sodium and liquid ammonia, and allowing the final oxidation step to V to occur by the action of air (*Schroth, Billig* and *G. Reinhold*, Angew. Chem. intern. Edn. Engl., 1967, **6**, 698):

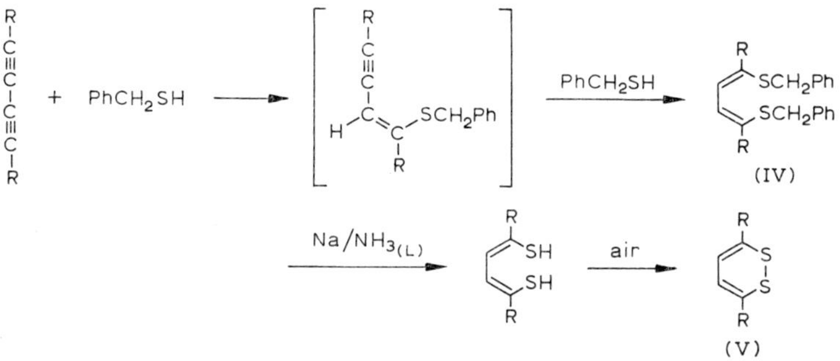

TABLE 3

SOME 3,6-DISUBSTITUTED-1,2-DITHIINS (V)

R	m.p. (°C)	λ_{max}(log ε)(nm)
Ph	143	468 (3.12)
4-MeC$_6$H$_4$	162	470 (3.46)
4-MeOC$_6$H$_4$	179	467 (4.43)
2-thienyl	123	481 (3.76)

PPP–HMO Calculations indicate that the heterocyclic structure for dithiin is correct, rather than the valence-tautomeric bisthioaldehyde representation VI. In particular, the calculated $\pi \to \pi^*$ transition (λ_{max} 439 nm) agrees closely with that experimentally observed (λ_{max} 451 nm, cyclohexane) (*J. Fabian et al.*, Tetrahedron, 1970, **26**, 3227). For the thioaldehyde VI, λ_{max} for the $\pi \to \pi^*$ transition is calculated to be at 376 nm. Sodium in liquid ammonia, or sodium tetrahydridoborate, effect reductive cleavage of the S–S bond in 1,2-dithiins, and pyridazines are formed by reaction with hydrazine.

When heated or photolysed, 1,2-dithiins undergo desulphurization and ring contraction to thiophenes. *Schroth et al.* (1967, *loc. cit.*) consider this to implicate maleic bisthioaldehyde (VI) as an intermediate in the reaction:

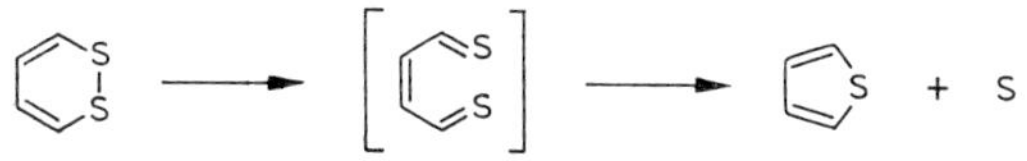

3,6-**Dihydro**-1,2-**dithiin** (VII) is an unstable compound, obtained by *Schöberl* and *Gräffe (loc. cit.)* in poor yield by oxidising *cis*-but-2-ene-1,4-dithiol with ferric chloride:

(VII)

On standing at room temperature, it rapidly forms a dark grey polymer (*N. A. Rosenthal*, U.S.P. 3,284,466/1966; C.A., 1968, **68**, 13015n). 2-Methyl-1,3-pentadiene in acetone containing hydrogen sulphide and traces of hydroquinone and water at 140°, with sulphur dioxide under pressure, yields 3,5-dimethyl-3,6-dihydro-1,2-dithiin. Under similar conditions 4-methyl-3,6-dihydro-1,2-dithiin is obtained from 3-methyl-3-sulpholene and isoprene (*G. B. Payne*, U.S.P. 3,366,644/1968; C.A., 1968, **69**, 19167f).

(iii) Benzodithiins

1,4-**Dihydrobenzo**[*d*]-1,2-**dithiin**(1*H*,4*H*-2,3-**dithianaphthalene**)(VIII), m.p. 80°, 2-*oxide*, m.p. 143°, 2,2-*dioxide*, m.p. 108°, is prepared by the oxidation of 1,2-bis(mercaptomethyl)benzene with ferric chloride (*A. Lüttringhaus* and *K. Hägele*, Angew. Chem., 1955, **67**, 304):

(VIII)

A second route to 1,4-dihydrobenzo[*d*]-1,2-dithiin consists in heating the Bunte salt IX with thiourea in hot dilute hydrochloric acid (*B. Milligan* and *J. M. Swan*, J. chem. Soc., 1965, 2901):

(IX)

It may also be synthesised by reacting *o*-xylylene bromide (X, R = H) with sulphur and sodium sulphide (*M. G. Voronkov et al.*, Khim. geterot. Soedin., 1970, 885; C.A., 1971, **74**, 42337y). 1,4-*Dihydro-1-methylbenzo*[*d*]-1,2-*dithiin* (XI, R = Me) is formed similarly by the action of potassium disulphide upon (1-(1′-bromoethyl)-2-bromomethylbenzene (X, R = Me). It has m.p. 40°, b.p. 145–150°/13 mm, and is volatile in steam (*J. von Braun* and *K. Weissbach*, Ber., 1929, **62**, 2420):

(X)   (XI)

Oxidation of biphenyl-2,2'-dithiol with alcoholic ferric chloride yields **dibenzo[c,e]-1,2-dithiin** (XII), m.p. 113°; when heated with copper, this product undergoes ring contraction and loss of sulphur to give dibenzo[b,d]-thiophene (XIII) (C.C.C., 2nd Edn., Vol. IV A, p. 302) (*H. J. Barber* and *S. Smiles*, J. chem. Soc., 1928, 1142):

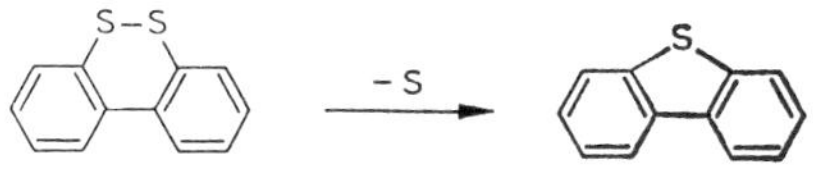

(XII)   −S   (XIII)

Similarly, *dinaphtho*[2,1-c:1',2'-e]-1,2-*dithiin* (1,1'-*binaphthalene* 2,2'-*disulphide*) (XIV) (formed by oxidation of 1,1'-binaphthyl-2,2'-dithiol), a yellow solid with m.p. 214°, when heated with copper at 250° yields *dinaphtho*[2,1-b:1',2'-d]*thiophene* (XV), m.p. 202°:

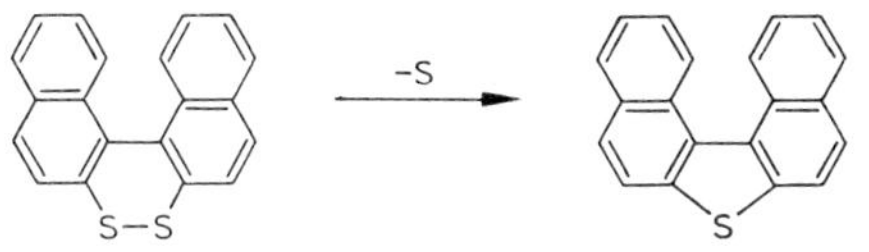

(XIV)   −S   (XV)

3,8-Dinitrodibenzo[c,e]-1,2-dithiin and 2,9-dinitrodibenzo[c,e]-1,2-dithiin were obtained by treating the corresponding biphenyldisulphonyl chlorides with hydriodic acid in acetic acid (*A. Ya. Zheltov et al.*, Zhur. Vses. Khim. obshcheĭ, 1968, **13**, 228; C.A., 1968, **69**, 106635a):

HI/AcOH

(b) 1,3-Dithianes, 2,4-dihydro-1,3-dithiins and their benzo derivatives

(i) 1,3-Dithianes

1,3-**Dithiane** (m-*dithiane*), m.p. 54°, the parent compound of this group, may be synthesised in poor yield by the action of formaldehyde upon propane-1,3-dithiol (*E. E. Campaigne* and *G. Schaefer*, Bol. Col. Quím. Puerto Rico, 1952, **9**, 25; see also *W. Autenrieth* and *K. Wolff*, Ber., 1899, **32**, 1375). An alternative preparation requires the slow addition (8 h) of propane-1,3-dithiol and methylal to boron trifluoride etherate in chloroform and acetic acid at reflux (*E. J. Corey* and *D. Seebach*, Org. Synth., 1970, **50**, 72).

Other 2-substituted and 2,2-disubstituted members of this group may be prepared by the addition of aldehydes or ketones to propane-1,3-dithiols; normally yields are much better than is the case with formaldehyde. Many 2-substituted 1,3-dithianes have been obtained from 1,3-dithiane by deprotonation with butyl-lithium and treatment with an alkyl halide:

(XVI) $\xrightarrow[\text{THF}]{\text{Bu}^n\text{Li}}$ $\xrightarrow{\text{R}^1\text{X}}$ (XVII)

Physical constants for a number of 2-substituted 1,3-dithianes appear in Table 4.

TABLE 4

SOME 2-SUBSTITUTED-1,3-DITHIANES (XVI)

(*R. C. Carlson* and *P. M. Helquist*, Tetrahedron Letters, 1969, 173)

R	m.p. (°C)	b.p. (°C/mm)
Ph–	70–71	
Bu^n-		97/0.7
Bu^t–		113/15
Bu$^\beta$–		135/12

Similarly, 2-substituted-1,3-dithianes may be deprotonated by organo-lithium compounds in tetrahydrofuran and the carbanions, so formed, again reacted with alkyl halides (R^{1}X) to yield 2,2-disubstituted products (*Corey*, Angew. Chem. intern. Edn., Engl., 1965, **4**, 1075).

A number of such derivatives are listed in Table 5.

TABLE 5

SOME 2,2-DISUBSTITUTED-1,3-DITHIANES (XVII)

R	R^1	b.p. (°C) (bath temp.)/mm
Me–	Pr$^\beta$–	60/0.3
Me–	–CH$_2$Ph	113/0.08
C$_5$H$_{11}{}^n$–	C$_5$H$_{11}{}^n$–	100/0.05
C$_5$H$_{11}{}^n$–	Pr$^\beta$–	84/0.03
Ph–	Pr$_n{}^\beta$–	m.p. 50.5–51.5
Bu^n–	C$_5$H$_{11}{}^n$–	100/0.05
Bu^n–	–CH$_2$Ph	210/15

2-Acyl-1,3-dithianes may be synthesised by the interaction of enamines with 1,3-bis(toluene-4'-sulphonylthio)propane, prepared in turn from 1,3-dibromopropane and S-potassium toluene-4-thiosulphonate. The method is illustrated by the formation of 2-*phenacyl*-1,3-*dithiane*, m.p. 94° (*R. B. Woodward, I. J. Pachter* and *M. L. Scheinbaum*, J. org. Chem., 1971, **36**, 1137):

$$\begin{array}{ccc}
\text{CH}_2\text{S Tosyl} & & \text{CH}_3 \\
| & & | \\
\text{CH}_2 & + & \text{C=CHPh} \\
| & & | \\
\text{CH}_2\text{S Tosyl} & & \text{N} \\
& & R \quad R^1
\end{array} \xrightarrow[\text{H}_2\text{O}]{-2\ \text{Tosyl}^{\ominus}} \quad \text{[dithiane]}-\text{CO}\cdot\text{CH}_2\text{Ph}$$

Numerous other substituted 1,3-dithianes have been made (see for example, *D. S. Breslow* and *H. Skolnik* in "The Chemistry of Heterocyclic Compounds", Vol. 21, Part II: "Multi-Sulfur and Sulfur and Oxygen Five- and Six-membered Heterocycles", *A. Weissberger*, Ed., Interscience, New York, 1966, Chapter 12, pp. 979–1028 and refs. cited therein).

The basic structure of the 1,3-dithiane system has been determined by an X-ray crystallographic analysis of 2-phenyl-1,3-dithiane (*H. T. Kalff* and *C. Romers*, Acta Crystallogr., 1966, **20**, 490). A résumé of this study is shown below:

Structure with values: 115.5°, 115°, 1.82 Å, 1.49 Å, 1.80 Å, 116.5°, 100°, Ph

Torsional angles

CS–CS 57°

CS–CC 55°

SC–CC 61.5°

E. L. Eliel and his co-workers have conducted a determination of the conformational preferences of alkyl groups in the 2-, 4- and 5-positions of the 1,3-dithiane system (Accounts of Chemical Research, 1970, **3**, 1; see also *Eliel* and *R. O. Hutchins*, J. Amer. chem. Soc., 1969, **91**, 2703).

Thus, for example, conformational energies for the 2-position of *cis*-4,6-dimethyl-dithianes were obtained by the equilibration procedure shown (XVIII ⇌ XIX) and the results are compared with cyclohexane values in Table 6.

TABLE 6

CONFORMATIONAL ENERGIES IN
2-ALKYL-*CIS*-4,6-DIMETHYL-1,3-DITHIANES

R–	$-\Delta G°(kJ \cdot mol^{-1})(69°)$	$-\Delta G°(kJ \cdot mol^{-1})$ (cyclohexane)
Me	7.43	7.14
Et	6.47	7.35
Pr^β	8.19	9.03
Bu^t	11.42	⩾ 20.5

(XVIII)   (XIX)

It is noteworthy that similar values are obtained for the substituents in both series, although there is a disparity in the case of the tertiary butyl group.

Additional data, based upon ^{1}H-n.m.r. studies has been presented by *M. Anteunis, G. Swaelens* and *J. Gelan* (Tetrahedron, 1971, **27**, 1917), and an analysis of the ultraviolet spectrum is given by *D. R. Williams et al.* (J. chem. Soc., B, 1968, 1132), and by *L. K. Dyall* and *S. Winstein* (Spectrochim. Acta, Part A, 1971, **27**, 1619).

S. A. Khan et al. (J. Amer. chem. Soc., 1975, **97**, 1468) have shown that 1,3-dithiane-1-oxide, prepared by the method of *Carlson* and *Helquist* (J. org. Chem., 1968, **33**, 2596), exists as two isomers in the ratio 21 : 4 at $-81.5°$, with the equatorial form predominating:

Although the 1,1-*dioxide* (m.p. 139–140 °C) and the *trans*-1,3-dioxide are able to undergo ring reversal, by symmetry, they have an equilibrium constant of unity; the *cis*-1,3-dioxide on the other hand has a choice of two conformations, but the di-equatorial isomer is stabilized by dimer formation:

The mass-spectral fragmentation patterns of 2-aryl-1,3-dithianes have been deduced by deuterium labelling experiments (*J. H. Bowie* and *P. Y. White*, Org. Mass Spectrometry, 1969, **2**, 611). In the case of 2-phenyl-1,3-dithiane two major fragmentations are established as shown below:

Reduction of 1,3-dithianes with calcium in liquid ammonia affords the corresponding α,γ-dithiols (*B. C. Newman* and *Eliel*, J. org. Chem., 1970, **35**, 3641).

Considerable interest in 1,3-dithianes in recent years has attended the utilisation of the system in aldehyde and ketone synthesis. *Corey* and co-workers have shown that hydrolysis of 2-mono- or 2,2-di-substituted 1,3-dithianes yield aldehydes and ketones respectively (*Corey* and *Seebach*, Angew. Chem. intern. Edn., Engl., 1965, **4**, 1075, 1077):

It is reported, however, by *Corey* and *B. W. Erickson* (J. org. Chem., 1971, **36**, 3553) that the usual reagents ($HgCl_2$ + HgO or $CdCO_3$) give low yields of aldehydes in the hydrolysis of 2-alkyl-1,3-dithianes. Better results are obtained using red mercuric oxide and boron trifluoride etherate in aqueous tetrahydrofuran (*E. Vedejs* and *P. L. Fuchs*, ibid., 1971, **36**, 366). A method for the chain extension of aldehydes utilizes the reaction of aldehydes with the ylide obtained from 1,3-dithian-2-thione and trimethyl phosphite, thus affording ketene dithioacetals which may then be reduced and hydrolysed (*F. A. Carey* and *J. R. Neergaard*, ibid., 1971, **36**, 2731):

(XX)

(XXI a)

In this reaction the carbonium ion **XXIa** formed by the protonation of **XX** with trifluoracetic acid is stabilized by the flanking sulphur atoms.

Carbanions of the type **XXIb** readily react with 1,2-oxides, thus providing access to β-hydroxyaldehydes or β-hydroxyketones which normally dehydrate to the corresponding olefins:

(XXIb)

$$RCO \cdot CH_2 \cdot \underset{\underset{OH}{|}}{CHR^1} \xrightarrow{-H_2O} RCO \cdot CH = CHR^1$$

Oxetanes react similarly with γ-hydroxy-aldehydes and -ketones (*Corey* and *Seebach*, ibid., 1975, **40**, 231). With carbon dioxide, at −70°, 1,3-dithianyl derivatives of α-oxocarboxylic acids are formed and, with nitriles, acyl derivatives are obtained (*Corey* and *Seebach*, 1965, *loc. cit.*):

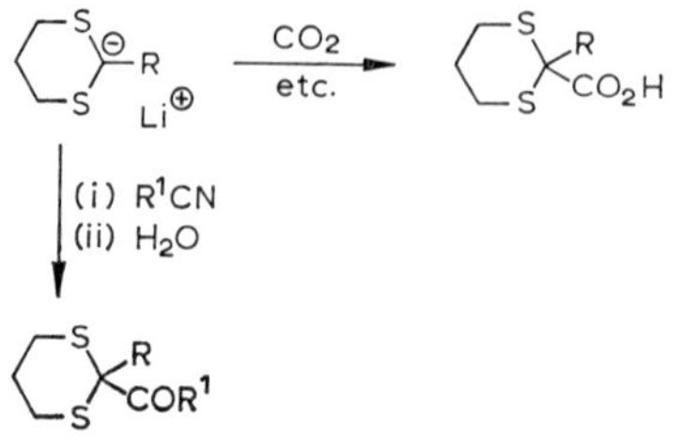

A simple synthesis of esters *via* the nucleophilic reaction of 2-lithio-2-methylthio-1,3-dithiane with aldehydes, ketones and alkyl halides has been described (*W. D. Woessner*, Chem. Letters, 1976, 43). 2-Lithio-1,3-dithiane undergoes 1,2-addition to α,β-unsaturated ketones, but a 1,4-addition has been noted with nitrostyrenes (*Seebach* and *H. F. Leitz*, Angew. Chem., 1969, **81**, 1047).

2-Lithio-2-phenyl-1,3-dithiane when treated with nitrobenzene or bromo-2,4-dinitrobenzene gives *2,2'-diphenyl-2,2'-bi(1,3-dithiane)* (XXII), m.p. 204°, whereas 2-bromopyridine yields 2-phenyl-2-(2'-pyridyl)-1,3-dithiane (XXIII). Neither 3-bromopyridine nor 4-bromopyridine react with this salt (*W. H. Baarischers* and *T. L. Loh*, Tetrahedron Letters, 1971, **37**, 3483):

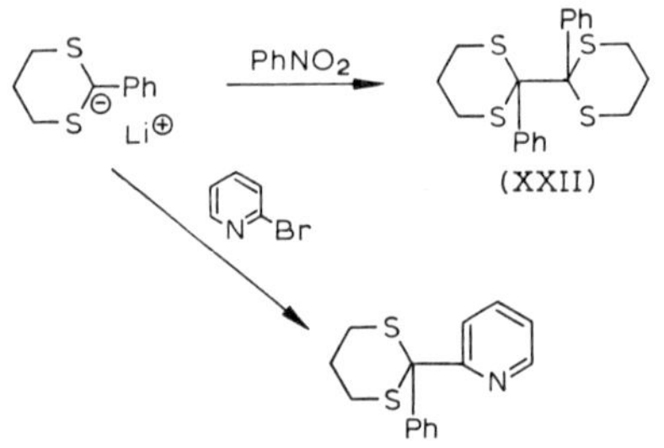

The *spirocycle* XXIV, m.p. 116°, has been prepared by the interaction of tetramethyl orthothiocarbonate and propane-1,3-dithiol (*D. L. Coffen*, J. heterocyclic Chem., 1970, **7**, 201; *cf. T. P. Johnston, C. R. Strongfellow* and *A. Gallagher*, J. org. Chem., 1962, **27**, 4068):

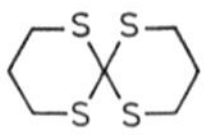

Other related products have also been synthesised (*G. Zinner* and *R. Vollrath*, Chem. Ber., 1970, **103**, 766; *M. P. Mertes* and *A. A. Ramsay*, J. med. Chem., 1969, **12**, 342).

For other applications of 1,3-dithianes in synthesis see, for example, *E. J. Corey et al.* (J. Amer. chem. Soc., 1967, **89**, 434; 1968, **90**, 3245; J. org. Chem., 1968, **33**, 298; Org. Synth., 1970, **50**, 72), *A. G. Brook et al.* (J. Amer. chem. Soc., 1967, **89**, 431), *C. A. Reece et al.* (Tetrahedron, 1968, **24**, 4249), *T. Hylton* and *V. Boekelheide* (J. Amer. chem. Soc., 1968, **90**, 6887), *D. Seebach* (Synthesis, 1969, **1**, 17), *Seebach* and *Corey* (J. org. Chem., 1975, **40**, 231) and *A. I. Meyers* (Heterocycles in Organic Synthesis, Wiley, New York, 1974, pp. 192–197).

(ii) 2,4-Dihydro-1,3-dithiins

A few compounds of this class are known, thus *A. Lüttringhaus* and *H. Prinzbach* (Ann., 1959, **624**, 79), *Lüttringhaus, M. Mohr* and *N. Engelhard* (*ibid.*, 1963, **661**, 84) have prepared 2,5-diphenyl-2,4-dihydro-1,3-dithiin and 2-(4-methoxyphenyl)-5-phenyl-2,4-dihydro-1,3-dithiin by the following reaction sequence:

$$RCHO \; + \; 2\,HSCH_2 \cdot CO_2Et \; \longrightarrow$$

(R = Ph or [4] MeOC₆H₄)

2,2-Dibenzyl-4,6-diphenyl-2,4-dihydro-1,3-dithiin, is obtained from 1,3-diphenyl-propane-2,2-dithiol and chalcone in ethanolic hydrogen chloride (*E. Campaigne* and *B. E. Edwards*, J. org. Chem., 1962, **27**, 4488); whereas 2,2-propanedithiol affords 2,2-dimethyl-4,6-diphenyl-2,4-dihydro-1,3-dithiin (*C. Demuynck et al.*, Bull. Soc. chim. Fr., 1966, **6**, 1920):

(R = CH₂Ph or Me)

Aromatic aldehydes and acetylenic esters together with hydrogen sulphide and boron trifluoride afford substituted 2,4-dihydro-1,3-dithiins of type XXV (*U. Eisner* and *T. Krishnamurthy*, Tetrahedron, 1971, **27**, 5763):

$$HC\equiv C \cdot CO_2Me \; + \; ArCHO \; \xrightarrow[\;BF_3\;]{\;H_2S\;}$$

(Ar = Ph, m.p. 134–135°; [4] MeC₆H₄, m.p. 176–177°; [4] ClC₆H₄, m.p. 135–136°)

(XXV)

The benzo derivative 2-*phenylbenzo*[d]-2,4-*dihydro*-1,3-*dithiin* (XXVI, R=H, R′ = Ph), m.p. 113–114°, is obtained from 2-mercaptomethylbenzenethiol and benzaldehyde (*A. Lüttringhaus et al.*, Angew. Chem., 1955, **67**, 304); acetone yields 2,2-*dimethylbenzo*[d]-2,4-*dihydro*-1,3-*dithiin* (XXVI, R = R′ = Me), b.p. 140–141°:

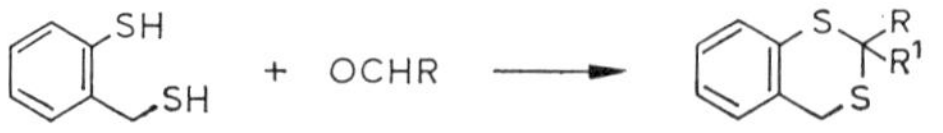

When treated with bromine, conversion to the corresponding benzo-1,3-dithiolium cation occurs, which then undergoes ring-opening and subsequent contraction. This appears to be a general reaction of 2-aryl-2,4-dihydro-1,3-dithiins (*Lüttringhaus et al.*, 1963, *loc. cit.*):

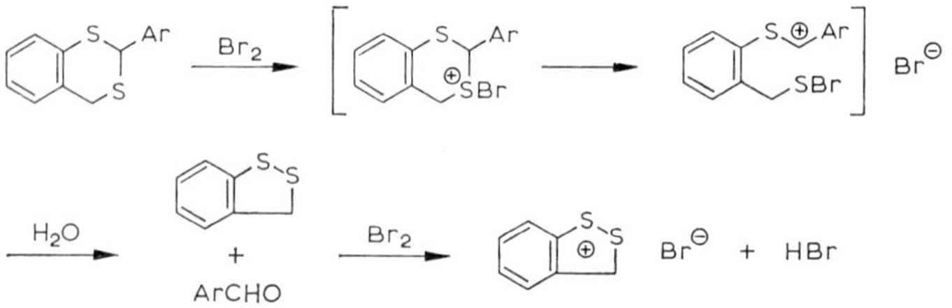

W. B. Price and *S. Smiles* (J. chem. Soc., 1928, 2372) prepared 2-*phenylnaphtho*-[1,8-de]-2,4-*dihydro*-1,3-*dithiin* (2-*phenyl*-peri-*naphtha*-1,3-*dithiane*), m.p. 116°, by reacting naphthalene-1,8-dithiol with benzaldehyde:

(c) *1,4-Dithiane, 1,4-dithiin and their benzo derivatives*

(i) *1,4-Dithiane and simple derivatives*

1,4-Dithiane may be synthesised from 1,2-dibromoethane and potassium sulphide (*J. M. Crafts*, Compt. rend., 1862, **54**, 1277), or from 1,2-dibromo-ethane by the action of dimethyl sulphide (*O. Masson*, J. chem. Soc., 1886, **49**, 234); 1,2-ethanedithiol and sodium ethoxide have also been used (*R. G. Gillis* and *A. B. Lacey*, Org. Synth. Coll., Vol. 4, 1963, p. 396):

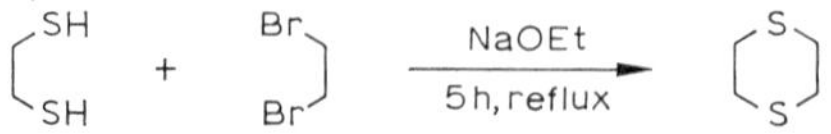

1,2-Ethanedithiol and palladium chloride in carbon monoxide under pressure give 1,4-dithiane (*J. A. Scheben* and *I. L. Mador*, U.S.P. 3,480,632/1969; see also *W. T. Eisfeld* and *E. D. Weil*, U.S.P. 3,503,991/1970).

Thermal degradation of polyethylene sulphide gives 1,4-dithiane as a major product, while polypropylene sulphide yields 2,5-dimethyl-1,4-dithiane among other products (*C. Brown* and *D. R. Hogg*, J. chem. Soc., B, 1969, 1054).

Cyclohexene when photolysed with thiobenzophenone gives 2,2,3,3-tetraphenylperhydrobenzo-1,4-dithiin (XXVII), by *trans* addition (*A. Ohno*, *Y. Ohnishi* and *G. Tsuchihashi*, J. Amer. chem. Soc., 1969, **91,** 5038):

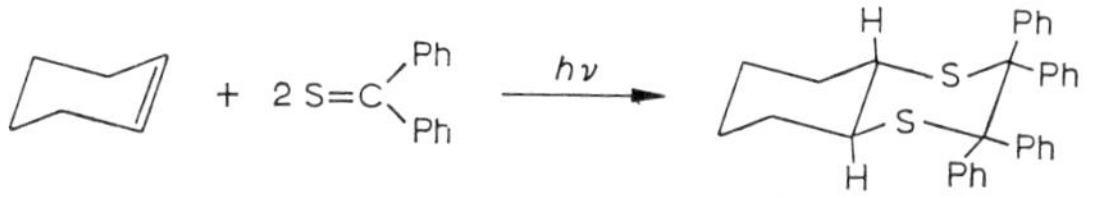

(XXVII)

Similarly thiobenzophenone and *tert*-butyl vinyl ether yield XXVIII; *tert*-butyl vinyl sulphide on the other hand gives *cis*-5-*tert*-butylthio-2,2,3,3-tetraphenyl-1,4-dithiane (XXIX):

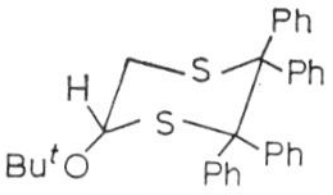

(XXVIII) 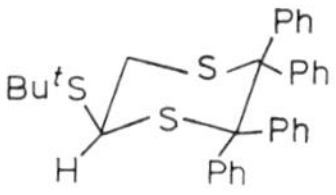   (XXIX)

Allyl disulphide and methallyl disulphide react with sulphuryl chloride yielding a mixture of *cis* and *trans* isomers of 2,5-bis(chloromethyl)-1,4-dithiane (XXX, R = H) and 2,5-bis(chloromethyl)-3,5-dimethyl-1,4-dithiane (XXX, R = Me), respectively (*W. A. Thaler* and *P. E. Butler*, J. org. Chem., 1969, **34,** 3389):

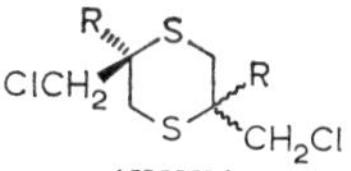

(XXX)

The addition of acetylene or but-2-yne to ethane-1,2-disulphenyl chloride gives trans-2,3-*dichloro*-1,4-*dithiane*, m.p. 127–129°, and trans-2,3-*dichloro*-2,3-*dimethyl*-1,4-*dithiane*, m.p. 86–89°, respectively (*W. H. Mueller* and *M. Dines*, J. heterocyclic Chem., 1969, **6,** 627). With propyne the product is 2-*chloro*-3-*methyl*-5,6-*dihydro*-1,4-*dithiin* (XXXI, R = Me), b.p. 83°/0.08 mm, the formation of which was taken as evidence that the addition involves carbonium rather than episulphonium ion intermediates:

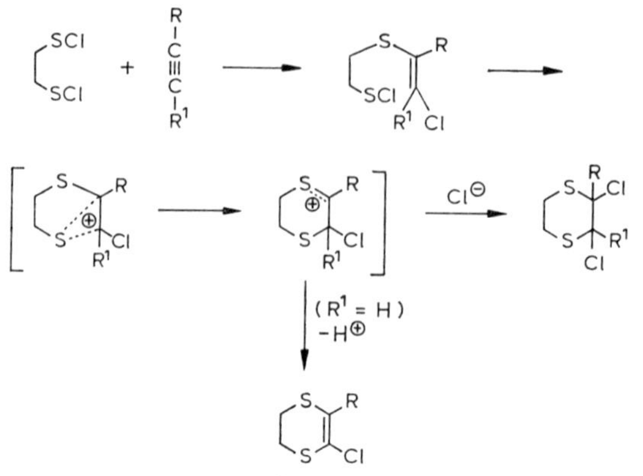

(XXXI)

Ethane-1,2-dithiol reacts with oxalyl chloride to give the *dispiro* derivative (XXXII), m.p. 210–213° (*Coffen et al.*, J. Amer. chem. Soc., 1971, **93**, 2258):

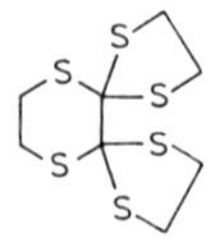

(XXXII)

1,4-Dithiane is a colourless solid, m.p. 111–112°, b.p. 199–200°/760 mm, which sublimes at ordinary temperatures and is steam-volatile. The ^{1}H-n.m.r. spectrum of 1,4-dithiane and some of its derivatives has been determined (*N. E. Alexandrou* and *P. M. Hadjimihalakis*, Org. Magn. Res., 1969, 401); a significant effect on the chemical shift position of the signals is induced by aromatic solvents. A mass spectrometric comparison of 1,4-dioxane, 1,4-oxathiane (p. 417) and 1,4-dithiane (*G. Conde-Caprace* and *J. E. Collin*, Org. Mass Spectra, 1969, **2**, 1277), reveals that the last two give rise to similar fragmentation patterns, rather different to the first mentioned compound.

The *monoxide*, m.p. 125°, which can be obtained along with the 1,4-*dioxide* by careful oxidation of 1,4-dithiane with hydrogen peroxide, forms two isomeric 4-*tolylsulphilimines* (XXXIII), m.p. 230–234° (decomp.) and 176–177° (decomp.), by reaction with chloramine-T:

$$O = S \quad SNSO_2C_6H_4Me[4]$$

(XXXIII)

The water-soluble *dioxide* also exists in two forms, α-, m.p. 263° (main product) and β-, decomp. 235–250°. The monoxide is also formed by the action of potassium sulphide on β,β'-dichlorodiethyl sulphoxide, whilst a similar reaction with the corresponding sulphone gives 1,4-*dithiane* 1,1-*dioxide*, m.p. 203° (*E. V. Bell* and *G. M. Bennett*, J. chem. Soc., 1927, 1798; 1928, 86).

Absorption spectral data for a number of 1,4-dithianes and their 1-oxides and various of their metal complexes are consistent with them adopting a chair conformation (*J. A. W. Dalziel, M. J. Hitch* and *S. D. Ross*, Spectrochim. Acta, 1969, **25A**, 1055, 1061; *P. Klaboe, ibid.*, 1969, **25A**, 1437; *Hitch* and *Ross, ibid.*, p. 1041; *J. Reedijk, A. H. M. Fleur* and *N. L. Groeneveld*, Rec. Trav. chim., 1969, **88**, 1115; *J. D. Donaldson* and *D. G. Nicholson*, J. chem. Soc., A, 1970, 145).

2,5-*Diethoxy*-1,4-*dithiane*, m.p. 91°, has been obtained by the slow spontaneous self-condensation of 2,2-diethoxyethanethiol; this has been hydrolysed to 2,5-*dihydroxy*-1,4-*dithiane* which assumes two forms giving distinct *acetates*, m.p. 173° and 109° (*G. Hesse* and *I. Jörder*, Chem. Ber., 1952, **85**, 924) (see p. 384).

The physical constants for a number of 1,4-dithianes are collected into Table 7.

TABLE 7

SOME SUBSTITUTED 1,4-DITHIANES

Substituent	m.p. (°C)	b.p. (°C/mm)	n_D^{20}	Ref.
2-Me	20	209–210/760	—	1
2,5-Me$_2$	—	85–87/12	1.5420	2,3
2,6-Me$_2$	—	80–100/15	1.5324	3,4
2,5-Pr$_2$^{n}	—	145–155/20	1.5255	2
2,5-Ph$_2$	—	190–195/30	1.6060	2
2,3,5,6-Me$_4$	—	145–150/35	—	5

References

1   *J. R. Meadow* and *E. E. Reid*, J. Amer. chem. Soc., 1934, **56**, 2177.
2   *F. J. Glavis, L. L. Ryden* and *C. S. Marvel, ibid.*, 1937, **59**, 707.
3   *Marvel* and *E. D. Weil, ibid.*, 1954, **76**, 61.
4   *D. Harman* and *W. E. Vaughan*, U.S.P. 2,562,145/1951.
5   *W. J. Pope* and *J. L. B. Smith*, J. chem. Soc., 1921, 396.

(*ii*) 1,4-Dithiins

2,2-Dimethoxyethanethiol when treated with acid affords 2,5-dimethoxy-1,4-dithiane, which with phosphorus pentoxide at 100–120° yields a mixture of **3-methoxy-2,3-dihydro-1,4-dithiin** (XXXIV) and **1,4-dithiin** (XXXV); at 180–200° only the latter is formed (*W. E. Parham, H. Wynberg* and *F. L. Ramp*, J. Amer. chem. Soc., 1953, **75**, 2065):

These authors were unable to repeat the experiment of *L. E. Levi* (Chem. News, 1890, **62**, 216) in which thiodiglycollic acid and phosphorus tri-sulphide were claimed to yield 1,4-dithiin.

1,4-*Dithiin* has m.p. 99–100°, sublimes at 120°/0.6 mm, and has b.p. 77°/17 mm. The corresponding 1,1,4,4-*tetraoxide* has m.p. 241.5–242.5° (decomp.).

W. Schroth et al. (Z. Chem., 1970, **10**, 147) have shown that 1,4-dithiin, its 2-phenyl, 2,5-diphenyl and 2,3,5,6-tetraphenyl derivatives undergo two-step electro-oxidation at a platinum anode in acetonitrile solution, forming first the corresponding monocation and then the dication.

Diaryl-1,4-dithiins are of some interest since they undergo substitution (nitration and bromination) in the heterocycle and, in some reactions, ring contraction to thiophenes.

2,5-Diphenyl-1,4-**dithiin**, m.p. 115–117°, is formed when sodium phenacyl thiosulphate is boiled with alcoholic hydrochloric acid (*R. H. Baker* and *C. Barkenbus*, J. Amer. chem. Soc., 1936, **58**, 262). Its 3,6-*dibromo* deriv., m.p. 165–166°, when oxidised in glacial acetic acid with hydrogen peroxide is converted into 3,5-dibromo-2,4-diphenylthiophene (*Parham et al., ibid.*, 1954, **76**, 4960; 1955, **77**, 68; 1956, **78**, 850; *H. H. Szmant* and *I. M. Alfonso, ibid.*, 1956, **78**, 1064).

2,5-Di(methoxycarbonyl)-1,4-dithiins (XXVII, R = 1- or 2-naphthyl) are prepared from the corresponding dithiolones (XXXVI) in methanolic alkali (*A. Marei* and *M. M. A. El Sukkary*, U.A.R. J. Chem., 1971, **14**, 101; C.A., 1972, **77**, 126, 473):

XXXVI   (XXXVII)

Photolysis of diaryl and benzo-1,2,3-thiadiazoles is considered to produce biradicals XXXVIII, which then dimerize to 1,4-dithiins (*K. Zeller, J. Meier* and *E. Müller*, Tetrahedron Letters, 1971, 537):

(XXXVIII)

$$\left(\text{R} = \text{R}^1 = \text{Ph} \quad \text{or} \quad \text{R}, \text{R}^1 = \right)$$

The 2,3-dihydro-1,4-dithiin system is readily synthesised by reacting ethane-1,2-dithiol with α-bromoketones (*H. Rubinstein* and *M. Wuerthele*, J. org. Chem., 1969, **34**, 2763) and 2,3-**dihydro**-1,4-**dithiin** (XXXIX, R = R′ = H), b.p. 101°/29 mm (*Parham et al.*, J. Amer. chem. Soc., 1955, **77**, 1169) can be obtained from bromoacetaldehyde diethylacetal and ethane-1,2-dithiol in the presence of toluene-4-sulphonic acid.

(XXXIX)

Other 2,3-dihydro-1,4-dithiins (see Table 8) are available through the interaction of thiiranes and α-mercaptoketones (*F. Asinger et al.*, Ann., 1971, **753**, 151), or *via* a thermal ring-expansion of 1,3-dithiolane 1-oxides (*C. H. Chen*, Tetrahedron Letters, 1976, 25; *Chen* and *B. A. Donatelli*, J. org. Chem., 1976, **41**, 3053).

TABLE 8

2,3-DIHYDRO-1,4-DITHIINS

2.3-Dihydro-1,4-dithiin	*b.p. (°C)/mm*	n_D^{20}
5-Methyl-	83–84/1	—
6-Ethyl-5-methyl-	69/0.8	1.5742
5-*tert*-Butyl-	m.p. 63.6–65.5	—
5,6-Dimethyl-	66–67/0.9	1.5880
5-Methyl-6-phenyl-	103/0.03	1.6508
5,6-Trimethylene-	64/0.2	1.6115
6-Ethyl-2,5-dimethyl-	74/0.8	1.5535
2,5-Dimethyl-6-phenyl-	111/0.13	1.6282
2-Methyl-5,6-trimethylene-	80–81/0.5	1.5890
6-Ethyl-5-methyl-2,3-tetramethylene-	97/0.15	1.5625
2,3-Tetramethylene-5,6-trimethylene-	100/0.05 (m.p. 39)	
2-Diethylaminomethyl-6-ethyl-5-methyl-	121/0.2	1.4850
2-Diethylaminomethyl-5-methyl-6-phenyl-	144/0.1	1.5870
6-Ethyl-2-methoxymethyl-5-methyl-	85/0.15	1.5448
2-Methoxymethyl-5-methyl-6-phenyl-	116.5/0.04	1.6099
2-Methoxymethyl-5,6-trimethylene-	87–88/0.1	1.5705
5-Anisyl-	m.p. 99–101	—
5-(4-Nitrophenyl)-	m.p. 142–143	—
6-Benzyl-5-phenyl-	m.p. 132–133	—

$(CH_2)_{n-1}$ (ring structure with two S atoms)

$n = 4$	71–72/0.4	
$= 5$	109–110/1.4	
$= 6$	88–89/0.3	
$= 11$	169–170/0.1	
$= 12$	186–189/0.8–0.9	

1,4-Dithiin and its 2,3-dihydro derivative form the sulphonium salts XL and XLI, respectively, when treated with triethyloxonium tetrafluoroborate (*Schroth* and *M. Hassfeld*, Z. Chem., 1970, **10**, 296).

Since the chemical shifts for the 2-(δ 8.36, $J_{2,3} = 10$ Hz) and 3-(δ 6.82)proton signals in the ^{1}H-n.m.r. spectrum of XL compare closely with those of the olefinic protons in XLI (δ 8.03, 6.76, $J_{2,3} = 10$ Hz) the alternative representation XLII, requiring dπ–pπ interaction and valence-shell expansion at sulphur, for the first compound is discounted:

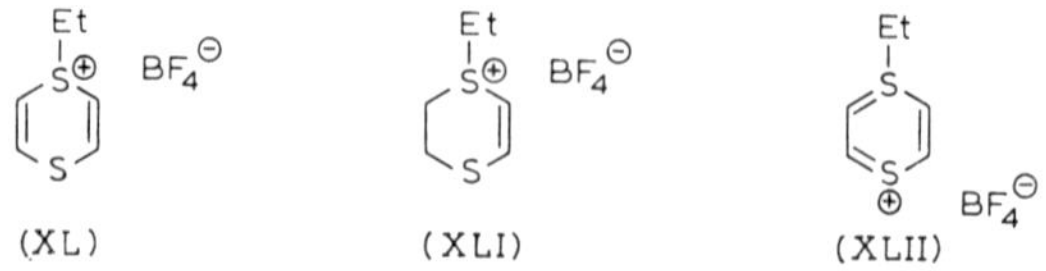

(XL)   (XLI)   (XLII)

It is to be anticipated that the last representation XLII should be aromatic and sustain a diamagnetic ring-current; clearly this effect is not observed in the spectrum of the salt from 1,4-dithiin.

(iii) Benzo-1,4-dithiins

Benzene-1,2-dithiol and 1,2-dibromoethane afford 2,3-**dihydrobenzo**-1,4-**dithiin** (XLIII), b.p. 82.5–85°, n_D^{25} 1.6713, *disulphone*, m.p. 269° (*Parham, T. M. Roder* and *W. R. Hasek*, J. Amer. chem. Soc., 1953, **75**, 1647):

(XLIII)

Condensation of benzene-1,2-dithiol with diethyl bromoacetal in ethanolic potassium hydroxide solution gives 2,2-diethoxyethyl 2-mercaptophenyl sulphide (XLIV), which may be cyclised, by treatment with hydrogen chloride, to 2-*ethoxy*-2,3-*dihydrobenzo*-1,4-*dithiin* (XLV), b.p. 124–125°/0.9 mm; this by heating with phosphorus pentoxide at 170–175° affords **benzo 1,4-dithiin** (XLVI):

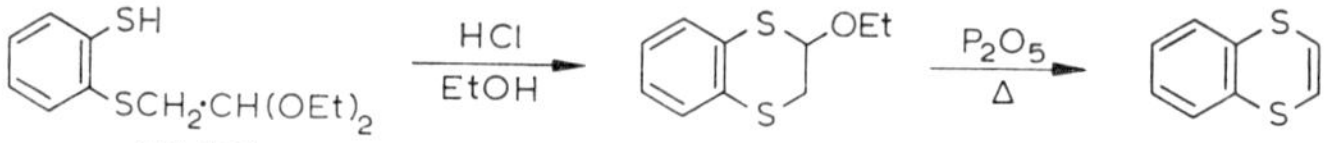

(XLIV)   (XLV)   (XLVI)

An improved method of preparation for benzo-1,4-dithiin using benzenethiol and 1,2-dibromoethane has been described (*ä. H. Verheijen* and *H. Kloosterziel*, Synthesis, 1975, 451); the intermediate 2,3-dihydrobenzo-1,4-dithiin is dehydrogenated by the action of sulphuryl chloride in quinoline.

Benzo-1,4-*dithiin* is a pale yellow-green oil, b.p. 67–70°/0.1 mm. It may be oxidised to a *tetraoxide (disulphone)*, m.p. 221.5–222.5°. With electrophilic reagents (nitration, acylation by Friedel–Crafts reaction) attack takes place at position 2 *(Parham et al., loc. cit.).* K. K. *Deb*, J. E. *Bloor* and T. C. *Cole* (Org. Magn. Resonance, 1970, **2**, 431) have examined the high resolution ^{1}H-n.m.r. spectrum of this compound and the e.s.r. spectrum of the corresponding radical cation in aluminium trichloride–nitromethane has been recorded by P. D. *Sullivan* (J. Amer. chem. Soc., 1968, **90**, 3618).

Benzo-1,4-dithiin behaves in a similar manner to 1,4-dithiin (p. 403) on anodic oxidation, affording first a monocationic radical then the dication (*Schroth et al.*, 1970, *loc. cit.*).

2,3-*Dihydro-6-methyl*-1,4-*benzodithiin*-2,3-*dione dioxime*, XLVII, m.p. 216–217° (decomp.), is formed *via* the addition of cyanogen di-*N*-oxide to toluene-3,4-thiol (*N. E. Alexandrou* and *D. N. Nicolaides*, J. chem. Soc., C, 1969, 2319):

(XLVII)

The benzo-1,4-dithiins (XLVIII, R = H) and (XLVIII, R = Me) may be oxidised with hydrogen peroxide to mono-*S*-oxides (sulphoxides) (XLIX, R = H) and (XLIX, R = Me). On further oxidation partial desulphurization and ring contraction occurs affording L (*K. Fickentscher*, Chem. Ber., 1970, **103**, 3000):

(XLVIII)   (XLIX)   (L)

(iv) Thianthrene and its derivatives

Thianthrene (LIII), m.p. 160°, is obtained by reaction of benzene with sulphur monochloride in presence of aluminium chloride (*K. Fries* and *W. Vogt*, Ann., 1911, **382**, 320). 1,2,3-Benzothiazole (LI) when reacted with phenyl radicals yields three products: diphenyl sulphide, the bibenzothiophene (LIV) and thianthrene. The last product arises from the intermediate LII. Evidence for the intermediate consists in its isolation, as a carbon disulphide adduct, from the reaction medium (*L. Benati, A. Tundo* and *G. Zanardi*, Chem. Comm., 1952, 590; *Benati et al.*, J. chem. Soc., Perkin I, 1974, 1276):

(LI)   (LII)

Ph₂S +   (LIV)   (LIII)

1,2,3-Benzothiadiazole with phenylthio radicals affords dibenzo[*c,e*]-1,2-dithiin (see p. 392), thianthrene and 2-(phenylthio)diphenyl disulphide (*Benat et al.*, J. org. Chem., 1976, **41**, 1331):

Thianthrene is also formed, in good yield, when 2,2′-diiododiphenyl sulphide is heated with copper at 180° (*Barber* and *Smiles*, J. chem. Soc., 1928, 1149). Diphenyl sulphide and aluminium chloride, at 170–180°, give thianthrene in 68% yield (*A. V. Tolstaya* and *A. V. Petrakov*, Izvest. Akad. Nauk, S.S.S.R. Ser. Khim., 1970, 2752; C.A., 1971, **75**, 5819g).

Numerous other syntheses have been described (see for example, *Breslow* and *Skolnik*, *loc. cit.*, pp. 1156–1161 and refs. cited therein).

The mass spectrum has been analysed (*J. Heiss, K. P. Zeller* and *B. Zeeh*, Tetrahedron, 1968, **24**, 3255; *G. Saintruf, J. Servoinsidoine* and *J. P. Coic*, J. heterocyclic Chem., 1974, **11**, 287). The principal mode of fragmentation is loss of sulphur from the molecular ion to give the dibenzothiophene ion.

The ^{1}H-n.m.r. spectrum of the dioxide is discussed by *K. F. Purcell* and *J. R. Berschied Jr.* (J. Amer. chem. Soc., 1967, **89**, 1579).

Crystal structure studies indicate that thianthrene is folded about the axis of the sulphur atoms, the planes of the benzene rings being inclined to one another at an angle of 128° (*R. G. Wood* and *J. E. Crackston*, Phil. Mag., 1941, **31**, 62; *H. Lynton* and *E. G. Cox*, J. chem. Soc., 1956, 4886; *I. Rowe* and *B. Post*, Acta Cryst., 1956, **9**, 827).

Thianthrene gives rise by oxidation, for example by nitric acid, to two stereoisomeric 5,10-*dioxides*; the α-form has m.p. 284°, dipole moment 1.7D, the β-form, m.p. 249°, dipole moment 4.2D (*E. Bergmann* and *M. Tschudnovski*, Ber., 1932, **65**, 457). Crystallographic studies (*S. Hosoya* and *R. G. Wood*, Chem. and Ind., 1957, 1042) have confirmed a suggestion of *T. W. J. Taylor* (J. chem. Soc., 1935, 625) that the two forms have the configurations shown below, the reverse of that which was originally suggested by *Bergmann* and *Tschudnovski*:

The 5,5,10,10-*tetraoxide*, m.p. 337.5–340° (*H. V. Drushel* and *J. F. Miller*, Anal. Chem., 1958, **30**, 1271) may be formed by the oxidation of thianthrene with chromic oxide in acetic acid, or with potassium dichromate in sulphuric acid.

A number of simple thianthrene derivatives are listed in Table 9.

In the last decade or so there has been interest in the chemistry of the **thianthrene (thianthrenium) radical cation.** *H. J. Shine* and co-workers (J. Amer. chem. Soc., 1962, **84**, 4798; J. org. Chem., 1964, **29**, 21) have observed the formation of the cation in concentrated sulphuric acid, and

TABLE 9

SOME SUBSTITUTED THIANTHRENES

Substituent	m.p. (°C)	Ref.
2,7-Me$_2$	116	1,2,3
2,3,7,8-Me$_4$	125–128	4
1,3,6,8-Me$_4$	118	1,2
1,4,6,9-Me$_4$	242	5
1-Cl	85–85.5	6
1-Br	145	6
1-I	187.5–188.5	7
2-Cl	84	8
2-Br	88–89	9
2-NO$_2$	128	10
1-NH$_2$	120–121	6,11
2-NH$_2$	160.5–161	9
1-OH	117–118	12
2-OH	145	13
2,7-(OH)$_2$	233	14
1-OMe	81–81.5	15
2,7-(OMe)$_2$	131	14

References

1   *P. Jacobson* and *E. Ney*, Ber., 1889, **22**, 904.
2   *Idem*, Ann., 1893, **277**, 232.
3   *I. N. Tits-Skvortsova et al.*, C.A., 1955, **49**, 921.
4   *A. F. Damanski* and *K. D. Kostić*, Bull. chim. Belgrade, 1947, **12**, 243; C.A., 1952, **46**, 5051.
5   *M. Sen* and *J. N. Ray*, J. chem. Soc., 1926, 1139.
6   *H. Gilman* and *D. R. Swayampati*, J. Amer. chem. Soc., 1957, **79**, 108.
7   *G. A. Martin Jr.*, Iowa State Coll. J., 1946, **21**, 38; C.A., 1947, **41**, 952.
8   *K. Fries* and *W. Vogt*, Ann., 1911, **381**, 312.
9   *Gilman* and *Swayampati*, J. Amer. chem. Soc., 1955, **77**, 5944.
10 *S. Krishna*, J. chem. Soc., 1923, **123**, 156.
11 *Gilman* and *C. G. Stuckwisch*, J. Amer. chem. Soc., 1943, **65**, 1461.
12 *Gilman* and *D. L. Esmay*, ibid., 1954, **76**, 5787.
13 *B. Pützer* and *F. Muth*, G.P. 606,350/1935; C.A., 1935, **29**, 1434; B.P. 427,816/1935; C.A., 1935, **29**, 6608.
14 *Fries* and *E. Engelbertz*, Ann., 1915, **407**, 194.
15 *Gilman* and *Swayampati*, J. Amer. chem. Soc., 1957, **79**, 991.

the e.s.r. spectrum in an aluminium chloride–nitromethane medium has been recorded by *Sullivan (loc. cit.)*. The perchlorate, an explosive solid, may be obtained either from thianthrene by treatment with perchloric acid in carbon tetrachloride (not chloroform as stated in the reference) (*E. A. C. Lucken*, J. chem. Soc., 1962, 4963; see also *Shine et al.*, J. Amer. chem. Soc.,

1969, **91,** 1872), or by the disproportionation of thianthrene and thianthrene-5-oxide in perchloric acid (*W. Rundell* and *K. S. Scheffler*, Tetrahedron Letters, 1963, 993). Besides the perchlorate, the pentachloroantimonate (*M. Kinoshita*, Bull. chem. Soc., Japan, 1962, **35,** 1137), the tetrafluoroborate and the trichlorodiiodide (*Shine et al.*, J. org. Chem., 1969, **34,** 3368; J. Amer. chem. Soc., 1972, **94,** 11026; Tetrahedron Letters, 1974, 99) are known.

The perchlorate when treated with dry ammonia affords LV (*idem.* J. org. Chem., 1969, *loc. cit.*), whereas primary alkyl amines yield *N*-protonated *N*-alkylsulphilimine perchlorates (*B. K. Bandlish et al., ibid.*, 1975, **40,** 2590):

β-Oxoalkylsulphonium perchlorates (LVI) are formed with methyl ketones (*K. Kim* and *Shine*, Tetrahedron Letters, 1974, 4413):

LVI (a) R = Me
(b) R = Bu^t
(c) R = Ph
(d) R = 2-naphthyl

In the presence of triethylamine, LVIa and LVIb yield the ylides LVIIa and LVIIb, respectively:

LVIIa, R = Me
LVIIb, R   Bu^t

Various amine or phenol derivatives react with the perchlorate affording *S*-aryl derivatives LVIII (*Shine et al.*, J. org. Chem., 1974, 2534, 2537):

X = NHAc, OH, OH, OH, NMe$_2$, NH$_2$
Y = H,   H, Cl, Bu^t, H,   H,
m.p. 136.7 256.5- 229- 205- 115- –
(°C)   257   230 206 116   –

3. Oxathianes and oxathiins

The title compounds contain one oxygen atom and one sulphur atom in a six-membered ring, they are thus mono thio analogues of the dioxanes and dioxiins, respectively.

(a) 1,2-Oxathianes and oxathiins

(i) 1,2-Oxathianes

Interest in 1,2-oxathianes centres almost entirely upon the 2,2-dioxides (δ-sultones), thus 1,2-oxathiane-2,2-dioxide (1,4-butane sultone) is conveniently prepared by treating the dimesylate of 1,4-butandiol with *n*-butyl-lithium in tetrahydrofuran (*T. Durst* and *K.-C. Tin*, Canad. J. Chem., 1970, **48**, 845):

It may also be prepared from 1-acetoxy-4-chlorobutane by the following sequence (*J. H. Helberger* and *H. Lantermann*, Ann., 1954, **586**, 158; *W. E. Truce* and *F. D. Hoerger*, J. Amer. chem. Soc., 1954, **76**, 5357):

$$Cl(CH_2)_4OCO{\cdot}CH_3 \xrightarrow{Na_2SO_3} NaO_3S(CH_2)_4OCO{\cdot}CH_3 \xrightarrow{HCl/MeOH} HO_3S(CH_2)_4OH \longrightarrow$$

Alkenes react with sulphur trioxide in dioxane at $-70°$ to form 4,6-disubstituted-1,2-oxathiane-2,2-dioxides (*F. G. Bordwell et al.*, ibid., 1959, **81**, 2000, 2002; Org. Synth., 1954, **34**, 85):

1,2-Oxathiane 2,2-dioxide has m.p. 15°, b.p. 153°/14 mm (*J. Willems*, Bull. Soc. chim. Belges, 1955, **80**, 5418). 4-*Methyl-*, b.p. 155–157°/15 mm, 6-*methyl-*, b.p. 121–123°/2 mm, 6-*propyl-*, b.p. 126°/0.4 mm, 4,4,6,6-*tetramethyl-*, m.p. 99–100°, 4,4,6,6-*tetraphenyl-1,2-oxathiane 2,2-dioxide*, m.p. 94–95°.

The parent compound is a colourless, odourless liquid, insoluble in cold, but partly soluble in hot water. It is miscible with most organic solvents, but not with carbon tetrachloride. It has a weak carcinogenic effect upon the rat (*H. Druckrey et al.*, Z. Krebsforsch., 1970, **75**, 69).

When heated with primary or secondary amines ring-fission occurs affording derivatives of 4-aminobutanesulphonic acid (*B. Helferich* and *V. Böllert*, Ann., 1961, **647**, 37):

$$\overset{RNHR^1}{\longrightarrow} \quad \overset{R}{\underset{R^1}{>}} N CH_2 \cdot CH_2 \cdot CH_2 \cdot CH_2 SO_3 H$$

6-Lithio derivatives are formed when δ-sultones are reacted with butyl-lithium; in these compounds, which exist in a chair conformation, the lithium atom shows a strong tendency to assume an equatorial orientation (*Durst*, Tetrahedron Letters, 1971, 4171). 1,2-**Oxathiane 2-oxide** (I), b.p. 58–64°/0.2 mm, is obtained, together with thiolane 1,1-dioxide (II), when 1,2-dithiane 2,2-dioxide is treated with tri(diethylamino)phosphorus (*D. N. Harpp, J. G. Gleason* and *D. K. Ash*, J. org. Chem., 1971, **36**, 322):

$$\xrightarrow{(Et_2N)_3P}$$

(90 %)   +   (10 %)
(I)   (II)

The monoxide I adopts a chair conformation with the oxygen substituent in an axial orientation (*Harpp* and *Gleason, ibid.*, p. 1314; *cf.* p. 390).

(ii) Dihydro 1,2-oxathiins and 1,2-oxathiins

3,6-Dihydro-1,2-oxathiin 2,2-dioxides are made by reacting 1,3-butadienes with sulphur trioxide in dioxane (*Bordwell et al.*, J. Amer. chem. Soc., 1959, **81**, 2002).

A similar $(4 + 2)\pi$ cycloaddition using sulphur dioxide does not occur because of the more favourable process leading to 2,5-dihydrothiophene-1,1-dioxides. Indeed, 3,6-dihydro-1,2-oxathiin 2-oxide, prepared as shown below, is thermally unstable and readily undergoes a retro Diels–Alder change (*F. Jung et al., ibid.*, 1974, **96**, 935).

$$\xrightarrow{0°} \quad \left[\right] \longrightarrow \quad + \; SO_2$$

K. S. Dhami (Chem. and Ind., 1968, 1004) claims that 3,6-dihydro-4-methyloxathiin 2-oxide is obtained by the treatment of 4-chloro-4-methyl-oxathiane 2-oxide with diethylamine. The last compound arises through the action of thionyl chloride and pyridine upon 4-hydroxy-2-methyl-1-butene:

Some 1,2-oxathiin-2,2-dioxides may be synthesised by treatment of unsaturated aliphatic ketones containing either of the units III or IV, with acetic anhydride/sulphuric acid.

$$-CH_2-\underset{\underset{Alkyl}{|}}{C}=C-\overset{\overset{O}{\|}}{C}-Alkyl \qquad -CH=\underset{\underset{Alkyl}{|}}{C}-CH-\overset{\overset{O}{\|}}{C}-Alkyl$$

(III)   (IV)

Thus mesityl oxide gives 4,6-dimethyl-1,2-oxathiin 2,2-dioxide, but 3-penten-2-one, 3-decen-2-one and 3-methyl-3-penten-2-one fail to give the appropriate cyclisation products. Physical constants for a number of substituted 1,2-oxathiin 2,2-dioxides are summarised in Table 10.

TABLE 10

SOME 1,2-OXATHIIN 2,2-DIOXIDES

Substituents	m.p. (°C)	b.p. (°C/mm)	Ref.
4,5-Me$_2$-	85.6		1
4,6-Me$_2$-	70.5–71		1,2
6-Et-4-Me-		136–138/3.5	1
4-Me-6-Ph-	83.4		1
3,4,6-Me$_3$-	68.5–69		1
4-Et-3,6-Me$_2$-		152–156/11	1
4,6-Er$_2$-3,5-Me$_2$-		135–139/3	1

References

1   *T. Morel* and *P. E. Verkade*, Rec. Trav. chim., 1949, **68**, 619.
2   *R. H. Eastman* and *D. Gallup*, J. Amer. chem. Soc., 1948, **70**, 864.

Eastman and *Gallup* obtained from 4,6-dimethyl-1,2-oxathiin 2,2-dioxide, a monobromo derivative in which the bromine atom was regarded as being attached to position 5 of the ring. In fact, however, the product is *3-bromo-4,6-dimethyl-1,2-oxathiin 2,2-dioxide*, m.p. 77–78° (*W. E. Barnett* and *J. McCormack*, Tetrahedron Letters, 1969, 651). *5-Bromo-4,6-dimethyl-1,2-oxathiin 2,2-dioxide* has m.p. 62–63°, and is prepared unambiguously as follows:

(*b*) *1,3-Oxathianes*

Until quite recently few simple 1,3-oxathianes were known, although *B. Sjoberg* (Ber., 1942, **75**, 13) had prepared 2,2-dimethyl-1,3-oxathiane by heating 3-mercaptopropanol and acetone with phosphorus pentoxide three decades ago:

$$\text{(OH, SH)} \xrightarrow[\text{P}_2\text{O}_5]{\text{MeCOMe}} \text{(O, S)}\text{CMe}_2$$

The reaction was modified by *C. Djerassi* and *M. Gorman* (J. Amer. chem. Soc., 1953, **75**, 3704; 1958, **80**, 4723) by replacing phosphorus pentoxide

TABLE 11

SOME 1,3-OXATHIANES

Substituent	b.p. (°C/mm)	n_D^{20}
None	48/11	1.5026
6-Me	52–54/13	1.4920
2-Me	66–70/24	1.4922
2-Et	73–75/19	1.4856
2-Pr$^\beta$	75–78/16	1.4853
2,2-Me$_2$	65–68/17	1.4890
cis-2,6-Me$_2$	54–56/12	1.4827
cis-2-Et-6-Me	66–69/20	1.4771
cis-6-Me-2-Pr$^\beta$	69–72/13	1.4747
2-Et-2-Me	61–63/10	1.4936
2-Me-2-Pr$^\beta$	64–67/10	1.4923
2-Bu^t-2-Me	94–97/27	1.4942
2,2,6-Me$_3$	55–57/12	1.4789
2-Et-2,6-Me$_2$[a]	61–64/12	1.4812
trans-2-Et-2,6-Me$_2$[a]		1.4804
cis-2-Et-2,6-Me$_2$[a]		1.4827
2,6-Me$_2$-2-Pr$^\beta$ (*trans*)	68–70/12	1.4819
2-Bu^t-2,6-Me$_2$ (*trans*)	68–70/11	1.4827
6-Me-2-Ph[b]	111–113/0.45	1.5673
2-Bu^t-6-Me[b]	46–48/0.05	—
2-Bu^t-4-Ph[b]	m.p. 45.8	—
4-Ph[b]	m.p. 51–53	—
2,4-Ph$_2$[b]	m.p. 83.5	—

[a] mixture of stereoisomers.
[b] *Y. Allingham et al.* (Org. Mag. Resonance, 1971, **3**, 37).

with toluene-4-sulphonic acid and distilling off the water produced as the benzene azeotrope.

K. Pihlaja and *P. Pasanen* (Acta Chem. Scand., 1970, **24**, 2257) employ dichloromethane as solvent and have extended the reaction to aldehydes, as well as ketones (see Table 11).

Primary interest in 1,3-oxathianes at this time focusses upon their stereo-chemistry and many papers on this subject have been published, see for example *N. de Wolf* and *H. R. Buys* (Tetrahedron Letters, 1970, 551), *M. Anteunis et al.* (Tetrahedron, 1971, **27**, 1917), *Allingham et al.* (Org. Mag. Resonance, 1971, **3**, 37), *Pasanen* (Suom. Kemistilehti B., 1972, **45**, 363; C.A., 1973, **78**, 83682h), *Pasanen* and *Pihlaja* (Tetrahedron, 1972, **28**, 2617).

N. de Wolf et al. (Acta Crystallogr., Sect. B, 1972, **28**, 2424) have examined the crystal structure of 2-(4-nitrophenyl)-1,3-oxathiane and shown that the aryl substituent occupies an equatorial position with respect to a chair-shaped ring; but an analysis of the thermal motions point to positional disorder of several atoms in the ring and in the nitro group, indicating that the observed result is an average of at least two conformations.

The mass spectra of 1,3-oxathianes have also been analysed (*Pihlaja* and *Pasanen*, Org. Mass Spectrom., 1971, **5**, 763; *J. H. Bowie* and *A. C. Ho*, Austral. J. Chem., 1973, **26**, 2009).

2-Oxo-1,3-oxathiane, b.p. 134°/0.001 mm, is obtained by the reaction of ethyl carbonate and thallic nitrate upon 3-mercapto propanol (*D. L. Johnson*, Fr. P. 1,445,849/1966; C.A., 1967, **66**, 55065).

(c) *1,4-Oxathianes, 2,4-dihydro-1,4-oxathiins and 1,4-oxathins*

(i) *1,4-Oxathianes*

1,4-Oxathiane (thioxane, *p*-oxathiane) was prepared for the first time by *H. T. Clarke* (J. chem. Soc., 1912, 1788) by heating bis(2-iodoethyl) ether and alcoholic potassium sulphide. Since then, many other methods have been described and thioxane is commercially available in quantity. One process employs the acid-catalysed cyclisation of bis(2-hydroxyethyl) sulphide:

Yields ranging from 80–86% are claimed by reacting bis(β-chloroethyl) ether and sodium sulphide nonahydrate under reflux in 30–70% aqueous diethylene glycol for 28–34 hours (*J. M. Whisenhunt*, U.S.P. 2,894,956/1959):

The condensation of 2-mercaptoethanol and 1,2-ethanedithiol in the presence of silicone rubber or grease is also recommended (*C. H. Mihm*, U.S.P. 3,578,681/1968; C.A., 1971, **75**, 36064m):

Substituted 1,4-oxathianes may be obtained by adaptations of the above synthesis, thus *M. Hunt* and *C. S. Marvel* (J. Amer. chem. Soc., 1935, **57**, 1691) prepared 2,6-dimethyl-1,4-oxathiane (see Table 12) from bis(2-chloropropyl) ether and sodium sulphide in alcohol under reflux. *Clarke* and *Smiles* (J. chem. Soc., 1909, **95**, 992) record the synthesis of 2,6-di-ethoxy-1,4-oxathiane from di(2,2-diethoxyethyl) sulphide in dry ether or ethanol by the action of dry hydrogen chloride. The same product results from ethyl vinyl ether and sulphur monochloride in carbon tetrachloride at $-20°$, followed by the addition of water:

Some other derivatives can be prepared by the following route which appears to offer general applicability (*W. E. Parnham*, J. Amer. chem. Soc., 1947, **69**, 2449):

Physical constants for a number of representative 1,4-oxathianes are summarized in Table 12.

TABLE 12

SOME 1,4-OXATHIANES

Substituent	m.p. (°C)	b.p. (°C/mm)	n_D/°C	Ref.
None		145.5–146.4/740	1.5072/20	1,2
2-OMe		57/5	1.4911/23	3
2-OEt		81–82/4	1.4840/20	4
2,6-Me₂		162/760	1.4733/20	5
3,5-Me₂		113–114/160	1.4850/20	6,7
2,3-Ph₂	95–95.5		—	8
2,6-(OEt)₂	101		—	9

References

1 *P. A. Akishin et al.*, C.A., 1954, **48**, 10436.
2 *H. T. Clarke*, J. chem. Soc., 1912, **101**, 1788.
3 *W. E. Parham, I. Gordon* and *J. D. Swalen*, J. Amer. chem. Soc., 1952, **74**, 1824.
4 *Parham*, ibid., 1947, **69**, 2449.
5 *M. Hunt* and *C. S. Marvel*, J. Amer. chem. Soc., 1935, **57**, 1691.
6 *D. Harman* and *W. E. Vaughan*, ibid., 1950, **72**, 631.
7 *Harman* and *Vaughan*, U.S.P. 2,562,145/1951.
8 *A. H. Haubein*, J. Amer. chem. Soc., 1959, **81**, 144.
9 *Clarke* and *S. Smiles*, J. chem. Soc., 1909, **95**, 992.

Alkyl halides react with 1,4-oxathianes to form **oxathianium salts,** thus with ethyl bromide 4-*ethyloxathianium bromide*, m.p. 85°, is produced (*E. Fromm* and *B. Ungar*, Ber., 1923, **56B**, 2286):

1,4-Oxathiane in dry carbon tetrachloride solution, at $-20°$, reacts with chlorine to afford the 3-chloro derivative; attempts to isolate this product are prevented because it easily dehydrohalogenates to 2,3-dihydro-1,4-oxathiin, but when chlorine is passed into the solution at 30°, a mixture containing 40% of trans-2,3-*dichloro*-1,4-*oxathiane*, m.p. 37–38°, and 30% of 3,3-*dichloro*-1,4-*oxathianes* is obtained. Other products are the cis-3,5-*deriv.* and 2,3,3-*trichloro*-1,4-*oxathiane*, m.p. 53–54° (*N. de Wolff, P. W. Henniger* and *E. Havinga*, Rec. Trav. chim., 1967, **86**, 1227).

The fluorination of 1,4-oxathiane over potassium tetrafluoro-cobaltate is similarly complex (*J. Burdon* and *I. W. Parsons*, Tetrahedron, 1971, **27**, 4533). Electrofluorination in anhydrous hydrogen fluoride gives **perfluoro**-1,4-**oxathiane** (*R. D. Dresdner* and *J. A. Young*, J. Amer. chem. Soc., 1959, **81**, 574):

In common with the 1,3-isomers, conformational aspects of 1,4-oxathiane chemistry have been carefully analysed by ^{1}H n.m.r. studies (*De Wolff, Henniger* and *Havinga, loc. cit.*). Normally they exist in a conformational equilibrium with two chair forms predominating (see also *N. S. Zefirov et al.*, Tetrahedron, 1971, **27**, 3111).

The mass spectrometric fragmentation patterns of 1,4-oxathianes are very similar to those of 1,4-dithianes, but different to 1,4-dioxanes (*G. Conde Caprace* and *J. E. Collin*, Org. Mass Spectrom., 1969, **2**, 1277; see p. 402).

1,4-Oxathiane 4,4-dioxide (sulphone) is formed by the action of aqueous sodium hydroxide upon bis(2-chloroethyl) sulphone. It is described as an extremely stable solid, m.p. 129°, soluble in water, alcohol and acetone, but not in ether. The 4-**monoxide** is obtained similarly from bis(2-chloroethyl) sulphoxide (*A. E. Cashmore*, J. chem. Soc., 1923, 1738).

2,6-*Dihydroxy*-1,4-*oxathiane*, m.p. 73°, is a cyclic form of thiodiacetaldehyde in solution (*R. D. Coghill*, J. Amer. chem. Soc., 1952, **74**, 1824).

2,6-**Dioxo**-1,4-**oxathianes** can be prepared by heating thiodiacetic acids and acetic anhydride (*L. C. Lappas* and *G. L. Jenkins*, J. Amer. pharm. Assoc. Sci. Ed., 1952, **41**, 257):

$$S(CHR \cdot CO_2H)_2 \xrightarrow[\Delta]{Ac_2O}$$

R = H   b.p. 158–159°/12 mm
R = Me b.p. 133–137°/14 mm
R = Et b.p. 149–150°/15 mm

3,3-*Diphenyl*-1,4-*oxathiane*-2,6-*dione*, m.p. 108–109°, was synthesised as follows (*G. S. Skinner* and *J. B. Bicking*, J. Amer. chem. Soc., 1954, **76**, 2776):

(ii) Dihydro-1,4-oxathiins and 1,4-oxathiins

2,3-Dihydro-1,4-oxathiins (V) (see Table 13) are available through the acid-catalysed cyclisation of α-(2′-hydroxyethylthioketones (*F. Asinger et al.*, Ann., 1971, **753**, 151; see also *H. Schubert, H.-J. Dietz* and *P. Göhmann*, Z. Chem., 1976, 272):

(V)

The parent compound, 2,3-**dihydro**-1,4-**oxathiin** was first obtained by *W. E. Parham, L. Gordon* and *J. D. Swalen* (J. Amer. chem. Soc., 1952, **74**, 1824) by heating 3-methoxy-1,4-oxathiane with phosphorus pentoxide at 160°; it has b.p. 47–48°/19 mm, $n/°C$ 1.5209/20, and slowly polymerises upon standing to a colourless product, m.p. 200–220°; it is stable, however, if stored free from acids or peroxides:

Methods to this system, potentially of general application, have been described by *J. R. Marshall* and *H. A. Stevenson* (J. chem. Soc., 1959, 2360). Thus 2,3-*dihydro*-5,6-*diphenyl*-1,4-*oxathiin* (VI), m.p. 63–65°, has been prepared by the following sequence:

TABLE 13

2,3-DIHYDRO-1,4-OXATHIINS (V)

R	R^1	R^2	R^3	b.p. ($°C/mm$)	n_D^{20}
H	H	Me	Et	44/1.5	1.5132
H	H	Me	Ph	96/0.3	1.6152
Me	H	Me	Et	52/2.5	1.4990
Me	H	Me	Ph	97–98/0.4	1.5955
-(CH$_2$)$_4$-			-(CH$_2$)$_4$-	100.5/0.5	—
H	H		-(CH$_2$)$_3$-	64.5/1.5	1.5518
H	H		(-CH$_2$)$_4$-	73–74/1.4	1.5528
Me	H		-(CH$_2$)$_3$-	62.5/1.5	—
Me	H		-(CH$_2$)$_4$-	70.5/1.0	1.5360
CH$_2$Cl	H	Me	Et	60–61/0.1	1.5139
CH$_2$Cl	H	Me	Ph	118–119/0.2	1.6055
CH$_2$Cl	H		-(CH$_2$)$_3$-	74/0.3	1.5525
CH$_2$Cl	H		-(CH$_2$)$_4$-	90/0.5	1.5530
-(CH$_2$)$_4$-		Me	Et	75/0.4	1.5200
-(CH$_2$)$_4$-		Me	Ph	130–132/0.2	—
CH$_2$OH	H	Me	Et	57–58/0.3	1.5020
CH$_2$OMe	H	Me	Et	73.4/0.7	1.4989
CH$_2$OPh	H	Me	Et	127–128/0.3	1.5560
CH$_2$SEt	H	Me	Et	90/0.9	1.5320
CH$_2$NEt$_2$	H	Me	Et	121/0.2	1.4850
CO$_2$Et	Me	Me	Et	84/0.2	1.4912
-(CH$_2$)$_3$-		Me	Et	58–59/0.3	1.5250
H	H	H	Ph	126/0.7	—
H	H	H	C$_6$H$_4$Br[4]	m.p. 95–97	—
II	H	Ph	Ph	131/0.35; m.p. 63–65	—
H	H	H	C$_6$H$_4$NO$_2$[3]	m.p. 53–57	—
H	H	H	C$_6$H$_4$NO$_2$[4]	m.p. 143–144	—
H	H	H	C$_6$H$_4$Cl[4]	m.p. 70–72	—

while 2,3-*dihydro-6-methyl-1,4-oxathiin* (VII), b.p. 56°/10 mm, is obtained from 2-hydroxyethanethiol and chloroacetone by heating in methanolic sodium hydroxide solution:

Although simple 2,3-dihydro-1,4-oxathiins may be very sensitive to hydrogen peroxide, 5-carbamoyl derivatives are not and such treatment affords sulphoxides and sulphones, many of which are systemic fungicides and bacteriocides. *M. Kulka, D. S. Thiara* and *W. A. Harrison* (Fr. P. 1,477,062/ 1967; C.A., 1968, **68**, P95830m; Neth. Appl. 6,605,527/1967; C.A., 1968, **68**, 78294x) describe nearly 100 2,3-dihydro-5-carbamoyl-6-methyl-1,4-oxathiin sulphones and sulphoxides of the type:

in which R is most commonly a substituted phenyl group and R^1, hydrogen.

(d) Benzoxathiins

Few examples of this class are known and most representatives are dihydro derivatives (sometimes, but incorrectly, called benzoxathianes) of the parent ring systems.

(i) 1,2-Benzoxathiin derivatives

3,4-**Dihydro**-1,2-**benzoxathiin** 2-oxide (VIII), b.p. 120–140°/0.05 mm, is obtained by the chlorination of 2-(2′-hydroxyphenyl)ethanethiol in acetic acid at room temperature:

(VIII)

In hydrolytic solvents (aqueous acetic acid, methanol, tertiary butanol) chlorination of the thiol gives 6-*chloro*-, m.p. 133–135°, 8-*chloro*-, m.p. 153–155°, and 6,8-*dichloro*-3,4-*dihydro*-1,2-*benzoxathiin* 2,2-dioxides (*E. N. Givens* and *L. E. Hamilton*, J. org. Chem., 1967, **32**, 2857). The structure of VIII was proved by its oxidation using hydrogen peroxide to the 2,2-*dioxide* IX, m.p. 109–112°; this sultone was prepared originally by *W. E. Truce* and *F. D. Hoerger* (J. Amer. chem. Soc., 1954, **76**, 5357) *via* ring-closure of 2(2′-hydroxyphenyl)ethanesulphonic acid:

(IX)

Molecular parameters for the sultone are summarized in formula X (*E. B. Fleischer et al.,* Chem. Comm., 1967, 197):

(X)

4-*Methyl*-1,2-*benzoxathiin*-2,2-*dioxide* (XI), m.p. 86–87°, is synthesised in high yield by treating 2-acetylphenylmethanesulphonate with potassium hydroxide in pyridine (*E. M. Philbin et al.,* J. chem. Soc., 1956, 4414):

(XI)

In the mass spectrometer, 1,2-benzoxathiin-2,2-dioxides undergo expulsion of sulphur dioxide and the elimination of sulphene (*J. M. Clancy, A. G. Esmonde* and *R. F. Timoney,* Int. J. Sulphur Chemistry, Part A, 1972, **2**, 249).

(ii) 2,1-Benzoxathiin derivatives

$3H$-2,1-Benzoxathiole 1,1-dioxide (XII) when treated with sodium cyanide gives sodium phenylacetonitrile-2-sulphonate (IX) which with hydrochloric acid and subsequent treatment with thionyl chloride affords 3,4-*dihydro*-2,1-*benzoxathiin*-3-*one* 1,1-*dioxide* (XIV), m.p. 107,5–108.5°, a rare example of this class of ring system (*J. F. King et al.,* Canad. J. Chem., 1971, **49**, 936):

(XII)   (XIII)   (XIV)

(iii) 2,3-Benzoxathiin derivatives

Cathodic reduction of sulphur dioxide in aprotic solvents affords an anion radical which combines with *o*-xylylene dibromide to give 1,4-**dihydro**-2,3-**benzoxathiin 3-oxide.** This readily undergoes oxidation to give the 3,3-**dioxide** (*D. Knuttel* and *B. Kastening,* J. appl. Electrochem., 1973, **3**, 291):

(iv) Dihydro-1,3-benzoxathiins

2,4-Dihydro-1,3-benzoxathiin (1,3-benzoxathiane) (XV), is formed in low yield when phenol is reacted with dimethyl sulphoxide, dicyclohexylcarbodiimide and phosphoric acid (*K. E. Pfitzner, J. P. Marino* and *R. A. Olofson,* J. Amer. chem. Soc., 1965, **87,** 4658). The *naphto analogue* XVI, m.p. 65–66°, is similarly prepared from 2-naphthol:

(XV)   (XVI)   (XVII)

In a similar way, *J. Burdon* and *J. G. Moffat* (*ibid.,* 1966, **88,** 5855) obtained 2,4-*dihydro-6-nitro-1,3-benzoxathiin,* m.p. 135–136°, from 4-nitrophenol and 2,4-*dihydro-8-methyl-1,3-benzoxathiin,* b.p. 60°/0.001 mm, from 2-cresol. 1-Naphthol affords XVII, m.p. 60–62°. Probable mechanisms for these and related reactions are discussed in a later paper (*idem, ibid.,* 1967, **89,** 4725).

In a similar reaction XI is obtained from phenol by the action of dimethyl sulphoxide and phosphorus pentoxide, the yield was 14% and the product has b.p. 104–105°/5 mm (*Y. Hayashi* and *R. Oda,* J. org. Chem., 1967, **32,** 457). 4-Nitrophenol gives 2,4-*dihydro-6-nitro-1,3-benzoxathiin,* m.p. 136–137°.

(v) 2,4-Dihydro-3,1-benzoxathiin derivatives

2-Mercaptobenzoic acid when heated with benzal diacetate in the presence of sulphuric acid affords 4-*oxo-2-phenyl-2,4-dihydro-3,1-benzoxathiin* (XVIII), m.p. 90° (*D. T. Mowry et al.,* J. Amer. chem. Soc., 1947, **69,** 2358):

(XVIII)   (XIX)

(XX)   +   (XXI)

Low temperature photolysis of XVIII affords 2-oxobenzothiethane (XIX), which at −40° decomposes to XX (*O. L. Chapman* and *C. L. McIntosh,* ibid., 1970, **92,** 7007) not XXI, as was suggested by *A. O. Peterson et al.* (Tetrahedron, 1970, **26,** 1157).

When the diazonium salt of anthranilic acid is allowed to react with thiobenzophenone in the presence of propylene oxide, 2,2-*diphenyl-4-oxo-*

2,4-*dihydro*-3,1-*benzoxathiin* (XXII), m.p. 185°, is formed (*D. C. Dittmer* and *E. S. Whitman*, J. org. Chem., 1969, **34**, 2004).

This product on treatment with sodium hydroxide gives the dianion XXIII, which on oxidation and protonation yields 2,2′-dicarboxydiphenyl disulphide:

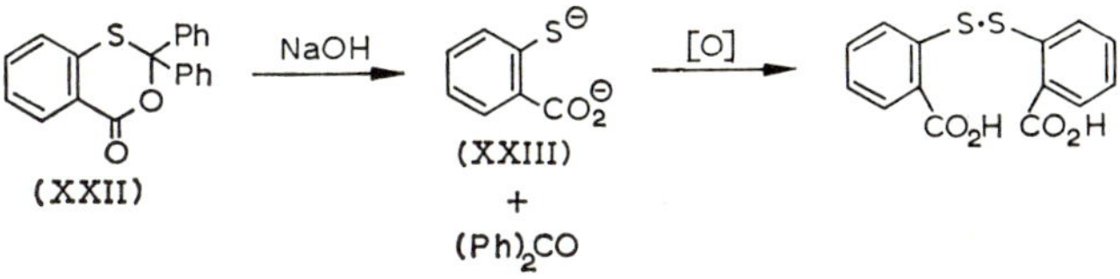

(vi) 1,4-Benzoxathiin

1,4-**Benzoxathiin** (XXIV), b.p. 49°/0.08 mm, is synthesised, in low yield, from 2-*ethoxy*-2,3-*dihydro*-1,4-*benzoxathiin*, b.p. 129–136°/9 mm, by heating with phosphorus pentoxide (*Parham* and *J. D. Jones*, J. Amer. chem. Soc., 1954, **76**, 1068):

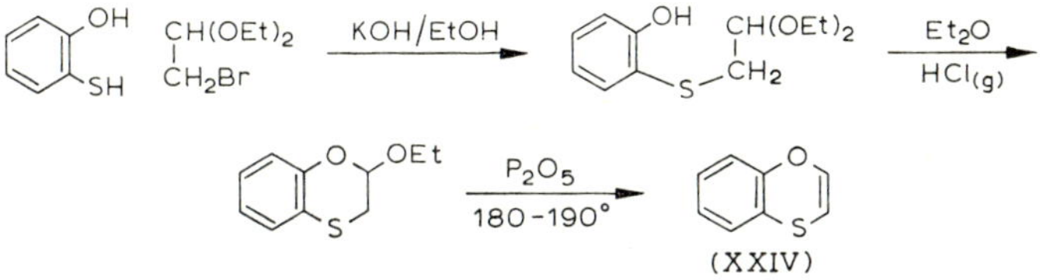

It is more advantageous, however, to convert the ether (XXV, R = OEt) into the corresponding alcohol (XXV, R = OH), which as the acetoxy derivative (XXV, R = Ac) affords 1,4-benzoxathiin in 76% yield on pyrolysis:

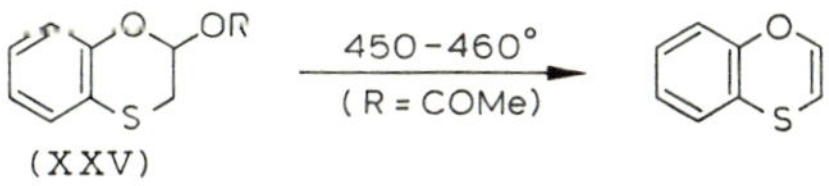

Another route (*J. H. Verheijen* and *H. Kloosterziel*, Synthesis, 1975, 45) employs the following sequence:

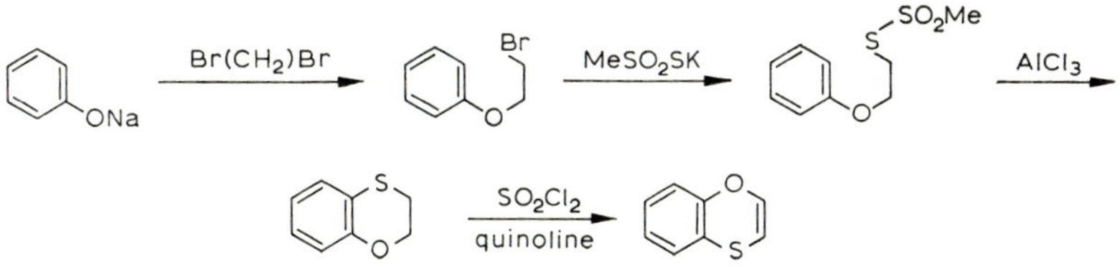

This method is claimed to be more efficient than the earlier approaches and, using benzenethiol in place of phenol, can be adapted to the synthesis of benzo-1,4-dithiin (p. 406).

Marshall and *Stevenson (loc. cit.)* have employed the following route to obtain 2,3-*diphenyl*-1,4-*benzoxathiin*, m.p. 108–109°:

1,4-Benzoxathiin is polymerized on exposure to air (*Parham* and *Jones, loc. cit.*), but, with hydrogen peroxide in glacial acetic acid at reflux, the 4,4-*dioxide*, m.p. 154–155°, is produced. Bromine in carbon tetrachloride at 0° affords 2,3-*dibromo-2,3-dihydro*-1,4-*benzoxathiin*, m.p. 105–106°.

2,3-**Dihydro**-1,4-**benzoxathiin** (XXVI), b.p. 90°/2 mm, is formed from 2-mercaptophenol and 1,2-dibromoethane in the presence of sodium ethoxide (*D. Greenwood* and *Stevenson*, J. chem. Soc., 1953, 1514):

(XXVI)

Dilute potassium permanganate at 70° oxidises XXVI to the 4,4-*dioxide*, m.p. 82°, whereas hydrogen peroxide at room temperature in acetic acid solution affords the 4-*monoxide*, m.p. 85°.

(vii) Phenoxathiin (phenothioxin or dibenzo-1,4-oxathiin)

Phenoxathiin (XXVII) colourless needles, m.p. 57–58° (ethanol) (*C. M. Suter* and *F. O. Green*, Org. Synth., 1938, **18**, 64) can be obtained by the reaction of diphenyl ether with sulphur in the presence of aluminium chloride (*F. Ackerman*, G.P. 234,743/1910; Chem. Zentr., 1911, **2**, 1768; see also *Suter* and *C. E. Maxwell*, Org. Synth., Coll. Vol. **2**, 1943, p. 485):

(XXVII)

The 10-*monoxide*, m.p. 152–153°, may be prepared by oxidation of phenoxathiin with 30% hydrogen peroxide in ethanol at reflux for 12 h; but, if the solvent used is glacial acetic acid and the reflux time reduced to 1 h, after standing overnight at room temperature, the 10,10-*dioxide*, m.p. 147–148°, is obtained (*H. Gilman* and *D. R. Swayampati*, J. Amer. chem. Soc., 1956, **78**, 2163).

Phenoxathiin and some of its derivatives show insecticidal, bacteriocidal and anthelmintic activities; in view of this, synthetic work in this area has been extensive and many analogues are known (for a review see *D. S. Breslow* and *H. Skolnik*, in "The Chemistry of heterocyclic Compounds", Vol. 21, Part II, Chap. 12, pp. 868–951, ed. *A. Weissberger*, Interscience, New York, 1966).

4. Miscellaneous compounds containing selenium or tellurium atoms

1,4-Oxaselenane (**1,4-selenoxane**) results from the action of sodium selenide on β,β'-dichlorodiethyl ether in boiling water, being conveniently isolated as the 4,4-*dibromide*, m.p. 132°, which is readily reduced by sodium bisulphite. It is a liquid of penetrating odour, b.p. 167.5–168.5°/763 mm, m.p. $-21.5°$ (*C. S. Gibson* and *J. D. A. Johnson*, J. chem. Soc., 1931, 266).

1,4-**Oxatellurane** is a foul smelling yellow solid, m.p. 6°; it may be prepared in a similar manner to 1,4-oxaselenane using sodium telluride in place of sodium selenide (*W. V. Farrar* and *J. M. Gulland, ibid.*, 1945, 11):

$$\text{ClCH}_2\text{CH}_2\text{—O—CH}_2\text{CH}_2\text{Cl} + \text{Na}_2\text{X} \longrightarrow \quad X = \text{Se or Te}$$

Phenoxaselenin, m.p. 88°, is obtained from **phenoxatellurin,** m.p. 78–79°, by heating with selenium (*H. D. K. Drew, ibid.*, 1928, 511). The latter compound may be formed by heating tellurium tetrachloride with diphenyl ether; this affords *phenoxatellurin-*10,10-*dichloride*, m.p. 265°, which may be reduced to phenoxatellurin (*idem, ibid.*, 1926, 223). The reaction proceeds *via* the intermediate phenoxyphenyltelluritrichloride (I):

2-Halogeno derivatives of phenoxatellurin are formed in a similar way from 4-halogenophenyl phenyl ethers (*A. Gioaba et al.*, Rev. Roum. Chim., 1976, **21**, 739). The mass spectrum of phenoxaselenin has been interpreted (*I. C. Calder et al.*, Org. Mass Spectrum, 1970, **4**, 121) and it, and related compounds, form isolable complexes with trinitrobenzene (*B. Hetnarski* and *A. Grabowski*, Bull. Acad. Pol. Sci. Ser. Sci. Chim., 1969, **17**, 333; C.A., 1970, **72**, 78957y).

Phenothiatellurin (III), light yellow needles, m.p. 123–124°, is obtained by reducing *phenothiatellurin dichloride* (II), m.p. 264–266° (decomp.) with either sodium sulphide (*N. Patragnani*, Tetrahedron, 1960, **11,** 15) or sodium bisulphite (*A. Gioaba* and *O. Maior*, Rev. Roum. Chim., 1970, **15,** 1967).

The starting material is prepared by heating tellurium tetrachloride and diphenyl sulphide at 190–200°:

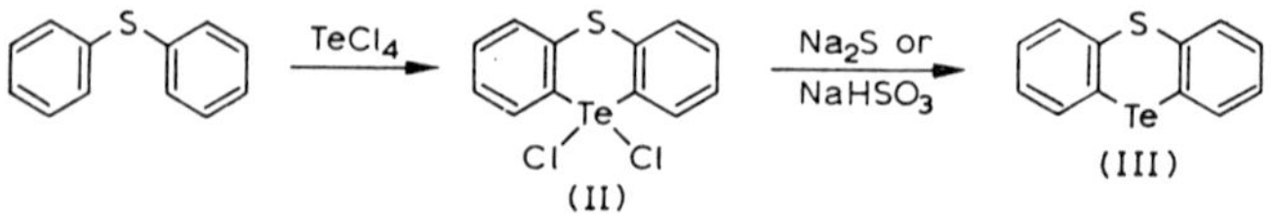

Chapter 41

Compounds Containing a Six-Membered Ring with Two Hetero Atoms from Groups V and VI, respectively of the Periodic Table. Oxazines, Thiazines and their Analogues

MALCOLM SAINSBURY

1. Oxazines

Nomenclature. Oxazines are six-membered monocyclic structures of the general molecular formula C_4H_5NO. Depending upon the relative orientation of the oxygen and nitrogen atoms and the disposition of the double bonds in the heterocycle nine isomeric forms are possible; these are illustrated below:

2*H*-1,2-Oxazine   4*H*-1,2-Oxazine   6 *H*-1,2-Oxazine

2*H*-1,3-Oxazine   4 *H*-1,3-Oxazine   6*H*-1,3-Oxazine

2*H*-1,4-Oxazine   4*H*-1,4-Oxazine   6*H*-1,4-Oxazine

Of these possible structures some are known, but others are recognised only as derivatives, most often the dihydro- or tetrahydro- compounds.

(*a*) *1,2-Oxazines*

(*i*) *6H-1,2-Oxazines*

A few aryl-substituted 6*H*-1,2-oxazines have been prepared; thus, for example, aryl methyl ketoximes may be converted by the action of ethylmagnesium bromide into the corresponding 3,5-diaryl-6*H*-1,2-oxazines (*L. W. Deady*, Tetrahedron, 1967, **23**, 3505):

In the presence of concentrated halogen acids these products rearrange to form 2,4-diarylpyrroles.

5-Hydroxy-6-methoxy-3,4,6-triphenyl-6H-1,2,-oxazine (I), m.p. 173°, with reddening and decomposition, gives an O-*benzoate*, m.p. 163°, and an N-*methyl* derivative, m.p. 147°. It is formed by the action of a weak acid upon the monoxime of 1,3,4-triphenylbutane-1,2,4-trione:

On boiling with water the oxime is regenerated.

With phenylmagnesium bromide, the methoxyl group is replaced and the resulting 5-*hydroxy*-3,4,6,6-*tetraphenyl-6H*-1,2-*oxazine*, m.p. 220° (decomp.), may be alkylated, e.g. with ethyl bromide, to give 2-*ethyl-5-oxo*-3,4,6,6-*tetraphenyl-5,6-dihydro-6H*-1,2-*oxazine* (II), m.p. 167–168° (*E. P. Kohler*, J. Amer. chem. Soc., 1926, **48**, 754).

3,6-**Diphenyl-6H-1,2-oxazine**, m.p. 213–214°, is claimed as the product obtained by refluxing hydroxylamine and 1,4-diphenyl-3-nitro-3-buten-1-one in pyridine–ethanol. It decolourizes potassium permanganate solution and reacts with bromine by substitution (*L. I. Smith* and *R. E. Kelly, ibid.*, 1952, **74**, 3300):

6-Hydroxy-3,5,6-triphenyl-6H-1,2-oxazine (III, R = H), m.p. 159–160°, from hydroxylamine and *cis*-1,2-dibenzoyl-1-phenylethene forms an O-methyl derivative (III, R = Me) which may also be obtained by ring expansion from the nitrone IV on pyrolysis (*A. H. Blatt*, J. org. Chem., 1938, **3**, 91; 1950, **15**, 869):

The oxime of mucobromic acid on heating gives 4,5-**dibromo-6-oxo-1,2-oxazine** (V, R = Br), m.p. 125° (*A. Bistrzycki* and *H. Simonis*, Ber., 1899, **32**, 536), and the corresponding 4,5-*dichloro-6-oxo-1,2-oxazine* (V, R = Cl), m.p. 77°. Some monohalogeno compounds have been prepared in similar manner from appropriate starting materials. On heating, these rearrange to the corresponding imides and, on hydrolysis, afford mono ammonium salts of substituted maleic acids (*H. B. Hill* and *E. T. Allen*, Amer. Chem. J., 1897, **19**, 650):

3,5-Dimethyl-6-oxo-1,2-oxazine (VI), m.p. 109–110°, may be prepared in the same way, and also from 2,4-dimethylpyrrole by nitrosation and rearrangement (*T. Ajello* and *S. Cusmano*, Gazz., 1940, **70**, 512, 755):

Some other derivatives include the corresponding 3,5-*diphenyl-*, m.p. 153° (*Hill* and *Allen*, *loc. cit.*) and 3-(4-*chlorophenyl*)-5-*phenyl*-6-*oxo*-1,2-*oxazine*, m.p. 164° (*Kohler* and *J. B. Shohan*, J. Amer. chem. Soc., 1926, **48**, 2425); the latter is prepared from nitromethane and 1-(4-chlorophenyl)-3-phenylprop-2-en-1-one, followed by bromination and treatment of the product with potassium acetate in methanol.

3,5-**Diphenyl-6-oxo-1,2-oxazine** with Meerwein's reagent (Et$_3$O$^{\oplus}$B$^{\ominus}$F$_4$) in dichloromethane, followed by the addition of ferric chloride and concentrated hydrochloric acid, gives the 1,2-oxazin-1-ium salt (R = Et) (*O. P. Shelyapin, I. V. Samartseva* and *L. A. Pavolova*, Zhur. org. Khim., 1973, **9**, 1987):

When treated with dimethyl sulphate the corresponding methyl ether (R = Me) is formed.

(*ii*) 2H- *and* 4H-*1,2-oxazines*

Rather fewer representatives of these structures have been described in the chemical literature; 3-(4-chlorophenyl)-4-hydroxy-6-methoxy-5-phenyl-4*H*-1,2-oxazine is an example, prepared as shown:

2-(4-**Dimethylaminophenyl**)-3,4,5,6-**tetraphenyl-2***H***-1,2-oxazine** (VIII), m.p. 212–213°, is obtained from tetracyclone and 4-nitrosodimethylaniline (*W. Dilthey et al.*, J. pr. Chem., 1940, [ii], **156**, 27):

(iii) Dihydro- and tetrahydro-1,2-oxazines

3,6-**Dihydro**-2*H*-1,2-**oxazine** (IX) can be prepared by the action of aqueous alkali upon the adduct formed by the cycloaddition of 2-chloro-2-nitroso-propane and butadiene (*Yu. A. Arbuzov* and *A. Markovskaya*, Izvest. Akad. Nauk S.S.S.R. Otdel. khim. Nauk, 1952, 363; C.A., 1953, **47**, 15104):

Other dienes may be employed (see *J. Hamer* and *M. Ahmad*, in "1,4-Cycloaddition Reactions", ed. *Hamer*, Academic Press, New York, 1967, p. 422 and refs. cited therein).

Thus, for example, chloroprene (2-chloro-1,3-butadiene) affords 5-chloro-3,6-dihydro-2*H*-1,2-oxazine, whereas *cis*-1-phenyl-1,3-butadiene gives 6-phenyl-3,6-dihydro-2*H*-1,2-oxazine. Nitrosobenzene and butadiene give *2-phenyl-3,6-dihydro-2H-1,2-oxazine* (X), m.p. 53°, b.p. 108–109°/4 mm (*O. Wichterle* and *J. Vogel*, Coll. Czech. chem. Comm., 1949, **14**, 209; *Arbuzov*, Doklady Akad. Nauk, S.S.S.R., 1948, **60**, 993; C.A., 1949, **45**, 650):

Hydrogenation of X in cyclohexane yields 2-**phenyltetrahydro**-1,2-**oxazine**, b.p. 104–105°/4 mm, which is cleaved by treatment with zinc dust and sodium hydroxide and by high-pressure hydrogenation with the formation of 4-phenylamino-1-butanol (*H. E. Winberg*, U.S.P. 2,628,978/1953). From 2,3-dimethylbutadiene and nitrosobenzene 4,5-*dimethyl-2-phenyl-3,6-dihydro-2H-1,2-oxazine*, m.p. 41°, is obtained and this is reduced by zinc and acetic acid to an open-chain compound, 2,3-dimethyl-4-phenylamino-2-buten-1-ol, PhNHCH$_2$ · CMe = CMe · CH$_2$OH (*S. Kojima*, J. chem. Soc., Japan, 1954, **57**, 371; B.P. 696,471/1953/.

4-Nitrosotoluene and butadiene in chloroform at 0° give 2-(4-*methylphenyl*)-3,6-*dihydro*-2H-1,2-*oxazine*, m.p. 47–48°, in excellent yield (*Arbuzov*, Doklady Akad. Nauk,

S.S.S.R., 1951, **78,** 59; C.A., 1952, **46,** 993). Nitrosyl chloride and silver cyanide react with butadiene to yield 2-cyano-3,6-dihydro-2*H*-1,2-oxazine (*P. Horsewood* and *G. W. Kirby*, Chem. Comm., 1971, 1139); this substance may be used subsequently as a source of nitrosyl cyanide:

Chloroprene and 1-chloro-1-nitrosocyclohexane form 5-*chloro*-2-*cyclohexylidene*-3,6-*dihydro*-1,2-*oxazonium chloride*, m.p. 164°, which with alkali gives 5-*chloro*-3,6-*dihydro*-2H-1,2-*oxazine*, b.p. 67°/11 mm; *picrate*, m.p. 139.5–140°; *benzoyl* derivative, m.p. 66.5° (*O. Wichterle* and *M. Kolínský*, Chem. Listy, 1953, **47,** 1787; C.A., 1955, **49,** 201). Other applications of the cycloaddition of nitroso compounds and 1,3-dienes will be found in the following studies: *Arbuzov et al.* (Izvest. Akad. Nauk, S.S.S.R., Otdel Khim. Nauk, 1952, 566; Bull. Acad. Sci., U.S.S.R., Div. chem. Sci., 1952, 539; C.A., 1953, **47,** 4342; 1954, **48,** 5125; *Arbuzov* and *T. A. Pisha*, Doklady Akad. Nauk, S.S.S.R., 1957, **116,** 71; C.A., 1958, **52,** 6357); *Hamer, Ahmad* and *R. E. Holliday* (J. org. Chem., 1963, **28,** 3034); *S. Kojima*, Kôgyô Kagaku Zasshi, 1956, **59,** 951; C.A., 1958, **52,** 12870); *Hamer* and *R. E. Bernard* (Rec. Trav. chim., 1962, **81,** 734); *G. Kresze* and *J. Firl* (Angew. Chem. intern. Edn. Engl., 1964, **3,** 382); *Kresze et al.* (*ibid.*, 1963, **2,** 321); *T. Sasaki et al.* (J. org. Chem., 1971, **36,** 2886); *Kresze et al.* (Tetrahedron, 1971, **27,** 1941).

Whereas alkyl halides convert ethyl *N*-hydroxycarbamates to ethyl *O*-alkylhydroxy-carbamates (*L. W. Jones* and *R. E. Oesper*, J. Amer. chem. Soc., 1914, **36,** 730, 2208; *L. Neuffer* and *A. L. Hoffman*, *ibid.*, 1925, **47,** 1685) the use of tri- and tetramethylene bromides yields the cyclic compounds, *N*-ethoxycarbonyliso-oxazolidine and 2-*ethoxy-carbonyltetrahydro*-1,2-*oxazine*, b.p. 113–116°/12 mm, respectively.

Hydrolysis of the above oxazine by acid yields **tetrahydro**-1,2-**oxazine** isolated as its *hydrochloride*, m.p. 141–143°. Methyl iodide upon the free base gives 2-*methyltetrahydro-*1,2-*oxazine methiodide*, m.p. 176° (*H. King*, J. chem. Soc., 1942, 432). A series of 2-ethoxy-carbonyltetrahydro-1,2-oxazines have been prepared recently by similar means (*F. G. Riddell* and *D. A. R. Williams*, Tetrahedron, 1974, **30,** 1083).

An alternative preparation of tetrahydro-1,2-oxazine by the reduction of 5-chloro-3,6-dihydro-2*H*-1,2-oxazine in the presence of a platinum catalyst, afforded the *base*, b.p. 50–51°/30 mm, *picrate*, m.p. 150–151° (*Wichterle* and *Kolínský*, *loc. cit.*), and not m.p. 160–161°, as recorded by *King (loc. cit.)*.

Tetrahydro-1,2-oxazine is converted into 5,6-**dihydro-4***H*-1,2-**oxazine** by oxidation with lead tetra-acetate (*R. O. C. Norman, R. Purchase* and *C. B. Thomas*, J. chem. Soc., 1972, 1701); on standing the product slowly trimerizes:

5,6-Dihydro-3,6,6-trimethyl-4H-1,2-oxazine (XI) is one of the products of a Beckmann rearrangement of isobutylideneacetone oxime (*Nguyen Thoai, Nguyen Ngoc Chieu* and *J. Wiemann*, Ann. Chim., 1971, **6**, 235):

(XI)

2-Nitro-5-bromopentane with sodium methoxide in methanol yields 5,6-dihydro-3-methyl-4H-1,2-oxazine N-oxide (XII, R = Me); when 1-chloro-4-nitrobutane is used, 5,6-**dihydro-4H-1,2-oxazine** N-**oxide** (XII, R = H) is obtained and this with styrene affords the adduct XIII. Similar reactions with XII (R = Me) and the appropriate alkenes afford XIV (R = CO$_2$Me, Ph or CH$_2$Cl) (*I. E. Chlenov et al.*, Izvest. Akad. Nauk, S.S.S.R., Ser. Khim., 1970, 2641; C.A., 1971, **74**, 141662g):

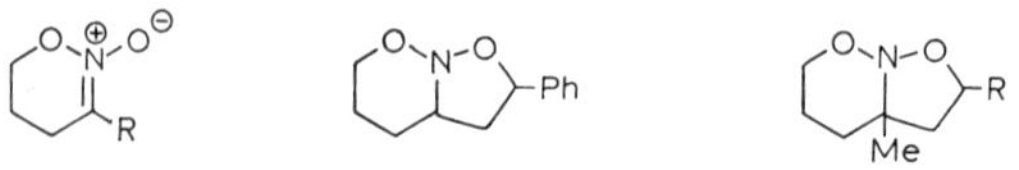

(XII)   (XIII)   (XIV)

Nicotine N-oxide is converted, by heating for 1 min at 190–200°/1 mm, into 2-**methyl-6-(3-pyridyl)tetrahydro**-1,2-**oxazine** (XV), b.p. 90°/1 mm; *dihydrochloride*, m.p. 187–188°, *dipicrate*, m.p. 194–195° (*C. H. Rayburn et al.*, J. Amer. chem. Soc., 1950, **72**, 1721): it is identical with a base obtained by *A. Pinner* (Ber., 1895, **28**, 456). The structure is proved by its conversion into 4-methylamino-1-(3-pyridyl)butan-1-one by hydrolysis with hydrochloric acid and also by its reduction to 4-methylamino-1-(3-pyridyl)-1-butanol:

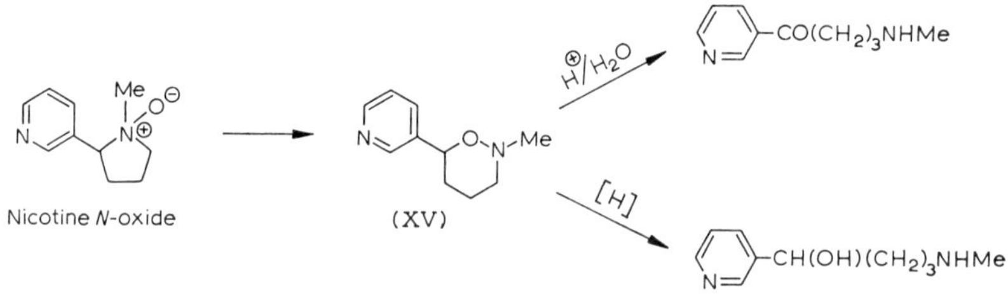

Nicotine N-oxide   (XV)

The mass spectra of a number of 1,2-tetrahydro-oxazines have been recorded and analysed (*R. A. W. Johnstone, B. J. Millard* and *E. J. Wise*, J. chem. Soc., 1967, 307), and the conformational equilibria of 2,4- and 2,5-dimethyltetrahydro-1,2-oxazines have been studied (*F. G. Riddell*, Tetrahedron, 1975, **31**, 523).

(b) 1,3-Oxazines

Known monocyclic 1,3-oxazines (metoxazines) are mostly derivatives of 4H-1,3-oxazine, rather than of the 2H-1,3- and 6H-1,3-forms. The hydroxy-1,3-oxazines are tautomeric with the corresponding oxodihydro compounds.

At one time 4H-1,3-oxazines were named pentoxazoles, which emphasises their resemblance to oxazoles, both in structure and in methods of synthesis.

(i) 4H-*1,3-oxazines*

4H-1,3-oxazines are formed by the cyclodehydration of β-acylamino derivatives of aldehydes (or their acetals), ketones or esters. β-Acylamino-ketones, for example, give 2,6-disubstituted compounds XVI in which the substituents are alkyl or aryl (*S. Gabriel*, Ber., 1910, **43**, 134, 1283; Ann., 1915, **409**, 305). β-Acylamino-esters give 6-alkoxy compounds (*P. Karrer* and *E. Miyamichi*, Helv., 1926, **9**, 336):

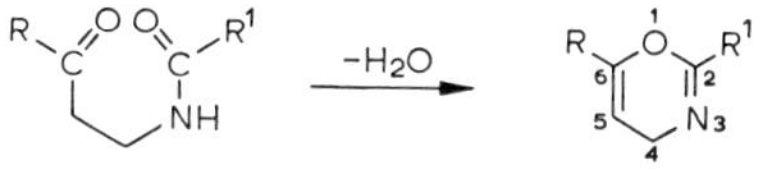

(XVI)

2-Phenyl-4H-1,3-oxazine (XVI, R = H; R^1 = Ph), m.p. 171°, is prepared from β-benz-amidopropionaldehyde diethyl acetal by the action of aqueous oxalic acid at 25° (*A. Wohl*, Ber., 1901, **34**, 1914). In the case of acylaminoketones and their acetals phosphorus pentachloride is employed as, for example, in the synthesis of 2,6-*diphenyl-4H-1,3-oxazine* (XVI, R = R^1 = Ph), m.p. 86–87°, and 2,4,4,6-*tetramethyl-4H-1,3-oxazine*, b.p. 144°/754 mm (*Gabriel*, Ann., 1915, **409**, 305).

Similarly, ethyl β-benzamidopropionate is cyclised with phosphorus pentoxide to 6-*ethoxy*-2-*phenyl-4H-1,3-oxazine* (XVI, R^1 = Ph, R = OEt), m.p. 74°; *picrate*, m.p. 103° (*Karrer* and *Miyamichi, loc. cit.*); such oxazines lack the stability of the oxazoles and are readily hydrolysed by aqueous acid to an acylated β-aminoketone or β-carboxylic acid.

5-Hydroxy-2-phenyl-4H-1,3-oxazine, m.p. 91°, is prepared from hippuryl chloride and diazomethane and, on hydrolysis, gives hippuric acid and formaldehyde (*Karrer* and *R. Widmer*, Helv., 1925, **8**, 204). 6-*Hydroxy-4-methyl-2-phenyl-4H-1,3-oxazine**, m.p. 95°, from β-benzamidocrotonic acid and acetic anhydride, gives 6-hydroxy-4-methyl-2-phenyl-pyrimidine on treatment with ammonia (*C. C. Barker*, J. chem. Soc., 1954, 317).

Suitable β-chlorocarbonyl derivatives on treatment with stannic chloride, yield intermediate carbonium ions which with nitriles undergo cycloaddition to form 4H-1,3-oxazin-3-ium salts (XVII). Deprotonation affords the corresponding free bases:

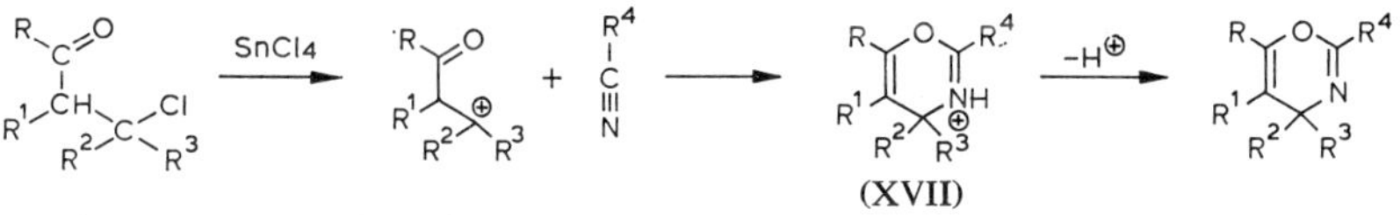

(XVII)

N-Chloroalkylamides similarly form ions of the type XVIII with stannic chloride; these react with alkynes providing an alternative synthesis of

---

* Presumably this and related structures are in tautomeric equilibrium with the corresponding oxo-forms.

4H-1,3-oxazin-3-ium salts (*R. R. Schmidt*, Chem. Ber., 1965, **98**, 334, 3892; *Schmidt, D. Schwille* and *S. Sommer*, Ann., 1969, **723**, 111; *M. Lora-Tamayo et al.*, Chem. Ber., 1964, **97**, 2234):

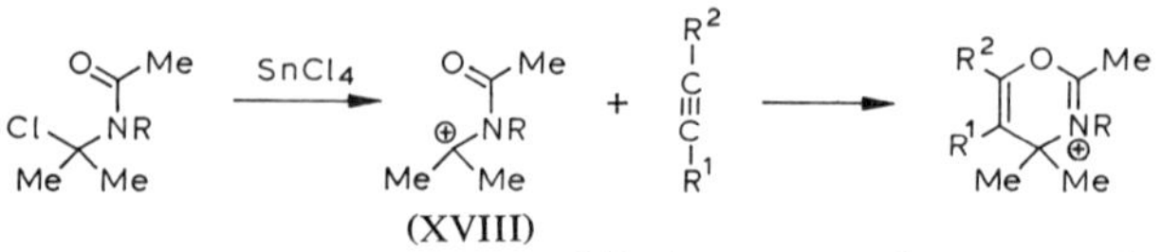

(XVIII)

In a related sequence the aldimine XIX when treated with carboxylic acid chlorides in the presence of stannic chloride yields cations which with phenylacetylene give the corresponding 4H-1,3-oxazinium salts (*Schmidt*, Angew. Chem., 1965, **77**, 218; Angew. Chem., internat. Edn., 1965, **4**, 241):

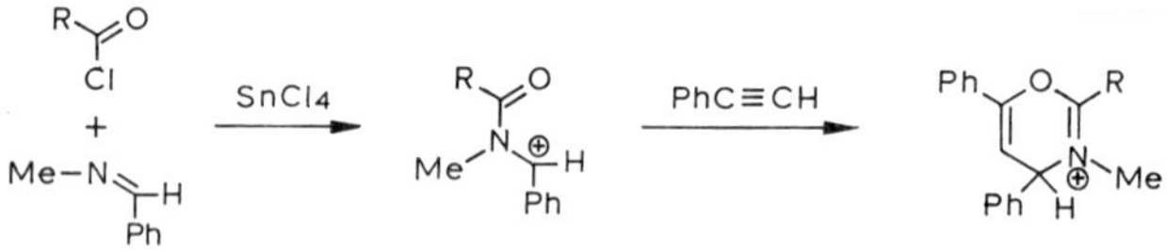

$$R = Me;\ Ph;\ [4]MeC_6H_4\ \ or\ \ [4]ClC_6H_4$$

(ii) 1,3-Oxazin-1-ium salts

S. *Hünig* and *K. Hübner* (Chem. Ber., 1962, **95**, 937) were the first to prepare a 1,3-oxazin-1-ium salt; *e.g.* from the 4-thioxo-1,3-oxazine XX* by alkylation with methyl iodide:

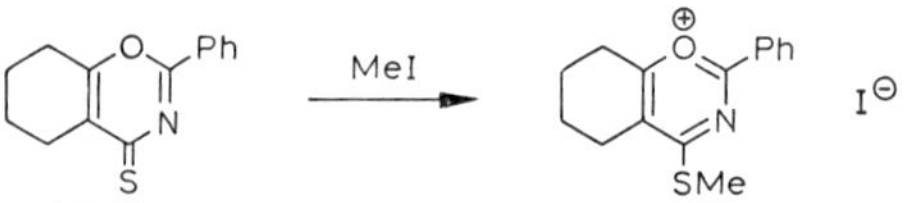

(XX)

β-Acylaminoalkenes of the type XXI (where $R^1 = Ph$, $[4]ClC_6H_4$, $[4]MeC_6H_4$ or Bu^t; $R^2 = Me$, OEt or Ph and $R^3 = OEt$, Ph or NMe_2) cyclise to 2,4,6-trisubstituted 1,3-oxazin-1-ium cations (XXII) in the presence of absolute perchloric acid. The starting alkenes are themselves obtained by treating the appropriate 2,4,6-trisubstituted 4H-1,3-oxazines (see p. 433) with 1,4-benzoquinone followed by hydrolysis (*Schmidt, Schwille* and *Sommer, loc. cit.*). The reaction proceeds *via* an isolable 1:1 adduct of structure XXIII. In an analogous reaction to the formation of 4H-1,3-oxazin-3-ium salts, from β-chlorocarbonyl compounds and nitriles, referred to on p. 433, 1,3-oxazin-1-ium salts may be made directly by the 1,4-cyclo-addition of α,β-unsaturated β-chlorocarbonyl compounds and nitriles in

---

* See I.U.P.A.C. Rule 532.3; the grouping $\mathord{>}C = S$ is designated as "thioxo" in place of "thiono" used in some languages.

the presence of stannic chloride (*Schmidt*, Habilitationsschrift, Universität Stuttgart, 1968; Synthesis, 1972, **7**, 333). Similarly *N*-acylimidoyl chlorides and alkynes may also be employed:

When 4*H*-1,3-oxazines bearing a hydrogen atom at C-4 are reacted with equimolar proportions of trityl perchlorate in absolute acetonitrile, 1,3-oxazin-1-ium perchlorates are formed in excellent yields (*K. Wohl*, Ber., 1901, **34**, 1914).

The salts react with nucleophiles in a number of ways: thus hydrolysis affords a quantitative conversion to β-acylaminoalkenes, a reaction is probably initiated by attack of a water molecule at C-6. Ammonia or ammonia derivatives, such as hydrazine, hydroxylamine and ureas, cause ring scission and recyclisation. Ammonia, for example, yields pyrimidines (*Hünig* and *Hühner*, loc. cit.), whereas hydrazine gives 1,2-imidazoles and hydroxylamine 1,2-oxazoles (*Schmidt*, Chem. Ber., 1965, **98**, 334):

Carbanions, such as that generated from malonodinitrile with triethylamine in anhydrous acetonitrile, give a mixture of 1,3-butadiene isomers. The reaction proceeds *via* the intermediate 6*H*-1,3-oxazine (XXIV); such intermediates may, in fact, be isolated when alkyl malonodinitriles are employed. Here, however, diene formation is prevented since the substituent in the 6-position no longer bears an acidic α-hydrogen atom (*Schmidt, Synthesis, loc. cit.*):

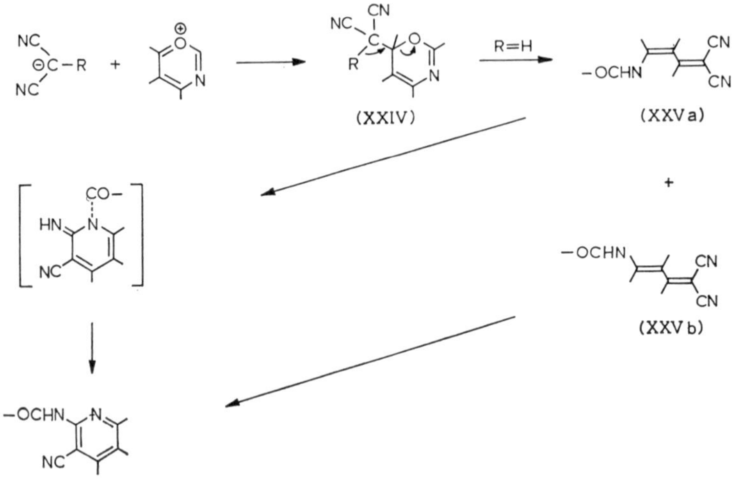

The isomers XXVa readily cyclise to *N*-acyl-2,4-dihydropyridine derivatives, which in turn rearrange to pyridines. Although the configuration of the isomers XXVb is opposed to cyclisation, they too yield the same pyridines when heated above the melting point.

Phenate ions, generated under anhydrous conditions, normally react at position-6 (*Schmidt, Schwille* and *H. Wolf*, Chem. Ber., 1970, **103**, 2760); for example, 2,6-di-*tert*-butylphenol and 2,4,6-triphenyl-1,3-oxazin-1-ium perchlorate yield the tautomers XXVIa and XXVIb.

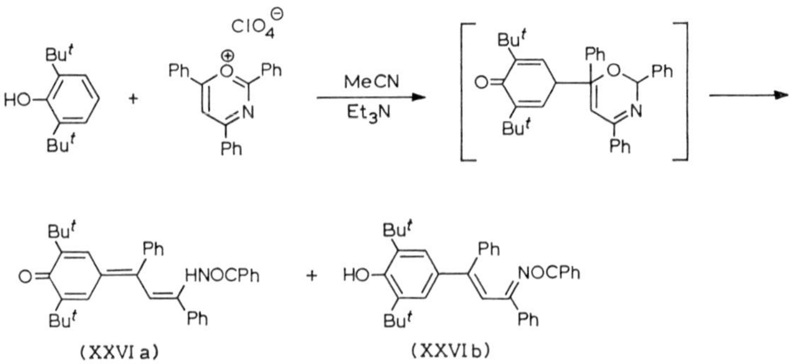

Likewise with enamines, such as morpholinocyclohexene, isomeric products *e.g.* XXVIIa and XXVIIb result:

(XXVIIa)   (XXVIIb)

The corresponding 1,5,6,7-*tetrahydroquinoline*, XXVIII, m.p. 165°, is readily formed from XXVIIa by the action of hydrochloric acid (*Schmidt, ibid.*, 1965, *loc. cit.*):

(XXVIII)   (XXIX)

Benzylmagnesium bromide and 2,4,6-triphenyl-1,3-oxazin-1-ium perchlorate yield 6-*benzyl*-2,4,6-*triphenyl*-6H-1,3-*oxazine* (XXIX), m.p. 124°, which when heated affords 2,3,4,6-tetraphenylpyridine (*Schmidt*, Habilitations-schrift, *loc. cit.*).

When 2-(4-chlorophenyl)-4,6-dimethyl-1,3-oxazin-1-ium perchlorate (XXX) was treated with triethylamine in the absence of air the 4-*methylene* derivative, m.p. 74°, was produced (*Schmidt, Schwille* and *Wolf, loc. cit.*). In view of the reactivity of the C-6 position in oxazinium salts, it is perhaps surprising that this product was formed in preference to the 6-methylene isomer; the yield, however, was only 48%:

(XXX)

(iii) Hydro-1,3-oxazines

Compounds with one double bond are dihydro-1,3-oxazines. The following structures are possible:

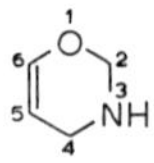

3,4-Dihydro-2*H*-1,3-oxazine

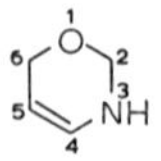

3,6-Dihydro-2*H*-1,3-oxazine

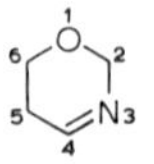

5,6-Dihydro-2*H*-1,3-oxazine

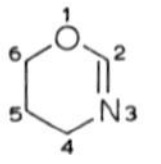

5,6-Dihydro-4*H*-1,3-oxazine

(*1*) *3,4-Dihydro-2H-1,3-oxazines.* Although condensations employing diketene and amidines or thioureas usually form pyrimidine derivatives, N,N^1-diphenylguanidine and *S*-methyl-*N*-phenylthiourea yield **4-oxo-3,4-dihydro**-1,3-**oxazines** (XXXI) which rearrange to pyrimidones (XXXII) on boiling with hydrochloric acid (*R. N. Lacey*, J. chem. Soc., 1954, 839):

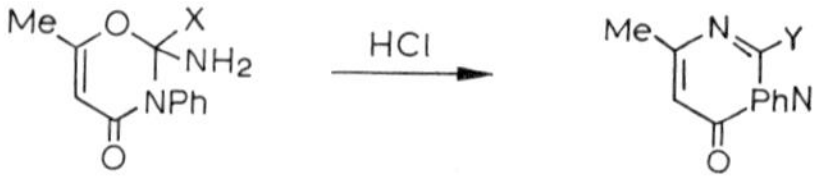

(XXXI, X = NHPh
or SMe)

(XXXII, Y = NHPh
or OH)

Similarly, *N,N'*-diethyl- and *N,N'*-diphenyl-*S*-methylthioureas react with diketene, with the elimination of methanethiol, giving substituted 2-imino-4-oxo(3*H*)-1,3-oxazines (XXXIII) [3-*ethyl-2-ethylimino*-, (XXXIII, R = Et), m.p. 30°, and 6-*methyl-4-oxo-3-phenyl-2-phenylimino*-2H-1,3-*oxazine* (XXXIII, R = Ph), m.p. 184–185°] which are readily hydrolysed by hydrochloric acid to 2,4-dioxo compounds XXXIV (3-*ethyl-*, m.p. 69–70°, and 6-*methyl-2,4-dioxo-3-phenyl*-1,3-*oxazine*, m.p. 170°, respectively). Confirmation of the structures assigned to these compounds is obtained by their reaction with amines to give known substituted uracils XXXV. Imino-oxazines like XXXIII are hydrolysed by sodium hydroxide to *N,N'*-disubstituted ureas *(Lacey, loc. cit.).* Related work shows that diketene and carbodi-imides in boiling benzene yield similar compounds to XXXIII (*Lacey* and *W. R. Ward*, J. chem. Soc., 1958, 2134).

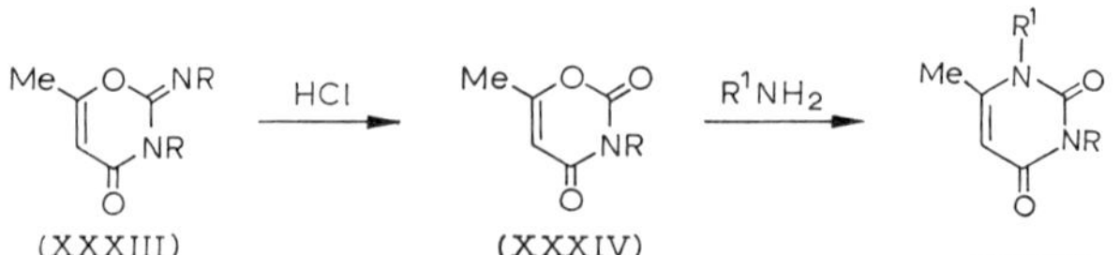

(XXXIII)

(XXXIV)

(XXXV)

Potassium cyanate and diketene in acetic acid react exothermically to give 6-*methyl-2,4-dioxo(3H)*-1,3-*oxazine* (XXXVI, R^1 = H, R^2 = Me), m.p. 230–231°, (*V. I. Gunar et al.*, Khim. geterot. Soedin., 1967, 48; C.A., 1967, **67**, 54097p):

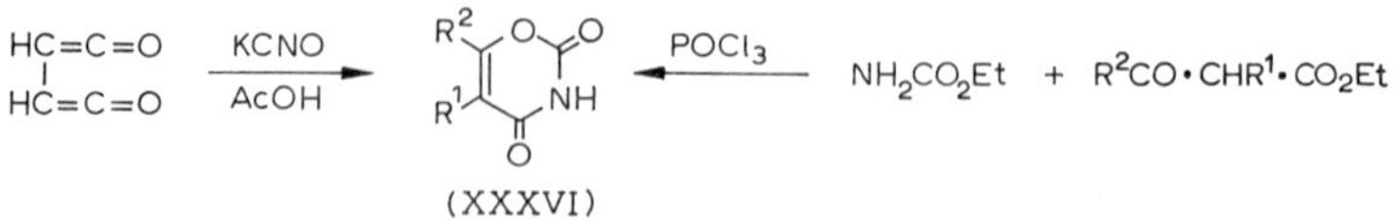

(XXXVI)

Other derivatives are available from the reaction of β-oxo esters and ethyl-glycinate in the presence of phosphoryl chloride (*S. S. Washburne* and *K. K. Park*, Tetrahedron Letters, 1976, 243).

Phenyliminocyclohexane reacts with malonic acid in acetic anhydride at 20° to afford the *spiro derivative*, m.p. 130°, shown below (*E. Ziegler, K. Belegratis* and *G. Brus*, Monatsh., 1967, **98**, 555):

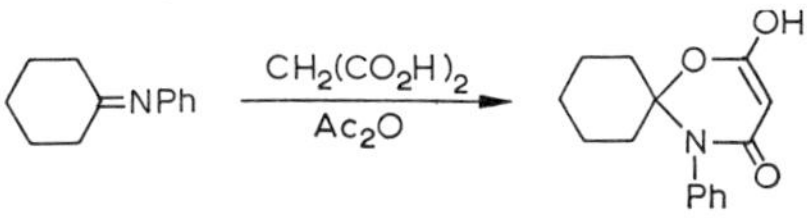

At low temperatures chloro- and fluoro-sulphonyl isocyanates react with α,β-unsaturated ketones to yield 2-**oxo**-3,4-**dihydro**-1,3-**oxazines** (*K. Claus, H. J. Friedrich* and *H. Jensen*, Ann., 1974, 561):

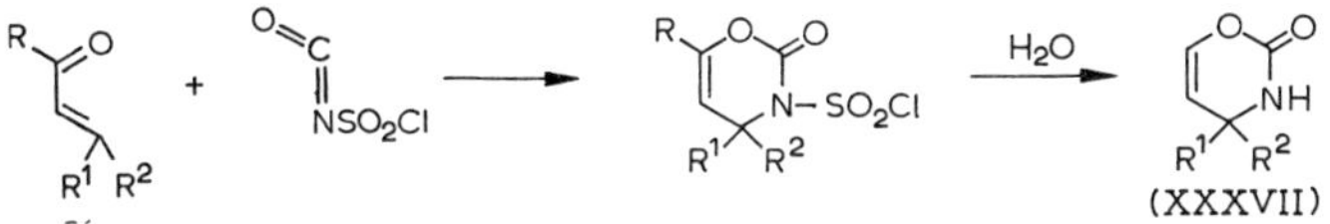

The physical constants for a number of the compounds prepared are summarized in Table 1.

TABLE 1

2-OXO-3,4-DIHYDRO-2H-1,3-OXAZINES (XXXVII)

R	R^2	R^3	m.p.($°C$)
Me	Me	Me	110
Ph	Me	Me	161
Ph	H	Ph	189–190
Mc	H	[4]MeOC$_6$H$_4$	119
Ph	H	[4]ClC$_6$H$_4$	141
CMe$_0$	H	Ph	159
PhCH $=$ CH	H	Ph	204

Partial hydrolysis of the pyrimidine XXXVIII, prepared from ethyl α-acetyl-β-phenylaminohydrocinnamate and benzalaniline, affords ethyl 6-methyl-2,3,4-triphenyl-3,4-dihydro-2H-1,3-oxazine-5-carboxylate, a rare example of an unoxygenated derivative of the parent 3,4-dihydro-2H-1,3-oxazine system (*J. G. Erickson*, J. Amer. chem. Soc., 1945, **67**, 1382).

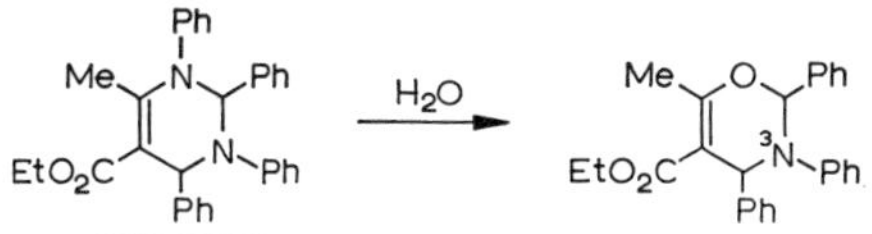

(*2*) 5,6-*Dihydro*-4H-1,3-*oxazines*. *Preparation*: The parent compound 5,6-**dihydro**-4*H*-1,3-**oxazine** (XXXIX, R = H), has been prepared by the cyclisation of *N*-formyl-2-hydroxyethylamine on silica gel (*F. Becke* and *P. Paessler*, Ger. Öffen., 1,923,022/1970). Derivatives were first prepared by *S. Gabriel* and *P. Elfelt* (Ber., 1891, **24**, 3213) by treating acyl derivatives of γ-bromopropylamine with alkali:

(XXXIX)

2-*Phenyl*-5,6-*dihydro*-4H-1,3-*oxazine* (XXXIX, R = Ph), *picrate*, m.p. 151°, and the corresponding 2-amino- and 2-nitro-phenyl compounds (XXXIX, R = [2], [3] or [4] $NH_2C_6H_4$ or $NO_2C_6H_4$) are typical products of this route (*A. Novelli* and *R. Adams*, J. Amer. chem. Soc., 1937, **59**, 2259). 6-*Methyl-2-phenyl*-, (oil), *picrate*, m.p. 146–148°, and 4,4,6-*trimethyl-2-phenyl*-5,6-*dihydro*-4*H*-1,3-*oxazine*, m.p. 32°, are obtained from 3-chlorobutylbenzamide and *N*-(3-bromo-1,1-dimethylbutyl)benzamide, respectively (*A. Luchmann*, Ber., 1896, **29**, 1428; *M. Kahan, ibid*., 1897, **30**, 1319).

Alternatively, 5,6-dihydro-4*H*-1,3-oxazines may be synthesised from *O*-acyl derivatives of 3-propanolamines (*Gabriel* and *H. Ohle*, Ber., 1917, **50**, 819):

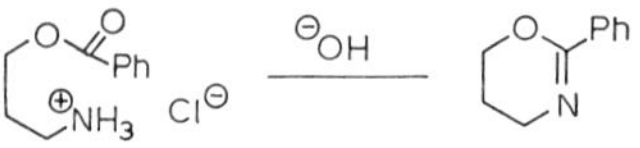

Acetylenic ethers (Neth. P. 81,868/1958; Chem. Zentr., 1959, 3980; U.S.P. 2,813,862/1958; Chem. Zentr., 1959, 17309; B.P. 785,373/1958; Chem. Zentr., 1959, 17309), esters of unsaturated acids (*I. J. Rinkes*, Rec. Trav. chim., 1927, **46**, 268; *Amer. Cyanamid Co.*, B.P. 773,011/1957; C.A., 1957, **51**, 15605F; *C. J. Schmidle*, G.P. 953,524/1951; Chem. Zentr., 1958, 1143), salts of amidines (*P. Oxley* and *W. F. Short*, B.P. 615,006/1948; C.A., 1949, **43**, 7512; J. chem. Soc., 1950, 1100) and lactic esters (*W. P. Ratchford*, J. Amer. chem. Soc., 1950, **72**, 3297) all react with 3-propanolamines to form 5,6-dihydro-4*H*-1,3-oxazines.

Aromatic aldehydes react with 3-azidopropanol to form 2-aryl derivatives (*J. H. Boyer* and *J. Hamer, ibid*., 1955, **77**, 951; *Boyer et al., ibid*., 1956, **78**, 325):

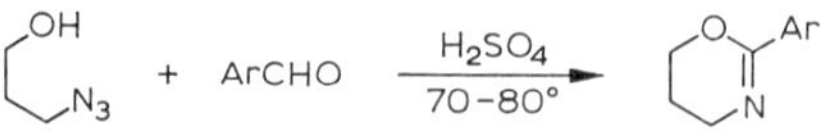

Dihydro-oxazines are formed when nitriles are reacted with 1,3-propandiols, in the presence of concentrated sulphuric acid, thus, for example, 1,1,3-trimethyl-1,3-propandiol and acetonitrile yield 2,4,4,6-**tetramethyl**-5,6-

dihydro-4*H*-1,3-oxazine (XL) (*J. W. Lynn*, J. org. Chem., 1959, **24**, 711; *E. J. Tillmanns* and *J. J. Ritter, ibid.*, 1957, **22**, 839; *A. I. Meyers, ibid.*, 1960, **25**, 145). The reaction proceeds *via* the most stable carbonium ion which undergoes cycloaddition with the nitrile.

A further method, of wide application, involves a cycloaddition between *N*-acylimines and electron-rich alkenes such as enamines (*Schmidt* and *E. Schlipf*, Chem. Ber., 1970, **103**, 3783). Less reactive alkenes may be employed if the electrophilic nature of the *N*-acylimine is increased by *N*-protonation or alkylation. The latter are formed *in situ* from *N*-(1-halogenoalkyl)carboxamides (XLI, X = hal.) or *N*-(1-hydroxyalkyl)carbox-amides (XLI, X = OH) (*H. E. Zaugg*, Synthesis, 1970, 49; *H. Hellmann*, Angew. Chem., 1957, **69**, 463; *Hellmann* and *G. Opitz*, "α-Aminoalkylierung", Verlag Chemie, Weinheim, 1960, pp. 64–79; *Schmidt*, Chem. Ber., 1965, **98**, 334):

C. Giordano, G. Ribaldone and *G. Borsotti* (Synthesis, 1971, 92) report that a similar process occurs when carboxamides, alkenes and formaldehyde (or formaldehyde derivatives) are reacted in acetic acid containing catalytic amounts of sulphuric acid, hydrochloric acid or toluene-4-sulphonic acid (see also *Schmidt*, Chem. Ber., 1970, **103**, 3242). α-Isocyano-γ-butyrolactone, prepared from α-amino-γ-butyrolactone or α-hydroxyimino-γ-butyrolactone, when reacted with ethoxide ion in ethanol undergoes lactone cleavage and rearrangement to yield 4-ethoxycarbonyl-5,6-dihydro-4*H*-1,3-oxazine (*U. Kraatz, H. Wamhoff* and *F. Korte*, Ann., 1971, **794**, 33). 2-Amino-5,6-dihydro-4*H*-1,3-oxazines may be prepared by the action of aryl isothiocyanates upon 3-aminopropanols followed by treatment of the resultant thioureas with methyl iodide and then with base (*L. A. Ignatova et al.*, Khim. geterot. Soedin., 1974, 354). 2-Amino-5,6-dihydro-4*H*-1,3-oxazines (XLIIb) are tautomeric with the 2-imino forms (XLIIa), but if the latter are treated consecutively with methyl iodide and base 5,6-dihydro-4*H*-1,3-oxazines containing a secondary amino group in the 2-position are obtained:

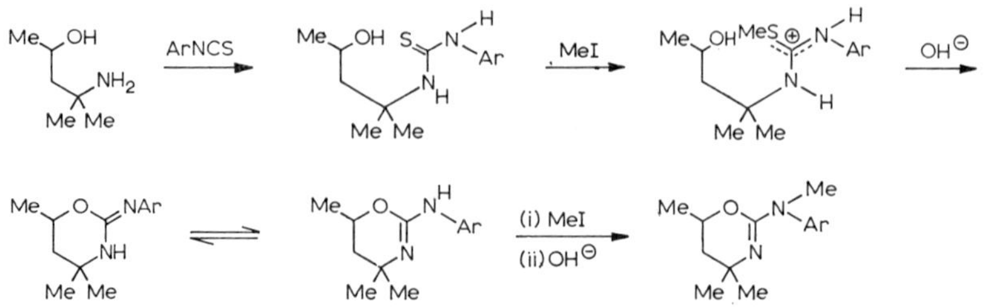

Ar = Ph; [4]BrC$_6$H$_4$; [3]ClC$_6$H$_4$; [4]MeOC$_6$H$_4$;
[4]EtOC$_6$H$_4$; [4]MeC$_6$H$_4$; [3]MeC$_6$H$_4$

When the corresponding oxo derivatives are employed instead of the alcohols, 2-amino-4H-1,3-oxazines are formed (*idem, ibid.*, p. 764):

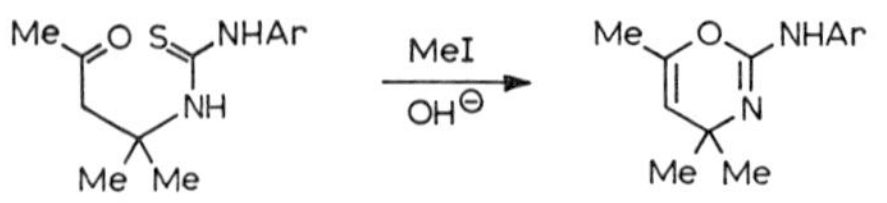

Properties: 5,6-Dihydro-4H-1,3-oxazines hydrolyse in dilute acid solution to salts of the corresponding 3-aminopropyl esters which, on basification, then undergo rearrangement to 3-hydroxypropylamides (*S. Gabriel*, Ann., 1915, **409**, 305):

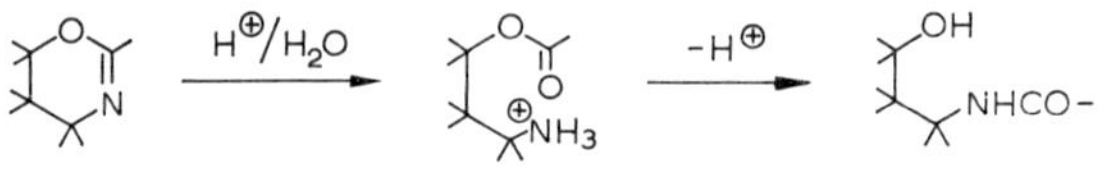

Schmidt (Chem. Ber. 1970, **103**, 3242) notes that certain derivatives *e.g.* the 6-(4-anisyl)-2-phenyl derivative, undergo a stereospecific elimination affording *trans-N*-(Δ^2-alkenyl) benzamides:

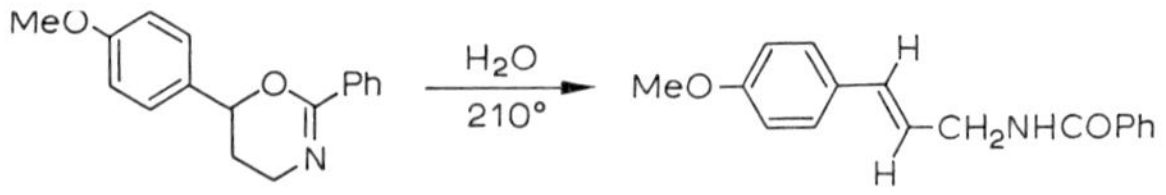

With oxiranes an addition reaction occurs to give dicyclic derivatives (*W. Seeliger et al.*, Angew. Chem., 1966, **78**, 913; internat. Edn., 1966, **5**, 875):

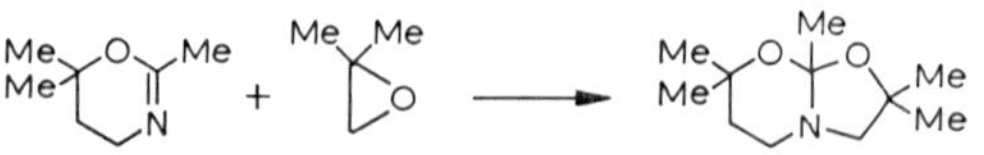

2-Alkyl-5,6-dihydro-4H-1,3-oxazines can be reduced by sodium tetrahydridoborate to the corresponding tetrahydro-1,3-oxazines (*Meyers et al.*, J. Amer. chem. Soc., 1969, **91**, 764, 765, 2155, 5886, 5887; 1969, **92**, 1084; 1970, **93**, 2314; 1972, **94**, 3243; Tetrahedron Letters,

1969, 1783, 4809; 1970, 3715, 4355, 5151; "Heterocycles in Organic Synthesis", Wiley, New York, 1974, pp. 201–208. On acid hydrolysis these products yield aldehydes and propanolamine salts:

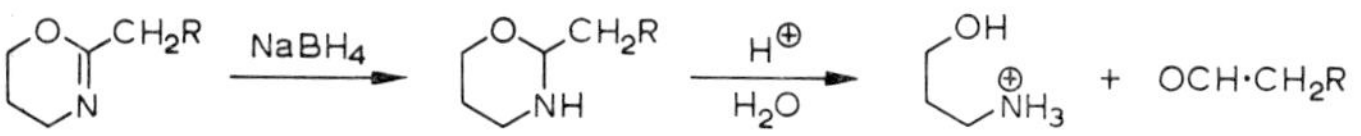

Although Grignard reagents do not react with simple 2-alkyl-5,6-dihydro-4H-1,3-oxazines, n-butyl-lithium in tetrahydrofuran at low temperature affords carbanions (*e.g.* XLIII) which with electrophiles extend the scope of the above reaction.

Thus with alkyl halides (R^{1}X), followed by reduction and hydrolysis, numerous substituted aldehydes may be prepared, and, by using oxiranes, γ-hydroxyaldehydes are formed. Similarly, if aldehydes or ketones are used as the electrophiles α,β-unsaturated aldehydes result:

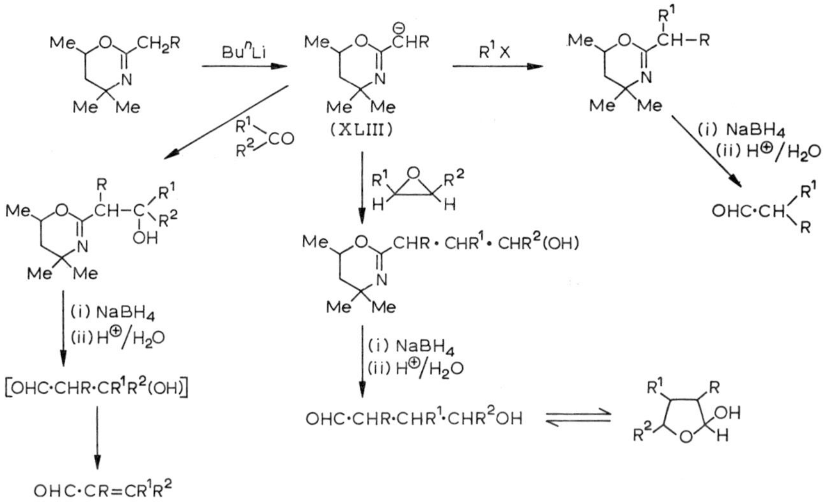

When 4,4,6-trimethyl-5,6-dihydro-4H-1,3-oxazine, or its 2-bromo derivative, is treated with n-butyl-lithium in tetrahydrofuran solution the carbanion XLIV is formed. This readily rearranges to the isonitrile salt XLV, neutralization and pyrolysis of which reforms the starting material (*Meyers* and *A. W. Adickes*, Tetrahedron Letters, 1969, 5151):

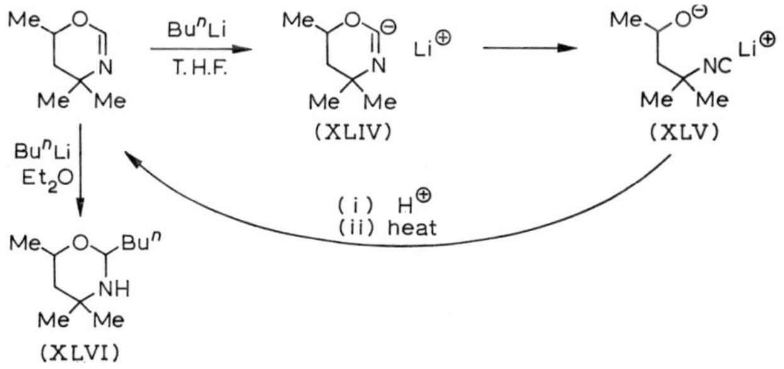

If the solvent for the initial reaction is changed from tetrahydrofuran to diethyl ether the tetrahydro-1,3-oxazine XLVI is formed. 2-Aryl-5,6-dihydro-4*H*-1,3-oxazines yield the corresponding 3-methyl-1,3-oxazolinium salts when treated with dimethyl sulphate, these may now react with alkyl Grignard reagents by addition to the imminium bond. The product *N*-methyltetrahydrooxazoles, afford ketones upon hydrolysis (*Meyers* and *E. M. Smith*, J. Amer. chem. Soc., 1970, **92**, 1084):

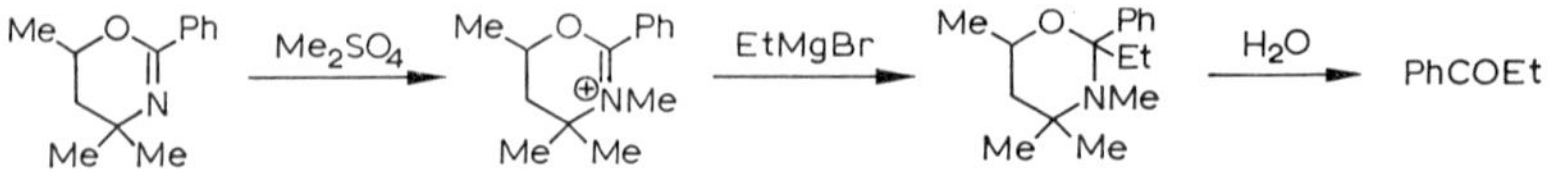

(*3*) *3,6-Dihydro-2H-1,3-oxazines.* Only oxo-derivatives are known: thus *S. L. Shapiro*, *V. Bandurco* and *L. Freedman* (J. org. Chem., 1961, **26**, 3710) obtained some 3-aryl-3,6-dihydro-2*H*-1,3-oxazines by cyclising dialkyl ethynyl arylcarbamates:

3,6-**Dihydro**-2,6-**dioxo**-2*H*-1,3-**oxazine** is prepared by reacting maleic imide with sodium hypochlorite (*I. J. Rinkes*, Rec. Trav. chim., 1927, **46**, 268):

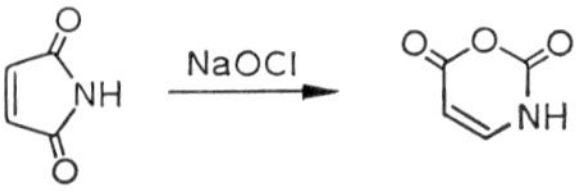

By heating amido esters of the type shown below at 270° 2-alkyl-3,6-dihydro, 4-methyl-1,3-oxazin-6-ones are formed (*W. Stroglich et al.*, Angew. Chem.-**86**, 596):

(*4*) *Tetrahydro-1,3-oxazines.* Compounds of this class may be easily obtained from 5,6-dihydro-4*H*-1,3-oxazines by reduction with sodium tetrahydrido-borate (see above) or by the condensation of aldehydes and ketones with 3-aminopropan-1-ols in basic solution. Thus the parent heterocycle, **tetrahydro**-1,3-**oxazine,** is synthesised by the condensation of 3-aminopropanol with formaldehyde (*E. D. Bergmann* and *A. Kaluszyner*, Rec. Trav. chim., 1959, **78**, 315; *J. S. Eden*, U.S.P. 2,911,294/1959; C.A., 1960, **54**, 3842). Excess of the carbonyl compound leads to the methane derivative XLVII:

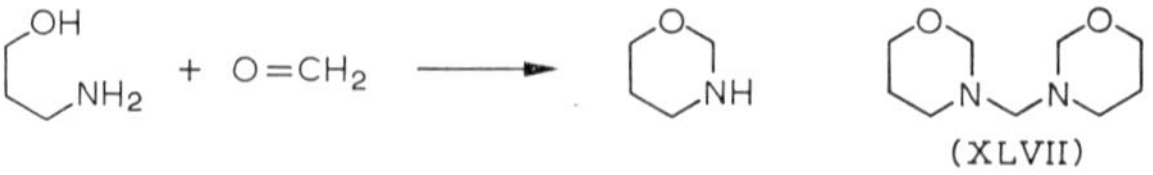

(XLVII)

Similarly, 2-*methyltetrahydro*-1,3-*oxazine*, b.p. 68–70°/100 mm, $n_D^{25°}$ 1.4389, is prepared from 3-aminopropanol and acetaldehyde (*W. H. Watanabe* and *L. E. Conlon*, J. Amer. chem. Soc., 1957, **79**, 2825). This reaction forms the basis of a general route (*M. Kohn*, Ber., 1916, **49**, 250 and earlier papers): 3,4-*dimethyl*-, b.p. 40–45°/20 mm, *hydrochloride*, m.p. 175° (*C. Mannich* and *K. Roth*, Arch. Pharm., 1937, **274**, 527; C.A., 1937, **31**, 2216) and 3,5,5-*trimethyltetrahydro*-1,3-*oxazine*, b.p. 40–42°/12 mm, *hydrochloride*, m.p. 203° (*Mannich* and *H. Wieder*, Ber., 1932, **65**, 385).

2-Methyltetrahydro-1,3-oxazine may be synthesised also by the interaction of 3-amino-propanol and ethyne at 150° under pressure without a catalyst. Because the product is not stable under these conditions the reaction is stopped after 4 h with 10–30% conversion; under similar conditions ethanolamine yields 2-methyloxazolidine (*Watanabe* and *Conlon*, loc. cit.).

Several 5-nitro-3,5-dialkyltetrahydro-1,3-oxazines are prepared from β-nitropropane-diols, primary amines and at least two molecular equivalents of formaldehyde (*M. Senkus*, C.A., 1949, **43**, 1068). Formaldehyde, nitromethane and benzylamine yield 3-*benzyl-5-hydroxymethyl-5-nitrotetrahydro*-1,3-*oxazine* (XLVIII, R = CH₂OH), m.p. 140–142°, oxidised by alkaline hydrogen peroxide to 3-*benzyl-5-nitrotetrahydro*-1,3-*oxazine* (XLVIII, R = H), *hydrochloride*, m.p. 210–212°:

$$O_2N{-}CR(OH)(CH_2OH) \xrightarrow[\text{HCHO}]{\text{MeNH}_2} \text{(XLVIII)}$$

(XLVIII)

which is reduced to the corresponding 5-amino compound (*T. Urbański* and *D. Gürne*, Roczniki Chem., 1954, **28**, 175; C.A., 1955, **49**, 8826). Other nitro-alkanes behave similarly yielding the 3-*benzyl-5-methyl*- and 3-*benzyl-5-ethyl-tetrahydro*-1,3-*oxazines* (XLVIII, R = Me and Et), m.p. 66–68° and 68–70°, respectively. These oxazines form nitroso and acyl derivatives but the ring is cleaved by mineral acid (*idem*, Bull. Acad. Polon. Sci., Classe III, 1955, **3**, 175; C.A., 1955, **49**, 13998). 1-Nitrobutane or 1-nitroisobutane, formaldehyde and ammonia yield 5-*nitro-5-propyltetrahydro*-1,3-*oxazine*, *hydrochloride*, m.p. 190–192°; *picrate*, m.p. 163–164°, or the corresponding 5-*isopropyl* isomer, *hydrochloride*, m.p. 190°, *picrate*, m.p. 167–168° (*Urbański et al.*, Bull. Acad. Polon. Sci., Classe III, 1955, **3**, 179; C.A., 1956, **50**, 4152; Roczniki Chem., 1955, **29**, 379; C.A., 1956, **50**, 4967; J. chem. Soc., 1959, 1912; Tetrahedron, 1961, **16**, 30; Synthesis, 1974, 613).

Most of one hundred and ten 5-nitrotetrahydro-1,3-oxazines prepared for anticancer screening show some activity against Ehrlich ascites carcinoma and Amytal ascites sarcoma *in vitro* (*M. Mordarski, B. Chylinska* and *Urbański*, Arch. Immunol. Ther. Exp., 1970, **18**, 679).

2-**Phenyltetrahydro**-1,3-**oxazine**, b.p. 175–176°, *picrate*, m.p. 131°, is prepared from benzaldehyde, 3-aminopropanol and potassium cyanide (*A. I. Kiprianov* and *B. A. Rash-kovan*, J. gen. Chem., U.S.S.R., 1937, **7**, 1026; C.A., 1937, **31**, 5356).

3-*Phenyltetrahydro*-1,3-*oxazine*, b.p. 94°/0.4 mm, is available from *N*-γ-hydroxy-propylaniline and formaldehyde (*G. A. R. Kon* and *J. J. Roberts*, J. chem. Soc., 1950, 978) and appropriately for 3-*methyl-6,6-diphenyltetrahydro*-1,3-*oxazine*, m.p. 83–85° (*A. L. Morrison* and *H. Rinderknecht*, ibid., p. 1510); 6-*phenyltetrahydro*-1,3-*oxazine*, b.p. 94°/1 mm, from styrene, ammonium chloride and formaldehyde (3-*acetyl* deriv., m.p. 83–83.5°) (*H. D. Hartough, J. Dickert* and *S. L. Meisel*, U.S.P. 2,647,117–8/1953; C.A., 1954, **48**, 7645, 8265). The ultraviolet and infrared spectra of tetrahydro-1,3-oxazines have been discussed (see *Z. Eckstein* and *T. Urbański*, "Advances in Heterocyclic Chemistry", Vol. 2, Academic Press, New York, 1963, pp. 311–341, and refs. cited therein).

Tetrahydro-1,3-oxazines usually possess the chair conformation (*R. Lukeš, K. Blaha* and *J. Kovař*, Collect. Czech. chem. Comm., 1958, **23**, 306), and dipole moment and V_{NH} first overtone infrared measurements indicate that there is a strong preference for the N–H bond to be axial (*M. J. Cook et al.*, J. chem. Soc., Perkin II, 1973, 325).

Similarly, measurements of CH–NH coupling constants, at low temperature, in the ^{1}H-n.m.r. spectra show that the axial orientation for the NH bond is the dominant or sole conformation. Nuclear Oberhauser effects also show partial axial character for the 3-methyl substituent in 3,4,4,6-tetramethyltetrahydro-1,3-oxazine (*H. Booth* and *R. U. Lemieux*, Canad. J. Chem., 1971, **49**, 777; see also *A. R. Katritsky et al.*, J. chem. Soc., B, 1971, 1320; *Y. Allingham, R. C. Cookson* and *T. A. Crabb*, Tetrahedron, 1968, **24**, 4625; *M. Anteunis, G. Swaelens* and *J. Gelan*, ibid., 1971, **27**, 1917; *Yu. Yu. Samitov et al.*, Zhur. org. Khim., 1973, **9**, 193, 201; C.A., 1973, **78**, 110394z, 110429q).

(5) *Oxotetrahydro*-1,3-*oxazines*. The **2-oxotetrahydro**-1,3-**oxazines** are prepared from γ-amino alcohols and ethyl chloroformate or ethyl chlorocarbonate, *e.g.* 2-*oxo*-, m.p. 77°, *nitroso* deriv., m.p. 54° (*R. Delaby et al.*, Compt. rend., 1954, **239**, 674), 2-*oxo*-3,5,5-*trimethyl*-, b.p. 133°/12 mm (*Mannich* and *Wieder*, Ber., 1932, **65**, 385) and 2-*oxo*-3,4,4,6-*tetramethyltetrahydro*-1,3-*oxazine*, m.p. 84–87° (*M. Kohn*, Monatsh., 1905, **26**, 939).

With carbon disulphide, 4-amino-4-methylpentan-2-ol (diacetonealkamine) yields 2-*thio*-4,4,6-*trimethyltetrahydro*-1,3-*oxazine*, m.p. 210–211° (*H. L. Fischer*, U.S.P. 2,326,732/1944; C.A., 1944, **38**, 894). **2-Imino-5-methoxytetrahydro**-1,3-oxazine, *picrate*, m.p. 196–197°, is reported by *D. E. Pearson* and *M. V. Sigal* (J. org. Chem., 1950, **15**, 1055). 2,4-**Dioxotetrahydro**-1,3-**oxazines** are formed when β-hydroxy acids react with sodium cyanate and the resultant carbamates are then treated with thionyl chloride (*R. N. Lacey*, J. chem. Soc., 1954, 845; *E. Testa*, G.P. (D.A.S.), 1,074,585/1954; B.P. 853,954/1961; C.A., 1961, **55**, 12342c):

$$
\underset{R^1}{\overset{R}{>}}C\underset{CO_2H}{\overset{CH_2OH}{<}} \quad \xrightarrow[\text{HCl}]{\text{NaCNO}} \quad \underset{R^1}{\overset{R}{>}}C\underset{CO_2H}{\overset{CH_2OCONH_2}{<}} \quad \xrightarrow{\text{SOCl}_2} \quad
$$

2,6-Dioxotetrahydro-1,3-oxazines, from β-amino acids and phosgene, give polyamides and carbon dioxide on thermal decomposition: 5,5-*dimethyl*-2,6-*dioxotetrahydro*-1,3-*oxazine*, m.p. 126–127° (*H. E. Winberg*, U.S.P. 2,600,596/1962; C.A., 1953, **47**, 7536).

(c) 1,4-Oxazines and related compounds

(i) 1,4-Oxazines

Simple examples of this type of system are rare. Ethyl α-bromoacet- or α-bromopropionamidoacetoacetate (XLIX, R = CO_2Et, X = H or Me) give a substituted 4*H*-1,4-oxazine (L, X = H or Me) on treatment with liquid ammonia, or preferably sodium ethoxide:

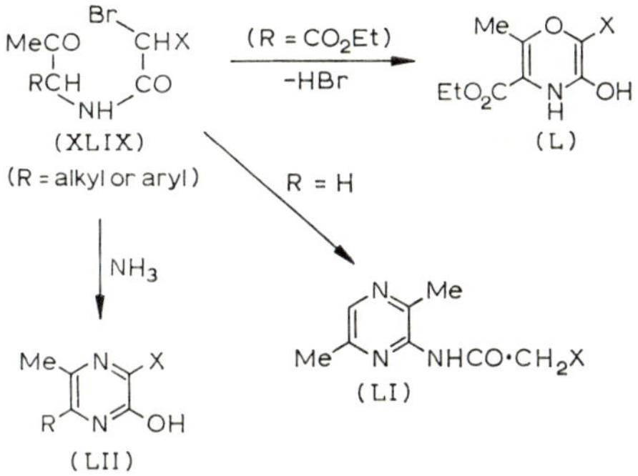

The effect of variation of the substituent R is remarkable since, when R = H, XLIX undergoes self-condensation leading to acylaminopyrazines (LI), but when R = alkyl or aryl, reaction with ammonia gives hydroxypyrazines (LII) instead.

Ethyl 5-hydroxy-2-methyl-4H-1,4-oxazine-3-carboxylate (L, X = H), m.p. 112°, *acid*, m.p. 226° (decomp.) and the corresponding 2,6-*dimethyl* compound (L, X = Me), m.p. 96°, *acid*, m.p. 214° (decomp.). Decarboxylation of the acids gives 5-*hydroxy-2-methyl-*, m.p. 54–55° and 5-*hydroxy-2,6-dimethyl-4H-1,4-oxazine*, m.p. 73°, respectively, both of which are hygroscopic (*G. T. Newbold, F. S. Spring* and *W. Sweeney*, J. chem. Soc., 1950, 909).

(ii) Dihydro-1,4-oxazines

The hydro-1,4-oxazines are almost exclusively tetrahydro compounds, although 5,6-dihydro-2*H*-1,4-oxazine *N*-oxide (LIII), stable only in solution, may be obtained from 4-hydroxymorpholine (see below) by the action of a variety of oxidising agents (*J. F. Elsworth* and *M. Lamchen*, J. chem. Soc., C, 1968, 2423). When the solvent (water or chloroform) is removed, however, polymers believed to be chain structures (*e.g.* as shown) are formed:

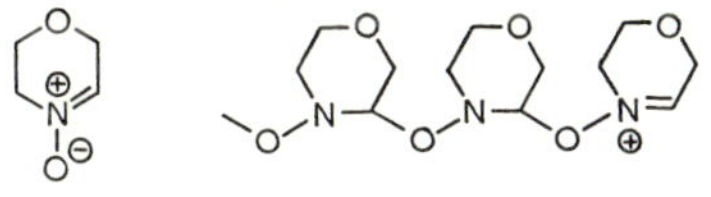

The work of *R. E. Lutz et al.*, establishes ring-chain tautomerism for some α-(β-hydroxyethylamino)deoxybenzoins (LIV, R^1 = Ph) and also for ω-(β-hydroxyethylamino)acetophenone (LIV, R^1 = H). Such compounds are tautomeric with cyclic 2-hydroxymorpholines (LV, R^1 = Ph and H,

respectively) forming various alkoxy derivatives (LV, R = Me, Et or
CH$_2$Ph) and on dehydration yield 5,6-dihydro-1,4-oxazines (LVI, X = Me,
Et or CH$_2$Ph); 2-*phenyl-*, m.p. 163.5–165.5°, and 4-*ethyl-2,3-diphenyl-5,6-
dihydro*-4H-1,4-*oxazine*, m.p. 89–90°, are typical (*Lutz, J. A. Freek* and
R. S. Murphey, J. Amer. chem. Soc., 1948, **70**, 2015; see also *N. H. Cromwell*
and *K. Tsou, ibid.*, 1949, **71**, 993).

4-*Benzoyl-2-(3,4-dihydroxyphenyl)*5,6-*dihydro*-4H-1,4-*oxazine*, m.p. 222–225°, is reported
by *R. Hill* and *G. Powell, ibid.*, 1945, **67**, 1462):

$$\text{(LIV)} \quad\rightleftharpoons\quad \text{(LV)} \quad\longrightarrow\quad \text{(LVI)}$$

(*iii*) *Tetrahydro-1,4-oxazines*: *morpholines*

Tetrahydro-1,4-oxazine or morpholine, so called by its discoverer, *L.
Knorr* (Ber., 1889, **22**, 2081; Ann., 1898, **301**, 1) on account of a supposed
relationship to morphine, is one of the best known oxazine derivatives. It
is widely employed in synthesis, as an organic base and as a solvent.
Morpholine is stable at elevated temperatures and pressures and is commercially useful in boilers as a corrosion inhibitor by restricting carbon dioxide
contamination (*J. J. Maguire*, Ind. Eng. Chem., 1954, **46**, 994). It is used
to form wax-emulsions giving a high gloss and for the preparation of rubber
accelerators (*A. Albert*, "Heterocyclic Chemistry", 1959, Athlone Press,
p. 29).

*Methods of synthesis**. (*1*) Diethanolamine is heated with concentrated
hydrochloric acid or 70% sulphuric acid at 160° for about 12 hours. After
making the reaction product alkaline, morpholine is isolated by steam
distillation. The route, although poor for the preparation of the parent,
gives good yields for many *N*-substituted derivatives. Improvements are
obtained by heating in the presence of zinc salts (*L. W. Jones* and *G. R.
Burns*, J. Amer. chem. Soc., 1925, **47**, 2966), or by heating the amine
hydrochloride at 200–210° for 15 hours without an excess of mineral acid
(*B. L. Hampton* and *C. B. Pollard, ibid.*, 1936, **58**, 2338). The use of concentrated sulphuric acid at 175–180° is claimed to give a nearly quantitative
yield (*L. Médard*, Bull. Soc. chim. Fr., 1936, [v], **3**, 1338).

---

* For a review see *S. F. Potnis* and *K. S. Chitnis* (Chem. Process Eng. (Bombay), 1969,
3, 27).

(*2*) 2,2′-Dihalogenodiethyl ethers are reacted with ammonia or a primary amine. *N*-Substituted derivatives of morpholine are usually obtained in good yield especially with three molecular equivalents of amine to one of ether, or with added sodium or potassium hydroxide (*H. T. Clarke*, J. chem. Soc., 1912, **101**, 1808). Sulphonamides have also been reacted with the halogeno ethers and the resulting *N*-sulphonyl derivatives may be hydrolysed to morpholine (*J. Sand*, Ber., 1901, **34**, 2906).

Both the above general methods have been widely applied with the method of choice decided by the availability of the appropriate amine and dihalogeno-ether.

(*3*) Oxiranes are reacted with ammonia or an aminoacetal. Ethene oxide and β-aminoacetal yield 2-*ethoxymorpholine*, b.p. 253–255° (*Knorr, ibid.*, 1899, **32**, 729). Styrene oxide and ammonia at 90–125° for 8 hours yield 2,6-*diphenylmorpholine*, b.p. 170–185°/3 mm (*W. S. Emerson*, J. Amer. chem. Soc., 1945, **67**, 516).

Morpholine (LVII), b.p. 128.3°, *d* 1.0016 (*A. L. Wilson*, Ind. Eng. Chem., 1935, **27**, 867) is hygroscopic, volatile in ether and steam, and soluble in water in all proportions but not in strong alkali. It is a weaker base (pK_a 8.7) than piperidine (pK_a 11.2) which it may, however, replace as a condensing agent: derivatives include the *hydrochloride*, m.p. 175–176°, *aurichloride*, m.p. 240° (decomp.), *picrate*, m.p. 146–148°, and *picrolonate*, m.p. 255° (decomp.). *Morpholine thiosulphate*, m.p. 152–153°, is prepared from morpholine and sulphur (*E. M. Peters* and *W. T. Smith*, Proc. Iowa Acad. Sci., 1950, **57**, 211; C.A., 1952, **46**, 993). The dipole moment at 20° in benzene is 1.48D (piperidine, 1.17D) in line with a structure of the Sachse *Z*-form (*J. R. Partington* and *D. I. Coomber*, Nature, 1938, **141**, 918).

Morpholine is prepared by the general methods and by patent procedures from dichloro-diethyl ether and ammonia at 50° and 1750 p.s.i. pressure for 24 hours with either nitrogen (*A. W. Campbell*, U.S.P. 2,034,427/1936; C.A., 1936, **30**, 2986) or benzene as diluent (*P. H. Groggins* and *A. J. Stirton*, Ind. Eng. Chem., 1937, **29**, 1353). It is purified by fractionation and then by refluxing over metallic sodium for the removal of water. Morpholine and water do not form a constant boiling mixture.

With oleic and stearic acids, morpholine forms soaps which provide stable oil-in-water emulsions (*Wilson, loc. cit.*). Morpholinium alkyl sulphates are useful detergents of good comptability and germicidal activity (Ann. Rep. appl. Chem., 1951, **36**, 227). Morpholine with ammonium carbonate and refined sugar gives good yields of citric acid by deep fermentation (*ibid.*, 1953, **38**, 720).

The morpholine ring is stable to hydrolysis by water at 200°, 10% sodium hydroxide or concentrated hydrochloric acid at 160° and to oxidation by permanganate (*L. Knorr*, Ann., 1898, **301**, 1) however, ruthenium (VIII) oxide affords 3-oxomorpholines (see p. 453). The ring is opened by cyanogen bromide (*J. von Braun* and *Z. Köhler*, Ber., 1918, **51**, 255).

N-*Methylmorpholine methiodide*, m.p. ∼246°, in the Hofmann degradation gives the expected vinyl β-dimethylaminoethyl ether. This on further treatment forms very little divinyl ether but rather a polymer which on heating gives ethyne (*Knorr* and *H. Matthes*, *ibid.*, 1899, **32,** 736).

A comparison of ring stability in the spiro compounds LVIII and LIX when subjected to the Hofmann degradation shows that not the morpholine ring in LVIII but the adjoining six-membered ring in LIX is cleaved (*von Braun* and *Köhler*, *loc. cit.*):

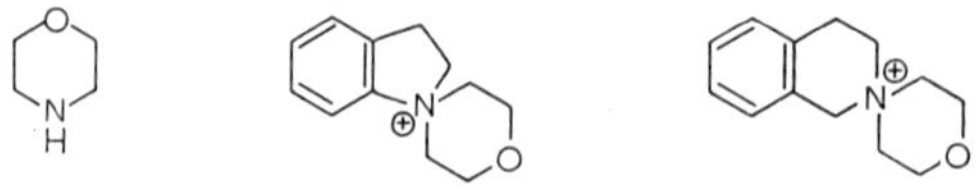

(LVII)   (LVIII)   (LIX)

The use of morpholine in the Willgerodt reaction as modified by *K. Kindler* (Ann., 1923, **431**, 193, 222; Arch. Pharm., 1927, **265,** 389) is exemplified by the preparation of 4-methoxyphenylacetic acid from 4-methoxyacetophenone (*D. A. Shirley*, "Preparation of Organic Intermediates", Wiley, New York, 1951, p. 192) as follows:

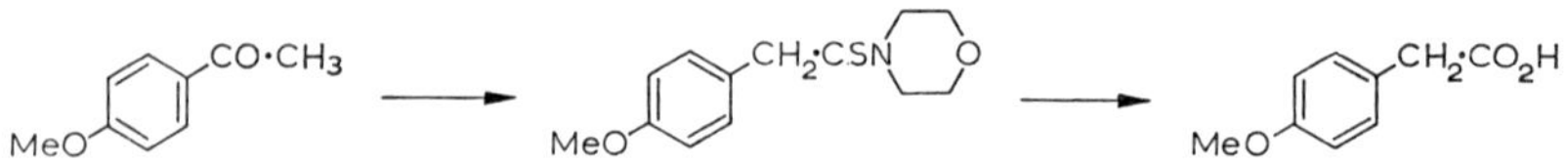

Morpholine may be used as the amine in the Mannich reaction (C.C.C., Vol. I C, p. 44) Acetone, morpholine hydrochloride and paraformaldehyde yield 4-(N-*morpholinyl*)*butane-2-one hydrochloride*, m.p. 149° (free *base*, b.p. 116°/20 mm; *picrate*, m.p. 114°), which is reduced by aluminium amalgam to the corresponding *alcohol*, b.p. 95–100°/2.5 mm, *picrate*, m.p. 142–144° (*R. H. Harradence* and *F. Lions*, J. Proc. roy. Soc., N.S.W., 1939, **72**, 233; C.A., 1939, **33,** 5855).

The reaction of morpholine, cupric nitrate and phenol yields 4,5-(N,N′-*dimorpholinyl*)-1,2-*benzoquinone* as red needles, m.p. 204°, readily converted with *o*-phenylenediamine to 2,3-*dimorpholinylphenazine*, m.p. 221.5°. This on hydrolysis with acid and characterisation of the product as 2,3-diacetoxyphenazine proves the 4,5-positions for the morpholinyl groups in the benzoquinone (*W. Brackman* and *E. Havinga*, Rec. Trav. chim., 1955, **74,** 937).

The oxidation of morpholine by hydrogen peroxide yields 4-*hydroxymorpholine*, b.p. 100–102°/16 mm, which rapidly reduces silver nitrate solution (*R. A. Henry* and *W. M. Dehn*, J. Amer. chem. Soc., 1950, **72**, 2280).

2,6-*Dihydroxymorpholine hydrochloride*, m.p. 121–124°, prepared by hydrolysing diacetalylamine, bis(diethoxyethyl)amine (C.C.C. Vol. I D, p. 57) with hydrochloric acid, is a strong reducing agent (*L. Wolff* and *R. Marburg*, Ann., 1908, **363,** 169).

C-Alkylmorpholines. Cyclisation of the product from an appropriate alkylene oxide and ethanolamine gives morpholine homologues, *e.g.*, propene oxide and ethanolamine yield 2-**methylmorpholine**, b.p. 135–136°, *tosyl* deriv., m.p. 88–89°, and from similar reactions 2-*ethyl-*, b.p. 38–40°/8 mm, *tosyl* deriv., m.p. 116–117°, 3-*ethyl-*, b.p. 156.5–157.5°, 3,3-*dimethyl-*, b.p. 143°, 5-*ethyl-2-methyl-morpholine*, b.p. 164.5–165°, are obtained (*D. L. Cottle et al.*, J. org. Chem., 1946, **11,** 286). Diallylamine with mercuric acetate forms 2,6-bis(chloromercuri)morpholine which on reduction with sodium amalgam gives 2,6-*dimethylmorpholine*, b.p. 140–145° (*A. N. Nesmeyanov* and *I. F. Lutsenko*, Bull. Acad. Sci., U.R.S.S., Classe Sci. chim., 1943, 296; C.A., 1944, **38**, 5498).

N-*Substituted derivatives of morpholine*. The general methods have indicated routes to N-substituted derivatives which are also obtained from morpholine by substitution on the secondary amino group. 4-**Methylmorpholine**, b.p. 115.4°, *d* 0.9051 (both constants are lower than the corresponding values for morpholine), *hydrochloride*, m.p. 205°; *picrate*, m.p. 189–190°, and 4-*benzylmorpholine*, b.p. 128–129°/13 mm, *picrate*, m.p. 184–185°, are typical.

4-**Phenylmorpholine**, m.p. 58°, is prepared from β-hydroxyethylaniline and sulphuric acid or phosphorus pentoxide, or from β,β'-dichlorodiethyl ether and aniline (*H. Adkins* and *R. M. Simington*, J. Amer. chem. Soc., 1925, **47**, 1687) and is a useful intermediate of the dimethylaniline type. It nitrosates readily and the resulting (4-*nitrosophenyl*)*morpholine*, m.p. 100° forms morpholine on alkaline hydrolysis (*O. Kamm* and *J. H. Waldo*, ibid., 1921, **43**, 2223). 4-(4-*Tolyl*)-, m.p. 51°; 4-(1-*naphthyl*)-, m.p. 83° and 4-(2-*naphthyl*)*morpholine*, m.p. 90° (*L. H. Cretcher et al.*, ibid., 1925, **47**, 163, 1173).

Some 2-substituted 4-methylmorpholines are known, *e.g.*, 4-*methyl-2,2-diphenylmorpholine* (LX), m.p. 76–77°, the cyclic form of benadryl, dimethyl(β-diphenylmethoxy)-ethylamine, a widely used antihistamine drug introduced in 1945 (*H. Gilman* and *C. C. Wanser*, ibid., 1951, **73**, 4030; *T. A. Geissman et al.*, ibid., p. 5874; *A. L. Morrison et al.*, J. chem. Soc., 1951, 952):

Crystalline products are obtained by condensing morpholine (2 mol.) with aromatic aldehydes (1 mol.) (also furfural) (*Henry* and *Dehn*, J. Amer. chem. Soc., 1949, **71**, 2271).

4-**Chloromorpholine**, b.p. 52–53°/17 mm, is prepared from morpholine and sodium hypochlorite in the presence of an acetate or phosphate buffer (*W. S. Metcalf*, J. chem. Soc., 1942, 148; the paper reports absorption spectral data for chloroamines). It is a lachrymator and rapidly darkens with the deposition of morpholine hydrochloride after about eight hours. The corresponding 4-bromomorpholine is similarly prepared but the resulting bright yellow solid is not well characterised and decomposes completely in ten hours giving some *morpholine hydrobromide*, m.p. 204°, together with an unstable unidentified material, m.p. 117–119°. Phenyl isocyanate with 4-chloromorpholine gives (4-*chlorophenyl*)*carbamoylmorpholine*, m.p. 196°, formed by rearrangement of the intermediate N-chlorourea (*cf.* N-chloroacetanilide) (*Henry* and *Dehn*, J. Amer. chem. Soc., 1950, **72**, 2280):

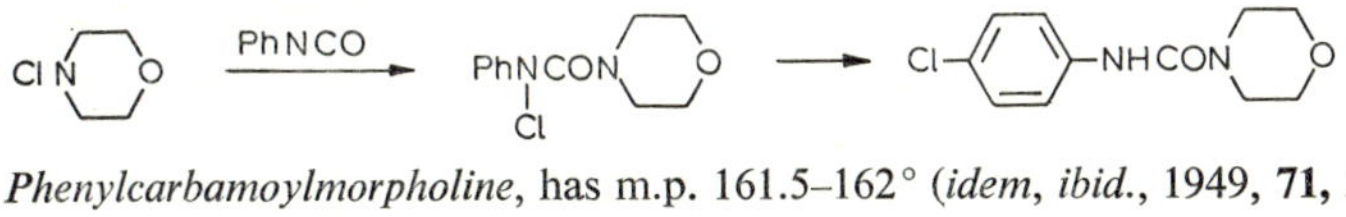

Phenylcarbamoylmorpholine, has m.p. 161.5–162° (*idem, ibid.*, 1949, **71**, 2297).

4-**Aminomorpholine**, b.p. 168°, *hydrochloride*, m.p. 164°, is prepared from hydrazine hydrate and ethylene oxide or by reduction of 4-*nitrosomorpholine*, m.p. 29° (*Knorr* and *H. W. Brownsdon*, Ber., 1902, **35**, 4474). It reacts with potassium cyanate giving *morpho-*

linylsemicarbazide, m.p. 218° (decomp.) (*idem, ibid.*, p. 4477), and with benzaldehyde forming benzylideneaminomorpholine.

Arenediazonium chlorides react with morpholine in the 4-position, giving, *e.g.* colourless 4-*phenylazo-*, m.p. 29–30°, yellow 4-(4-*nitrophenyl*)*azo-*, m.p. 137.5–138.5°, and orange 4-(2-*naphthylazo*)-morpholine, m.p. 99.5–100.5° (*Henry* and *Dehn*, J. Amer. chem. Soc., 1943, **65**, 479).

2-(4-**Morpholinyl**)**ethanol** (*β-morpholinoethanol*), b.p. 227°, 118–120°/24 mm, is obtained from morpholine and ethylene oxide (*Knorr*, Ann., 1898, **301**, 1) or from triethanolamine hydrochloride on heating (*J. H. Gardner* and *E. O. Haenni*, J. Amer. chem. Soc., 1931, **53**, 2763). The cyclisation of triethanolamine by pyrophosphoric acid may be accompanied by phosphorylation (*E. Cherbuliez* and *J. Rabinowitz*, Helv., 1958, **41**, 1168).

3-(4-*Morpholinyl*)*propan-2-ol*, b.p. 97–98°/13 mm, from propylene oxide, gives a hygroscopic *hydrochloride*, m.p. 125–126° (*D. J. Brown et al.*, J. chem. Soc., 1949, S111). Morpholinyl alcohols of the type $R(CH_2)_nOH$ (together with the corresponding dimorpholinylalkanes), in which $n = 4$ and up to 10, are prepared from the appropriate glycol and morpholine (*G. W. Anderson* and *C. B. Pollard*, J. Amer. chem. Soc., 1939, **61**, 3440).

2-(4-**Morpholinyl**)**ethylamine**, b.p. 202°, prepared from the corresponding hydroxy compound *via* the halogeno compound, is a useful reagent for the characterisation of esters. Most of the resulting amides are crystalline but in the few cases in which oils are formed, these are readily converted into crystalline methiodides. The following derivatives are formed by heating the reagent and ester for several hours with the addition of ethylene glycol for volatile esters: *formyl-*, oil (*methiodide*, m.p. 134°), *acetyl-*, m.p. 95.2°, *propionyl-*, m.p. 85°, *benzoyl-2-(4-morpholinyl)ethylamine*, m.p. 123.4° (*R. W. Bost* and *L. V. Mullen, ibid.*, 1951, **73**, 1967).

Morpholine readily forms a 4-*formyl-*, m.p. 17.5°, -*acetyl-*, m.p. 14.5° (*L. Médard*, Bull. Soc. chim. Fr., 1936, [v], **3**, 1338), -*benzoyl-*, m.p. 74–75° and -4-*tosyl-* derivative, m.p. 147° and with ethyl chloroformate gives *ethyl morpholine-4-carboxylate*, b.p. 220–221°, *amide*, m.p. 110–113° (*Knorr*, Ann., 1898, **301**, 8).

4-**Cyanomorpholine**, b.p. 123–124°/18 mm, is prepared from the 4-chloro compound and potassium cyanide (*Henry* and *Dehn, loc. cit.*) or from morpholine and cyanogen bromide in ether (*G. F. D'Alelio* and *J. J. Pyle*, U.S.P. 2,375,628/1945).

4-**Morpholinylacetonitrile**, b.p. 232°, m.p. 61–62°, is prepared from morpholine, sodium bisulphite, formaldehyde and aqueous potassium cyanide whilst in this reaction, acetaldehyde and acetone yield the α-*methyl-*, b.p. 97–98°/5 mm, and the α,α-*dimethyl-*, b.p. 123–124°/20 mm, analogues, respectively. The last two compounds from crude picrates which on crystallisation decompose into morpholine picrate (*H. R. Henze et al.*, J. Amer. chem. Soc., 1957, **79**, 6230).

Morpholine has been widely used in the synthesis of products of potential chemotherapeutic value, generally replacing conventional secondary amines. *Ethyl 4-morpholinyl-2-phenylbutyrate*, b.p. 135°/1.5 mm, *hydrochloride*, m.p. 169°, shows enhanced analgesic activity when the morpholino replaces a dimethylamino group (*R. M. Anker* and *A. H. Cook*, J. chem. Soc., 1948, 806). The pharmaceutical value of the morpholinyl group in other substituted phenylpropylamines is supported by other studies (*P. A. J. Janssen* and *J. C. Janssen*, J. Amer. chem. Soc., 1956, **78**, 3862).

2-Morpholinylethylmorphine (pholcodine) is a cough suppressant (Ann. Rept. appl. Chem., 1953, **38**, 143).

Some morpholinylalkyl esters of 2-alkoxycinchoninic acids, as hydrochlorides, are local anaesthetics (*J. H. Gardner* and *W. M. Hammel*, J. Amer. chem. Soc., 1936, **58**, 1360), as are esters of 2-(4-morpholinyl)-ethanol and -propanol and the 2-*naphthylcarbamate*, m.p. 96.5°, of the former alcohol (*Gardner, D. V. Clarke* and *J. Semb, ibid.*, 1933, **55**, 2999).

(iv) Oxomorpholines (morpholones)

2-**Oxomorpholines** (2-**morpholones**) are cyclic lactones and are formed by the spontaneous cyclisation of the appropriate 5-hydroxycarboxylic acids, especially *N*-substituted derivatives:

The following are typical: 4-*methyl-*, b.p. 233°, *picrate*, m.p. 190–192°, *methiodide*, m.p. 228° (decomp.) (*Knorr*, Ann., 1899, **307**, 199); 4-*benzenesulphonyl-6-methyl-*, m.p. 128–128.5° and 4-(4-*toluenesulphonyl*) deriv., m.p. 92–93°; 6,6-*dimethyl-* (same derivs., m.p. 134–135° and 133–134°, respectively) (*W. Cocker*, J. chem. Soc., 1943, 373); 4-*phenyl-*, m.p. 75°; 4-(2-*hydroxyethyl*)-2-*morpholone*, b.p. 175–176°/5 mm (*A. Kiprianov*, C.A., 1928, **22**, 3134).

Ethylene oxide and its analogues lead to homologues of the 2-morpholones, *e.g.* α-ethylaminopropionic acid with the former yields 3-*methyl-*, b.p. 100–110°/6 mm, *picrate*, m.p. 156–157°, whilst isobutylene oxide gives 4-*ethyl-3,6,6-trimethyl-2-morpholone*, b.p. 121.5–123°/16 mm, *picrate*, m.p. 155° (*Kiprianov* and *D. M. Krasinskaya*, C.A., 1931, **25**, 5143).

4-*Methyl-3,3-diphenyl-2-morpholone* has m.p. 135–136°, *picrate*, m.p. 174–175° (*H. S. Mosher et al.*, J. Amer. chem. Soc., 1953, **75**, 5326). 4-*Methoxycarbonyl-2-morpholone*, m.p. 185° (decomp.), is prepared from ethanolamine, sodium cyanide and formaldehyde (*L. W. Ziemlak et al.*, J. org. Chem., 1950, **15**, 255).

The 2-oxomorpholines show normal basic properties and the lactone group is readily hydrolysed.

3-**Oxomorpholine** (3-**morpholone**), m.p. 105°, b.p. 145°, with a carbonyl group adjoining the nitrogen atom is a cyclic amide and is formed from 2-aminoethoxyacetic acid. This acid has a tendency to form a linear crystalline polyamide but the monomer does not polymerise when heated for 24 hours at 140° (*R. Leimu* and *J. C. Jansson*, Suomen. Kemi, 1945, **18B**, 40; C.A., 1947, **41**, 769). Ruthenium (VIII) oxide and sodium metaperiodate in a two-phase system (carbon tetrachloride/water) oxidize 4-substituted morpholines to the corresponding 3-oxo derivatives (*R. Perrone, G. Bettoni* and *V. Tortorella*, Synthesis, 1976, 598). Several *gem*-diphenyl compounds have been examined as analgetics. 4-*Methyl-2,2-diphenyl-3-morpholone*, m.p. 97–98°, is prepared by heating sodium 2-(2-dimethylaminoethoxy)-2,2-diphenylacetate with thionyl chloride (*A. L. Morrison et al.*, J. chem. Soc., 1950, 2887; see also *Mosher et al., loc. cit.*):

The diphenylmethylmorpholone is reduced by lithium tetrahydridoaluminate to 4-methyl-2,2-diphenylmorpholine with some *3-hydroxy-4-methyl-2,2-diphenylmorpholine*, m.p. 97–99°, *hydrochloride*, m.p. 194–195°; *picrate*, m.p. 164–165°. The latter with nitromethane yields the 4-*methyl-3-nitromethyl-2,2-diphenylmorpholine*, m.p. 144–145°, and on reduction gives the corresponding 3-*aminomethyl* derivative (*dihydrochloride*, m.p. 200–203°) (*idem, ibid.*, 1951, 952).

Dioxomorpholines. There are three possible dioxomorpholines, 2,5-, 2,6- and 3,5-dioxomorpholine; representative members of each type are known.

2,5-**Dioxomorpholines** combine the characteristics of lactone and amide groups. *N*-Aryl compounds are prepared by the lactonisation of appropriate *N*-glycoloylglycines: 4-*phenyl*, m.p. 169°, 4-(2-*tolyl*)-2,5-*dioxomorpholine*, m.p. 108–109° (*P. W. Abenius*, J. pr. Chem., 1889, [ii], **40**, 500).

(LXI)

Enniatin A (Lateritiin-I) $C_{36}H_{63}O_9N_3$, m.p. 121–122°, $[\alpha]_D^{20°}$ $-95.6°$, is a peptolide antibiotic from *Fusarium lateritium*; the structure of this molecule and related metabolites is discussed in a review by *E. von Schröder* and *K. Lübke* (Experientia, 1963, **19**, 57). On hydrolysis with aqueous sodium hydroxide, it gives α-hydroxyisovalerylvaline, which lactonises to 3,6-*diisopropyl-4-methyl-2,5-dioxomorpholine*, m.p. 82°, in which the *trans* configuration LXI is preferred (*A. H. Cook et al.*, J. chem. Soc., 1949, 1022).

2,5-Dioxomorpholines distil unchanged and are not much affected by concentrated nitric acid, at 100° for 1 hour, by potassium permanganate or by bromine in chloroform.

2,6-**Dioxomorpholines** are the anhydrides of iminodiacetic acid (diglycollamidic acid) but in the enolic form may be regarded as 2,6-dihydroxy-1,4-oxazines:

2,6-*Dioxo-4-phenylmorpholine*, m.p. 148°, from aniline-*N,N*-diacetic acid (*C. A. Bischoff* and *A. Hausdörfer*, Ber., 1892, **25**, 2272) and 1-(2,6-*dioxo-3,3,5,5-tetramethylmorpholinyl*) *phenylacetic acid*, m.p. 180–181° (decomp.), are known (*G. L. Stadnikow*, J. Russ. phys. chem. Soc., 1911, **43**, 1235; Chem. Ztbl., 1912, I, 1621). The former is readily hydrolysed by boiling water and cleaved by aniline giving the appropriate dicarboxylic acid and acid-anilide, respectively.

3,5-Dioxomorpholines. 3,5-*Dioxomorpholine*, m.p. 142°, the cyclic imide of diglycollic acid (C.C.C. Vol. I D, p. 94), is prepared from diglycollic anhydride and ammonia (*W. Heintz*, Ann., 1863, **128**, 134; *M. A. Wurtz*, Ann., 1863, [iii], **69**, 345). Other primary amines readily yield *N*-substituted derivatives.

4-*Methyl-*, m.p. 78°, -*ethyl-*, b.p. 95°/15 mm (hygroscopic), -*phenyl-*, m.p. 195°, -4-*tolyl-*, m.p. 182°, -(1-*naphthyl*)-, m.p. 174–176°, -(2-*naphthyl*)-3,5-*dioxomorpholin* , m.p. 172.5° (*M. Sido*, Ber. pharm. Ges., 1921, **31**, 48; C.A., 1922, **16**, 900; *R. Anschütz* and *S. Jaeger*, Ber., 1922, **55**, 675). The alkyl compounds especially 4-(n-*propyl*)- and 4-(n-*butyl*)-3,5-*dioxomorpholine*, both oils, have a very sweet taste (not observed for the parent or aryl compounds) but they are too readily hydrolysed to be of use in this connection.

Diglycollic esters, from α-bromo esters and the sodium derivative of an α-hydroxy ester, are converted into the imide *via* the diamide:

$$\underset{\substack{|\\ CO_2Et}}{RCH}\!\!-\!ONa \;\; + \;\; \underset{\substack{|\\ CO_2Et}}{Br\!-\!CHR^1} \;\longrightarrow\; \underset{CO_2Et\;\;CO_2Et}{R\!-\!CH\!-\!O\!-\!CH\!-\!R^1} \;\longrightarrow\; \underset{CONH_2\;CONH_2}{R\!-\!CH\!-\!O\!-\!CH\!-\!R^1} \;\longrightarrow\; \text{(LXII)}$$

Dilactyldiamide (LXII, R = R^1 = Me) in aqueous solution undergoes spontaneous resolution and on heating at 160–170° forms the imide 3,5-*dioxo*-2,6-*dimethylmorpholine*, m.p. 122°, but this is always optically inactive (*P. Vielès*, Ann., 1935, [ii], **3**, 143; *E. Jungfleisch* and *M. Godchot*, Compt. rend., 1907, **145**, 72).

As cyclic imides the 3,5-dioxomorpholines are readily hydrolysed to the corresponding acid amides by barium hydroxide and to the diglycollic acids by potassium hydroxide.

2. Benzoxazines

Nominally there are nine possible benzoxazines, illustrated in the formulae I–IX, some are readily interconverted by tautomerism.

The numbering system starts from the apex of the heterocycle selecting a heteroatom where possible and ensuring that oxygen has the lowest number. The chemistry of some of the above systems remains relatively unexplored, whilst that of others is well established.

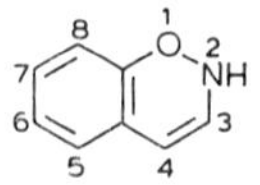

(I)
2*H*-1,2-Benzoxazine

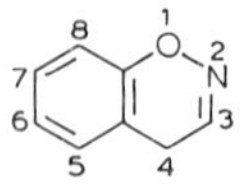

(II)
4*H*-1,2-Benzoxazine

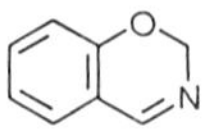

(III)
2*H*-1,3-Benzoxazine

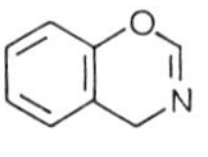

(IV)
4*H*-1,3-Benzoxazine

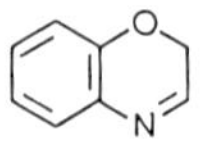

(V)
2*H*-1,4-Benzoxazine

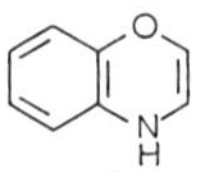

(VI)
4*H*-1,4-Benzoxazine

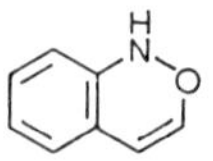

(VII)
1*H*-2,1-Benzoxazine

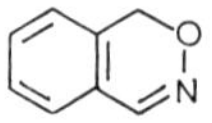

(VIII)
1*H*-2,3-Benzoxazine

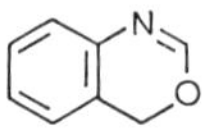

(IX)
4*H*-3,1-Benzoxazine

(a) *1,2-Benzoxazines*

The tautomeric 2*H*- and 4*H*-1,2-benzoxazines are poorly represented in the chemical literature, although 7-*nitro*-4-*oxo*-3-*phenyl*-1,2-*benzoxazine*, m.p. 169°, has been known for many years (*G. Bishop* and *O. L. Brady*, J. chem. Soc., 1926, 810).

(b) *1,3-Benzoxazines and their hydro derivatives*

(i) *2H-1,3-Benzoxazines*

4-Phenyl-2*H*-**1,3-benzoxazine** (XI, R = H), m.p. 49–50°, is obtained by treatment of 2-methyl-3-phenyl-1,2-benzisoxazolium tetrachloroferrate (X, R = H) with sodium hydroxide; the reaction probably proceeds through an ylid intermediate (*J. F. King* and *R. Durst*, Canad. J. Chem., 1962, **40**, 882):

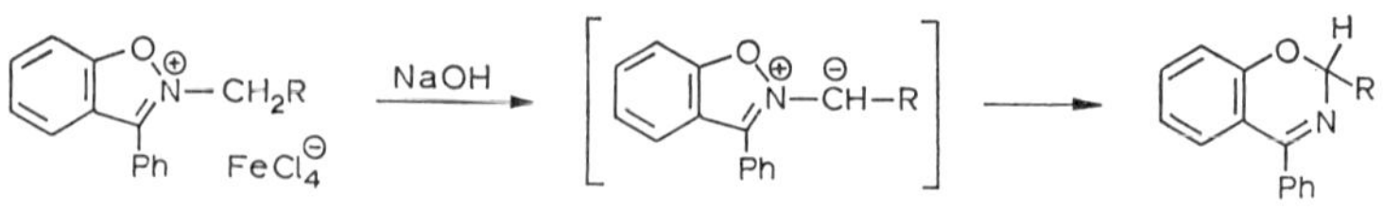

(X) NaOH (XI)

2,4-*Diphenyl*-2H-1,3-*benzoxazine* (XI, R = Ph), m.p. 97–99°, is prepared similarly from 2-benzyl-3-phenyl-1,2-benzisoxazolium perchlorate (X, R = Ph).

2-*Methyl*-4-*phenyl*-2H-1,3-*benzoxazine*, m.p. 33–35°, b.p. 175–177°/11 mm, *picrate*, m.p. 128–129° (its preparation involves several steps) is readily hydrolysed to 2-hydroxybenzophenone, acetaldehyde and ammonia (*E. P. Kohler* and *W. F. Bruce*, J. Amer. chem. Soc., 1931, **53**, 644).

2-Hydroxy-6-nitrobenzaldehyde with ammonium acetate forms an aldehyde–ammonia type of compound, which cyclises by loss of water forming 2-(2-*hydroxy-6-nitrophenyl*)-5-*nitro*-2H-1,3-*benzoxazine* (XII), m.p. 204–205°, *acetate*, m.p. 164° (*R. J. S. Beer et al.*, J. chem. Soc., 1948, 1605):

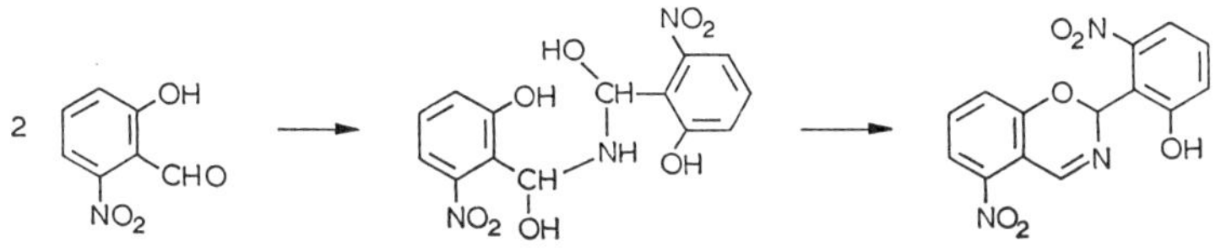

2-Methyl-2,3-dihydro-1,3-benzoxazin-4-one, m.p. 146°, is prepared from salicylamide, paraldehyde and hydrogen chloride (*W. L. Hicks, ibid.*, 1910, **97**, 1032) and in small yield from this amide and vinyl acetate (*D. T. Mowry et al.*, J. Amer. chem. Soc., 1947, **69**, 2358). On oxidation by chromic acid the oxazine gives *N*-acetylsalicylamide whilst hydrolysis with alkali yields ethylidenesalicylamide:

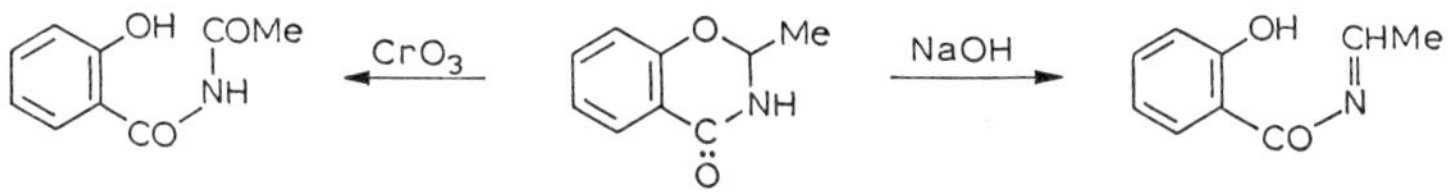

2,2-*Dimethyl-4-methoxy-2H-1,3-benzoxazine*, b.p. 108–110°/13 mm, *acetyl* deriv., m.p. 30–32°, is prepared by the methylation of 2,2-*dimethyl-2,3-dihydro-1,3-benzoxazine-4-one*, m.p. 137°, from salicylamide and acetone (*H. O. L. Fischer et al.*, Ber., 1932, **65**, 1032):

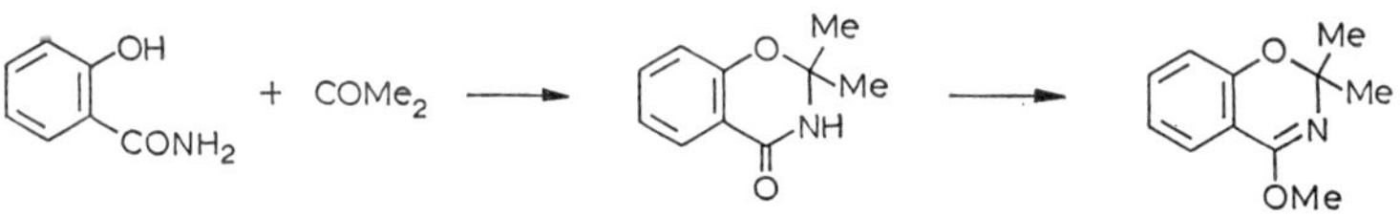

3,4-*Dihydro-1,3-benzoxazin-2-one*, m.p. 188°, results from the cyclisation of 2-hydroxyphenylacetyl azide (*H. Lindemann* and *W. Schultheis*, Ann., 1928, **464**, 246):

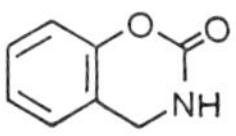

A general synthesis of 3,4-*dihydro*-1,3-*benzoxazin*-2-*ones* has been developed by *L. Bernardi et al.* (Experientia, 1968, **24**, 774): an appropriate amino ester is reductively alkylated by hydrogenation in presence of palladium on charcoal in ethanol solution containing a molecular equivalent of a 2-acylphenol. The resultant 2-hydroxybenzylamines are then treated with phosgene, thus affording 3-alkoxycarbonylalkyl-3,4-dihydro-1,3-benzoxazin-2-ones, alkaline hydrolysis of which gives the corresponding carboxylic acid. Finally, the latter are reacted with ammonia, amines, hydroxylamine or hydrazine in the presence of dicyclohexylcarbodiimide and *N*-hydroxysuccinimide to give 3-substituted 3,4-dihydro-1,3-benzoxazin-2-ones. Some compounds prepared in this way are listed in Table 2.

TABLE 2

SOME 3,4-DIHYDRO-1,3-BENZOXAZIN-2-ONE-DERIVATIVES

R^1	R^2	R^3	R^4	m.p.(°C)
$CH_2 \cdot CH_2 \cdot CONH_2$	H	H	H	65–66
$CH_2 \cdot CONH_2$	H	H	H	203–205
$CONH_2$	H	H	H	181–183
$CH_2 \cdot CH_2 \cdot CO_2H$	H	H	H	130–131
$CH_2 \cdot CO_2H$	H	H	H	169–171
$CH_2 \cdot CH_2 \cdot CO_2Et$	H	H	H	42
$CH_2 \cdot CONHMe$	H	H	H	186–188
$CH_2 \cdot CON(Et)_2$	H	H	H	90–92
$CH_2 \cdot CONHCH_2 \cdot CH_2OH$	H	H	H	183–184
$CH_2 \cdot CONH \cdot NH_2$	H	H	H	180–183
CH_2CN	H	H	H	136–137
$CH_2 \cdot CONHOH$	H	H	H	187–188
$CHMe \cdot CONH_2$	H	H	H	175–176
$CH(CH_2Ph) \cdot CONH_2$	H	H	H	165–168
$CH_2 \cdot CONH_2$	H	H	Me	144–145
$CH_2 \cdot CONH_2$	H	H	Et	185–186
$CH_2 \cdot CONH_2$	OMe	H	H	219–220
$CH_2 \cdot CONH_2$	Br	Br	H	240–243
$CH_2 \cdot CONH_2$	NO_2	NO_2	H	195–197
$CH_2 \cdot CSNH_2$	H	H	H	180–182

Similarly *E. L. May* (J. med. Chem., 1967, **10**, 505) reports the preparation of 3-cyclo-hexyl-2-cyclohexylimino-1,3-benzoxazin-4-one and the naphthyl analogue by the action of dicyclohexylcarbodiimide upon salicylic acid and 1-hydroxynaphthalene-2-carboxylic acid, respectively:

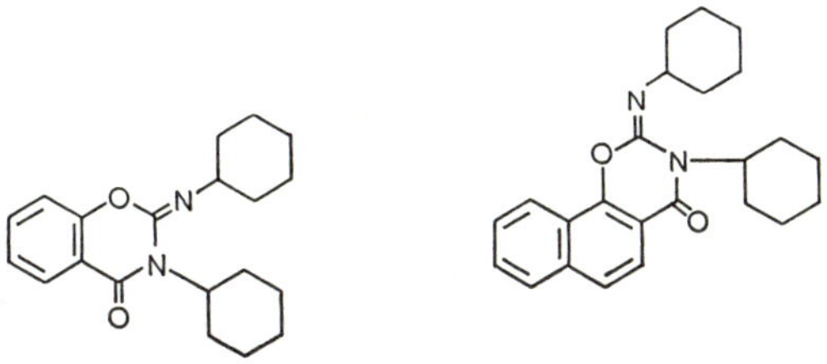

Phosgene condenses with salicyloylamide in pyridine to give 1,3-*benzoxazine*-2,4(3H)-*dione*, m.p. 227°, which forms crystalline metal salts: its *benzoyl* derivative has m.p. 172°; N-*methyl* derivative, m.p. 146° (*A. Einhorn* and *C. Mettler*, Ber., 1902, **35**, 3647). The compound can also be obtained using ethyl chloroformate or urea in place of phosgene and it is also formed by rearrangement of the 2-monoxime of 2,3-dioxo-2,3-dihydro-benzofuran:

N-Aryl derivatives of 1,3-benzoxazine-2,4(3*H*)-dione can be obtained by condensing salicylamide or a salicylic ester with the *S*-methyl derivative of a diarylthiourea, MeSC(NHAr) = NAr (*J. F. Deck* and *F. B. Dains*, J. Amer. chem. Soc., 1933, **55**, 4986). *Kemp* and *Woodward* (Tetrahedron, 1965, **21**, 3019) utilize the reaction of cyanate or thiocyanate ions with 2-ethyl-1,2-benzisoxazolium tetrafluoroborate to prepare 4-*ethylimino*-1,3-*benzoxazin*-2(3H)-*one*, m.p. 193.5–194°, and 4-*ethylimino*-1,3-*benzoxazine*-2(3H)-*thione*, m.p. 207.8–208°:

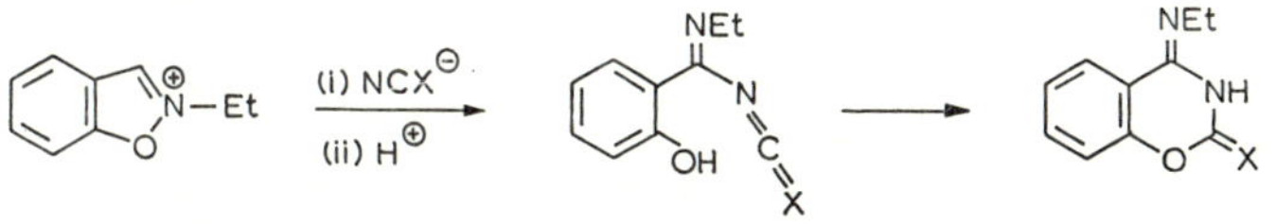

X = O or S

3,4-Dihydro-2*H*-1,3-benzoxazines may be obtained by Mannich type reactions upon phenols (*W. J. Burke*, J. Amer. chem. Soc., 1949, **71**, 609; *P. B. Talukar, S. K. Sengupta* and *A. K. Dutta*, J. Indian chem. Soc., 1970, **47**, 187) and a number of 3-acyl derivatives show activity against pathogenic fungi affecting plants (*H. Huber-Emden*, G.P. 2,018,625/1970; C.A., 1971, **74**, 22856).

2-Hydroxybenzylamines react with α-dicarbonyl compounds to yield 3,3′,4,4′-tetrahydro-2,2′-bi-2H-1,3-benzoxazines (*H. Kanatomi* and *I. Murase*, Bull. chem. Soc. Japan, 1970, **43**, 226):

R = H or Me

With formaldehyde, 2-hydroxybenzylamine condenses to give the methylene derivative XIII, see p. 444 (*F. W. Holly* and *A. C. Cope*, J. Amer. chem. Soc., 1944, **66**, 1877).

(XIII)

(*ii*) 4H-*1,3-Benzoxazines*

Starting materials for the synthesis of 4H-1,3-benzoxazines are derivatives of 2-hydroxybenzylamine or of salicylic acid thus: O,N-diacetyl-2-hydroxy-benzylamine when heated in a current of hydrogen chloride forms 2-**methyl-**4H-1,3-**benzoxazine** (*S. Gabriel*, Ann., 1915, **409**, 325). The base is unstable and is isolated as the *picrate*, m.p. 187° (decomp.).

Derivatives of salicylic acid have been used in a variety of ways to synthesise 4H-1,3-benzoxazines or their dihydro derivatives. A simple example is the cyclisation by dehydration of O-benzoylsalicylamide (or of the N-benzoyl compound) to give the 4-oxo derivative XIV. The same product is formed by the interaction of phenyl salicylate and N-phenylbenzamidine, as shown:

(XIV)

2-Alkyl derivatives are available through the condensation of salicylamides and ketones (*G. N. Dorofeeko*, Khim. gerot. Soedin, 1975, 460).

4-*Oxo-2-phenyl-1,3-benzoxazine* has m.p. 106°; acids hydrolyse it to N-benzoylsalicylamide (*A. W. Titherley*, J. chem. Soc., 1910, **97**, 200). 2-Hydroxyacetophenone imines *e.g.* XV (where R = H or Ph), when treated with ethyl orthoformate followed by distillation of ethanol, afford 4-methylene derivatives of 4H-1,3-benzoxazine (*R. R. Schmidt, D. Schwille* and *H. Wolf*, Chem. Ber., 1970, **103**, 2760):

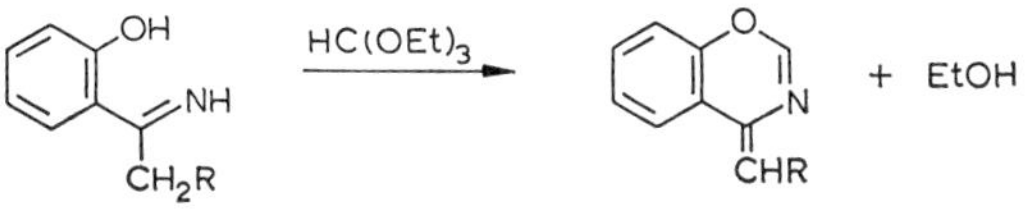

(XV)

V. A. Zagorevskii, K. I. Lopalma and *S. M. Klynev* describe the preparation of 2,4,4-trisubstituted-4*H*-1,3-benzoxazines from 2-hydroxyphenylmethanol and nitriles in the presence of perchloric and sulphuric acids (U.S.S.R. P. 225,197/1968; C.A., 1969, **70**, 28928d).

(c) *1,4-Benzoxazines and their hydro derivatives*

(i) *1,4-Benzoxazines*

The 2*H*-1,4-benzoxazine and 4*H*-1,4-benzoxazine systems are tautomeric and only a few representative structures are known, although recently *A. McKillop* and *T. S. B. Sayer* (J. org. Chem., 1976, **41**, 1079) report that bis-copper(II) complexes derived from 2-nitrosophenols react with dimethyl acetylenedicarboxylate by [4 + 2]cycloaddition to give the corresponding 4-hydroxy-4*H*-1,4-benzoxazines:

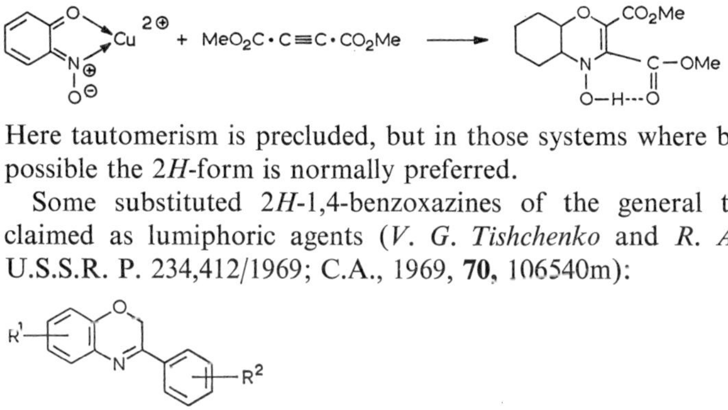

Here tautomerism is precluded, but in those systems where both forms are possible the 2*H*-form is normally preferred.

Some substituted 2*H*-1,4-benzoxazines of the general type XVI are claimed as lumiphoric agents (*V. G. Tishchenko* and *R. A. Semenenko*, U.S.S.R. P. 234,412/1969; C.A., 1969, **70**, 106540m):

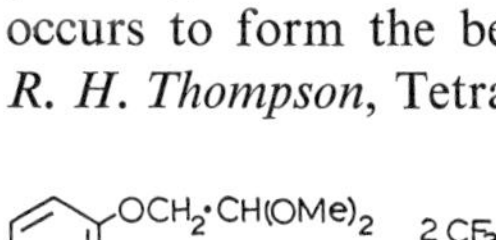

(XVI)

2*H*-1,4-**Benzoxazine** (XVII) itself can be generated *in situ* from 1-(2-aminophenoxy)-2,2′-dimethoxyethane in trifluoroacetic acid, further reaction then occurs to form the benzoxazine dye XVIII (*F. Chiroccara, G. Prota* and *R. H. Thompson*, Tetrahedron, 1976, **32**, 1407):

(XVII)   (XVIII)

The mechanism by which XVIII is produced is uncertain.

3-Phenyl-2*H*-1,4-benzoxazine gives 2-hydroxy-3-phenyl-2*H*-1,4-benzoxazine, 3-phenyl-1,4-benzoxazin-2-one and 2-phenylbenzoxazole through autoxidation in air. A suggested mechanism is as follows (*F. Chioccara et al.*, Tetrahedron, 1976, **32**, 2033):

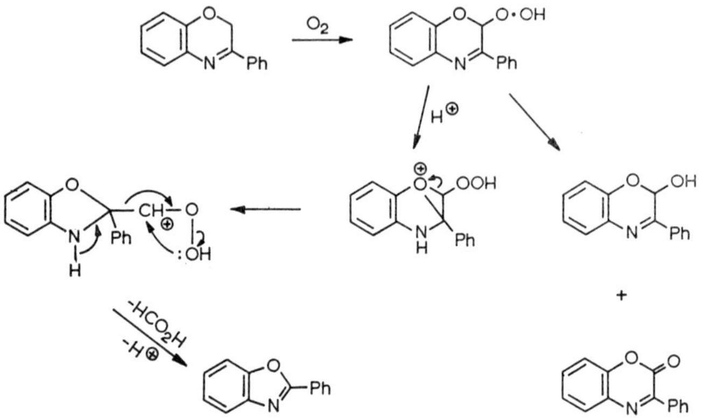

(ii) Dihydro-1,4-benzoxazinones

The most familiar examples of the class can also be formulated as 3-hydroxy-4*H*-1,4-benzoxazines, although their chemistry suggests that they are best considered as 2*H*-1,4-benzoxazin-3(4*H*)-ones (see *Mme. Ramart-Lucas* and *M. V. Vantu*, Bull. Soc. chim. Fr., 1936, [v], **3**, 1165). Such compounds may be prepared from 2-chloroacetamidophenols by treatment with sodium hydroxide (*K. von Auwers* and *E. Frese*, Ber., 1926, **59**, 539; *E. Puxeddu* and *G. Sanna*, Gazz., 1931, **61**, 158), or from the corresponding *O*-methyl ethers by demethylation with aluminium chloride (*5. W. Cook et al.*, J. chem. Soc., 1952, 3904; *J. D. Loudon* and *J. Ogg*, ibid., 1955, 739):

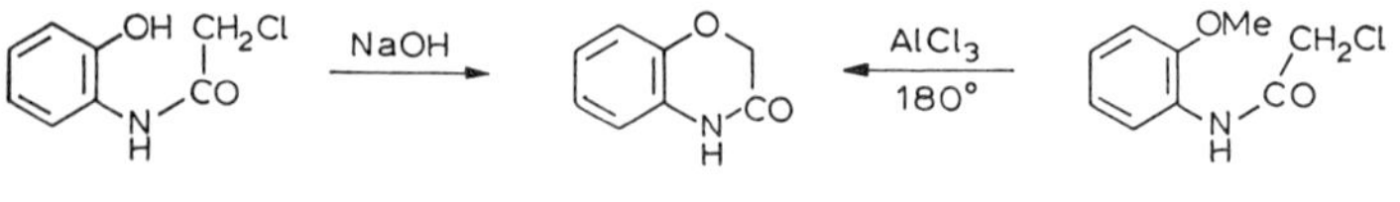

An alternative route involves the reduction of 2-nitrophenoxyacetic acids (*G. Newbery* and *M. A. Phillips*, ibid., 1928, 3046; *W. G. Christiansen*, J. Amer. chem. Soc., 1926, **48**, 460):

The parent, 2*H*-1,4-**benzoxazin**-3(4*H*)-**one**, m.p. 173° (*W. A. Jacobs* and *M. Heidelberger*, *ibid.*, 1917, **39**, 2188), is not acetylated by acetic anhydride but the silver salt with acetyl chloride gives an *acetyl* deriv., m.p. 77° (*H. L. Wheeler* and *B. Barnes*, Amer. Chem. J., 1898, **20**, 565). It is best methylated by methyl iodide and alkali giving an N-*methyl* deriv., m.p. 58°, which is reduced by lithium tetrahydridoaluminate to 4-methylphenomorpholine (see p. 464). Both the parent and *N*-methyl compound form benzylidene derivatives at the reactive 2-position: 2-*benzylidenephenomorphol*-3-*one*, m.p. 260–261°, and its 4-*methyl* deriv., m.p. 155–156°.

2-*Methyl*-2H-1,4-*benzoxazin*-3(4H)-*one*, m.p. 143–144°, from *N*-(2-chloropropionyl)-2-methoxyaniline and aluminium chloride, gives on methylation the 2,4-*dimethyl* deriv., m.p. 49.5–50°. Other compounds include 5-*methyl*-, m.p. 188–190° and 6-*methyl*-2H-1,4-*benzoxazin*-3(4H)-*one*, m.p. 229–230°, 3,6-*dihydroxy*-, m.p. 249–250°, 7-*hydroxy*-2H-1,4-*benzoxazin*-3(4H)-*one*, m.p. 208–209° and their 4-*methyl* derivs., m.p. 86–87°, not given, 209° and 180–182° respectively (*Loudon* and *Ogg*, loc. cit.; *Puxeddu* and *Sanna*, loc. cit.). Some derivatives of 2*H*-1,4-benzoxazin-3(4*H*)-one substituted with aminoalkyl and hydroxyalkyl have been prepared by Madame *Thuillier* and her co-workers (Eur. J. Med. Chem. Chim. Therap., 1975, **10**, 37). In this reference a number of synthetic routes are described.

From the appropriate 2-aminonitrophenols all four nitro-2*H*-1,4-benzoxazin-3(4*H*)-ones having the nitro group in the aromatic ring are prepared and the amines are obtained by reduction: 5-*nitro*-, m.p. 115–116°, 6-*nitro*-, m.p. 233–234°, 7-*nitro*-, m.p. 232° and 8-*nitro*-2H-1,4-*benzoxazin*-3(4H)-*one*, m.p. 255°, and the corresponding *amines*, m.p.'s 236, 255, 220 and 272°, respectively (*Newbery* and *Phillips*, loc. cit.). Direct nitration of 2*H*-1,4-benzoxazin-3(4*H*)-one gives the 6-nitro compound together with a second product, not identical with a *Bz*-nitro compound, considered to be 2-*nitro*-2H-1,4-*benzoxazin*-3(4H)-*one*, m.p. 155–157°. On reduction with stannous chloride in hydrochloric acid it reforms the parent, but with iron and water (drop of acetic acid) a base (*acetyl* deriv., m.p. 283°) is obtained.

Nitration of 7-*acetamido*-2H-1,4-*benzoxazin*-3(4H)-*one*, m.p. 276°, gives only the 6-nitro compound which is hydrolysed to 7-*amino*-6-*nitro*-, m.p. 317° (decomp.), and deaminated to 6-nitro-2*H*-1,4-benzoxazin-3(4*H*)-one, thereby proving the position of entry of the nitro group. The 8-*acetamido* derivative, m p. 257°, was similarly examined and gave a mixture of 6- and 7-nitro compounds (*I. E. Balaban*, J. chem. Soc., 1929, 2607).

2,3-**Dioxo**-4H-1,4-**benzoxazine** (2,3-*dioxophenomorpholine**), m.p. 258–259°, is readily prepared from 2-aminophenol and ethyl oxalate or oxalyl chloride and similarly for the 6-*methyl* homologue, m.p. 272–273° (*R. E. Meijer* and *A. Seeliger*, Ber., 1896, **29**, 2643; *Puxeddu* and *Sanna*, loc. cit.; Gazz., 1932, **62**, 558).

(iii) *Dihydro-1,4-benzoxazines*

3,4-Dihydro-2*H*-1,4-benzoxazine* is sometimes known as phenomorpholine (a term which will be used for convenience here) and incorrectly

---

* In this account these compounds are referred to as 3,4-dihydro-2*H*-1,4-benzoxazines rather than as derivatives of 2,3-dihydro-4*H*-1,4-benzoxazines.

as tetrahydro-1,4-benzoxazine or benzomorpholine. Compounds of this group may be synthesised in the following ways:

(*1*) Phenomorpholine is formed by heating 2-(2-chloroethylamino)phenol with alkali (*L. Knorr*, Ber., 1889, **22**, 2085) or by heating the corresponding 2-(2-hydroxyethylamino)phenol *(idem, ibid.)* or hydroxyethyl-2-anisidine (*J. von Braun* and *J. Seemann, ibid.*, 1922, **55**, 3818) with hydrochloric acid at 160–180°.

(*2*) Reduction of 2-nitrophenyl 2-(2-hydroxyethyl) ethers is accompanied by cyclisation; thus 6-amino-3,4-dihydro-2*H*-benzo-1,4-oxazine has been prepared (*A. Fairbourne* and *H. Toms*, J. chem. Soc., 1921, **119**, 2076). Similarly, alkyl or aryl 2-nitrophenoxymethyl ketones give 3-substituted dihydrobenzo-1,4-oxazines.

(*3*) Oxiranes *e.g.* XIX may be used, thus *S. P. Gupta et al.* (Synthesis, 1974, 660) have prepared a series of 3-methylamino-3,4-dihydro-2*H*-1,4-benzoxazines by first heating XIX with an amine, then oxidising the resultant alcohol with Jones' reagent (CrO_3/H_2SO_4/acetone) and finally submitting the product to reductive cyclisation by hydrogenation over Raney nickel:

(XIX)

(XX)

(XX   R = R^1 = NH$_2$ m.p. 81 – 83°)

(XX   R = COPh ; R^1 = Bz   m.p. 149 – 150°)

(XX   R,R^1 = [1,2] CO·C$_6$H$_4$·CO   m.p. 132 – 134°)

(XX   R,R^1 = CO·CH$_2$·CH$_2$·CO   m.p. 81 – 83°)

Phenomorpholine is a mobile oil, b.p. 268° or 127–128°/12 mm, forming a hygroscopic *hydrochloride*, m.p. ~ 120°. With methyl iodide it forms a quaternary iodide which on distillation gives 4-*methylphenomorpholine*, b.p. 124°/12.5 mm, *picrate*, m.p. 144° (149–150° has been recorded). The 4-methyl compound has also been obtained from 4-methyl-2*H*-1,4-benzoxazin-3(4*H*)-one by reduction with lithium tetrahydridoaluminate (*Loudon* and *Ogg*, J. chem. Soc., 1955, 739). With caustic soda the quaternary compound gives 2-vinyloxydimethylaniline (*Knorr*, Ber., 1899, **32**, 734). It is concluded from degradation studies that the hetero-ring of phenomorpholine is at least as stable as that of tetrahydroquinoline (*J. von Braun* and *J. Seemann, ibid.*, 1922, **55**, 3818; see also Vol. IV F, p. 351).

Cyanogen bromide converts 4-methyl- into 4-cyano-phenomorpholine. Phenomorpholine forms diazoamino compounds with aromatic diazonium salts.

6-*Aminophenomorpholine*, from 2,4-dinitrophenoxyethanol by reduction, has m.p. 80°.

Other homologues of phenomorpholine include: 3-*methyl*-, b.p. 254–256°, 150–152°/24 mm, *picrate*, m.p. 141°, *acetyl*-, m.p. 87° and *benzoyl*- deriv., m.p. 126° (*R. Stoermer* and *H. Brockerhof*, Ber., 1897, **30**, 1631). 3,4-*Dimethylphenomorpholine*, b.p. 259–261°, *hydrochloride*, m.p. 170°, *picrate*, m.p. 136° (*Stoermer* and *M. Franke*, *ibid.*, 1898, **31**, 752).

(*iv*) *1*H-*2,3-Benzoxazines*

The parent base, 1*H*-2,3-**benzoxazine**, m.p. 53–55°, *picrate*, 222° (decomp.), is obtained by the action of phosphorus pentachloride upon 1*H*-2,3-benzoxazin-4(3*H*)-one (*G. Pifferi, P. Consonni* and *E. Testa*, Tetrahedron, 1968, **24**, 4923). The authors of this paper also describe the synthesis of a number of 4-amino derivatives (see also *Pifferi* and *Testa*, B.P. 1,111,184/1966; *A. Omodei-Sale, Pifferi* and *Consonni*, Ger. Offen. 1,235,544/1971) which show hypotensive, myrorelaxant, anti-inflammatory and sedative properties. Similarly obtained 4-hydrazino-1*H*-2,3-benzoxazines (XXI) react with aldehydes in acid solution to form 2-aryl-*s*-triazoles (XXII) (*Pifferi* and *Consonni*, J. heterocycl. Chem., 1972, 581):

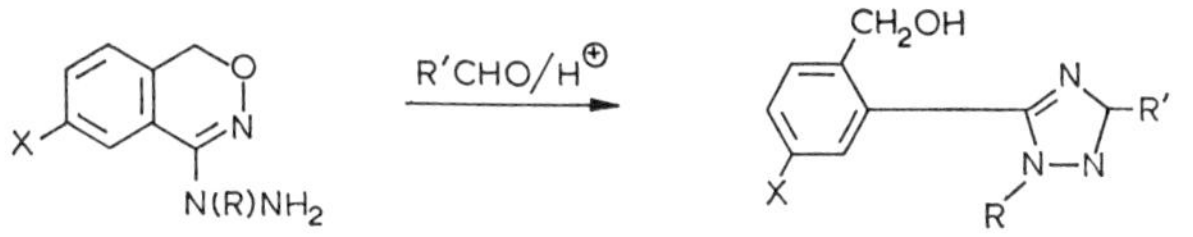

$$\text{R'CHO/H}^{\oplus}$$

(**XXI** ; R = H , Me ;　X = H , Cl)

(**XXII** ; R′ = Me, Ph , 4-NO$_2$C$_6$H$_4$)

Phthalaldehyde gives a monoxime which is a tautomeric mixture of the normal oxime and 1-hydroxy-1*H*-2,3-benzoxazine (XXIII).

A second pair of tautomerides XXIV and XXV is also suggested by the isolation of a compound behaving as 1-*hydroxyisoindole 2-oxide* (XXV), N-*hydroxyphthalimidine* (XXVI), m.p. 181°, from preparations carried out at 60° or in the presence of alkali:

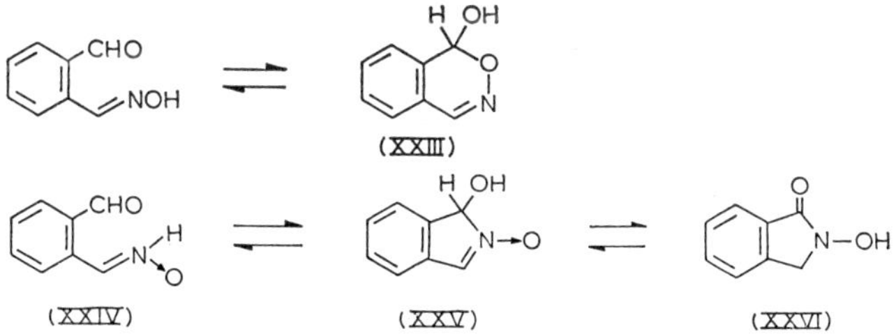

(**XXIII**)

(**XXIV**)　　　　(**XXV**)　　　　(**XXVI**)

The ring form XXIII, 1-*hydroxy*-1H-2,3-*benzoxazine*, m.p. 114–115°, with methanol yields an O-*methyl* derivative XXVII (R = Me), m.p. 77°, and with ethanol the corresponding *O*-ethyl derivative XXVII (R = Et) (m.p. not recorded). These are devoid of aldehydic properties and are hydrolysed back to the parent substance by a small quantity of mineral acid. The formation of *phthalaldehyde*-1-*oxime* 2-*semicarbazone*, m.p. 235° (decomp.), derived from the normal monoxime of phthaldehyde proceeds only slowly and suggests the need for a tautomeric change from XXIII before reaction with the semicarbazide. On oxidation with neutral permanganate the main product from XXIII is 1-*oxo*-2,3-*benzoxazine* (XXIX) the anhydride of XXVIII, m.p. 120°, derived from the normal oxime. Such observations suggest that the ring form XXIII is the more stable:

(XXVII)   (XXVIII)   (XXIX)

Evidence for the formation of XXV follows from its reduction to phthalimidine and oxidation to phthalimide. It gives N-*methoxyphthalimidine*, m.p. 74°, (from XXVII) with sodium methoxide and methyl iodide and is characterised with ferric chloride by an intense violet colour not given by XXIII.

Both XXIII and XXV reduce Fehling's solution. The latter, XXV, is formed when XXIII is kept in a solvent (not alcohols or ketones) or heated with aqueous sodium carbonate (*J. P. Griffiths* and *C. K. Ingold*, J. chem. Soc., 1925, **127**, 1698).

2-Carboxy-aldehydes and -ketones behave similarly to phthalaldehyde giving oximes XXVIII or oxime anhydrides. The anhydrides are 1-oxo-2,3-benzoxazines *e.g.* XXIX. 4-*Methyl*-, m.p. 158–159°, is obtained from acetophenone-2-carboxylic acid or 2-cyanoacetophenone (*J. H. Helberger* and *A. v. Rebay*, Ann., 1937, **531**, 279); 4-*ethyl*-, m.p. 115–117° (*R. L. Frank et al.*, J. Amer. chem. Soc., 1944, **66**, 1) and 7,8-*dimethoxy-1-oxo-2,3-benzoxazine*, m.p. 114–115°, from opianic acid (XXX). The last compound decomposes on heating to hemipinimide (dimethoxyphthalimide) (*W. H. Perkin*, J. chem. Soc., 1890, **57**, 1070; *O. L. Brady et al., ibid.*, 1928, 529):

(XXX)

3,4-Dihydro-1*H***-2,3-benzoxazine** (XXXI) has been prepared (*Pifferi, Consonni* and *Testa*, Gazz., 1966, **96**, 1671) and a number of 3-acetyl-3,4-dihydro derivatives have been synthesised and tested as potential anti-inflammatory drugs (*Pifferi* and *G. Schiatti*, S. African P. 6,808,449/1969; C.A., 1970, **72**, 79064s):

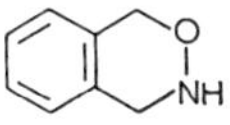

(XXXI)

(*v*) *4H-3,1-Benzoxazines and related compounds*

Synthetic methods for the 4*H*-3,1-benzoxazines resemble those for the 4*H*-1,3-benzoxazines, the starting materials being 2-aminobenzyl alcohol and anthranilic acid or their equivalents.

By warming the hydrobromide of 2-aminobenzyl bromide with glacial acetic acid *S. Gabriel* and *Th. Posner* obtained 2-**methyl-4***H***-3,1-benzoxazine** as its *hydrobromide*, m.p. 170–172°. The base is steam-volatile but is decomposed by water; the *picrate* has m.p. 146–149° (Ber., 1894, **27**, 3515).

The dihydro compounds are more stable towards hydrolysing agents than the 3,1-benzoxazines themselves. They can be prepared by condensing 2-aminobenzyl alcohol with aldehydes or ketones, e.g. in boiling benzene, sometimes with a little glacial acetic acid as catalyst (*F. Eiden et al.*, Arch. Pharm., 1975, **308**, 622 and refs. cited therein).

Formaldehyde in this way gives 1,2-*dihydro*-4H-3,1-*benzoxazine*, m.p. 43–44° and acetone gives the 2,2-*dimethyl* derivative, m.p. 123–124°. The cyclic structure of these compounds is supported by their molecular refractions and ultraviolet absorptions. Benzaldehyde under these conditions does not give a cyclic compound but only benzylidene-2-aminobenzyl alcohol (*F. W. Holly* and *A. C. Cope*, J. Amer. chem. Soc., 1944, **66**, 1875). Compounds of this series were earlier obtained by *O. Widmann* (Ber., 1883, **16**, 2576) by heating 3-amino-4-[2-(2-hydroxypropyl)]benzoic acid with acyl anhydrides or chlorides; acetic anhydride gave the so-called methylcoumazonic acid (XXXII), 2,4,4-trimethyl-4*H*-3,1-benzoxazine-7-carboxylic acid:

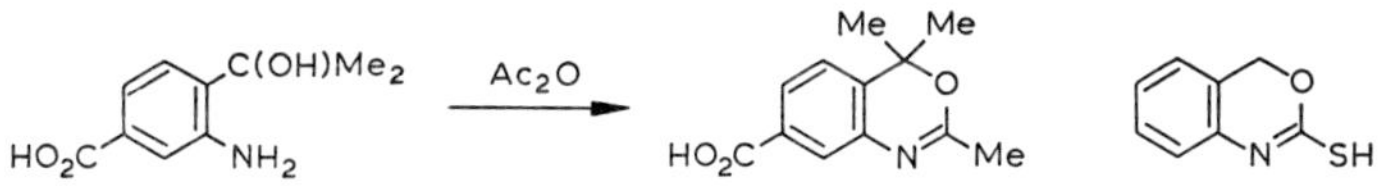

(XXXII)   (XXXIII)

When 2-aminobenzyl alcohol reacts with carbon disulphide in ethanol the so-called **thiocoumazone**, XXXIII, 2-*mercapto*-4H-3,1-*benzoxazine* (or its tautomer, 4H-3,1-*benzoxazine*-2(1H)-*thione*) is formed. It is a stable substance, m.p. 142° (decomp.), soluble in alkalis. Corresponding 2-arylimino derivatives can be obtained by cyclising 2-hydroxymethyl-diarylureas. Both the 2-thioxo- and 2-arylimino-compounds are converted into quinazolines when heated with aromatic amines (*C. Paal et al.*, Ber., 1892, **25**, 2978; 1894, **27**, 1866, 2424).

Certain 1,2-dihydro-4*H*-3,1-benzoxazines are cleaved, depending on the experimental conditions, into anilides, acridans or (alkyl) imidates (*Eiden et al., loc. cit.*):

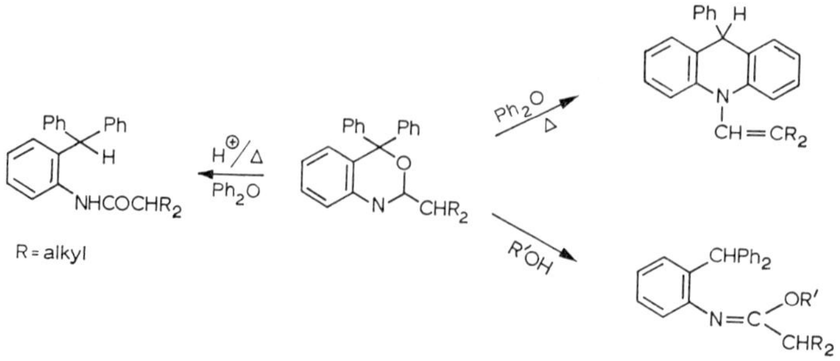

3,1-Benzoxazin-4-ones ("acylanthranils") are formed by the acylation of 2,1-benzisoxazoles (anthranils) with acid chlorides (*I. R. Gambir* and *S. S. Joshki*, J. Indian chem. Soc., 1974, **41**, 47) or anhydrides (*E. Erdman*, Ber., 1899, **32**, 2158). A similar reaction occurs between anthranil (XXXIV) and *N*-phenylbenzimidoyl chloride (*O. Mumm* and *H. Hesse, ibid.*, 1910, **43**, 2508).

It is probable that the change involves the following mechanism (see *K. H. Wünsch* and *A. J. Boulton*, Adv. in heterocycl. Chem., Academic Press, N.Y., 1967, Vol. 8, pp. 277–343):

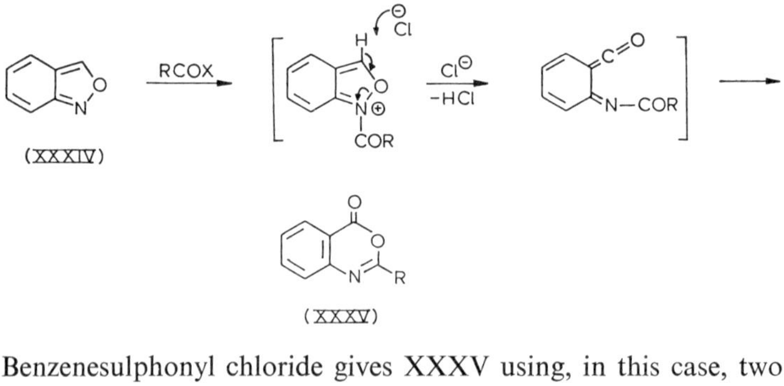

Benzenesulphonyl chloride gives XXXV using, in this case, two molecular equivalents of anthranil (*G. Schroeter*, Ber., 1907, **40**, 2628):

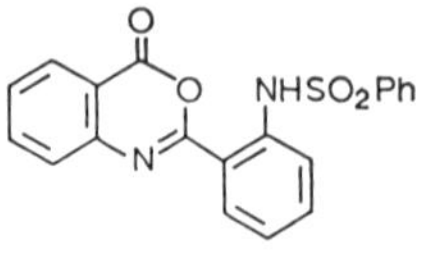

(XXXVI)

3,1-Benzoxazin-4-ones may be prepared by dehydrating *N*-acylanthranilic acids with acetic anhydride. The simplest member of the series (XXXV, R = H), m.p. 43–44°, is very sensitive to hydrolysis, as are the 2-*methyl* derivative (XXXV, R = Me), m.p. 80–81° and other alkyl derivatives; the 2-*phenyl* compound, m.p. 123–124° is more stable. Heated with ammonia they form quinazolines (*D. T. Zentmyer* and *E. C. Wagner*, J. org. Chem., 1949, **14**, 967) and when allowed to react with primary amines then afford 2-acylamidobenzanilides and/or quinazolones, the latter being formed *via* the intermediacy of *N*-(2-carbonyloxyphenyl)-*N'*-arylacylamidines as shown (*L. Errede*, J. org. Chem., 1976, **41**, 1763):

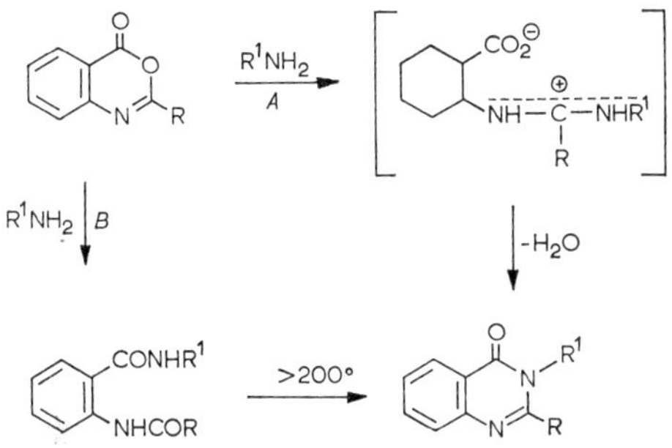

Simple primary amines such as ethylamine or aniline follow pathway *A*, whereas secondary amines and those primary amines with substituents on the α-carbon atom, such as isopropylamine, follow route *B*. The rate of conversion to the various products varies directly with the p*K*a but inversely with the size of the amine substituents (*Errede, J. J. McBrady* and *H. T. Oien*, ibid., 1977, **42**, 656).

   2-**Phenyl**-3,1-**benzoxazin-4-one** is also obtained when 3-benzoyl-2,1-benzisoxazole (XXXVII) is heated. The best conditions for this reaction require pyridine/acetic anhydride (16:5) as solvent and 120–122° for 20 h (*J. L. Pinkus, H. A. Jessup* and *T. Cohen*, J. chem. Soc., C, 1970, 242):

(XXXVII)

Pyrolysis of saccharin leads to 2-(2-aminophenyl)-3,1-benzoxazin-4-one (*A. J. Barker* and *R. K. Smalley*, Tetrahedron Letters, 1971, 4629); sulphonylamino derivatives of which (*e.g.* XXXVII) are lumiphors producing a green-yellow green light (*B. M. Bolotin, D. A. Drapkina* and *V. G. Brudz*, Proc. int. Conf. Lumin., 1966, **1**, 626; C.A., 1969, **71**, 4481w).

(XXXVIII; R = [2]- or [4]- MeC$_6$H$_4$
    or 2-C$_{10}$H$_7$)

1,4-Dihydro-3,1-benzoxazin-2-one, m.p. 119–120°, is formed by heating 2-hydroxymethylbenzoyl azide (*H. Lindemann* and *W. Schulthesis*, Ann., 1928, **464**, 246; see p. 457):

4-Alkyl derivatives are known (*W. E. Savage*, Austral. J. Chem., 1975, **28**, 2275).

1,2-Dihydro-3,1-benzoxazin-4-one (XXXIX), m.p. 145–150°, is prepared from anthranilic acid and formaldehyde in aqueous or ethereal suspension (*B.A.S.F.*, G.P. 155,628/1903; Chem. Ztbl., 1904, II, 1444):

(XXXIX)

Substituted anthranilic acids similarly give 1-*phenyl*, m.p. 89° (*P. Friedländer* and *K. Kunz*, Ber., 1922, **55**, 1597), 5,8-*dichloro*, m.p. 159–161° and 5,6,7,8-tetrachloro-4-oxo-1,2-dihydro-3,1-benzoxazine (*V. Villiger* and *L. Blangey*, *ibid.*, 1909, **42**, 3529, 3549).

Benzylideneanthranilic acid is tautomeric with 1,2-dihydro-2-phenyl-3,1-benzoxazin-4-one. The N-*acetyl* compound, m.p. 131–132°, and N-*phenyl-ureide*, m.p. 171°, are described (*H. R. Snyder et al.*, J. Amer. chem. Soc., 1938, **60**, 2025).

Isatoic anhydride (2*H*-3,1-**benzoxazine-2,4(1***H***)-dione** is the anhydride of phenylglycine-2-carboxylic acid. It is prepared (*E. C. Wagner* and *M. F. Fegley*, Org. Synth., Coll. Vol. 3, 1955, p. 488) by passing phosgene into a solution of anthranilic acid in aqueous hydrochloric acid. It is soluble in alkali with a blue fluorescence and is hydrolysed by acids to anthranilic acid. Heated with excess of ammonia it forms 2,4-dihydroxyquinazoline; for a study of this and other reactions see *Wagner et al.* (J. org. Chem., 1948, **13**, 347; 1943, **8**, 239; 1944, **9**, 55).

Isatoic anhydride combines with 3-hydroxyisothiazole (XL) at 80–95°, with loss of water and carbon dioxide, to give 3-anthranoyloxyisothiazole (XLII) and isothiazole[2,3b]quinazolin-9-one (XLIII). The formation of the second product can readily be interpreted *via* a dehydrative cyclisation of the initial *N*-acyl product XLI, but its competitor XLII probably requires an $N{\rightarrow}O$ acyl migration (*A. W. K. Chan* and *W. D. Crow*, Austral. J. Chem., 1969, **22**, 2497):

(XL)   (XLI)

(XLII)   (XLIII)

Numerous *N*-substituted 2*H*-3,1-benzoxazine-2,4(1*H*)diones have been prepared (*G. E. Hardtmann, G. Koletar* and *O. R. Pfister*, J. heterocyclic Chem., 1975, **12**, 565) from 2-chloro- and 2-nitro-benzoic acids and *N*-unsubstituted isatoic anhydrides as starting materials. In a later paper (*G. M. Coppola, Hardtmann* and *Pfister*, J. org. Chem., 1976, **41**, 825) the authors discuss the reactions of these compounds with various nucleophiles such as thiopseudoureas and carban ions.

3. Phenoxazines

Phenoxazine (I) was first synthesised in 1887 by *Bernthsen* (Ber., 1887, **20**, 942), subsequently it has aroused interest because some derivatives are highly coloured and have been used as dyes and because certain phenoxazines show pharmacological activity and yet others have been isolated as natural products. For earlier reviews of phenoxazine chemistry see *D. E. Pearson* ("Heterocyclic Compounds", *R. C. Elderfield*, ed., Vol. 6, Ch. 14, p. 685, New York, 1957; *M. Ionescu* and *H. Mantsch*, Advances in Heterocyclic Chemistry, Academic Press, London and New York, 1967, Vol. 8, pp. 83–112).

(I)

Formula I adopts the numbering system recommended by the I.U.P.A.C. Rules of Organic Nomenclature (Nomenclature of Organic Chemistry, Sections A and B, 2nd Edn., Butterworths, London, 1966, p. 54) and replaces the alternative system proposed by *Bernthsen*.

(a) Methods of synthesis

(*1*) Phenoxazine itself is formed by heating a mixture of 2-aminophenol and catechol (functioning as an acid catalyst) *(A. Bernthsen, loc. cit.)*, or by heating 2-aminophenol and its hydrochloride in equimolar proportions at 240° (*F. Kehrmann* and *A. A. Neil*, Ber., 1914, **47**, 3102; *S. Granik, L. Michaelis* and *M. P. Schubert*, J. Amer. chem. Soc., 1940, **62**, 1802; *H. Gilman* and *L. O. Moore*, J. Amer. chem. Soc., 1957, **79**, 3485). The reaction, which proceeds *via* the intermediacy of a 2,2′-dihydroxydiphenylamine, may be extended to the preparation of symmetrically substituted dialkylphenoxazines from alkyl-substituted 2-aminophenols (*J. de Antoni*, Comp. rend., 1961, **252**, 3274; Bull. Soc. chim. Fr., 1963, 2871):

Similarly, in the presence of iodine 2-aminophenols afford intermediate 2,2′-dihydroxydiphenylamines which yield phenoxazines on thermal cyclohydration (*P. Müller, N. P. Buu-Höi* and *R. Rips*, J. org. Chem., 1959, **24**, 37; *V. G. Samolovova, T. V. Gortinskaya* and *M. N. Shchukina*, Zhur. obshcheĭ Khim., 1961, **31**, 1942) and catechols when heated with ammonia also give phenoxazines (*K. Ley*, Angew. Chem., 1962, **74**, 871).

(*2*) 2-Aminophenols with picryl chloride in alkaline solution give 2-nitrophenoxazines *via* 2-hydroxy-2′,4′-dinitrodiphenylamines. Mechanistically the reaction requires an intramolecular nucleophilic displacement of the 2′-nitro group by the phenoxide function at position-2 and works best when the diphenylamine bears a substituent at C-6′:

$$(R = NO_2 ; SO_3H ; CO_2H ; CH_3 ; OCH_3)$$

(*G. S. Turpin*, J. chem. Soc., 1891, **59**, 714; *B. Boothroyd* and *E. R. Clark*, *ibid.*, 1953, 1499, 1504; *A. B. Sen* and *R. C. Sharma*, J. Indian chem. Soc., 1957, **34**, 877; *G. E. Bonvicino, L. H. Yogodzinski* and *R. A. Hardy Jr.*, J. org. Chem., 1961, **26**, 2797; *H. Musso* and *P. Wagner*, Chem. Ber., 1961, **94**, 2551; *S. P. Gupta* and *S. S. Joshi*, J. Indian chem. Soc., 1963, **40**, 400; *F. Ullmann*, Ann., 1909, **366**, 79; *O. L. Brady* and *C. Waller*, J. chem. Soc., 1930, 1218). *Musso* (Chem. Ber., 1963, **96**, 1927) has shown, however, that if the sodium salt of the intermediate diphenylamine is heated in either dimethyl sulphoxide or dimethylformamide better yields are obtained.

In a related sequence *J. Gruz* and *Z. Stransky*, C.A., 1969, **71**, 70553x) condensed picryl chloride and 2-acetamido-3-nitrophenols in ethanol containing sodium acetate to form 1,3,9-trinitrophenoxazines.

(*3*) 2,2'-Dimethoxydiphenylamines can be *O*-demethylated and cyclo-dehydrated by heating with pyridine hydrochloride (*H. Beecken* and *Musso*, Chem. Ber., 1961, **94**, 601).

(*4*) 2-Chloronitrobenzenes serve as the starting materials for the preparation of 2,2'-disubstituted diphenyl ethers which may then be cyclised to phenoxazines. Thus when heated with potassium hydroxide and 2-methoxyphenols, 2-methoxy-2'-nitrodiphenyl ethers are prepared. These may be reduced catalytically and demethylated to yield phenoxazines (*N. M. Cullinane, H. G. Davey* and *H. J. H. Padfield*, J. chem. Soc., 1934, 716; *M. P. Olmsted et al.*, J. org. Chem., 1961, **26**, 1901):

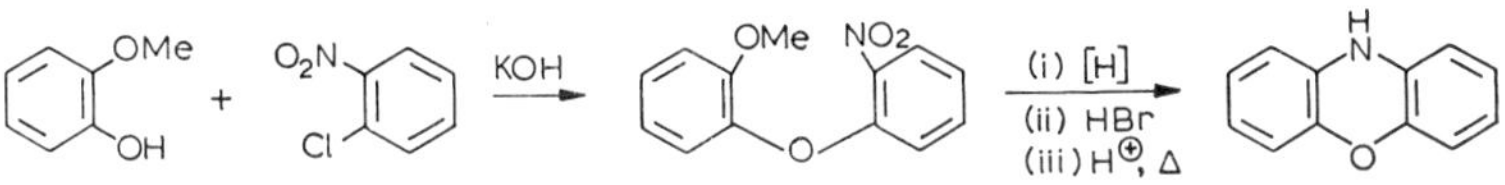

In place of 2-methoxyphenols, 2-nitrophenols may be used (*R. Higginbottom* and *H. Suschitzky*, J. chem. Soc., 1962, 2367), or 2-bromophenols. In the last case the reduction step is achieved by the action of stannous chloride or iron fillings in acetic acid, and ring closure by heating in dimethylformamide solution in the presence of copper and potassium carbonate (*Bonvicino et al., loc. cit.; Olmsted et al., loc. cit.*; see also *Bonvicino, Yogodzinski* and *Hardy Jr.*, J. org. Chem., 1962, **27**, 4272).

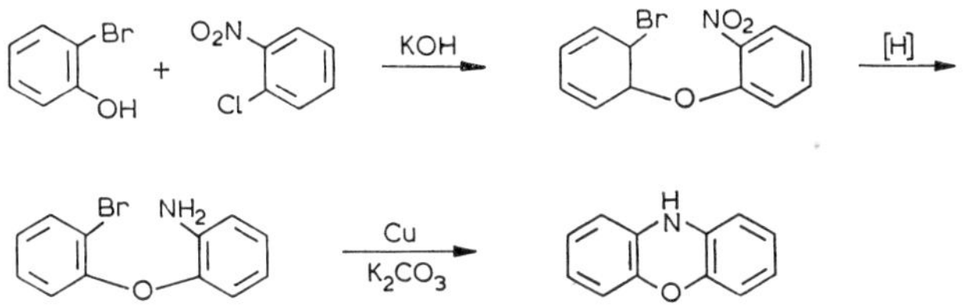

(b) Phenoxazine and its simple derivatives

Phenoxazine, m.p. 158–159°, dissolves in organic solvents and in concentrated sulphuric acid, the solutions showing a violet or violet-red fluorescence; the alcohol solution gives a blue-green colour with ferric chloride. The *acetyl* deriv. has m.p. 142°. *N*-Alkylation of phenoxazine is difficult but with sodamide in ether and the requisite halide the 10-*methyl*-, m.p. 35–37°, *picrate*, m.p. 101–102°, 10-*ethyl*-, m.p. 46–47°, *picrate*, m.p. 93.5–94°, and 10-*benzyl*-, m.p. 127.5–128°, phenoxazines have been obtained, as well as the 10-*phenyl* compound, m.p. 138–139° (*Gilman* and *Moore*, *loc. cit.*).

3-*Methylphenoxazine*, m.p. 124°, is obtained by heating catechol with 4-amino-3-hydroxytoluene; 3,7-*dimethylphenoxazine*, m.p. 205°, is formed together with the 3,8-dimethyl isomer by heating 4-methylcatechol with the same amine (*Kehrmann*, Ber., 1901, **34**, 1623; Ann., 1902, **322**, 17).

TABLE 3

NITRO DERIVATIVES OF PHENOXAZINE

Derivative	m.p.(°C)	Description	Ref.
1-Nitro	66	violet	1
1-Nitro-3-carboxylic acid	320–321	purple	1,3
3-Benzoyl-1-nitro	220	red	1
1-Nitro-3-sulphonic acid	—	*aniline salt* 285° (decomp.)	1
3-Methyl-1-nitro	165	black needles	2
4-Methyl-1-nitro	169	dark violet	2
3,9-Dimethyl-1-nitro	162	violet	2
8-Chloro-1-nitro	240–241 (decomp.)	—	2
1-Methyl-3-nitro	212–213 (decomp.)	red needles	3
8-Chloro-1-methyl-3-nitro	271–273 (decomp.)	red needles	3
3-Nitro-1-carboxylic acid	295 (decomp.)	orange-red	1
1,4-Dimethoxy-3-nitro	173	crimson	2
10-Benzyl-3-nitro	175–175.5	red needles	3
10-Benzyl-8-chloro-3-nitro	226.5–227	orange red needles	3
1,3-Dinitro	213	purple-brown	4
10-Benzyl-2,3-dinitro	238–240	brownish red plates	3

References

1 *F. Ullmann*, Ann., 1909, **366**, 90.
2 *O. L. Brady* and *C. Waller*, J. chem. Soc., 1930, 1218.
3 *B. Boothroyd* and *E. R. Clark*, *ibid.*, 1953, 1499, 1506.
4 *G. S. Turpin*, *ibid.*, 1891, **59**, 723.

(i) Nitrophenoxazines

Nitrophenoxazines can be obtained by nitrating *N*-acylphenoxazines or through Turpin's method (method 2, p. 472) by cyclising certain 2′-hydroxy-2-nitrodiphenylamines. Phenoxazine itself reacts very violently with nitric acid but the acetyl derivative can be nitrated in acetic anhydride to a mixture of acetyl derivatives of 3,7-di- and 1,3,7,9-tetra-nitrophenoxazines, separable by crystallisation from benzene. 3,7-**Dinitrophenoxazine** forms dark red needles from acetic acid solution and decomposes above 200°; *acetyl* deriv., m.p. 192° (*Kehrmann* and *A. Saager*, Ber., 1903, **36**, 475).

Many phenoxazines containing nitro groups have been synthesised by the Turpin process. Table 3 lists some representative examples.

Unsymmetrically substituted halonitrobenzenes which have two nitro groups *ortho* to the reactive halogen atom produce two isomeric phenoxazines although one isomer usually predominates (*Gupta* and *Joshi, loc. cit.*).

Thus when 2′-hydroxy-2,6-dinitro-3-methyldiphenylamine is cyclised only 1-nitro-4-methylphenoxazine (II) is isolated; the other isomer, 1-nitro-2-methylphenoxazine (III) is not detected. The identity of the product was shown by an independent synthesis from 2-chloro-2′-hydroxy-6-nitro-3-methyldiphenylamine:

(II)   (III)

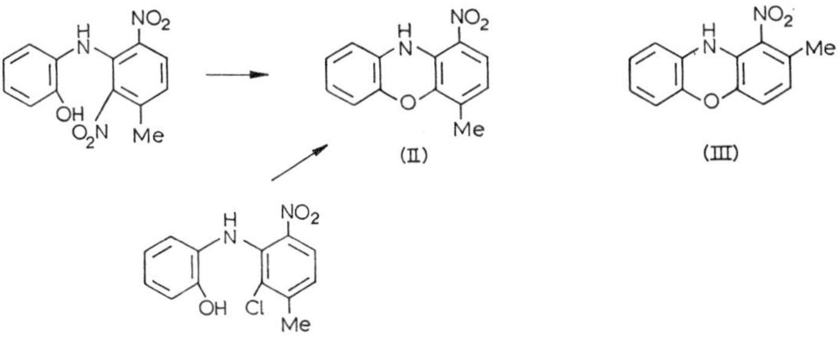

(ii) Aminophenoxazines

1-**Aminophenoxazine** is obtained by reduction of the nitro compound with stannous chloride; it forms a crystalline *hydrochloride*, m.p. 249–250°, and an *acetyl* deriv., m.p. 197° (*Kehrmann* and *L. Löwy*, Ber., 1911, **44**, 3006). 1-*Amino-8-chlorophenoxazine hydrochloride* has m.p. 228–230° (decomp.) (*Boothroyd* and *Clark*, J. chem. Soc., 1953, 1507). 1-*Amino-phenoxazine-3-carboxylic acid hydrochloride*, pale green needles, m.p. above 340°.

3-**Aminophenoxazine** is difficult to prepare, because of the ready oxidisability of 3-aminophenazines to phenazimes and beyond. *Boothroyd* and *Clark*, after much study, obtained it best by catalytic reduction (using palladium-charcoal) in *n*-butyl alcohol of 10-benzyl-3-nitrophenoxazine, the benzyl group being removed during reduction. The base crystallises from alcohol containing sodium dithionate in fawn-coloured plates, m.p. 172–173° (decomp.); *acetyl* deriv., m.p. 171.5–172.5° (decomp.). Other catalytic methods of reduction afford some phenoxazine, characterised as *picrate*, m.p. 186–187° (decomp.). 1,3-**Diaminophenoxazine**, prepared by catalytic reduction of the dinitro compound, is characterised as *sulphate*, m.p. 221–223° (decomp.).

(iii) 3-Methoxyphenoxazines

A number of substituted 3-methoxyphenoxazines (see Table 4) have been prepared from the corresponding 3*H*-phenoxazin-3(3*H*)-ones (see p. 479) by reductive methylation with dimethyl sulphate and sodium dithionate in aqueous acetone solution (*W. Schaefer* and *I. Geyer*, Tetrahedron, 1972, **28**, 5261).

TABLE 4

SUBSTITUTED 3-METHOXYPHENOXAZINES

R^1	R^2	R^3	m.p.(°C)
COMe	OMe	H	74
COMe	OMe	[8]COMe	155
COMe	NH$_2$	[8]COMe	116
COMe	NHMe	[8]COMe	139
COMe	N(Me)$_2$	[8]COMe	138
COMe	Cl	[8]COMe	181
COMe	OMe	[8]CH $=$ CH $\cdot$ COMe	157
COMe	OMe	[6]CH $=$ CH $\cdot$ COMe	123
COMe	OMe	[8]CO$_2$Me	131
COMe	N(Me)$_2$	[8]CO$_2$Me	171
COMe	NHPh	[6]COMe	224
COMe	[1,4]NHC$_6$H$_4$OMe	H	208
CO$_2$Me	OMe	H	115
CO$_2$Me	OMe	[7]Me	127
CO$_2$Me	OMe	[6]COMe	123
CO$_2$Me	OMe	[5]COMe	191
CN	OMe	H	178

2-Azidophenyl-2,6-dimethoxyphenyl ether in decalin at 160° gives 4-methoxyphenoxazine (35%) and 1,2-dimethoxyphenoxazine (15%) (*J. I. G. Cadogan* and *P. K. K. Lim*, Chem. Comm., 1971, 1431).

(iv) Phenoxazinecarboxylic acids

Phenoxazine-1-carboxylic acids may be obtained from 1-halogeno-2-nitrobenzenes and 2-chloro-3-hydroxybenzoic acids in four steps (see p. 473) (*B. Blank* and *L. L. Baxter*, J. med. Chem., 1968, **11**, 807) and some α-(10-methylphenoxazin-2-yl)alkanoic acids and their esters have been prepared similarly (*D. Farge, G. Jeanmart* and *M. N. Messer*, F.P. 1,516,746/1966). Phenoxazine is metalated by *n*-butyl-lithium in ether and the product when treated with carbon dioxide yields **phenoxazine-4-carboxylic acid**, m.p. 244–245°, *methyl ester*, m.p. 112.5–114° (*H. Gilman* and *L. O. Moore*, J. Amer. chem. Soc., 1958, **80**, 2195).

10-*Ethylphenoxazine-4-carboxylic acid*, m.p. 163.5–165°, is obtained similarly, but the proof of structure is incomplete. Bromination of 10-ethylphenoxazine gives the 3-bromo derivative which is converted by the Grignard method into 10-*ethylphenoxazine*-3-, *carboxylic acid*, m.p. 207.5–210.5° *(idem, ibid.)*.

(*v*) *Halogenophenoxazines*

The action of bromine in benzene solution upon phenoxazine gives a mixture of 3-**bromophenoxazine** and 3,7-**dibromophenoxazine** together with brominated polyphenoxazines (*H. Musso*, Chem. Ber., 1959, **92**, 2862). 1,3,7,9-**Tetrabromophenoxazine** is available through bromination of *N*-acetylphenoxazine (*Kehrmann* and *Saager*, Ber., 1903, **36**, 475). The mono- and di-bromophenoxazines may be *N*-acetylated but the tetrabromo compound is inert even under vigorous conditions of acetylation.

Chlorination of phenoxazine with thionyl chloride yields 1,3,7,9-**tetrachlorophenoxazine** (*G. S. Predvoditeleva* and *M. N. Shchukina*, Zhur. obshcheĭ Khim., 1960, **30**, 1893).

6-*Chloro-*, m.p. 283–285°, 7-*chloro-*, m.p. 290–291°, 8-*chloro-*, m.p. 276–277°, 7-*fluoro-*, m.p. 248–250° and 8-*fluoro-*, m.p. 264–265°, -*phenoxazine-1-carboxylic acids* have been prepared (*Blank* and *Baxter*, *loc. cit.*) *via* the appropriate 1-halogeno-2-nitrobenzenes and 2-chloro-3-hydroxybenzoic acid:

(*vi*) *N-Alkylphenoxazines and N-arylphenoxazines*

Table 5 lists a number of *N*-alkylphenoxazines, which are prepared by the action of alkyl halides, oxiranes or acrylonitrile upon phenoxazine in the presence of a strong base. In general, sodamide in liquid ammonia or an aromatic hydrocarbon solvent, is used as the base since the N-atom of phenoxazine is insufficiently basic to react directly with the alkylating agent.

10-**Arylphenoxazines** can be prepared either by the action of an aryl halide and sodamide upon phenoxazine or by the reaction between phenoxazine and an aryl bromide or iodide in the presence of anhydrous potassium carbonate with copper bronze as catalyst. 10-*Phenylphenoxazine*, m.p. 138–139° (*Gilman* and *Moore*, *loc. cit.*; see also *F. Sparatore* and *V. Boido*, Ann. Chim., Rome, 1964, **54**, 591).

The e.s.r. spectra of the cation radicals derived from 10-arylphenoxazines shows the spin distribution to be influenced inductively for substituents other than 4-alkylamino (*D. Clarke*, *B. C. Gilbert* and *P. Hanson*, J. chem. Soc., Perkin II, 1976, 114).

TABLE 5

N-ALKYLPHENOXAZINES

Alkyl group	m.p.(°C)	Ref.
Me	35–37	1
Et	46–47	1
CH_2Ph	127–128	1
$CH_2 \cdot CH_2Cl$	62	2
$CH_2 \cdot CH_2OH$	109–110	3,4
$CH_2 \cdot CH_2 \cdot CH_2Cl$	54–55	5,6
$CH_2 \cdot CH_2 \cdot CH_2OH$	68	7
$CH_2 \cdot CH(OH)Me$	95–98	5
$CH_2 \cdot CH_2 \cdot CH_2NH_2$	173–174	8
$CH_2 \cdot CH_2 \cdot CO_2H$	138–139	7,8,9
$CH_2 \cdot CH_2 \cdot CN$	123–124	7,8,9

References

1 *H. Gilman* and *L. O. Moore*, J. Amer. chem. Soc., 1957, **79**, 3485.
2 *D. A. Shirley, K. Sen* and *J. C. Gilmer*, J. org. Chem., 1961, **26**, 3587.
3 *H. Linde*, Arch. Pharm., 1960, **293/295**, 537.
4 *V. G. Samolovova, T. V. Gortinskaya* and *M. N. Shchukina*, Zhur. obshcheĭ Khim., 1962, **32**, 1085.
5 *M. Vanderhaeghe* and *L. Verlooy*, J. org. Chem., 1961, **26**, 3827.
6 *A. E. Gal* and *S. Avakian*, J. med. pharm. Chem., 1963, **6**, 809.
7 *G. Frangatos, G. Kohan* and *F. L. Chubb*, Canad. J. Chem., 1960, **38**, 1021.
8 *Samolovova, Gortinskaya* and *Shchukina*, Zhur. obshcheĭ Khim., 1960, **30**, 1516.
9 *P. Müller, N. P. Buu Höi* and *R. Rips*, J. org. Chem., 1959, **24**, 1699.

TABLE 6

N-ACYLPHENOXAZINES

N-Acyl group	m.p.(°C)	Ref.
COMe	142	1
COCl	142–144	2,3,4
CO_2Me	119–120	5
CO_2Et	74–75	2

References

1 *N. M. Cullinane, H. G. Davey* and *H. J. H. Padfield*, J. chem. Soc., 1934, 716
2 *A. E. Gal* and *S. Avakian*, J. med. pharm. Chem., 1963, **6**, 809.
3 *V. G. Samolovova, T. V. Gorkinskaya* and *M. N. Shchukina*, Zhur. obshcheĭ Khim., 1960, **30**, 1516.
4 *M. Claesen* and *H. Vanderhaeghe*, J. org. Chem., 1961, **26**, 4130.
5 *Samolovova, Gortinskaya* and *Shchukina*, Zhur. obshcheĭ Khim., 1962, **32**, 1085.

(vii) *Acylphenoxazines*

Acetyl chloride or acetic anhydride react with phenoxazine to yield 10-acetylphenoxazine; similarly phosgene affords phenoxazine-10-carbonyl chloride from which a number of esters have been obtained by reacting with alcohols. Some examples of *N*-acylphenoxazines appear in Table 6.

C-Acylphenoxazines. *N*-Acetylphenoxazine combines with acetyl chloride in carbon disulphide containing aluminium trichloride to give 2,8-**diacetylphenoxazine** (*J. de Antoni,* Bull. Soc. chim. Fr., 1963, 2874). 10-Ethylphenoxazine, however, gives 3,7-*diacetyl*-10-*ethylphenoxazine*, m.p. 178–180° (*H. Vanderhaeghe,* J. org. Chem., 1960, **25,** 747). Phenoxazine itself can be acetylated with acetyl chloride and aluminium chloride to yield 2-*acetylphenoxazine*, m.p. 211–213°, not the 3-acetyl derivative as first reported (*Müller, Buu Höi* and *Rips, loc. cit.*).

10-*Ethyl-3-formylphenoxazine*, m.p. 130–132°, is prepared by a Vilsmeier–Haack reaction (*N*-methylformamide/POCl$_3$) upon 10-ethylphenoxazine *(Vanderhaeghe, loc. cit)*.

1-*Formylphenoxazine*, m.p. 80–81°, has been obtained by an indirect route from phenoxazine (*H. Harfenist,* J. org. Chem., 1962, **27,** 4326).

(c) *Oxidation products of phenoxazines*

(i) *Phenoxazinones and phenoxazonium salts*

In acid solution phenoxazine may be oxidised by a variety of reagents (ferric chloride, potassium permanganate, bromine, hydrogen peroxide, etc.) to form a radical cation (IV), the structure of which has been studied by electron spin resonance spectroscopy (*L. D. Tuck* and *D. W. Schieser,* J. phys. Chem., 1962, **66,** 937). On neutralisation, or basification, deprotonation is effected yielding the radical V which either polymerises to polyphenoxazines (linked through positions 3 and 10) or oxidises to phenoxazine-3(3*H*)-one (VI).

The relative yields of the two possible products is dependent on pH, alkaline conditions favouring polymerisation (*H. Musso,* Chem. Ber., 1959, **92,** 2862, 2873):

Phenoxazin-3(3*H*)-**one**, which crystallizes in golden brown leaves, **m.p.** 216–217° (decomp.) (*Kehrmann* and *Saager*, Ber., 1902, **35**, 341), may also be obtained by the oxidation of phenoxazine with nitrous acid at 5° (*G. W. K. Cavill, P. S. Clezy* and *F. B. Whitfield*, Tetrahedron, 1961, **12**, 139). When phenoxazine is oxidised with ferric chloride in the presence of anilinium chloride, the red oxonium salt VII (R = H) is formed. Further treatment of the salt with anilinium chloride and oxidant affords VII (R = NHPh) (*Kehrmann* and *Saager, loc. cit.*). 3-(N-*Phenylimino*-3H-*phenoxazine* (VIII) forms blood red crystals, m.p. 196–198°.

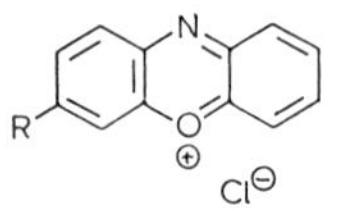

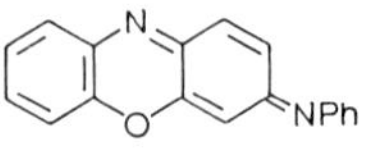

Aminophenoxazines readily oxidise to the corresponding imines and care is needed in their preparation (see p. 472).

Oxidation of 2-aminophenols gives phenoxazin-3(3*H*)-ones. For example, 2-aminophenol when oxidised with silver oxide gives 2-**aminophenoxazin**-3(3*H*)-**one** (IX), m.p. 249° (*O. Fischer* and *O. Jona*, Ber., 1894, **27**, 2784). However, 2-amino-5-dimethylaminophenol, oxidised by air in weak acetic acid solution, yields the phenoxazine equivalent of methylene blue—the 3,7-bis(dimethylamino)phenoxazonium salt (X) (see p. 531) (*Kehrmann* and *W. Poplawski*, Ber., 1909, **42**, 1275):

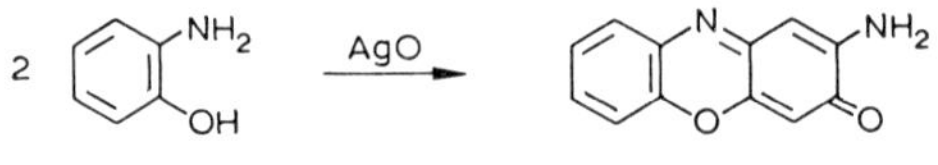

2-Hydroxy-1,4-quinones with aminophenols yield phenoxazin-3(3*H*)-ones (*Kehrmann*, Ber., 1895, **28**, 353; *Kehrmann* and *J. Messinger*, ibid., 1893, **26**, 2375).

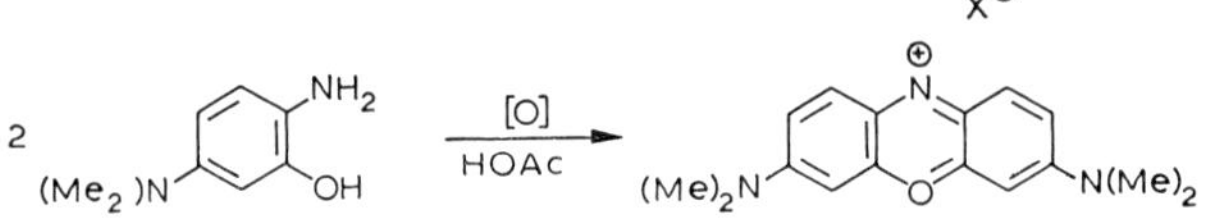

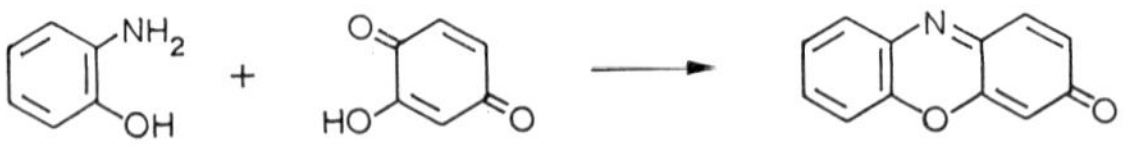

Resorcinol with nitrous acid gives 7-hydroxyphenoxazin-3(3*H*)-one (resorufin); this compound was originally obtained when resorcinol was treated with nitric acid adulterated with nitrous acid. It was contaminated with another product, resazurin, shown later to be the *N*-oxide of resorufin (*R. Nietski et al.*, Ber., 1889, **22**, 3020) (see p. 481).

Autoxidation of resorcinol derivatives, *e.g.* orcinol in the presence of ammonia gives phenoxazin-3(3*H*)-ones (see p. 479).

Ultraviolet and visible spectra of some fifty substituted phenoxazin-3(3H)-ones have been recorded (*W. Schaefer* and *I. Geyer*, Tetrahedron, 1972, **28**, 5261).

See *Spanke* and *Walter* (Ber., 1967, **100**, 1436) for an interpretation of the infrared spectra of phenoxazin-3(3H)-ones, and also *Musso* (Chem. Ber., 1957, **90**, 1814).

Resorufin (**7-hydroxyphenoxazin-3(3H)-one**) (*acetyl* derivative, m.p. 223°, *methyl ether*, m.p. 247°) was first obtained by *P. Weselsky* (Ann., 1872, **162**, 273); its constitution and the mechanism of formation were determined by *Nietzki et al.* (Ber., 1889, **22**, 3020; 1891, **24**, 3366; see above). Resorufin dissolves in alkalis with a rose red colour showing a splendid cinnaber-red fluorescence. Orcinol gives a similar compound, 1,9-dimethyl-resorufin, with nitrous–nitric acid. Another homologue, 7-hydroxy-2,8-dimethylphenoxazin-3(3H)-one, has been obtained from 5-amino-2,4-dihydroxytoluene, which is autoxidised in faintly alkaline solution to an indophenol; this is converted by acid or alkalies to the phenoxazinone, the *acetyl* derivative of this last product has m.p. 216–218° (*F. Heinrich* and *F. Götz*, Ber., 1925, **58**, 1055):

Resorufin

Resazurin

Resazurin, resorufin-N-oxide, is formed by oxidation of resorufin with hydrogen peroxide. It forms green iridescent prisms which dissolve in alkalis to blue solutions and in alcohol with a brick-red fluorescence. The *acetyl* deriv. has m.p. 222° and the *methyl ether* m.p. 252° (decomp.). By reduction both resazurin and resorufin give 3,7-dihydroxy-phenoxazine, which forms a *triacetyl* deriv., m.p. 216°.

Resorcinol with benzoquinone dichloroimide gives a small yield of 7-aminophenoxazin-3(3H)-one but with orcinol (5-methylresorcinol) a much better yield of 7-amino-1-methyl-phenoxazine-3(3H)one (XI) is obtained. The formula shown is preferred to the tautomeric hydroxyphenoxazine formula since the substance is insoluble in alkalis. It gives red salts with acids and an orange-red fluorescent solution with alkalis in alcoholic solution (*Nietzki* and *H. Mäckler*, Ber., 1890, **23**, 723).

(XI)

Orceins. The long-known violet colouring matter formed by the combined action of air and ammonia on certain lichens, *Rocella* and *Lecanora* species, known as orseille or orcein, was shown by *H. Robiquet* in 1835 to owe its appearance to the presence of orcinol in the plant (C.C.C. 2nd Edn. Vol. III A, p. 401). The researches of *H. Musso* and colleagues (Ber., 1956, **89**, 1659; 1957, **90**, 1814; 1958, **91**, 2001; Angew. Chem., 1957, **69**, 178) have revealed that orcein is a complex mixture of phenoxazinones. It has been separated chromatographically into fourteen components, the structures of several of which have been identified. These are derivatives either of 7-hydroxy- or of 7-aminophenoxazin-3(3*H*)-one, having methyl groups in the 1 and 9 positions and a 4,6-dihydroxy-2-methylphenyl group in the 8 or in both 2 and 8 positions.

(XII)   (XIII)

α-*Aminoorcein* is given formula XII (R = NH_2); it forms a red *triacetate*, m.p. 218–219.5°. The corresponding α-hydroxyorcein (XII, R = OH) forms two isomeric *triacetates* (owing to tautomerism between positions 3 and 7), both orange crystals, m.p. 189–191° and 211°, respectively. Both can be reduced to the same reoxidisable *leuco* form (dihydrophenoxazine) which gives a *tetra-acetate*, m.p. 210–212° (decomp.). Compounds of general formula XIII (R = OH or NH_2), can exist in *cis*- and *trans*-forms due to restricted rotation of the large substituent groups in positions 2 and 8. The *β*- and *γ*-amino- and *β*- and *γ*-hydroxy-orceins are racemic *cis*- and *trans*-compounds of this type.

(*d*) *Naturally occurring phenoxazine derivatives*

Actinomycins represent a large group of antibiotics which contain a 2-amino-4,6-dimethylphenoxazin-3(3*H*)-one-1,9-dicarboxylic acid nucleus bound through peptide links at the carboxyl groups to pentapeptide side chains. Actinomycin A was the first of these antibiotics to be isolated (*S. A. Waksman* and *H. B. Woodruff*, Proc. Soc. exptl. Biol. Med., 1940, **45**, 609; C.A., 1941, **35**, 1081).

The various structures of this group differ in the amino acid sequences in the peptide chains. Some have been synthesised (*H. Brockmann* and *H. Lackner*, Naturwiss., 1960, **47**, 230; 1964, **51**, 384; *Brockmann*, Angew. Chem., 1960, **72**, 939) and analogues have also been made (*B. Weinstein et al.*, J. org. Chem., 1962, **27**, 1389; *M. T. Wu* and *R. E. Lyle*, J. heterocyclic Chem., 1971, **8**, 989):

Actinomycins

Some aspects of their biosynthesis have been studied recently (see, for example, *R. B. Herbert*, Tetrahedron Letters, 1974, 4525).

Cinnabarine (polystictin) is a fungal metabolite isolated from the Australian wood rotting fungus *Coriolus sanguineus*, abundant on decaying pines and eucalypts, and identical with cinnabarin isolated from *Trametes cinnabarina*, has the composition $C_{14}H_{10}O_5N_2$.

On chemical, degradative and spectroscopic evidence it is given the structure XIV. By zinc dust distillation it gives a small quantity of a product probably identical with phenoxazin-3(3H)-one:

(XIV)

Cinnabarine crystallises in light red-brown needles or prisms from dioxane decomp. above 300°. It forms an O-*methyl* deriv., variable m.p. above 200° and an O-*acetyl* deriv., red needles, m.p. 250–252° (decomp.) (*G. W. K. Cavill et al.*, J. chem. Soc., 1953, 525; 1957, 2646; Proc. Chem. Soc., 1957, 346).

Cinnabaric acid (XV) and **tramesanguine** (VI) are related structures found as natural products in fungi (*J. Gripenberg*, Acta Chem. Scand., 1958, **12**, 603; 1963, **17**, 703; *Cavill et al.*, Tetrahedron, 1959, **5**, 275; 1961, **12**, 139).

(XV, R = CO$_2$H)
(XVI, R = CHO)

It is interesting to note that 2-aminophenols are oxidised by human and mouse melanoma to cinnabaric acid (*L. R. Moran et al.*, Cancer Res., 1967, **27**, 2395). Several 2-amino-phenoxazin-3(3H)-ones, including cinnabarine, have been obtained from the appropriate 2-aminophenols by oxidative dimerization on thin-layer chromatographic plates (*N. N. Gerber*, Canad. J. Chem., 1968, **46**, 790). Cinnabarine has also been synthesised by a selective method (*W. Schaefer* and *H. Schlude*, Tetrahedron Letters, 1968, 2161).

Questiomycin A (2-Aminophenoxazin-3(3H)-one) has been isolated from *Streptomyces fungicidus* (*M. Matsuoka*, J. Antibiotics, Tokyo, Ser. A., 1960, **13**, 121) and, together with its 2-acetyl derivative, from members of the *Waksmania* genus (*Gerber* and *M. P. Lechevalier*, Biochemistry, 1964, **3**, 598). Enzymes from various natural sources catalyse the conversion of 2-aminophenol into 2-aminophenoxazin-3(3H)-one (*H. R. Gutman* and *H. T. Nagasawa*, J. biol. Chem., 1959, **234**, 1493; *E. Katz* and *H. Weissbach*, ibid., 1962, **237**, 882; *P. M. Nair* and *C. S. Vaidyanathan*, Biochim. Biophys. Acta, 1964, **81**, 507) (see p. 480 for alternative synthesis).

Xanthommatin, rhodommatin and **ommatin D** are obtained from the butterfly *Vanessa urticae*. Collectively these compounds and related structures are called ommochromes (*A. Butenandt* and *W. Schäfer*, in "Recent Progress in the Chemistry of Natural and Synthetic Colouring Matters", Academic Press, New York, 1962, p. 13). The structure of xanthommatin is confirmed by its synthesis firstly by ferric chloride oxidation of hydroxykynurenine (XVII) and secondly by condensation of the latter with the 4,6-dihydroxyquinolinequinone-2-carboxylic acid (*Butenandt et al.*, Ann., 1954, **588**, 106; **590**, 75):

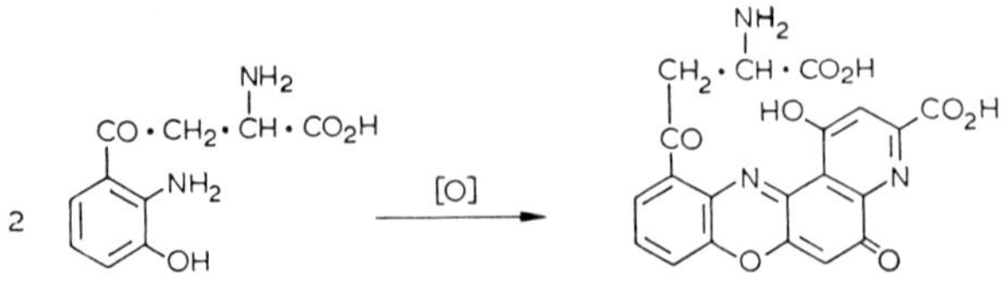

(XVII) Xanthommatin

Rhodommatin is represented by structure XVIII and ommatin D by XIX:

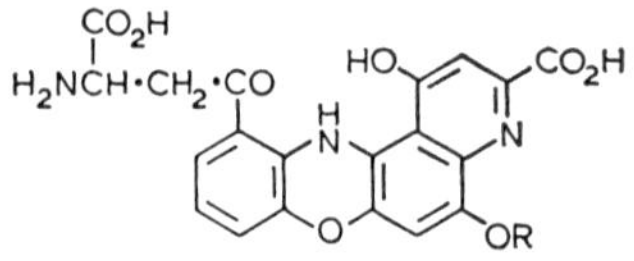

(XVIII, R = β-glucosyl)
(XIX, R = SO_3H)

(*e*) *Benzophenoxazines and benzophenoxazinones*

When 1,2-dihydroxynaphthalene is heated with 2-aminophenol the main product is 7H-*benzo*[c]*phenoxazine* (XX)*, yellow crystals, m.p. 127–128° (sealed tube). The isomeric product, 12H-*benzo[a]phenoxazine* (XXI), a yellow unstable solid, m.p. 107° (decomp.), is formed at the same time in small amount and is also obtained by heating 1-amino-2-naphthol with 2-hydroxyanilinium chloride (*H. Goldstein* and *Z-Ludwig Semelitch*, Helv., 1919, **2**, 655):

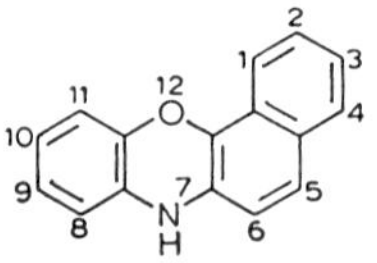

(XX)

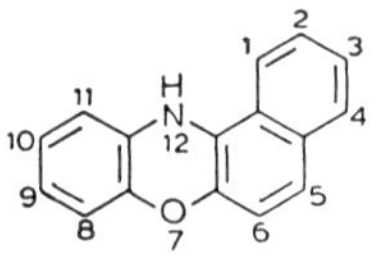

(XXI)

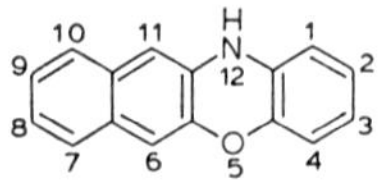

(XXII)

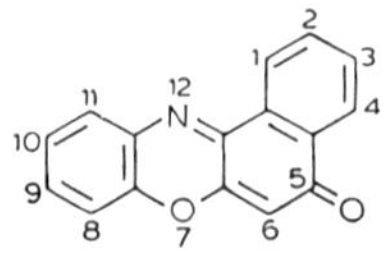

(XXIII)

* The nomenclature follows *A. M. Patterson*, *L. I. Capell* and *D. F. Walker* (The Ring Index 2nd Edn., J. Amer. chem. Soc., 1959, p. 665).

The third isomer 12H-*benzo*[b]*phenoxazine* (XXII), m.p. 302°, is prepared from 2,3-dihydroxynaphthalene (*Kehrmann* and *A. A. Neil*, Ber., 1914, **47**, 3102).

Benzo[a]*phenoxazin*-5(5H)-*one* (XXIII), m.p. 191°, is obtained by hydrolysis of its 3-*phenylimino* derivative, m.p. 215°, which is formed by oxidation, *e.g.* by air, of 1,4-dianilino-2-naphthol. This derivative can be reduced to 5-(N-*phenylamino*)-12H-*benzo*-[a]*phenoxazine*, m.p. 200°, which is reoxidised by air (*R. Lantz*, Ann., 1934, [xi], **2**, 101). In the last quoted paper other benzophenoxazine derivatives are described.

Benzo[a]*phenoxazin*-9(9H)-*one* has been obtained from β-naphthol and 4-nitrosophenol in the presence of zinc chloride and acetic acid. It has m.p. 211° (softening at 200°), is soluble in ether and benzene and gives a blue-green colour in concentrated sulphuric acid changing through violet to brown on dilution (*O. Fischer* and *E. Hepp*, Ber., 1903, **36**, 1807).

Meldola's Blue. The dyestuff Meldola's Blue was discovered by *R. Meldola* (J. chem. Soc., 1881, **39**, 37) through the reaction between 4-nitrosodimethylaniline and β-naphthol:

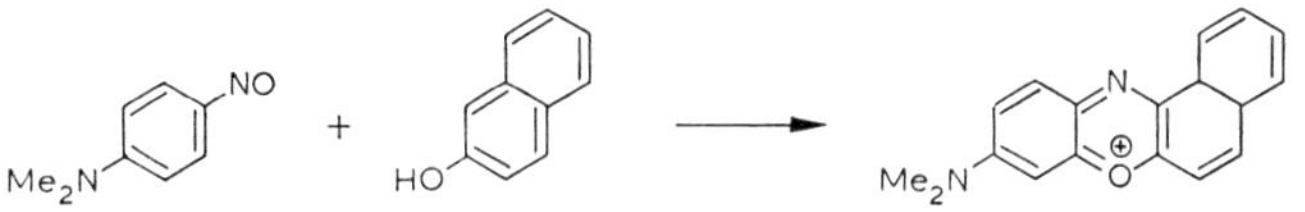

Meldola supposed the dye to be an indophenol, but later *Nietzki* and *Otto* (Ber., 1888, **21**, 1744) deduced the correct benzophenoxazonium structure. The indophenol first formed undergoes oxidation to the benzoxazine, then to the dye salt, at the expense of excess 4-nitrosodimethylaniline which is reduced to N,N-dimethyl-1,4-phenylenediamine. Variants of this method use quinone-di-imines and -dichloroimines, or aminoazo compounds in place of 4-nitrosophenol.

A number of benzo[*c*]phenoxazinones have been prepared by *Kehrmann et al.* (Helv., 1926, **9**, 866) by reacting β-naphthaquinone and its 4-hydroxy-derivative with 2-aminophenols. It is of interest that β-naphthaquinone gives, with 2-amino-5-dimethylaminophenol the isomer of Meldola's Blue, XXIV, not Meldola's Blue itself:

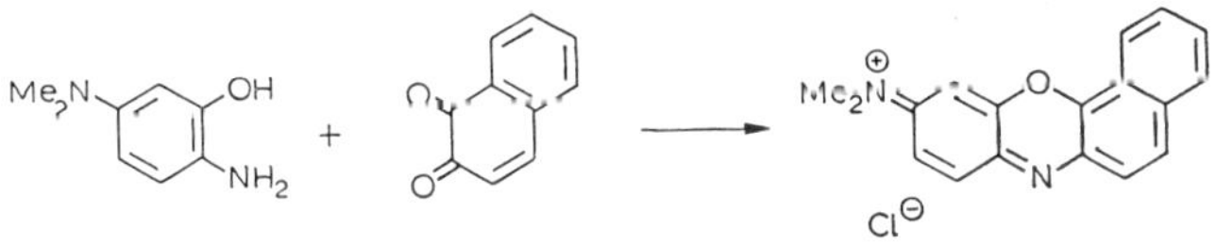

(XXIV)

Benzo[c]*phenoxazin*-10(10H)-*one* (XXV) has been obtained by the condensation of α-naphthol with 4-nitrosophenol in acetic acid containing zinc chloride the yield, however, was low (*K. I. Pashevich, G. B. Afanaséva* and *I. Ya. Postovskii*, Khim. geterot. Soedin, 1971, **7**, 746; C.A., 1972, **76**, 25208f):

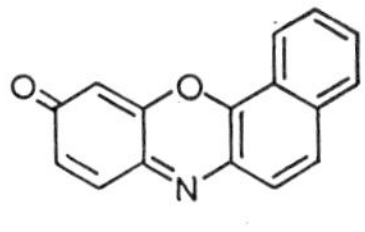

(XXV)

Some 5-arylimino-9-dialkylaminobenzo[*a*]phenoxazinones have been tested for anti-tubercular activity (*R. C. Clapp et al.*, J. Amer. chem. Soc., 1952, **74**, 1989) and the electronic spectra of benzophenoxaxines have been studied (*V. Stuzka* and *V. Simanek*, Coll. Czech. chem. Comm., 1972, **37**, 1121; *Pashkevich et al., loc. cit.*).

Phenanthroxazine, *tetrabenzo*[a,c,d,f]*phenoxazine* (XXVI), m.p. 350–355°, brown crystals with a green lustre, is obtained from 9,10-dihydroxyphenanthrene and ammonia (*E. Bamberger* and *J. Grob*, Ber., 1901, **34**, 535) and by prolonged exposure to sunlight of either phenanthraquinonemonoxime or monoimine (*A. Schönberg et al.*, J. chem. Soc., 1950, 374); dinitro compounds are reported (*J. Schmidt* and *E. Aeckerle*, Ber., 1924, **57**, 363):

(XXVI)

4. Thiazines and related compounds

The isomeric forms of thiazines, having one sulphur and one nitrogen atom in a six-membered ring, and their hydro- and benzo- derivatives correspond with structures in the oxazine series and are named and numbered in the same way. Oxidised forms, having no parallel among oxazines, exist as cyclic sulphoxides or sulphones; these are not usually obtained by oxidation of the parent thiazine, but rather by synthesis from intermediates in which the sulphur atom is already at the correct oxidation level.

(a) 1,2-Thiazines and their hydro derivatives

Few simple examples of this class are known, however, a number of 3,6-**dihydro**-2*H*-1,2-**thiazine** 1-**oxides** (III) are available through hydrolysis of their 2-trichloroethoxycarbonyl derivatives II, which can be prepared by a cycloaddition reaction of the appropriate diene and trichloroethyl *N*-sulphinylcarbamate (I) (*L. Wald* and *W. Wucherpfennig*, Ann., 1971, **746**, 28). Cleavage of the 2-trichloroethoxycarbonyl function in II is effected by the action of zinc dust in boiling *tert*-butyl alcohol:

R^1, R^2 = H, Me or Cl

Wucherpfennig (*ibid.*, 1971, **746**, 16) has observed that 2-sulphonyl derivatives of III are readily attacked by nucleophiles such as hydroxyl ions thus undergoing ring opening to give 4-sulphonylamino-*cis*-but-2-enesulphinates:

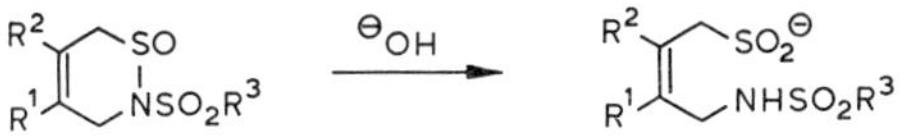

2-Substituted-3,6-dihydro-2*H*-1,2-thiazine 1-oxides are prepared by the addition of *N*-sulphinylamines, R–N = S → O (R = aryl), to butadiene and its derivatives (*E. G. Kataev* and *V. V. Plemenkov*, Zhur. obshcheĭ Khim., 1962, **32**, 3817; C.A., 1963, **58**, 12544f; *O. Wichterle* and *J. Roček*, Chem. Listy, 1953, **47**, 1768; Coll. Czech. Chem. Comm., 1954, **19**, 282). Similarly, *N*-sulphinylsulphonamides afford 2-sulphonyl-3,6-dihydro-2*H*-1,2-thiazine 1-oxides (*G. Kresze et al.*, Angew. Chem. intern. Edn., Engl., 1962, **2**, 89; *E. S. Levchenko* and *A. V. Kirsanov*, Zhur. obshcheĭ Khim., 1962, **32**, 161; C.A., 1962, **57**, 12365i; *Levchenko et al.*, *ibid.*, 1963, **33**, 1579; C.A., 1963, **59**, 12801a).

2-Aryl-3,6-dihydro-2*H*-1,2-thiazine 1-oxides slowly yield unsaturated sultams when treated with hydrogen peroxide in dilute aqueous sodium hydroxide (*Wichterle* and *Roček, loc. cit.*):

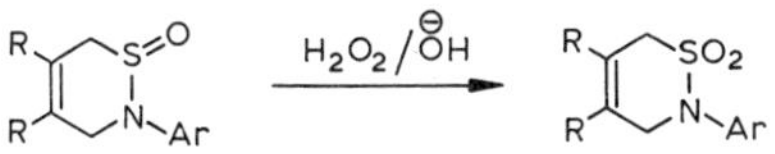

When perbenzoic acid is employed epoxides result through attack at the 4,5-double bond:

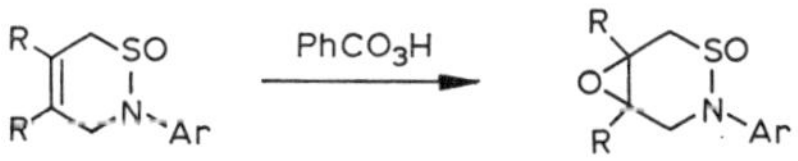

5,6-**Dihydro-2***H***-1,2-thiazine** 1,1-**dioxides** (VI) result from a cycloaddition of benzoylsulphene (V), formed *in situ* from triethylamine and benzoyl-methanesulphonyl chloride, and cinnamylidenamines (IV):

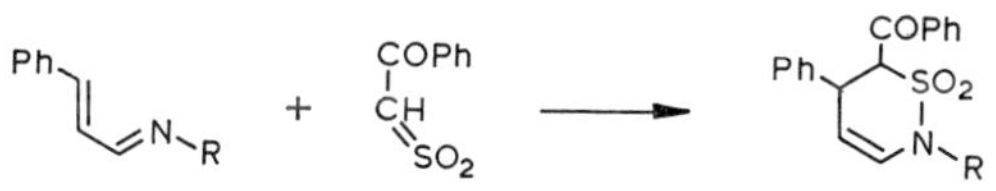

(IV, R = alkyl or aryl)   (V)   (VI)

In these products (VI) the benzoyl and phenyl groups are *quasi*-axial and *quasi*-equatorial respectively, but when treated with sodium methoxide in methanol, or silica gel in chloroform, epimerisation occurs affording isomers in which both substituents occupy *quasi*-equatorial positions (*O. Tsuge* and *S. Iwanami*, Bull. chem. Soc. Japan, 1971, **44**, 2750).

(b) 1,3-Thiazines and related compounds

(i) 4H-1,3-Thiazines

4H-1,3-**Thiazines** are available by the action of α-thioamidoalkylating agents upon arylacetylenes (*C. Giordano*, Gazz., 1974, **104**, 849):

(R and R^1 = Me , Ph or subst.Ph ; R^2 = H or Ph)

Ethyl N-cyanoacetyldithiocarbamate (VII) with triethyl orthoformate yields 5-**cyano**-2-**ethylthio**-1,3-**thiazin**-4(4H)-**one** (VIII) (*M. R. Atkinson et al.*, **J. chem. Soc.**, 1956, 3847):

(VIII)

Similarly, 2-ethylthio-5-methyl-1,3-thiazin-4-one has been obtained from ethyl N-propionyldithiocarbamate (*E. N. Cain* and *R. N. Warrener*, Austral. J. Chem., 1970, **23**, 51).

1,2,3,4-Tetrahydropyrimidine-2-thiones (*e.g.* IX) undergo acid-catalysed ring-opening and recyclisation to yield the isomeric 2-amino-4H-1,3-thiazines Xa and Xb.

(IX)

(Xa)   (Xb)

The products appear to exist principally in the amino forms (Xb) (*L. A. Ignatova, P. L. Ovechkin* and *B. V. Unkovskii*, Zhur. Vsesoyuz. Khim. obshcheĭ im. *D. I. Mendeleeva*, 1970, **15**, 238; C.A., 1970, **73**, 25408j; *Unkovskii et al.*, Khim. geterotsikl. Soedin., 1970, 1690; C.A., 1971, **74**, 99962c; *Ovechkin, Ignatova* and *Unkovskii, ibid.*, 1971, 946; C.A., 1971, **75**, 151129y).

(*ii*) *1,3-Thiazinium salts*

The 1,3-thiazinium salts XI illustrated in Table 7 have been prepared from the corresponding 1,3-oxazinium salts by the action of hydrogen sulphide in anhydrous acetonitrile followed by treatment of the resultant 3-acyl-amidoprop-2-ene-thiones with anhydrous perchloric acid (*R. R. Schmidt* and *D. Schwille*, Chem. Ber., 1969, **102**, 269):

$$R^2\text{-}C(\!=\!S)\text{-}CH\!=\!C(R^3)\text{-}NHCOR^1 \xrightarrow{HClO_4} (XI)$$

(XI)

TABLE 7

SOME 1,3-THIAZINIUM PERCHLORATES (XI)

R^1	R^2	R^3	m.p.(°C)
Ph-	Ph	Ph	280–282
[4]MeC$_6$H$_4$-	Ph	Ph	238–241
[4]ClC$_6$H$_4$-	Ph	Ph	206–208
Ph-	Ph	[4]ClC$_6$H$_4$	257–260
Ph-	[4]ClC$_6$H$_4$	Ph	253–255
Ph-	OEt	Me	112
[4]ClC$_6$H$_4$-	OEt	Me	125
[4]MeC$_6$H$_4$-	OEt	Me	123

In general these compounds react with nucleophiles in a similar manner to 1,3-oxazinium salts (p. 435). Thus with alkalis ring-opening occurs to give thioamides (XII) and the action of hydrogen sulphide leads to XIII. Treatment of the latter with perchloric acid refurbishes the parent salt XI, whereas XII with perchloric acid yields XI or a mixture of XI and the 1,3-oxazinium salt XIV, depending upon the substituents present:

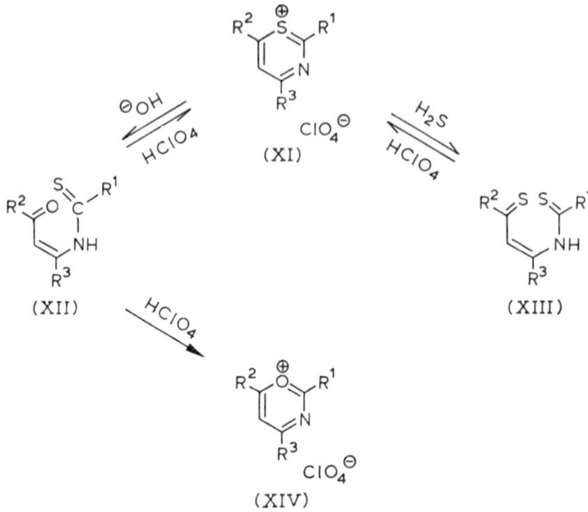

1,3-Thiazinium salts may also combine with carbanions and their potential as synthetic intermediaries is considerable (see for example, *Schmidt, Synthesis*, 1972, 333).

(iii) Dihydro-2H-1,3-thiazines

A number of dihydro-1,3-thiazines are made by combining dithiocarbamic acids and β-propiolactones (*M. Tisler*, Arch. Pharm., 1961, **294**, 348), β-chloropropionic acid (*A. P. Grishchuk* and *I. R. Barilyak*, Zhur. obshcheĭ Khim., 1963, **33**, 3972; C.A., 1964, **60**, 9273g) or acrylic acids (*J. E. Jansen* and *R. A. Mathes*, J. Amer. chem. Soc., 1955, **77**, 2866).

Potentially the reaction between dithiocarbamic acid and propiolic acid derivatives may yield a number of products (see *Cain* and *Warrener, loc. cit.*), but at low temperature under mildly acidic conditions, attack by one of the sulphur atoms at the β-acetylenic position occurs affording products of type XV.

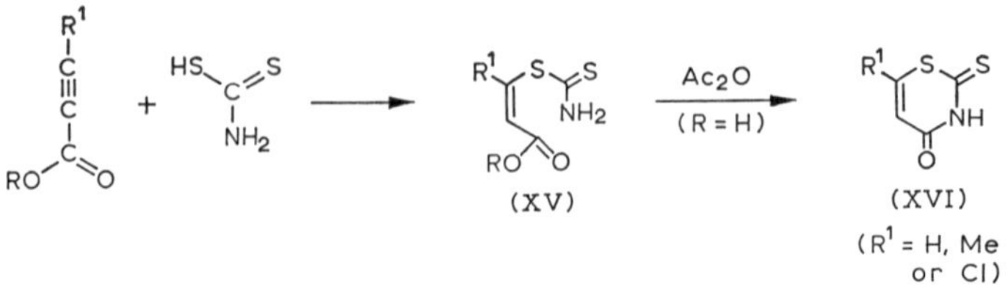

When R $=$ H, cyclisation to the corresponding 2-**thioxo**-1,3-**thiazine**-4 3(*H*)-**ones** (XVI) is easily effected by treatment with hot acetic anhydride.

Esters, *e.g.* XV (R $=$ Et), are much more resistant and, in such cases, the product are *N*-acetyl derivatives of the acyclic system XV.

Dithiocarbamic acid (prepared *in situ* from its ammonium salt) adds readily to activated alkenes, but whilst dithiocarbamoyl compounds are formed from crotonaldehyde, mesityl oxide and methyl acrylate, only the products from the last two compounds can be cyclised to 4,6,6-**trimethyl**-3,6-**dihydro**-1,3-**thiazine**-2-**thione** (XVII), m.p. 96–97°, and 4-**oxo**-2-**thioxo**-**tetrahydro**-1,3-**thiazine** (XVIII), m.p. 119–120°, respectively:

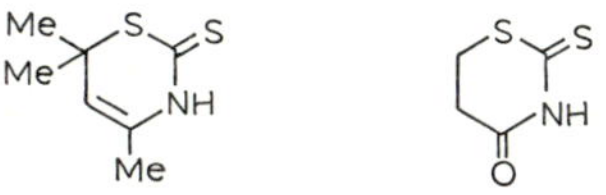

(XVII)

(XVIII)

Phenylpropiolic acid fails to react with dithiocarbamic acid, but ethyl phenylpropiolate reacts with thiourea in refluxing acetone to afford 2-**imino**-6-**phenyl**-1,3-**thiazin**-4(3*H*)-**one** (XIX) (*Cain* and *Warrener*, *loc. cit.*; *H. Tanaka* and *A. Yokoyama*, Chem. pharm. Bull. Tokyo, 1962, **10**, 19; C.A., 1963, **58**, 3347b):

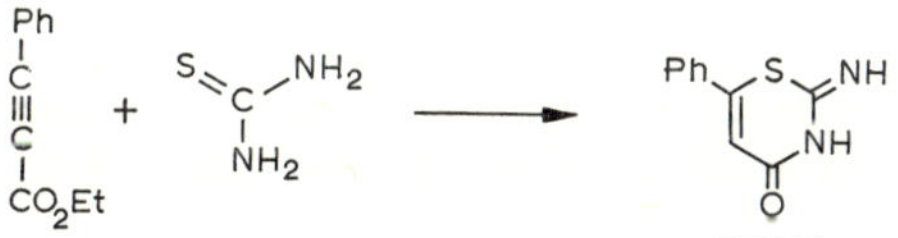

(XIX)

Yokoyama (Bull. chem. Soc. Japan, 1971, **44**, 1610) reports that 2,3-di-hydro-1,3-thiazin-4-ones (XX) are available from the condensation of 3-alkylthio-2-cyano-3-mercaptoacrylamides with ketones or aldehydes in the presence of sulphuric acid:

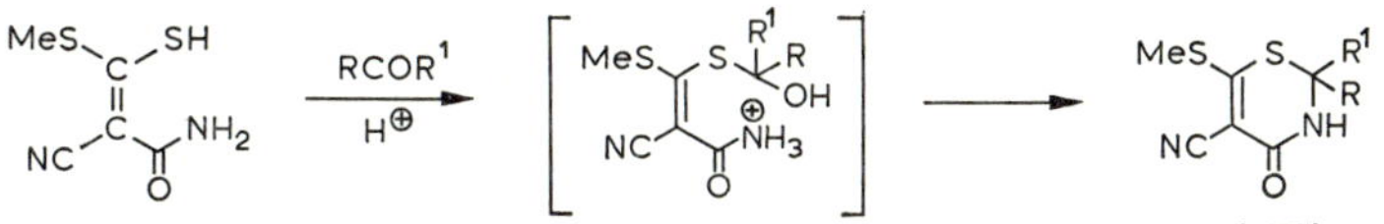

(XX)

Two routes to 4,4,6-**trimethyl**-3,4-**dihydro**-1,3-**thiazine**-2-**thione** (XXI), m.p. 120.5–121.5° are described by *J. E. Jansen* and *R. A. Mathes* (J. Amer. chem. Soc., 1955, **77**, 5431). The same intermediate is obtained from di-acetoneamine and carbon disulphide and from the addition of ammonium dithiocarbamate to mesityl oxide and hydrochloric acid:

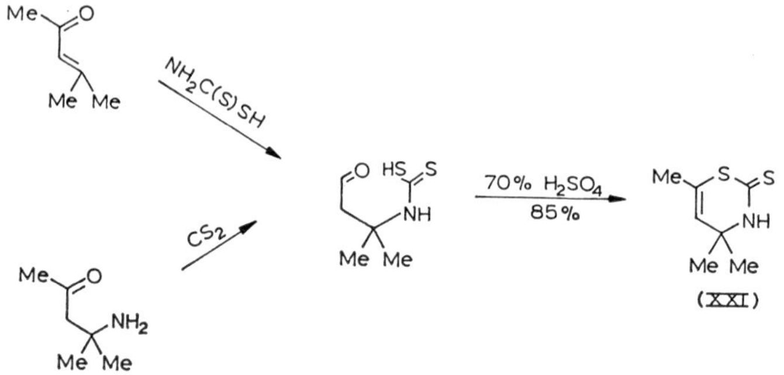

There are numerous syntheses of dihydro-2*H*-1,3-thiazines which employ the condensation of an appropriate C_3 unit with a molecule containing a N–C–S fragment. Often the latter is thiourea, some examples have already been cited (for others see *H. Behringer* and *P. Zillikens*, Ann., 1951, **574**, 140; *B. H. Chase* and *J. Walker*, J. chem. Soc., 1955, 4443; *R. Zimmermann*, Angew. Chem. intern. Edn., 1962, **1**, 663; *T. B. Johnson*, J. Amer. chem. Soc., 1914, **36**, 1891; *E. V. Vladimirskaya* and *N. M. Tirkevich*, C.A., 1965, **63**, 2967h; *Cain* and *Warrener*, Austral. J. Chem., 1970, **23**, 51). In some cases, however, the reaction takes an alternative path and 2-thiopyrimidines are produced (see *D. J. Brown*, "The Pyrimidines", Interscience, Amsterdam, 1962, p. 272).

Claims for the synthesis of 3,4-dihydro-2*H*-1,3-thiazines (XXII) by the addition of thioureas to dimethyl acetylenedicarboxylate (*J. W. Lown* and *J. C. N. Ma*, Canad. J. Chem., 1967, **45**, 939; *E. Winterfeldt* and *J. M. Nelke*, Chem. Ber., 1967, **100**, 3671; *Y. Kishida* and *A. Terada*, Chem. and Pharm. Bull., Japan, 1968, **16**, 1351) are incorrect. The condensation products are in fact 1,3-thiazolinones of type XXIII (*L. K. Mushkalo* and *G. Y. Yangol*, Ukrain. Khim. Zhur., 1955, **21**, 732; C.A., 1956, **50**, 16751a; *J. B. Hendrickson*, *R. Rees* and *J. F. Templeton*, J. Amer. chem. Soc., 1964, **86**, 107; *H. Sasaki*, *H. Sakata* and *Y. Iwanami*, J. chem. Soc. Japan, 1964, **85**, 704; C.A., 1965, **62**, 14678f; *A. F. Cameron et al.*, Chem. Comm., 1970, 890; *Warrener* and *Cain*, Austral. J. Chem., 1970, **23**, 51; *ibid.*, 1971, **24**, 785; *J. F. B. Mercer et al.*, Synthetic Comm., 1972, **2**, 35):

(iv) *5,6-Dihydro-4H-1,3-thiazines and tetrahydro-1,3-thiazines*

5,6-Dihydro-4H-1,3-thiazines are prepared by the condensation of thio-amides and 1,3-dihalogenopropanes or γ-halogenoamines. Other methods include the cyclisation of N-acyl-γ-bromoamines (XXIV, Y = Br) by treatment with phosphorus pentasulphide and of N-acyl-γ-mercaptoamines

X = Br
or NH$_2$

(XXIV)

TABLE 8

ALKYL- AND ARYL-5,6-DIHYDRO-4H-1,3-THIAZINES

Compound	m.p.(°C)	Picrate, m.p.(°C)	Method	Ref.
5,6,-Dihydro-4H-1,3-thiazine	b.p. 174	130–132	HS(CH$_2$)$_3$NHCHO and P$_2$O$_5$	1
2-Methyl-	b.p. 175	138	MeCSNH$_2$ and Cl(CH$_2$)$_3$Br	1,2
2-Benzyl-	oil	—	PhCH$_2$ · CSNH$_2$ and Cl(CH$_2$)$_3$Br	2
2-Phenyl-	44–45	—	{ PhCSNH$_2$ and Cl(CH$_2$)$_3$Br [PhCONH(CH$_2$)$_3$]$_2$S and PCl$_5$	3
4,4,6-Trimethyl-2-phenyl-	34	152	PhCSNH$_2$ and BrCHMe · CH$_2$ · CMe$_2$NH$_2$	4
2-p-Tolyl-	52–53	—	MeC$_6$H$_4$CSNH$_2$ and Cl(CH$_2$)$_3$Br	2
2-(2-Naphthyl)-5,6-dihydro-4H-1,3-thiazine	82	169	[2]C$_{10}$H$_7$CSNH$_2$ and Cl(CH$_2$)$_3$Br	5

References

1   S. Gabriel, Ber., 1916, **49**, 1114.
2   G. Pinkus, ibid., 1893, **26**, 1077.
3   M. Lehmann, ibid., 1894, **27**, 2173.
4   M. Kahan, ibid., 1897, **30**, 1318.
5   F. Saulmann, ibid., 1900, **33**, 2634.

(XXIV, Y = SH) with acyl anhydrides. Some examples are included in Table 8.

Spectroscopic evidence shows that compounds formed by the reaction of 1-arylthioureas with 1,3-dihalogenopropanes (*A. C. Glasser* and *R. M. Doughty*, J. pharm. Sci., 1965, **54**, 1005) and formulated as 2-arylamino-5,6-dihydro-4*H*-1,3-thiazines (XXV) actually have the isomeric structures XXVI (*L. Toldy* and *P. Sohár*, Tetrahedron Letters, 1970, 181; *Toldy et al.*, ibid., p. 2167).

(XXV)   (XXVI)

2-Amino-5,6-**dihydro**-4*H*-1,3-**thiazines** may be prepared from γ-hydroxy- or γ-halogeno-amines and an isothiocyanate:

(XXVII)   (XXVIII)

X = OH or Cl ; R = H or Ph

Typical products include 2-*amino*-, oil (*hydrobromide*, m.p. 135–136°, *picrate*, m.p. 128°) (*S. Gabriel* and *W. E. Lauer*, Ber., 1890, **23**, 93); 2-*sulphanilamido*-, m.p. 88° (*G. W. Raiziss*, J. Amer. chem. Soc., 1941, **63**, 2739); 2-*anilino-6-methyl*-, m.p. 106.5°, *picrate*, m.p. 163–164° (*A. Luchmann*, Ber., 1896, **29**, 1429); 2-*ethylamino-4,4,6-trimethyl*-, m.p. 190–191° (decomp.), *picrate*, m.p. 156–157°, and 2-*anilino-4,4,6-trimethyl-5,6-dihydro-4H-1,3-thiazine*, m.p. 147–148° (*M. Kahan*, Ber., 1897, **30**, 1318). It is not certain, however, whether these compounds should be represented as 2-amino-5,6-dihydro-4*H*-1,3-thiazines (XXVIII) or whether they are better described as the 2-iminotetrahydro tautomers (XXVII).

2-Hydroxy-5,6-**dihydro**-4*H*-1,3-**thiazine** (2-oxotetrahydro-1,3-thiazine) is the probable product from trimethylene chlorobromide and xanthamide (*G. Pinkus*, Ber., 1893, **26**, 1077):

2-Mercapto-5,6-**dihydro**-4*H*-1,3-**thiazines,** from carbon disulphide and γ-halogenoamines, may exist preferably in the tautomeric thione form, 2-**thioxotetrahydro**-1,3-**thiazines,** on the evidence of ultraviolet absorption (*F. M. Hamer* and *R. J. Rathbone*, J. chem. Soc., 1943, 243): 2-*mercapto*-, m.p. 129–131°, 2-*mercapto-6-methyl*-, m.p. 131° *(Luchmann, loc. cit.)* and 2-*mercapto-4,4,6-trimethyl-5,6-dihydro*-4H-1,3-*thiazine*, m.p. 180° (ref. 4, Table 8). These on alkylation yield *S*-alkyl compounds, *e.g.* 2-*methylthio*-, m.p. 155–160°/50 mm, *methiodide*, m.p. 132° and 2-*ethylthio*-, b.p. 145–150°/40 mm, *ethiodide*, m.p. 99°, which on heating in pyridine rearrange to *N*-alkyl compounds, 4-*methyl*-, m.p. 88°, and 4-*ethyl-2-thioxotetrahydro*-1,3-*thiazine*, m.p. 68°, respectively (*Hamer* and *Rathbone, loc. cit.*).

2,4-**Dioxo**-, m.p. 159° and 2-**methylimino-4-oxo-tetrahydro**-1,3-**thiazine**, m.p. 210° (decomp.), are prepared from β-iodopropionic acid and xanthamide or *N*-methylthiourea, respectively: 2-*phenylimino-4-oxo* compound, m.p. 157° (*N. A. Langlet*, Ber., 1891, **24**, 3852; 1894, **27**, 247).

4-**Oxo**-2-**thioxotetrahydro**-1,3-**thiazine** (XXIX, R = H) (see p. 491), is prepared from β-propiolactone and dithiocarbamic acid (from its ammonium salt) by treatment with acetic anhydride (with a drop of concentrated sulphuric acid) and similarly for the 3-*ethyl*- deriv. (XXIX, R = Et), m.p. 65–66° (*T. L. Gresham et al.*, J. Amer. chem. Soc., 1948, **70**, 1001; see also p. 491).

(XXIX)

Cephalosporins. The cephalosporins (XXX) exemplified by cephalosporine C (R = –(CH$_2$)$_3$·CH(NH$_2$) · CO$_2$H contain a dihydro-1,3-thiazine nucleus. They are an important group of antibiotics and their chemistry, synthesis and pharmacological activity has been reviewed (see for example *R. B. Morin* and *B. G. Jackson*, Fortschr. Chem. org. Naturstoffe, 1970, **28**, 343), and their inter-relationship with penicillins has also been examined and discussed (*D. H. R. Barton* and *P. G. Sammes*, Proc. roy. Soc., 1971, Δ**179**, 1971):

(XXX)

In view of the occurrence of the dihydro-1,3-thiazine moiety in the cephalosporins, novel syntheses to dihydro-1,3-thiazines related in structure to the antibiotics have been developed (see for example *J. E. Dolfini, J. Schwartz* and *F. Weisenborn*, J. org. Chem., 1969, **34**, 1582 and Table 9, p. 497).

(c) *1,4-Thiazines and their hydro derivatives*

(i) *Thiazines*

Thiodiglycollimide (XXXI), m.p. 128°, may be formulated also as 3,5-**dihydroxy-2***H*-1,4-**thiazine** or as the 4*H*-tautomer. When it is passed over alumina on pumice at 450°, it gives a 13 % yield of 2*H*-1,4-**thiazine** (XXXIII) (*C. Barkenbus* and *P. S. Landis*, J. Amer. chem. Soc., 1948, **70**, 684). The fact that the product, b.p. 76.5–77°, *picrate*, m.p. 158–159°, *chloroplatinate*, m.p. 236–238° and *hydrochloride* (hygroscopic), m.p. 74–75°, fails to form a sulphonamide is regarded as evidence in favour of its formulation as

2*H*-1,4-thiazine (XXXII) rather than as the 4*H*-isomer XXXIII, but this is hardly conclusive:

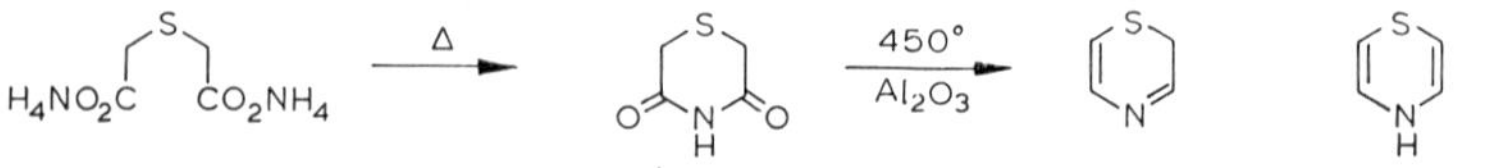

(XXXI)   (XXXII)   (XXXIII)

When heated in ethanol solution containing picric acid or nitrobenzene, 3,5-disubstituted-2*H*-1,4-thiazines undergo oxidative coupling at position 2, giving 2,2'-bithiazines (*D. Sica, C. Santacrocea* and *G. Prota*, J. heterocyclic Chem., 1970, **7**, 1143):

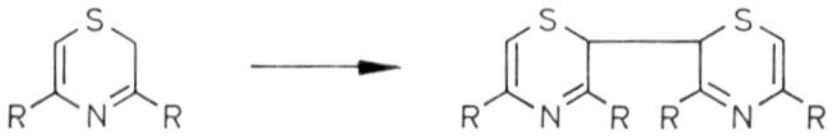

(R = Ph or [4]CH$_3$·C$_6$H$_4$)

This reaction has relevance to the biosynthesis of the trichosiderin pigments present in red hair and feathers (see p. 510).

C. R. Johnson and *C. B. Thanawalla* (*ibid.*, 1969, **6**, 247) have prepared 3,4-*dimethyl*-5-*methylthio*-4H-1,4-*thiazine* (XXXIV) and its *methiodide*, m.p. 125–126°, by the sequence shown below:

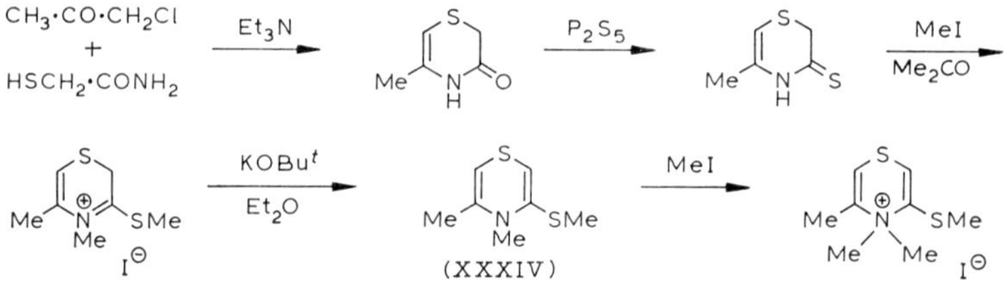

(XXXIV)

The *sulphone* XXXV, m.p. 270–272°, and its *anion* XXXVI have also been prepared (*Johnson* and *I. Sataty*, J. med. Chem., 1967, **10**, 501; *Sataty*, J. org. Chem., 1969, **34**, 250); the former is remarkably stable remaining unchanged after two weeks in boiling triethylamine (see also *V. Baliah* and *T. Rangarajan*, *ibid.*, 1961, **26**, 970):

(XXXV)   (XXXVI)

(ii) Dihydro-1,4-thiazines

Thioglycollamide from ammonia and methylthioglycollate with chloroacetone gives 5-**methyl**-3,4-**dihydro**-2*H*-1,4-**thiazine**-3-**one** (XXXVIII), m.p.

144°, which is hydrolysed by an acidic solution of 2,4-dinitrophenyl-hydrazine and forms the *dinitrophenyl hydrazone*, m.p. 159–161°, of the intermediate *acetonylcarbamidomethyl sulphide* (XXXVII) (*H. Sokol* and *J. J. Ritter*, J. Amer. chem. Soc., 1948, **70**, 3517):

(XXXVII)   (XXXVIII)

3,4-Dihydro-2*H***-1,4-thiazine-5,6-dicarboxylates** (XXXIX) have been synthesised by the reaction of 2-mercaptoethylamine with 2,3-dibromoacrylic esters (*5. Alexander et al.*, J. chem. Soc., Perkin I, 1974, 2092):

(XXXIX)

A number of derivatives useful in relation to analogues of the cephalosporins (p. 495) have been made (see Table 9).

TABLE 9

3,4-DIHYDRO-2H-1,4-THIAZINE-5,6-DICARBOXYLIC ACID
DERIVATIVES (XXXIX)

R^1	R^2	m.p.(°C)
Me	CO_2H	100–102
Et	CO_2Et	75–77
CH_2Ph	CO_2Bu^t	118
CH_2Ph	CO_2Me	97–98

Sulphur and aziridine (ethyleneimine) (see Vol. IV A, p. 15 *et seq.*) with aldehydes in the presence of dimethylformamide or potassium carbonate yield 6-alkyl-3,4-dihydro-2*H*-1,4-thiazines together with 2-alkylthiazolidines (*F. Asinger et al.*, Monatsh., 1970, **101**, 1281, 1295; 1971, **102**, 321). Ketones under the same conditions afford a mixture of 5-alkyl- and 5,6-dialkyl-3,4-dihydro-2*H*-1,4-thiazines plus 2,2-dialkylthiazolidines:

(R^1 = H or alkyl)

2-Alkynylcysteine-S-oxide and S,S-dioxides react with ammonium hydroxide to give 3,4-dihydro-2H-1,4-thiazine 1-oxides or 1,1-dioxides (*J. F. Carson* and *L. E. Boggs*, J. org. Chem., 1971, **36**, 611). Thus S-(2-propynyl)-L-cysteine S,S-dioxide yields *compound* XL, decomp. 185°, which adopts a half-chair conformation:

$(+)$-S-(2-Propynyl)-L-cysteine S-oxide gives the analogous 1-*oxide*, decomp. 182–184°, but the $(-)$-enantiomer fails to react.

(iii) Tetrahydro-1,4-thiazines; 1,4-thiazanes

N-substituted tetrahydro-1,4-thiazines (1,4-thiazanes) or thiomorpholines have attracted much attention and are known in great variety.

Methods of synthesis. (*1*) Primary amines combine with mustard gas, β,β'-dichlorodiethyl sulphide, giving N-substituted thiomorpholines (*H. T. Clarke*, J. chem. Soc., 1912, **101**, 1583). Alcoholic ammonia at 60° in this reaction gives the parent XLI (R = H) (*W. Davies*, ibid., 1920, **117**, 297).

The corresponding β,β'-dichlorodiethyl sulphoxide and sulphone behave similarly (*W. E. Lawson* and *E. E. Reid*, J. Amer. chem. Soc., 1925, **47**, 2821).

(*2*) Sodium sulphide reacts with N-phenyl-β,β'-dichlorodiethylamine (XLII, R = Ph) giving 4-phenylthiomorpholine (*V. V. Korshak* and *Yu. A. Strepikheev*, C.A., 1945, **39**, 3790):

(*3*) Ethylene dibromide reacts with β-aminoethanethiol in alcoholic potassium hydroxide; thiomorpholine was first prepared in this way (*N. A. Langlet*, Beilstein, 1937, Vol. 27, p. 9).

(*4*) Divinyl sulphone and amines give 4-substituted thiomorpholine 1-dioxides (*A. H. Ford-Moore et al.*, J. chem. Soc., 1946, 819; *M. A. Stahmann et al.*, J. org. Chem., 1946, **11**, 719).

Thiomorpholine, (1,4-thiazane), b.p. 169°, a colourless liquid with an unpleasant pyridine-like odour, is miscible with water and organic solvents. It is a strong base which rapidly absorbs atmospheric carbon dioxide, forming a non-deliquescent carbonate: *hydrochloride*, m.p. 163°, *picrate*, m.p. 198° (decomp.), and *picrolonate*, m.p. 242° (decomp.). Some

N-substituted compounds are included in Table 10. The *N*-alkylthiomorpholines are colourless, water-soluble oils but the higher members become insoluble. Their 1-oxides and 1,1-dioxides cannot be distilled without decomposition even under reduced pressure (*Lawson* and *Reid, loc. cit.*).

TABLE 10

N-ALKYLTHIOMORPHOLINES AND DERIVATIVES

N-deriv.	b.p.(*°C/mm*)	HCl salt m.p.(*°C*)	1-Oxide m.p.(*°C*)	HCl salt m.p.(*°C*) (decomp.)	1,1-Dioxide m.p.(*°C*)	HCl salt m.p.(*°C*)
Methyl	163–164	239	52	241	82	> 280
Ethyl	184	188	oil	177.5	oil	248–249
Benzyl	154/13	225	oil	224	76.5	237–239

4-*Cetylthiomorpholine*, m.p. 78°, *hydrochloride*, m.p. 162°, *picrate*, m.p. 112°, *methiodide*, m.p. 244° (*W. F. Hart* and *J. B. Niederl.* J. Amer. chem. Soc., 1944, **66**, 1610) and other higher alkyl compounds, including oxides and dioxides, are also known (*idem, ibid.*, 1946, **68**, 714).

4-**Phenylthiomorpholine**, b.p. 160–162°/18 mm, m.p. 32.3–32.6°, *hydrochloride*, m.p. 162°, *picrate*, m.p. 141–142°, with nitrous acid yields 4-(4-*nitrosophenyl*)*thiomorpholine*, m.p. 60° (*Korshak* and *Strepikheev, loc. cit.*). A product, m.p. 108–111°, from β,β'-dichlorodiethyl sulphide and aniline is also reported as being this 4-phenyl compound (*O. B. Helfrich* and *Reid*, J. Amer. chem. Soc., 1920, **42**, 1226).

4-(**2-Hydroxyethyl**)**thiomorpholine**, m.p. 35.5°, *hydrochloride*, m.p. 162–163°, 1-*oxide*, m.p. 173–174°, 1,1-*dioxide*, m.p. 175.5°, and many of its esters have been described (*L. A. Burrows* and *Reid, ibid.*, 1934, **56**, 1720). 4-(*2-Chloroethyl*)*thiomorpholine* 1,1-*dioxide*, m.p. 188–189°, is prepared *via* divinyl sulphone and ethanolamine *(Ford-Moore et al., loc. cit.)*.

Thiomorpholine reacts with 2-chloro-5-nitropyridine and with 4-bromo-3-nitrophenyl-arsonic acid giving 2-(4-*thiomorpholino*)-5-*nitropyridine*, m.p. 132.1–132.6°, and 3-*nitro*-4-(4-*thiomorpholino*)*phenylarsonic acid*, m.p. 187–189°, respectively (*E. J. Cragoe* and *C. S. Hamilton*, J. Amer. chem. Soc., 1945, **67**, 536).

β,β'-Dichlorodiethyl sulphide and glycine ester give the non-cyclic compound di-β,β'-(ethoxycarbonylmethylamino)ethyl sulphide, $S(CH_2 \cdot CH_2NHCH_2 \cdot CO_2Et)_2$, but the corresponding sulphone gives the cyclic 4-*methoxycarbonylmethylthiomorpholine*-1,1-*dioxide*, m.p. 68.5°, which is hydrolysed to the *acid*, m.p. 177° (*A. E. Cashmore* and *H. McCombie*, J. chem. Soc., 1923, **123**, 2888) also obtained directly from the sulphone and glycine (*Boursnell et al.*, Biochem. J., 1946, **40**, 734 and later papers).

Diethyl or dimethyl sulphonyldiacetate (XLIII, R = CO_2Et or CO_2Me), when condensed with aromatic (not aliphatic) aldehydes and ammonia or aliphatic (not aromatic) primary amines in alcohol with ammonium acetate, give substituted thiomorpholines of type XLIV. Sulphonyldiacetic acid behaves similarly in acetic acid but decarboxylation of the product occurs. 3,5-*Diphenylthiomorpholine* 1,1-*dioxide* (XLIV, R = R^2 = H,

R^1 = Ph), m.p. 205–206°, *hydrochloride*, m.p. 275–278° (decomp.) and the 2,6-*bisethoxy-carbonyl* derivative (XLIV, R = CO₂Et, R^1 = Ph, R^2 = H), m.p. 184–185°, are typical (*V. Baliah* and *T. Rangarajan*, J. chem. Soc., 1954, 3068).

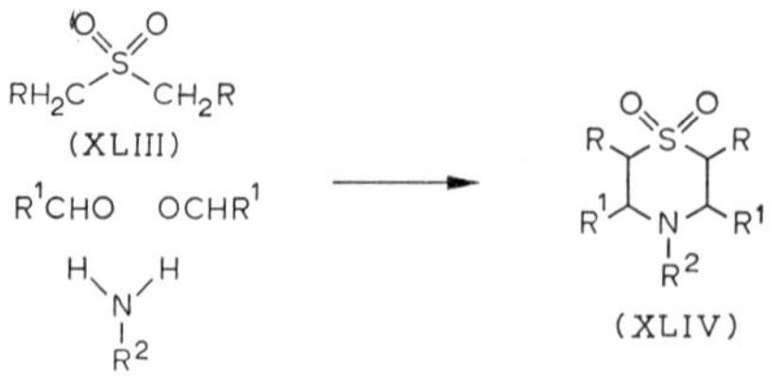

Thiomorpholine-3,5-dicarboxylic acid (XLVII, R = H), m.p. 253–254°, is prepared from thiodiacetaldehyde (XLV) by a Strecker reaction. The product from bromoacetal and potassium sulphide is hydrolysed by dilute hydrochloric acid and the intermediate XLVI is treated with hydrocyanic acid and ammonia or a primary amine, but not aniline.

At 0–10° the product from ammonia is 3-*cyanothiomorpholine-5-carboxylic acid*, m.p. 192° (decomp.), 4-*methyl-*, m.p. 208° (decomp.), and 4-*ethyl-* deriv., m.p. 177° (decomp.), but at 65–70°, 3,5-*dicyanothiomorpholine*, m.p. 214° (decomp.), 4-*methyl-*, m.p. 178°, and 4-*ethyl-* deriv., m.p. 137°, is formed:

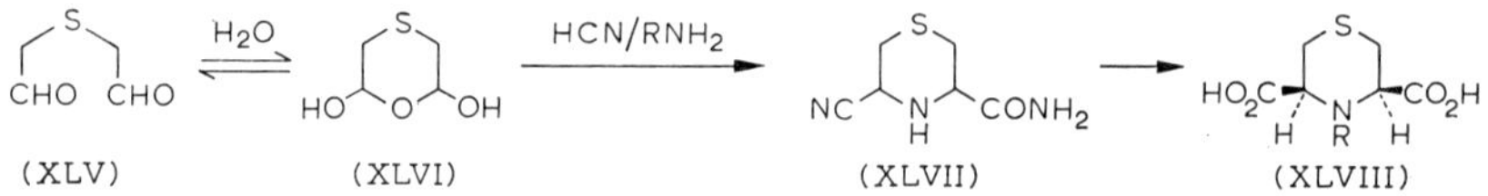

On attempted hydrolysis with hydrochloric acid these products tend to decompose but 4-*methyl-3-cyanothiomorpholine-5-carboxylic acid*, m.p. 184–185°, and *thiomorpholine-3,5-dicarboxylic acid, diethyl ester*, b.p. 154–156°/3 mm, *ester hydrochloride*, m.p. 148°, are obtained from the appropriate amide XLVII. The dicarboxylic acid XLVIII (R = H) may be in the *cis*-form, as shown, since on boiling with acetic anhydride it gives 4-*acetyl-thiomorpholine*-3,5-*dicarboxylic anhydride*, m.p. 143°. With hydrogen peroxide the dicarboxylic acid gives a 1-*mono-oxide*, m.p. 242° (decomp.) (*R. D. Coghill et al.*, J. Amer. chem. Soc., 1937, **59**, 801; 1940, **62,** 1613, 1615).

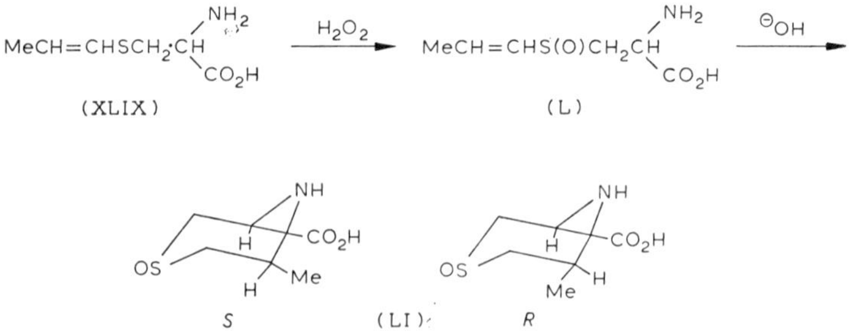

cis-S-(1-Propenyl)L-cysteine (XLIX) may be oxidised with aqueous hydrogen peroxide to a mixture of (+)- and (−)-*cis-S*(1-propenyl)L-cysteine sulphoxides (L). Reaction of either in aqueous base may then form cycloalliin (3-methyl-1,4-thiazane-5-carboxylic acid 1-oxide) (LI). Alliin (*trans* isomer of XLIX, is a component of onion *(Allium cepa)* aroma (*Carson* and *Boggs*, J. org. Chem., 1966, **31**, 2862; see p. 498):

5. Benzothiazines

The numbering systems and nomenclature of the benzothiazines follow the same patterns as those described for the benzoxazines (p. 456).

(a) 1,2-, 2,1- and 2,3-Benzothiazines and their hydro derivatives

(i) 2H-1,2-Benzothiazines

1,1-**Dioxo**-2*H*-1,2-**benzothiazin**-3(4*H*)-**one** (I) is prepared by the cyclisation of 2-sulphamoylphenylacetic acid with polyphosphoric acid (*E. Sianesi, I. Setnikar* and *E. Massarani*, Ger. Offen. 2,022,694/1970).

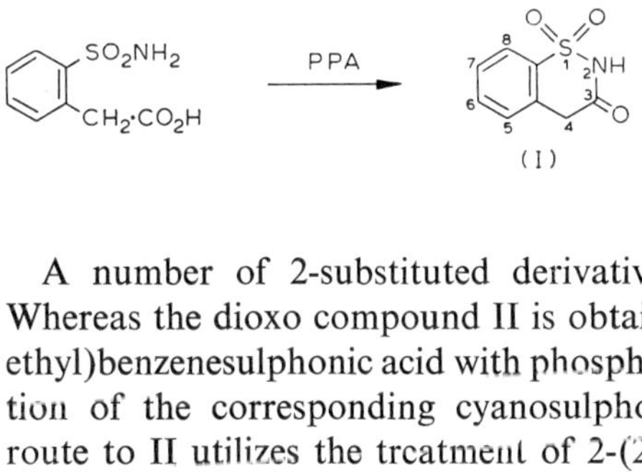

A number of 2-substituted derivatives have been prepared similarly. Whereas the dioxo compound II is obtained by dehydration of 2-(2-aminoethyl)benzenesulphonic acid with phosphoryl chloride or through hydrogenation of the corresponding cyanosulphonamide in acid solution, a third route to II utilizes the treatment of 2-(2-chloroethyl)benzenesulphonamide with alkalis (*Sianesi et al.*, Chem. Ber., 1971, **104**, 1880).

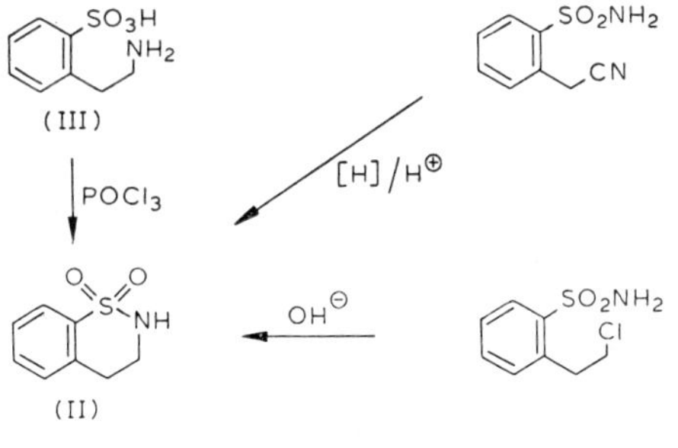

The oxidation of α-tetralinsulphonamide gives 3-sulphonamidophthalonic acid and its *sultam* III, m.p. 261°, which corresponds to the 1,1-dioxide of 2*H*-1,2-**benzothiazine**-3,4-**dione**-5-**carboxylic acid** (*J. Braun* and *G. Hahn*, Ber., 1923, **56**, 2343):

Substituted 2*H*-1,2-benzothiazine 1,1-dioxides (V) are prepared from sulphonamides, *e.g.* IV by cyclodehydration with sulphuric acid (*H. Watanabe et al.*, J. org. Chem., 1969, **34**, 919):

Benzyne, generated *in situ* from lead tetra-acetate and 1-aminobenzotriazole, reacts with thionylaniline to give 6*H*-dibenzo[*c,e*]-1,2-thiazine 5,5-dioxide (VI), m.p. 196° (*C. D. Campbell* and *C. W. Rees*, J. chem. Soc. C, 1969, 748):

The same product is obtained by thermolysis of biphenyl-2-sulphonyl azide (VII) at 150° in dodecane (*R. A. Abramovitch, C. I. Azogu* and *I. T. Mc-Master*, J. Amer. chem. Soc., 1969, **91**, 1219). The reaction probably involves the nitrene VIIa.

(ii) 1H-2,1-Benzothiazines

1*H*-2,1-**Benzothiazine** 2,2-**dioxide** (**sulphostyril**) (IX), m.p. 153–155°, can be synthesised by a Bamford–Stevens procedure (decomposition of the tosylhydrazone) from 1*H*-2,1-benzthiazine-4(3*H*)-one 2,2-dioxide (VIII) (*B. Loer, M. F. Kormendy* and *K. M. Snader*, J. org. Chem., 1966, **31**, 3531). Starting materials for this, and for derivatives substituted in the benzenoid ring, are sulphonamides of the general type $ArNHSO_2CH_2 \cdot CO_2Me$, which are cyclised by the action of polyphosphoric acid:

$$\text{(VIII)} \xrightarrow[-\text{MeOH}]{PPA} \text{(IX)}$$

Sulphostyrils are more stable to ultraviolet irradiation than carbostyrils which dimerize under prolonged exposure (*Loer* and *Snader*, J. heterocycl. Chem., 1967, **4**, 407).

3,4-Dihydro-1*H*-2,1-benzthiazine 2,2-dioxide (X) is known (*Sianesi et al., loc. cit.*). It may be prepared either by pyrolysis of 2-(2-aminophenyl)-ethanesulphonamide or the sodium salt of the corresponding sulphonic acid:

$$\xrightarrow{\Delta} \text{(X)} \xleftarrow{\Delta}$$

(iii) 1H-2,3-Benzothiazine derivatives

3,4-*Dihydro*-1H-2,3-*benzothiazine* 2,2-*dioxide*, m.p. 142–143°, is formed by the action of phenylmethanesulphonamide (1 mol.) and trioxane (0.33 mol.) in a mixture of methanesulphonic acid and acetic acid during 3 h. If the reaction time is reduced to 2 min, 1,3,5-tribenzylsulphonylhexahydro-1,3,5-cyclotriazine—the kinetic product—is obtained (*O. O. Orazi* and *R. A. Corral*, Chem. Comm., 1976, 470):

Several 1*H*-2,3-benzothiazin-4(3*H*)-one 2,2-dioxides have been prepared (*idem*, Chem. Ber., 1970, **103**, 1922; Ger. Öffen., 2,022,694/1970) by the following route:

(b) 1,3- and 3,1-Benzothiazines and their hydro derivatives

(i) 1,3-Benzothiazines

4*H*-1,3-**Benzothiazine** (XI, R = H) may be prepared in 35% yield by treating the *N,S*-dimagnesium dibromide of 2-mercaptobenzylamine with bromoform (*D. Bourgoin-Legay* and *R. Boudet*, Compt. rend., 1971, **273**, C, 372):

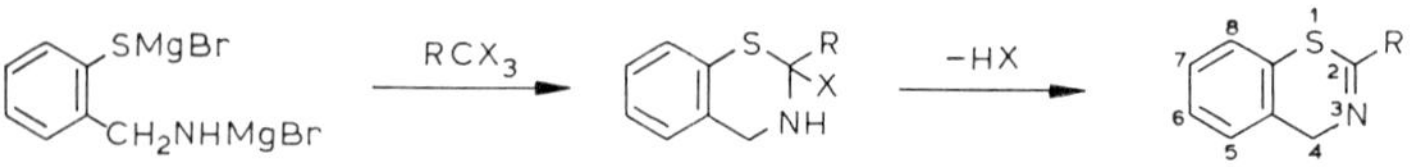

When benzotrichloride is employed, instead of bromoform, 2-*phenyl*-4H-1,3-*benzothiazine* (XI, R = Ph), m.p. 41.9–42°, is obtained, whereas benzal chloride affords 3,4-dihydro-2-phenyl-2*H*-1,3-benzothiazine. 2-Alkyl- or 2-aryl- substituted 4*H*-1,3-thiazines are also available through cyclisation of *N,S*-diacyl-2-mercaptobenzylamines with phosphoryl chloride (*idem*, Bull. Soc. chim. Fr., 1967, 4441).

2-*Methyl*-, b.p. 71–72°/0.3 mm; 2-n-*propyl*-, b.p. 85°/0.2 mm; 2-*isopropyl*-, b.p. 79°/0.2 mm; 2-*ethyl*-, b.p. 79–80°/0.35 mm; 2-(4-*anisyl*)-, m.p. 90°; 2-(4-*nitrophenyl*)-, m.p. 112°; 2-(4-*bromophenyl*)-, m.p. 131°; 3-(4-*nitrophenyl*)-4H-1,3-*benzothiazine*, m.p. 112°.

J. Szabó (Kem. Kozlem., 1966, **26**, 21; C.A., 1967, **66**, 463840) reports the synthesis of 6,7-dialkoxyl-2-phenyl-4*H*-1,3-benzothiazines. These are easily converted into the corresponding quaternary salts (XII) by heating with alkyl halides in a sealed tube (*Szabó et al.*, Acta Chim. Acad. Sci. Hung., 1971, **69**, 459; **70**, 71; C.A. 1971, **75**, 140778h, 151745w). These salts are susceptible to nucleophilic attack, thus in water or alkali ring-opening occurs to give an *S*-(2-ammoniomethylphenyl)thiobenzoate halide as the primary reaction product:

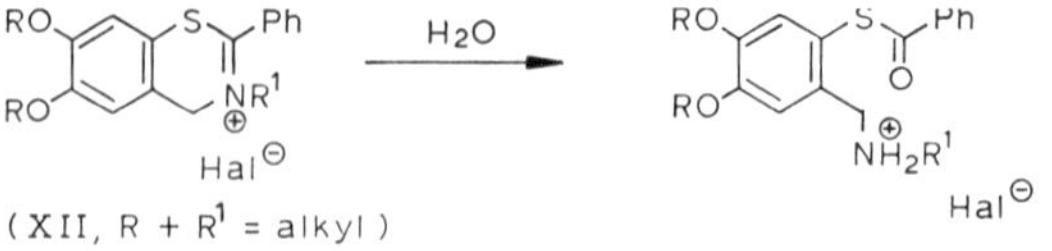

2-Substituted-4,4-diethyl-4*H*-1,3-benzothiazines can be made by reacting the appropriate 2-alkyl-, aryl- or aralkyl-thiobenzyl alcohols with nitriles in the presence of acid (*V. A. Zagorevskii et al.*, Khim. geterot. Soedin., 1974, 1437):

(R or R^1 = Me, Ph, Bz)

Although the products are contaminated with the tricyclic 1,1,2,2-tetra-ethyl-5,6-dithia[3,4-7,8]dibenzo-octatetraene, the sequence appears to have general application.

Bourgoin-Legay and *Boudet* (Bull. Soc. chim. Fr., 1969, 2524) report that 2-alkyl- and 2-aryl-4*H*-1,3-benzothiazines are fairly resistant to ring-opening, thus the 2-phenyl derivative is unaffected by boiling alkali and concentrated acids. Hot dilute acids, on the other hand, yields 2-aminomethylbenzenethiol:

Depending upon the reaction conditions oxidation may occur at $C_{(4)}$ yielding a benzo-1,3-thiazin-4(4*H*)-one or at sulphur giving a sulphoxide.

3-Chloro-2-methyl-1,2-benzisothiazolium chloride (XIII) reacts with *N*-methylformamide in pyridine solution at 35–40° to give predominantly **3-methyl-4-methylimino-3,4-dihydro-2*H*-1,3-benzthiazin-2-one** (XIV); at 100° in 1,2-dichlorobenzene **3-methyl-2-methylimino-3,4-dihydro-2*H*-1,3-benzthiazin-4-one** (XV) is obtained; *N*-phenylformamide, instead of *N*-methyl-formamide, yields the 2-phenylimino-analogue under both sets of reaction conditions (*H. Böshagen et al.*, Chem. Ber., 1971, **104**, 3757):

(XIII)   (XIV)   (XV)

3-Amino-2-imino-1,3-benzothiazine-4(4*H*)-ones can be prepared from cyanogen and 2-mercaptobenzhydrazides (*N. D. Heindel* and *L. A. Schaeffer*, J. pharm. Sci., 1975, **64**, 1425):

(R = H, Cl, Me)

2-Chlorobenzo-1,3-thiazin-4(4*H*)-ones are formed by reacting hydrogen chloride and 2-thiocyanatobenzoyl chlorides (R = Cl) or phosphorus pentachloride and the corresponding 2-thiocyanobenzoic acids (R = OH) (*G. Simchen* and *J. Wenzelburger*, Chem. Ber., 1970, **103**, 413), *e.g.*:

(R = Cl, OH)   (m.p. 115–117°)

The 2-chloro substituent is readily displaced by nucleophiles. 2-Phenylbenzo-1,3-thiazin-4(4*H*)-ones can be prepared by the acid-catalysed rearrangement and ring-closure of *N*-(arylthiomethyl)benzamides; the resultant products react with alkyl or aryl Grignard reagents to give 2-substituted 2-phenyl-3,4-dihydrobenzothiazin-4(2*H*)-ones and 4-substituted 2-phenyl-4*H*-benzothiazin-4-ols (*I. Varga et al., ibid.*, 1975, **108**, 2523). 2-**Methyl-1-thia-azaphenalene, naphtho**[1,8-*de*]-**1,3-thiazine** (2-**methylperinaphthothiazine**) (XVI), m.p. 96.5–97.5°, has been obtained by heating 1-acetamidonaphthalene-8-thiol (as its stannic chloride addition compound) with sodium acetate in glacial acetic acid; it condenses with benzaldehyde to form a 2-*styryl*- derivative, m.p. 132.5° (*M. T. Bogert* and *J. H. Bartlett*, J. Amer. chem. Soc., 1931, **53**, 4046):

(XVI)

It has been used as a component of cyanine dyes (*F. M. Hamer* and *R. J. Rathbone*, J. chem. Soc., 1943, 487). The *phenyl* analogue has m.p. 102–103° (*A. Reissert*, Ber., 1922, **55**, 858).

(ii) 4H-3,1-Benzothiazine derivatives

These may be obtained by: (*1*) the action of phosphorus pentasulphide on 2-acylaminobenzyl alcohol; (*2*) condensing a 2-aminobenzyl halide with a thioamide:

(*3*) the action of phosphorus pentasulphide on the corresponding benzoxazine; (*4*) from a 2-acylaminophenylmethanethiol by treatment with phosphorus pentachloride.

4*H*-3,1-**Benzothiazine**, b.p. 116–118°/6 mm, is formed by condensing 2-aminobenzyl chloride with thioformamide; its *picrate* has m.p. 181.5–181.6° (*S. Gabriel*, Ber., 1916, **49**, 1115).

2-*Methyl*-4H-3,1-*benzothiazine*, m.p. 46°, has been obtained from 2-acetamidobenzyl alcohol by the first method and has also been prepared by the third and fourth methods (*Gabriel* and *Th. Posner*, *ibid.*, 1894, **27**, 3509). The *picrate* has m.p. 178°, darkening from 170°. The substance has also been made by method (*2*) from thioacetamide (*B. Beilenson* and *Hamer*, J. chem. Soc., 1942, 98). The base is stable in boiling acid solution but at high temperatures is hydrolysed by strong acids to 2-aminophenylmethanethiol. 2-*Ethyl-*, *picrate*, m.p. 135–136°, 2-p-*tolyl-*, m.p. 109–110°, *picrate*, m.p. 156–157° and 2-(4-*methoxyphenyl*-4H-3,1-*benzothiazine*, *hydrochloride*, m.p. 212° (*H. Kippenburg*, Ber., 1897, **30**, 1141) have been prepared by the thioamide process.

2-*Amino*-4H-3,1-*benzothiazine*, m.p. 135–137°, is formed when 2-aminobenzyl chloride hydrochloride is heated with thiourea at 160–170°, but the yield is very variable; it forms an *ethiodide*, m.p. 220° (decomp.). This compound, and also 2-methyl-4*H*-3,1-benzo-thiazine, have been used as components of cyanine dyes (*Beilenson* and *Hamar*, *loc. cit.*).

Although 2-aminobenzyl alcohol reacts with carbon disulphide in alcohol to form 2-mercapto-4*H*-3,1-benzoxazine, when the two interact in alcoholic potassium hydroxide there is formed 2-*mercapto*-4H-3,1-*benzothiazine* (or the tautomeric 2-thioxodihydro compound), yellow needles, m.p. 166°. The substance is stable in boiling aqueous alkali but when boiled with aniline is converted into 3-phenyl-2-thioxotetrahydroquinazoline (*C. Paal* and *O. Commerell*, Ber., 1894, **27**, 2427).

2-**Methyl**-3,1-**benzothiazin**-4(4*H*)-**one** (XVII) is formed from anthranilic acid in thio-acetic acid at 100° (*G. C. Barrett*, *J. R. Chapman* and *A. K. Khokhar*, Chem. Comm., 1969, 818):

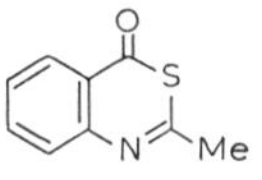

(XVII)

D. Lednicer and *D. E. Emmert* (J. hctcrocycl. Chem., 1971, **8**, 903) report that when 2-alkylthio-4*H*-3,1-benzothiazines (*e.g.* XVIII) are treated with potassium amide in liquid ammonia ring contraction with expulsion of sulphur occurs yielding the corre-sponding 2-alkylthio-indole (XX). It is probable that an episulphide intermediate (XIX) participates in this reaction:

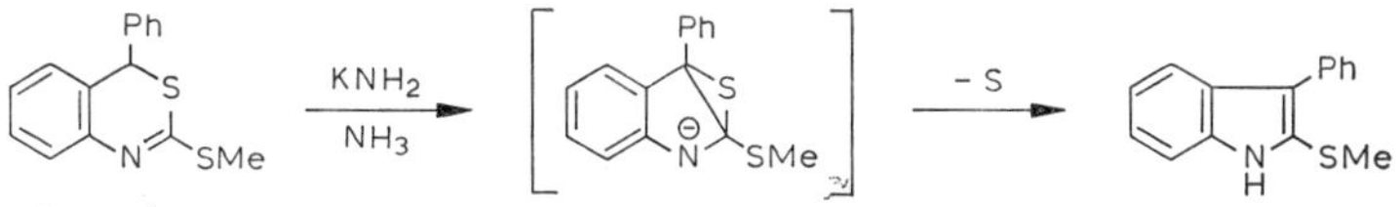

(XVIII)   (XIX)   (XX)

2-**Phenyl**-3,1-**benzothiazine**-4(4*H*)-**thione** (XXI) reacts with diazomethane at room tempera-ture to form the isomeric bi(1,3-dithiolanes) XXII and XXIII, whereas with diazoaceto-phenone at ∼ 150° the product is XXIV (*M. Ebel* and *N. Lozac'h*, Bull. Soc. chim. Fr., 1971, 180, 183, 187):

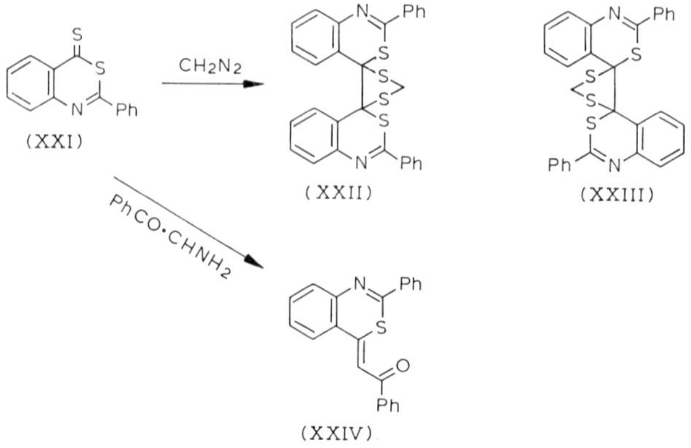
S
S
N
Ph
(XXI)
CH2N2
N
Ph
S
S
S
S
N
Ph
(XXII)
N
Ph
S
S
S
S
Ph
N
(XXIII)
PhCO·CHNH2
N
Ph
S
O
Ph
(XXIV)

The reaction of 2-aryl-3,1-benzothiazine-4(4*H*)thiones with secondary amines in benzene solution typically yields thioamides (*e.g.* XXV → XXVI) (*C. Denis-Garez*, *L. Legrand* and *N. Lozac'h*, *ibid.*, 1969, 3727). In ethanol, very surprisingly, the ring sulphur atom is replaced by a NMe function, thus with dimethylamine the quinazoline-4-thione (XXVII) is formed. This last change was confirmed since methylamine under the same conditions also gives XXVII:

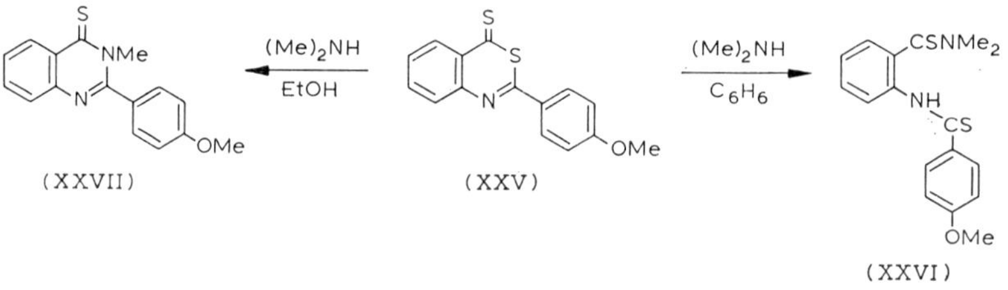
S
NMe
N
OMe
(XXVII)
(Me)2NH
EtOH
S
S
N
OMe
(XXV)
(Me)2NH
C6H6
CSNMe2
NH
CS
OMe
(XXVI)

Similar reactions are reported by *A. Sammour et al.* (Indian J. Chem., 1973, **11,** 437). Thus aliphatic primary amines react at room temperature to give 2-phenyl-3-alkyl-4-thioxoquinazolines, aromatic primary amines require a temperature of 170°. Two equivalents of alkyl Grignard reagents afford 1-(1-hydroxyalkyl)-2-thiobenzamidobenzenes, and aromatic hydrocarbons in the presence of aluminium trichloride yield 1-arylthiocarbonyl-2-diarylmethyliminobenzenes:

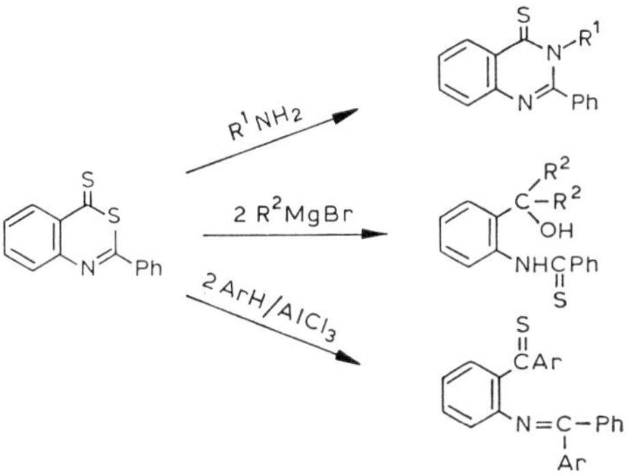

2,4-**Dihydro-2,4-dithioxo-1*H*-benzo-3,1-thiazine (trithioisatic anhydride)** reacts with hydrazine at room temperature to give 3-amino-2,4-dithioxoquinazoline (XXVIII), which with hot formic acid yields 1,3,4-thiadiazolo[2,3-*b*]-3,4-dihydroquinazoline-4-thione (XXIX) (*G. Wagner* and *S. Leistner*, Z. Chem., 1971, **11,** 65):

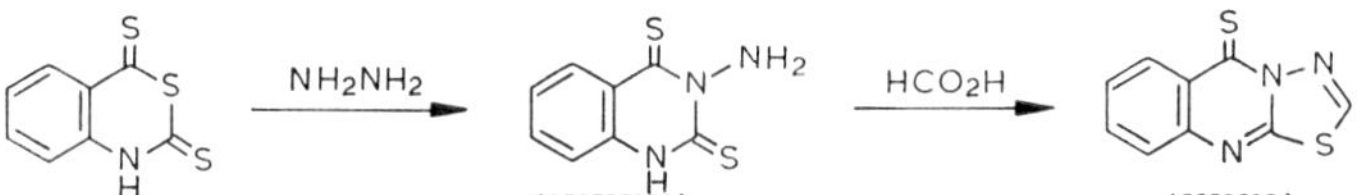

(XXVIII) (XXIX)

The parent substance can be converted into 2-*thiocyanato*-4-*thioxo*-4H-*benzo*-3,1-*triazine* (XXX), m.p. 153–154.5°, which may be oxidised to the *disulphide* XXXI, m.p. 203–204° and the *sulphide* XXXII, m.p. 222–223° (*Leistner* and *Wagner*, Pharmazie, 1975, **30,** 643):

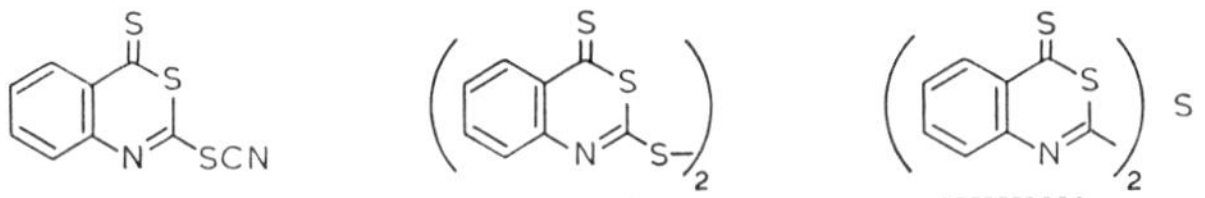

(XXX) (XXXI) (XXXII)

(*c*) *1,4-Benzothiazines and their hydro derivatives*

(*i*) *2H-1,4-Benzothiazines*

It was claimed that phenacyl bromide and 2-aminobenzenethiol gave 3-phenyl-2*H*-1,4-benzothiazine (XXXIII) (*O. Unger*, Ber., 1897, **30,** 607) and similarly that 2-amino-4-chlorobenzenethiol gave 6-*chloro*-3-*phenyl*-2H-1,4-*benzothiazine*, m.p. 64° (decomp. on keeping). However, ^{1}H n.m.r. and u.v. studies now show that the first product is really a mixture of the 2*H*- (80%) and 4*H*- (XXXIV) (20%) tautomers:

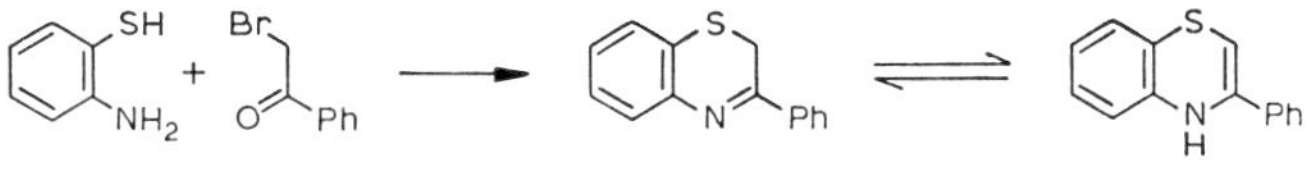

(XXXIII) (XXXIV)

Ethyl 6-chloro-3-methyl-2H-1,4-benzothiazine-2-carboxylate, m.p. 177–178°, is reported by *Zincke* and *Baeumer* (Ann., 1918, **416**, 108), and *Santacrocea et al.* (Gazz., 1968, **98**, 85), describe the preparation of 3-(4-*chlorophenyl*)-, 3-(4-*methylphenyl*)-, m.p. 53–55°, 2,3-*diphenyl*-, m.p. 89–90°, 2-ethyl-3(4-methoxyphenyl)- and 7-methoxy-3-(4-methylphenyl)-2H-1,4-benzothiazines.

The claim by *Langlet* (Bihang. till Svenska Vet. Akad. Hanglinger 22II, N.I.S. 20; *Beilstein*, 1937, **27**, 44) to have prepared 2H-1,4-benzothiazine by condensing 2-aminophenol and 1,2-dibromoethane is incorrect, the true product is 2-methylbenzothiazole (*Santacrocea et al., loc. cit.*).

The parent heterocycle, 2H-1,4-**benzothiazine** (XXXIII), is unstable, but may be prepared in solution by treatment of 1-(2-aminophenylthio)-2,2-diethoxyethane with methanolic hydrogen chloride. In aqueous hydrochloric acid 2H-1,4-benzothiazine exists in equilibrium with various aldolization products, one of which, a *trimer*, m.p. 222–224°, may be isolated (*F. Chioccara, Prota* and *R. H. Thompson*, Tetrahedron, 1976, **32**, 1407; *Chioccara et al.*, Chem. Comm., 1977, 50). Aerial oxidation of 2H-1,4-benzothiazine rapidly affords $\Delta^{2,2'}$-**bi(2H-1,4-benzothiazine)** (XXXVI) the parent ring system of a group of amino acid pigments (trichosiderins or trichochromes) found in mammalian red hair, in the feathers of many birds (*Prota, E. Ponsiglione* and *R. Ruggiero*, Tetrahedron, 1974, **30**, 2781) and also in the urine of humans suffering malignant melanoma metastases (*Prota et al.*, Experientia, 1976, **32**, 1122).

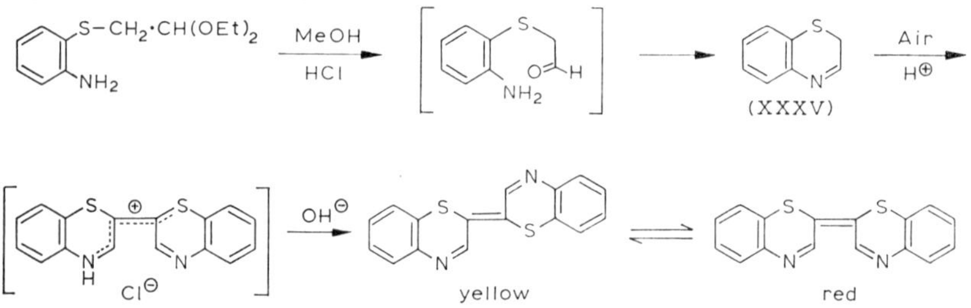

A convenient source of these pigments are the feathers of New Hampshire Chickens, from which three homogenous **trichosiderins (C, E** and **F)** have been isolated and characterized as:

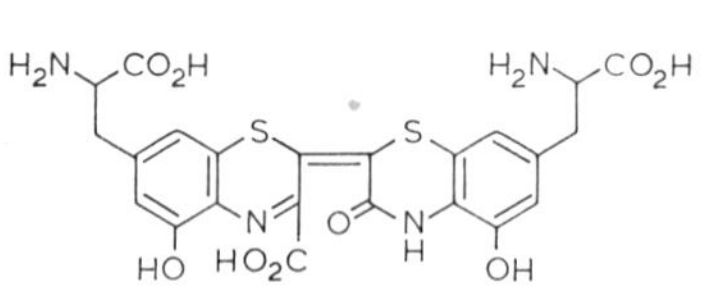

Trichosiderin C
(Trichochrome C)

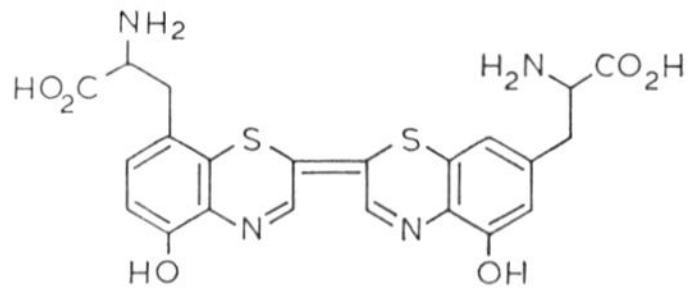

Trichosiderin E
(Trichochrome E)

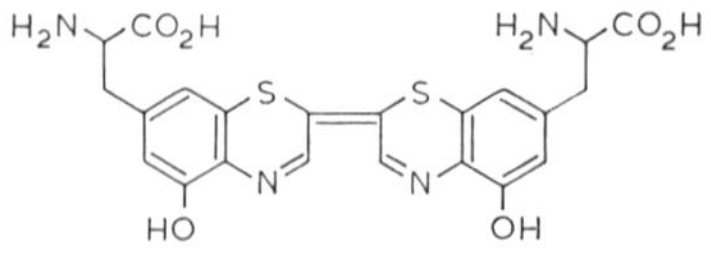

Trichosiderin F
(Trichochrome F)

Other related structures are known (see *Prota et al.*, Gazz., 1969, **99**, 323; Experientia, 1971, **27**, 1381; *Prota*, in "Pigmentation: its genesis and biological control", ed. *V. Riley*, pp. 615–630, Appleton-Century-Crofts, New York, 1972). *B. L. Kaul* (Helv., 1974, **57**, 2664) describes a simple trichosiderin synthesis and a number of "biogenetic" type syntheses of model trichosiderins have been published by *Prota* and his co-workers (Chim. Ind., Milan, 1975, **57**, 392 and papers already cited) thus, 3,4-dihydroxytoluene with cysteine and potassium ferricyanide form an intermediate XXXVII which with acid yields XXXVIII:

3,3-**Diphenyl**-7,7-**bi**(2*H*-1,4-**benzothiazine**), m.p. 325°, has been prepared by the condensation of phenacyl bromide and 3,3′-dimercaptobenzidine (*P. Bottex, B. Sillion* and *G. de Gaudemaris*, Compt. rend., 1968, **267**, 711):

(ii) 2H-1,4-Benzothiazinium salts

2-Aminobenzenethiols of the general formula $C_6H_4(NHR)SH[1,2]$ and α-epoxyketones react to form 2-substituted methylene salts of type XXXIX (*L. G. Kovalenko, L. K. Mushkalo* and *V. A. Chiuguk*, Ukr. Khim. Zhur., 1969, **35**, 1278; C.A., 1970, **12**, 78963x). Some examples are collected in Table 11.

TABLE 11

SOME 2*H*-1,4-BENZOTHIAZINIUM SALTS
OF GENERAL FORMULA XXXIX

R	R^1	R^2	R^3	x^-	m.p.(°C)
H	Me	Me	Me	Cl	115–117
H	Me	Me	H	ClO_4	156–158
H	Me	Et	H	ClO_4	160
H	Me	Pr^i	H	ClO_4	185
H	Me	Bu^i	H	ClO_4	185–186
H	Me	Ph	H	Cl	145–146
H	Ph	Ph	H	Br	220–221
Me	Me	Bu^i	H	ClO_4	184–186

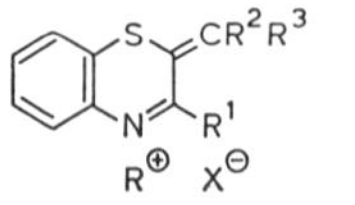

(XXXIX) 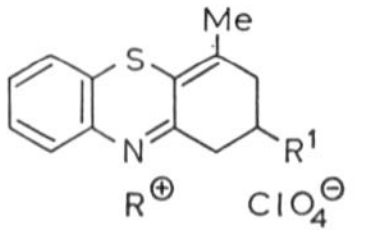   (XL)

Similarly, 2,3-epoxy-3-methylcyclohexanones and 2-aminobenzenethiol afford derivatives of the type XL (R = H; R^1 = *Ph*, m.p. 192°; R = H; R^1 = *Me*, m.p. 203°; R = R^1 = H, m.p. 174°; R = R^1 = *Me*, m.p. 190–191°; R = *Me*; R^1 = *Ph*, m.p. 199–200°).

(iii) 2H-1,4-Benzothiazin-3(4H)-ones

2*H*-1,4-**Benzothiazin**-3(4*H*)-**one** (XLI), m.p. 179°, is prepared from 2-aminobenzenethiol and bromoacetic acid (*O. Unger* and *G. Graaf*, Ber., 1897, **30,** 608, 2389) and by the reduction of 2-nitrophenylthioglycollic acid, being the sultam of 2-aminophenylthioglycollic acid:

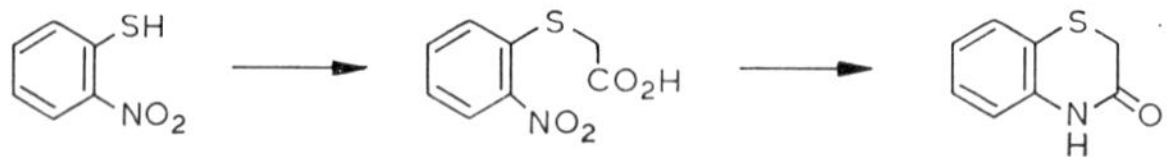

(XLI)

2-Aminobenzenethiol reacts with glycidic esters at room temperature in the presence of alcoholic potassium hydroxide as shown below. The ester group, not the hydroxyl group, is involved in the subsequent cyclisation and the resulting products XLII are readily cleaved by aqueous alkali to the same 3-oxo compound XLI. Cyclisation does not occur with any product which has suffered hydrolysis but the salt XLIII on acidification cyclises spontaneously. Condensations with ethyl phenylglycidate (R = Ph, R′ = H) and ethyl dimethylglycidate (R = R′ = Me) are described (*C. C. J. Culvenor et al.*, J. chem. Soc., 1949, 278):

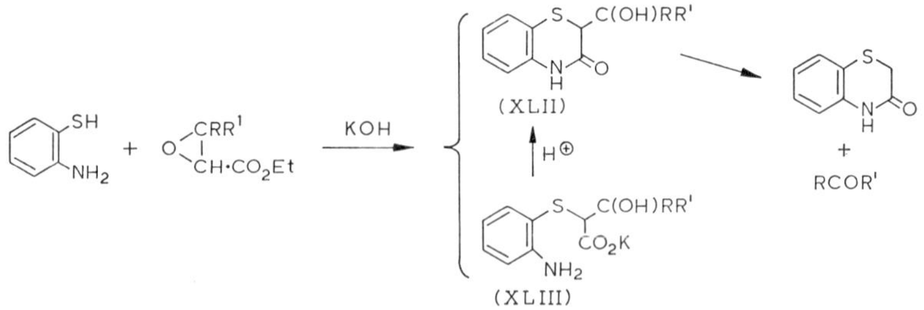

Ethyl cinnamate, cinnamic acid and cinnamoyl chloride react quite differently towards 2-aminobenzenethiol, the products being 2-*benzyl*-2H-1,4-*benzothiazin*-3(4H)-*one* (XLIV, R = CH₂Ph), m.p. 159–160°, the seven-membered ring *compound* (XLV, R = Ph), m.p. 177° and 2-styrylbenzthiazole (XLVI), respectively. Acrylic and crotonic acids yield XLV (R = H), m.p. 215–216°, and XLII (R = Me), m.p. 205–206°, whilst maleic acid

or monobromosuccinic acid with 2-aminobenzenethiol gives 3-*oxo*-2H-1,4-*benzothiazine*, 2-*acetic acid* (XLI, R = CH$_2$ · CO$_2$H), m.p. 195–196° (*W. H. Mills* and *J. B. Whitworth* ibid., 1927, 2738):

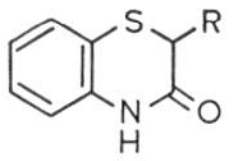

(XLIV) 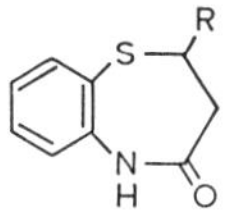   (XLV) 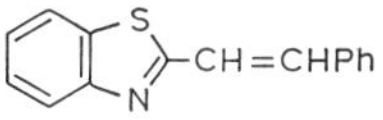   (XLVI)

2*H*-1,4-Benzothiazin-3(4*H*)-one (XLI) has a reactive methylene group in the 2-position, the activity of which is increased by phosphorylation *(J. W. Worley, K. W. Ratts* and *K. L. Cammack,* J. org. Chem., 1975, **40**, 1731), and a typical amide group capable of alkaline hydrolysis. On chlorination with sulphuryl chloride, it forms 2-*chloro*-, m.p. 215° (decomp.) and then 2,2-*dichloro*-2H-1,4-*benzothiazin*-3(4H)-*one*, m.p. 195–196°; the latter with concentrated sulphuric acid yields 2,3-*dioxo*-4H-1,4-*benzothiazine*, m.p. 250° (decomp.) and with aniline forms its 2-*phenylimide*, m.p. 254–256° (*K. Zahn,* Ber., 1923, **56**, 581). The monochloro compound and its N-*methyl* derivative, m.p. 50–53°, react with triethyl phosphite to give the 2-phosphonates which, in turn, combine with aldehydes and ketones to yield 2-alkylidene derivatives *(idem, ibid.).*

6-*Chloro*-, m.p. 204° (*J. Pollak et al.,* Monatsh., 1928, **49**, 219), 7-*chloro*-, m.p. 206° and 5,7-*dichloro*-2H-1,4-*benzothiazin*-3(4H)-*one*, m.p. 164° (*R. Herz,* G.P. 364,822/1914; Chem. Ztbl., 1923, II, 918) are prepared from the corresponding 2-aminophenylthioglycollic acid.

The 3-oxo compound XLI with amyl nitrite yields 2-*hydroxyimino*-2H-1,4-*benzothiazin*-3(4H)-*one*, m.p. 267°, but the reactivity in the 2-position is enhanced in 2H-1,4-*benzothiazine*-3(4H)-*one* 1,1-*dioxide*, m.p. 207–208°, which is readily available by the reduction of the *S*-dioxide of 2-nitrophenylthioglycollic acid and by oxidation of the 3-oxo compound with permanganate. Direct oxidation of a thiazine to the corresponding sulphone is rarely possible.

The 1,1-dioxide of XLI couples with diazonium salts in the 2-position and undergoes electrophilic substitution in the 7-position: 7-*nitro*-2H-1,4-*benzothiazin*-3(4H)-*one* 1,1-*dioxide*, m.p. 219–220° (decomp.) (*M. Claasz,* Ber., 1912, **45**, 747; 1916, **49**, 350).

4-Hydroxy-1,4-benzothiazin-3(4*H*)-one-2-acetic acid (XLVII) when heated under reflux with sodium hydroxide in water yields 2*H*-1,4-benzthiazin-3(4*H*)-one-2-acetic acid and 2-carboxymethylene-1,4-benzothiazin-3(4*H*)-one (XLIX) (*R. T. Coutts et al.,* Canad. J. Chem., 1970, **48**, 1859, 3727):

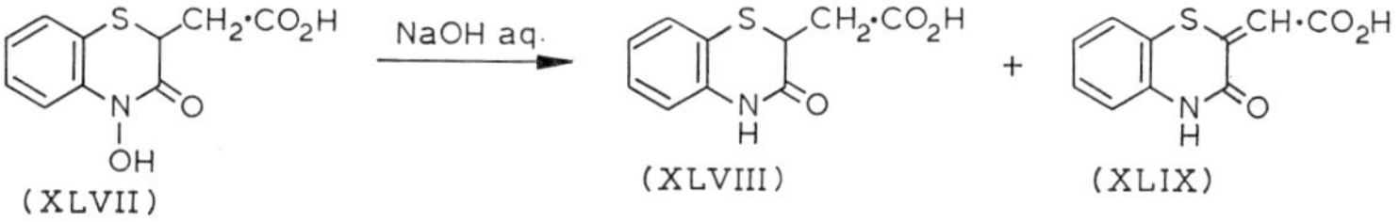

(XLVII)   NaOH aq. →   (XLVIII)   +   (XLIX)

The formation of the latter molecule probably requires the operation of the following mechanism:

(iv) Phenothiomorpholines

3,4-**Dihydro-2***H*-1,4-**benzothiazine, phenothiomorpholine** (thiophenomorpholine), b.p. 144°/1.5 mm, m.p. 38°, *hydrochloride* hygroscopic; *picrolonate*, m.p. 168° (decomp.); *thiocarbanilide*, m.p. 129°, is prepared (*a*) from 2-aminobenzenethiol and either ethylene dibromide (*N. A. Langlet*, Beilstein, 1937, Vol. 27, p. 34) or ethylene oxide (*Culvenor et al.*, J. chem. Soc., 1949, 278) and (*b*) by condensation of 1,2-chloronitrobenzene and *β*-hydroxyethanethiol and reduction of the products *S*-(2-hydroxyethyl)- or the corresponding *S*-(2-chloroethyl)-2-nitrobenzenethiol by tin and hydrochloric acid.

Phenothiomorpholine is not very stable; alkaline benzoylation or treatment with nitrous acid opens the heterocyclic ring, giving an *O,N*-dibenzoyl derivative or a diazonium salt, respectively:

Phenothiomorpholine 1,1-*dioxide*, m.p. 144°, is prepared by the reduction of *β*-chloroethyl-2-nitrophenyl sulphone by stannous chloride but not by direct oxidation of the parent.

3-*Methyl-*, b.p. 140°/1.5 mm and 3-*phenyl-phenothiomorpholine*, m.p. 103°, are the products from 2-aminobenzenethiol and propylene and styrene oxides, respectively (*R. Fusco* and *G. Palazzo*, Gazz., 1951, **81**, 735).

(v) 4H-1,4-Benzothiazines

These products are tautomeric with 2*H*-1,4-benzothiazines, but in the absence of stabilizing factors the 4*H*-tautomer is preferred. For example, when 2-(*N*-methylamino)benzenethiol is reacted with phenacyl bromides, 4-methyl-4*H*-1,4-benzothiazines are formed (*G. Santacroce, D. Sica* and *R. A. Nicolaus*, *ibid.*, 1968, **98**, 85):

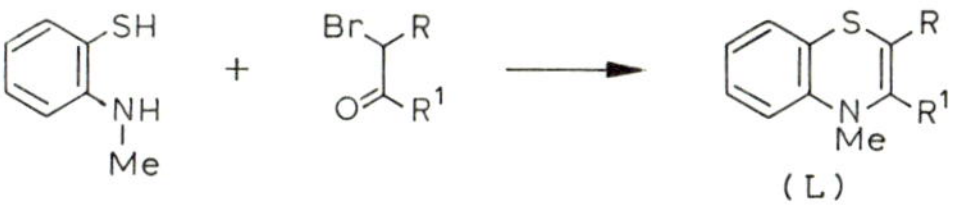

(L, R = H; R^1 = [4]BrC$_6$H$_4$), m.p. 79–80°; (L, R = H; R^1 = [4]ClC$_6$H$_4$), m.p. 68–70°; (L, R = R^1 = Ph), m.p. 178–179°.

2-Aminobenzenethiol normally reacts with β-ketones and β-oxo-esters to give dihydro-benzothiazoles or bisdihydrobenzothiazoles (*A. I. Kiprianov* and *V. A. Portnyagina*, Zhur. obshcheĭ Khim., 1955, **25**, 2257), but with β-diketones in dimethyl sulphoxide solution, 4H-1,4-benzothiazines are produced (*S. Miyano et al.*, J. chem. Soc. Perkin I, 1976, 1146):

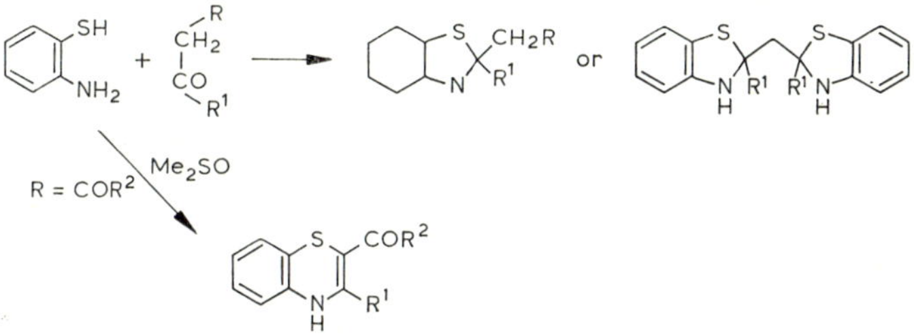

R^1 = R^2 = Me, m.p. 194–196°
R^1 = Me; R^2 = Ph, m.p. 188–190°
R^1 = Me; R^2 = OEt, m.p. 144–145°

On treatment with sulphuryl chloride in methylene chloride at room temperature, 3-acetylbenzothiazolines undergo ring-expansion to give the corresponding 4H-1,4-benzo-thiazines (*Chioccara et al.*, Chim. Ind., Milan, 1976, **58**, 546):

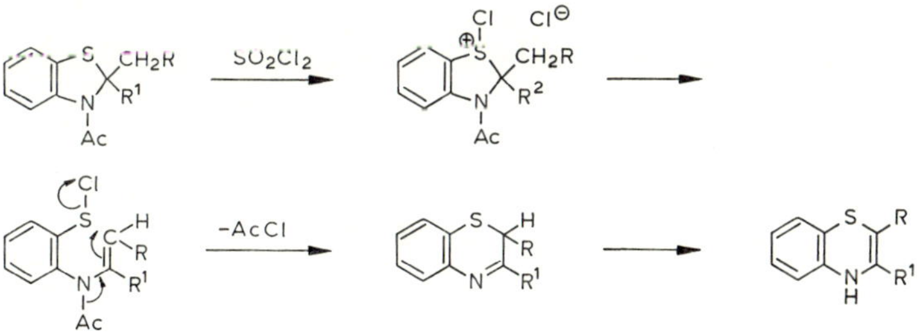

Ethyl 4H-1,4-benzothiazine-2-carboxylates are available by the action of ethyl 2-halogeno-3-oxopropionates upon 2-aminobenzenethiol (*F. Duro et al.*, Ann. Chim., Rome, 1970, **6**, 383).

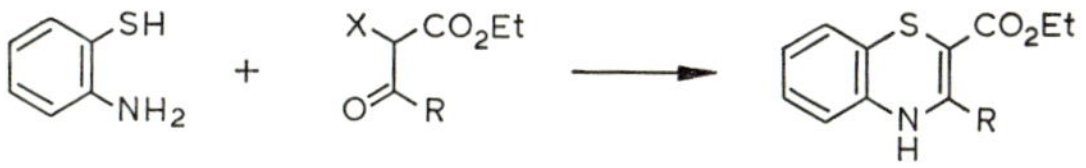

(R = H, Me or CO$_2$Et)

2-Aminobenzenethiol and ethyl bromopyruvate yield ethyl 4H-1,4-benzothiazine-3-carboxylate (LI) (*Santacroce et al.*, *loc. cit.*; *Páppalaro, Condorelli* and *M. Raspagliesi*, Gazz., 1966, **96**, 1147). In ethanolic solution, LI is oxidised by air affording the dehydro-dimers LII and LIII in the ratio 3:2 (*Duro et al.*, Ann. Chim., Rome, 1971, **7**, 351); however, when ethyl azodicarboxylate is used as the oxidant LII alone is obtained.

Sodium 2-nitrobenzenesulphinate and allyl bromide yield allyl 2-nitrophenyl sulphone which, when ozonized and the ozonide hydrogenated over palladium on charcoal, gives 4H-1,4-**benzothiazine** 1,1-**dioxide** (LIV), m.p. 195–196° (*G. Pagani* and *S. B. Pagani*, Tetrahedron Letters, 1968, 1041):

This dioxide is said to undergo bromination (Br_2/$CHCl_3$/$CaCO_3$), Vilsmeir formylation and a Mannich type reaction (morpholine/HCHO) all at position 2.

6. Phenothiazines

Nomenclature

The unqualified term phenothiazines normally refers to 10H-phenothiazine derivatives, since these are by far the most important members of this group, but isomerisation is possible and representatives of 3H-phenothiazines are known in which both hydrogen atoms at C-3 are substituted.

Phenothiazine
(10 H–Phenothiazine)

3 H–Phenothiazine

Phenothiazines represent an important class of drug substances which show activity against numerous disorders; the parent structure has been known for many years, but early interest in this group was directed mainly towards derivatives which were used as dyestuffs (see Chapter 47).

Since their original discovery by *A. Bernthsen* (Ber., 1833, **18**, 2896) much work has been published concerning the chemistry and biological properties of phenothiazines and these studies have been extensively reviewed. (*i*) **Chemistry.** *S. P. Massie*, Chem. Reviews, 1954, **54**, 797; *D. E. Pearson*, in

"Heterocyclic Compounds", ed. *R. C. Elderfield*, Vol. 6, Chapter XIV, p. 624, Wiley, New York, 1957; *C. Bodea* and *I. Silberg*, Advances in Heterocyclic Compounds, Vol. 9; *A. R. Katritsky* and *A. J. Boulton*, Eds., Academic Press, New York, 1968, p. 321. (*ii*) **Pharmacology.** *E. Schenker* and *H. Herbst*, Progr. Drug Res., 1963, **5**, 269; *F. Mietzsch*, Angew. Chem., 1954, **66**, 363; *M. Gordon, P. N. Craig* and *C. L. Zirkle*, Adv. Chem. Ser., 1964, **45**, 140; *L. Valzelli* and *S. Garattini*, Princ. Psychopharmacol., *W. G. Clark*, ed., Academic Press, New York, 1970, p. 255–258; *H. J. Meneses*, Medicina (Mexico City), 1972, **52**, 301.

(*a*) *Methods of synthesis*

(*1*) An arylamine is heated with sulphur, alone at 180–230° or in a solvent such as xylene or 1,2-dichlorobenzene (*Massie* and *P. K. Kadaba*, J. org. Chem., 1956, **21**, 347). In place of sulphur, sulphur monochloride has been used, *e.g.* with 2-naphthylamine (*O. Kym*, Ber., 1888, **21**, 2807), and sulphur dichloride in benzene is also recommended (*Sandoz Ltd.*, B.P. 890,912/19; C.A., 1963, **58**, 1472g).

The time of heating may be short or extended to several hours, and improved yields follow the addition of 1 % of iodine or aluminium chloride (*F. Kehrmann* and *J. H. Dardel*, Ber., 1922, **55**, 2348). Although many substituted diphenylamines fail to thionate (*A. Mackie* and *A. A. Cutler*, J. chem. Soc., 1954, 2577; *A. Roe et al.*, J. org. Chem., 1956, **21**, 28) the route is of wide application. Typical examples include 3-methyl- and 3,7-dihydroxy-phenothiazine from 4-methyl- (*H. Gilman* and *D. A. Shirley*, J. Amer. chem. Soc., 1944, **66**, 888) and 4,4'-dihydroxy-diphenylamine, respectively (*D. F. Houston et al.*, *ibid.*, 1949, **71**, 3816).

From *meta*-substituted diphenylamines, both 2- and 4-substituted phenothiazines are possible and proof of structure is not easy to establish. The main product is, however, considered to be the 2-compound since desulphurisation by heating with copper yields some 2-substituted carbazole (*A. Goske*, Ber., 1887, **20**, 232):

(*2*) Certain diphenyl sulphides from 2-chloronitrobenzene and 2-amino-benzenethiol rearrange in alkali giving 2-nitro-2'-mercaptodiphenylamines which readily cyclise (*cf.* formation of phenoxazines, Turpin reaction, p. 000). The preparation of 1,3-dinitrophenothiazine is typical (*Kehrmann* and *J. Steinberg*, *ibid.*, 1911, **44**, 3011):

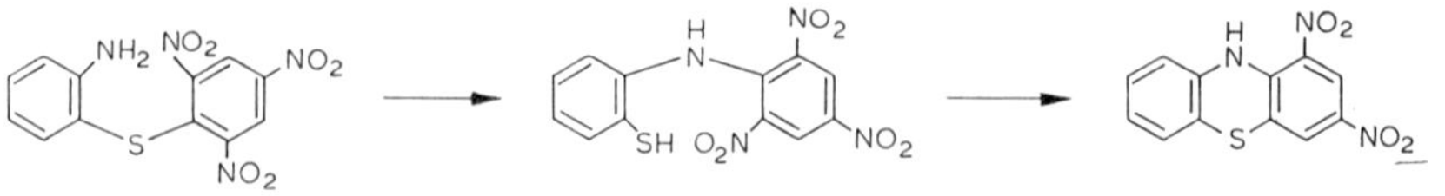

Later developments show that what has become to be known as the Smiles rearrangement (*S. Smiles et al.*, J. chem. Soc., 1931, 914) depends upon considerable deactivation of the ring B by electronegative groups (NO_2 etc.) at the positions marked *; no further substituents in ring A need be present:

XH = OH, SH, NHR, CONHR or SO_2NHR
Y = S, SO, SO_2 or O

The rearrangement is interpreted as intramolecular nucleophilic aromatic substitution (*J. F. Bunnett* and *R. E. Zahler*, Chem. Reviews, 1951, **49**, 273).

A thermal version of the Smiles rearrangement is known also (*F. A. Davis* and *R. B. Wetzel*, Tetrahedron Letters, 1969, 4483). Thus when 1'-amino-4-methyl-2-nitrodiphenyl sulphide is heated at 190°, alone or in *N,N*-di-methylacetamide, 3-methylphenothiazine (III) together with 2-methyl-dibenzothiophene (V) are formed; it is believed that the intermediates I and II are involved in the formation of the first product:

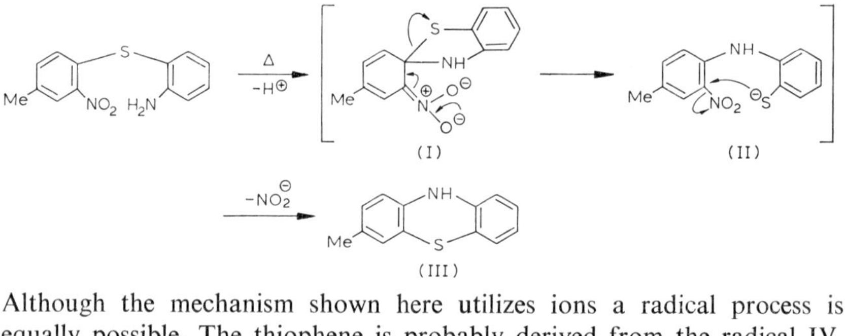

Although the mechanism shown here utilizes ions a radical process is equally possible. The thiophene is probably derived from the radical IV, which may be formed from the starting material in a number of ways:

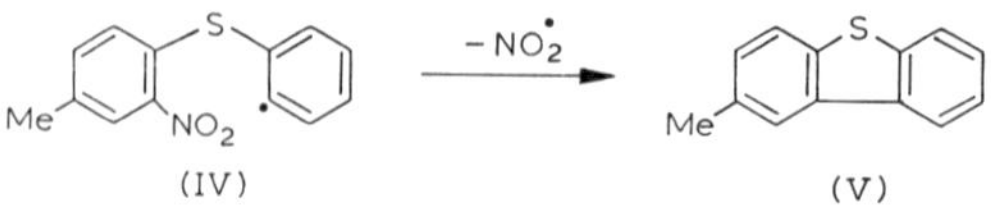

Were it not for this complication the thermal reaction would be attractive, principally because it avoids the need for an electron-withdrawing substituent attached to ring B.

(*3*) 2-Amino-2′-halogenodiphenyl sulphides may eliminate the appropriate hydrogen halide to form phenothiazines; thus, for example, 2′-amino-2-iodo-4,4′-dinitrodiphenyl sulphide gives 2,8-dinitrophenothiazine (*J. G. Michels* and *E. D. Amstutz*, J. Amer. chem. Soc., 1950, **72**, 888). This represents an Ulmann-type cyclisation, but a Smiles type rearrangement is also possible in which a 2,7-disubstituted phenothiazine would result:

Indeed, contrary to this early report indicating only an Ulmann type reaction, evidence has been produced to show that in some cases a Smiles rearrangement does actually occur (*G. Páppalardo* and *G. Scapini*, Bol. Sci. Fac. chim. Ind. Bologna, 1963, **21**, 143; C.A., 1963, **59**, 13972e; *E. A. Nodiff* and *M. Hausman*, J. org. Chem., 1964, **29**, 2453).

(*4*) Some 2,2′-diaminodiphenyl sulphides (or sulphones) cyclise by loss of ammonia but many unsuccessful attempts are reported also (*H. H. Hodgson et al.*, J. chem. Soc., 1948, 1104). This failure is in marked contrast to the ready formation of carbazole from 2,2′-diaminobiphenyls (C.C.C., Vol. IV A, p. 490). *J. I. G. Cadogan* and his co-workers (Chem. Comm., 1966, 491) have shown that phenothiazines may be prepared from 2-nitro-diphenyl sulphides by reduction with triethyl phosphite. The reaction probably involves a spiro intermediate VI since when 2-nitro-2′-diphenyl sulphide (R = Me) is reduced, the product is 1-methylphenothiazine (VII, R = Me) (*Cadogan et al.*, J. chem. Soc., C, 1970, 2437, see also Chem. Comm., 1971, 1431, 2621; J. chem. Soc., Perkin I, 1975, 2376, 2396):

(*5*) When 4-nitrodiphenylamine-2-sulphinic acid is dissolved in concentrated sulphuric acid and the solution is immediately diluted, the product is 3-nitrophenothiazine 5-oxide. If the solution in acid is allowed to stand 30 min before dilution, the product is 3-nitrophenothiazine (*S. Krishna* and *M. S. Jain*, Proc. 15th Indian Sci. Cong., 1928, 153; C.A., 1931, **25**, 3001).

In a similar way, diphenylamine with oleum leads to a phenothiazine

5-dioxide tetrasulphonic acid (*I. G. Farbenind.*, B.P. 420,444/1934) as follows:

(b) *Physical properties*

Phenothiazine, yellow plates, m.p. 180–181°, which turn green on exposure to air is obtained by heating diphenylamine with sulphur, by deamination of Lauth's violet (p. 531) and also by heating together catechol and 2-aminobenzenethiol at 220°. It sublimes at 130°/1 mm and then has m.p. 185.1° but not above 184.2° when further crystallised from benzene or alcohol (*L. E. Smith* and *O. A. Nelson*, J. Amer. chem. Soc., 1942, **64,** 461). Phenothiazine is only slightly soluble in water (1 part in 800,000) but soluble in most organic solvents. It is of some value as an insecticide and as a veterinary anthelmintic for which particle size is a factor in determining efficiency (Ann. Rep. appl. Chem., 1956, **41,** 522). A review of the chemotherapy of helminthiasis is given by *T. I. Watkins* (J. Pharm. and Pharmacol. 1958, **10,** 209).

The configuration of phenothiazine and some of its derivatives has been the subject of some discussion (see *Bodea* and *Silberg, loc. cit.*, and refs. cited therein, and also *D. Simov, L. Kamenov* and *St. Stoyanov,* Proc. 2nd Conf. appl. phys. Chem., 1971, **1,** 387; C.A., 1972, **76,** 58805b). X-Ray crystallographic studies show that phenothiazine does not consist of two planar halves symmetrically folded along the N–S axis, But the N and S atoms deviate from the planes of the two aromatic rings by significant distances (N 0.03Å–0.01Å; S 0.18Å–0.17Å). The dihedral angle between the planes is 153.3° (*J. D. Bell et al.,* Chem. Comm., 1968, 1656). The bonds about the N atom assume a flattened tetrahedron, with the NH bond having a *quasi* equatorial orientation with respect to the heterocyclic ring.

The ultraviolet spectrum of phenothiazine shows maxima at $\sim \lambda$ 240, 280, 300 and 340 nm and correlations between the ultraviolet spectra of phenothiazine derivatives and their structures have been made by several authors (see for example *R. J. Warren et al.,* J. pharm. Sci., 1966, **55,** 144 and *Simov et al.,* God. Sofii, Univ. Khim. Fak., 1968–1969; C.A., 1972, **76,** 112284w).

There is a considerable literature on the infrared spectroscopy of phenothiazines (see *Bodea* and *Silberg, loc. cit.* and refs. cited therein). *Warren et al. (loc. cit.)* have surveyed the infrared spectra of phenothiazine drugs and observed four characteristic bands in the fingerprint region, at 928–915, 870–840, 800–785 and 755–730 cm^{-1}, which may be used for analytical purposes. The –NH band of phenothiazine in the solid state appears at $\sim$ 3340 cm^{-1} and the aromatic system gives rise to bands at 1560 and 1590 cm^{-1}.

^{1}H n.m.r. spectroscopic analysis of substituted phenothiazines has been reported (see *J. Cymerman-Craig et al.,* J. med. Chem., 1965, **8,** 392; *N. S. Angerman* and *S. S. Danyluk,* Org. Magn. Resonance, 1972, **4,** 897) and the mass spectra of some derivatives has been recorded (*R. A. Coombs,* Arch. Mass Spectral. Data, 1971, **2,** 754; *H. M. Fales et al., ibid.,* p. 692).

(*c*) *Chemical properties*

(*i*) *Oxidation, phenothiazinyl radical cation and phenothiazonium (phenazathionium) salts*

Phenothiazine and its derivatives are easily oxidized and numerous chemical oxidants have been employed. The reactions involved are complex and a variety of end-products are possible. It would seem, however, that when first oxidized phenothiazine forms a radical cation VIII which may deprotonate in the presence of a base to the phenothiazinyl radical (see below). Alternatively, further oxidation and deprotonation may occur to give the phenothiazonium cation (IX). In strongly acidic solution the dication (X) may form (see *E. R. Biehl et al.*, J. heterocycl. Chem., 1975, **12**, 397):

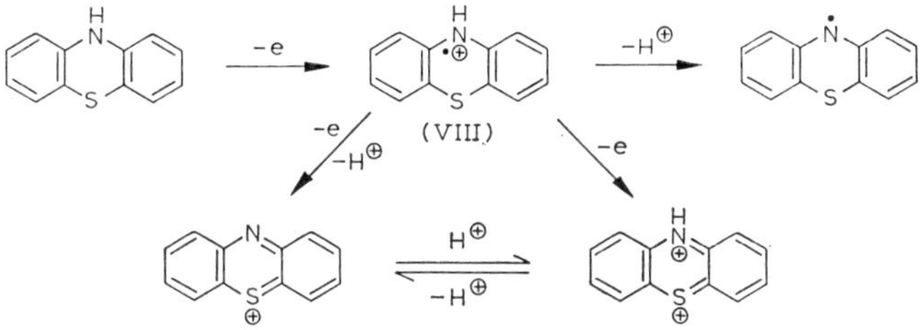

The structure of the radical cation VIII has been studied by e.s.r. spectroscopy (*J. P. Billon et al.*, Bull. Soc. chim. Fr., 1960, 2062; Compt. rend., 1961, **253**, 1593; J. Chim. phys., 1964, **61**, 374; *T. Constantienescu et al.*, Farmacia Bucharest, 1971, 731; C.A., 1972, **76**, 140679x) and it has been shown that this species is an obligatory intermediate in the formation of phenothiazonium salts from phenathiazine (*cf. L. Michaelis, S. Granick* and *M. P. Schubert*, J. Amer. chem. Soc., 1941, **63**, 351). Thus, for example, the phenothiazine radical cation is formed when phenothiazine reacts with iodine in dimethyl sulphoxide (*Y. Tsujino*, Tetrahedron Letters, 1969, 763), subsequently the *dimer* XI, m.p. 199–200°, is produced (see p. 523) together with phenothiazine-3(3*H*)-one (XII) and 3-iodophenothiazine:

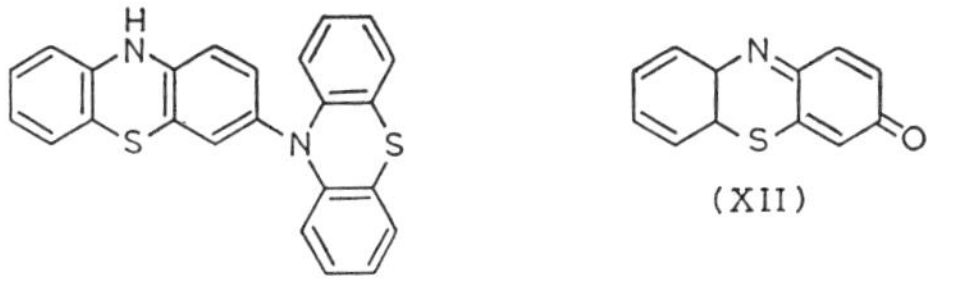

When phenothiazine is dissolved in concentrated sulphuric acid the golden coloured solution first obtained contains the radical cation. On standing, the colour changes to green; this is due to the formation of the protonated phenothiazonium cation (X) (*H. J. Shine* and *E. E. Mach*, J. org. Chem., 1965, **301**, 2130). The green solution becomes red-brown on dilution, and it is considered that IX and other products are re-formed; further dilution yields a solution of the radical cation (see *F. Kehrmann, J. Speitel* and *E. Grandmougin*, Ber., 1914, **47**, 2976; *Kehrmann* and *M. Sandoz*, ibid., 1917, **50**, 1673, for an early study of this phenomenon).

Phenothiazine 5-oxide in sulphuric acid behaves in a similar manner to phenothiazine itself (*Shine* and *Mach, loc. cit.*) and oxidized species corresponding to VIII and X may be obtained by chemical or electrochemical oxidation of *C*-substituted phenothiazines (*Billon et al.*, Compt. rend., *loc. cit.*; Ann. Chim., Paris, 1962, **7**, 183; *5. M.* L*hoste* and *F. Tonnard*, J. Chim. phys., 1966, **63**, 678).

Phenothiazine 5-oxide and **5,5-dioxide.** Phenothiazine readily forms the 5-oxide, m.p. ~ 250° (decomp.), on oxidation with hydrogen peroxide in acetone containing some sodium ethoxide (*E. de B. Barnett* and *Smiles*, J. chem. Soc., 1909, **95**, 1265) and the 5,5-*dioxide*, m.p. 257–258°, is best obtained by alkaline hydrolysis of 10-ethoxycarbonylphenothiazine 5,5-dioxide (*H. I. Bernstein* and *L. R. Rothstein*, J. Amer. chem. Soc., 1944, **66**, 1886). Many reagents oxidise phenothiazine to a mixture of these products and it has been shown that phenothiazine 5-oxides are formed when phenothiazonium cations, formed electrochemically, are treated with aqueous base (*Billon*, 1961, *loc. cit.*):

3*H*-Phenothiazin-3-ones (phenothiazones). Oxidation of phenothiazine in aqueous acid affords phenothiazin-3(3*H*)-one (XV) (*S. P. Massie*, Chem. Reviews, 1954, **54**, 797 and refs. cited therein). It is envisaged that the mesomeric phenothiazonium cation (XIII), which is first formed, undergoes nucleophilic attack by a water molecule at C-3 to give the intermediate XIV. Further oxidation and deprotonation then yields XV (*Shine* and *Mach, loc. cit.*). The reaction leading to phenothiazin-3(3*H*)-one is obviously concurrent with that producing phenothiazine 5-oxide (see above), however, the pH of the solution is the controlling factor. It should be noted that 5-oxide formation, unlike that of the 3-one, is reversible in the presence of acid and thus acid conditions direct the reaction in favour of the production of phenothiazinone. Reagents commonly used for the preparation of phenothiazinones are ferric chloride or potassium dichromate.

(XIII)   (XIV)

(XV)

Phenothiazinyl radicals. Phenothiazine is readily oxidised, by such reagents as silver oxide, lead oxide, lead tetra-acetate or potassium permanganate, to the phenothiazinyl radical (*C. Jackson* and *N. K. D. Patel*, Tetrahedron Letters, 1967, 2255). Photochemical oxidation of phenothiazine derivatives is also facile, so that problems may arise during storage of drug preparations. Similarly, patients treated with phenothiazine derivatives at high dose rates may suffer skin reactions; these are thought to result from photolysis of the drugs accumulated in or near the cuticle (*M. S. Blois Jr.*, J. Invest. Dermatol., 1965, **44**, 475).

The structure of the phenothiazinyl radical, as well as those derived from pharmacologically active phenothiazines, has been studied by various techniques including electron-spin resonance spectroscopy (*C. Lagercrantz* and *M. Yhland*, Acta Chim. Scand., 1962, **16**, 508; *Lagercrantz*, Psychopharmacol. Serv. Center Bull., 1962, **2**, 53; *Shine* and *Mach*, *loc. cit.*; *B. C. Gilbert et al.*, Chem. Comm., 1966, 161; *Jackson* and *Patel*, *loc. cit.*).

The greatest spin density in the phenothiazinyl radical (XVI) is located on the nitrogen and sulphur atoms and at C-3 and C-7. Although coupling through the sulphur atoms does not occur, 10,10′-dimers result, when phenothiazines are reacted with mercuric salts (*L. Pesci*, Gazz., 1916, **46**,I, 103; *C. Finzi*, Giorn. Chim. Ind. appl., 1927, **9**, 176). Phosphoric acid chlorides cause a similar reaction (*B. A. Arbuzov* and *D. Kh. Yarmukhametov*, Isvest. Akad. Nauk, S.S.S.R. Otdel. Khim. Nauk, 1962, 1405; C.A., 1963, **58**, 2468f).

N–C Coupling is also possible (see p. 521) and it is considered that the "green product" obtained in solution by several workers (see for example *R. Pummerer* and *S. Gassner*, Ber., 1913, **46**, 2310) during the conversion of phenothiazonium cations into phenothiazinones has the structure XVII, and may arise as indicated in Scheme 1 (*R. Foster* and *P. Hanson*, Biochim. Biophys. Acta, 1966, **112**, 482; *Hanson* and *R. O. C. Norman*, J. chem. Soc., 1973, 264; *Jackson* and *Patel*, *loc. cit.*):

(XIV)

(XV)

C–C Coupling occurs when the N-atoms are substituted; thus *S. Fujisawa* and his co-workers (C.A., 1965, **62,** 7750c; J. chem. Soc. Japan, 1966, **86,** 514; C.A., 1966, **65,** 10585f) suggest that bimethopromazine has the structure:

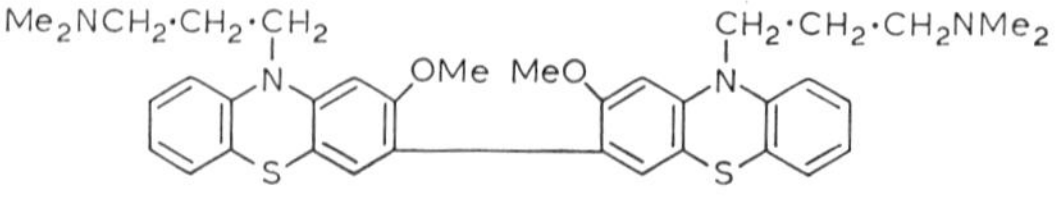

$Me_2NCH_2 \cdot CH_2 \cdot CH_2$ $CH_2 \cdot CH_2 \cdot CH_2NMe_2$

OMe   MeO

Bimethopromazine

It is formed from an aqueous solution of the parent drug methopromazine by irradiation with white fluorescent light.

(*ii*) *Electrophilic substitution*

Molecular-orbital calculations using the SCF–PPP approximation (*H. H. Mantsch* and *J. Dehler*, Canad. J. Chem., 1969, **47,** 3173) establish the π-electron charge densities (localization energies) in electron volts summarized in formula XVI for electrophilic substitution (see also *N. Tyutyulkov et al.,* Compt. rend. Acad. Bulg. Sci., 1970, **23,** 1095; C.A., 1971, **27,** 5453).

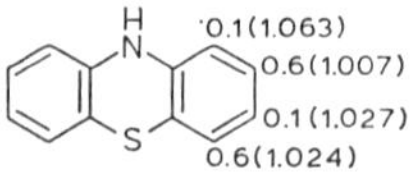

H
N
·0.1(1.063)
0.6(1.007)
0.1(1.027)
0.6(1.024)

(XVI)

These results support the fact that position 3 is about, or slightly more, reactive than position 1, but because of the ease of oxidation electrophilic substitution of phenothiazine and its derivatives is often complicated by side reactions.

(*1*) *Halogenation.* Chlorine in glacial acetic acid yields a mixture of di- and tetra-chlorophenothiazines from which 3,7-**dichlorophenothiazine,** m.p. 222°, was obtained (37%) (*C. Bodea* and *M. Raileanu,* Studii Cercetari. Chim., Cluj, 1958, **9,** 159; C.A., 1961, **55,** 15497b). In dimethyl sulphoxide solution at 40°, phenothiazine affords 3,7-dichlorophenothiazine in 84% yield (*T. Tsujino, N. Naito* and *J. Sugita,* J. chem. Soc., Japan, 1970, **91,** 1075; C.A., 1971, **74,** 141,663h). In nitrobenzene solution poly-chlorinated products are obtained (*Bodena* and *Silberg,* Rev. Roumaine Chim., 1964, **9,** 425), but if cupric chloride is employed as the chlorinating agent, 1,7-*dichlorophenothiazine,* m.p. 101°, is obtained. An even better yield is afforded if *meta*-cresol is employed as the reaction medium. Cupric chloride converts 3,7-dichlorophenothiazine into 1,3,7-*trichlorophenothiazine,* m.p. 176°, whereas thionyl chloride converts 1,7-dichlorophenothiazine into 1,3,7,9-*tetrachlorophenothiazine,* m.p. 235° (*Silberg, V. Farcasan* and *M. Dindea,* J. pr. Chem., 1976, **318,** 353). Phenothiazonium perbromide (XIX) is said to be obtained by the action of bromine (*Kehrmann* and *O. Vessily*, Ann., 1902, **322**, 34; *R. Pummerer* and *S. Gassner*, Ber., 1913, **46**, 2321; *Kehrmann* and *L. Dieserens*, *ibid.*, 1915, **48**, 318). This product is now considered to be charge-transfer complex, it is unstable and is converted into poly-brominated phenothiazines (*Bodea* and *Răileanu*, Ann. Chim., 1960, **631**, 194):

(XIX) $Br_3^{\ominus}$

Although charge transfer complexes are also formed with iodine, these do not yield ring-iodinated products when reacted further.

(2) *Nitration*. Although there are claims for the preparation of simple *C*-nitrophenothiazines by the action of nitric acid, it seems that using this reagent the sulphur atom is always attacked giving nitro-substituted 5-oxides or 5,5-dioxides. Phenothiazine gives 3-**nitrophenothiazine** 5-**oxide** and 1,3,7,9-**tetranitrophenothiazine** 5-**oxide** (*Ch. Monard, H. Ficheroulle* and *R. Fournier*, Mem. Poudres, 1952, **34**, 179; *K. Toei*, Nippon Kagaku Zashi, 1955, **76**, 1083; C.A., 1957, **51**, 11913d; *Bodea* and *Răileanu*, C.A., 1960, **54**, 22657g; *cf. Kehrmann* and *P. Zybs*, Ber., 1919, **52**, 132).

(3) *Sulphonation*. Sulphuric acid causes oxidation, but chlorosulphonic acid effects sulphonation at positions 3- and 7- (*M. Fujimoto*, Bull. chem. Soc., Japan, 1959, **32**, 483).

(4) *Acylation*. The secondary amino function of phenothiazine is easily acylated. Acetyl chloride, for example, is said to yield first 10-acetyl-phenothiazine and then, with difficulty, 2,10-diacetyl- and 2,8,10-triacetyl-phenothiazines depending upon the amount of reagent employed in a Friedel–Crafts reaction with carbon disulphide as solvent. Two molecular equivalents of aluminium chloride are required because of complex formation with the nitrogen atom of the substrate. *J. J. Lafferty et al.* (J. org. Chem., 1962, **27**, 1346) suggest, however, that both *N*- and *C*-substitution occur under these conditions.

2,10-**Diacetylphenothiazine** has m.p. 105–106° and 2,8,10-*triacetylphenothiazine*, m.p. 199–202° (*Michels* and *Amstutz*, loc. cit.). The former on hydrolysis with hydrochloric acid gives 2-*acetylphenothiazine*, yellow-orange needles, m.p. 192–193°. Chloroacetyl chloride and 10-acetylphenothiazine in a Friedel–Crafts type reaction yield 2-*chloro-acetyl-10-acetyl-*, m.p. 171–172.5°, hydrolysed by hydrochloric acid to 2-*chloroacetyl-phenothiazine*, red needles, m.p. 198–199° (decomp.) (*A. Burger* and *J. B. Clements*, J. org. Chem., 1954, **19**, 1113; see also *A. Georgiev*, Comp. rend. Acad. Bulgare Sci., 1964, **17**, 267; C.A., 1964, **61**, 8303b).

Acetic anhydride and aluminium chloride give a diacetyl product of unknown structure

(*S. P. Massie, I. Cooke* and *W. A. Hills*, J. org. Chem., 1956, **21**, 1006; see also *H. Braeuniger* and *K. F. Ahrend*, Pharmazie, 1966, **21**, 645).

N-Alkylphenothiazines upon acylation normally afford the corresponding 3,7-diacyl-phenothiazines (*G. Cauquil* and *A. Casadevall*, Bull. Soc. chim. Fr., 1955, 1061).

(5) *Alkylation and arylation.* The hydrogen of the secondary amino-group of phenothiazine is readily substituted. Many such substituted compounds have been examined for chemotherapeutic activity and are remarkable for the widely differing pharmacological effects produced by small changes of structure in the *N*-substituent.

N-Alkyl groups are introduced by heating with the appropriate alcohol and hydrogen chloride under pressure. Alternatively condensation with alkyl halides is best achieved by sodamide in liquid ammonia (*H. Gilman et al.*, J. Amer. chem. Soc., 1952, **74**, 4205). Methyl sulphate in dioxane may be used for methylation but ethyl sulphate fails.

N-Alkylphenothiazines are obtained simply by treating phenothiazine and alkyl chlorides in benzene solution (*A. Smith*, G.P. 922,467/1958; C.A., 1958, **52**, 5485h).

Certain esters of phenothiazine-10-carboxylic acid (see below) may undergo thermal decomposition to afford the corresponding 10-alkyl derivatives (*J. Schmitt et al.*, Bull. Soc. chim. Fr., 1957, 938). Similarly, esters of phenothiazine-10-thiocarboxylic acid eliminate carbonyl sulphide but here conditions are more severe and yields tend to be lower (*idem, ibid.*, p. 1474).

Arylation proceeds smoothly by heating phenothiazine with iodobenzene, copper powder and sodium carbonate. Some alkyl and aryl compounds are given in Table 12.

TABLE 12

10-ALKYL- AND 10-ARYL PHENOTHIAZINES

Substituent	m.p.(°C)	Ref.
10-Methyl-	99.5	1
10-Ethyl-	103–104.5	2
3,10-Dimethyl-	148	3
10-Phenyl-	94.5	4
10-*p*-Tolyl-	135–136	5

References

1   *H. Gilman* and *R. D. Nelson*, J. Amer. chem. Soc., 1953, **75**, 5422.
2   *Gilman et al.*, J. org. Chem., 1954, **19**, 560.
3   *Ng. Ph. Buu-Höi* and *N. G. Hoán*, J. chem. Soc., 1951, 1834.
4   *Gilman, P. R. van Ess* and *D. A. Shirley*, J. Amer. chem. Soc., 1944, **66**, 1216; *Shirley*, Prep. of Organic Intermediates, 1951, p. 256.
5   *Gilman et al.*, J. Amer. chem. Soc. 1952, **74**, 4205.

Phenothiazine with diphenyliodotetrafluoroborate, Ph_2IBF_4 at 210° gives the *S*-phenylphenothiazine cation (XX), which deprotonates to *S*-**phenylphenothiazine**. When this product is heated in toluene at 110°, 10-phenylphenothiazine is obtained (*A. N. Nesmeyanov et al.*, Izvest. Akad. Nauk, S.S.S.R., Ser. Khim., 1973, 1678; C.A., 1973, **79**, 105167u):

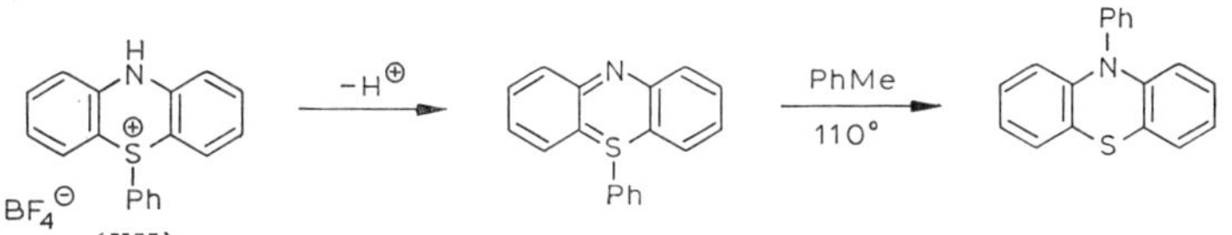

The e.s.r. spectrum of the radical cation of this compound has been studied recently (*D. Clarke, B. C. Gilbert* and *P. Hanson*, J. chem. Soc., Perkin II, 1975, 1078; 1976, 114).

(*6*) *Reactions with nucleophiles.* Phenothiazine is deprotonated by bases to give the corresponding anion; reagents such as butyl-lithium are effective and the resultant salt may then be carbonated to give phenothiazine-1-carboxylic acid (*H. Gilman et al.*, J. Amer. chem. Soc., 1944, **66**, 626; *ibid.*, p. 1216; J. org. Chem., 1962, **27**, 1260; *G. Cauquil et al.*, Compt. rend., 1955, **240**, 1784; *ibid.*, 1956, **243**, 590; Bull. Soc. chim. Fr., 1960, 1049; *ibid.*, 1964, 590).

Phenothiazine-1-carboxylic acid has m.p. 264–265°, *methyl ester*, m.p. 113–114°, 2-Acetyl-, 2,10-diacetyl- and 2-chloroacetyl-phenothiazine are converted by aqueous sodium hypochlorite to *phenothiazine-2-carboxylic acid*, m.p. 276–277°, *methyl ester*, m.p. 166–167° (*Baltzly et al.*, J. Amer. chem. Soc., 1946, **68**, 2673). *Phenothiazine-2,8-dicarboxylic acid*, decomp. over 360°, is similarly prepared from 2,8,10-triacetylphenothiazine.

The acid chloride, from phenothiazine-2-carboxylic acid and technical thionyl chloride, is converted to the amide when added to ammonia. The product is thereby characterised as a *tetrachlorophenothiazine-2-carboxamide*, m.p. 266–267°, and it is converted by phosphoryl chloride to the corresponding *nitrile*, m.p. 225.5–227.5° (*Burger* and *Clements, loc. cit.*).

(*d*) *Phenothiazine derivatives*

(*i*) *Phenothiazine homologues*

1-, 2-, 3- and 4-**Methylphenothiazine**, m.p.s. 137.5–138.5, 187–188, 166–168 and 114–118°, respectively, are obtained from the appropriate diphenylamine by the thionation route (*Massie* and *Kadaba, loc. cit.*; *P. Charpentier et al.*, Compt. rend., 1952, **235**, 59; *Gilman* and *Shirley, loc. cit.*). Similarly prepared are 3-*ethyl-*, m.p. 131° and 3,7-*dimethyl-*, m.p. 228° (*G. Cauquil, A. Casadevall* and *E. Casadevall*, Compt. rend., 1953, **236**, 1363) whilst 2-*ethyl-phenothiazine*, m.p. 135–136°, results from a Wolff–Kishner reduction on the 2-acetyl compound (*Burger* and *Clements, loc. cit.*).

10-Substituted compounds are very important as a group and are dealt with separately (p. 529).

(ii) *Halogenophenothiazines* (see pp. 524, 525)

Simple monohalogenophenothiazines are not available by direct halogenation (see above) but can be synthesised by method (*1*). The 1-chloro- and 1-fluoro-compound are each accompanied by phenothiazine after the thionation. Chloro- and bromo-compounds are also prepared by the action of hydrochloric acid, hydrobromic acid or molecular bromine upon phenothiazine 5-oxides. Phenothiazine 5-oxide and hydrochloric acid, for example, yield a mixture of 3-chloro- and 3,7-dichlorophenothiazine (*H. J. Page* and *S. Smiles*, J. chem. Soc., 1910, **97**, 1112). The mechanism of the reductive halogenation reaction appears complex, see for example *Bodea* and *Silberg*, loc. cit., pp. 384–389 and refs. cited therein. 3,7-Dichloro-10-methylphenothiazine 5-oxide and hydrochloric acid give 1,3,7,9-tetrachlorophenothiazine, the chlorination being accompanied by demethylation (*Schmalz* and *Burger*, ref. 4, Table 13).

TABLE 13

MONOHALOGENOPHENOTHIAZINES

Substituent position	m.p.(°C)			
	1	2	3	4
Fluoro-phenothiazine	81.5–82[1]	199[1]	178–179[1]	—
Chloro-phenothiazine	92–93[2]	196–197[3]	199[4]	116[3]

References

1 *A. Roe* and *W. F. Little*, J. org. Chem., 1955, **20**, 1577.
2 *S. P. Massie* and *P. K. Kadaba*, ibid., 1956, **21**, 347.
3 *Soc. des Usines chim. Rhone Poulenc*, B.P. 716,205–6/1954; C.A., 1956, **50**, 1929, 1931.
4 *W. J. Evans* and *S. Smiles*, J. chem. Soc., 1935, 1263; *A. C. Schmalz* and *A. Burger*, J. Amer. chem. Soc., 1954, **76**, 5455.

This tetrachloro compound is also prepared by heating phenothiazine in hydrochloric acid containing hydrogen peroxide (*Page* and *Smiles*, loc. cit.). Reductive fluorination or iodination of phenothiazine with either hydrofluoric or hydroiodic acid fails (*Schmalz* and *Burger*, loc. cit.; *H. Gilman* and *J. Eisch*, J. Amer. chem. Soc., 1955, **77**, 3862). Iodine converts phenothiazine into the hydriodide of phenothiazonium iodide (*R. Pummerer* and *S. Gassner*, Ber., 1913, **46**, 2310):

2,7-, 2,8- and 3,7-**Difluorophenothiazine** have m.p.s 153–155°, 198° (decomp.) and 165–167°, respectively (*Roe* and *Little*, loc. cit.). The trifluoromethyl group is present in some 10-aminoalkylphenothiazines which have pronounced pharmacological activity.

2- and 4-**Trifluoromethylphenothiazine,** m.p. 188–189° and 72–73°, respectively, are prepared simultaneously from 3-trifluoromethyldiphenylamine in the ratio of 3:2 (*H. L. Yale et al.,* J. Amer. chem. Soc., 1957, **79**, 4375), but the proportion of 4-isomer is also given as less than 5% by *P. N. Craig et al.* (J. org. Chem., 1957, **22**, 709). 3-*Trifluoromethylphenothiazine,* m.p. 217–218°, is obtained in good yield (*Roe* and *Little, loc. cit.*).

(*iii*) *Nitrophenothiazines* (see p. 525)

Direct nitration is not an important route to nitrophenothiazines. With fuming nitric acid at 0°, 3,7-dinitrophenothiazine 5-oxide is one of two products and on reduction it gives Lauth's violet (*A. Bernthsen,* Ann., 1885, **230**, 100). Other preparations are covered by general methods *2, 3* and *4*.

The nitro compounds are insoluble in alkali but the corresponding sulphoxides with a nitro group in the 1- or 3-position form red to violet solutions of the sodium salt of the pseudo-acid, *e.g.*:

1- and 3-**Nitro-,** 1,3-, 2,8- and 3,7-**dinitro-** and 1,3,7-**trinitrophenothiazine** have m.p. 111°, 218°, 188°, 355–360°, 276° and 214°, respectively (*Kehrmann et al.,* Ber., 1913, **46**, 2809, 3014; *Evans* and *Smiles,* J. chem. Soc., 1935, 181). The 1,3-dinitro compound is obtained from picryl chloride and 2-aminobenzenethiol (*cf.* 1,3-dinitrophenoxazine: *Kehrmann* and *J. Steinberg,* Ber., 1911, **44**, 3011).

(*iv*) *N-Alkyl- and N-acyl- derivatives* (see pp. 525, 526)

Substituted *N*-alkyl compounds are of considerable importance. With ethylene and propylene oxides, phenothiazine gives 10-(2-**hydroxyethyl**)-, b.p. 170–175°/0.2 mm and 10-(2-**hydroxypropyl**)-**phenothiazine,** b.p. 185–190°/0.3 mm, respectively (*R. Dahlbom,* Swed. P. 129,843/1950; C.A., 1951, **45**, 5193) and cyanoethylation yields 10-(2-*cyanoethyl*)-*phenothiazine,* m.p. 158–159° (*N. L. Smith,* J. org. Chem., 1950, **15**, 1125). Direct condensations occur with dialkylaminoalkyl halides in an inert solvent using sodamide as condensing agent (*Craig et al., loc. cit.*) but some rearrangement in the side-chain is not unusual with an alkaline condensation agent (*P. Charpentier,* Compt. rend., 1947, **225**, 306). A procedure involving fewer secondary effects is to add the dialkylaminoalkyl halide in xylene to molten phenothiazine at 200° (*Aktiebolaget Recip.,* B.P. 681,410/1952).

A series of 10-dialkylaminoalkylphenothiazines (*Gilman* and *Shirley,* J. Amer. chem. Soc., 1944, **66**, 888) lack antimalarial activity but some are useful antihistamines of low toxicity (*B. N. Halpern* and *R. Ducrot,* Compt. rend. Soc. Biol., 1946, **140**, 361) especially 10-(2-dimethylaminoethyl)- (also a good local anaesthetic) 10-(2-dimethylamino-1-propyl)- (**Phenergan**), 10-β-diethylaminoethyl-phenothiazine (**Diparcol**) (suggested for Parkinson's Disease) (*D. Bovet et al., ibid.,* 1950, **144**, 514), **chloropromazine** (*S. Courvoisier et al.,* Arch. intern. Pharmacodynamie, 1953, **92**, 305; C.A., 1953, **47**, 10103) and **pyrrolazote** (*M. J. Vander Brook et al.,* J. Pharmacol. exptl. Therap., 1948, **94**, 209; C.A., 1949, **43**, 759):

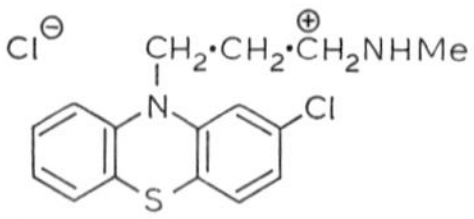

10-(3-*Dimethylaminopropyl*)-2-*trifluoromethylphenothiazine* (XXI), b.p. 162–164°/0.4 mm, as the *hydrochloride* (**Vesprin**), m.p. 172–174° (decomp.), is used as a tranquillizer and is prepared as follows (*Yale et al.*, see p. 529):

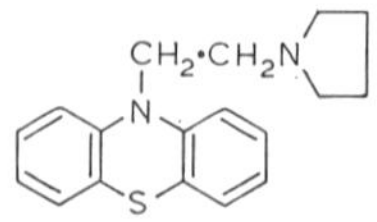

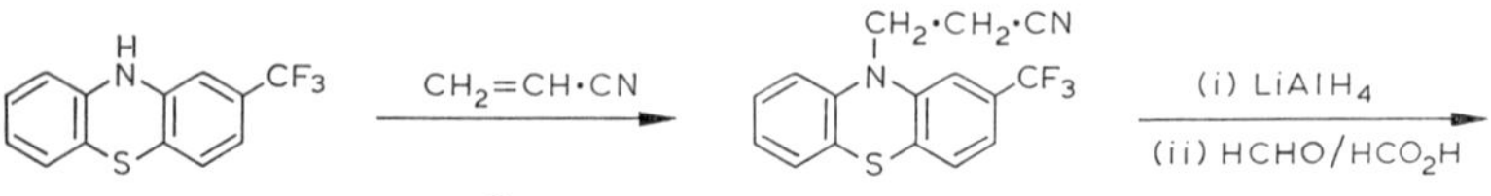

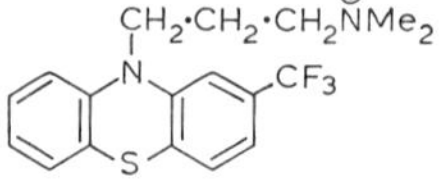

Related compounds are described and also by *Craig et al.* (p. 529).

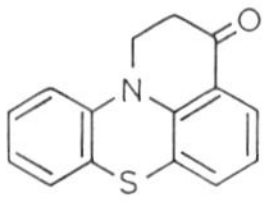

Ethyl bromoacetate and phenothiazine in the presence of potassium carbonate and copper bronze give *ethyl phenothiazine-*10-*acetate*, two allotropic forms, m.p. 102° and 112°, from which the corresponding *acid*, m.p. 210°, is obtained. This, by boiling in acetic acid, is decarboxylated smoothly to 10-methylphenothiazine. The corresponding *propionic ester*, m.p. 64°, can be prepared similarly but in this case the resulting *acid*, m.p. 161°, is not decarboxylated (*G. Cauquil* and *A. Casadevall*, Compt. rend., 1947, **225**, 578) but with phosphorus pentoxide cyclises to the ketone XXII, m.p. 112–113° (*N. L. Smith*, J. org. Chem., 1950, **15**, 1125; *A. Mackie* and *A. L. Misra*, J. chem. Soc., 1955, 1281).

For the acylation of phenothiazine the acid chloride is employed in a solvent, preferably toluene (*R. Dahlbom* and *T. Ekstrand*, Acta Chem. Scand., 1952, **6**, 1285); 10-*acetyl* derivative, m.p. 197–198°. Reduction of 10-aminoacylphenothiazines with lithium tetrahydridoaluminate results in cleavage to give phenothiazine and the corresponding amino-alcohol (*idem, ibid.*, 1951, **5**, 102).

Aroyl chlorides in either toluene or acetic acid yield 10-aroylphenothiazines: N-*benzoyl-phenothiazine*, m.p. 177–178° (*Mackie et al.*, J. chem. Soc., 1954, 2577; 1955, 1281). Phosgene and phenothiazine give a good yield of the 10-*carbonyl chloride*, m.p. 167.5° (*S. Paschkowezky*, Ber., 1891, **24**, 2905; 10-*ethoxycarbonylphenothiazine*, m.p. 109–110° (*N. Fraenkal, ibid.*, 1885, **18**, 1845). Acid chlorides and a suspension of 10-phenothiazinyl-magnesium chloride in ether form 10-acyl compounds (*G. Cauquil* and *A. Casadevall*, Compt. rend., 1953, **236**, 1569). Phenothiazine is re-formed (together with 2-amino-4-hydroxythiazole) when its 10-*chloroacetyl* deriv., m.p. 115–116.5° (*Ekstrand*, Acta Chem. Scand., 1949, **3**, 1281) and thiourea are boiled together in ethanol (*Mackie* and *Misra, loc. cit.*). 10-Chloroacetylphenothiazine shows activity against ascites sarcoma-180 in mice, and in this study twenty-two other halogenoacetylphenothiazines were also tested (*F. Kanzawa, A. Akio* and *K. Kuretani*, Gann, 1972, **63**, 225; C.A., 1972, **77**, 70249b).

Sulphonyl derivatives are prepared in pyridine solution: 10-*benzenesulphonyl*-, m.p. 170–170.5° (*S. E. Hazlet* and *E. E. Roderuck*, J. Amer. chem. Soc., 1945, **67**, 496); p-

toluenesulphonyl-phenothiazine, m.p. 155–156° (*H. I. Bernstein* and *L. R. Rothstein, ibid.*, 1944, **66**, 1886).

10-Ethyl- or 10-phenyl-phenothiazine with butyllithium followed by carbonation forms a 4-carboxylic acid. If the lithium compound from 10-ethylphenothiazine is treated with iodine, 10-*ethyl-4-iodophenothiazine*, m.p. 188–189°, is formed. When this, as its 5-dioxide, is treated with sodamide in liquid ammonia, a rearrangement gives 3-amino-10-ethylphenothiazine-5,5-dioxide (*H. Gilman et al.*, J. org. Chem., 1954, **19**, 560).

On treatment with mercuric acetate in ethanol, 10-methylphenothiazine yields a mixture of 3-*acetoxy*-, m.p. 165°, and 2,7-diacetoxy-mercuri compounds, from which the corresponding chlorides, bromides and iodides may be prepared; the 10-ethyl compound behaves similarly. Acetoxymercuri groups are replaced by nitro groups on treatment with concentrated nitric acid with simultaneous formation of the 5-oxide (*C. Finzi*, Gazz., 1932, **62**, 175; *Gilman et al., loc. cit.*). The product from the 2,7-compound is identical with 10-methyl-2,7-dinitrophenothiazine 5-oxide from the direct nitration of 10-methylphenothiazine (*F. Kehrmann* and *P. Zybs*, Ber., 1919, **52**, 130).

(v) *Aminophenothiazines*

Phenothiazines with amino groups in the 3- and 7-position, and especially in both, are very susceptible to oxidation. The dyestuffs, Lauth's violet (XXIII, R = H) and methylene blue (C.I. Basic Blue 9) (XXIII, R = Me) are typical products (see Chap. 47):

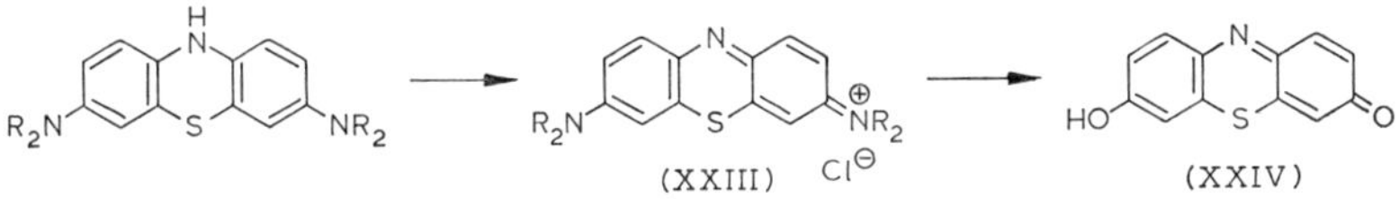

(XXIII)   (XXIV)

The amino groups are readily hydrolysed to 7-hydroxyphenothiazin-3(3*H*)-one (thionol) (XXIV) and methylene violet (7-dimethylaminophenothiazin-3(3*H*)-one) is an inter mediate from methylene blue.

3-**Aminophenothiazine** is prepared from 4-aminodiphenylamine and sulphur (*A. Bernthsen*, Ann., 1885, **230**, 106) and by reduction of 3-nitrophenothiazine 5-oxide. It is readily oxidised by ferric chloride in acid solution to **phenothiazin-3(3*H*)imine** (XXV, R = H), bright red plates from ether which blacken above 130°. This on reduction re-forms 3-aminophenothiazine and on hydrolysis with dilute sodium hydroxide gives phenothiazin-3(3*H*)-one (XXVI):

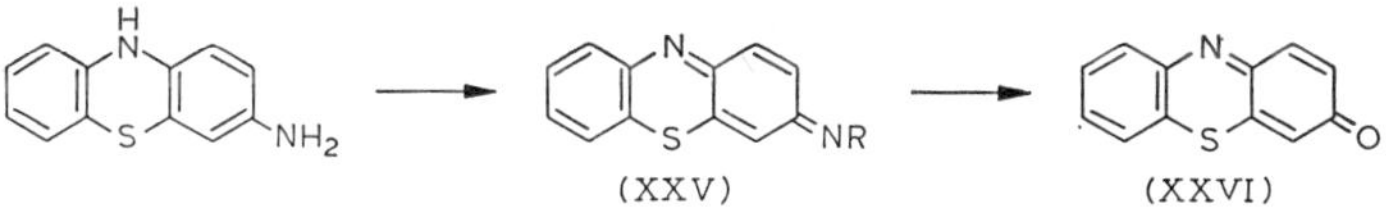

(XXV)   (XXVI)

3-*Acetamidophenothiazine*, bright yellow needles, m.p. 208°, is oxidised by ferric chloride to 2-acetamidophenothiazonium chloride.

Phenothiazine, aniline hydrochloride and ferric chloride in ethanolic solution yield 3-*phenylimino-3H-phenothiazine* (XXV, R = Ph), dark red plates, m.p. 150°.

3,7-**Diaminophenothiazine** is prepared from 4,4′-diaminodiphenylamine and sulphur, and by reduction of 3,7-dinitrophenothiazine 5-oxide. It is oxidised by ferric chloride in aqueous sulphuric acid to thionin, 7-aminophenothiazin-3(3*H*)imine (XXV, R = H and with NH₂ at position 7) and the 3,7-*diacetyl* derivative, needles, m.p. 280°, is oxidised by ferric chloride to 3,7-diacetamidophenothiazonium chloride (*Kehrmann*, Ber., 1906, **39**, 916).

2,8-Diaminophenothiazine 5,5-dioxide has been prepared from the corresponding dicarboxylic acid (*J. G. Michels* and *E. D. Amstutz*, J. Amer. chem. Soc., 1950, **72**, 888).

(*vi*) *Hydroxyphenothiazines*

3-Hydroxyphenothiazine, from 4-hydroxydiphenylamine and sulphur is readily oxidised by ferric chloride in aqueous alcohol at 90° to phenothiazin-3(3*H*)-one (XXVI, m.p. 160–161°, together with a black polymer (dimer). The latter is also formed by passing oxygen through an alkaline solution of the hydroxy compound (*Pummerer* and *Gassner*, loc. cit.; *S. C. J. Olivier* and *W. P. Combé*, Rec. Trav. chim., 1950, **69**, 526), from phenothiazine by oxidation at 200° by air and, together with phenothiazine 5-oxide, by irradiation at room temperature in alcoholic solution (*G. P. Brown et al.*, J. org. Chem., 1955, **20**, 1772).

Further oxidation of phenothiazin-3(3*H*)-one or oxidation of phenothiazine by hydrogen peroxide in hydrochloric acid (*F. De Eds* and *C. W. Eddy*, J. Amer. chem. Soc., 1938, **60**, 1446) or by 90% sulphuric acid at 150–160° for 25 hours, gives thionol, 7-hydroxyphenothiazin-3(3*H*)-one (XXIV). Thionol occurs as red-brown needles and gives red aqueous and purple alcoholic solutions. It is reduced by zinc dust and ammonia to **leucothionol**, 3,7-dihydroxyphenothiazine, needles from ether, which is not very soluble in water but soluble in acid. With acetic anhydride, thionol is simultaneously reduced and acetylated to *diacetyl-leucothionol*, m.p. 212° (*Kehrmann*, loc. cit.; *De Eds* and *Eddy*, loc. cit.).

3-Methoxydiphenylamine and sulphur give only 2-**methoxyphenothiazine**, m.p. 179–180°, and not a mixture of 2- and 4-monosubstituted phenothiazines as occurs in the methyl- and chloro-series (*P. Charpentier et al.*, Compt. rend., 1952, **235**, 59). 3-*Methoxyphenothiazine*, m.p. 160–161°, is similarly obtained from 4-methoxydiphenylamine (*H. Gilman* and *D. A. Shirley*, J. Amer. chem. Soc., 1944, **66**, 888; *R. Baltzly et al.*, ibid., 1946, **68**, 2673). Pyridine hydrochloride has been employed to demethylate 1-, 2- and 3-methoxyphenothiazines (*C. Bodea, V. Farcasan* and *T. Panea*, Rev. Roumaine Chim., 1967, **12**, 697) and sodium in isoamyl alcohol effects the reductive fission of benzyloxy substituents affording the corresponding hydroxy compounds (*Chao Ho Ts'ao et al.*, Yao. Hsueh. Hsueh. Pao., 1963, **10**, 394; C.A., 1963, **59**, 13971g).

(*vii*) *3H-Phenothiazine derivatives*

Familiar examples of this class of compound include **phenothiazin-3(3*H*)-one**, already mentioned many times previously and methylene blue (p. 531). Other examples will also be found in the foregoing text.

Oxidation of phenothiazonium salts forms 7-hydroxyphenothiazin-3(3*H*)-one (*F. De Eds* and *C. W. Eddy*, J. Amer. chem. Soc., 1938, **60**, 1446) or, in the presence of amines, imino compounds (*Kehrmann et al.*, Ann., 1902, **322**, 38; Ber., 1917, **50**, 1662, 1663):

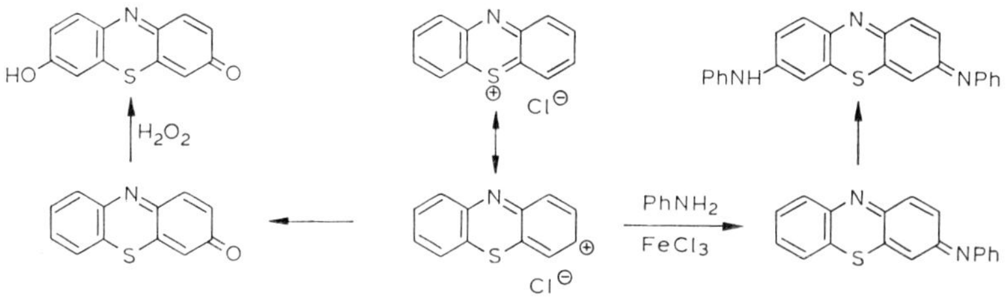

Similarly derivatives of general formula XXVII, where X = 4,4-dimethyl-2,6-dioxocyclohexylidene or 4,4-dimethyl-3,5-dioxocyclohexylidine are prepared by reacting phenothiazine with ferric chloride in methanol containing potassium acetate and then adding dimedone or Meldrum's acid respectively (*J. Daneke et al.*, Tetrahedron Letters, 1970, 1271; Ann., 1970, **740**, 52):

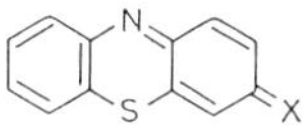

(XXVII)

Of special interest is the preparation of methylene blue from dimethylamine, the oxidation being achieved by utilising the charge-transfer complex formed from phenothiazine and bromine (*Pummerer* and *Gassner, loc. cit.*; *Kehrmann* and *R. Speitel*, Ber., 1916, **49**, 53).

On reduction the amino- and hydroxy-phenothiazonium salts form leuco compounds which, are readily reoxidised.

Some tumour-inhibiting compounds of the type XXVIII where X = OH and Y = OH or various organic or inorganic acid anions are prepared by oxidising phenothiazine with bromine (0.6 mol) in the presence of di(2-hydroxyethyl)amine (*J. Korosi* and *G. Csaba*, Ger. Offen. 1,906,527/1969). When treated with thionyl chloride the compound XXVIII (X = Cl, Y = Cl⁻) is obtained. This product shows 49 % inhibition of the Ehrlich ascites tumour in mice:

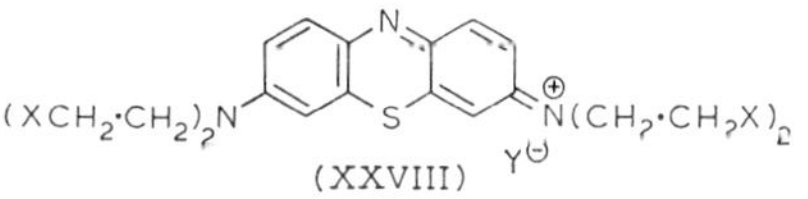

(XXVIII)

It is suggested that alkaline treatment of 3-nitrophenothiazine yields a salt XXIX (*J. S. Driscole* and *R. N. Nealey*, J. heterocyclic Chem., 1965, **2**, 272); 3,7-dinitrophenothiazine behaves similarly and the intense blue colours which result from the reaction in alcohol serve to distinguish 3-nitrophenothiazines from 2-nitrophenothiazines. The latter give red-violet colours (*G. Páppalardo, L. Amoretti* and *G. P. Gardini*, Ann. Chim., Rome, 1965, **55**, 196):

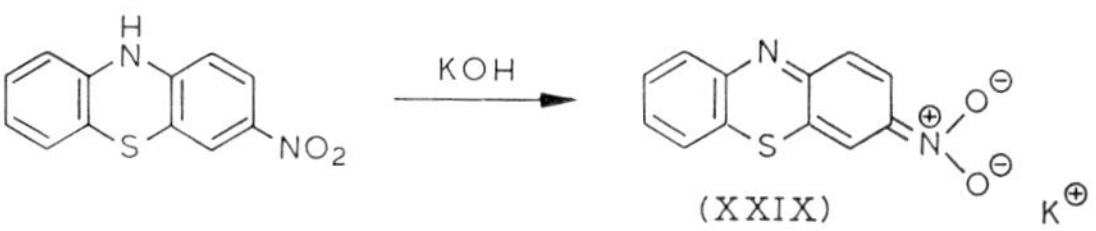

(XXIX)

7-Substituted-2-(or 1-)-methoxyphenothiazin-3(3*H*)-ones may be prepared directly by the condensation of 2-amino-5-substituted benzenethiols and 5-chloro-2-methoxy- and

6-chloro-2-**methoxy**-1,4-benzoquinones (*J. K. Jain* and *R. L. Mital*, Ind. J. Chem., 1974, **12**, 780):

(R = Cl, Br, Me, OMe or OEt)

All seven possible isomers of monomethyl-phenothiazin-3(3*H*)-one have been prepared; the 2-, 4- and 6-isomers by oxidation of the corresponding methylphenothiazines with ferric chloride, the 1- and 2-isomers by condensation of zinc 2-aminobenzenethiolate with methyl or halogenomethyl-1,4-benzoquinones, and the 6-, 7-, 8- and 9-isomers by thionation of 4-hydroxymethyldiphenylamines (*M. Terdic*, Ann., 1971, **746**, 200).

(*viii*) *Desulphurization of phenothiazines*

Before the advent of spectroscopy the elimination of sulphur from substituted phenothiazines was of value in elucidating structure, since they form easily recognizable carbazoles on heating with copper powder (p. 517); 2- and 4-methylphenothiazine were characterised in this way (*P. Charpentier et al.*, Compt. rend., 1952, **235**, 59). By the action of lithium on *N*-ethylphenothiazine in tetrahydrofuran, *N*-ethylcarbazole is formed in 27% yield (*Gilman* and *J. J. Dietrich*, J. Amer. chem. Soc., 1958, **80**, 380).

Desulphurisation of phenothiazine by Raney nickel gives diphenylamine (*K. H. Shah, B. D. Tilak* and *K. Venkataraman*, Proc. Indian Acad. Sci., 1948, **28A**, 142; C.A., 1950, **44**, 3958). The position of the substituent in 2-acetylphenothiazine is established by removal of the sulphur from the corresponding acid by Raney nickel. Diphenylamine-3-carboxylic acid is obtained and the acetyl group is in either the 2- or 4-position. The former is favoured since on *N*-ethylation of the acid the resulting 10-ethylphenothiazinecarboxylic acid is not identical with the known 4-carboxylic acid (*R. Baltzly et al.*, loc. cit.). Other related studies include the work of *H. Brauniger* and *K. F. Ahrend* (Arch. Pharm., 1965, 298, 627), *Skorodumov et al.* (Zhur. obshcheĭ Khim., 1960, **30**, 1680) and *Cauquil et al.* (Bull. Soc. chim. Fr., 1960, 1049).

(*ix*) *Azaphenothiazines*

1-**Azaphenothiazine** (pyrido[3,2-*b*]-1,4-benzothiazine) (XXX, R = H), m.p. 112-114°, 10-*acetyl* deriv., m.p. 171–172°, is prepared *via* a Smiles rearrangement of 2-aminophenyl 2-nitro-3-pyridyl sulphide, obtained by condensation of 2-chloro-3-nitropyridine and 2-aminobenzenethiol:

(XXX)

1-Azaphenothiazine is also formed by the thionation of *N*-(2-pyridyl)aniline (*A. R. Gennaro*, J. org. Chem., 1959, **24**, 1156).

The 8-*chloro*-1-*azaphenothiazine* (XXX, R = Cl), m.p. 207–208°, 10-*acetyl* deriv., m.p. 189–190°, is prepared similarly from 2-amino-4-chlorobenzenethiol (*H. L. Yale* and *F. Sowinski*, J. Amer. chem. Soc., 1958, **80**, 1651).

1-*Nitro*-3-*azaphenothiazine*, (5-*nitropyrido*[4,3-b]*benzo*-1,4-*thiazine*), m.p. 141–142°, is obtained from 4-chloro-3,5-dinitropyridine and aminobenzenethiol. It is reduced to 1-*amino*-3-*azaphenothiazine*, m.p. 310–311°, *acetyl* deriv., m.p. 230–231°) and on nitration gives 1,7-*dinitro*-3-*azaphenothiazine* 5-*oxide*, m.p. 282°, which on reduction yields 1,7-*diamino*-3-*azaphenothiazine, dihydrochloride*, m.p. > 300°) (*V. A. Petrow* and *E. L. Rewald*, J. chem. Soc., 1945, 591).

3-*Ethoxy*-7,9-*dinitro*-4-*azaphenothiazine*, m.p. 212° (*T. Takahashi* and *E. Yoshii*, Pharm. Bull. Japan, 1954, **2**, 382; C.A., 1956, **50**, 13032).

Although 4,4-dipyridylamine yields XXXI on thionation (*E. Kopp* and *M. Strell*, Arch. Pharm., Weinheim, 1962, 295, **99**, 561), *N*-(4-pyridyl)aniline fails to react (*Petrow* and *Rewald, loc. cit.*):

(XXXI)

2,3-**Diazaphenothiazine** (XXXII) has been synthesised (*G. Páppalardo, F. Duro* and *G. Scapini*, Ann. Chim., Ital., 1971, **58**, 280), as well as derivatives of 2,4- (*V. G. Granik* and *R. G. Glushkov*, Khim. Farm. Zhur., 1971, **5**, 10; C.A., 1971, **75**, 63,723g) and 3,4-diaza-phenothiazines (*D. E. Ames* and *N. D. Griffiths*, J. chem. Soc., C, 1971, 2672):

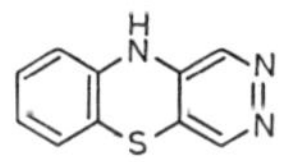

(XXXII)

Phenoselenazine, yellow plates, m.p. 195°, from diphenylamine and selenium chloride in benzene, on nitration gives 3-nitrophenoselenazine and 3,7-dinitrophenoselenazine 5-oxide and on treatment with iron or zinc gives carbazole (*W. Cornelius*, J. pr. Chem., 1913, [ii], **88**, 403; *H. Bauer*, Ber., 1914, **47**, 1873). The alkylation reactions of this molecule have been studied (*D. Simov* and *N. Khristova*, God. Sofi. Univ. Khim. Fak., 1964, **237**, 59).

This index is constructed in a similar manner to the volume indexes of the first edition of the Chemistry of Carbon Compounds. However, to make the index easier to use, more descriptive entries have been made for the commonly occurring individual, and groups of chemicals.

The indexes cover primarily the chemical compounds mentioned in the text, and also include reactions and techniques, where named, and some sources of chemical compounds such as plant and animal species, oils, etc.

Chemical compounds have been indexed alphabetically under the names used by authors, editing being restricted to ensuring uniformity of entries under the same heading. In view of the alternative nomenclature that can often be used, a limited amount of cross-referencing has been done where it is considered to be helpful, but attention is particularly drawn to Convention 2 below.

For this and the succeeding volumes, the indexing conventions listed below have been adopted.

1. *Alphabetisation*

(a) The following prefixes have not been counted for alphabetising:

n-	*o-*	*as-*	*meso-*	D	*C-*
sec-	*m-*	*sym-*	*cis-*	DL-	*O-*
tert-	*p-*	*gem-*	*trans-*	L	*N-*
	vic-				*S-*
		lin-			*Bz-*
					Py-

Some prefixes and numbering have been omitted in the index, where they do not usefully contribute to the reference.

(b) The following prefixes have been alphabetised:

Allo	Epi	Neo
Anti	Hetero	Nor
Cyclo	Homo	Pseudo
	Iso	

(c) A letter by letter alphabetical sequence is followed for entries, firstly for the main entry, followed by the descriptive entry. The only exception to this sequence is the placing of plural entries in front of the corresponding individual entries to prevent these being overlooked by a strict alphabetical sequence which could lead to a considerable separation of plural from individual entries. Thus "butanes" will come before *n*-butane, "butenes" before 1-butene, and 2-butene, etc.

2. *Cross references*

In view of the many alternative trivial and systematic names for chemical compounds, the indexes should be searched under any alternative names which may be indicated in the main body of the text. Only a limited amount of cross-referencing has been carried out, where it is considered that it would be helpful to the user.

3. *Esters*

In the case of lower alcohols esters are indexed only under the acid, *e.g.* propionic methyl ester, not methyl propionate. Ethyl is normally omitted *e.g.* acetic ester.

4. *Derivatives*

Simple derivatives are not normally indexed if they follow in the same short section of the text.

5. *Collective and plural entries*

In place of "— derivatives" or "— compounds" the plural entry has normally been used. Plural entries have occasionally been used where compounds of the same name but differing numbering appear in the same section of the text.

6. *Main entries*

The main entry of the more common individual compounds is indicated by heavy type. Where entries relate to sections of three pages or more, the page number is followed by "ff".

Index

1,2-Benzoxazines, 456
1,3-Benzoxazines, 456
4*H*-1,3-Benzoxazines, 460
–, trisubstituted, 461
1,4-Benzoxazines, 461
2*H*-1,4-Benzoxazine, **461**
1*H*-2,3-Benzoxazines, 465
1*H*-2,3-Benzoxazine, **465**
4*H*-3,1-Benzoxazines, 467
1,3-Benzoxazine-2,4(3*H*)-dione, 459
2*H*-3,1-Benzoxazine-2,4-(1*H*)-diones, 471
2*H*-3,1-Benzoxazine-2,4-(1*H*)-dione, 470
Benzoxazine dyes, 461
4*H*-3,1-Benzoxazine-2(1*H*)-thione, 467
2*H*-1,4-Benzoxazin-3(4*H*)-one, **463**
1*H*-2,3-Benzoxazin-4(3*H*)-one, 465
3,1-Benzoxazin-4-ones, **468**, 469
3-Benzoyl-2,1-benzisoxazole, 469
4-Benzoyl-2-(3,4-dihydroxyphenyl)-5,6-dihydro-1,4-oxazine, 448
1-Benzoyl-2,5-dimethyl-4-azaindole, 357
Benzoylmethanesulphonyl chloride, 487
4-Benzoylmorpholine, 452
Benzoyl-2-(4-morpholinyl) ethylamine, 452
N-Benzoylphenothiazine, 530
N-Benzoylsalicylamide, 460
O-Benzoylsalicylamide, 460
Benzoylsulphene, 487
N-Benzoyltrimethylcolchinic acid, 237
1,2-Benzphenanthridine, 173
Benzylamine, 293, 445
Benzylaminoacetals, 17, 140
2-Benzyl-2*H*-1,4-benzothiazin-3(4*H*)-one, 512
Benzylcyclobutanes, 212
1-Benzyl-3,4-dihydroisoquinolines, 131, 133
3-Benzyl-5-ethyltetrahydro-1,3-oxazine, 445
1-Benzyl-6′-halogenoisoquinolines, 89
1-Benzyl-4-hydroxyisoquinoline alkaloids, 55
3-Benzyl-5-hydroxymethyl-5-nitrotetrahydro-1,3-oxazine, 445
1-Benzyl-4-hydroxytetrahydroisoquinoline, 17, 55
Benzylic acid rearrangement, 239
Benzylidene-2-aminobenzyl alcohol, 467

Benzylideneaminomorpholine, 452
Benzylideneanthranilic acid, 470
2-Benzylidenephenomorphol-3-one, 463
1-Benzylisoquinolines, 1, 10, 64, 84, 115, 145, 199, 254
1-Benzylisoquinoline alkaloids, biosynthesis, 24
N-Benzylisoquinolinium salts, 136
Benzylmagnesium bromide, 437
3-Benzyl-5-methyltetrahydro-1,3-oxazine, 445
4-Benzylmorpholine, 451
N-Benzylnarcotine, 154
Benzylnitrophenoxazines, 474
10-Benzyl-3-nitrophenoxazine, 475
3-Benzyl-5-nitrotetrahydro-1,3-oxazine, 445
2-(3-Benzyloxy-4,5-dimethoxyphenyl)ethylamine, 4
10-Benzylphenoxazine, 474
2-Benzyl-3-phenyl-1,2-benzisoxazolium perchlorate, 456
6-Benzyl-5-phenyl-2,3-dihydro-1,4-dithiin, 405
1-Benzyl-4-piperidone phenylhydrazone, 360
2-Benzylpyridine, 262
1-Benzyltetrahydroisoquinolines, 56, 60, 68, 84, 91, 121, 128, 129, 154
–, biosynthesis, 141
–, synthesis, 20
Benzylthiomorpholine, 499
6-Benzyl-2,4,6-triphenyl-6*H*-1,3-oxazine, 437
Benzyne, 502
Berbamine, 64
Berbamine-*O*-methyl ether, 64
Berberal, 112
Berberastine, 55, 111, 112, 124, 141, 144
Berberidaceae sp., 111, 145, 152, 286
Berberilic acid, 112
Berberine, 110, 111, 112, 143, 191, 193
–, oxidative degradation, 112
–, structure, 113
–, synthesis, 133
Berberine acetone, 126
Berberine alkaloids, 110
Berberine phenol, 126, 127
Berberine phenol betaine, 126, 210
Berberinium salts, 124
Berberis vulgaris, 111

Roemrefine, 48
Romneine, 30
Rubiaceae sp., 216
Rubremetine, **227**
Rutaceae sp., 111, 145, 365

Saccharin, 469
Salicylamides, 460
Salicylamide, 457, 459
Salicylic acid, 459, 460
Salicylic esters, 459
Salicyloylamide, 459
Saligenin, 380
Saligenin methylene ether, 380
Salsolidine, 5
Salutaridine, 242
Sanguilutine, **173**, 175
Sanguinaria canadensis, 178
Sanguinarine, **173**, 175, 176, 178, **179**, 187, 195
–, biosynthesis, 197
–, from protopine, 191
–, synthesis, 189, 192
Sanguinarine pseudocyanide, 185
Sanguirubine, **173**, 175
Sarodesmine, 309
Sarothamnus catalaunicus, 309
Schefferine, 115
Scoulerine, 120, 121, 130, 132, 143, 144, 152, 162, 196
Secologanin, **232**
Selenium heterocyclic compounds, 425
1,4-Selenoxane, **425**
Sendeverine, 23
Sibiricine, **198**
Silybin, **387**
Sinactine, 147
Skimmia japonica, 351
Skimmianal, 351
Skimmianic acid, 351
Skimmianine, 350, **351**
Skimmis repens, 350
Skraup reaction, 344, 346, 371, 372
Smiles rearrangement, 388, 518, 519, 534
Sodium phenacyl thiosulphate, 404
Solanaceae sp., 286
Sophocarpine, **325**, 326, 327
Sophora sp., 319
Sophora chrysophylla, 290
Sophora flavescens, 324, 325, 326

Sophoramine, 324, **325**, 326, 327
Sophoranol, **326**, 327
Sophora pachycarpa, 324, 325
Sophoridine, **327**
Spartalupine, 314
Sparteilene, **311**
Sparteine, 300, 302, 305, 310, **311**, 312, 313, 315, 316, 317, 318, 319, 337
Sparteine alkaloids, 299
Sparteine-like alkaloids, 311
Sparteine *N*-oxide, 313
Spathelia sorbifolia, 366
Spirobenzylisoquinolines, 1, 160, 168, 170, 197, 198
–, biosynthesis, 209
–, chemical synthesis, 205, 206
–, from protoberberines, 209
–, u.v. spectral characteristics, 203
–, synthesis, 212
Stearic acid, 449
Stephanine, 97
Stepharotine, 111, 132
Steporphine, 90
Stevens rearrangement, 21, 37, 47
Strecker reaction, 500
Streptomyces fungicidus, 483
Stylopine, 143, 144, 152, 195, 196, 197
Stylopine metho salts, 151
Styrene, 377, 432, 445
Styrene oxide, 449, 514
2-Styrylbenzthiazole, 512
2-Styrylpyridine, 274
Substance CC-12, 242
Succinoyl peroxide, 376
2-Sulphamoylphenylacetic acid, 501
2-Sulphanilamido-5,6-dihydro-4*H*-1,3-thiazine, 494
N-Sulphinylamines, 487
N-Sulphinylsulphonamides, 487
Sulphonamides, 449, 502
3-Sulphonamidophthalonic acid, 502
Sulphonium salts, 405
4-Sulphonylamino-*cis*-but-2-enesulphinates, 487
Sulphonyldiacetic acid, 499
Sulphonyldiacetic ester, 499
2-Sulphonyl-3,6-dihydro-2*H*-1,3-thiazine 1-oxides, 487
Sulphostyril, **503**
δ-Sultones, 411